THE WILEY
ENGINEER'S
DESK REFERENCE

THE WILEY ENGINEER'S DESK REFERENCE

A CONCISE GUIDE FOR
THE PROFESSIONAL ENGINEER

SECOND EDITION

SANFORD I. HEISLER, P.E.
Foster City, California

JOHN WILEY & SONS, INC.

New York · Chichester · Weinheim · Brisbane · Singapore · Toronto

Copyright © 1998 by John Wiley & Sons, Inc.

Library of Congress Cataloging-in-Publication Data
Heisler, Sanford, I.
    The Wiley engineer's desk reference : a concise guide for the
professional engineer / Sanford I. Heisler.—2nd ed.
        p.   cm.
    Includes index.
    ISBN 0-471-16827-0 (cloth : alk. paper)
    1. Engineering—Handbooks, manuals, etc.   I. John Wiley & Sons.
II. Title.
TA151.H425   1998
620—dc21   97-30199

Printed in the United States of America

10 9 8 7 6 5 4 3 2

CONTENTS

6 Electricity and Electronics   **291**

J. M. Shulman

Fellow Engineer (Retired)
Westinghouse Electric Corporation

PREFACE

Slightly more than a decade has passed since the first edition of the *Wiley Engineer's Desk Reference* was written. In that time both the breadth of the body of engineering knowledge and its depth have expanded almost explosively. Engineers today must be able to deal not only with the classical engineering topics, such as hydraulics, structures, and electrical engineering, but with newer ones such as composite structures, lasers, and electro-optics. This has led to a blurring of the distinctions between the classical engineering disciplines, but interestingly enough, it closely parallels what typically occurs in one's career.

This second edition has been prepared not only to update the earlier edition but also to provide data on some of these newer areas. Significant expansion and re-ordering of the existing work includes new complete chapters on structures and the design process, as well as expanded information on the characteristics and application of electric motors, corrosion information, conversion factors, codes, the electromagnetic spectrum, ISO 9000, engineering operations, writing for publication, and intellectual property. The newer areas also include composite materials, plastics, lasers, automatic data collection, reliability, finite element analysis, statistical process control, nondestructive examination, good manufacturing practices (for medical devices), and computer chip packaging. Much of the existing material has been updated to reflect the current state of the art and preferred practices.

Over a span of years engineers accumulate all manner of useful data, often in summary or graphical form. Some of the material in this volume is of this type: which while perhaps conservative, has stood the test of time and proven extremely useful.

Comments from users of the first edition of the *Desk Reference* has asked that future editions include more application data, so considerable stress has been placed on this and more of this type of information has been introduced. Although not intended to replace a more fundamental approach, for many purposes application data provide a shortcut to an economical and practical first approach to a particular problem. More references have been included to assist the working engineer who wishes to dig more deeply. The material herein is not intended to cover all areas of engineering activity but concentrates on those felt by the author to be of particular interest and importance.

The period between the two editions of this book has seen a virtual revolution in the development and adaptation of the digital computer to problem solution in the engineering profession. Because of its domination of the computational and modeling aspects of the work, it is necessary to consider its impact on the methodology of engineering problem solution.

The application of digital computers has two major aspects. The first and I believe more important aspect of the growth of computer utilization is the importance of the logic of the program(s) that manipulates the data. This logic is fundamentally dependent on the formulas and data provided in this volume. Thus while the power and computational ability of computers have a major impact on the mechanics of problem solution, the underlying logic and methodology remain dependent on the engineering and physical relationships expressed herein.

The second aspect represents a highly developed computational tool to perform mathematical operations leading to a numerical solution. Compounding this is the time dependency of computer-related technical data. The growth in the speed and capability of microprocessors and in the number and sophistication of programs for engineering applications seemingly has been exponential. Each day sees improvements and extensions in methodology and scope. As a result, we are faced with material that is highly time dependent. A desk reference must necessarily present data that are more fundamental and not subject to significant change with time; thus the data herein are intended to support and be used in conjunction with currently available computer programs. A subset to that are the data and machine control aspects of computer utilization that permitted the development of computer-aided design, drafting, and manufacturing, and similar techniques. These are covered in the book in a way that provides an overview and permits practicing engineers to adopt and utilize these tools effectively.

In addition to the expansion of the body of knowledge, increasing interaction with the public requires that the practice of engineering now include methodology in design work that is more rigorously defined than previously. Consideration of this often involves reviewing alternatives or lists of possibilities. It is not wise or possible to eliminate the use of judgment and reduce engineering to a list of steps to be followed rigorously. To assist the engineer, however, checklists and similar material have been added to act as an *aide memoire* in assuring that the design process considers all factors in an orderly way.

Keeping the book of manageable size has also been a goal and has required considerable selectivity, resulting in the necessary omission of some material in the first edition which while meritorious was not felt to be of sufficiently wide use. The widespread availability and use of hand-held computers, particularly the scientific type that most engineers utilize, has permitted a reduction in the number of tables in this edition. The tables that have been omitted are those commonly found within these computers and permit the utilization of that space in this volume for other information.

In preparing this edition, many persons have provided valuable information and given freely of their time. Without their active assistance, this work would not have been possible. I wish to thank each of them for their help and for the active role they played in making this book possible. Mr. Michael Chessman of Varian Associates has been of great assistance not only with advice on the section on the design process, but also with general help regarding the thrust of major portions of the work; Mrs. Samar Adranly of ICF Kaiser Engineers has contributed significantly to the new chapter on structures; Mr. Don Pastor of the Westinghouse Motor Company, on electric motors; Dr. Jim Leslie of Advanced Composite Products and Technologies, on composite materials, and Mr. Rick Stevenson of Excel Control Laser, on lasers. The staff at Wiley is to be thanked for their work in turning the manuscript into a finished work. Finally, I would like to thank my wife, Lois, for putting up with what at times became an all-consuming effort in preparing this edition.

It is hoped that you will find this edition useful; we welcome comments and suggestions for future editions.

SANFORD I. HEISLER

Foster City, CA
May 1997

THE WILEY
ENGINEER'S
DESK REFERENCE

CHAPTER 1
MATHEMATICS

1.1 SYMBOLS AND ABBREVIATIONS

$+$	Plus Positive	$\angle$	Angle
		$\perp$	Perpendicular to
$-$	Minus Negative	$\parallel$	Parallel to
		$(\ )$	Parentheses
		$[\ ]$	Brackets
$\pm$	Plus or minus Positive or negative	$\{\ \}$	Braces
		$-$	Vinculum
$\mp$	Minus or plus Negative or positive	a°	a degrees (angle)
		a'	a minutes (angle) a prime
$\times$ or $\cdot$	Multiplied by	a''	a seconds (angle) a second a double prime
$\div$ or $:$	Divided by		
$=$, or $::$	Equals, as		
$\neq$	Does not equal	a'''	a third a triple prime
$\approx$	Equals approximately		
$>$	Greater than	a_n	a sub n
$<$	Less than	sin	Sine
$\geqq$	Greater than or equal to	cos	Cosine
$\leqq$	Less than or equal to	tan	Tangent
$\equiv$	Is identical to	cot	Cotangent
$\rightarrow$ or $\doteq$	Approaches as a limit	sec	Secant
$\propto$	Varies directly as	csc	Cosecant
$\therefore$	Therefore	vers	Versed sine
$\sqrt{\ }$	Square root	covers	Coversed sine
$\sqrt[n]{\ }$	nth root	exsec	Exsecant
a^n	nth power of a		
$n!$	$1 \cdot 2 \cdot 3 \cdots n$	$\sin^{-1} a$	Anti-sine a Angle whose sine is a Inverse sine a
log	Common logarithm Briggsian logarithm	sinh	Hyperbolic sine
		cosh	Hyperbolic cosine
		tanh	Hyperbolic tangent
ln or $\log_e$	Natural logarithm Hyberbolic logarithm Napierian logarithm	$\sinh^{-1} a$	Anti-hyperbolic sine a Angle whose hyperbolic sine is a
e or ϵ	Base (2.718) of natural system of logarithms		
π	Pi (3.1416)		
		$P(x,y)$	Rectangular coordinate of point P

$P(r,\theta)$	Polar coordinate of point P	$\dfrac{\partial z}{\partial x}$	Partial derivative of z with respect to x
$f(x)$, $F(x)$, or $\phi(x)$	Function of x		
		$\dfrac{\partial^2 z}{\partial x\,\partial y}$	Second partial derivative of z with respect to y and x
Δy	Increment of y		
Σ	Summation of		
∞	Infinity	$\displaystyle\int$	Integral of
dy	Differential of y		
$\dfrac{dy}{dx}$ or $f'(x)$	Derivative of $y = f(x)$ with respect to x	$\displaystyle\int_a^b$	Integral between the limits a and b
$\dfrac{d^2y}{dx^2}$ or $f''(x)$	Second derivative of $y = f(x)$ with respect to x	j	Imaginary quantity $(\sqrt{-1})$
		$\dot{x} = a + jb$	Symbolic vector notation
$\dfrac{d^n y}{dx^n}$ or $f^{(n)}(x)$	nth derivative of $y = f(x)$ with respect to x		

1.2 ALGEBRA

Powers and Roots (1)

$$a^n = a \cdot a \cdot a \cdots \text{to } n \text{ factors.} \qquad a^{-n} = \frac{1}{a^n}.$$

$$a^m \cdot a^n = a^{m+n}; \quad \frac{a^m}{a^n} = a^{m-n}. \qquad (ab)^n = a^n b^n; \qquad \left(\frac{a}{b}\right)^n = \frac{a^n}{b^n}.$$

$$(a^m)^n = (a^n)^m = a^{mn}. \qquad (\sqrt[n]{a})^n = a.$$

$$a^{1/n} = \sqrt[n]{a}; \qquad a^{m/n} = \sqrt[n]{a^m}. \qquad \sqrt[n]{ab} = \sqrt[n]{a}\sqrt[n]{b}; \qquad \sqrt[n]{\frac{a}{b}} = \frac{\sqrt[n]{a}}{\sqrt[n]{b}}.$$

$$\sqrt[n]{\sqrt[m]{a}} = \sqrt[mn]{a}.$$

Operations with Zero and Infinity (2)

$$a \cdot 0 = 0; \; a \cdot \infty = \infty; \; 0 \cdot \infty \text{ is indeterminate}$$

$$\frac{0}{a} = 0; \qquad \frac{a}{0} = \infty; \qquad \frac{0}{0} \text{ is indeterminate}$$

$$\frac{\infty}{a} = \infty; \qquad \frac{a}{\infty} = 0; \qquad \frac{\infty}{\infty} \text{ is indeterminate}$$

$$a^0 = 1; \qquad 0^a = 0; \qquad 0^0 \text{ is indeterminate}$$

$$\infty^a = \infty; \qquad\qquad\qquad \infty^0 \text{ is indeterminate}$$

see Section 1.15.

$a^\infty = \infty$, if $a^2 > 1$; $a^\infty = 0$, if $a^2 < 1$; $a^\infty = 1$, if $a^2 = 1$
$a^{-\infty} = 0$, if $a^2 > 1$; $a^{-\infty} = \infty$, if $a^2 < 1$; $a^{-\infty} = 1$, if $a^2 = \Big\}$ see Section 1.15.
$a - a = 0$; $\infty - a = \infty$; $\infty - \infty$ is indeterminate

Binomial Expansions (3)

$(a \pm b)^2 = a^2 \pm 2ab + b^2.$
$(a \pm b)^3 = a^3 \pm 3a^2b + 3ab^2 \pm b^3.$
$(a \pm b)^4 = a^4 \pm 4a^3b + 6a^2b^2 \pm 4ab^3 + b^4.$
$$(a \pm b)^n = a^n \pm \frac{n}{1} a^{n-1}b + \frac{n(n-1)}{1\cdot2} a^{n-2}b^2 \pm \frac{n(n-1)(n-2)}{1\cdot2\cdot3} a^{n-3}b^3 + \cdots$$

NOTE. n may be positive or negative, integral or fractional. When n is a positive integer, the series has $(n + 1)$ terms; otherwise, the number of terms is infinite.

Logarithms (4)

Definition.   If b is a finite positive number other than 1, and $b^x = N$, then x is the logarithm of N to the base b, or $\log_b N = x$. If $\log_b N = x$, then $b^x = N$.

Properties of Logarithms

$$\log_b b = 1; \quad \log_b 1 = 0; \quad \log_b 0 = \begin{cases} +\infty, \text{ when } b \text{ lies between 0 and 1,} \\ -\infty, \text{ when } b \text{ lies between 1 and } \infty. \end{cases}$$

$\log_b M \cdot N = \log_b M + \log_b N.$ $\qquad \log_b \dfrac{M}{N} = \log_b M - \log_b N.$

$\log_b N^p = p \log_b N.$ $\qquad \log_b \sqrt[r]{N^p} = \dfrac{p}{r} \log_b N.$

$\log_b N = \dfrac{\log_a N}{\log_a b}.$ $\qquad \log_b b^N = N; \; b^{\log_b N} = N.$

Systems of Logarithms

- *Common* (Briggsian): base 10
- *Natural* (Napierian or hyperbolic): base 2.7183 (designated by e or ϵ)

NOTE.   The abbreviation of *common logarithm* is *log* and the abbreviation of *natural logarithm* is *ln*.

Characteristic or Integral Part (c) of the Common Logarithm of a Number (N).   If N is not less than 1, c equals the number of integral figures in N, minus 1. If N is less than 1, c equals 9 minus the number of zeros between the decimal point and the first significant figure, minus 10 (the -10 being written after the mantissa).

Mantissa or Decimal Part (m) of the Common Logarithm of a Number N.   If N has not more than three figures, find the mantissa directly from your calculator or from a table of common logarithms. If N has four figures, $m = m_1 +$

$(f/10)(m_2 - m_1)$, where m_1 is the mantissa corresponding to the first three figures of N, m_2 is the next larger mantissa in the table, and f is the fourth figure of N.

Number (N) Corresponding to a Common Logarithm That Has a Characteristic (c) and a Mantissa (m). If N is desired to three figures, find the mantissa nearest to m and the corresponding number is N.

If N is desired to four figures, find the next smaller mantissa, m_1, and the next larger mantissa, m_2. The first three figures of N correspond to m_1 and the fourth figure equals the nearest whole number to $10[(m - m_1)/(m_2 - m_1)]$.

NOTE. If c is positive, the number of integral figures in N equals c plus 1. If c is negative (e.g., $9 - 10$, or -1), write numeric c minus 1 zeros between the decimal point and the first significant figure of N.

Natural Logarithm (ln) of a Number (N). Any number, N, can be written $N = N_1 \times 10^{\pm p}$, where N_1 lies between 1 and 1000. Then $\ln N = \ln N_1 \pm p \ln 10$. If N_1 has not more than three figures, find $\ln N_1$ directly from a table of natural logarithms. If N_1 has four figures, N_2 is the number composed of the first three figures of N_1, and f is the fourth figure of N_1, then

$$\ln N_1 = \ln N_2 + \frac{f}{10}[\ln(N_2 + 1) - \ln N_2].$$

Number (N) Corresponding to a Natural Logarithm, ln N. Any logarithm, $\ln N$, can be written $\ln N = \ln N_1 \pm p \ln 10$, where $\ln N_1$ lies between $4.6052 = \ln 100$ and $6.9078 = \ln 1000$. Then $N = N_1 \times 10^{\pm p}$. The first three figures of N_1 correspond to the next smaller logarithm, $\ln N_2$, and the fourth figure, f, of N_1 equals the nearest whole number to $10\{(\ln N_1 - \ln N_2)/[\ln(N_2 + 1) - \ln N_2]\}$.

Solution of Algebraic Equations (5)

The Quadratic Equation. If

$$ax^2 + bx + c = 0,$$

then

$$x = \frac{-b \pm \sqrt{b^2 - 4ac}}{2a} = \frac{2c}{-b \mp \sqrt{b^2 - 4ac}}.$$

If $b^2 - 4ac \begin{cases} > \\ = 0 \\ < \end{cases}$ $\begin{cases} \text{the roots are real and unequal,} \\ \text{the roots are real and equal,} \\ \text{the roots are imaginary.} \end{cases}$ The second equation serves best when the two values of x are nearly equal.

The Cubic Equation. Any cubic equation, $y^3 + py^2 + qy + r = 0$ may be reduced to the form $x^3 + ax + b = 0$ by substituting for y the value $[x - (p/3)]$. Here $a = \frac{1}{3}(3q - p^2)$, $b = \frac{1}{27}(2p^3 - 9pq + 27r)$.

Algebraic Solution of $x^3 + ax + b = 0$. Let

$$A = \sqrt[3]{-\frac{b}{2} + \sqrt{\frac{b^2}{4} + \frac{a^3}{27}}}, \qquad B = \sqrt[3]{-\frac{b}{2} - \sqrt{\frac{b^2}{4} + \frac{a^3}{27}}},$$

then

$$x = A + B, \quad -\frac{A+B}{2} + \frac{A-B}{2}\sqrt{-3}, \quad -\frac{A+B}{2} - \frac{A-B}{2}\sqrt{-3}.$$

If $\dfrac{b^2}{4} + \dfrac{a^3}{27} = 0$ $\begin{cases} > \\ = \\ < \end{cases}$ $\begin{cases} \text{1 real root, 2 conjugate imaginary roots,} \\ \text{3 real roots of which 2 are equal,} \\ \text{3 real and unequal roots.} \end{cases}$

Trigonometric Solution of $x^3 + ax + b = 0$. In the case where $(b^2/4) + (a^3/27) < 0$, the formulas above give the roots in a form impractical for numerical computation. In this case, a is negative. Compute the value of the angle ϕ from $\cos \phi = \sqrt{(b^2/4) \div (-a^3/27)}$, then

$$x = \mp 2 \sqrt{-\frac{a}{3}} \cos \frac{\phi}{3}, \quad \pm 2 \sqrt{-\frac{a}{3}} \cos\left(\frac{\phi}{3} + 120°\right), \quad \mp 2 \sqrt{-\frac{a}{3}} \cos\left(\frac{\phi}{3} + 240°\right),$$

where the upper or lower signs are to be used according as b is positive or negative.

In the case where $(b^2/4) + (a^3/27) > 0$, compute the values of the angles ψ and ϕ from $\cot 2\psi = [(b^2/4) \div (a^3/27)]^{1/2}$, $\tan \phi = (\tan \psi)^{1/3}$; then the real root of the equation is

$$x = \pm 2 \sqrt{\frac{a}{3}} \cot 2\phi,$$

where the upper or lower sign is to be used according as b is positive or negative.

In the case where $(b^2/4) + (a^3/27) = 0$, the roots are

$$x = \mp 2 \sqrt{-\frac{a}{3}}, \quad \pm \sqrt{-\frac{a}{3}}, \quad \pm \sqrt{-\frac{a}{3}},$$

where the upper or lower signs are to be used according as b is positive or negative.

Graphical Solution of the Cubic Equations. To find the real roots of the cubic equation

$$x^3 + ax + b = 0,$$

draw the parabola (Section 1.5) $y^2 = 2x$, and the circle (Section 1.5) the coordinates of whose center are $x = (4 - a)/4$, $y = -b/8$ and which passes through the vertex of the parabola. Measure the ordinates of the points of intersection; these give the real roots of the equation.

The Binomial Equation. If $x^n = a$, the n roots of this equation are: (a) if a is *positive*,

$$x = \sqrt[n]{a}\left(\cos\frac{2k\pi}{n} + \sqrt{-1}\sin\frac{2k\pi}{n}\right);$$

(b) if a is *negative*,

$$x = \sqrt[n]{-a}\left[\cos\frac{(2k+1)\pi}{n} + \sqrt{-1}\sin\frac{(2k+1)\pi}{n}\right],$$

where k takes in succession the values $0, 1, 2, 3, \ldots, n-1$.

1.3   TRIGONOMETRY

Definition of Angle (6)

An angle is the amount of rotation (in a fixed plane) by which a straight line may be changed from one direction to any other direction. If the rotation is counterclockwise the angle is said to be positive, if clockwise, negative.

Measure of Angle (7)

A **degree** is $\frac{1}{360}$ of the plane angle about a point. A **radian** is the angle subtended at the center of a circle by an arc equal in length to the radius.

Trigonometric Functions of an Angle (8)

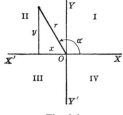

Fig. 1.1

sine (sin) α $= \dfrac{y}{r}$.

cosine (cos) α $= \dfrac{x}{r}$.

tangent (tan) α $= \dfrac{y}{x}$.

cotangent (cot) α $= \dfrac{x}{y}$.

secant (sec) α $= \dfrac{r}{x}$.

cosecant (csc) α $= \dfrac{r}{y}$.

exsecant (exsec) α $= \sec\alpha - 1$
versine (vers) α $= 1 - \cos\alpha$.
coversine (covers) $\alpha = 1 - \sin\alpha$.

NOTE. *x* is positive when measured along *OX* and negative along *OX'*; *y* is positive when measured parallel to *OY* and negative parallel to *OY'*.

Signs of the Functions (9)

Quadrant	sin	cos	tan	cot	sec	csc
I	+	+	+	+	+	+
II	+	−	−	−	−	+
III	−	−	+	+	−	−
IV	−	+	−	−	+	−

Functions of 0°, 30°, 45°, 60°, 90°, 180°, 270°, 360° (10)

	0°	30°	45°	60°	90°	180°	270°	360°
sin	0	$\dfrac{1}{2}$	$\dfrac{\sqrt{2}}{2}$	$\dfrac{\sqrt{3}}{2}$	1	0	-1	0
cos	1	$\dfrac{\sqrt{3}}{2}$	$\dfrac{\sqrt{2}}{2}$	$\dfrac{1}{2}$	0	-1	0	1
tan	0	$\dfrac{\sqrt{3}}{3}$	1	$\sqrt{3}$	∞	0	∞	0
cot	∞	$\sqrt{3}$	1	$\dfrac{\sqrt{3}}{3}$	0	∞	0	∞
sec	1	$\dfrac{2\sqrt{3}}{3}$	$\sqrt{2}$	2	∞	-1	∞	1
csc	∞	2	$\sqrt{2}$	$\dfrac{2\sqrt{3}}{3}$	1	∞	-1	∞

Fundamental Relations Among the Functions (11)

$$\sin \alpha = \frac{1}{\csc \alpha}; \qquad \cos \alpha = \frac{1}{\sec \alpha}; \qquad \tan \alpha = \frac{1}{\cot \alpha} = \frac{\sin \alpha}{\cos \alpha}.$$

$$\csc \alpha = \frac{1}{\sin \alpha}; \qquad \sec \alpha = \frac{1}{\cos \alpha}; \qquad \cot \alpha = \frac{1}{\tan \alpha} = \frac{\cos \alpha}{\sin \alpha}.$$

$$\sin^2\alpha + \cos^2\alpha = 1; \qquad \sec^2\alpha - \tan^2\alpha = 1; \qquad \csc^2\alpha - \cot^2\alpha = 1.$$

Functions of Multiple Angles (12)

$$\sin 2\alpha = 2 \sin \alpha \cos \alpha;$$
$$\cos 2\alpha = 2 \cos^2\alpha - 1 = 1 - 2 \sin^2\alpha = \cos^2\alpha - \sin^2\alpha.$$
$$\sin 3\alpha = 3 \sin \alpha - 4 \sin^3\alpha;$$

$\cos 3\alpha = 4 \cos^3\alpha - 3 \cos \alpha.$
$\sin 4\alpha = 4 \sin \alpha \cos \alpha - 8 \sin^3\alpha \cos \alpha;$
$\cos 4\alpha = 8 \cos^4\alpha - 8 \cos^2\alpha + 1.$
$\sin n\alpha = 2 \sin(n - 1)\alpha \cos \alpha - \sin(n - 2)\alpha;$
$\cos n\alpha = 2 \cos(n - 1)\alpha \cos \alpha - \cos(n - 2)\alpha.$

Functions of Half-Angles (13)

$$\sin \frac{\alpha}{2} = \sqrt{\frac{1 - \cos \alpha}{2}}; \qquad \cos \frac{1}{2} \alpha = \sqrt{\frac{1 + \cos \alpha}{2}}.$$

$$\tan \frac{1}{2} \alpha = \frac{1 - \cos \alpha}{\sin \alpha} = \frac{\sin \alpha}{1 + \cos \alpha} = \sqrt{\frac{1 - \cos \alpha}{1 + \cos \alpha}}.$$

Powers of Functions (14)

$\sin^2\alpha = \frac{1}{2}(1 - \cos 2\alpha);$ $\cos^2\alpha = \frac{1}{2}(1 + \cos 2\alpha).$

$\sin^3\alpha = \frac{1}{4}(3 \sin \alpha - \sin 3\alpha);$ $\cos^3\alpha = \frac{1}{4}(\cos 3\alpha + 3 \cos \alpha).$

$\sin^4\alpha = \frac{1}{8}(\cos 4\alpha - 4 \cos 2\alpha + 3);$ $\cos^4\alpha = \frac{1}{8}(\cos 4\alpha + 4 \cos 2\alpha + 3).$

$$\sin^n\alpha = \frac{1}{(2\sqrt{-1})^n} \left(y - \frac{1}{y}\right)^n; \qquad \cos^n\alpha = \frac{1}{(2)^n} \left(y + \frac{1}{y}\right)^n.$$

Functions of Sum or Difference of Two Angles (15)

$\sin(\alpha \pm \beta) = \sin \alpha \cos \beta \pm \cos \alpha \sin \beta.$
$\cos(\alpha \pm \beta) = \cos \alpha \cos \beta \mp \sin \alpha \sin \beta.$

$$\tan(\alpha \pm \beta) = \frac{\tan \alpha \pm \tan \beta}{1 \mp \tan \alpha \tan \beta}.$$

Sums, Differences, and Products of Two Functions (16)

$\sin \alpha \pm \sin \beta = 2 \sin \frac{1}{2}(\alpha \pm \beta) \cos \frac{1}{2}(\alpha \mp \beta).$
$\cos \alpha + \cos \beta = 2 \cos \frac{1}{2}(\alpha + \beta) \cos \frac{1}{2}(\alpha - \beta).$
$\cos \alpha - \cos \beta = -2 \sin \frac{1}{2}(\alpha + \beta) \sin \frac{1}{2}(\alpha - \beta).$

$$\tan \alpha \pm \tan \beta = \frac{\sin(\alpha \pm \beta)}{\cos \alpha \cos \beta}.$$

$\sin^2\alpha - \sin^2\beta = \sin(\alpha + \beta) \sin(\alpha - \beta).$
$\cos^2\alpha - \cos^2\beta = -\sin(\alpha + \beta) \sin(\alpha - \beta).$
$\cos^2\alpha - \sin^2\beta = \cos(\alpha + \beta) \cos(\alpha - \beta).$

$\sin \alpha \sin \beta = \frac{1}{2} \cos(\alpha - \beta) - \frac{1}{2} \cos(\alpha + \beta).$
$\cos \alpha \cos \beta = \frac{1}{2} \cos(\alpha - \beta) + \frac{1}{2} \cos(\alpha + \beta).$
$\sin \alpha \cos \beta = \frac{1}{2} \sin(\alpha + \beta) + \frac{1}{2} \sin(\alpha - \beta).$

Equivalent Expressions for sin α, cos α, and tan α (17)

$$\sin \alpha = \sqrt{1 - \cos^2\alpha} = \frac{\tan \alpha}{\sqrt{1 + \tan^2\alpha}} = \frac{1}{\sqrt{1 + \cot^2\alpha}} = \frac{\sqrt{\sec^2\alpha - 1}}{\sec \alpha} = \frac{1}{\csc \alpha}$$

$$= \cos \alpha \tan \alpha = \frac{\cos \alpha}{\cot \alpha} = \frac{\tan \alpha}{\sec \alpha} = \frac{\sin 2\alpha}{2 \cos \alpha} = \sqrt{\frac{1}{2}(1 - \cos 2\alpha)}$$

$$= 2 \sin \frac{\alpha}{2} \cos \frac{\sigma}{2}.$$

$$\cos \alpha = \sqrt{1 - \sin^2\alpha} = \frac{1}{\sqrt{1 + \tan^2\alpha}} = \frac{\cot \alpha}{\sqrt{1 + \cot^2\alpha}} = \frac{1}{\sec \alpha} = \frac{\sqrt{\sec^2\alpha - 1}}{\csc \alpha}$$

$$= \sin \alpha \cot \alpha = \frac{\sin \alpha}{\tan \alpha} = \frac{\cot \alpha}{\csc \alpha} = \frac{\sin 2\alpha}{2 \sin \alpha} = \sqrt{\frac{1}{2}(1 + \cos 2\alpha)}$$

$$= \cos^2 \frac{\alpha}{2} - \sin^2 \frac{\alpha}{2} = 1 - 2 \sin^2 \frac{\alpha}{2} = 2 \cos^2 \frac{\alpha}{2} - 1.$$

$$\tan \alpha = \frac{\sin \alpha}{\sqrt{1 - \sin^2\alpha}} = \frac{\sqrt{1 - \cos^2\alpha}}{\cos \alpha} = \frac{1}{\cot \alpha} = \sqrt{\sec^2\alpha - 1} = \frac{1}{\sqrt{\csc^2\alpha - 1}}$$

$$= \frac{\sin \alpha}{\cos \alpha} = \frac{\sec \alpha}{\csc \alpha} = \frac{\sin 2\alpha}{1 + \cos 2\alpha} = \frac{1 - \cos 2\alpha}{\sin 2\alpha} = \frac{2 \tan(\alpha/2)}{1 - \tan^2(\alpha/2)}.$$

Inverse or Antifunctions (18)

$\text{Sin}^{-1}a$, is defined as the angle whose sine is a, and has an infinite number of values. If α is the value of $\sin^{-1}a$ that lies between -90 and $+90°$ ($-\pi/2$ and $+\pi/2$ radians), and if n is any integer,

$$\sin^{-1}a = (-1)^n\alpha + n \cdot 180° = (-1)^n\alpha + n\pi \text{ (similarly for csc}^{-1}a).$$

$\text{Cos}^{-1}a$, is defined as the angle whose cosine is a, and has an infinite number of values. If α is the value of $\cos^{-1}a$ that lies between 0 and $180°$ (0 and π radians), and if n is any integer,

$$\cos^{-1}a = \pm\alpha + n \cdot 360° = \pm\alpha + 2n\pi \text{ (similarly for sec}^{-1}a).$$

$\text{Tan}^{-1}a$, is defined as the angle whose tangent is a, and has an infinite number of values. If α is the value of $\tan^{-1}a$ that lies between 0 and $180°$ (0 and π radians), and if n is any integer,

$$\tan^{-1}a = \alpha + n \cdot 180° = \alpha + n\pi \text{ (similarly for cot}^{-1}a).$$

Some Relations Among Inverse Functions (19)

$$\sin^{-1}a = \cos^{-1}\sqrt{1 - a^2} = \tan^{-1}\frac{a}{\sqrt{1 - a^2}} = \cot^{-1}\frac{\sqrt{1 - a^2}}{a}$$

$$= \sec^{-1}\frac{1}{\sqrt{1 - a^2}} = \csc^{-1}\frac{1}{a}.$$

$$\cos^{-1}a = \sin^{-1}\sqrt{1 - a^2} = \tan^{-1}\frac{\sqrt{1 - a^2}}{a} = \cot^{-1}\frac{a}{\sqrt{1 - a^2}}$$

$$= \sec^{-1}\frac{1}{a} = \csc^{-1}\frac{1}{\sqrt{1 - a^2}}.$$

$$\tan^{-1}a = \sin^{-1}\frac{a}{\sqrt{1 + a^2}} = \cos^{-1}\frac{1}{\sqrt{1 + a^2}} = \cot^{-1}\frac{1}{a} = \sec^{-1}\sqrt{1 + a^2}$$

$$= \csc^{-1}\frac{\sqrt{1 + a^2}}{a}.$$

$$\cot^{-1}a = \tan^{-1}\frac{1}{a}; \sec^{-1}a = \cos^{-1}\frac{1}{a}; \csc^{-1}a = \sin^{-1}\frac{1}{a}.$$

$$\text{vers}^{-1}a = \cos^{-1}(1 - a); \text{covers}^{-1}a = \sin^{-1}(1 - a); \text{exsec}^{-1}a = \sec^{-1}(1 + a).$$

$$\sin^{-1}a \pm \sin^{-1}b = \sin^{-1}(a\sqrt{1 - b^2} \pm b\sqrt{1 - a^2}).$$

$$\cos^{-1}a \pm \cos^{-1}b = \cos^{-1}(ab \mp \sqrt{1 - a^2}\sqrt{1 - b^2}).$$

$$\tan^{-1}a \pm \tan^{-1}b = \tan^{-1}\frac{a \pm b}{1 \mp ab}.$$

$$\sin^{-1}a + \cos^{-1}a = 90°; \tan^{-1}a + \cot^{-1}a = 90°; \sec^{-1}a + \csc^{-1}a = 90°,$$
if $\sin^{-1}a, \tan^{-1}a, \csc^{-1}a$ lie between -90 and $+90°$
and $\cos^{-1}a, \cot^{-1}a, \sec^{-1}a$ lie between 0 and 180°.

Properties of Plane Triangles (20)

Notation: α, β, γ = angles; a, b, c = sides.
A = area; h_b = altitude on b; $s = \frac{1}{2}(a + b + c)$.
r = radius of inscribed circle; R = radius of circumscribed circle.
$\alpha + \beta + \gamma = 180° = \pi$ radians

$$\frac{a}{\sin \alpha} = \frac{b}{\sin \beta} = \frac{c}{\sin \gamma}.$$

$$\frac{a + b}{a - b} = \frac{\tan \frac{1}{2}(\alpha + \beta)}{\tan \frac{1}{2}(\alpha - \beta)}.$$

$$a^2 = b^2 + c^2 - 2bc \cos \alpha, \quad a = b \cos \gamma + c \cos \beta.$$

$$\cos \alpha = \frac{b^2 + c^2 - a^2}{2bc}, \quad \sin \alpha = \frac{2}{bc}\sqrt{s(s - a)(s - b)(s - c)}.$$

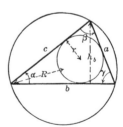

Fig. 1.2

$$\sin\frac{\alpha}{2} = \sqrt{\frac{(s-b)(s-c)}{bc}} \, , \quad \cos\frac{\alpha}{2} = \sqrt{\frac{s(s-a)}{bc}} \, ,$$

$$\tan\frac{\alpha}{2} = \sqrt{\frac{(s-b)(s-c)}{s(s-a)}} = \frac{r}{s-a} \, .$$

$$h_b = c \sin\alpha = a \sin\gamma = \frac{2}{b} \sqrt{s(s-a)(s-b)(s-c)}.$$

$$r = \sqrt{\frac{(s-a)(s-b)(s-c)}{s}} = (s-a) \tan\frac{\alpha}{2} \, .$$

$$R = \frac{a}{2 \sin\alpha} = \frac{abc}{4A} \, .$$

$$A = \frac{1}{2} bh_b = \frac{1}{2} ab \sin\gamma = \frac{a^2 \sin\beta \sin\gamma}{2 \sin\alpha} = \sqrt{s(s-a)(s-b)(s-c)} = rs.$$

Solution of the Right Triangle (21)

Given any two sides, or one side and any acute angle α, to find the remaining parts.

$$\sin\alpha = \frac{a}{c}, \, \cos\alpha = \frac{b}{c}, \, \tan\alpha = \frac{a}{b}, \, \beta = 90° - \alpha.$$

$$a = \sqrt{(c+b)(c-b)} = c \sin\alpha = b \tan\alpha.$$

$$b = \sqrt{(c+a)(c-a)} = c \cos\alpha = \frac{a}{\tan\alpha} \, .$$

$$c = \frac{a}{\sin\alpha} = \frac{b}{\cos\alpha} = \sqrt{a^2 + b^2}.$$

$$A = \frac{1}{2} ab = \frac{a^2}{2 \tan\alpha} = \frac{b^2 \tan\alpha}{2} = \frac{c^2 \sin 2\alpha}{4} \, .$$

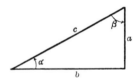

Fig. 1.3

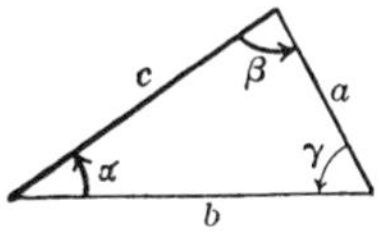

Fig. 1.4

Solution of Oblique Triangles (22)

CASE I.   Given any two angles α and β and any side c,

$$\gamma = 180° - (\alpha + \beta); \quad a = \frac{c \sin\alpha}{\sin\gamma} \, ; \quad b = \frac{c \sin\beta}{\sin\gamma} \, .$$

CASE II.   Given any two sides a and c and an angle opposite one of these, say α,

$$\sin\gamma = \frac{c \sin\alpha}{a} \, , \quad \beta = 180° - (\alpha + \gamma), \quad b = \frac{a \sin\beta}{\sin\alpha} \, .$$

NOTE. γ may have two values, $\gamma_1 < 90°$ and $\gamma_2 = 180° - \gamma_1 > 90°$. If $\alpha + \gamma_2 > 180°$, use only γ_1.

CASE III. Given any two sides b and c and their included angle α. Use any one of the following sets of formulas:

1. $\dfrac{1}{2}(\beta + \gamma) = 90° - \dfrac{1}{2}\alpha;$ $\tan\dfrac{1}{2}(\beta - \gamma) = \dfrac{b - c}{b + c}\tan\dfrac{1}{2}(\beta + \gamma);$

$\beta = \dfrac{1}{2}(\beta + \gamma) + \dfrac{1}{2}(\beta - \gamma);$ $\gamma = \dfrac{1}{2}(\beta + \gamma) - \dfrac{1}{2}(\beta - \gamma);$ $a = \dfrac{b\sin\alpha}{\sin\beta}.$

2. $a = \sqrt{b^2 + c^2 - 2bc\cos\alpha};$ $\sin\beta = \dfrac{b\sin\alpha}{a};$ $\gamma = 180° - (\alpha + \beta).$

3. $\tan\gamma = \dfrac{c\sin\alpha}{b - c\cos\alpha};$ $\beta = 180° - (\alpha + \gamma);$ $a = \dfrac{c\sin\alpha}{\sin\gamma}.$

CASE IV. Given the three sides a, b, and c. Use either of the following sets of formulas.

1. $s = \dfrac{1}{2}(a + b + c);$ $r = \sqrt{\dfrac{(s - a)(s - b)(s - c)}{s}}.$

$\tan\dfrac{1}{2}\alpha = \dfrac{r}{s - a};$ $\tan\dfrac{1}{2}\beta = \dfrac{r}{s - b};$ $\tan\dfrac{1}{2}\gamma = \dfrac{r}{s - c}.$

2. $\cos\alpha = \dfrac{b^2 + c^2 - a^2}{2bc};$ $\cos\beta = \dfrac{c^2 + a^2 - b^2}{2ca};$ $\gamma = 180° - (\alpha + \beta).$

1.4   MENSURATION: LENGTHS, AREAS, VOLUMES

Notation: a, b, c, d, s denote lengths; A denotes area; V denotes volume.

Right Triangle (23) (Fig. 1.5)

$A = \frac{1}{2}ab.$ [For other formulas, see (21).]
$c = \sqrt{a^2 + b^2}, \quad a = \sqrt{c^2 - b^2}, \quad b = \sqrt{c^2 - a^2}.$

Fig. 1.5

Oblique Triangle (24) (Fig. 1.6)

$A = \frac{1}{2}bh.$ [For other formulas, see (22).]

Fig. 1.6

Equilateral Triangle (25) (Fig. 1.7)

$A = \dfrac{1}{2}ah = \dfrac{1}{4}a^2\sqrt{3}, \quad r_1 = \dfrac{a}{2\sqrt{3}}$

$h = \dfrac{1}{2}a\sqrt{3}; \quad r_2 = \dfrac{a}{\sqrt{3}}$

Fig. 1.7

Fig. 1.8

Fig. 1.9

Fig. 1.10

Square (26) (Fig. 1.8)

$A = a^2; d = a\sqrt{2}$.

Rectangle (27) (Fig. 1.9)

$A = ab; d = \sqrt{a^2 + b^2}$.

Parallelogram (Opposite Sides Parallel) (28) (Fig. 1.10)

$A = ah = ab \sin \alpha$.
$d_1 = \sqrt{a^2 + b^2 - 2ab \cos \alpha}$;
$d_2 = \sqrt{a^2 + b^2 + 2ab \cos \alpha}$.

Trapezoid (One Pair of Opposite Sides Parallel) (29) (Fig. 1.11)

$A = \frac{1}{2}h(a + b)$.

Trapezium (No Sides Parallel) (30) (Fig. 1.12)

$A = \frac{1}{2}(ah_1 + bh_2) = $ sum of areas of two triangles.

Regular Polygon of n Sides (All Sides Equal, All Angles Equal) (31) (Fig. 1.13)

$\beta = \frac{n - 2}{n} 180° = \frac{n - 2}{n} \pi$ radians.

$\alpha = \frac{360°}{n} = \frac{2\pi}{n}$ radians.

Fig. 1.11

Fig. 1.12

Fig. 1.13

Characteristics of Regular Polygons

n	a	r	R	A
3	$2r\sqrt{3} = R\sqrt{3}$	$\frac{1}{6}a\sqrt{3}$	$\frac{1}{3}a\sqrt{3}$	$\frac{1}{4}a^2\sqrt{3} = 3r^2\sqrt{3} = \frac{3}{4}R^2\sqrt{3}$
4	$2r = R\sqrt{2}$	$\frac{1}{2}a$	$\frac{1}{2}a\sqrt{2}$	$a^2 = 4r^2 = 2R^2$
6	$\frac{2}{3}r\sqrt{3} = R$	$\frac{1}{2}a\sqrt{3}$	a	$\frac{3}{2}a^2\sqrt{3} = 2r^2\sqrt{3} = \frac{3}{2}R^2\sqrt{3}$
8	$2r(\sqrt{2}-1)$ $= R\sqrt{2-\sqrt{2}}$	$\frac{1}{2}a(\sqrt{2}+1)$	$\frac{1}{2}a\sqrt{4+2\sqrt{2}}$	$2a^2(\sqrt{2}+1) = 8r^2(\sqrt{2}-1)$ $= 2R^2\sqrt{2}$
n	$2r\tan\frac{\alpha}{2} = 2R\sin\frac{\alpha}{2}$	$\frac{a}{2}\cot\frac{\alpha}{2}$	$\frac{a}{2}\csc\frac{\alpha}{2}$	$\frac{na^2}{4}\cot\frac{\alpha}{2} = nr^2\tan\frac{\alpha}{2}$ $= \frac{nR^2}{2}\sin\alpha$

Circle (32)

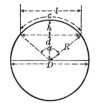

Fig. 1.14

Notation: C = circumference, α = central angle in radians.

$C = \pi D = 2\pi R.$

$c = R\alpha = \dfrac{1}{2} D\alpha = D \cos^{-1} \dfrac{d}{R} = D \tan^{-1} \dfrac{1}{2d}.$

$l = 2\sqrt{R^2 - d^2} = 2R \sin \dfrac{\alpha}{2} = 2d \tan \dfrac{\alpha}{2} = 2d \tan \dfrac{c}{D}.$

$d = \dfrac{1}{2}\sqrt{4R^2 - l^2} = \dfrac{1}{2}\sqrt{D^2 - l^2} = R \cos \dfrac{\alpha}{2} = \dfrac{1}{2} l \cot \dfrac{\alpha}{2} = \dfrac{1}{2} l \cot \dfrac{c}{D}.$

$h = R - d.$

$\alpha = \dfrac{c}{R} = \dfrac{2c}{D} = 2 \cos^{-1} \dfrac{d}{R} = 2 \tan^{-1} \dfrac{l}{2d} = 2 \sin^{-1} \dfrac{l}{D}.$

$A(\text{circle}) = \pi R^2 = \frac{1}{4}\pi D^2 = \frac{1}{2} RC = \frac{1}{4} DC.$

$A(\text{sector}) = \frac{1}{2} Rc = \frac{1}{2} R^2\alpha = \frac{1}{8} D^2\alpha.$

$A(\text{segment}) = A(\text{sector}) - A(\text{triangle}) = \dfrac{1}{2} R^2(\alpha - \sin \alpha) = \dfrac{1}{2} R \left(c - R \sin \dfrac{c}{R} \right)$

$\qquad = R^2 \sin^{-1} \dfrac{l}{2R} - \dfrac{1}{4} l\sqrt{4R^2 - l^2} = R^2 \cos^{-1} \dfrac{d}{R} - d\sqrt{R^2 - d^2}$

$\qquad = R^2 \cos^{-1} \dfrac{R - h}{R} - (R - h)\sqrt{2Rh - h^2}.$

Ellipse* (33)

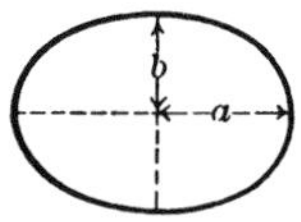

Fig. 1.15

$$A = \pi ab.$$

Perimeter (s)

$= \pi(a + b) \left[1 + \dfrac{1}{4}\left(\dfrac{a - b}{a + b}\right)^2 + \dfrac{1}{64}\left(\dfrac{a - b}{a + b}\right)^4 + \dfrac{1}{256}\left(\dfrac{a - b}{a + b}\right)^6 + \cdots \right]$

$\approx \pi \dfrac{a + b}{4} \left[3(1 + \lambda) + \dfrac{1}{1 - \lambda} \right], \lambda = \left[\dfrac{a - b}{2(a + b)}\right]^2$

Parabola* (34)

$$A = \tfrac{2}{3} ld.$$

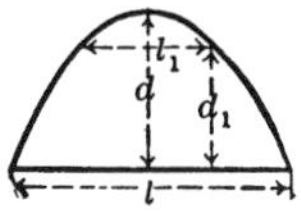

Length of arc (s) $= \dfrac{1}{2}\sqrt{16d^2 + l^2} + \dfrac{l^2}{8d} \ln\left(\dfrac{4d + \sqrt{16d^2 + l^2}}{l}\right)$

$\qquad = l\left[1 + \dfrac{2}{3}\left(\dfrac{2d}{l}\right)^2 - \dfrac{2}{5}\left(\dfrac{2d}{l}\right)^4 + \cdots \right].$

Fig. 1.16

Height of segment $(d_1) = \dfrac{d}{l^2}(l^2 - l_1^2).$

---

*For a definition and equation, see Section 1.5.

Width of segment $(l_1) = l \sqrt{\dfrac{d - d_1}{d}}$.

Cycloid* (35)

Notation: r = radius of generating circle.

Fig. 1.17

$A = 3\pi r^2$.
Length of arc $(s) = 8r$.

Catenary* (36)

Length of arc $(s) = 1\left[1 + \dfrac{2}{3}\left(\dfrac{2d}{l}\right)^2\right]$

Fig. 1.18

approximately, if d is small in comparison with l.

Area by Approximation (37) (Fig. 1.19)

Let y_0, y_1, y_2, . . . , y_n be the measured lengths of a series of equidistant parallel chords, and let h be their distance apart; then the area enclosed by any boundary is given approximately by one of the following rules.

- Trapezoidal rule

  $A_T = h[\frac{1}{2}(y_0 + y_n) + y_1 + y_2 + \cdots + y_{n-1}]$

- Durand's rule

  $A_D = h[0.4(y_0 + y_n) + 1.1(y_1 + y_{n-1}) + y_2 + y_3 + \cdots + y_{n-2}]$

- Simpson's rule, where n is even

  $A_s = \frac{1}{3}h[(y_0 + y_n) + 4(y_1 + y_3 + \cdots + y_{n-1}) + 2(y_2 + y_4 + \cdots + y_{n-2})]$

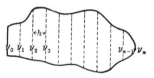

Fig. 1.19

Fig. 1.20

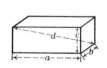

Fig. 1.21

*For a definition and equation, see Section 1.5.

The larger the value of n, the greater is the accuracy of approximation. In general, for the same number of chords, A_s gives the most accurate, A_T the least accurate approximation.

Cube (38) (Fig. 1.20)

$V = a^3$; $d = a\sqrt{3}$.
Total surface $= 6a^2$.

Rectangular Parallelopiped (39) (Fig. 1.21)

$V = abc$; $d = \sqrt{a^2 + b^2 + c^2}$.
Total surface $= 2(ab + bc + ca)$.

Prism or Cylinder (40) (Fig. 1.22)

$V = $ (area of base) $\times$ (altitude).
Lateral area $= $ (perimeter of right section) $\times$ (lateral edge).

Pyramid or Cone (41) (Fig. 1.23)

$V = \frac{1}{3}$(area of base) $\times$ (altitude).
Lateral area of regular figure $= \frac{1}{2}$(perimeter of base) $\times$ (slant height).

Frustum of Pyramid or Cone (42) (Fig. 1.24)

$$V = \frac{1}{3}(A_1 + A_2 + \sqrt{A_1 \times A_2})h,$$

where A_1 and A_2 are areas of bases, and h is altitude.

Lateral area of regular figure $= \frac{1}{2}$(sum of perimeters of bases) $\times$ (slant height).

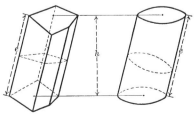

Fig. 1.22

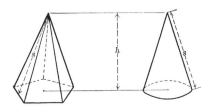

Fig. 1.23

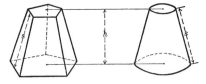

Fig. 1.24

Fig. 1.25

Prismatoid (Bases Are in Parallel Planes, Lateral Faces Are Triangles or Trapezoids) (43) (Fig. 1.25)

$V = \frac{1}{6}(A_1 + A_2 + 4A_m)h,$

where A_1 and A_2 are areas of bases, A_m is area of midsection, and h is altitude.

Sphere (44)

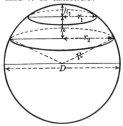

$$A(\text{sphere}) = 4\pi R^2 = \pi D^2.$$
$$A(\text{zone}) = 2\pi Rh = \pi Dh.$$
$$V(\text{sphere}) = \frac{4}{3}\pi R^3 = \frac{1}{6}\pi D^3.$$
$$V(\text{spherical sector}) = \frac{2}{3}\pi R^2 h = \frac{1}{6}\pi D^2 h.$$
$$V(\text{spherical segment of one base}) = \frac{1}{6}\pi h_1(3r_1^2 + h_1^2)$$
$$V(\text{spherical segment of two bases}) = \frac{1}{6}\pi h(3r_1^2 + 3r_2^2 + h^2).$$

Fig. 1.26

Ellipsoid (45)

$V = \frac{4}{3}\pi abc.$

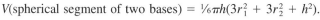

Fig. 1.27

Paraboloidal Segment (46)

$V(\text{segment of one base}) = \frac{1}{2}\pi r_1^2 h.$
$V(\text{segment of two bases}) = \frac{1}{2}\pi d(r_1^2 + r_2^2).$

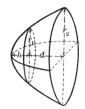

Torus (47) (Fig. 1.29)

Fig. 1.28

$V = 2\pi^2 Rr^2.$
Surface $(S) = 4\pi^2 Rr.$

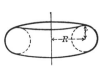

Fig. 1.29

Fig. 1.30

Solid (V) or Surface (S) of Revolution (48) (Fig. 1.30)

Generated by revolving any plane area (A) or arc (s) about an axis in its plane, and not crossing the area or arc.

$$V = 2\pi RA; \quad S = 2\pi Rs,$$

where R is the distance of center of gravity (G) of area or arc from axis.

1.5   ANALYTIC GEOMETRY

1.5.1   Plane

Rectangular (or Cartesian) Coordinates (49)

Let two perpendicular lines, $X'X$ (x-axis) and $Y'Y$ (y-axis) meet in a point O (origin). The position of any point $P(x,y)$ is fixed by the distances x (abscissa) and y (ordinate) from $Y'Y$ and $X'X$, respectively, to P (Fig. 1.31).

NOTE. x is + to the right and − to the left of $Y'Y$, y is + above and − below $X'X$.

Polar Coordinates (50)

Let O (origin or pole) be a point in the plane and OX (initial line) be any line through O. The position of any point $P(r,\theta)$ is fixed by the distance r (radius vector) from O to the point and the angle θ (vectorial angle) measured from OX to OP (Fig. 1.31).

NOTE. r is + measured along terminal side of θ, r is − measured along the opposite side of θ; θ is + measured counterclockwise, θ is − measured clockwise.

Relations Connecting Rectangular and Polar Coordinates (51)

$$x = r \cos \theta, \quad y = r \sin \theta.$$

$$r = \sqrt{x^2 + y^2}, \quad \theta = \tan^{-1} \frac{y}{x}, \quad \sin \theta = \frac{y}{\sqrt{x^2 + y^2}}, \quad \cos \theta = \frac{x}{\sqrt{x^2 + y^2}}, \quad \tan \theta = \frac{y}{x}.$$

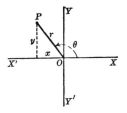

Fig. 1.31

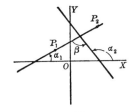

Fig. 1.32

Points and Slopes (52)

Let $P_1(x_1,y_1)$ and $P_2(x_2,y_2)$ be any two points, and let α be the angle from OX to P_1P_2, measured counterclockwise (Fig. 1.32).

$$P_1P_2 = d = \sqrt{(x_2 - x_1)^2 + (y_2 - y_1)^2}.$$

Midpoint of P_1P_2 is $\left(\dfrac{x_1 + x_2}{2}, \dfrac{y_1 + y_2}{2} \right).$

Point that divides P_1P_2 in the ratio $m_1{:}m_2$ is $\left(\dfrac{m_1x_2 + m_2x_1}{m_1 + m_2}, \dfrac{m_1y_2 + m_2y_1}{m_1 + m_2} \right).$

Slope of $P_1P_2 = \tan \alpha = m = \dfrac{y_2 - y_1}{x_2 - x_1}.$

Angle between two lines of slopes m_1 and m_2 is $\beta = \tan^{-1} \dfrac{m_2 - m_1}{1 + m_1m_2}.$

Two lines of slopes m_1 and m_2 are perpendicular if $m_2 = -\dfrac{1}{m_1}.$

Locus and Equation (53)

The collection of all points that satisfy a given condition is called the **locus** of that condition; the condition expressed by means of the variable coordinates of any point on the locus is called the **equation** of the locus.

The locus may be represented by equations of three kinds:

1. A **rectangular equation** involves the rectangular coordinates (x,y).
2. A **polar equation** involves the polar coordinates (r,θ).
3. **Parametric equations** express x and y or r and θ in terms of a third independent variable called a *parameter.*

The following equations are given in the system in which they are most simply expressed; sometimes, several forms of the equation in one or more systems are given.

Straight Line (54)

Figure 1.33:

$Ax + By + C = 0.$ $(-A \div B = $ slope.$)$
$y = mx + b.$ $(m = $ slope, $b = $ intercept on $OY.)$
$y - y_1 = m(x - x_1).$ $[m = $ slope, $P_1(x_1,y_1)$ is a known point on the line.$]$
$d = \dfrac{Ax_2 + By_2 + C}{\pm\sqrt{A^2 + B^2}}.$ $[d = $ distance from a point $P_2(x_2,y_2)$ to the line $Ax +$
$\qquad By + C = 0.]$

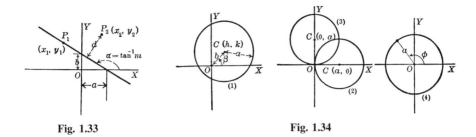

Fig. 1.33   Fig. 1.34

Circle (55)

Locus of a point at a constant distance (radius) from a fixed point C (center). [For mensuration of a circle, see (32).]

Circle (1) in Fig. 1.34:

$$(x - h)^2 + (y - k)^2 = a^2, \qquad C(h,k), \text{ rad.} = a.$$
$$r^2 + b^2 - 2br \cos (\theta - \beta) = a^2, \quad C(b,\beta), \text{ rad.} = a.$$

Circle (2) in Fig. 1.34:

$$x^2 + y^2 = 2ax, \quad C(a,0), \text{ rad.} = a.$$
$$r = 2a \cos \theta, \quad C(a,0), \text{ rad.} = a.$$

Circle (3) in Fig. 1.34:

$$x^2 + y^2 = 2ay, \quad C(0,a), \text{ rad.} = a.$$
$$r = 2a \sin \theta, \quad C\left(a, \frac{\pi}{2}\right), \text{ rad.} = a.$$

Circle (4) in Fig. 1.34:

$$x^2 + y^2 = a^2, \qquad C(0,0), \text{ rad.} = a.$$
$$r = a, \qquad C(0,0), \text{ rad.} = a.$$
$$x = a \cos \phi, \; y = a \sin \phi, \quad C(0,0), \text{ rad.} = a, \; \phi = \text{angle from } OX \text{ to radius.}$$

Conic (56)

Locus of a point whose distance from a fixed point (focus) is in a constant ratio, e (called **eccentricity**), to its distance from a fixed straight line (**directrix**) (Fig. 1.35).

$$x^2 + y^2 = e^2(d + x)^2. \; (d = \text{distance from focus to directrix.})$$
$$r = \frac{de}{1 - e \cos \theta}.$$

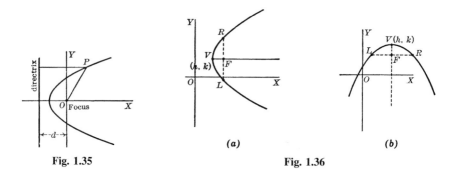

Fig. 1.35   Fig. 1.36

The conic is called a **parabola** when $e = 1$, an **ellipse** when $e < 1$, and a **hyperbola** when $e > 1$.

Parabola (57)

Conic where $e = 1$. [For mensuration of a parabola, see (34).]

Figure 1.36*a*:

$(y - k)^2 = a(x - h)$.   vertex (h,k), axis $\parallel OX$.
$y^2 = ax$,   vertex $(0,0)$, axis along OX.

Figure 1.36*b*:

$(x - h)^2 = a(y - k)$.   vertex (h,k), axis $\parallel OY$.
$x^2 = ay$,   vertex $(0,0)$, axis along OY.

Distance from vertex to focus $= VF = \frac{1}{4}a$.
Latus rectum $= LR = a$.

Ellipse (58)

Conic where $e < 1$. [For mensuration of an ellipse, see (33).]

$\dfrac{(x - h)^2}{a^2} + \dfrac{(y + k)^2}{b^2} = 1$,   center (h,k), axes $\parallel OX, OY$.

$\dfrac{x^2}{a^2} + \dfrac{y^2}{b^2} = 1$,   center $(0,0)$, axes along OX, OY.

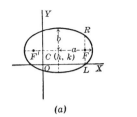
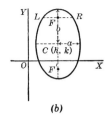

(a)
(b)

Fig. 1.37

	$a > b$ Fig. 1.37a	$b > a$ Fig. 1.37b
Major axis	$2a$	$2b$
Minor axis	$2b$	$2a$
Distance from center to either focus	$\sqrt{a^2 - b^2}$	$\sqrt{b^2 - a^2}$
Latus rectum	$\dfrac{2b^2}{a}$	$\dfrac{2a^2}{b}$
Eccentricity, e	$\dfrac{\sqrt{a^2 - b^2}}{a}$	$\dfrac{\sqrt{b^2 - a^2}}{b}$
Sum of distances of any point from the foci, $PF' + PF$	$2a$	$2b$

Hyperbola (59)

Conic where $e > 1$.

Figure 1.38a:

$$\frac{(x - h)^2}{a^2} - \frac{(y - k)^2}{b^2} = 1, \; C(h,k), \qquad \text{transverse axis} \parallel OX.$$

$$\frac{x^2}{a^2} - \frac{y^2}{b^2} = 1, \; C(0,0), \qquad \text{transverse axis along } OX.$$

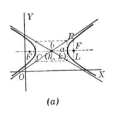
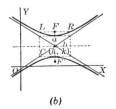
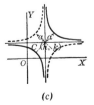

(a)
(b)
(c)

Fig. 1.38

Figure 1.38*b*:

$$\frac{(y - k)^2}{a^2} - \frac{(x - h)^2}{b^2} = 1, \ C(h,k), \qquad \text{transverse axis} \parallel OY.$$

$$\frac{y^2}{a^2} - \frac{x^2}{b^2} = 1, \ C(0,0), \qquad \text{transverse along } OY.$$

Transverse axis = 2*a*; conjugate axis = 2*b*.
Distance from center to either focus = $\sqrt{a^2 + b^2}$.

Latus rectum = $\dfrac{2b^2}{a}$.

Eccentricity, $e = \dfrac{\sqrt{a^2 + b^2}}{a}$.

Difference of distances of any point from the foci = 2*a*.

   Asymptotes are two lines through the center to which the branches of the hyperbola approach indefinitely near; their slopes are $\pm b/a$ (Fig. 1.38*a*) or $\pm a/b$ (Fig. 1.38*b*). Rectangular (equilateral) hyperbola, *b* = *a*. The asymptotes are perpendicular.

Figure 1.38*c*:

$$(x - h)(y - k) = \pm\frac{a^2}{2}, \qquad \text{center } (h,k), \text{ asymptotes} \parallel OX, OY.$$

$$xy = \pm\frac{a^2}{2}, \qquad \text{center } (0,0), \text{ asymptotes along } OX, OY.$$

where the + sign gives the smooth curve in Fig. 1.38*c* and where the − sign gives the dashed curve in Fig. 1.38*c*.

Sine Wave (60)

Figure 1.39:

$$y = a \sin(bx + c).$$

$$y = a \cos(bx + c') = a \sin(bx + c), \text{ where } c = c' + \frac{\pi}{2}.$$

$$y = m \sin bx + n \cos bx = a \sin(bx + c), \text{ where } a = \sqrt{m^2 + n^2}, c = \tan^{-1}\frac{n}{m}.$$

   The curve consists of a succession of waves, where

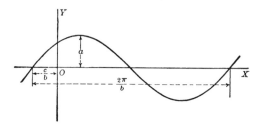

Fig. 1.39

a = amplitude = maximum height of wave.

$\dfrac{2\pi}{b}$ = wavelength = distance from any point on wave to the corresponding point on the next wave.

$x = -\dfrac{c}{b}$ (called the *phase*) marks a point on OX from which the positive half of the wave starts.

Tangent and Cotangent Curves (61)

Figure 1.40:

   1. $y = a \tan bx.$
   2. $y = a \cot bx.$

Secant and Cosecant Curves (62)

Figure 1.41:

   1. $y = a \sec bx.$
   2. $y = a \csc bx.$

Exponential or Logarithmic Curves (63)

Figure 1.42:

   1. $y = ab^x$ or $x = \log_b \dfrac{y}{a}$.

   2. $y = ab^{-x}$ or $x = -\log_b \dfrac{y}{a}$.

   3. $x = ab^y$ or $y = \log_b \dfrac{x}{a}$.

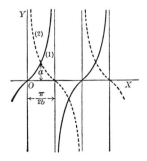

Fig. 1.40

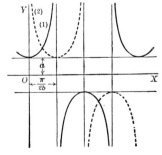

Fig. 1.41

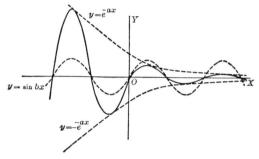

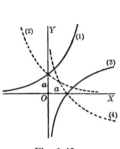

Fig. 1.42   Fig. 1.43

4. $x = ab^{-y}$ or $y = -\log_b \dfrac{x}{a}$.

The equations $y = ae^{\pm nx}$ and $x = ae^{\pm ny}$ are special cases of the above.

Oscillatory Wave of Decreasing Amplitude (64)

Figure 1.43:

$$y = e^{-ax} \sin bx.$$

NOTE.   The curve oscillates between $y = e^{-ax}$ and $y = -e^{-ax}$.

Catenary (65)

Curve made by a chain or cord of uniform weight suspended freely between two points at the same level (Fig. 1.44). [For mensuration of a catenary, see (36).]

$$y = \frac{a}{2} (e^{x/a} + e^{-x/a}).$$

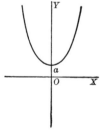

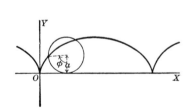

Fig. 1.44   Fig. 1.45

Cycloid (66)

Curve described by a point on a circle which rolls along a fixed straight line (Fig. 1.45).

$x = a(\phi - \sin \phi)$.
$y = a(1 - \cos \phi)$.

Epicycloid (67)

Curve described by a point on a circle that rolls along the outside of a fixed circle (Fig. 1.46).

$$x = (a + b) \cos \phi - b \cos\left(\frac{a + b}{b} \phi\right).$$

$$y = (a + b) \sin \phi - b \sin\left(\frac{a + b}{b} \phi\right).$$

Hypocycloid (68)

Curve described by a point on a circle that rolls along the inside of a fixed circle.

$$x = (a - b) \cos \phi + b \cos\left(\frac{a - b}{b} \phi\right).$$

$$y = (a - b) \sin \phi - b \sin\left(\frac{a - b}{b} \phi\right).$$

(For a hypocycloid of n cusps, radius of fixed circle $= nx$ radius of rolling circle; Fig. 1.47).

Involute of the Circle (69)

Curve described by the end of a string that is kept taut while being unwound from a circle (Fig. 1.48).

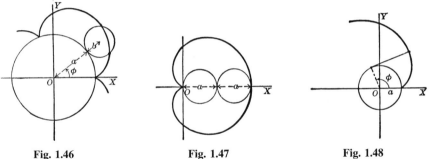

Fig. 1.46   Fig. 1.47   Fig. 1.48

$$x = a \cos \phi + a \phi \sin \phi.$$
$$y = a \sin \phi - a \phi \cos \phi.$$

1.5.2 Solid

Coordinates (70)

Let three mutually perpendicular planes, *XOY*, *YOZ*, *ZOX* (coordinate planes) meet in a point *O* (origin).

Rectangular System. The position of a point $P(x,y,z)$ in space is fixed by its three distances x, y, and z from the three coordinate planes.

Cylindrical System. The position of any point $P(r,\theta,z)$ is fixed by z, its distance from the *XOY* plane, and by (r,θ), the polar coordinates of the projection of *P* in the *XOY* plane.

Relations connecting rectangular and cylindrical coordinates are the same as those given in (51).

Points, Lines, and Planes (71)

The distance (d) between two points $P_1(x_1,y_1,z_1)$ and $P_2(x_2,y_2,z_2)$,

$$d = \sqrt{(x_2 - x_1)^2 + (y_2 - y_1)^2 + (z_2 - z_1)^2}.$$

The direction cosines of a line (cosines of the angles α, β, γ which the line or any parallel line makes with the coordinate axes) are related by

$$\cos^2\alpha + \cos^2\beta + \cos^2\gamma = 1.$$

If $\cos \alpha{:}\cos \beta{:}\cos \gamma = a{:}b{:}c$, then

$$\cos \alpha = \frac{a}{\sqrt{a^2 + b^2 + c^2}}, \quad \cos \beta = \frac{b}{\sqrt{a^2 + b^2 + c^2}}, \quad \cos \gamma = \frac{c}{\sqrt{a^2 + b^2 + c^2}}.$$

The direction cosines of the line joining $P_1(x_1,y_1,x_1)$ and $P_2(x_2,y_2,z_2)$,

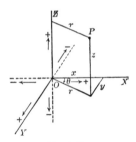

Fig. 1.49

$$\cos \alpha : \cos \beta : \cos \gamma = x_2 - x_1 : y_2 - y_1 : z_2 - z_1.$$

The angle (θ) between two lines, whose direction angles are α_1, β_1, γ_1 and α_2, β_2, γ_2,

$$\cos \theta = \cos \alpha_1 \cos \alpha_2 + \cos \beta_1 \cos \beta_2 + \cos \gamma_1 \cos \gamma_2.$$

The equation of a plane is of the first degree in x, y, and z,

$$Ax + By + Cz + D = 0,$$

where A, B, C are proportional to the direction cosines of a normal or perpendicular to the plane.

The angle between two planes is the angle between their normals.

The equations of a straight line are two equations of the first degree,

$$A_1x + B_1y + C_1z + D_1 = 0, \qquad A_2x + B_2y + C_2z + D_2 = 0.$$

The equations of a straight line through the point $P_1(x_1,y_1,z_1)$ with direction cosines proportional to a, b, and c,

$$\frac{x - x_1}{a} = \frac{y - y_1}{b} = \frac{z - z_1}{c}.$$

Cylindrical Surfaces (72)

The locus in space of an equation containing only two of the coordinates x, y, z is a cylindrical surface with its elements perpendicular to the plane of the two coordinates. Considered as a plane geometry equation, the equation represents the curve of intersection of the cylinder with the plane of the two coordinates.

Circular cylinders [For mensuration, see (40).]

Figure 1.50a:

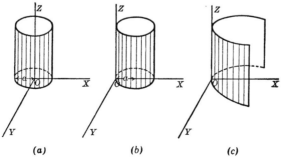

(a)   (b)   (c)

Fig. 1.50

$x^2 + y^2 = a^2.$
$r = a.$

Figure 1.50*b*:

$x^2 + y^2 = 2ax.$
$r = 2a \cos \theta.$

Figure 1.50*c*:

Parabolic cylinder $y^2 = ax.$

Surfaces of Revolution (73)

The equation of the surface of revolution obtained by revolving the plane curve $y = f(x)$ or $z = f(x)$ about OX,

$$y^2 + z^2 = [f(x)]^2.$$

Sphere (revolve circle $x^2 + y^2 = a^2$ about OX):

$$x^2 + y^2 + z^2 = a^2. \text{ [For mensuration of a sphere, see (44).]}$$

Spheroid [revolve ellipse $(x^2/a^2) + (y^2/b^2) = 1$ about OX]:

$$\frac{x^2}{a^2} + \frac{y^2 + z^2}{b^2} = 1 \text{ (prolate if } a > b, \text{ oblate if } b > a).$$

[For mensuration of an ellipsoid, see (45).]

Cone (revolve line $y = mx$ about OX):

$$y^2 + z^2 = m^2x^2. \text{ [For mensuration of a cone, see (41).]}$$

Paraboloid (revolve parabola $y^2 = ax$ about OX):

$$y^2 + z^2 = ax. \text{ [For mensuration of a paraboloid, see (46).]}$$

Space Curves (74)

A curve in space may be represented by two equations connecting the coordinates x, y, z of any point on the curve, or by three equations expressing the coordinates x, y, z in terms of a fourth variable or parameter.

Helix.   A curve generated by a point moving on a cylinder so that the distance traversed parallel to the axis of the cylinder is proportional to the angle of rotation about the axis (Fig. 1.51):

$$x = a \cos \theta, \quad y = a \sin \theta, \quad z = k\theta,$$

where a is the radius of cylinder and $2\pi k$ is the pitch.

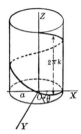

1.6 DIFFERENTIAL CALCULUS

Definition of Function: Notation (75)

Fig. 1.51

A variable y is said to be a function of another variable x, if when x is given, y is determined.

The symbols $f(x)$, $F(x)$, $\phi(x)$, and so on, represent various functions of x.

The symbol $f(a)$ represents the value of $f(x)$ when $x = a$.

Definition of Derivative: Notation (76)

Let $y = f(x)$. If Δx is any increment (increase or decrease) given to x, and Δy is the corresponding increment in y, then the derivative of y with respect to x is the limit of the ratio of Δy to Δx as Δx approaches zero; that is:

$$\frac{dy}{dx} = \lim_{\Delta x \to 0} \frac{\Delta y}{\Delta x} = \lim_{\Delta x \to 0} \frac{f(x + \Delta x) - f(x)}{\Delta x} = f'(x),$$

$$\frac{d^2 y}{dx^2} = \frac{d}{dx}\left(\frac{dy}{dx}\right) = \frac{d}{dx} f'(x) = f''(x), \quad \text{(second derivative)}$$

$$\frac{d^n y}{dx^n} = \frac{d}{dx}\left(\frac{d^{n-1} y}{dx^{n-1}}\right) = \frac{d}{dx} f^{(n-1)}(x) = f^{(n)}(x). \quad \text{(nth derivative)}$$

The symbols $f'(a)$, $f''(a)$, ..., $f^{(n)}(a)$ represent the values of $f'(x)$, $f''(x)$, ..., $f^{(n)}(x)$, respectively, when $x = a$.

Some Relations Among Derivatives (77)

If $x = f(y)$, then $\dfrac{dy}{dx} = 1 \div \dfrac{dx}{dy}$.

If $x = f(t)$, and $y = F(t)$, then $\dfrac{dy}{dx} = \dfrac{dy}{dt} \div \dfrac{dx}{dt}$.

If $y = f(u)$, and $u = F(x)$, then $\dfrac{dy}{dx} = \dfrac{dy}{du} \cdot \dfrac{du}{dx}$.

List of Derivatives (78)

Functions of x are represented by u and v; constants are represented by a, n, and e.

$$\frac{d}{dx}(x) = 1. \quad (79)$$

$$\frac{d}{dx}(a) = 0. \quad (80)$$

$$\frac{d}{dx}(u \pm v \pm \cdots) = \frac{du}{dx} \pm \frac{dv}{dx} \pm \cdots . \quad (81)$$

$$\frac{d}{dx}(au) = a\frac{du}{dx}. \quad (82)$$

$$\frac{d}{dx}(uv) = u\frac{dv}{dx} + v\frac{du}{dx}. \quad (83)$$

$$\frac{d}{dx}\left(\frac{u}{v}\right) = \frac{v\frac{du}{dx} - u\frac{dv}{dx}}{v^2}. \quad (84)$$

$$\frac{d}{dx}(u^n) = nu^{n-1}\frac{du}{dx}. \quad (85)$$

$$\frac{d}{dx}\log_a u = \frac{\log_a e}{u}\frac{du}{dx}. \quad (86)$$

$$\frac{d}{dx}\ln u = \frac{1}{u}\frac{du}{dx}. \quad (87)$$

$$\frac{d}{dx}a^u = a^u \ln a \frac{du}{dx}. \quad (88)$$

$$\frac{d}{dx}e^u = e^u \frac{du}{dx}. \quad (89)$$

$$\frac{d}{dx}u^v = vu^{v-1}\frac{du}{dx} + u^v \ln u \frac{dv}{dx}. \quad (90)$$

$$\frac{d}{dx}\sin u = \cos u \frac{du}{dx}. \quad (91)$$

$$\frac{d}{dx}\cos u = -\sin u \frac{du}{dx}. \quad (92)$$

$$\frac{d}{dx}\tan u = \sec^2 u \frac{du}{dx}. \quad (93)$$

$$\frac{d}{dx}\cot u = -\csc^2 u \frac{du}{dx}. \quad (94)$$

$$\frac{d}{dx}\sin^{-1} u = \frac{1}{\sqrt{1 - u^2}}\frac{du}{dx} \left(\text{where } \sin^{-1} u \text{ lies between } -\frac{\pi}{2} \text{ and } +\frac{\pi}{2}\right). \quad (95)$$

$$\frac{d}{dx}\cos^{-1} u = -\frac{1}{\sqrt{1 - u^2}}\frac{du}{dx} \,(\text{where } \cos^{-1} u \text{ lies between } 0 \text{ and } \pi). \quad (96)$$

$$\frac{d}{dx}\tan^{-1} u = \frac{1}{1 + u^2}\frac{du}{dx}. \quad (97)$$

$$\frac{d}{dx}\cot^{-1} u = -\frac{1}{1 + u^2}\frac{du}{dx}. \quad (98)$$

$$\frac{d}{dx}\sec^{-1} u = \frac{1}{u\sqrt{u^2 - 1}}\frac{du}{dx} \,(\text{where } \sec^{-1} u \text{ lies between } 0 \text{ and } \pi). \quad (99)$$

$$\frac{d}{dx}\csc^{-1} u = -\frac{1}{u\sqrt{u^2 - 1}}\frac{du}{dx} \,\left(\text{where } \csc^{-1} u \text{ lies between } -\frac{\pi}{2} \text{ and } +\frac{\pi}{2}\right). \quad (100)$$

$$\frac{d}{dx}\text{vers}^{-1} u = \frac{1}{\sqrt{2u - u^2}}\frac{du}{dx} \,(\text{where } \text{vers}^{-1} u \text{ lies between } 0 \text{ and } \pi). \quad (101)$$

The nth Derivative of Certain Functions (102)

$$\frac{d^n}{dx^n}e^{ax} = a^n e^{ax}. \quad (103)$$

$$\frac{d^n}{dx^n}a^x = (\ln a)^n a^x. \quad (104)$$

$$\frac{d^n}{dx^n} \ln x = \frac{(-1)^{n-1}\lfloor n-1}{x^n}, \quad \lfloor n-1 = 1 \cdot 2 \cdot 3 \cdots (n-1). \quad (105)$$

$$\frac{d^n}{dx^n} \sin ax = a^n \sin\left(ax + \frac{n\pi}{2}\right). \quad (106)$$

$$\frac{d^n}{dx^n} \cos ax = a^n \cos\left(ax + \frac{n\pi}{2}\right). \quad (107)$$

Slope of a Curve: Tangent and Normal (108)

The slope of the curve (slope of the tangent line to the curve) whose equation is $y = f(x)$ is

$$\text{slope} = m = \tan \phi = \frac{dy}{dx} = f'(x).$$

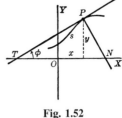

Fig. 1.52

The slope at $x = x_1$ is $m_1 = f'(x_1)$.
 The equation of the tangent line at $P_1(x_1, y_1)$ is

$$y - y_1 = m_1(x - x_1).$$

The equation of the normal at $P_1(x_1, y_1)$ is

$$y - y_1 = -\frac{1}{m_1}(x - x_1).$$

The angle (β) of intersection of two curves whose slopes at a common point are m_1 and m_2 is

$$\beta = \tan^{-1} \frac{m_2 - m_1}{1 + m_1 m_2}.$$

Derivative of Length of Arc: Radius of Curvature (109)

If s is the length of arc measured along the curve $y = f(x)$ from some fixed point to any point $P(x,y)$, and ϕ is the inclination of the tangent line at P to OX, then (Fig. 1.52)

$$\frac{dx}{ds} = \cos \phi = \frac{1}{\sqrt{1 + (dy/dx)^2}}, \quad \frac{dy}{ds} = \sin \phi = \frac{1}{\sqrt{1 + (dx/dy)^2}},$$

$$\left(\frac{dx}{ds}\right)^2 + \left(\frac{dy}{ds}\right)^2 = 1$$

The radius of curvature (ρ) at any point of the curve $y = f(x)$ or $r = f(\theta)$.

$$\rho = \frac{ds}{d\phi} = \frac{[1 + (dy/dx)^2]^{3/2}}{d^2y/dx^2} = \frac{\{1 + [f'(x)]^2\}^{3/2}}{f''(x)} = \frac{[1^2 + (dr/d\theta)^2]^{3/2}}{1^2 + 2(dr/d\theta)^2 - r(d^2r/d\theta^2)}.$$

ρ at $x = a$ is $\{1 + [f'(a)]^2\}^{3/2}/f''(a)$.

The curvature (k) at any point is $k = 1/\rho$.

Maximum and Minimum Values of a Function (110)

The maximum (minimum) value of a function $f(x)$ in an interval $x = a$ to $x = b$ is the value of the function that is larger (smaller) than the values of the function in its immediate vicinity. Thus in Fig. 1.53, the value of the function at M_1 and M_2 is a maximum, its value at m_1 and m_2 is a minimum.

Test for a maximum at $x = x_1$: $f'(x_1) = 0$ or ∞, and $f''(x_1) < 0$.
Test for a minimum at $x = x_1$: $f'(x_1) = 0$ or ∞, and $f''(x_1) > 0$.

If $f'(x_1) = 0$ or ∞, then for a maximum, $f'''(x_1) = 0$ or ∞, and $f^{(iv)}(x_1) < 0$, for a minimum, $f'''(x_1) = 0$ or ∞, and $f^{(iv)}(x_1) > 0$, and similarly if $f^{(iv)}(x_1) = 0$ or ∞, and so on.

In a practical problem which suggests that the function $f(x)$ has a maximum or has a minimum in an interval $x = a$ to $x = b$, merely equate $f'(x)$ to zero and solve for the required value of x. To find the largest or smallest values of a function $f(x)$ in an interval $x = a$ to $x = b$, also find the values $f(a)$ and $f(b)$, for (see Fig. 1.53 at L and S) these may be the largest and smallest values, although they are not maximum or minimum values.

Points of Inflection of a Curve (111)

Wherever $f''(x) < 0$, the curve is concave down. Wherever $f''(x) > 0$, the curve is concave up.

The curve is said to have a **point of inflection** at $x = x_1$ if $f''(x_1) = 0$ or ∞ and the curve is concave up on one side of $x = x_1$ and concave down on the other (see points I_1 and I_2 in Fig. 1.54).

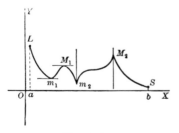

Fig. 1.53

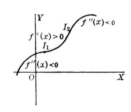

Fig. 1.54

Taylor's and Maclaurin's Theorems (112)

Any $f(x)$ may, in general, be expanded into a Taylor's series:

$$f(x) = f(a) + f'(a)\frac{x-a}{1} + f''(a)\frac{(x-a)^2}{2!} + f'''(a)\frac{(x-a)^3}{3!} + \cdots$$

$$+ f^{(n-1)}(a)\frac{(x-a)^{n-1}}{(n-1)!} + R_n,$$

where a is any quantity whatever so chosen that none of the expressions $f(a)$, $f'(a)$, $f''(a)$, . . . become infinite. If the series is to be used for the purpose of computing the approximate value of $f(x)$ for a given value of x, a should be chosen so that $(x - a)$ is numerically very small, and thus only a few terms of the series need be used. $R_n = f^{(n)}(x_1)[(x - a)^n / n!]$, where x_1 lies between a and x, is the remainder after n terms, and gives the limits between which the error lies in using n terms of the series for the value of the function. $n! = 1 \cdot 2 \cdot 3 \cdots n$.

If $a = 0$, the foregoing series becomes Maclaurin's series:

$$f(x) = f(0) + f'(0)\frac{x}{1} + f''(0)\frac{x^2}{2!} + f'''(0)\frac{x^3}{3!} + \cdots + f^{(n-1)}(0)\frac{x^{n-1}}{(n-1)!} + R_n.$$

This series may be used for purposes of computation when x is numerically very small.

Differential of a Function (113)

If $y = f(x)$ and Δx = increment in x, then the differential of x equals the increment of x, or $dx = \Delta x$; and the differential of y is the derivative of y multiplied by the differential of x, thus

$$dy = \frac{dy}{dx}dx = \frac{df(x)}{dx}dx = f'(x)\,dx \quad \text{and} \quad \frac{dy}{dx} = dy \div dx.$$

If $x = f_1(t)$ and $y = f_2(t)$, then $dx = f_1'(t)\,dt$, $dy = f_2'(t)\,dt$.

Every derivative formula has a corresponding differential formula; thus from the list of derivatives (78), we have, for example,

$$d(uv) = u\,dv + v\,du; \quad d(\sin u) = \cos u\,du; \quad d(\tan^{-1} u) = \frac{du}{1 + u^2},$$

and so on.

Functions of Several Variables: Partial Derivatives Differentials (114)

Let z be a function of two variables, $z = f(x, y)$, then its partial derivatives are

$$\frac{\partial z}{\partial x} = \frac{dz}{dx} \text{ when } y \text{ is kept constant.}$$

$$\frac{\partial z}{\partial y} = \frac{dz}{dy} \text{ when } x \text{ is kept constant.}$$

$$\frac{\partial^2 z}{\partial x^2} = \frac{\partial}{\partial x}\left(\frac{\partial z}{\partial x}\right) ; \frac{\partial^2 z}{\partial y^2} = \frac{\partial}{\partial y}\left(\frac{\partial z}{\partial y}\right) ; \frac{\partial^2 z}{\partial x\,\partial y} = \frac{\partial}{\partial x}\left(\frac{\partial z}{\partial y}\right) = \frac{\partial}{\partial y}\left(\frac{\partial z}{\partial x}\right) = \frac{\partial^2 z}{\partial y\,\partial x}.$$

Similarly, if $z = f(x, y, u, \ldots)$, then, for example,

$$\frac{\partial z}{\partial x} = \frac{dz}{dx} \qquad \text{when } y, u, \ldots \text{ are kept constant.}$$

If $z = f(x, y, \ldots)$ and $x, y, \ldots$ are functions of a single variable, t,

$$\frac{dz}{dt} = \frac{\partial z}{\partial x}\frac{dx}{dt} + \frac{\partial z}{\partial y}\frac{dy}{dt} + \cdots.$$

If $z = f(x,y,\ldots)$,

$$dz = \frac{\partial z}{\partial x}\,dx + \frac{\partial z}{\partial y}\,dy + \cdots.$$

If $F(x,y,z,\ldots) = 0$,

$$\frac{\partial F}{\partial x}\,dx + \frac{\partial F}{\partial y}\,dy + \frac{\partial F}{\partial z}\,dz + \cdots = 0.$$

If $f(x,y) = 0$,

$$\frac{dy}{dx} = -\frac{\partial f}{\partial x} \div \frac{\partial f}{\partial y}.$$

Maxima and Minima of Functions of Two Variables (115)

If $u = f(x,y)$, the values of x and y that make u a maximum or a minimum must satisfy the conditions

$$\frac{\partial u}{\partial x} = 0, \frac{\partial u}{\partial y} = 0, \left(\frac{\partial^2 u}{\partial x\,\partial y}\right)^2 < \left(\frac{\partial^2 u}{\partial x^2}\right)\left(\frac{\partial^2 u}{dy^2}\right).$$

A maximum (minimum) also requires both $\partial^2 u / \partial x^2$ and $\partial^2 u / \partial y^2$ to be negative (positive).

Space Curves: Surfaces (116) (see Section 1.5)

Let $x = f_1(t)$, $y = f_2(t)$, $z = f_3(t)$ be the equations of any space curve. The direction cosines of the tangent line to the curve at any point are proportional to dx, dy, and dz, or to dx/dt, dy/dt, and dz/dt.

Equations of tangent line at a point (x_1, y_1, z_1) are

$$\frac{x - x_1}{(dx)_1} = \frac{y - y_1}{(dy)_1} = \frac{z - z_1}{(dz)_1},$$

where $(dx)_1$ = value of dx at (x_1, y_1, z_1), and so on.

Angle between two space curves is the angle between their tangent lines (see Section 1.5).

Let $F(x, y, z) = 0$ be the equation of a surface.

Direction cosines of the normal to the surface at any point are proportional to $\partial F/\partial x$, $\partial F/\partial y$, and $\partial F/\partial z$.

Equations of the normal at any point (x_1, y_1, z_1) are

$$\frac{x - x_1}{(\partial F/\partial x)_1} = \frac{y - y_1}{(\partial F/\partial y)_1} = \frac{z - z_1}{(\partial F/\partial z)_1}.$$

Equation of the tangent plane at any point (x_1, y_1, z_1) is

$$(x - x_1)\left(\frac{\partial F}{\partial x}\right)_1 + (y - y_1)\left(\frac{\partial F}{\partial y}\right)_1 + (z - z_1)\left(\frac{\partial F}{\partial z}\right)_1 = 0,$$

where $(\partial F/\partial x)_1$ is the value of $(\partial F/\partial x)$ at the point (x_1, y_1, z_1), and so on.

Angle between two surfaces is the angle between their normals.

1.7 INTEGRAL CALCULUS

Definition of Integral (117)

$F(x)$ is said to be the integral of $f(x)$ if the derivative of $F(x)$ is $f(x)$, or the differential of $F(x)$ is $f(x)\, dx$; in symbols:

$$F(x) = \int f(x)\, dx \text{ if } \frac{dF(x)}{dx} = f(x), \text{ or } dF(x) = f(x)\, dx.$$

In general: $\int f(x)\, dx = F(x) + C$, where C is an arbitrary constant.

Fundamental Theorems on Integrals (118)

$$\int df(x) = f(x) + C.$$

$$d \int f(x)\, dx = f(x)\, dx.$$

$$\int [f_1(x) \pm f_2(x) \pm \cdots]\, dx = \int f_1(x)\, dx \pm \int f_2(x)\, dx \pm \cdots.$$

$$\int af(x)\, dx = a \int f(x)\, dx, \text{ where } a \text{ is any constant.}$$

$$\int u^n\, du = \frac{u^{n+1}}{n+1} + C \ (n \neq -1); \ u \text{ is any function of } x.$$

$$\int \frac{du}{u} = \ln u + C; \ u \text{ is any function of } x.$$

$$\int u\, dv = uv - \int v\, du; \ u \text{ and } v \text{ are any functions of } x.$$

1.7.1 Selected Integrals

In the following list, the constant of integration (c) is omitted but should be added to the result of every integration. The letter x represents any variable; the letter u represents any function of x; all other letters represent constants which may have any finite value unless otherwise indicated; $\ln = \log_e$; all angles are in radians.

Functions Containing $ax + b$ (119)

$$\int (ax + b)^n\, dx = \frac{1}{a(n+1)} (ax + b)^{n+1}.\ (n \neq -1)\ (120)$$

$$\int \frac{dx}{ax + b} = \frac{1}{a} \ln (ax + b).\ (121)$$

$$\int x(ax + b)^n\, dx = \frac{1}{a^2(n+2)} (ax + b)^{n+2} - \frac{b}{a^2(n+1)} (ax + b)^{n+1}.\ (122)$$
$$(n \neq -1, -2)$$

$$\int \frac{x\, dx}{ax + b} = \frac{x}{a} - \frac{b}{a^2} \ln (ax + b).\ (123)$$

$$\int \frac{x\, dx}{(ax + b)^2} = \frac{b}{a^2(ax + b)} + \frac{1}{a^2} \ln (ax + b).\ (124)$$

$$\int \frac{dx}{x(ax + b)} = \frac{1}{b} \ln \frac{x}{ax + b}.\ (125)$$

$$\int \frac{dx}{x^2(ax + b)} = -\frac{1}{bx} + \frac{a}{b^2} \ln \frac{ax + b}{x}.\ (126)$$

$$\int \frac{dx}{x(ax + b)^2} = \frac{1}{b(ax + b)} - \frac{1}{b^2} \ln \frac{ax + b}{x}.\ (127)$$

$$\int \frac{dx}{x^2(ax + b)^2} = -\frac{b + 2ax}{b^2x(ax + b)} + \frac{2a}{b^3} \ln \frac{ax + b}{x}.\ (128)$$

$$\int \frac{dx}{x\sqrt{ax + b}} = \frac{1}{\sqrt{b}} \ln \frac{\sqrt{ax + b} - \sqrt{b}}{\sqrt{ax + b} + \sqrt{b}}.\ (b \text{ pos.})\ (129)$$

$$\int \frac{dx}{x\sqrt{ax + b}} = \frac{2}{\sqrt{-b}} \tan^{-1} \sqrt{\frac{ax + b}{-b}}.\ (b \text{ neg.})\ (130)$$

$$\int \frac{\sqrt{ax+b}}{x}\,dx = 2\sqrt{ax+b} + \sqrt{b}\,\ln\frac{\sqrt{ax+b}-\sqrt{b}}{\sqrt{ax+b}+\sqrt{b}}. \quad (b \text{ pos.}) \ (131)$$

$$\int \frac{\sqrt{ax+b}}{x}\,dx = 2\sqrt{ax+b} - 2\sqrt{-b}\,\tan^{-1}\sqrt{\frac{ax+b}{-b}}. \quad (b \text{ neg.}) \ (132)$$

$$\int \frac{px+q}{\sqrt{ax+b}}\,dx = \frac{2}{3a^2}\,(3aq - 2bp + apx)\sqrt{ax+b}. \ (133)$$

$$\int \frac{\sqrt{ax+b}}{px+q}\,dx = \frac{2\sqrt{ax+b}}{p} - \frac{2}{p}\sqrt{\frac{aq-bp}{p}}\,\tan^{-1}\sqrt{\frac{p(ax+b)}{aq-bp}}. \ (134)$$
$$(p \text{ pos.}, \ aq > bp)$$

$$\int \frac{\sqrt{ax+b}}{px+q}\,dx = \frac{2\sqrt{ax+b}}{p}$$
$$+ \frac{1}{p}\sqrt{\frac{bp-aq}{p}}\,\ln\frac{\sqrt{p(ax+b)}-\sqrt{bp-aq}}{\sqrt{p(ax+b)}+\sqrt{bp-aq}}. \ (135)$$
$$(p \text{ pos.}, \ bp > aq)$$

Functions Containing $ax^2 + b$ (136)

$$\int \frac{dx}{ax^2+b} = \frac{1}{\sqrt{ab}}\,\tan^{-1}\left(x\sqrt{\frac{a}{b}}\right). \quad (a \text{ and } b \text{ pos.}) \ (137)$$

$$\int \frac{dx}{ax^2+b} = \frac{1}{2\sqrt{-ab}}\,\ln\frac{x\sqrt{a}-\sqrt{-b}}{x\sqrt{a}+\sqrt{-b}}. \quad (a \text{ pos.}, b \text{ neg.})$$
$$= \frac{1}{2\sqrt{-ab}}\,\ln\frac{\sqrt{b}+x\sqrt{-a}}{\sqrt{b}-x\sqrt{-a}}. \quad (a \text{ neg.}, b \text{ pos.}) \ (138)$$

$$\int \frac{dx}{(ax^2+b)^n} = \frac{1}{2(n-1)b}\frac{x}{(ax^2+b)^{n-1}} + \frac{2n-3}{2(n-1)b}\int \frac{dx}{(ax^2+b)^{n-1}} \ (139)$$
$$(n \text{ integ.} > 1)$$

$$\int (ax^2+b)^n x\,dx = \frac{1}{2a}\frac{(ax^2+b)^{n+1}}{n+1}. \quad (n \ne -1) \ (140)$$

$$\int \frac{x\,dx}{ax^2+b} = \frac{1}{2a}\,\ln(ax^2+b). \ (141)$$

$$\int \frac{dx}{x(ax^2+b)} = \frac{1}{2b}\,\ln\frac{x^2}{ax^2+b}. \ (142)$$

$$\int \frac{x^2\,dx}{ax^2+b} = \frac{x}{a} - \frac{b}{a}\int \frac{dx}{ax^2+b}. \ (143)$$

$$\int \frac{x^2\,dx}{(ax^2+b)^n} = -\frac{1}{2(n-1)a}\frac{x}{(ax^2+b)^{n-1}} + \frac{1}{2(n-1)a}\int \frac{dx}{(ax^2+b)^{n-1}}. \ (144)$$
$$(n \text{ integ.} > 1)$$

$$\int \frac{dx}{x^2(ax^2+b)^n} = \frac{1}{b}\int \frac{dx}{x^2(ax^2+b)^{n-1}} - \frac{a}{b}\int \frac{dx}{(ax^2+b)^n}. \quad (n \text{ pos. integ.}) \ (145)$$

$$\int \sqrt{ax^2+b}\,dx = \frac{x}{2}\sqrt{ax^2+b} + \frac{b}{2\sqrt{a}}\,\ln(x\sqrt{a}+\sqrt{ax^2+b}). \quad (a \text{ pos.}) \ (146)$$

$$\int \sqrt{ax^2+b}\,dx = \frac{x}{2}\sqrt{ax^2+b} + \frac{b}{2\sqrt{-a}}\,\sin^{-1}\left(x\sqrt{-\frac{a}{b}}\right). \quad (a \text{ neg.}) \ (147)$$

$$\int \frac{dx}{\sqrt{ax^2 + b}} = \frac{1}{\sqrt{a}} \ln (x\sqrt{a} + \sqrt{ax^2 + b}). \quad (a \text{ pos.}) \quad (148)$$

$$\int \frac{dx}{\sqrt{ax^2 + b}} = \frac{1}{\sqrt{-a}} \sin^{-1} \left(x\sqrt{-\frac{a}{b}} \right). \quad (a \text{ neg.}) \quad (149)$$

$$\int \sqrt{ax^2 + b} \; x \; dx = \frac{1}{3a} (ax^2 + b)^{3/2}. \quad (150)$$

$$\int \frac{x \; dx}{\sqrt{ax^2 + b}} = \frac{1}{a} \sqrt{ax^2 + b}. \quad (151)$$

$$\int \frac{\sqrt{ax^2 + b}}{x} \; dx = \sqrt{ax^2 + b} + \sqrt{b} \ln \frac{\sqrt{ax^2 + b} - \sqrt{b}}{x}. \quad (b \text{ pos.}) \quad (152)$$

$$\int \frac{\sqrt{ax^2 + b}}{x} \; dx = \sqrt{ax^2 + b} - \sqrt{-b} \tan^{-1} \frac{\sqrt{ax^2 + b}}{\sqrt{-b}}. \quad (b \text{ neg.}) \quad (153)$$

$$\int \frac{dx}{x\sqrt{ax^2 + b}} = \frac{1}{\sqrt{b}} \ln \frac{\sqrt{ax^2 + b} - \sqrt{b}}{x}. \quad (b \text{ pos.}) \quad (154)$$

$$\int \frac{dx}{x\sqrt{ax^2 + b}} = \frac{1}{\sqrt{-b}} \sec^{-1} \left(x\sqrt{-\frac{a}{b}} \right). \quad (b \text{ neg.}) \quad (155)$$

$$\int \frac{\sqrt{ax^2 + b} \; dx}{x^n} = -\frac{(ax^2 + b)^{3/2}}{b(n - 1)x^{n-1}} - \frac{(n - 4)a}{(n - 1)b} \int \frac{\sqrt{ax^2 + b}}{x^{n-2}} \; dx. \quad (n > 1) \quad (156)$$

$$\int \frac{dx}{x^n\sqrt{ax^2 + b}} = -\frac{\sqrt{ax^2 + b}}{b(n - 1)x^{n-1}} - \frac{(n - 2)a}{(n - 1)b} \int \frac{dx}{x^{n-2}\sqrt{ax^2 + b}}. \quad (n > 1) \quad (157)$$

$$\int \frac{dx}{x(ax^n + b)} = \frac{1}{bn} \ln \frac{x^n}{ax^n + b}. \quad (158)$$

$$\int \frac{dx}{x\sqrt{ax^n + b}} = \frac{1}{n\sqrt{b}} \ln \frac{\sqrt{ax^n + b} - \sqrt{b}}{\sqrt{ax^n + b} + \sqrt{b}}. \quad (b \text{ pos.}) \quad (159)$$

$$\int \frac{dx}{x\sqrt{ax^n + b}} = \frac{2}{n\sqrt{-b}} \sec^{-1} \sqrt{\frac{-ax^n}{b}}. \quad (b \text{ neg.}) \quad (160)$$

Functions Containing $ax^2 + bx + c$ (161)

$$\int \frac{dx}{ax^2 + bx + c} = \frac{1}{\sqrt{b^2 - 4ac}} \ln \frac{2ax + b - \sqrt{b^2 - 4ac}}{2ax + b + \sqrt{b^2 - 4ac}}. \quad (b^2 > 4ac) \quad (162)$$

$$\int \frac{dx}{ax^2 + bx + c} = \frac{2}{\sqrt{4ac - b^2}} \tan^{-1} \frac{2ax + b}{\sqrt{4ac + b^2}}. \quad (b^2 < 4ac) \quad (163)$$

$$\int \frac{dx}{ax^2 + bx + c} = -\frac{2}{2ax + b}. \quad (b^2 = 4ac) \quad (164)$$

$$\int \frac{x \; dx}{ax^2 + bx + c} = \frac{1}{2a} \ln (ax^2 + bx + c) - \frac{b}{2a} \int \frac{dx}{ax^2 + bx + c}. \quad (165)$$

$$\int \frac{x^2 \; dx}{ax^2 + bx + c} = \frac{x}{a} - \frac{b}{2a^2} \ln (ax^2 + bx + c) + \frac{b^2 - 2ac}{2a^2} \int \frac{dx}{ax^2 + bx + c}.$$

$$(166)$$

$$\int \frac{dx}{\sqrt{ax^2 + bx + c}} = \frac{1}{\sqrt{a}} \ln \left(2ax + b + 2\sqrt{a}\sqrt{ax^2 + bx + c}\right). \ (a \text{ pos.}) \ (167)$$

$$\int \frac{dx}{\sqrt{ax^2 + bx + c}} = \frac{1}{\sqrt{-a}} \sin^{-1} \frac{-2ax - b}{\sqrt{b^2 - 4ac}}. \ (a \text{ neg.}) \ (168)$$

$$\int \sqrt{ax^2 + bx + c} \ dx = \frac{2ax + b}{4a} \sqrt{ax^2 + bx + c}$$
$$+ \frac{4ac - b^2}{8a} \int \frac{dx}{\sqrt{ax^2 + bx + c}}. \ (169)$$

$$\int \frac{x \ dx}{\sqrt{ax^2 + bx + c}} = \frac{\sqrt{ax^2 + bx + c}}{a} - \frac{b}{2a} \int \frac{dx}{\sqrt{ax^2 + bx + c}}. \ (170)$$

$$\int \sqrt{ax^2 + bx + c} \ x \ dx = \frac{(ax^2 + bx + c)^{3/2}}{3a} - \frac{b}{2a} \int \sqrt{ax^2 + bx + c} \ dx. \quad (171)$$

$$\int \frac{dx}{x\sqrt{ax^2 + bx + c}} = -\frac{1}{\sqrt{c}} \ln \left(\frac{\sqrt{ax^2 + bx + c} + \sqrt{c}}{x} + \frac{b}{2\sqrt{c}} \right). \ (c \text{ pos.}) \ (172)$$

$$\int \frac{dx}{x\sqrt{ax^2 + bx + c}} = \frac{1}{\sqrt{-c}} \sin^{-1} \frac{bx + 2c}{x\sqrt{b^2 - 4ac}}. \ (c \text{ neg.}) \ (173)$$

$$\int \frac{dx}{x\sqrt{ax^2 + bx}} = -\frac{2}{bx} \sqrt{ax^2 + bx}. \ (174)$$

Functions Containing sin ax (175)

$$\int \sin u \ du = -\cos u. \ (u \text{ is any function of } x) \ (176)$$

$$\int \sin ax \ dx = -\frac{1}{a} \cos ax. \ (177)$$

$$\int \sin^2 ax \ dx = \frac{x}{2} - \frac{\sin 2ax}{4a}. \ (178)$$

$$\int \sin^n ax \ dx = -\frac{\sin^{n-1} ax \cos ax}{na} + \frac{n-1}{n} \int \sin^{n-2} ax \ dx. \ (n \text{ pos. integ.}) \ (179)$$

$$\int \frac{dx}{\sin ax} = \frac{1}{a} \ln \tan \frac{ax}{2} = \frac{1}{a} \ln (\csc ax - \cot ax). \ (180)$$

$$\int \frac{dx}{\sin^2 ax} = -\frac{1}{a} \cot ax. \ (181)$$

$$\int \frac{dx}{\sin^n ax} = -\frac{1}{a(n-1)} \frac{\cos ax}{\sin^{n-1} ax} + \frac{n-2}{n-1} \int \frac{dx}{\sin^{n-2} ax}. \ (n \text{ integ.} > 1) \ (182)$$

$$\int \frac{dx}{1 + \sin ax} = -\frac{1}{a} \tan \left(\frac{\pi}{4} - \frac{ax}{2} \right). \ (183)$$

$$\int \frac{dx}{1 - \sin ax} = \frac{1}{a} \cot \left(\frac{\pi}{4} - \frac{ax}{2} \right). \ (184)$$

Functions Containing cos ax (185)

$$\int \cos u \ du = \sin u. \ (u \text{ is any function of } x) \ (186)$$

$$\int \cos ax \, dx = \frac{1}{a} \sin ax. \ (187)$$

$$\int \cos^2 ax \, dx = \frac{x}{2} + \frac{\sin 2ax}{4a}. \ (188)$$

$$\int \cos^n ax \, dx = \frac{\cos^{n-1} ax \sin ax}{na} + \frac{n-1}{n} \int \cos^{n-2} ax \, dx. \ (n \text{ pos. integ.}) \ (189)$$

$$\int \frac{dx}{\cos ax} = \frac{1}{a} \ln \tan \left(\frac{ax}{2} + \frac{\pi}{4} \right) = \frac{1}{a} \ln (\tan ax + \sec ax). \ (190)$$

$$\int \frac{dx}{\cos^2 ax} = \frac{1}{a} \tan ax. \ (191)$$

$$\int \frac{dx}{\cos^n ax} = \frac{1}{a(n-1)} \frac{\sin ax}{\cos^{n-1} ax} + \frac{n-2}{n-1} \int \frac{dx}{\cos^{n-2} ax}. \ (n \text{ integ.} > 1) \ (192)$$

$$\int \frac{dx}{1 + \cos ax} = \frac{1}{a} \tan \frac{ax}{2}. \ (193)$$

$$\int \frac{dx}{1 - \cos ax} = -\frac{1}{a} \cot \frac{ax}{2}. \ (194)$$

$$\int \sqrt{1 - \cos x} \, dx = \sqrt{2} \int \sin \frac{x}{2} \, dx. \ (195)$$

$$\int \sqrt{1 + \cos x} \, dx = \sqrt{2} \int \cos \frac{x}{2} \, dx. \ (196)$$

Functions Containing sin ax and cos ax (197)

$$\int \sin ax \cos bx \, dx = -\frac{1}{2} \left[\frac{\cos(a-b)x}{a-b} + \frac{\cos(a+b)x}{a+b} \right]. \ (a^2 \neq b^2) \ (198)$$

$$\int \sin^n ax \cos ax \, dx = \frac{1}{a(n+1)} \sin^{n+1} ax. \ (n \neq -1). \ (199)$$

$$\int \frac{\cos ax}{\sin ax} \, dx = \frac{1}{a} \ln \sin ax. \ (200)$$

$$\int (b + c \sin ax)^n \cos ax \, dx = \frac{1}{ac(n+1)} (b + c \sin ax)^{n+1}. \ (n \neq -1) \ (201)$$

$$\int \frac{\cos ax \, dx}{b + c \sin ax} = \frac{1}{ac} \ln (b + c \sin ax). \ (202)$$

$$\int \cos^n ax \sin ax \, dx = -\frac{1}{a(n+1)} \cos^{n+1} ax. \ (n \neq -1). \ (203)$$

$$\int \frac{\sin ax}{\cos ax} \, dx = -\frac{1}{a} \ln \cos ax. \ (204)$$

$$\int (b + c \cos ax)^n \sin ax \, dx = -\frac{1}{ac(n+1)} (b + c \cos ax)^{n+1}. \ (n \neq -1) \ (205)$$

$$\int \frac{\sin ax}{b + c \cos ax} \, dx = -\frac{1}{ac} \ln (b + c \cos ax). \ (206)$$

$$\int \frac{dx}{b \sin ax + c \cos ax} = \frac{1}{a\sqrt{b^2 + c^2}} \ln \left[\tan \frac{1}{2} \left(ax + \tan^{-1} \frac{c}{b} \right) \right]. \ (207)$$

$$\int \sin^2 ax \cos^2 ax \, dx = \frac{x}{8} - \frac{\sin 4ax}{32a}. \ (208)$$

$$\int \frac{dx}{\sin ax \cos ax} = \frac{1}{a} \ln \tan ax. \quad (209)$$

$$\int \frac{dx}{\sin^2 ax \cos^2 ax} = \frac{1}{a} (\tan ax - \cot ax). \quad (210)$$

$$\int \frac{\sin^2 ax}{\cos ax} dx = \frac{1}{a} \left[-\sin ax + \ln \tan \left(\frac{ax}{2} + \frac{\pi}{4} \right) \right]. \quad (211)$$

$$\int \frac{\cos^2 ax}{\sin ax} dx = \frac{1}{a} \left[\cos ax + \ln \tan \frac{ax}{2} \right]. \quad (212)$$

Functions Containing tan *ax* $\left(= \dfrac{1}{\cot ax} \right)$ **or cot *ax*** $\left(= \dfrac{1}{\tan ax} \right)$ **(213)**

$$\int \tan u \, du = -\ln \cos u. \ (u \text{ is any function of } x) \ (214)$$

$$\int \tan ax \, dx = -\frac{1}{a} \ln \cos ax. \quad (215)$$

$$\int \tan^2 ax \, dx = \frac{1}{a} \tan ax - x. \quad (216)$$

$$\int \cot u \, du = \ln \sin u. \ (u \text{ is any function of } x) \ (217)$$

$$\int \cot ax \, dx = \int \frac{dx}{\tan ax} = \frac{1}{a} \ln \sin ax. \quad (218)$$

$$\int \cot^2 ax \, dx = \int \frac{dx}{\tan^2 ax} = -\frac{1}{a} \cot ax - x. \quad (219)$$

$$\int \frac{dx}{b + c \tan ax} = \int \frac{\cot ax \, dx}{b \cot ax + c}$$

$$= \frac{1}{b^2 + c^2} \left[bx + \frac{c}{a} \ln (b \cos ax + c \sin ax) \right]. \quad (220)$$

$$\int \frac{dx}{b + c \cot ax} = \int \frac{\tan ax \, dx}{b \tan ax + c}$$

$$= \frac{1}{b^2 + c^2} \left[bx - \frac{c}{a} \ln (c \cos ax + b \sin ax) \right]. \quad (221)$$

$$\int \frac{dx}{\sqrt{1 + \tan^2 ax}} = \frac{1}{a} \sin ax. \quad (222)$$

$$\int \frac{dx}{\sqrt{b + c \tan^2 ax}} = \frac{1}{a\sqrt{b - c}} \sin^{-1} \left(\sqrt{\frac{b - c}{b}} \sin ax \right). \ (b \text{ pos.}, b^2 > c^2) \ (223)$$

Functions Containing sec *ax* $\left(= \dfrac{1}{\cos ax} \right)$ **or csc *ax*** $\left(= \dfrac{1}{\sin ax} \right)$ **(224)**

$$\int \sec u \, du = \ln (\sec u + \tan u) = \ln \tan \left(\frac{u}{2} + \frac{\pi}{4} \right). \ (u \text{ is any function of } x) \ (225)$$

$$\int \sec ax \, dx = \frac{1}{a} \ln \tan \left(\frac{ax}{2} + \frac{\pi}{4} \right). \quad (226)$$

$$\int \sec^2 ax \ dx = \frac{1}{a} \tan ax. \ (227)$$

Functions Containing tan ax and sec ax or cot ax and csc ax (228)

$$\int \tan u \sec u \ du = \sec u. \ (u \text{ is any function of } x) \ (229)$$

$$\int \tan ax \sec ax \ dx = \frac{1}{a} \sec ax. \ (230)$$

$$\int \frac{\sec^2 ax \ dx}{\tan ax} = \frac{1}{a} \ln \tan ax. \ (231)$$

Inverse Trigonometric Functions (232)

$$\int \sin^{-1} ax \ dx = x \sin^{-1} ax + \frac{1}{a} \sqrt{1 - a^2 x^2}. \ (233)$$

$$\int \cos^{-1} ax \ dx = x \cos^{-1} ax - \frac{1}{a} \sqrt{1 - a^2 x^2}. \ (234)$$

$$\int \tan^{-1} ax \ dx = x \tan^{-1} ax - \frac{1}{2a} \ln (1 + a^2 x^2). \ (235)$$

$$\int \cot^{-1} ax \ dx = x \cot^{-1} ax + \frac{1}{2a} \ln (1 + a^2 x^2). \ (236)$$

Algebraic and Trigonometric Functions (237)

$$\int x \sin ax \ dx = \frac{1}{a^2} \sin ax - \frac{1}{a} x \cos ax. \ (238)$$

$$\int x^n \sin ax \ dx = -\frac{1}{a} x^n \cos ax + \frac{n}{a} \int x^{n-1} \cos ax \ dx. \ (n \text{ pos.}) \ (239)$$

$$\int \frac{\sin ax \ dx}{x} = ax - \frac{(ax)^3}{3 \cdot 3!} + \frac{(ax)^5}{5 \cdot 5!} - \cdots. \ (240)$$

$$\int x \cos ax \ dx = \frac{1}{a^2} \cos ax + \frac{1}{a} x \sin ax. \ (241)$$

$$\int \frac{\cos ax \ dx}{x} = \ln ax - \frac{(ax)^2}{2 \cdot 2!} + \frac{(ax)^4}{4 \cdot 4!} - \cdots. \ (242)$$

Exponential, Algebraic, Trigonometric, and Logarithmic Functions (243)

$$\int b^u \ du = \frac{b^u}{\ln b}. \ (u \text{ is any function of } x) \ (244)$$

$$\int e^u \ du = e^u. \ (u \text{ is any function of } x) \ (245)$$

$$\int b^{ax} \ dx = \frac{b^{ax}}{a \ln b}. \ (246)$$

$$\int e^{ax}\, dx = \frac{1}{a}\, e^{ax}. \ (247)$$

$$\int \frac{dx}{b + ce^{ax}} = \frac{1}{ab}\, [ax - \ln (b + ce^{ax})]. \ (248)$$

$$\int \frac{e^{ax}\, dx}{b + ce^{ax}} = \frac{1}{ac}\, \ln (b + ce^{ax}). \ (249)$$

$$\int \frac{dx}{be^{ax} + ce^{-ax}} = \frac{1}{a\sqrt{bc}}\, \tan^{-1} \left(e^{ax}\, \sqrt{\frac{b}{c}} \right). \ (b \text{ and } c \text{ pos.}) \ (250)$$

$$\int xb^{ax}\, dx = \frac{xb^{ax}}{a \ln b} - \frac{b^{ax}}{a^2(\ln b)^2}. \ (251)$$

$$\int xe^{ax}\, dx = \frac{e^{ax}}{a^2}\, (ax - 1). \ (252)$$

$$\int x^n b^{ax}\, dx = \frac{x^n b^{ax}}{a \ln b} - \frac{n}{a \ln b} \int x^{n-1} b^{ax}\, dx. \ (n \text{ pos.}) \ (253)$$

$$\int x^n e^{ax}\, dx = \frac{1}{a}\, x^n e^{ax} - \frac{n}{a} \int x^{n-1} e^{ax}\, dx. \ (n \text{ pos.}) \ (254)$$

$$\int \frac{e^{ax}}{x}\, dx = \ln x + ax + \frac{(ax)^2}{2 \cdot 2!} + \frac{(ax)^3}{3 \cdot 3!} + \cdots . \ (255)$$

$$\int \frac{e^{ax}}{x^n}\, dx = \frac{1}{n-1} \left(-\frac{e^{ax}}{x^{n-1}} + a \int \frac{e^{ax}}{x^{n-1}}\, dx \right). \ (n \text{ integ.} > 1) \ (256)$$

$$\int e^{ax} \ln x\, dx = \frac{1}{a}\, e^{ax} \ln x - \frac{1}{a} \int \frac{e^{ax}}{x}\, dx. \ (257)$$

$$\int e^{ax} \sin bx\, dx = \frac{e^{ax}}{a^2 + b^2}\, (a \sin bx - b \cos bx). \ (258)$$

$$\int e^{ax} \cos bx\, dx = \frac{e^{ax}}{a^2 + b^2}\, (a \cos bx + b \sin bx). \ (259)$$

$$\int \ln ax\, dx = x \ln ax - x. \ (260)$$

$$\int (\ln ax)^n\, dx = x(\ln ax)^n - n \int (\ln ax)^{n-1}\, dx. \ (n \text{ pos.}) \ (261)$$

$$\int x^n \ln ax\, dx = x^{n+1} \left[\frac{\ln ax}{n + 1} - \frac{1}{(n + 1)^2} \right]. \ (n \neq -1) \ (262)$$

$$\int \frac{(\ln ax)^n}{x}\, dx = \frac{(\ln ax)^{n+1}}{n + 1}. \ (n \neq -1) \ (263)$$

$$\int \frac{dx}{x \ln ax} = \ln (\ln ax). \ (264)$$

$$\int \frac{dx}{\ln ax} = \frac{1}{a} \left[\ln (\ln ax) + \ln ax + \frac{(\ln ax)^2}{2 \cdot 2!} + \frac{(\ln ax)^3}{3 \cdot 3!} + \cdots \right]. \ (265)$$

$$\int \sin (\ln ax)\, dx = \frac{x}{2}\, [\sin (\ln ax) - \cos (\ln ax)]. \ (266)$$

$$\int \cos (\ln ax)\, dx = \frac{x}{2}\, [\sin (\ln ax) + \cos (\ln ax)]. \ (267)$$

Some Definite Integrals (268)

$$\int_0^a \sqrt{a^2 - x^2}\, dx = \frac{\pi a^2}{4}. \ (269)$$

$$\int_0^a \sqrt{2ax - x^2} \, dx = \frac{\pi a^2}{4}. \quad (270)$$

$$\int_0^\infty \frac{dx}{ax^2 + b} = \frac{\pi}{2\sqrt{ab}}. \quad (a \text{ and } b \text{ pos.}) \ (271)$$

$$\int_0^{\pi/2} \sin^n ax \, dx = \int_0^{\pi/2} \cos^n ax \, dx = \frac{1 \cdot 3 \cdot 5 \cdot \ldots \cdot (n-1)}{2 \cdot 4 \cdot 6 \cdot \ldots \cdot n} \frac{\pi}{2a}.$$

$$(n \text{ pos. even integ.}) \ (272)$$

$$\int_0^{\pi/2} \sin^n ax \, dx = \int_0^{\pi/2} \cos^n ax \, dx = \frac{2 \cdot 4 \cdot 6 \cdot \ldots \cdot (n-1)}{1 \cdot 3 \cdot 5 \cdot \ldots \cdot n} \frac{1}{a}.$$

$$(n \text{ pos. odd integ.}) \ (273)$$

$$\int_0^\pi \sin^2 ax \, dx = \int_0^\pi \cos^2 ax \, dx = \frac{\pi}{2}. \quad (274)$$

$$\int_0^\infty e^{-ax^2} \, dx = \frac{1}{2}\sqrt{\frac{\pi}{a}}. \quad (275)$$

$$\int_0^\infty x^n e^{-ax} \, dx = \frac{n!}{a^{n+1}}. \quad (n \text{ pos. integ.}) \ (276)$$

1.7.2 Definite Integrals

Definition and Approximate Value of the Definite Integral (277)

If $f(x)$ is continuous from $x = a$ to $x = b$ inclusive, and this interval is divided into n equal parts by the points $a, x_1, x_2, \ldots, x_{n-1}, b$ such that $\Delta x = (b - a) \div n$, then the definite integral of $f(x) \, dx$ between the limits $x = a$ to $x = b$ is

$$\int_a^b f(x) \, dx = \lim_{n \doteq \infty} [f(a) \, \Delta x + f(x_1) \, \Delta x + f(x_2) \, \Delta x + \cdots + f(x_{n-1}) \, \Delta x]$$

$$= \left[\int f(x) \, dx\right]_a^b = [F(x)]_a^b = F(b) - F(a).$$

If $y_0, y_1, y_2, \ldots, y_{n-1}, y_n$ are the values of $f(x)$ when $x = a, x_1, x_2, \ldots, x_{n-1}, b$, respectively, and if $h = (b - a) \div n$, then approximate values of this definite integral are given by the trapezoidal, Durand's, and Simpson's rules (37).

Some Fundamental Theorems on Definite Integrals (278)

$$\int_a^b [f_1(x) + f_2(x) + \cdots] \, dx = \int_a^b f_1(x) \, dx + \int_a^b f_2(x) \, dx + \cdots. \quad (279)$$

$$\int_a^b kf(x) \, dx = k \int_a^b f(x) \, dx. \quad (k \text{ is any constant}) \ (280)$$

$$\int_a^b f(x) \, dx = -\int_b^a f(x) \, dx. \quad (281)$$

$$\int_a^b f(x) \, dx = \int_a^c f(x) \, dx + \int_c^b f(x) \, dx. \quad (282)$$

$$\int_a^b f(x)\, dx = (b - a)f(x_1), \text{ where } x_1 \text{ lies between } a \text{ and } b. \quad (283)$$

$$\int_a^\infty f(x)\, dx = \lim_{b \pm \infty} \int_a^b f(x)\, dx. \quad (284)$$

Some Applications of the Definite Integral

Plane Area (285). Area (A) bounded by the curve $y = f(x)$, the axis OX, and the ordinates $x = a$, $x = b$ (Fig. 1.55a).

$$dA = y\, dx, \qquad A = \int_a^b f(x)\, dx.$$

Area (A) bounded by the curve $x = f(y)$, the axis OY, and the abscissas $y = c$, $y = d$ (Fig. 1.55b).

$$dA = x\, dy, \qquad A = \int_c^d f(y)\, dy.$$

Area (A) bounded by the curve $x = f_1(t)$, $y = f_2(t)$, the axis OX, and $t = a$, $t = b$.

$$dA = y\, dx, \qquad A = \int_a^b f_2(t)f_1(t)\, dt.$$

Area (A) bounded by the curve $r = f(\theta)$ and two radii $\theta = \alpha$, $\theta = \beta$ (Fig. 1.55c).

$$dA = \tfrac{1}{2}r^2\, d\theta, \qquad A = \tfrac{1}{2}\int_\alpha^\beta [f(\theta)]^2\, d\theta.$$

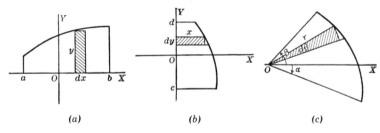

(a)   (b)   (c)

Fig. 1.55

Length of Arc (286).   Length (s) of arc of curve $f(x,y) = 0$ from the point (a,c) to the point (b,d) (Fig. 1.56a).

$$ds = \sqrt{(dx)^2 + (dy)^2}, \qquad s = \int_a^b \sqrt{1 + \left(\frac{dy}{dx}\right)^2}\, dx$$

$$= \int_c^d \sqrt{1 + \left(\frac{dx}{dy}\right)^2}\, dy.$$

Length (s) of arc of curve $x = f_1(t)$, $y = f_2(t)$ from $t = a$ to $t = b$.

$$ds = \sqrt{(dx)^2 + (dy)^2}, \qquad s = \int_a^b \sqrt{\left(\frac{dx}{dt}\right)^2 + \left(\frac{dy}{dt}\right)^2}\, dt$$

Length (s) of arc of curve $r = f(\theta)$ from $\theta = \alpha$ to $\theta = \beta$ (Fig. 1.56b).

$$ds = \sqrt{(dr)^2 + (r\, d\theta)^2}, \qquad s = \int_\alpha^\beta \sqrt{r^2 + \left(\frac{dr}{d\theta}\right)^2}\, d\theta.$$

Length (s) of arc of space curve $x = f_1(t)$, $y = f_2(t)$, $z = f_3(t)$ from $t = a$ to $t = b$.

$$ds = \sqrt{(dx)^2 + (dy)^2 + (dz)^2}, \qquad s = \int_a^b \sqrt{\left(\frac{dx}{dt}\right)^2 + \left(\frac{dy}{dt}\right)^2 + \left(\frac{dz}{dt}\right)^2}\, dt.$$

Volume of Revolution (287).   Volume (V) of revolution generated by revolving about the line $y = k$ the area enclosed by the curve $y = f(x)$, the ordinates $x = a$, $x = b$, and the line $y = k$ (Fig. 1.57).

$$dV = \pi R^2\, dx = \pi(y - k)^2\, dx,$$

$$V = \pi \int_a^b [f(x) - k]^2\, dx.$$

Volume (V) of revolution generated by revolving about the line $x = k$ the area enclosed by the curve $x = f(y)$, the abscissas $y = c$, $y = d$, and the line $x = k$.

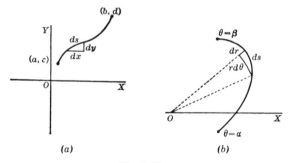

(a)   (b)

Fig. 1.56

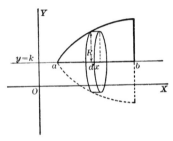

Fig. 1.57

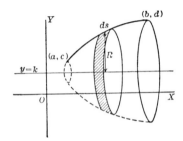

Fig. 1.58

$$dV = \pi R^2 \, dy = \pi(x - k)^2 \, dy, \qquad V = \pi \int_c^d [f(y) - k]^2 \, dy.$$

Area of Surface of Revolution (288). Area (S) of surface of revolution generated by revolving the arc of the curve $f(x,y) = 0$ from the point (a,c) to the point (b,d) (Fig. 1.58).

About $y = k$: $dS = 2\pi R \, ds$,

$$S = 2\pi \int_a^b (y - k) \sqrt{1 + \left(\frac{dy}{dx}\right)^2} \, dx.$$

About $x = k$: $dS = 2\pi R \, ds$,

$$S = 2\pi \int_c^d (x - k) \sqrt{1 + \left(\frac{dx}{dy}\right)^2} \, dy.$$

Area (S) of surface of revolution generated by revolving the arc of the curve $r = f(\theta)$ from $\theta = \alpha$ to $\theta = \beta$.

About OX: $dS = 2\pi R \, ds$, $\qquad S = 2\pi \int_\alpha^\beta r \sin \theta \sqrt{r^2 + \left(\frac{dr}{d\theta}\right)^2} \, d\theta.$

About OY: $dS = 2\pi R \, ds$, $\qquad S = 2\pi \int_\alpha^\beta r \cos \theta \sqrt{r^2 + \left(\frac{dr}{d\theta}\right)^2} \, d\theta.$

Volume by Parallel Sections (289). Volume (V) of a solid generated by moving a plane section of area A_x perpendicular to OX from $x = a$ to $x = b$ (Fig. 1.59).

$$dV = A_x \, dx, \qquad V = \int_a^b A_x \, dx,$$

where A_x must be expressed as a function of x.

Mass (209). Mass (m) constant or variable density (δ).

$$dm = \delta \, dA \quad \text{or} \quad \delta \, ds \quad \text{or} \quad \delta \, dV \quad \text{or} \quad \delta \, dS, \qquad m = \int dm,$$

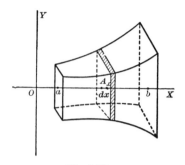

Fig. 1.59

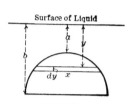

Fig. 1.60

where dA, ds, dV, dS are the elements of area, length, volume, surface in (281) to (285), and δ = mass per unit element.

Moment (291). Moment (M) of a mass (m).

$$\text{About } OX\text{: } M_x = \int y\, dm = \int r \sin \theta\, dm.$$

$$\text{About } OY\text{: } M_y = \int x\, dm = \int r \cos \theta\, dm.$$

$$\text{About } O\text{: } \quad M_0 = \int \sqrt{x^2 + y^2}\, dm = \int r\, dm.$$

Moment of Inertia (292). Moment of inertia (I) of a mass (m).

$$\text{About } OX\text{: } I_x = \int y^2\, dm = \int r^2 \sin^2 \theta\, dm.$$

$$\text{About } OY\text{: } I_y = \int x^2\, dm = \int r^2 \cos^2 \theta\, dm.$$

$$\text{About } O\text{: } \quad I_0 = \int (x^2 + y^2)\, dm = \int r^2\, dm.$$

Center of Gravity (293). Coordinates (x,y) of the center of gravity of a mass (m).

$$x = \frac{\int x\, dm}{\int dm}, \qquad y = \frac{\int y\, dm}{\int dm}.$$

NOTE. The center of gravity of the element of area may be taken at its midpoint. In the above equations x and y are the coordinates of the center of gravity of the element.

Work (294).   Work (W) done in moving a particle from $s = a$ to $s = b$ against a force whose component in the direction of motion is F_s.

$$dW = F_s \, ds, \qquad W = \int_a^b F_s \, ds,$$

where F_s must be expressed as a function of s.

Pressure (295).   Pressure (p) against an area vertical to the surface of the liquid and between depths a and b (Fig. 1.60).

$$dp = wyx \, dy, \qquad p = \int_a^b wyx \, dy,$$

where w is the weight of liquid per unit volume, y the depth beneath surface of liquid of a horizontal element of area, and x the length of horizontal element of area; x must be expressed in terms of y.

Center of Pressure (296).   The depth (y) of the center of pressure against an area vertical to the surface of the liquid and between depths a and b.

$$y = \frac{\displaystyle\int_a^b y \, dp}{\displaystyle\int_a^b dp} \; . \; [\text{For } dp, \text{ see (295).}]$$

1.8   DIFFERENTIAL EQUATIONS

Definitions and Notation

A **differential equation** is an equation involving differentials or derivatives.

The **order** of a differential equation is the same as that of the derivative of highest order which it contains.

The **degree** of a differential equation is the same as the power to which the derivative of highest order in the equation is raised, that derivative entering the equation free from radicals.

The **solution** of a differential equation is the relation involving only the variables (but not their derivatives) and arbitrary constants, consistent with the given differential equation.

The most **general solution** of a differential equation of the nth order contains n arbitrary constants. If particular values are assigned to these arbitrary constants, the solution is called a **particular solution.**

Notation: M and N denote functions of x and y; X denotes a function of x alone or a constant; Y denotes a function of y alone or a constant; $C, C_1, C_2, \ldots, C_n$ denote arbitrary constants of integration; $a, b, k, l, m, n, \ldots$ denote given constants.

Equations of First Order and First Degree. M dx + N dy = 0 (297)

Variables Separable: $X_1Y_1\,dx + X_2Y_2\,dy = 0$ (298)

Solution:

$$\int \frac{X_1}{X_2}\,dx + \int \frac{Y_2}{Y_1}\,dy = C.$$

Homogeneous Equation

$$dy - f\left(\frac{y}{x}\right) dx = 0. \quad (299)$$

Solution:

$$x = Ce^{\int [dv/f(v)-v]} \quad \text{and} \quad v = \frac{y}{x}.$$

NOTE. Here, $M \div N$ can be written in a form such that x and y occur only in the combination $y \div x$; this can always be done if every term in M and N is of the same degree in x and y.

Linear Equation

$$dy + (X_1 y - X_2)\,dx = 0. \quad (300)$$

Solution:

$$y = e^{-\int X_1 dx}\left(\int X_2 e^{\int X_1 dx}\,dx + C\right).$$

NOTE. A similar solution exists for $dx + (Y_1 x - Y_2)\,dy = 0$.

Exact Equation

$$M\,dx + N\,dy = 0, \text{ where } \frac{\partial M}{\partial y} = \frac{\partial N}{\partial x}. \quad (301)$$

Solution:

$$\int M\,dx + \int \left[N - \frac{\partial}{\partial y}\int M\,dx\right] dy = C,$$

where y is constant when integrating with respect to x.

Nonexact Equation

$$M\,dx + N\,dy = 0, \text{ where } \frac{\partial M}{\partial y} \neq \frac{\partial N}{\partial x}. \quad (302)$$

Solution:

The equation may be made exact by multiplying by an integrating factor $\mu(x,y)$. The form of this factor is readily recognized in a large number of cases. Then solve by (297).

Certain Special Equations of the Second Order

$$\frac{d^2y}{dx^2} = f\left(x, y, \frac{dy}{dx}\right). \quad (303)$$

Equation

$$\frac{d^2y}{dx^2} = X. \quad (304)$$

Solution:

$$y = x\int X\,dx - \int x X\,dx + C_1 x + C_2.$$

Equation

$$\frac{d^2y}{dx^2} = Y. \quad (305)$$

Solution:

$$x = \int \frac{dy}{\sqrt{2\int Y\,dy + C_1}} + C_2.$$

Equation

$$\frac{d^2y}{dx^2} = f\left(\frac{dy}{dx}\right). \quad (306)$$

Solution:

$$x = \int \frac{dp}{f(p)} + C_1 \quad \text{and} \quad y = \int \frac{p\,dp}{f(p)} + C_2.$$

From these two equations eliminate $p = dy/dx$ if necessary.

Equation

$$\frac{d^2y}{dx^2} = f\left(x, \frac{dy}{dx}\right). \quad (307)$$

Solution:

Place $dy/dx = p$ and $d^2y/dx^2 = dp/dx$, thus bringing the equation into the form $dp/dx = f(x,p)$. This is of the first order and may be solved for p by (297) to (301). Then replace p by dy/dx and integrate for y.

Equation

$$\frac{d^2y}{dx^2} = f\left(y, \frac{dy}{dx}\right). \quad (308)$$

Solution:

Place $dy/dx = p$ and $d^2y/dx^2 = p(dp/dy)$, thus bringing the equation into the form $p(dp/dy) = f(y,p)$. This is of the first order and may be solved for p by (297) to (301). Then replace p by dy/dx and integrate for y.

Linear Equations of Physics: Second Order with Constant Coefficients

$$\frac{d^2x}{dt^2} + 2l\frac{dx}{dt} \pm k^2x = f(t). \quad (309)$$

Equation

$$\frac{d^2x}{dt^2} - k^2x = 0. \quad (309a)$$

Solution:

$$x = C_1e^{kt} + C_2e^{-kt}.$$

Equation of Simple Harmonic Motion

$$\frac{d^2x}{dt^2} + k^2x = 0. \quad (310)$$

Solution:

This may be written in the following forms:

1. $x = C_1e^{kt\sqrt{-1}} + C_2e^{-kt\sqrt{-1}}$.
2. $x = C_1 \cos kt + C_2 \sin kt$.
3. $x = C_1 \sin(kt + C_2)$.
4. $x = C_1 \cos(kt + C_2)$.

Equation of Harmonic Motion with Constant Disturbing Force

$$\frac{d^2x}{dt^2} + k^2x = a. \quad (311)$$

Solution:

$$x = C_1 \cos kt + C_2 \sin kt + \frac{a}{k^2},$$

or
$$x = C_1 \sin(kt + C_2) + \frac{a}{k^2}.$$

Equation of Forced Vibration (312)

1. $\dfrac{d^2x}{dt^2} + k^2x = a \cos nt + b \sin nt$, where $n \neq k$.

Solution:

$$x = C_1 \cos kt + C_2 \sin kt + \frac{1}{k^2 - n^2}(a \cos nt + b \sin nt).$$

2. $\dfrac{d^2x}{dt^2} + k^2x = a \cos kt + b \sin kt$.

Solution:

$$x = C_1 \cos kt + C_2 \sin kt + \frac{t}{2k}(a \sin kt - b \cos kt).$$

Equation of Damped Vibration

$$\frac{d^2x}{dt^2} + 2l\frac{dx}{dt} + k^2x = 0. \quad (313)$$

Solution:

If $l^2 = k^2$, $\quad x = e^{-lt}(C_1 + C_2t)$.

If $l^2 > k^2$, $\quad x = e^{-lt}(C_1e^{\sqrt{l^2-k^2}\,t} + C_2e^{-\sqrt{l^2-k^2}\,t})$.

If $l^2 < k^2$, $\quad x = e^{-lt}(C_1 \cos \sqrt{k^2 - l^2}\,t + C_2 \sin \sqrt{k^2 - l^2}\,t)$

$\qquad$ or $\quad x = C_1e^{-lt} \sin(\sqrt{k^2 - l^2}\,t + C_2)$.

Equation of Damped Vibration with Constant Disturbing Force

$$\frac{d^2x}{dt^2} + 2l\frac{dx}{dt} + k^2x = a. \quad (314)$$

Solution:

$$x = x_1 + \frac{a}{k^2},$$

where x_1 is the solution of (309).

General Equation

$$\frac{d^2x}{dt^2} + 2l\frac{dx}{dt} + k^2x = f(t) = T. \quad (315)$$

Solution:

$$x = x_1 + I,$$

where x_1 is the solution of (313) and I is given by

CASE I. $l^2 = k^2$:

$$I = e^{-lt}\left[t\int e^{lt}T\, dt - \int e^{lt}Tt\, dt \right].$$

CASE II. $l^2 > k^2$:

$$I = \frac{1}{\alpha - \beta}\left[e^{\alpha t}\int e^{-\alpha t}T\, dt - e^{\beta t}\int e^{-\beta t}T\, dt \right],$$

where $\alpha = -1 + \sqrt{l^2 - k^2}$ and $\beta = -1 - \sqrt{l^2 - k^2}$.

CASE III. $l^2 < k^2$:

$$I = \frac{e^{\alpha t}}{\beta}\left[\sin\beta t\int e^{-\alpha t}\cos\beta t T\, dt - \cos\beta t\int e^{-\alpha t}\sin\beta t T\, dt \right],$$

where $\alpha = -l$ and $\beta = \sqrt{k^2 - l^2}$.

NOTE.   *I* may also be found by the method indicated in (317).

Linear Equations with Constant Coefficients: *n*th-Order Equation (316)

$$a_n\frac{d^nx}{dt^n} + a_{n-1}\frac{d^{n-1}x}{dt^{n-1}} + a_{n-2}\frac{d^{n-2}x}{dt^{n-2}} + \cdots + a_1\frac{dx}{dt} + a_0x = 0.$$

Solution:

Let $D = \alpha_1, \alpha_2, \alpha_3, \ldots, \alpha_n$ be the n roots of the auxiliary algebraic equation $a_nD^n + a_{n-1}D^{n-1} + a_{n-2}D^{n-2} + \cdots + a_1D + a_0 = 0$.
    If all roots are real and distinct,

$$x = C_1e^{\alpha_1 t} + C_2e^{\alpha_2 t} + \cdots + C_ne^{\alpha_n t}.$$

If two roots are equal: $\alpha_1 = \alpha_2$, the rest real and distinct,

$$x = e^{\alpha_1 t}(C_1 + C_2 t) + C_3 e^{\alpha_3 t} + \cdots + C_n e^{\alpha_n t}.$$

If p roots are equal: $\alpha_1 = \alpha_2 = \cdots = \alpha_p$, the rest real and distinct,

$$x = e^{\alpha_1 t}(C_1 + C_2 t + C_3 t^2 + \cdots + C_p t^{p-1}) + \cdots + C_n e^{\alpha_n t}.$$

If two roots are conjugate imaginary: $\alpha_1 = \beta + \gamma\sqrt{-1}$, $\alpha_2 = \beta - \gamma\sqrt{-1}$,

$$x = e^{\beta t}(C_1 \cos \gamma t + C_2 \sin \gamma t) + C_3 e^{\alpha_3 t} + \cdots + C_n e^{\alpha_n t}.$$

If there is a pair of conjugate imaginary double roots:

$$\alpha_1 = \beta + \gamma\sqrt{-1} = \alpha_2, \qquad \alpha_3 = \beta - \gamma\sqrt{-1} = \alpha_4,$$

$$x = e^{\beta t}[(C_1 + C_2 t) \cos \gamma t + (C_3 + C_4 t) \sin \gamma t] + \cdots + C_n e^{\alpha_n t}.$$

Equation (317)

$$a_n \frac{d^n x}{dt^n} + a_{n-1} \frac{d^{n-1} x}{dt^{n-1}} + \cdots + a_1 \frac{dx}{dt} + a_0 x = f(t). \quad (317)$$

Solution:

$$x = x_1 + I,$$

where x_1 is the solution of Equation (316) and where I may be found by the following method.

Let $f(t) = T_1 + T_2 + T_3 + \cdots$. Find the first, second, third, ... derivatives of these terms. If $\tau_1, \tau_2, \tau_3, \ldots, \tau_n$ are the resulting expressions that have different functional form (disregarding constant coefficients), assume that

$$I = A\tau_1 + B\tau_2 + C\tau_3 + \cdots + K\tau_k + \cdots + N\tau_n.$$

NOTE. Thus, if $T = a \sin nt + bt^2 e^{kt}$, all possible successive derivatives of $\sin nt$ and $t^2 e^{kt}$ give terms of the form: $\sin nt$, $\cos nt$, e^{kt}, te^{kt}, $t^2 e^{kt}$; hence assume that $I = A \sin nt + B \cos nt + C e^{kt} + D t e^{kt} + E t^2 e^{kt}$. Substitute this value of I for x in the given equation, expand, equate coefficients of like terms in the left and right members of the equation, and solve for $A, B, C, \ldots, N$.

NOTE. If a root, α_k, occurring m times, of the algebraic equation in D [see (316)] gives rise to a term of the form τ_k in x_1, the corresponding term in the assumed value of I is $Kt^m \tau_k$.

Simultaneous Equations (318)

$$a_n \frac{d^n x}{dt^n} + b_m \frac{d^m y}{dt^m} + \cdots + a_1 \frac{dx}{dt} + b_1 \frac{dy}{dt} + a_0 x + b_0 y = f_1(t).$$

$$c_k \frac{d^k x}{dt^k} + g_l \frac{d^l y}{dt^l} + \cdots + c_1 \frac{dx}{dt} + g_1 \frac{dy}{dt} + c_0 x + g_0 y = f_2(t).$$

Solution:

Write the equations in the form:

$$(a_n D^n + \cdots + a_1 D + a_0)x + (b_m D^m + \cdots + b_1 D + b_0)y = f_1(t),$$
$$(c_k D^k + \cdots + c_1 D + c_0)x + (g_l D^l + \cdots + g_1 D + g_0)y = f_2(t),$$

where

$$D = \frac{d}{dt}, \ldots, D^i = \frac{d^i}{dt^i}, \ldots.$$

Regarding this set of equations as a pair of simultaneous algebraic equations in x and y, eliminate y and x in turn, getting two linear differential equations of the form of (317) whose solutions are

$$x = x_1 + I_1, \qquad y = y_1 + I_2.$$

Substitute these values of x and y in the original equations, equate coefficients of like terms, and thus express the arbitrary constants in y_1, say, in terms of those in x_1.

Partial Differential Equations (319)

Equation of Oscillation (320)

$$\frac{\partial^2 y}{\partial t^2} = a^2 \frac{\partial^2 y}{\partial x^2}.$$

Solution:

$$y = \sum_{i=1}^{i=\infty} C_i e^{(x+at)\alpha_i} + \sum_{i=1}^{i=\infty} C_i' e^{(x-at)\alpha_i},$$

where C_i, C_i', and α_i are arbitrary constants.

Equation of Thermodynamics (321)

$$\frac{\partial u}{\partial t} = a^2 \frac{\partial^2 u}{\partial x^2}.$$

Solution:

$$u = \sum_{i=1}^{i=\infty} C_i e^{\alpha_i x} e^{a^2 \alpha_i^2 t},$$

where C_i and α_i are arbitrary constants.

Equation of Laplace or Condition of Continuity of Incompressible Liquids (322)

$$\frac{\partial^2 u}{\partial x^2} + \frac{\partial^2 u}{\partial y^2} = 0.$$

Solution:

$$u = \sum_{i=1}^{i=\infty} C_i e^{(x+y\sqrt{-1})\alpha_i} + \sum_{i=1}^{i=\infty} C_i' e^{(x-y\sqrt{-1})\alpha_i},$$

where C_i, C_i', and α_i are arbitrary constants.

1.9   COMPLEX QUANTITIES

Definition and Representation of a Complex Quantity (323)

If $z = x + jy$, where $j = \sqrt{-1}$ and x and y are real, z is called a complex quantity. z is completely determined by x and y.

If $P(x,y)$ is a point in the plane (Fig. 1.61), then the segment OP in magnitude and direction is said to represent the complex quantity $z = x + jy$.

If θ is the angle from OX to OP and r is the length of OP, then

$$z = x + jy = r(\cos\theta + j\sin\theta) = re^{j\theta},$$

where $\theta = \tan^{-1}(y/x)$, $r = +\sqrt{x^2 + y^2}$, and e is the base of natural logarithms. $x + jy$ and $x - jy$ are called **conjugate complex** quantities.

Properties of Complex Quantities (324)

Let z, z_1, and z_2 represent complex quantities; then:

Sum or difference: $z_1 \pm z_2 = (x_1 \pm x_2) + j(y_1 \pm y_2)$.

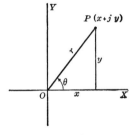

Fig. 1.61

Product: $z_1 \cdot z_2 = r_1 r_2 [\cos(\theta_1 + \theta_2) + j \sin(\theta_1 + \theta_2)]$
$$= r_1 r_2 e^{j(\theta_1 + \theta_2)} = (x_1 x_2 - y_1 y_2) + j(x_1 y_2 + x_2 y_1).$$

Quotient: $\dfrac{z_1}{z_2} = \dfrac{r_1}{r_2}[\cos(\theta_1 - \theta_2) + j \sin(\theta_1 - \theta_2)]$

$$= \frac{r_1}{r_2} e^{j(\theta_1 - \theta_2)} = \frac{x_1 x_2 + y_1 y_2}{x_2^2 + y_2^2} + j \frac{x_2 y_1 - x_1 y_2}{x_2^2 + y_2^2}.$$

Power: $z^n = r^n(\cos n\theta + j \sin n\theta) = r^n e^{jn\theta}.$

Root: $\sqrt[n]{z} = \sqrt[n]{r}\left(\cos\dfrac{\theta + 2k\pi}{n} + j \sin\dfrac{\theta + 2k\pi}{n}\right) = \sqrt[n]{r}\, e^{j(\theta + 2k\pi/n)},$

where k takes in succession the values 0, 1, 2, 3, . . . , $n - 1$.

Equation: If $z_1 = z_2$, then $x_1 = x_2$ and $y_1 = y_2$.

Periodicity: $z = r(\cos\theta + j \sin\theta) = r[\cos(\theta + 2k\pi) + j \sin(\theta + 2k\pi)],$

or $z = re^{j\theta} = re^{j(\theta + 2k\pi)}$ and $e^{j2k\pi} = 1$, where k is any integer.

Exponential-trigonometric relations:

$$e^{jz} = \cos z + j \sin z, \; e^{-jz} = \cos z - j \sin z,$$
$$\cos z = \frac{1}{2}(e^{jz} + e^{-jz}), \; \sin z = \frac{1}{2j}(e^{jz} - e^{-jz}).$$

Fig. 1.62

1.10 VECTORS

Definition and Graphical Representation of a Vector (325) (Fig. 1.62)

A vector (**V**) is a quantity that is completely specified by a magnitude and a direction. A scalar (s) is a quantity that is completely specified by a magnitude.

The vector (**V**) may be represented geometrically by the segment $\overrightarrow{OA}$, the length of OA signifying the magnitude of **V** and the arrow carried by OA signifying the direction **V**.

The segment $\overrightarrow{AO}$ represents the vector $-\mathbf{V}$.

Graphical Summation of Vectors (326)

If $\mathbf{V}_1$ and $\mathbf{V}_2$ are two vectors, their graphical sum, $\mathbf{V} = \mathbf{V}_1 + \mathbf{V}_2$, is formed by drawing the vector $\mathbf{V}_1 = \overrightarrow{OA}$ from any point O, and the vector $\mathbf{V}_2 = \overrightarrow{AB}$ from the end of $\mathbf{V}_1$, and joining O and B; then $\mathbf{V} = \overrightarrow{OB}$. Also $\mathbf{V}_1 + \mathbf{V}_2 = \mathbf{V}_2 + \mathbf{V}_1$ and $\mathbf{V}_1 + \mathbf{V}_2 - \mathbf{V} = 0$ (Fig. 1.63a).

Similarly, if $\mathbf{V}_1, \mathbf{V}_2, \mathbf{V}_3, \ldots, \mathbf{V}_n$ are any number of vectors drawn so that the initial point of one is the endpoint of the preceding one, then their graphical sum, $\mathbf{V} = \mathbf{V}_1 + \mathbf{V}_2 + \cdots + \mathbf{V}_n$, is the vector joining the initial point of $\mathbf{V}_1$ with the endpoint of $\mathbf{V}_n$ (Fig. 1.63b).

Components of a Vector, Analytic Representation (327)

A vector (**V**) considered as lying in the xy coordinate plane is completely determined by its horizontal and vertical components x and y. If **i** and **j** represent vectors of unit

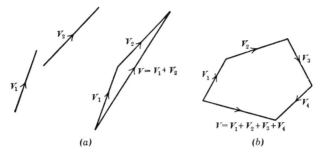

Fig. 1.63

magnitude along OX and OY, respectively, and a and b are the magnitudes of the components x and y, then $\mathbf{V}$ may be represented by $\mathbf{V} = a\mathbf{i} + b\mathbf{j}$, its magnitude by $|\mathbf{V}| = +\sqrt{a^2 + b^2}$, and its direction by $\alpha = \tan^{-1}(b/a)$.

A vector ($\mathbf{V}$) considered as lying in space is completely determined by its components x, y, and z along three mutually perpendicular lines OX, OY, and OZ, directed as in Fig. 1.64b. If $\mathbf{i}$, $\mathbf{j}$, $\mathbf{k}$ represent vectors of unit magnitude along OX, OY, OZ, respectively, and a, b, c are the magnitudes of the components x, y, z, respectively, then $\mathbf{V}$ may be represented by $\mathbf{V} = a\mathbf{i} + b\mathbf{j} + c\mathbf{k}$, its magnitude by $|\mathbf{V}| = +\sqrt{a^2 + b^2 + c^2}$, and its direction by $\cos \alpha : \cos \beta : \cos \gamma = a : b : c$.

Properties of Vectors (328)

$$\mathbf{V} = a\mathbf{i} + b\mathbf{j} \quad \text{or} \quad \mathbf{V} = a\mathbf{i} + b\mathbf{j} + c\mathbf{k}.$$

1. Vector sum ($\mathbf{V}$) of any number of vectors, $\mathbf{V}_1, \mathbf{V}_2, \mathbf{V}_3, \ldots$ (329)

$$\mathbf{V} = \mathbf{V}_1 + \mathbf{V}_2 + \mathbf{V}_3 + \cdots = (a_1 + a_2 + a_3 + \cdots)\mathbf{i}$$
$$+ (b_1 + b_2 + b_3 + \cdots)\mathbf{j} + (c_1 + c_2 + c_3 + \cdots)\mathbf{k}.$$

2. Product of a vector ($\mathbf{V}$) by a Scalar (s) (330)

$$s\mathbf{V} = (sa)\mathbf{i} + (sb)\mathbf{j} + (sc)\mathbf{k}.$$
$$(s_1 + s_2)\mathbf{V} = s_1\mathbf{V} + s_2\mathbf{V}; \quad (\mathbf{V}_1 + \mathbf{V}_2)s = \mathbf{V}_1 s + \mathbf{V}_2 s.$$

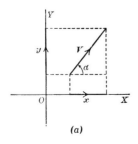

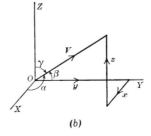

Fig. 1.64

NOTE. $s\mathbf{V}$ has the same direction as $\mathbf{V}$ and its magnitude is s times the magnitude of $\mathbf{V}$.

3. Scalar product of two vectors: $\mathbf{V}_1 \cdot \mathbf{V}_2$ (331)

$\mathbf{V}_1 \cdot \mathbf{V}_2 = |\mathbf{V}_1||\mathbf{V}_2| \cos \phi$, where ϕ is the angle between $\mathbf{V}_1$ and $\mathbf{V}_2$ (Fig. 1.65).

$\mathbf{V}_1 \cdot \mathbf{V}_2 = \mathbf{V}_2 \cdot \mathbf{V}_1; \quad \mathbf{V}_1 \cdot \mathbf{V}_1 = |\mathbf{V}_1|^2.$

$(\mathbf{V}_1 + \mathbf{V}_2) \cdot \mathbf{V}_3 = \mathbf{V}_1 \cdot \mathbf{V}_3 + \mathbf{V}_2 \cdot \mathbf{V}_3;$

$(\mathbf{V}_1 + \mathbf{V}_2) \cdot (\mathbf{V}_3 + \mathbf{V}_4) = \mathbf{V}_1 \cdot \mathbf{V}_3 + \mathbf{V}_1 \cdot \mathbf{V}_4 + \mathbf{V}_2 \cdot \mathbf{V}_3 + \mathbf{V}_2 \cdot \mathbf{V}_4.$

$\mathbf{i} \cdot \mathbf{i} = \mathbf{j} \cdot \mathbf{j} = \mathbf{k} \cdot \mathbf{k} = 1; \quad \mathbf{i} \cdot \mathbf{j} = \mathbf{j} \cdot \mathbf{k} = \mathbf{k} \cdot \mathbf{i} = 0.$

In plane: $\mathbf{V}_1 \cdot \mathbf{V}_2 = a_1 a_2 + b_1 b_2$; in space: $\mathbf{V}_1 \cdot \mathbf{V}_2 = a_1 a_2 + b_1 b_2 + c_1 c_2.$

NOTE. The scalar product of two vectors $\mathbf{V}_1 \cdot \mathbf{V}_2$ is a scalar quantity and may physically be represented by the work done by a constant force of magnitude $|\mathbf{V}_1|$ on a unit particle moving through a distance $|\mathbf{V}_2|$, where ϕ is the angle between the line of force and the direction of motion.

1.11   HYPERBOLIC FUNCTIONS

Definitions of Hyperbolic Functions (332)

Hyperbolic sine (sinh) $x = \dfrac{1}{2}(e^x - e^{-x})$; $\operatorname{csch} x = \dfrac{1}{\sinh x}$

Hyperbolic cosine (cosh) $x = \dfrac{1}{2}(e^x + e^{-x})$; $\operatorname{sech} x = \dfrac{1}{\cosh x}$

Hyperbolic tangent (tanh) $x = \dfrac{e^x - e^{-x}}{e^x + e^{-x}}$; $\operatorname{coth} x = \dfrac{1}{\tanh x}$

where e is the base of natural logarithms.

NOTE. The circular or ordinary trigonometric functions were defined with reference to a circle; in a similar manner, the hyperbolic functions may be defined with reference to a hyperbola. In the definitions above the hyperbolic functions are abbreviations for certain exponential functions.

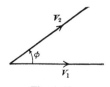

Fig. 1.65

Graphs of Hyperbolic Functions (333)

Figure 1.66

(a) $y = \sinh x$;
(b) $y = \cosh x$;
(c) $y = \tanh x$.

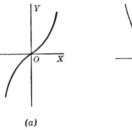
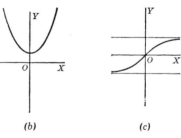

(a) (b) (c)

Fig. 1.66

Some Relations among Hyperbolic Functions (334)

$\sinh 0 = 0,$ $\cosh 0 = 1,$ $\tanh 0 = 0.$

$\sinh \infty = \infty,$ $\cosh \infty = \infty,$ $\tanh \infty = 1.$

$\sinh(-x) = -\sinh x,$ $\cosh(-x) = \cosh x,$ $\tanh(-x) = -\tanh x.$

$\cosh^2 x - \sinh^2 x = 1,$ $\mathrm{sech}^2 x + \tanh^2 x = 1,$ $\mathrm{csch}^2 x - \coth^2 x = -1.$

$\sinh 2x = 2\sinh x \cosh x,$ $\cosh 2x = \cosh^2 x + \sinh^2 x.$

$2\sinh^2 \dfrac{x}{2} = \cosh x - 1,$ $2\cosh^2 \dfrac{x}{2} = \cosh x + 1.$

$\sinh(x \pm y) = \sinh x \cosh y \pm \cosh x \sinh y.$

$\cosh(x \pm y) = \cosh x \cosh y \pm \sinh x \sinh y.$

$\tanh(x \pm y) = \dfrac{\tanh x \pm \tanh y}{1 \pm \tanh x \tanh y}.$

Hyperbolic Functions of Pure Imaginary and Complex Quantities (335)

$\sinh jy = j \sin y;$ $\cosh jy = \cos y;$ $\tanh jy = j \tan y.$

$\sinh(x + jy) = \sinh x \cos y + j \cosh x \sin y.$

$\cosh(x + jy) = \cosh x \cos y + j \sinh x \sin y.$

$\sinh(x + 2j\pi) = \sinh x;$ $\cosh(x + 2j\pi) = \cosh x.$

$\sinh(x + j\pi) = -\sinh x;$ $\cosh(x + j\pi) = -\cosh x.$

$\sinh(x + \tfrac{1}{2}j\pi) = j \cosh x;$ $\cosh(x + \tfrac{1}{2}j\pi) = j \sinh x.$

Inverse or Antihyperbolic Functions (336)

If $x = \sinh y$, then y is the antihyperbolic sine of x or $y = \sinh^{-1} x$.

$$\sinh^{-1} x = \ln(x + \sqrt{x^2 + 1});\qquad \mathrm{csch}^{-1} x = \sinh^{-1} \frac{1}{x}.$$

$$\cosh^{-1} x = \ln(x + \sqrt{x^2 - 1});\qquad \mathrm{sech}^{-1} x = \cosh^{-1} \frac{1}{x}.$$

$$\tanh^{-1} x = \frac{1}{2} \ln \frac{1 + x}{1 - x};\qquad \coth^{-1} x = \tanh^{-1} \frac{1}{x}.$$

Derivatives of Hyperbolic Functions (337)

$\dfrac{d}{dx} \sinh x = \cosh x;$ $\dfrac{d}{dx} \cosh x = \sinh x;$ $\dfrac{d}{dx} \tanh x = \mathrm{sech}^2 x.$

$$\frac{d}{dx} \coth x = -\mathrm{csch}^2 x; \qquad \frac{d}{dx} \mathrm{sech}\, x = -\mathrm{sech}\, x \tanh x;$$

$$\frac{d}{dx} \mathrm{csch}\, x = -\mathrm{csch}\, x \coth x.$$

$$\frac{d}{dx} \sinh^{-1} x = \frac{1}{\sqrt{x^2 + 1}}; \qquad \frac{d}{dx} \cosh^{-1} x = \frac{1}{\sqrt{x^2 - 1}}; \qquad \frac{d}{dx} \tanh^{-1} x = \frac{1}{1 - x^2}.$$

$$\frac{d}{dx} \coth^{-1} x = -\frac{1}{x^2 - 1}; \qquad \frac{d}{dx} \mathrm{sech}^{-1} x = -\frac{1}{x\sqrt{1 - x^2}};$$

$$\frac{d}{dx} \mathrm{csch}^{-1} x = -\frac{1}{x\sqrt{x^2 + 1}}.$$

Some Integrals Leading to Hyperbolic Functions (338)

$$\int \sinh x \, dx = \cosh x; \qquad \int \cosh x \, dx = \sinh x; \qquad \int \tanh x \, dx = \ln \cosh x.$$

$$\int \coth x \, dx = \ln \sinh x; \qquad \int \mathrm{sech}\, x \, dx = \sin^{-1}(\tanh x);$$

$$\int \mathrm{csch}\, x \, dx = \ln \tanh \frac{x}{2}.$$

$$\int \frac{dx}{\sqrt{x^2 + a^2}} = \sinh^{-1} \frac{x}{a}; \qquad \int \frac{dx}{\sqrt{x^2 - a^2}} = \cosh^{-1} \frac{x}{a};$$

$$\int \frac{dx}{a^2 - x^2} = \frac{1}{a} \tanh^{-1} \frac{x}{a}. \ (x < a)$$

$$\int \frac{dx}{x\sqrt{a^2 + x^2}} = -\frac{1}{a} \sinh^{-1} \frac{a}{x}; \qquad \int \frac{dx}{x\sqrt{a^2 - x^2}} = -\frac{1}{a} \cosh^{-1} \frac{a}{x};$$

$$\int \frac{dx}{x^2 - a^2} = -\frac{1}{a} \tanh^{-1} \frac{a}{x}. \ (x > a)$$

$$\int \sqrt{x^2 - a^2} \, dx = \frac{x}{2} \sqrt{x^2 - a^2} - \frac{a^2}{2} \cosh^{-1} \frac{x}{a}.$$

$$\int \sqrt{x^2 + a^2} \, dx = \frac{x}{2} \sqrt{x^2 + a^2} + \frac{a^2}{2} \sinh^{-1} \frac{x}{a}.$$

Expansions of Hyperbolic Functions into Series (339)

$$\sinh x = x + \frac{x^3}{3!} + \frac{x^5}{5!} + \cdots.$$

$$\cosh x = 1 + \frac{x^2}{2!} + \frac{x^4}{4!} + \cdots.$$

$$\tanh x = x - \frac{x^3}{3} + \frac{2x^5}{15} + \frac{17x^7}{315} + \cdots.$$

$$\sinh^{-1} x = x - \frac{1}{2} \frac{x^3}{3} + \frac{1 \cdot 3}{2 \cdot 4} \frac{x^5}{5} - \frac{1 \cdot 3 \cdot 5}{2 \cdot 4 \cdot 6} \frac{x^7}{7} + \cdots. \ (x < 1)$$

$$\sinh^{-1} x = \ln 2x + \frac{1}{2} \frac{1}{2x^2} - \frac{1 \cdot 3}{2 \cdot 4} \frac{1}{4x^4} + \frac{1 \cdot 3 \cdot 5}{2 \cdot 4 \cdot 6} \frac{1}{6x^6} - \cdots. \ (x > 1)$$

$$\cosh^{-1}x = \ln 2x - \frac{1}{2}\frac{1}{2x^2} - \frac{1 \cdot 3}{2 \cdot 4}\frac{1}{4x^4} - \frac{1 \cdot 3 \cdot 5}{2 \cdot 4 \cdot 6}\frac{1}{6x^6} - \cdots .$$

$$\tanh^{-1}x = x + \frac{x^3}{3} + \frac{x^5}{5} + \frac{x^7}{7} + \cdots .$$

The Catenary (340) (Fig. 1.67)

For a definition, see (36).

Equation

$$y = \frac{a}{2}(e^{x/a} + e^{-x/a}) = a \cosh \frac{x}{a} .$$

If the width of the span is l and the sag is d, then the length of the arc (s) is found by means of the equations

$$\cosh z = \frac{2d}{l}z + 1, \qquad s = \frac{l}{z}\sinh z,$$

where z is to be found approximately from the first of these equations and this value substituted in the second.

If s and l are known, d may be found similarly by means of

$$\sinh z = \frac{s}{l}z, \qquad d = \frac{l}{2z}(\cosh z - 1).$$

1.12   PROGRESSIONS

An **arithmetic progression** is a sequence of terms each of which differs from the preceding by the same number d, called the *common difference*. If $n =$ number of

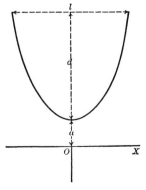

Fig. 1.67

terms, a = first term, l = last term, s = sum of n terms, then $l = a + (n - 1)d$, and $s = (n/2)(a + l)$. The arithmetic mean of two numbers is the number that placed between them would make them an arithmetic progression. Thus, the arithmetic mean of m, n is $(m + n)/2$.

$a, a + d, a + 2d, a + 3d, \ldots$, where d = common difference. (341)

The nth term, $t_n = a + (n - 1)d$.

The sum of n terms, $S_n = \dfrac{n}{2}[2a + (n - 1)d] = \dfrac{n}{2}(a + t_n)$.

The arithmetic mean of a and $b = \dfrac{a + b}{2}$.

A **geometric progression** is a sequence of terms each of which is obtained from the preceding by multiplying it by a fixed number r, called the *ratio*. If n = number of terms, a = first term, l = last term, s = sum of n terms, then $l = ar^{n-1}$, $s = (rl - a)/(r - 1) = a(1 - r^n)/(1 - r)$. The geometric mean of two numbers is the number which, placed between them, would make them a geometric progression. Thus, the geometric mean of m, n, is $\sqrt{m \cdot n}$.

$a, ar, ar^2, ar^3, \ldots$, where r = common ratio. (342)

The nth term, $t_n = ar^{n-1}$.

The sum of n terms, $S_n = a\left(\dfrac{r^n - 1}{r - 1}\right) = \dfrac{rt_n - a}{r - 1}$.

If $r^2 < 1$, S_n approaches a definite limit as n increases indefinitely, and

$$S_\infty = \frac{a}{1 - r}.$$

The geometric mean of a and $b = \sqrt{ab}$.

A **harmonic progression** is a sequence of terms whose reciprocals form an arithmetic progression. The harmonic mean of two numbers is the number which, placed between them, would make with them a harmonic progression. Thus, the harmonic mean of m, n, is $2mn/(m + n)$. (343)

The relation between **arithmetic, geometric,** and **harmonic means** of two numbers is expressed by the equality $G^2 = AH$, where G = geometric mean, A = arithmetic mean, and H = harmonic mean. (344)

1.13   SOME STANDARD SERIES

The following series are obtained through expansions of the functions by Taylor's or Maclaurin's theorems. The expression in parentheses following each series gives the region of convergence of the series, that is, the values of x for which the remainder, R_n, approaches zero as n increases, so that a number of terms of the series may be used for an approximation of the function. If the region of convergence is not indicated, it is to be understood that the series converges for all finite values of x. ($n! = 1 \cdot 2 \cdot 3 \cdots n$.)

Binomial Series (345)

$$(a + x)^n = a^n + na^{n-1}x + \frac{n(n-1)}{2!} a^{n-2}x^2 + \frac{n(n-1)(n-2)}{3!} a^{n-3}x^3 + \cdots .$$

$$(x^2 < a^2)$$

NOTE. The series consists of $(n + 1)$ terms when n is a positive integer; the number of terms is infinite when n is a negative or fractional number.

$$(a - bx)^{-1} = \frac{1}{a}\left(1 + \frac{bx}{a} + \frac{b^2x^2}{a^2} + \frac{b^3x^3}{a^3} + \cdots\right). \quad (b^2x^2 < a^2)$$

Exponential Series (346)

$$a^x = 1 + x \ln a + \frac{(x \ln a)^2}{2!} + \frac{(x \ln a)^3}{3!} + \cdots . \quad (347)$$

$$e^x = 1 + x + \frac{x^2}{2!} + \frac{x^3}{3!} + \cdots . \quad (348)$$

$$\frac{1}{2}(e^x + e^{-x}) = 1 + \frac{x^2}{2!} + \frac{x^4}{4!} + \cdots . \quad (349)$$

$$\frac{1}{2}(e^x - e^{-x}) = x + \frac{x^3}{3!} + \frac{x^5}{5!} + \cdots . \quad (350)$$

$$e^{-x^2} = 1 - x^2 + \frac{x^4}{2!} - \frac{x^6}{3!} + \frac{x^8}{4!} - \cdots . \quad (351)$$

Logarithmic Series (352)

$$\ln x = (x - 1) - \tfrac{1}{2}(x - 1)^2 + \tfrac{1}{3}(x - 1)^3 - \cdots . \quad (0 < x < 2) \quad (353)$$

$$\ln x = \frac{x - 1}{x} + \frac{1}{2}\left(\frac{x - 1}{x}\right)^2 + \frac{1}{3}\left(\frac{x - 1}{x}\right)^3 + \cdots . \quad (x > \tfrac{1}{2}) \quad (354)$$

$$\ln x = 2\left[\frac{x - 1}{x + 1} + \frac{1}{3}\left(\frac{x - 1}{x + 1}\right)^3 + \frac{1}{5}\left(\frac{x - 1}{x + 1}\right)^5 + \cdots\right]. \quad (x \text{ positive}) \quad (355)$$

$$\ln(1 + x) = x - \frac{x^2}{2} + \frac{x^3}{3} - \frac{x^4}{4} + \cdots . \quad (356)$$

$$\ln(a + x) = \ln a + 2\left[\frac{x}{2a + x} + \frac{1}{3}\left(\frac{x}{2a + x}\right)^3 + \frac{1}{5}\left(\frac{x}{2a + x}\right)^5 + \cdots\right].$$

$$\begin{pmatrix} a \text{ positive} \\ x \text{ between } -a \text{ and } +\infty \end{pmatrix} \quad (357)$$

$$\ln\left(\frac{1 + x}{1 - x}\right) = 2\left(x + \frac{x^3}{3} + \frac{x^5}{5} + \frac{x^7}{7} + \cdots\right). \quad (x^2 < 1) \quad (358)$$

$$\ln\left(\frac{x+1}{x-1}\right) = 2\left[\frac{1}{x} + \frac{1}{3}\left(\frac{1}{x^3}\right) + \frac{1}{5}\left(\frac{1}{x}\right)^5 + \frac{1}{7}\left(\frac{1}{x}\right)^7 + \cdots\right]. \quad (x^2 > 1) \quad (359)$$

$$\ln\left(\frac{x+1}{x}\right) = 2\left[\frac{1}{2x+1} + \frac{1}{3(2x+1)^3} + \frac{1}{5(2x+1)^5} + \cdots\right]. \quad (x \text{ positive}) \quad (360)$$

$$\ln(x + \sqrt{1+x^2}) = x - \frac{1}{2}\frac{x^3}{3} + \frac{1\cdot 3}{2\cdot 4}\frac{x^5}{5} - \frac{1\cdot 3\cdot 5}{2\cdot 4\cdot 6}\frac{x^7}{7} + \cdots. \quad (x^2 < 1) \quad (361)$$

Trigonometric Series (362)

$$\sin x = x - \frac{x^3}{3!} + \frac{x^5}{5!} - \frac{x^7}{7!} + \cdots. \quad (363)$$

$$\cos x = 1 - \frac{x^2}{2!} + \frac{x^4}{4!} - \frac{x^6}{6!} + \cdots. \quad (364)$$

$$\tan x = x + \frac{x^3}{3} + \frac{2x^5}{15} + \frac{17x^7}{315} + \frac{62x^9}{2835} + \cdots. \quad \left(x^2 < \frac{\pi^2}{4}\right) \quad (365)$$

$$\sin^{-1}x = x + \frac{1}{2}\frac{x^3}{3} + \frac{1\cdot 3}{2\cdot 4}\frac{x^5}{5} + \frac{1\cdot 3\cdot 5}{2\cdot 4\cdot 6}\frac{x^7}{7} + \cdots. \quad (x^2 < 1) \quad (366)$$

$$\tan^{-1}x = x - \frac{1}{3}x^3 + \frac{1}{5}x^5 - \frac{1}{7}x^7 + \cdots. \quad (x^2 \leqq 1) \quad (367)$$

Other Series

1. $1 + 2 + 3 + 4 + \cdots + (n - 1) + n = n(n + 1)/2.$ (368)

2. $p + (p + 1) + (p + 2) + \cdots + (q - 1) + q =$
                                             $(q + p)(q - p + 1)/2.$ (369)

3. $2 + 4 + 6 + 8 + \cdots + (2n - 2) + 2n = n(n + 1).$ (370)

4. $1 + 3 + 5 + 7 + \cdots + (2n - 3) + (2n - 1) = n^2.$ (371)

5. $1^2 + 2^2 + 3^2 + 4^2 + \cdots + (n - 1)^2 + n^2 = n(n + 1)(2n + 1)/6.$ (372)

6. $1^3 + 2^3 + 3^3 + 4^3 + \cdots + (n - 1)^3 + n^3 = n^2(n + 1)^2/4.$ (373)

7. $\dfrac{1 + 2 + 3 + 4 + 5 \cdots + n}{n^2} \rightarrow \dfrac{1}{2}$

8. $\dfrac{1 + 2^2 + 3^2 + 4^2 + \cdots + n^2}{n^3} \rightarrow \dfrac{1}{3}$ as $n \rightarrow \infty.$ (374)

9. $\dfrac{1 + 2^3 + 3^3 + 4^3 + \cdots + n^3}{n^4} \rightarrow \dfrac{1}{4}$

1.14   APPROXIMATIONS OF EXPRESSIONS CONTAINING SMALL TERMS

These may be derived from various infinite series given in Section 1.13. Some first approximations derived by neglecting all powers but the first of the small positive or negative quantity $x = s$ are given below. The expression in brackets gives the next term beyond that which is used and by means of it the accuracy of the approximation may be estimated.

$$\frac{1}{1 + s} = 1 - s. \qquad [+s^2] \quad (375)$$

$$(1 + s)^n = 1 + ns. \qquad \left[+\frac{n(n - 1)}{2} s^2\right] \quad (376)$$

$$e^s = 1 + s. \qquad \left[+\frac{s^2}{2}\right] \quad (377)$$

$$\ln (1 + s) = s. \qquad \left[-\frac{s^2}{2}\right] \quad (378)$$

$$\sin s = s. \qquad \left[-\frac{s^3}{6}\right] \quad (379)$$

$$\cos s = 1. \qquad \left[-\frac{s^2}{2}\right] \quad (380)$$

$$(1 + s_1)(1 + s_2) = (1 + s_1 + s_2). \quad [+s_1 s_2] \quad (381)$$

The following expressions are some that may be approximated by $1 + s$, where s is a small positive or negative quantity and n is any number.

$$\left(1 + \frac{s}{n}\right)^n. \quad (382) \quad e^s. \quad (385) \quad 1 + \ln \sqrt{\frac{1 + s}{1 - s}}. \quad (388)$$

$$\sqrt[n]{1 + ns}. \quad (383) \quad 2 - e^{-s}. \quad (386) \quad 1 + n \sin \frac{s}{n}. \quad (389)$$

$$\sqrt[n]{\frac{1 + \dfrac{ns}{2}}{1 - \dfrac{ns}{2}}}. \quad (384) \quad 1 + n \ln \left(1 + \frac{s}{n}\right). \quad (387) \quad \cos\sqrt{-2s}. \quad (390)$$

1.15   INDETERMINATE FORMS

Let $f(x)$ and $F(x)$ be two functions of x, and let a be a value of x.

If $\dfrac{f(a)}{F(a)} = \dfrac{0}{0}$ or $\dfrac{\infty}{\infty}$, use $\dfrac{f'(a)}{F'(a)}$ for the value of this fraction.   (391)

If $\dfrac{f'(a)}{F'(a)} = \dfrac{0}{0}$ or $\dfrac{\infty}{\infty}$, use $\dfrac{f''(a)}{F''(a)}$ for the value of this fraction, and so on.   (392)

If $f(a) \cdot F(a) = 0 \cdot \infty$ or if $f(a) - F(a) = \infty - \infty$, evaluate by changing the product or difference to the form $0/0$ or ∞/∞ and use (391) or (392).   (393)

If $f(a)^{F(a)} = 0^0$ or ∞^0 or 1^∞, then $f(a)^{F(a)} = e^{F(a) \cdot \ln f(a)}$, and the exponent, being of the form $0 \cdot \infty$, may be evaluated by (2).   (394)

1.16   MATRICES AND DETERMINANTS

Definitions

1.   A *matrix* is a system of mn quantities, called *elements,* arranged in a rectangular array of m rows and n columns.

$$A = \begin{pmatrix} a_{11} & a_{12} & \cdots & a_{1n} \\ a_{21} & a_{22} & \cdots & a_{2n} \\ \cdots\cdots & \ddots & \cdots \\ a_{m1} & a_{m2} & & a_{mn} \end{pmatrix} = \begin{Vmatrix} a_{11} & a_{12} & \cdots & a_{1n} \\ a_{21} & a_{22} & \cdots & a_{2n} \\ \cdots\cdots & \ddots & \cdots \\ a_{m1} & a_{m2} & & a_{mn} \end{Vmatrix} = (a_{ij}) = \|a_{ij}\| \begin{array}{l} i = 1, \dots, m \\ j = 1, \dots, n \end{array}$$

2. If $m = n$, then A is a *square* matrix of *order n*.

3. Two matrices are *equal* if and *only if* they have the same number of rows and of columns, and corresponding elements are equal.

4. Two matrices are *transposes* (sometimes called *conjugates*) of each other, if either is obtained from the other by interchanging rows and columns.

5. The *complex conjugate* of a matrix (a_{ij}) with complex elements is the matrix $(\bar{a}_{ij})$.

6. A matrix is *symmetric* if it is equal to its transpose, that is, if $a_{ij} = a_{ji}$; $i, j = 1, \dots, n$.

7. A matrix is *skew-symmetric,* or *antisymmetric,* if $a_{ij} = -a_{ji}$; $i, j = 1, \dots,$ n. The diagonal elements $a_n = 0$.

8. A matrix all of whose elements are zero is a *zero matrix.*

9. If the nondiagonal elements a_{ij}, $i \neq j$, of a square matrix A are all zero, then A is a *diagonal matrix.* If, furthermore, the diagonal elements are all equal, the matrix is a *scalar matrix;* if they are all 1, it is an *identity* or *unit matrix,* denoted by I.

10. The *determinant* $|A|$ of a square matrix (a_{ij}); $i, j = 1, \dots, n$, is the sum of the $n!$ products $a_{1r_1}a_{2r_2} \cdots a_{nr_n}$, in which $r_1, r_2, \dots, r_n$ is a permutation of $1, 2, \dots, n$, and the sign of each product is $+$ or $-$ according as the permutation is obtained from $1, 2, \dots, n$ by an even or an odd number of interchanges of two numbers.

    Symbols used are

$$|A| = \begin{vmatrix} a_{11} & a_{12} & \cdots & a_{1n} \\ a_{21} & a_{22} & \cdots & a_{2n} \\ \cdots & \cdots & \ddots & \cdots \\ a_{n1} & a_{n2} & & a_{nn} \end{vmatrix} = |a_{ij}| \quad i, j = 1, \dots, n$$

11. A square matrix (a_{ij}) is *singular* if its determinant $|a_{ij}|$ is zero.

12. The determinants of the square submatrices of any matrix A, obtained by striking out certain rows or columns, or both, are called the *determinants* or *minors* of A. A matrix is of *rank r* if it has at least one r-rowed determinant that is not zero, while all its determinants of order higher than r are zero. The *nullity d* of a square matrix of order n is $d = n - r$. The zero matrix is of rank 0.

13. The *minor* D_{ij} of the element a_{ij} of a square matrix is the determinant of the submatrix obtained by striking out the row and column in which a_{ij} lies. The *cofactor* A_{ij} of the element a_{ij} is $(-1)^{i+j}D_{ij}$. A *principal minor* is the minor obtained by striking out the same rows as columns.

14. The *inverse* of the square matrix A is

$$A^{-1} = \begin{pmatrix} \dfrac{A_{11}}{|A|} & \cdots & \dfrac{A_{n1}}{|A|} \\ \cdots & \ddots & \cdots \\ \dfrac{A_{1n}}{|A|} & & \dfrac{A_{nn}}{|A|} \end{pmatrix}$$

$$AA^{-1} = A^{-A} = I$$

15. The *adjoint* of A is

$$\text{adj } A = \begin{pmatrix} A_{11} & \cdots & A_{n1} \\ \cdots & \ddots & \cdots \\ A_{1n} & & A_{nn} \end{pmatrix}$$

16. *Elementary transformations* of a matrix are
  a. The interchange of two rows or of two columns.
  b. The addition to the elements of a row (or column) of any constant multiple of the corresponding elements of another row (or column).
  c. The multiplication of each element of a row (or column) by any nonzero constant.

17. Two $m \times n$ matrices A and B are *equivalent* if it is possible to pass from one to the other by a finite number of elementary transformations.
  a. The matrices A and B are equivalent if and only if there exist two non-singular square matrices E and F, having m and n rows, respectively, such that $EAF = B$.
  b. The matrices A and B are equivalent if and only if they have the same rank.

Matrix Operations

Addition and Subtraction. The sum or difference of two matrices (a_{ij}) and (b_{ij}) is the matrix $(a_{ij} \pm b_{ij})$, $i = 1, \ldots, m; j = 1, \ldots, n$.

Scalar Multiplication. The product of the scalar k and the matrix (a_{ij}) is the matrix (ka_{ij}).

Matrix Multiplication. The product (p_{ik}), $i = 1, \ldots, m; k = 1, \ldots, q$, of two matrices (a_{ij}), $i = 1, \ldots, m; j = 1, \ldots, n$, and (b_{jk}), $j = 1, \ldots, n; k = 1, \ldots, q$, is the matrix whose elements are

$$p_{ik} = \sum_{j=1}^{n} a_{ij}b_{jk} = a_{i1}b_{1k} + a_{i2}b_{2k} + \cdots + a_{in}b_{nk}$$

The element in the ith row and kth column of the product is the sum of the n products of the n elements of the ith row of (a_{ij}) by the corresponding n elements of the kth column of (b_{jk}).

Example

$$\begin{pmatrix} a_{11} & a_{12} \\ a_{21} & a_{22} \end{pmatrix} \begin{pmatrix} b_{11} & b_{12} & b_{13} \\ b_{21} & b_{22} & b_{23} \end{pmatrix} = \begin{pmatrix} a_{11}b_{11} + a_{12}b_{21} & a_{11}b_{12} + a_{12}b_{22} & a_{11}b_{13} + a_{12}b_{23} \\ a_{21}b_{11} + a_{22}b_{21} & a_{21}b_{12} + a_{22}b_{22} & a_{21}b_{13} + a_{22}b_{23} \end{pmatrix}$$

All the laws of ordinary algebra hold for the addition and subtraction of matrices and for scalar multiplication.

Multiplication of matrices is not in general commutative, but it is associative and distributive.

If the product of two or more matrices is zero, it does not follow that one of the factors is zero. The factors are *divisors of zero.*

Example

$$\begin{pmatrix} a & 0 \\ b & 0 \end{pmatrix} \begin{pmatrix} 0 & 0 \\ c & d \end{pmatrix} = \begin{pmatrix} 0 & 0 \\ 0 & 0 \end{pmatrix}$$

Linear Dependence

1. The quantities $l_1, l_2, \ldots, l_n$ are *linearly dependent* if there exist constants $c_1, c_2, \ldots, c_n$, not all zero, such that

    $$c_1 l_1 + c_2 l_2 + \cdots + c_n l_n = 0$$

    If no such constants exist, the quantities are *linearly independent.*
2. The linear functions

    $$l_i = a_{i1}x_1 + a_{i2}x_2 + \cdots + a_{in}x_n \qquad i = 1, 2, \ldots, m$$

    are *linearly dependent* if and only if the matrix of the coefficients is of rank $r < m$. Exactly r of the l_i form a linearly independent set.
3. For $m > n$, any set of m linear functions are linearly dependent.

Consistency of Equations

1. The system of homogeneous linear equations

    $$a_{i1}x_1 + a_{i2}x_2 + \cdots + a_{in}x_n = 0 \qquad i = 1, 2, \ldots, m$$

    has solutions not all zero if the rank r of the matrix (a_{ij}) is less than n.

    If $m < n$, there always exist solutions not all zero. If $m = n$, there exist solutions not all zero if $|a_{1j}| = 0$.

If r of the equations are so selected that their matrix is of rank r, they determine uniquely r of the variables as homogeneous linear functions of the remaining $n - r$ variables. A solution of the system is obtained by assigning arbitrary values to the $n - r$ variables and finding the corresponding values of the r variables.

2. The system of linear equations

$$a_{i1}x_1 + a_{i2}x_2 + \cdots + a_{in}x_n = k_i \qquad i = 1, 2, \ldots, m$$

is consistent if and only if the *augmented* matrix derived from (a_{ij}) by annexing the column $k_1, \ldots, k_m$ has the same rank r as (a_{ij}).

As in the case of a system of homogeneous linear equations, r of the variables can be expressed in terms of the remaining $n - r$ variables.

Linear Transformations

1. If a linear transformation

$$x'_i = a_{i1}x_1 + a_{i2}x_2 + \cdots + a_{in}x_n \qquad i = 1, 2, \ldots, n$$

with matrix (a_{ij}) transforms the variables x_i into the variables x'_i, and a linear transformation

$$x''_i = b_{i1}x'_1 + b_{i2}x'_2 + \cdots + b_{in}x'_n \qquad i = 1, 2, \ldots, n$$

with matrix (b_{ij}) transforms the variables x'_i into the variables x''_i, then the linear transformation with matrix $(b_{ij})(a_{ij})$ transforms the variables x_i into the variables x''_i directly.

2. A real *orthogonal* transformation is a linear transformation of the variables x_i into the variables x'_i such that

$$\sum_{i=1}^{n} x_i^2 = \sum_{i=1}^{n} x_i'^2$$

A transformation is orthogonal if and only if the transpose of its matrix is the inverse of its matrix.

3. A *unitary* transformation is a linear transformation of the variables x_i into the variables x'_i such that

$$\sum_{i=1}^{n} x_i \bar{x}_i = \sum_{i=1}^{n} x'_i \bar{x}'_i$$

A transformation is unitary if and only if the transpose of the conjugate of its matrix is the inverse of its matrix.

Quadratic Forms

A *quadratic form* in n variables is

$$\sum_{i,j=1}^{n} a_{ij}x_ix_j = a_{11}x_1^2 + a_{12}x_1x_2 + \cdots + a_{1n}x_1x_n$$

$$+ a_{21}x_2x_1 + a_{22}x_2^2 + \cdots + a_{2n}x_2x_n$$

$$\cdots$$

$$+ a_{n1}x_nx_1 + a_{n2}x_nx_2 + \cdots + a_{nn}x_n^2$$

in which $a_{ji} = a_{ij}$. The symmetric matrix (a_{ij}) of the coefficients is the *matrix* of the quadratic form and the rank of (a_{ij}) is the *rank* of the quadratic form.

A real quadratic form of rank r can be reduced by a real nonsingular linear transformations to the *normal form*

$$x_1^2 + \cdots + x_p^2 - x_{p+1}^2 - \cdots - x_p^2$$

in which the *index p* is uniquely determined.

If $p = r$, a quadratic form is *positive,* and, if $p = 0$, it is *negative.* If, furthermore, $r = n$, both are *definite.* A quadratic form is positive definite if and only if the determinant and all the principal minors of its matrix are positive.

A method of reducing a quadratic form to its normal form is illustrated.

Example

$$q = 3x^2 - 4y^2 - z^2 + 4xy - 2xz + 4yz$$

$$q = \begin{cases} 3x^2 + 2xy - xz \\ +2xy - 4y^2 + 2yz \\ -xz + 2yz - z^2 \end{cases}$$
$\quad\begin{aligned} &= \tfrac{1}{3}(3x + 2y - z)^2 + q_1, \text{ in which the quantity} \\ &\text{in parentheses is obtained by factoring } x \text{ out} \\ &\text{of the first row} \\ &= \tfrac{1}{3}(9x^2 + 4y^2 + z^2 + 12xy - 6xz - 4yz) + q_1 \end{aligned}$

$$q_1 = -\tfrac{4}{3}y^2 - \tfrac{1}{3}z^2 + \tfrac{4}{3}yz - 4y^2 + 4yz - z^2$$

$$= \begin{cases} -\tfrac{16}{3}y^2 + \tfrac{8}{3}yz \\ +\tfrac{8}{3}yz - \tfrac{4}{3}z^2 \end{cases} = -\tfrac{1}{16}(-\tfrac{16}{3}y + \tfrac{8}{3}z)^2 + q_2$$

$$q_2 = 0$$

The transformation

$$x' = 3x + 2y - z$$
$$y' = -\tfrac{16}{3}y + \tfrac{8}{3}z$$
$$z' = z$$

reduces q to $\tfrac{1}{3}x'^2 - \tfrac{3}{16}y'^2$.

The transformation

$$x'' = \sqrt{3}x'$$

$$y'' = \frac{4}{\sqrt{3}}y'$$

$$z'' = z'$$

further reduces q to the normal form $x''^2 - y''^2$ of rank 2 and index 1.

Expressing x, y, z in terms of x'', y'', z'', the real nonsingular linear transformation that reduces q to the normal form is

$$x = \frac{\sqrt{3}}{3}x'' + \frac{1}{2\sqrt{3}}y''$$

$$y = -\frac{\sqrt{3}}{4}y'' + \tfrac{1}{2}z''$$

$$z = z''$$

Hermitian Forms

A *Hermitian form* in n variables is

$$\sum_{i,j=1}^{n} a_{ij}x_i\bar{x}_j \qquad a_{ji} = \bar{a}_{ij}$$

The matrix (a_{ij}) is a *Hermitian matrix*. Its transpose is equal to its conjugate. The rank of (a_{ij}) is the *rank* of the Hermitian form.

A Hermitian form of rank r can be reduced by a nonsingular linear transformation to the *normal form*

$$x_1\bar{x}_1 + \cdots + x_p\bar{x}_p = x_{p+1}\bar{x}_{p+1} - \cdots - x_r\bar{x}_r$$

in which the *index* p is uniquely determined.

If $p = r$, the Hermitian form is *positive,* and, if $p = 0$, it is *negative.* If, furthermore, $r = n$, both are *definite.*

Determinants

Second- and third-order determinants are formed from their square symbols by taking diagonal products, down from left to right being positive and up negative.

$$\begin{vmatrix} a_{11} & a_{12} \\ a_{21} & a_{22} \end{vmatrix} = a_{11}a_{22} - a_{21}a_{12}$$

$$\begin{vmatrix} a_{11} & a_{12} & a_{13} \\ a_{21} & a_{22} & a_{23} \\ a_{31} & a_{32} & a_{33} \end{vmatrix} = a_{11}a_{22}a_{33} + a_{12}a_{23}a_{31} + a_{13}a_{32}a_{21}$$

$$- a_{31}a_{22}a_{13} - a_{32}a_{23}a_{11} - a_{33}a_{12}a_{21}$$

Third and higher order determinants are formed by selecting any row or column and taking the sum of the products of each element and its cofactor. This process is continued until second- or third-order cofactors are reached.

$$\begin{vmatrix} a_{11} & a_{12} & a_{13} \\ a_{21} & a_{22} & a_{23} \\ a_{31} & a_{32} & a_{33} \end{vmatrix} = a_{11} \begin{vmatrix} a_{22} & a_{23} \\ a_{32} & a_{33} \end{vmatrix} - a_{21} \begin{vmatrix} a_{12} & a_{13} \\ a_{32} & a_{33} \end{vmatrix} + a_{31} \begin{vmatrix} a_{12} & a_{13} \\ a_{22} & a_{23} \end{vmatrix}$$

Th determinant of a matrix A is

1. Zero, if two rows or two columns of A have proportional elements.
2. Unchanged, if
   a. The rows and columns of A are interchanged.
   b. To each element of a row or column of A is added a constant multiple of the corresponding element of another row or column.
3. Changed in sign, if two rows or two columns of A are interchanged.
4. Multiplied by c, if each element of any row or column of A is multiplied by c.
5. The sum of the determinants of two matrices B and C, if A, B, and C have all the same elements, except that in one row or column each element of A is the sum of the corresponding elements of B and C.

Example

$$\begin{vmatrix} 2 & 9 & 9 & 4 \\ 2 & -3 & 12 & 8 \\ 4 & 8 & 3 & -5 \\ 1 & 2 & 6 & 4 \end{vmatrix} = \begin{vmatrix} 2 & 5 & 9 & 4 \\ 2 & -7 & 12 & 8 \\ 4 & 0 & 3 & -5 \\ 1 & 0 & 6 & 4 \end{vmatrix} = 3 \begin{vmatrix} 2 & 5 & 3 & 4 \\ 2 & -7 & 4 & 8 \\ 4 & 0 & 1 & -5 \\ 1 & 0 & 2 & 4 \end{vmatrix}$$

Multiply 1st column   Factor 3 out of
by -2 and add to 2nd.   the 3rd column.

$$-3 \times (-5) \begin{vmatrix} 2 & 4 & 8 \\ 4 & 1 & -5 \\ 1 & 2 & 4 \end{vmatrix} + 3 \times (-7) \begin{vmatrix} 2 & 3 & 4 \\ 4 & 1 & -5 \\ 1 & 2 & 4 \end{vmatrix} = \qquad 0 \qquad -21 \begin{vmatrix} 1 & 1 & 0 \\ 4 & 1 & -5 \\ 1 & 2 & 4 \end{vmatrix}$$

Expand according to 2nd column.   1st and 3rd   Subtract 3rd
rows are proportional.   row from 1st.

$$= -21 \begin{vmatrix} 1 & -5 \\ 2 & 4 \end{vmatrix} - (-21) \begin{vmatrix} 4 & -5 \\ 1 & 4 \end{vmatrix} = -21[(4 + 10) - (16 + 5)] = +147$$

Expand according to 1st row.

1.17 PERMUTATIONS AND COMBINATIONS

Fundamental Principle. If in a sequence of s events the first event can occur in n_1 ways, the second in $n_2, \ldots$, the sth in n_s, then the number of different ways in which the sequence can occur is $n_1 n_2 \cdots n_r$.

A permutation of n objects taken r at a time is an arrangement of any r objects selected from the n objects. The number of permutations of n objects taken r at a time is

$$_nP_r = n(n-1)(n-2)\cdots(n-r+1) = \frac{n!}{(n-r)!}$$

In particular, $_nP_1 = n$, $_nP_n = n!$.
Cyclic permutations are

$$_nP_r^c = \frac{n!}{r(n-r)!} \qquad _nP_n^c = (n-1)!$$

If the n objects are divided into s sets each containing n_i objects that are alike, the distinguishable permutations are

$$n = n_1 + n_2 + \cdots + n_s, \qquad _nP_n = \frac{n!}{n_1!n_2!\cdots n_s!}$$

A combination of n objects taken r at a time is an unarranged selection of any r of the n objects. The number of combinations of n objects taken r at a time is

$$_nC_r = \frac{_nP_r}{r!} = \frac{n!}{r!(n-r)!} = _nC_{n-r}$$

In particular, $_nC_1 = n$, $_nC_n = 1$.
Combinations taken any number at a time, $_nC_1 + _nC_2 + \cdots + _nC_n = 2^n - 1$.

CHAPTER 2
MECHANICS/MATERIALS

2.1 KINEMATICS

2.1.1 Linear Motion

The **velocity** (v) of a particle that moves uniformly s feet in t seconds:

$$v = \frac{s}{t} \quad \text{feet/second.} \tag{2.1}$$

NOTE.   The velocity v of a moving particle at any instant equals ds/dt. The speed of a moving particle equals the magnitude of its velocity but has no direction.

The **acceleration** (a) of a particle whose velocity increases uniformly v feet per second in t seconds:

$$a = \frac{v}{t} \quad \text{feet/second}^2 \tag{2.2}$$

NOTE.   The acceleration a of a moving particle at any instant equals dv/dt or d^2s/dt^2. The acceleration g of a falling body *in vacuo* at sea level and latitude 45° equals 32.17 feet (9.805 m)/second^2.

The **velocity** (v_t) at the end of t seconds acquired by a particle having an initial velocity of v_0 feet per second and a uniform acceleration of a feet/second^2:

$$v_t = v_0 + at \quad \text{feet/second.} \tag{2.3}$$

NOTE. a is negative if the initial velocity and the acceleration act in opposite directions.

The **distance** (s) traversed in t seconds by a particle having an initial velocity of v_0 feet/second and a uniform acceleration of a feet/second^2:

$$s = v_0 t + \tfrac{1}{2}at^2 \quad \text{feet.} \tag{2.4}$$

The **distance** (s) required for a particle with an initial velocity of v_0 feet/second and a uniform acceleration of a feet/second^2 to reach a velocity of v_t feet/second:

$$s = \frac{v_t^2 - v_0^2}{2a} \quad \text{feet.} \tag{2.5}$$

The **velocity** (v_t) acquired, in traveling s feet, by a particle having an initial velocity of v_0 feet/second and a uniform acceleration of a feet/second^2:

$$v_t = \sqrt{v_0^2 + 2as} \quad \text{feet/second.} \tag{2.6}$$

The **time** (t) required for a particle having an initial velocity of v_0 feet/second and a uniform acceleration of a feet/second^2 to travel s feet:

$$t = \frac{-v_0 + \sqrt{v_0^2 + 2as}}{a} \quad \text{seconds.} \tag{2.7}$$

The **uniform acceleration** (a) required to move a particle, with an initial velocity of v_0 feet/second, s feet in t seconds:

$$a = \frac{2(s - v_0 t)}{t^2} \quad \text{feet/second}^2. \tag{2.8}$$

2.1.2   Circular Motion

The **angular velocity** (ω) of a particle moving uniformly through θ radians in t seconds:

$$\omega = \frac{\theta}{t} \quad \text{radians/second.} \tag{2.9}$$

NOTE.   The angular velocity (ω) of a moving particle at any instant equals $d\theta/dt^2$.

The **normal acceleration** (a) toward the center of its path of a particle moving uniformly with v feet/second tangential velocity and r feet radius of curvature of path:

$$a = \frac{v^2}{r} \quad \text{feet/second}^2. \tag{2.10}$$

NOTE.   The tangential acceleration of a particle moving with constant speed in a circular path is zero.

The **angular acceleration** (α) of a particle whose angular velocity increases uniformly ω radians/second in t seconds:

$$\alpha = \frac{\omega}{t} \quad \text{radians/second}^2. \tag{2.11}$$

NOTE.   The angular acceleration α of a moving particle at any instant equals $d\omega/dt$ or $d^2\theta/dt^2$.

The **angular velocity** (ω_t) at the end of t seconds acquired by a particle having an initial angular velocity of ω_0 radians/second and a uniform angular acceleration of α radians/second^2:

$$\omega_t = \omega_0 + \alpha t \quad \text{radians/second.} \tag{2.12}$$

The **angle** (θ) subtended in t seconds by a particle having an initial angular velocity of ω_0 radians/second and a uniform angular acceleration of α radians/second^2:

$$\theta = \omega_0 t + \tfrac{1}{2}\alpha t^2 \quad \text{radians.} \tag{2.13}$$

The **angle** (θ) subtended by a particle with an initial angular velocity of ω_0 radians/second and a uniform angular acceleration of α radians/second^2 in acquiring an angular velocity of ω_t radians/second:

$$\theta = \frac{\omega_t^2 - \omega_0^2}{2\alpha} \quad \text{radians.} \tag{2.14}$$

The **angular velocity** (ω_t) acquired in subtending θ radians by a particle having an initial angular velocity of ω_0 radians/second and a uniform angular acceleration of α radians/second^2:

$$\omega_t = \sqrt{\omega_0^2 + 2\alpha\theta} \quad \text{radians/second.} \tag{2.15}$$

The **time** (t) required for a particle having an initial angular velocity of ω_0 radians/second and a uniform angular acceleration of α radians/second^2 to subtend θ radians:

$$t = \frac{-\omega_0 + \sqrt{\omega_0^2 + 2\alpha\theta}}{\alpha} \quad \text{seconds.} \tag{2.16}$$

The **uniform angular acceleration** (α) required for a particle with an initial angular velocity of ω_0 radians/second to subtend θ radians in t seconds:

$$\alpha = \frac{2(\theta - \omega_0 t)}{t^2} \quad \text{radians/second}^2. \tag{2.17}$$

The **velocity** (v) of a particle r feet from the axis of rotation in a body making n revolutions/second:

$$v = 2\pi r n \quad \text{feet/second.} \tag{2.18}$$

The **velocity** (v) of a particle r feet from the axis of rotation in a body rotating with an angular velocity of ω radians/second:

$$v = \omega r \quad \text{feet/second.} \tag{2.19}$$

The **angular velocity** (ω) of a body making n revolutions/second:

$$\omega = 2\pi n \quad \text{radians/second.} \tag{2.20}$$

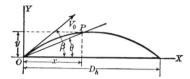

Fig. 2.1 Path of a projectile.

2.1.3   Path of a Projectile*

The **horizontal component of velocity** (v_x) of a particle having an initial velocity of v_0 feet/second in a direction making an angle of β degrees with the horizontal:

$$v_x = v_0 \cos \beta \quad \text{feet/second.} \tag{2.21}$$

The **horizontal distance** (x) traveled in t seconds by a particle having an initial velocity of v_0 feet/second at β degrees with the horizontal and a uniform downward acceleration of a feet/second^2:

$$x = v_0 t \cos \beta \quad \text{feet.} \tag{2.22}$$

The **vertical component of velocity** (v_y) at the end of t seconds of a particle having an initial velocity of v_0 feet/second at β degrees with the horizontal and a uniform downward acceleration of a feet/second^2:

$$v_y = v_0 \sin \beta - at \quad \text{feet/second.} \tag{2.23}$$

The **vertical distance** (y) traveled in t seconds by a particle having an initial velocity of v_0 feet/second at β degrees with the horizontal and a uniform downward acceleration of a feet/second^2:

$$y = v_0 t \sin \beta - \tfrac{1}{2} a t^2 \quad \text{feet.} \tag{2.24}$$

The **time** (t_v) to reach the highest point of the path of a particle having an initial velocity of v_0 feet/second at β degrees with the horizontal and a uniform downward acceleration of a feet/second^2:

$$t_v = \frac{v_0 \sin \beta}{a} \quad \text{seconds.} \tag{2.25}$$

The **vertical distance** (d_v) from the horizontal to the highest point of the path of a particle having an initial velocity of v_0 feet/second at β degrees with the horizontal and a uniform downward acceleration of a feet/second^2:

$$d_v = \frac{v_0^2 \sin^2 \beta}{2a} \quad \text{feet.} \tag{2.26}$$

*Friction of the air is neglected throughout.

The **velocity** (v) at the end of t seconds of a particle having an initial velocity of v_0 feet/second at β degrees with the horizontal and a uniform downward acceleration of a feet/second^2:

$$v = \sqrt{v_x^2 + v_y^2} = \sqrt{v_0^2 - 2v_0\, at \sin\beta + a^2 t^2} \quad \text{feet/second.} \quad (2.27)$$

The **time** (t_h) to reach the same horizontal as at the start for a particle having an initial velocity of v_0 feet/second at β degrees with the horizontal and a uniform downward acceleration of a feet/second^2:

$$t_h = \frac{2v_0 \sin\beta}{a} \quad \text{seconds.} \quad (2.28)$$

The **horizontal distance** (d_h) traveled by a particle having an initial velocity of v_0 feet/second at β degrees with the horizontal and a uniform downward acceleration of a feet/second^2 in returning to the same horizontal as at the start:

$$d_h = \frac{v_0^2 \sin 2\beta}{a} \quad \text{feet.} \quad (2.29)$$

The **time** (t) to reach any point P for a particle having an initial velocity of v_0 feet/second at β degrees with the horizontal and a uniform downward acceleration of a feet/second^2 if a line through P and the point of starting makes θ degrees with the horizontal:

$$t = \frac{2v_0 \sin(\beta - \theta)}{a \cos\theta} \quad \text{seconds.} \quad (2.30)$$

2.1.4 Harmonic Motion

Simple harmonic motion is the motion of the projection, on the diameter of a circle, of a particle moving with constant speed around the circumference of the circle. **Amplitude** is one-half the projection of the path of the particle or equal to the radius of the circle. **Frequency** is the number of complete oscillations per unit time.

The **displacement** (x) from the center t seconds after starting, of the projection on the diameter, of a particle moving with a uniform angular velocity of ω radians/second about a circle r feet in radius:

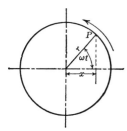

Fig. 2.2 Harmonic motion.

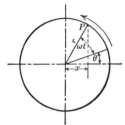

Fig. 2.3 Acceleration.

$$x = r \cos \omega t \quad \text{feet.} \tag{2.31}$$

The **velocity** (v), t seconds after starting, of the projection on the diameter, of a particle moving with a uniform angular velocity of ω radians/second about a circle r feet in radius:

$$v = -\omega r \sin \omega t \quad \text{feet/second.} \tag{2.32}$$

The **acceleration** (a), t seconds after starting, of the projection on the diameter, of a particle moving with a uniform angular velocity of ω radians/second about a circle r feet in radius:

$$a = -\omega^2 r \cos \omega t = -\omega^2 x \quad \text{feet/second}^2. \tag{2.33}$$

NOTE.   If the time t is reckoned from a position displaced by θ radians from the horizontal (called lead if positive and lag if negative) the formulas become $x = r \cos(\omega t + \theta)$ feet, $v = -\omega r \sin(\omega t + \theta)$ feet/second, and $a = -\omega^2 r \cos(\omega t + \theta)$ feet/second^2.

2.2   RELATIONS OF MASS AND SPACE

2.2.1   Mass

The **mass** (m) of a body weighing w pounds:

$$m = \frac{w}{g} \quad \text{pounds (mass).} \tag{2.34}$$

NOTE.   The mass m of a body may be measured by its weight w, designated "pounds (abs.)," or by its weight w divided by the acceleration due to gravity g, designated "pounds (mass)." In this book the latter unit is used throughout.

2.2.2   Center of Gravity

The **center of gravity** of a body or system of bodies is that point through which the resultant of the weights of the component particles passes, whatever position be given the body or system.

NOTE. The center of mass of a body is the same as the center of gravity. The center of gravity of a line, surface, or volume is obtained by considering it to be the center of gravity of a slender rod, thin plate, or homogeneous body and is often called the *centroid*.

The **moment** (M) of a body of weight w, or of mass m, about a plane if x is the horizontal distance from the center of gravity of the body to the plane:

$$M = wx \quad \text{or} \quad M = mx. \tag{2.35}$$

The **moment** (S) of an area A, about an axis X if x is the horizontal distance from the center of gravity of the area to the axis:

$$S = Ax. \tag{2.36}$$

NOTE. The moment of an area about an axis through its center of gravity is zero.

The **distances** (x_0, y_0, z_0) from each of three coordinate planes (X, Y, Z) to the center of gravity or mass of a system of bodies if Σw is the sum of their weights or Σm is the sum of their masses and Σwx, Σwy, Σwz, or Σmx, Σmy, Σmz are the algebraic sums of moments of the separate bodies about the X, Y, and Z planes:

$$
\begin{aligned}
x_0 &= \frac{\Sigma wx}{\Sigma w} = \frac{\Sigma mx}{\Sigma m}, \\
y_0 &= \frac{\Sigma wy}{\Sigma w} = \frac{\Sigma my}{\Sigma m}, \\
z_0 &= \frac{\Sigma wz}{\Sigma w} = \frac{\Sigma mz}{\Sigma m}
\end{aligned}
\tag{2.37}
$$

The **distances** (x_0, y_0, z_0) from each of three coordinate planes to the center of gravity of a volume if Σv is the sum of the component volumes and Σvx, Σvy, and Σvz are the algebraic sums of the moments of these component volumes about the X, Y, and Z planes:

$$x_0 = \frac{\Sigma vx}{\Sigma v}, \qquad y_0 = \frac{\Sigma vy}{\Sigma v}, \qquad z_0 = \frac{\Sigma vz}{\Sigma v}. \tag{2.38}$$

The **distances** (x_0, y_0) from each of two coordinate axes to the center of gravity of an area if ΣA is the sum of the component areas and ΣAx and ΣAy are the algebraic sums of the moments of these component areas about the X and Y axes:

$$x_0 = \frac{\Sigma Ax}{\Sigma A}, \qquad y_0 = \frac{\Sigma Ay}{\Sigma A}. \tag{2.39}$$

NOTE. The general method of finding the center of gravity of an irregular area is to divide it into component areas, the centers of gravity of which may be calculated or determined from Table 2.1, then find the sum of statical moments of the component areas about some convenient axis, and divide by the total area to obtain the distance from that

(*text continued on page 94*)

Table 2.1 Properties of Various Plane Sections

Section	Distance to Center of Gravity, x	Moment of Inertia, I^a	Radius of Gyration, K
Square	$x_a = x_b = \dfrac{b}{2}$ $x_d = \dfrac{b}{\sqrt{2}}$	$I_{AA} = I_{BB} = I_{DD} = \dfrac{b^4}{12}$ $I_{CC} = \dfrac{b^4}{3}$ $J = \dfrac{b^4}{6}$	$K_{AA} = K_{BB} = K_{DD} = \dfrac{b}{\sqrt{12}}$ $= 0.289b$ $K_{CC} = \dfrac{b}{\sqrt{3}} = 0.577b$
Hollow Square	$x_a = x_b = \dfrac{b}{2}$ $x_d = \dfrac{b}{\sqrt{2}}$	$I_{AA} = I_{BB} = I_{DD} = \dfrac{b^4 - b_1^4}{12}$ $I_{CC} = \dfrac{b^4}{3} - \dfrac{b_1^2(3b^2 + b_1^2)}{12}$ $J = \dfrac{b^4 - b_1^4}{6}$	$K_{AA} = K_{BB} = K_{DD} = \sqrt{\dfrac{b^2 + b_1^2}{12}}$ $= 0.289\sqrt{b^2 + b_1^2}$

$^a J$, polar moment of inertia, refers to an axis through the center of gravity.

Table 2.1 (*Continued*)

Section	Distance to Center of Gravity, x	Moment of Inertia, I^a	Radius of Gyration, K
Rectangle	$x_a = \dfrac{h}{2}$ $x_b = \dfrac{b}{2}$ $x_d = \dfrac{bh}{\sqrt{b^2 + h^2}}$	$I_{AA} = \dfrac{bh^3}{12}$ $I_{BB} = \dfrac{hb^3}{12}$ $I_{CC} = \dfrac{bh^3}{3}$ $I_{DD} = \dfrac{b^3 h^3}{6(b^2 + h^2)}$ $J = \dfrac{bh^3 + hb^3}{12}$	$K_{AA} = \dfrac{h}{\sqrt{12}} = 0.289h$ $K_{BB} = \dfrac{b}{\sqrt{12}} = 0.289b$ $K_{CC} = \dfrac{h}{\sqrt{3}} = 0.577h$ $K_{DD} = \dfrac{bh}{\sqrt{6(b^2 + h^2)}}$
Rectangle	$x = \dfrac{b \sin \alpha + h \cos \alpha}{2}$	$I_{AA} = \dfrac{bh(b^2 \sin^2 \alpha + h^2 \cos^2 \alpha)}{12}$	$K_{AA} = \sqrt{\dfrac{b^2 \sin^2 \alpha + h^2 \cos^2 \alpha}{12}}$

aJ, polar moment of inertia, refers to an axis through the center of gravity.

Table 2.1 (Continued)

Section	Distance to Center of Gravity, x	Moment of Inertia, I^a	Radius of Gyration, K
	$x_a = \dfrac{h}{2}$ $x_b = \dfrac{b}{2}$	$I_{AA} = \dfrac{bh^3 - b_1 h_1^3}{12}$ $I_{BB} = \dfrac{hb^3 - h_1 b_1^3}{12}$ $I_{CC} = \dfrac{bh^3}{3} - \dfrac{b_1 h_1}{12}(3h^2 + h_1^2)$	$K_{AA} = \sqrt{\dfrac{bh^3 - b_1 h_1^3}{12(bh - b_1 h_1)}}$ $K_{BB} = \sqrt{\dfrac{hb^3 - h_1 b_1^3}{12(hb - h_1 b_1)}}$
Triangle	$x_a = \dfrac{2}{3}h$	$I_{AA} = \dfrac{bh^3}{36}$ $I_{BB} = \dfrac{bh^3}{12}$	$K_{AA} = \dfrac{h}{\sqrt{18}} = 0.236h$ $K_{BB} = \dfrac{h}{\sqrt{6}} = 0.408h$
Trapezoid	$x_a = \dfrac{h(b_1 + 2b)}{3(b + b_1)}$ $x_b = \dfrac{h(b + 2b_1)}{3(b + b_1)}$	$I_{AA} = \dfrac{h^3(b^2 + 4bb_1 + b_1^2)}{36(b + b_1)}$ $I_{BB} = \dfrac{h^3(b + 3b_1)}{12}$	$K_{AA} = \dfrac{h\sqrt{2(b^2 + 4bb_1 + b_1^2)}}{6(b + b_1)}$ $K_{BB} = \dfrac{h\sqrt{b + 3b_1}}{\sqrt{6(b + b_1)}}$

$^a J$, polar moment of inertia, refers to an axis through the center of gravity.

Table 2.1 *(Continued)*

Section	Distance to Center of Gravity, x	Moment of Inertia, I^a	Radius of Gyration, K
Circle	$x_a = x_b = \dfrac{d}{2} = r$	$I_{AA} = \dfrac{\pi d^4}{64} = 0.0491 d^4$ $= \dfrac{\pi r^4}{4} = 0.7854 r^4$ $J = \dfrac{\pi r^4}{2}$	$K_{AA} = \dfrac{d}{4} = \dfrac{r}{2}$
Hollow circle	$x_a = x_b = \dfrac{d}{2} = r$	$I_{AA} = \dfrac{\pi(d^4 - d_1^4)}{64}$ $= 0.0491(d^4 - d_1^4)$ $= \dfrac{\pi(r^4 - r_1^4)}{4}$ $= 0.7854(r^4 - r_1^4)$ $J = \dfrac{\pi(r^4 - r_1^4)}{2}$	$K_{AA} = \dfrac{\sqrt{d^2 + d_1^2}}{4}$ $= \dfrac{\sqrt{r^2 + r_1^2}}{2}$
Semi-circle	$x_a = \dfrac{d(3\pi - 4)}{6\pi} = 0.288d$ $= 0.576r$ $x_b = \dfrac{2d}{3\pi} = 0.212d = \dfrac{4r}{3\pi} = 0.424r$	$I_{AA} = \dfrac{d^4(9\pi^2 - 64)}{1152\pi} = 0.00686 d^4$ $= 0.1098 r^4$	$K_{AA} = \dfrac{d}{12\pi}\sqrt{(9\pi^2 - 64)}$ $= 0.132d$

^{a}J, polar moment of inertia, refers to an axis through the center of gravity.

Table 2.1 (*Continued*)

Section	Distance to Center of Gravity, x	Moment of Inertia, I^a	Radius of Gyration, K
Hollow Half Circle	$x_b = \dfrac{2(d^3 - d_1^3)}{3\pi(d^2 - d_1^2)}$	$I_{AA} = \dfrac{\pi(d^4 - d_1^4)}{128} - \dfrac{4(d^3 - d_1^3)^2}{72\pi(d^2 - d_1^2)}$	$K_{AA} = \sqrt{\dfrac{(d^4 - d_1^4)}{16(d^2 - d_1^2)} - \dfrac{4(d^3 - d_1^3)^2}{9\pi^2(d^2 - d_1^2)^2}}$
Circular Segment	$x = \dfrac{2}{3}\dfrac{r^3 \sin^3\alpha}{A}$ $[A = \tfrac{1}{2}r^2(2\alpha - \sin 2\alpha)$ where first α is in radians]	$I_{AA} = \dfrac{1}{4}Ar^2\left(1 - \dfrac{2}{3}\dfrac{\sin^3\alpha\,\cos\alpha}{\alpha - \sin\alpha\,\cos\alpha}\right)$ $I_{BB} = \dfrac{1}{4}Ar^2\left(1 + \dfrac{2\sin^3\alpha\,\cos\alpha}{\alpha - \sin\alpha\,\cos\alpha}\right)$	$K = \sqrt{\dfrac{I}{A}}$
Circular Sector	$x = \dfrac{2}{3}\dfrac{r \sin\alpha}{\alpha}$	$I_{AA} = \dfrac{1}{4}Ar^2\left(1 - \dfrac{\sin\alpha\,\cos\alpha}{\alpha}\right)$ $I_{BB} = \dfrac{1}{4}Ar^2\left(1 + \dfrac{\sin\alpha\,\cos\alpha}{\alpha}\right)$	$K = \sqrt{\dfrac{I}{A}}$

$^a J$, polar moment of inertia, refers to an axis through the center of gravity.

Table 2.1 (*Continued*)

Section	Distance to Center of Gravity, x	Moment of Inertia, I^a	Radius of Gyration, K
Parabolic Segment	$x = \frac{3}{5}a$ (for half-segment, $y = \frac{3}{8}b$)	$I_{AA} = \frac{4}{15}ab^3$ $I_{BB} = \frac{4}{7}ba^3$	$K_{AA} = \frac{b}{\sqrt{5}} = 0.447b$ $K_{BB} = a\sqrt{\frac{3}{7}} = 0.654a$
Ellipse	$x_a = a$ $x_b = b$	$I_{AA} = \frac{\pi a^3 b}{4} = 0.7854a^3b$ $I_{BB} = \frac{\pi b^3 a}{4} = 0.7854ab^3$ $J = \frac{\pi ab(a^2 + b^2)}{4}$	$K_{AA} = \frac{a}{2}$ $K_{BB} = \frac{b}{2}$
Elliptical Ring	$x_a = a$ $x_b = b$	$I_{AA} = \frac{\pi}{4}(a^3b - a_1^3 b_1)$ $= 0.7854(a^3b - a_1^3 b_1)$ $I_{BB} = \frac{\pi}{4}(b^3a - b_1^3 a_1)$ $= 0.7854(b^3a - b_1^3 a_1)$	$K_{AA} = \frac{1}{2}\sqrt{\dfrac{a^3b - a_1^3 b_1}{ab - a_1 b_1}}$ $K_{BB} = \frac{1}{2}\sqrt{\dfrac{b^3a - b_1^3 a_1}{ba - b_1 a_1}}$

^aJ, polar moment of inertia, refers to an axis through the center of gravity.

Table 2.1 *(Continued)*

Section	Distance to Center of Gravity, x	Moment of Inertia, I^a	Radius of Gyration, K
Equal Angle	$x_a = x_b = \dfrac{a^2 + (a - t)t}{2(2a - t)}$ $[\alpha = 45°]$	$I_{AA} = \dfrac{t(a - x)^3 + ax^3 - a(x - t)^3}{3}$ $I_{BB} = I_{AA}$ $I_{CC} = \dfrac{bt^3 + b^3t + 3a^2bt + t^4}{12}$ $I_{DD} = \dfrac{bt^3 + b^3t + 3bt(a - 4x + 2t)^2 + t^4 + 6t^2(2x - t)^2}{12}$	$K = \sqrt{\dfrac{I}{A}}$
Unequal Angle	$x_a = \dfrac{t(b + 2c) + c^2}{2(b + c)}$ $x_b = \dfrac{t(2d + a) + d^2}{2(a + d)}$ $\tan 2d = \dfrac{t(2x_b - t)a(a - 2x_a) + d(2x_a - t)(b + t - 2x_b)}{2(I_{AA} - I_{BB})}$	$I_{AA} = \dfrac{t(a - x_a)^3 + bx_a^3 - d(x_a - t)^3}{3}$ $I_{BB} = \dfrac{t(b - x_b)^3 + ax_b^3 - c(x_b - t)^3}{3}$ $I_{CC} = \dfrac{I_{AA}\cos^2\alpha - I_{BB}\sin^2\alpha}{\cos 2\alpha}$ $I_{DD} = \dfrac{I_{BB}\cos^2\alpha - I_{AA}\sin^2\alpha}{\cos 2\alpha}$	$K = \sqrt{\dfrac{I}{A}}$
I-Beam	$x_a = \dfrac{d}{2}$ $x_b = \dfrac{b}{2}$	$I_{AA} = \dfrac{bd^3 - c^3(b - t)}{12}$ $I_{BB} = \dfrac{2mb^3 + ct^3}{12}$	$K_{AA} = \sqrt{\dfrac{bd^3 - c^3(b - t)}{12[bd - c(b - t)]}}$ $K_{BB} = \sqrt{\dfrac{2mb^3 + ct^3}{12[bd - c(b - t)]}}$

$^a J$, polar moment of inertia, refers to an axis through the center of gravity.

Table 2.1 *(Continued)*

Section	Distance to Center of Gravity, x	Moment of Inertia, I^a	Radius of Gyration, K
Channel	$x_a = \dfrac{d}{2}$ $x_b = \dfrac{(dt^2/2) + 2am(t + a/2)}{dt + 2am}$	$I_{AA} = \dfrac{bd^3 - ac^3}{12}$ $I_{BB} = \dfrac{dx_b^3 - d(x_b - t)^3 + 2m(b - x_b)^3}{3}$	$K_{AA} = \sqrt{\dfrac{bd^3 - ac^3}{12(bd - ac)}}$ $K_{BB} =$ $\sqrt{\dfrac{dx_b^3 - d(x_b - t)^3 + 2m(b - x_b)^3}{3(bd - ac)}}$
Tee	$x_a = \dfrac{(bm^2/2) + et\left(\dfrac{e}{2} + m\right)}{bm + et}$ $x_b = \dfrac{b}{2}$	$I_{AA} =$ $\dfrac{bx_a^3 + t(d - x_a)^3 - (b - t)(x_a - m)^3}{3}$ $I_{BB} = \dfrac{mb^3 + et^3}{12}$	$K_{AA} =$ $\sqrt{\dfrac{bx_a^3 + t(d - x_a)^3 - (b - t)(x_a - m)^3}{3(bm + et)}}$ $K_{BB} = \sqrt{\dfrac{mb^3 + et^3}{12(bm + et)}}$

[a] J, polar moment of inertia, refers to an axis through the center of gravity.

93

axis to the center of gravity of the whole area. In numerical problems it is often convenient to take the axis of reference through the center of gravity of one of the component areas, thereby eliminating the moment of that area and simplifying the numerical work.

2.2.3   Moment of Inertia

The **moment of inertia** (I) of an area about an axis is the sum of the products of the component areas into the square of their distances from the axis (ΣAx^2):

$$I = \Sigma Ax^2. \tag{2.40}$$

NOTE.   In general an expression for moment of inertia involves the use of calculus, the area being considered as divided into differential areas dA. $I_x = \int x^2\, dA$ and $I_y = \int y^2\, dA$. The unit of moment of inertia of an area is inches, feet, and so on, to the fourth power.

The **moment of inertia** (I_x) of an area A about any axis in terms of the moment of inertia I_0 about a parallel axis through the center of gravity of the area, if x_0 is the distance between the two axes:

$$I_x = I_0 + Ax_0^2. \tag{2.41}$$

The **radius of gyration** (K) of an area A from an axis about which the moment of inertia is I:

$$K = \sqrt{\frac{I}{A}}. \tag{2.42}$$

The **radius of gyration** (K_x) of an area A about any axis in terms of the radius of gyration K_0 about a parallel axis through the center of gravity of the area if x_0 is the distance between the two axes:

$$K_x^2 = K_0^2 + x_0^2. \tag{2.43}$$

The **product of inertia** (U) of an area with respect to two rectangular coordinate axes is the sum of the products of the component areas into the product of their distances from the two axes (ΣAxy):

$$U = \Sigma Axy. \tag{2.44}$$

NOTE.   The product of inertia, like the moment of inertia, is generally expressed by the use of calculus:

$$U = \int xy\, dA. \tag{2.45}$$

In case one of the areas is an axis of symmetry, the product of inertia is zero.

The **product of inertia** (U_{xy}) of an area A about any two rectangular axes in terms of the product of inertia U_0 about two parallel rectangular axes through the center of gravity of the area, if x_0 and y_0 are the distances between these two sets of axes:

$$U_{xy} = U_0 + Ax_0y_0. \tag{2.46}$$

The **moment of inertia** ($I_{x'}$ and $I_{y'}$) and **product of inertia** ($U_{x'y'}$) of an area A about each of two rectangular coordinate axes (X' and Y') in terms of the moments and product of inertia (I_x, I_y, U_{xy}) about two other rectangular coordinate axes making an angle α with X' and Y':

$$I_{x'} = I_y \sin^2\alpha + I_x \cos^2\alpha - 2U_{xy} \cos \alpha \sin \alpha. \tag{2.47}$$
$$I_{y'} = I_y \cos^2\alpha + I_y \sin^2\alpha + 2U_{xy} \cos \alpha \sin \alpha. \tag{2.48}$$
$$U_{x'y'} = (I_x - I_y) \cos \alpha \sin \alpha + U_{xy}(\cos^2\alpha - \sin^2\alpha). \tag{2.49}$$

The **principal axes of an area** are those axes, through any point, about one of which the moment of inertia is a maximum, the moment of inertia about the other being a minimum. The axes are at right angles to each other.

The **angle** (α) between the rectangular coordinate axes X and Y, about which the moments and products of inertia are I_x, I_y, and U_{xy}, and the principal axes through the point of intersection of X and Y:

$$\tan 2\alpha = \frac{2U_{xy}}{I_y - I_x}. \tag{2.50}$$

NOTE. An axis of symmetry is a principal axis. The product of inertia about principal axes is zero. If I_y and I_x are moments of inertia about principal axes the equations for the moments of inertia about rectangular axes making an angle α with these principal axes are: $I_{x'} = I_y \sin^2\alpha + I_x \cos^2\alpha$ and $I_{y'} = I_y \cos^2\alpha + I_x \sin^2\alpha$. The sum of the moments of inertia about rectangular coordinate axes is a constant for all pairs of axes intersecting at the same point, that is, $I_x + I_y = I_{x'} + I_{y'}$.

The **polar moment of inertia** (J) of an area is the moment of inertia about an axis perpendicular to the plane of the area and is equal to the sum of the products of the component areas into the squares of their distances from the axis (ΣAr^2):

$$J = \Sigma Ar^2. \tag{2.51}$$

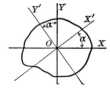

Fig. 2.4 Moment of inertia and product of inertia.

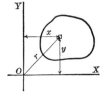

Fig. 2.5 Polar moment of inertia.

NOTE.   The polar moment of inertia is generally expressed by the use of calculus: $J = \int r^2 \, dA$.

The **polar moment of inertia** (J) of an area A in terms of the moments of inertia I_x and I_y about two rectangular coordinate axes intersecting on the polar axis:

$$J = I_x + I_y. \tag{2.52}$$

The **moment of inertia** (I_m) of a body about an axis, in terms of the mass, is the sum of the products of the component masses and the squares of their distances from the axis (Σmr^2):

$$I_m = \Sigma mr^2. \tag{2.53}$$

The **moment of inertia** (I) of a body about an axis, in terms of the weight, is the sum of the products of the component weights and the squares of their distances from the axis (Σwr^2):

$$I = \Sigma wr^2. \tag{2.54}$$

The **moment of inertia** (I_m) in terms of the mass for a case where the moment of inertia in terms of the weight is I:

$$I_m = \frac{I}{g}. \tag{2.55}$$

NOTE.   The moment of inertia of a body is generally expressed by calculus. $I_m = \int r^2 \, dm$. $I = \int r^2 \, dw$. The unit of moment of inertia of solid is pound-feet^2.

The **moment of inertia** (I_x) of a body of weight W about any axis in terms of the moment of inertia (I_o) about a parallel axis through the center of gravity of the body if x_0 is the distance between the axes:

$$I_x = I_o + Wx_0^2. \tag{2.56}$$

The **radius of gyration** (K) of a body of weight W from an axis about which the moment of inertia is I:

$$K = \sqrt{\frac{I}{W}}. \tag{2.57}$$

The **moment of inertia** (I_m), in terms of the mass, of a body of weight W about an axis for which the radius of gyration is K:

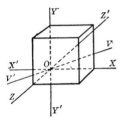

Fig. 2.6 Moment of inertia with respect to the axis $V'V$.

$$I_m = \frac{W}{g} K^2. \tag{2.58}$$

The **product of inertia** (U or U_m) of a body with respect to two coordinate planes is the sum of the products of the component weights (or masses) and the products of their distances from these planes (Σwxy or Σmxy):

$$U = \Sigma wxy, \qquad U_m = \Sigma mxy. \tag{2.59}$$

NOTE. The product of inertia of a body is generally expressed by calculus. $U = \int xy$ $dw. U_m = \int xy\ dm.$

The **moment of inertia** (I) with respect to the axis $V'V$ in terms of the moments of inertia I_x, I_y, and I_z with respect to the axes $X'X$, $Y'Y$, and $Z'Z$ and the products of inertia U_{xy}, U_{xz}, and U_{yz} with respect to the planes Y_{0y} and X_{0x}, the planes Y_{0z} and X_{0x}, and the planes X_{0z} and X_{0y}, respectively, where $V'V$ passes through the origin of these three axes and makes the angles α, β, and γ with the axes $X'X$, $Y'Y$, and $Z'Z$, respectively:

$$\begin{aligned}
I = I_x \cos^2\alpha + I_y \cos^2\beta + I_z \cos^2\gamma &- 2U_{xy} \cos\alpha \cos\beta \\
&- 2U_{xz} \cos\alpha \cos\gamma - 2U_{yz} \cos\beta \cos\gamma. \quad (2.60)
\end{aligned}$$

The **principal axes** of a body are those three rectangular axes through any point, about one of which the moment of inertia is a maximum and about another a minimum, the moment of inertia about the third axis being intermediate in value. The **principal planes** are the planes perpendicular to the principal axes. The products of inertia with respect to the principal planes are zero. (*Table 2.2 follows on page 98*)

2.3 KINETICS

2.3.1 Translation

Three laws of motion: (1) A body remains in a state of rest or of uniform motion except under the action of some unbalanced force. (2) A single force acting on a body causes it to move with accelerated motion in the direction of the force. The

Table 2.2 Properties of Various Solids[a]

Solid	Moment of Inertia, I	Radius of Gyration, K
Straight Rod	$I_{AA} = \frac{1}{12}Wl^2$ $I_{BB} = \frac{1}{3}Wl^2$ $I_{CC} = \frac{1}{3}Wl^2 \sin^2\alpha$	$K_{AA} = \dfrac{1}{\sqrt{12}}$ $K_{BB} = \dfrac{1}{\sqrt{3}}$ $K_{CC} = 1\sqrt{\dfrac{\sin\alpha}{3}}$
Rod bent into a Circular Arc	$I_{AA} =$ $\dfrac{1}{2}Wr^2\left(1 - \dfrac{\sin\alpha\cos\alpha}{\alpha}\right)$ $I_{BB} =$ $\dfrac{1}{2}Wr^2\left(1 + \dfrac{\sin\alpha\cos\alpha}{\alpha}\right)$	$K_{AA} =$ $r\sqrt{\dfrac{1}{2}\left(1 - \dfrac{\sin\alpha\cos\alpha}{\alpha}\right)}$ $K_{BB} =$ $r\sqrt{\dfrac{1}{2}\left(1 + \dfrac{\sin\alpha\cos\alpha}{\alpha}\right)}$
Cube	$I_{AA} = I_{BB} = \frac{1}{6}Wa^2$	$K_{AA} = K_{BB} = \dfrac{a}{\sqrt{6}}$
Rectangular Prism	$I_{AA} = \frac{1}{12}W(a^2 + b^2)$ $I_{BB} = \frac{1}{12}W(b^2 + c^2)$	$K_{AA} = \sqrt{\dfrac{a^2 + b^2}{12}}$ $K_{BB} = \sqrt{\dfrac{b^2 + c^2}{12}}$

[a]All axes pass through the center of gravity unless otherwise noted. $I_m = I/g$. W is the total weight of the body.

acceleration is directly proportional to the force and inversely proportional to the mass of the body. (3) To every action there is an equal and opposite reaction.

The **force** (F) imparting an acceleration of a feet/second^2 to a weight of m pounds (mass):

$$F = ma \quad \text{pounds.} \tag{2.61}$$

Table 2.2 (*Continued*)

Solid	Moment of Inertia, I	Radius of Gyration, K

Right Circular Cylinder

$$I_{AA} = \tfrac{1}{2}Wr^2$$
$$I_{BB} = \tfrac{1}{12}W(3r^2 + h^2)$$

$$K_{AA} = \frac{r}{\sqrt{2}}$$

$$K_{BB} = \sqrt{\frac{3r^2 + h^2}{12}}$$

Hollow Right Circular Cylinder

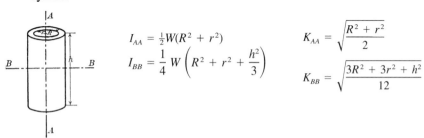

$$I_{AA} = \tfrac{1}{2}W(R^2 + r^2)$$
$$I_{BB} = \frac{1}{4}\,W\left(R^2 + r^2 + \frac{h^2}{3}\right)$$

$$K_{AA} = \sqrt{\frac{R^2 + r^2}{2}}$$

$$K_{BB} = \sqrt{\frac{3R^2 + 3r^2 + h^2}{12}}$$

Thin Hollow Cylinder

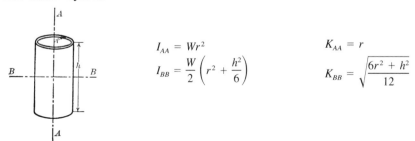

$$I_{AA} = Wr^2$$
$$I_{BB} = \frac{W}{2}\left(r^2 + \frac{h^2}{6}\right)$$

$$K_{AA} = r$$

$$K_{BB} = \sqrt{\frac{6r^2 + h^2}{12}}$$

*All axes pass through the center of gravity unless otherwise noted. $I_m = I/g$. W is the total weight of the body.

NOTE.   In terms of the weight w, $F = (w/g)a$.

The **impulse** (I) of a force of F pounds acting for t seconds:

$$I = Ft \quad \text{pound-seconds.} \tag{2.62}$$

The **momentum** ($\mathfrak{M}$) of a body of m pounds (mass) moving with a velocity of v feet/second:

Table 2.2 (*Continued*)

Solid	Moment of Inertia, I	Radius of Gyration, K

Elliptical Cylinder

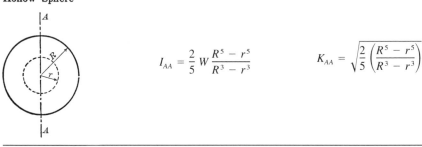

$$I_{AA} = \tfrac{1}{4}W(a^2 + b^2)$$
$$I_{BB} = \tfrac{1}{12}W(3b^2 + h^2)$$
$$I_{CC} = \tfrac{1}{12}W(3a^2 + h^2)$$

$$K_{AA} = \sqrt{\frac{a^2 + b^2}{2}}$$

$$K_{BB} = \sqrt{\frac{3b^2 + h^2}{12}}$$

$$K_{CC} = \sqrt{\frac{3a^2 + h^2}{12}}$$

Sphere

$$I_{AA} = \frac{2}{5}Wr^2$$

$$K_{AA} = \frac{2r}{\sqrt{10}}$$

Hollow Sphere

$$I_{AA} = \frac{2}{5}W\frac{R^5 - r^5}{R^3 - r^3}$$

$$K_{AA} = \sqrt{\frac{2}{5}\left(\frac{R^5 - r^5}{R^3 - r^3}\right)}$$

[a]All axes pass through the center of gravity unless otherwise noted. $I_m = I/g$. W is the total weight of the body.

$$\mathfrak{M} = mv \quad \text{pounds (mass)-feet/second.} \tag{2.63}$$

The **force** (F) required to change the velocity of m pounds (mass) from v_1 feet/second to v_2 feet/second in t seconds:

$$F = \frac{m(v_1 - v_2)}{t} \quad \text{pounds.} \tag{2.64}$$

Table 2.2 *(Continued)*

Solid	Moment of Inertia, I	Radius of Gyration, K

Thin Hollow Sphere

$$I_{AA} = \tfrac{2}{3}Wr^2$$

$$K_{AA} = \frac{2r}{\sqrt{6}}$$

Ellipsoid

$$I_{AA} = \tfrac{1}{5}W(b^2 + c^2)$$
$$I_{BB} = \tfrac{1}{5}W(a^2 + c^2)$$
$$I_{CC} = \tfrac{1}{5}W(a^2 + b^2)$$

$$K_{AA} = \sqrt{\frac{b^2 + c^2}{5}}$$

$$K_{BB} = \sqrt{\frac{a^2 + c^2}{5}}$$

$$K_{CC} = \sqrt{\frac{a^2 + b^2}{5}}$$

Torus

$$I_{AA} = W\left(R^2 + \tfrac{3}{4}r^2\right)$$
$$I_{BB} = W\left(\frac{R^2}{2} + \frac{5}{8}r^2\right)$$

$$K_{AA} = \tfrac{1}{2}\sqrt{4R^2 + 3r^2}$$

$$K_{BB} = \sqrt{\frac{4R^2 + 5r^2}{8}}$$

^{a}All axes pass through the center of gravity unless otherwise noted. $I_m = I/g$. W is the total weight of the body.

NOTE.   The change in momentum of a body during any time interval equals the impulse of the force acting on the body for that time.

The **work** (W) done by a force of F pounds acting through a distance of s feet:

$$W = Fs \quad \text{foot-pounds.} \tag{2.65}$$

NOTE.   If the force is variable, $W = \displaystyle\int_0^s F\, ds$.

The **power** (P) required to do W foot-pounds of work at a constant rate in t seconds:

Table 2.2 *(Continued)*

Solid	Distance to Center of Gravity, x	Moment of Inertia, I	Radius of Gyration, K
Right Rectangular Pyramid	$x = \dfrac{h}{4}$	$I_{AA} = \frac{1}{20}W(a^2 + b^2)$ $I_{BB} =$ $\dfrac{1}{20} W\left(b^2 + \dfrac{3h^2}{4}\right)$	$K_{AA} =$ $\sqrt{\dfrac{a^2 + b^2}{20}}$ $K_{BB} =$ $\sqrt{\frac{1}{80}(4b^2 + 3h^2)}$
Right Circular Cone	$x = \dfrac{h}{4}$	$I_{AA} = \frac{3}{10}Wr^2$ $I_{BB} =$ $\dfrac{3}{20} W\left(r^2 + \dfrac{h^2}{4}\right)$	$K_{AA} = \dfrac{3r}{\sqrt{30}}$ $K_{BB} =$ $\sqrt{\frac{3}{80}(4r^2 + h^2)}$
Frustum of a Cone	$x =$ $\dfrac{h(R^2 + 2Rr + 3r^2)}{4(R^2 + Rr + r^2)}$	$I_{AA} =$ $\dfrac{3}{10} W \dfrac{R^5 - r^5}{R^3 - r^3}$	$K_{AA} =$ $\sqrt{\dfrac{3}{10} \dfrac{R^5 - r^5}{R^3 - r^3}}$
Paraboloid	$x = \frac{1}{3}h$	$I_{AA} = \frac{1}{3}Wr^2$ $I_{BB} =$ $\frac{1}{18}W(3r^2 + h^2)$	$K_{AA} = \dfrac{r}{\sqrt{3}}$ $K_{BB} =$ $\sqrt{\frac{1}{18}(3r^2 + h^2)}$

^{a}All axes pass through the center of gravity unless otherwise noted. $I_m = I/g$. W = total weight of the body.

$$P = \frac{W}{t} \quad \text{foot-pounds/second.} \tag{2.66}$$

The **potential energy** (W), referred to a certain datum, of a body of w pounds weight and at an elevation of h feet above the datum:

$$W = wh \quad \text{foot-pounds.} \tag{2.67}$$

Table 2.2 *(Continued)*

Solid	Distance to Center of Gravity, x	Moment of Inertia, I	Radius of Gyration, K
Spherical Sector 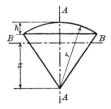	$x = \frac{3}{8}(2r - h)$	$I_{AA} = \frac{1}{5}W(3rh - h^2)$	$K_{AA} = \sqrt{\dfrac{3rh - h^2}{5}}$
Spherical Segment 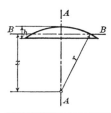	$x = \dfrac{3}{4}\dfrac{(2r - h)^2}{(3r - h)}$ For half-sphere $x = \frac{3}{8}r$	$I_{AA} = W\left(r^2 - \dfrac{3rh}{4} + \dfrac{3h^2}{20}\right)\dfrac{2h}{3r - h}$	$K_{AA} = \sqrt{\dfrac{I}{W}}$

*All axes pass through the center of gravity unless otherwise noted. $I_m = I/g$. W is the total weight of the body.

The **kinetic energy** (W) of a body of m pounds (mass) mass having a velocity of translation of v feet/second:

$$W = \frac{mv^2}{2} \quad \text{foot-pounds.} \tag{2.68}$$

The **force** (F) required to change the velocity of a mass of m pounds (mass) from v_1 feet/second to v_2 feet/second in s feet:

$$F = \frac{m(v_1^2 - v_2^2)}{2s} \quad \text{pounds.} \tag{2.69}$$

NOTE. The change in kinetic energy of the body equals the work done on the body.

The **force** (F) required to move m pounds (mass) in a circular path of r feet radius with a constant speed of v feet/second:

$$F = \frac{mv^2}{r} \quad \text{pounds.} \tag{2.70}$$

NOTE. This force acts along the normal to the path of the body toward the center of curvature and is called the *centripetal* or *deviating force*. The reaction to this force along the normal to the path of the body away from the center of curvature is called the *centrifugal force*.

2.3.2 Rotation

The **torque** or **moment** (T) about the axis of rotation imparting an angular acceleration of α radians/second^2 to a body with a mass moment of inertia I_m pound(mass)-feet squared about the axis of rotation:

$$T = I_m\alpha \quad \text{pound-feet.} \tag{2.71}$$

NOTE. In terms of the weight, w pounds, of the body and its radius of gyration, K feet, about the axis of rotation, $T = (w/g)K^2\alpha$ pound-feet.

The **angular impulse** (I_a) of a torque of T pound-feet acting for t seconds:

$$I_a = Tt \quad \text{pound-feet-seconds.} \tag{2.72}$$

The **angular momentum** ($\mathfrak{M}_a$) of a body with a mass moment of inertia of I_m pound (mass)-feet^2 about the axis of rotation and an angular velocity of ω radians/second:

$$\mathfrak{M}_a = I_m\omega \quad \text{pound(mass)-feet}^2/\text{second.} \tag{2.73}$$

NOTE. The angular momentum of a body is sometimes called its *moment of momentum.* The angular momentum of a body moving in a plane perpendicular to the axis of rotation is given by $\mathfrak{M}_a = \mathfrak{M}r$ pound(mass)-feet^2/second, where $\mathfrak{M}$ equals the momentum of the body in pound(mass)-feet/second and r equals the perpendicular distance in feet from the line of direction of the momentum to the axis of rotation.

The **torque** (T) required to change the angular velocity of a body of mass moment of inertia of I_m pound(mass)-feet^2 about the axis of rotation from ω_1 radians/second to ω_2 radians/second in t seconds:

$$T = \frac{I_m(\omega_1 - \omega_2)}{t} \quad \text{pound-feet.} \tag{2.74}$$

NOTE. The change in angular momentum of a body is equal to the angular impulse.

The **work** (W) done by a torque of T pound-feet acting through an angle of θ radians:

$$W = T\theta \quad \text{foot-pounds.} \tag{2.75}$$

NOTE. If the torque is variable, $W = \int_0^\theta T\,d\theta$. The work done by a torque of T pound-feet in N revolutions is given by $W = T \cdot 2\pi N$ foot-pounds.

The **kinetic energy** (W) of a body which has an angular velocity of ω radians per second and a mass moment of inertia of I_m pound(grav.)-feet mass squared about the axis of rotation:

$$W = \frac{I_m \omega^2}{2} \quad \text{foot-pounds.} \tag{2.76}$$

NOTE.   In terms of the weight, w pounds, of the body and its radius of gyration, K feet, about the axis of rotation, $W = (wK^2\omega^2)/2g$ foot-pounds.

The **torque** (T) required to change the angular velocity of a body of mass moment of inertia of I_m pound(mass)-feet^2 about the axis of rotation from ω_1 radians/second to ω_2 radians/second, the torque acting through an angle of θ radians:

$$T = \frac{I_m(\omega_1^2 - \omega_2^2)}{2\theta} \quad \text{pound-feet.} \tag{2.77}$$

NOTE.   The change in kinetic energy of a body equals the work done on the body.

The **center of percussion** with respect to the axis of rotation is the point through which the line of action of the resultant of all the external forces acting on the rotating body passes.

The **distance** (l) from the axis of rotation to the center of percussion of a body with a mass moment of inertia I_m pound(mass)-feet^2 about the axis of rotation, m pounds (mass), and x_0 feet between the axis and the center of gravity:

$$l = \frac{I_m}{x_0 m} \quad \text{feet.} \tag{2.78}$$

NOTE.   In terms of the radius of gyration K, $l = K^2/x_0$.

General Formulas for Rotation About a Fixed Axis

Assume a body AB rotating about the axis $Z'Z$. Let m = mass of the body; α = angular acceleration at any instant; ω = angular velocity at any instant, and x_0, y_0, and z_0 = the coordinates of the center of gravity of the body.

Considering the forces and motions of the small particles (as Δm) of which it may be composed, if ΣX, ΣY, ΣZ = the sums of the components of the forces parallel to the axes $X'X$, $Y'Y$, $Z'Z$, respectively; ΣT_x, ΣT_y, ΣT_z = the sums of the torques about the axes $X'X$, $Y'Y$, $Z'Z$, respectively; ΣI_m = the moment of inertia of the mass about the axis $Z'Z$; ΣU_{xzm}, ΣU_{yzm}, = the products of inertia of mass with respect to the planes YOZ and XOY and the planes XOZ and XOY, respectively.

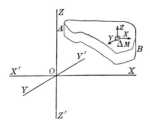

Fig. 2.7   Rotation about a fixed axis.

$$\Sigma X = -\alpha y_0 m - \omega^2 x_0 m \qquad \Sigma T_x = -\alpha U_{xz_m} + \omega^2 U_{yz_m}$$

$$\Sigma Y = +\alpha x_0 m - \omega^2 y_0 m \qquad \Sigma T_y = -\alpha U_{yz_m} - \omega^2 U_{xz_m}$$

$$\Sigma Z = 0, \qquad\qquad\qquad \Sigma T_z = \alpha I_m.$$

Table 2.3 Formulas for Translation and Rotation

Translation		Rotation	
Force	$F = ma$	Torque	$T = I_m\alpha$
Impulse	$I = Ft$	Angular impulse	$Ia = Tt$
Momentum	$\mathfrak{M} = mv$	Angular momentum	$\mathfrak{M}a = I_m\omega$
Change of momentum	$m(v_1 - v_0) = Ft$	Change of angular momentum	$I_m(\omega_1 - \omega_0) = Tt$
Work	$W = Fs$	Work	$W = T\theta$
Kinetic energy	$W = \frac{1}{2}mv^2$	Kinetic energy	$W = \frac{1}{2}I_m\omega^2$
Change of kinetic energy	$\frac{1}{2}m(v_1^2 - v_0^2) = F(s_1 - s_0)$	Change of kinetic energy	$\frac{1}{2}I_m(\omega_1^2 - \omega_1^2) = T(\theta_1 - \theta_0)$

2.3.3 Translation and Rotation

The **work** (W) done on a body by a force of F pounds having a torque of T pound-feet about the center of gravity of the body in moving the body s feet and causing it to rotate through an angle of θ radians:

$$W = Fs + T\theta. \tag{2.79}$$

The **kinetic energy** (W) of a body of m pounds (mass) with a mass moment of inertia of I_m pound(mass)-feet^2 about its center of gravity and having a velocity of translation of v feet/second and an angular velocity of ω radians/second:

$$W = \frac{1}{2}mv^2 + \frac{1}{2}\omega^2 I_m \quad \text{foot-pounds.} \tag{2.80}$$

NOTE. If the body weighs w pounds and has K feet radius of gyration about the center of gravity, $W = [\frac{1}{2}(w/g)v^2] + [\frac{1}{2}(w/g)K^2\omega^2]$.

The **kinetic energy** developed in a body during any displacement is equal to the external work done upon it:

$$Fs + T\theta = \frac{1}{2}mv^2 + \frac{1}{2}\omega^2 I_m \quad \text{foot-pounds.} \tag{2.81}$$

Instantaneous Axis. Any plane motion may be considered as a rotation about an axis which may be constantly changing to successive parallel positions. This axis at any instant is called the instantaneous axis.

NOTE. If the velocities, at any instant, of two points in a body are known, the instantaneous axis passes through the intersection of the perpendiculars to the lines of motion of these two points.

The **distance** (l) from the instantaneous axis to the center of percussion of a body of m pounds (mass) and mass moment of inertia of I_m pound(mass)-feet^2 about its center of gravity, for a position of the instantaneous axis of x_0 feet distance from the center of gravity of the body:

$$l = \frac{I_m}{x_0 m} + x_0.\qquad(2.82)$$

The **velocity of translation** (v_c) of the center of gravity of a body having an angular velocity of ω radians/second about the instantaneous axis which is x_0 feet from the center of gravity:

$$v_c = \omega x_0 \quad \text{feet/second.}\qquad(2.83)$$

The **kinetic energy** (W) of a body with a mass moment of inertia of I'_m pound(mass)-feet^2 about the instantaneous axis and an angular velocity of ω radians/second about the instantaneous axis:

$$W = \tfrac{1}{2}\omega^2 I'_m \quad \text{foot-pounds.}\qquad(2.84)$$

2.3.4 Pendulum

The imaginary pendulum conceived as a material point suspended by a weightless cord is called a **simple pendulum.** A real pendulum is called a **compound pendulum.**

The **period** (p) of oscillation (from a maximum deflection to the right to a maximum deflection to the left) of a simple pendulum l feet in length:

$$p = \pi \sqrt{\frac{l}{g}} \quad \text{seconds (for small oscillations).}\qquad(2.85)$$

NOTE. An approximate expression for all arcs is $t = \pi(l/g)^{1/2}[1 + (h/81)]$, where h is the vertical distance between the highest and lowest points of the path.

The **length** (l) of a simple seconds pendulum (one whose period of oscillation is 1 second):

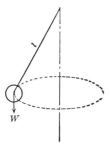

Fig. 2.8 Oscillation.

$$l = \frac{g}{\pi^2} \quad \text{feet.} \tag{2.86}$$

The **period** (p) of oscillation of a compound pendulum of K feet radius of gyration with respect to the axis of suspension and l feet length from the axis of suspension to the center of gravity of the pendulum:

$$p = \pi \sqrt{\frac{K^2}{lg}} \quad \text{(for small oscillations).} \tag{2.87}$$

The **distance** (d) from the center of suspension to the center of oscillation, of a compound pendulum, of K feet radius of gyration about the center of suspension, the distance from the center of suspension to the center of gravity being l feet:

$$d = \frac{K^2}{l} \quad \text{feet.} \tag{2.88}$$

NOTE. The period of oscillation, for a small oscillation, about an axis through the center of suspension is the same as that of a small oscillation about a parallel axis through the center of oscillation.

The **tension** (T) in the cord of a conical pendulum with a weight of W pounds and l feet length of cord, rotating with n revolutions/second:

$$T = \frac{Wl4\pi^2n^2}{g} \quad \text{pounds.} \tag{2.89}$$

NOTE. In terms of the angular velocity ω radians/second; $T = (Wl\omega^2)/g$ pounds.

The **period** (p) of oscillation of a simple cycloidal pendulum swinging on the arc of a cycloid described by a circle of r feet radius:

$$p = 2\pi \sqrt{\frac{r}{g}} \quad \text{seconds.} \tag{2.90}$$

2.3.5 Prony Brake

The **power** (P) indicated by a Prony brake when the perpendicular distance from the center of the pulley to the direction of a force of F pounds applied at the end of the brake arm is l feet and the pulley revolves at a speed of S revolutions/minute.

$$P = 1.903lSF \times 10^{-4} \quad \text{horsepower.} \tag{2.91}$$

NOTE. The torque of the pulley equals lF pound-feet. If l is made 5 feet 3 inches, $P = SF/1000$ horsepower.

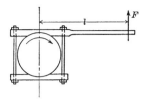

Fig. 2.9   Prony brake.

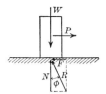

Fig. 2.10   Friction.

2.3.6   Friction

Static friction is the force, in addition to that overcoming inertia, required to set in motion one body in contact with another.

The **coefficient of static friction** (f) between two bodies, when N is the normal pressure between them and F is the corresponding static friction (N and F in the same units):

$$f = \frac{F}{N}. \tag{2.92}$$

The **resultant force** (R) between two bodies starting from relative rest with a normal pressure of N pounds and a static friction of F pounds between them:

$$R = \sqrt{F^2 + N^2} \text{ pounds}. \tag{2.93}$$

The **angle of static friction** (ϕ) for two surfaces with a normal pressure N and a static friction F between them (N and F in the same units):

$$\tan \phi = \frac{F}{N} = f. \tag{2.94}$$

NOTE.   The angle of repose is the angle of inclination of the surface of one body at which the other body will begin to slide along it, under the action of its own weight. The angle of repose (ϕ) is equal to the angle of static friction.

The **sliding friction** is the force, in addition to that overcoming inertia, required to maintain relative motion between two bodies.

NOTE.   (1) For moderate pressures the friction is proportional to the normal pressure between the surfaces. (2) For moderate pressures the friction is independent of the extent of the surface in contact. (3) At low velocities the friction is independent of the velocity of rubbing. The friction decreases as the velocity increases. (4) Sliding friction is usually less than static friction.

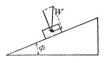

Fig. 2.11   Angle of static friction.

Table 2.4 Coefficients of Static and Sliding Friction

(Reference letters indicate the lubricant used; numbers in parentheses give the sources. See footnote)

Materials	Static		Sliding	
	Dry	Greasy	Dry	Greasy
Hard steel on hard steel	0.78 (1)	0.11 (1, a)	0.42 (2)	0.029 (5, h)
		0.23 (1, b)		0.081 (5, c)
		0.15 (1, c)		0.080 (5, i)
		0.11 (1, d)		0.058 (5, j)
		0.0075 (18, p)		0.084 (5, d)
		0.0052 (18, h)		0.105 (5, k)
				0.096 (5, l)
				0.108 (5, m)
				0.12 (5, a)
Mild steel on mild steel	0.74 (19)		0.57 (3)	0.09 (3, a)
				0.19 (3, u)
Hard steel on graphite	0.21 (1)	0.09 (1, a)		
Hard steel on babbitt (ASTM No. 1)	0.70 (11)	0.23 (1, b)	0.33 (6)	0.16 (1, b)
		0.15 (1, c)		0.06 (1, c)
		0.08 (1, d)		0.11 (1, d)
		0.085 (1, e)		
Hard steel on babbitt (ASTM No. 8)	0.42 (11)	0.17 (1, b)	0.35 (11)	0.14 (1, b)
		0.11 (1, c)		0.065 (1, c)
		0.09 (1, d)		0.07 (1, d)
		0.08 (1, e)		0.08 (11, h)
Hard steel on babbitt (ASTM No. 10)		0.25 (1, b)		0.13 (1, b)
		0.12 (1, c)		0.06 (1, c)
		0.10 (1, d)		0.055 (1, d)
		0.11 (1, e)		
Mild steel on cadmium silver				0.097 (2, f)
Mild steel on phosphor bronze			0.34 (3)	0.173 (2, f)
Mild steel on copper lead				0.145 (2, f)
Mild steel on cast iron		0.183 (15, c)	0.23 (6)	0.133 (2, f)
Mild steel on lead	0.95 (11)	0.5 (1, f)	0.95 (11)	0.3 (11, f)
Nickel on mild steel			0.64 (3)	0.178 (3, x)
Aluminum on mild steel	0.61 (8)		0.47 (3)	
Magnesium on mild steel			0.42 (3)	
Magnesium on magnesium	0.6 (22)	0.08 (22, y)		
Teflon on Teflon	0.04 (22)			0.04 (22, f)
Teflon on steel	0.04 (22)			0.04 (22, f)
Tungsten carbide on tungsten carbide	0.2 (22)	0.12 (22, a)		
Tungsten carbide on steel	0.5 (22)	0.08 (22, a)		
Tungsten carbide on copper	0.35 (23)			
Tungsten carbide on iron	0.8 (23)			
Bonded carbide on copper	0.35 (23)			
Bonded carbide on iron	0.8 (23)			
Cadmium on mild steel			0.46 (3)	
Copper on mild steel	0.53 (8)		0.36 (3)	0.18 (17, a)
Nickel on nickel	1.10 (16)		0.53 (3)	0.12 (3, w)
Brass on mild steel	0.51 (8)		0.44 (6)	
Brass on cast iron			0.30 (6)	
Zinc on cast iron	0.85 (16)		0.21 (7)	
Magnesium on cast iron			0.25 (7)	
Copper on cast iron	1.05 (16)		0.29 (7)	
Tin on cast iron			0.32 (7)	
Lead on cast iron			0.43 (7)	
Aluminum on aluminum	1.05 (16)		1.4 (3)	
Glass on glass	0.94 (8)	0.01 (10, p)	0.40 (3)	0.09 (3, a)
		0.005 (10, q)		0.116 (3, v)
Carbon on glass			0.18 (3)	
Garnet on mild steel			0.39 (3)	
Glass on nickel	0.78 (8)		0.56 (3)	
Copper on glass	0.68 (8)		0.53 (3)	
Cast iron on cast iron	1.10 (16)		0.15 (9)	0.070 (9, d)
				0.064 (9, n)
Bronze on cast iron			0.22 (9)	0.077 (9, n)
Oak on oak (parallel to grain)	0.62 (9)		0.48 (9)	0.164 (9, r)
				0.067 (9, s)
Oak on oak (perpendicular)	0.54 (9)		0.32 (9)	0.072 (9, s)
Leather on oak (parallel)	0.61 (9)		0.52 (9)	
Cast iron on oak			0.49 (9)	0.075 (9, n)
Leather on cast iron			0.56 (9)	0.36 (9, t)
				0.13 (9, n)
Laminated plastic on steel			0.35 (12)	0.05 (12, t)
Fluted rubber bearing on steel				0.05 (13, t)

(1) Campbell, *Trans. ASME*, 1939; (2) Clarke, Lincoln, and Sterrett, *Proc. API*, 1935; (3) Beare and Bowden, *Phil. Trans. Roy. Soc.*, 1935; (4) Dokos, *Trans. ASME*, 1946; (5) Boyd and Robertson, *Trans. ASME*, 1945; (6) Sachs, *Zeit. f. angew. Math. und Mech.*, 1924; (7) Honda and Yamada, *Jour. I of M*, 1925; (8) Tomlinson, *Phil. Mag.*, 1929; (9) Morin, *Acad. Roy. des Sciences*, 1838; (10) Claypoole, *Trans. ASME*, 1943; (11) Tabor, *Jour. Applied Phys.*, 1945; (12) Eyssen, General Discussion on Lubrication, *ASME*, 1937; (13) Brazier and Holland-Bowyer, General Discussion on Lubrication, *ASME*, 1937; (14) Burwell, *Jour. SAE*, 1942; (15) Stanton, "Friction," Longmans; (16) Ernst and Merchant, Conference on Friction and Surface Finish, M.I.T., 1940; (17) Gongwer, Conference on Friction and Surface Finish, M.I.T., 1940; (18) Hardy and Bircumshaw, *Proc. Roy. Soc.*, 1925; (19) Hardy and Hardy, *Phil. Mag.*, 1919; (20) Bowden and Young, *Proc. Roy. Soc.*, 1951; (21) Hardy and Doubleday, *Proc. Roy. Soc.*, 1923; (22) Bowden and Tabor, "The Friction and Lubrication of Solids," Oxford; (23) Shooter, *Research*, 4, 1951.

(a) Oleic acid; (b) Atlantic spindle oil (light mineral); (c) castor oil; (d) lard oil; (e) Atlantic spindle oil plus 2 percent oleic acid; (f) medium mineral oil; (g) medium mineral oil plus ½ percent oleic acid; (h) stearic acid; (i) grease (zinc oxide base); (j) graphite; (k) turbine oil plus 1 percent graphite; (l) turbine oil plus 1 percent stearic acid; (m) turbine oil (medium mineral); (n) olive oil; (p) palmitic acid; (q) ricinoleic acid; (r) dry soap; (s) lard; (t) water; (u) rape oil; (v) 3-in-1 oil; (w) octyl alcohol; (x) triolein; (y) 1 percent lauric acid in paraffin oil.

Source: Mark's Standard Handbook for Mechanical Engineers, 8th Edition, 1978. Used with permission from McGraw-Hill Book Co.

The **coefficient of sliding friction** (f) between two bodies when N is the normal pressure between them and F is the corresponding sliding friction (N and F in the same units):

$$f = \frac{F}{N}.$$ (2.95)

The **angle of sliding friction** (ϕ) for two surfaces with a normal pressure N and a sliding friction F between them (N and F in the same units).

NOTE. See Equation (2.94). The angle of sliding friction is the angle of inclination of the surface of one body, at which the motion of another body sliding upon it will be maintained. The angle of sliding friction is in general less than the angle of static friction.

Applications of Principles of Friction

Inclined Plane. Let W = weight in pounds of a body sliding on the plane, α = angle of inclination of plane, β = angle between force F and plane, ϕ = angle of repose, f = coefficient of friction (tan ϕ = f), and F = force applied to the body along the line of action indicated.

1. Force (F) to prevent slipping ($\alpha > \phi$):

$$F = W \frac{\sin(\alpha - \phi)}{\cos(\beta + \phi)} \quad \text{pounds.}$$ (2.96)

2. Force (F) to start the body up the plane ($\alpha > \phi$):

$$F = W \frac{\sin(\alpha + \phi)}{\cos(\beta - \phi)} \quad \text{pounds.}$$ (2.97)

3. Force (F) to start the body down the plane ($\alpha < \phi$):

$$F = W \frac{\sin(\phi - \alpha)}{\cos(\beta + \phi)} \quad \text{pounds.}$$ (2.98)

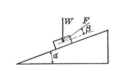

Fig. 2.12 Inclined plane.

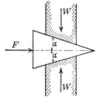

Fig. 2.13 Wedge.

Wedge. Let W = force in pounds opposing motion, α = angle of inclination of sides of wedge, ϕ = angle of friction, and F = force applied to wedge.

1. Force (F) to push wedge:

$$F = 2W \tan(\alpha + \phi) \quad \text{pounds.} \tag{2.99}$$

2. Force (F) to draw wedge ($\alpha > \phi$):

$$F = 2W \tan(\phi - \alpha) \quad \text{pounds.} \tag{2.100}$$

Square-Threaded Screw. Let r = mean radius of screw, p = pitch of screw, α = angle of pitch [$\tan \alpha = p/(2\pi r)$], F = force applied to screw at end of arm a, W = total weight in pounds to be moved, and ϕ = angle of friction (r and a in same units).

1. Force (F) to lower screw:

$$F = \frac{Wr(\tan \phi - \tan \alpha)}{a} \quad \text{pounds (approx.).} \tag{2.101}$$

2. Force (F) to raise screw:

$$F = \frac{Wr(\tan \phi + \tan \alpha)}{a} \quad \text{pounds (approx.).} \tag{2.102}$$

Sharp-Threaded Screw. Let r = mean radius of screw, α = angle of pitch, β = angle between faces of the screw, F = force in pounds applied to screw at end of arm a, W = total weight in pounds to move, and ϕ = angle of friction (r and a in same units).

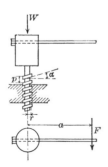

Fig. 2.14 Square-threaded screw.

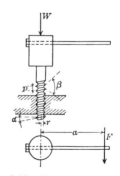

Fig. 2.15 Sharp-threaded screw.

1. Force (F) to lower screw:

$$F = \frac{Wr}{a} \left(\frac{\tan \phi \cos \alpha}{\cos(\beta/2)} - \tan \alpha \right) \text{ pounds (approx.)}. \qquad (2.103)$$

2. Force (F) to raise screw:

$$F = \frac{Wr}{a} \left(\frac{\tan \phi \cos \alpha}{\cos(\beta/2)} + \tan \alpha \right) \text{ pounds (approx.)}. \qquad (2.104)$$

The **coefficient of rolling friction** (c) of a wheel with a load of W pounds and with r inches radius, moved at a uniform speed by a force of F pounds applied at its center:

$$c = \frac{Fr}{W} \text{ inches.} \qquad (2.105)$$

NOTE. Coefficients of rolling friction:

Lignum vitæ roller on oak track	$c = 0.019$ in.
Elm roller on oak track	$c = 0.032$ in.
Iron on iron (and steel on steel)	$c = 0.020$ in.

Belt Friction

The **ratio** (F_1/F_2) of the pull F_1 on the driving side of a belt to the pull F_2 on the driven side of the belt, when slipping is impending, in terms of the coefficient of friction f and the angle of contact α, in radians ($\epsilon = 2.718$):

$$\frac{F_1}{F_2} = \epsilon^{f\alpha}. \qquad (2.106)$$

NOTE. Mean values of f are as follows:

Leather on wood (somewhat oily)	0.47	Hemp rope on wooden drum	0.40
Leather on cast iron (somewhat oily)	0.28	Hemp rope on polished drum	0.33
Leather on cast iron (moist)	0.38	Hemp rope on rough wood	0.50
Hemp rope on iron drum	0.25		

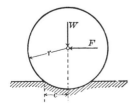

Fig. 2.16 Coefficient of rolling friction.

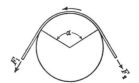

Fig. 2.17 Belt friction.

Table 2.5   Pivot Friction[a]

Type of Pivot	Torque T (lb-in.)	Power P Lost by Friction (ft-lb. per second)
Shafts and Journals		
(180° bearing)	$T = fWr$	$P = \dfrac{2\pi n}{12} fWr$
Flat Pivot	$T = \tfrac{2}{3} fWr$	$P = \dfrac{4\pi n}{3 \times 12} fWr$
Collar-bearing	$T = \dfrac{2}{3} fW \dfrac{R^3 - r^3}{R^2 - r^2}$	$P = \dfrac{4\pi n}{3 \times 12} fw \dfrac{R^3 - r^3}{R^2 - r^2}$
Conical Pivot	$T = \dfrac{2}{3} fW \dfrac{r}{\sin \alpha}$	$P = \dfrac{4\pi n fWr}{3 \times 12 \sin \alpha}$
Truncated-cone Pivot	$T = \dfrac{2}{3} fW \dfrac{R^3 - r^3}{(R^2 - r^2) \sin \alpha}$	$P = \dfrac{4\pi n fW(R^3 - r^3)}{3 \times 12(R^2 - r^2) \sin \alpha}$

[a] f, coefficient of friction; W, load in pounds; T, torque of friction about the axis of the shaft; r, radius in inches; n, revolutions per second.

Values of F_1/F_2 (slipping impending) are as follows:

$\frac{\alpha}{2\pi}$	f				$\frac{\alpha}{2\pi}$	f			
	0.25	0.33	0.40	0.50		0.25	0.33	0.40	0.50
0.1	1.17	1.23	1.29	1.37	0.6	2.57	3.47	4.52	6.59
0.2	1.37	1.51	1.65	1.87	0.7	3.00	4.27	5.81	9.00
0.3	1.60	1.86	2.13	2.57	0.8	3.51	5.25	7.47	12.34
0.4	1.87	2.29	2.73	3.51	0.9	4.11	6.46	9.60	16.90
0.425	1.95	2.41	2.91	3.80	1.0	4.81	7.95	12.35	23.14
0.45	2.03	2.54	3.10	4.11	1.5	10.55	22.42	43.38	111.2
0.475	2.11	2.68	3.30	4.45	2.0	23.14	63.23	152.4	535.5
0.5	2.19	2.82	3.51	4.81	2.5	50.75	178.5	535.5	2,576
0.525	2.28	2.97	3.74	5.20	3.0	111.3	502.9	1881	12,392
0.55	2.37	3.31	3.98	5.63	3.5	244.2	1418	6611	59,610

2.3.7 Impact*

The **common velocity** (v'), after direct central impact, of two inelastic bodies of mass m_1 and m_2 and initial velocities v_1 and v_2, respectively:

$$v' = \frac{m_1 v_1 + m_2 v_2}{m_1 + m_2} \tag{2.107}$$

The **final velocities,** v_1' and v_2', after direct central impact, of two perfectly elastic bodies of mass m_1 and m_2 and initial velocities v_1 and v_2, respectively:

$$v_1' = \frac{m_1 v_1 - m_2 v_1 + 2m_2 v_2}{m_1 + m_2},$$
$$v_2' = \frac{m_2 v_2 - m_1 v_2 + 2m_1 v_1}{m_1 + m_2}. \tag{2.108}$$

The **final velocities** v_1' and v_2', after direct central impact, of two partially but equally inelastic bodies of mass m_1 and m_2 and initial velocities v_1 and v_2, respectively, and constant e depending on the elasticity of bodies:

$$v_1' = \frac{m_1 v_1 + m_2 v_2 - e m_2 (v_1 - v_2)}{m_1 + m_2},$$
$$v_2' = \frac{m_1 v_1 + m_2 v_2 - e m_1 (v_2 - v_1)}{m_1 + m_2}. \tag{2.109}$$

NOTE. $e = (H/h)^{1/2}$ where H is the height of rebound of a sphere dropped from a height h onto a horizontal surface of a rigid mass. If the bodies are inelastic, $e = 0$, and if the bodies are perfectly elastic, $e = 1$.

*m_1 and m_2, v_1 and v_2 in the same units.

Fig. 2.18 Components of force F.

2.4 STATICS

2.4.1 Forces and Resultants

The **components of a force** F (F_x and F_y) parallel to two rectangular axes $X'X$ and $Y'Y$, the force F making an angle α with the axis $X'X$:

$$F_x = F \cos \alpha, \qquad F_y = F \sin \alpha. \tag{2.110}$$

The **moment** or **torque** (M) of a force of F pounds about a given point, the perpendicular distance from the point to the direction of the force being d feet.

$$M = Fd \quad \text{pound-feet.} \tag{2.111}$$

NOTE. A couple is formed by two equal, opposite, parallel forces acting in the same plane but not in the same straight line. The moment (M) of a couple of two forces, each of F pounds, with a perpendicular distance of d feet between them is Fd pound-feet. The moment, about any point, of the resultant of several forces, lying in the same plane, is the algebraic sum of the moments of the separate forces about that point.

The **resultant force** (R) of two forces, F_1 and F_2, which make an angle α with each other, the angle between the resultant force R and the force F_1 being θ.

$$R = \sqrt{F_1^2 + F_2^2 + 2F_1F_2 \cos \alpha}. \tag{2.112}$$

$$\tan \theta = \frac{F_2 \sin \alpha}{F_1 + F_2 \cos \alpha} \quad \text{or} \quad \sin \theta = \frac{F_1 \sin \alpha}{R}. \tag{2.113}$$

Parallelogram of Forces. The resultant force (R) of two forces F_1 and F_2 is represented in magnitude and direction by the diagonal lying between those two sides of a parallelogram which represent F_1 and F_2 in magnitude and direction.

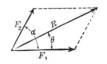

Fig. 2.19 Resultant force R.

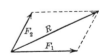

Fig. 2.20 Parallelogram of forces.

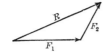

Fig. 2.21 Triangle of forces.

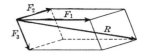

Fig. 2.22 Parallelopiped of forces.

Triangle of Forces. The resultant force (R) of two forces F_1 and F_2 is represented in magnitude and direction by the third side of a triangle in which the other two sides represent F_1 and F_2 in magnitude and direction.

The **resultant force** (R) of three forces F_1, F_2, and F_3 mutually at right angles to each other and not lying in the same plane, the angles between the resultant force R and the forces F_1, F_2, and F_3 being α, β, and γ, respectively.

$$R = \sqrt{F_1^2 + F_2^2 + F_3^2}. \tag{2.114}$$

$$\cos \alpha = \frac{F_1}{R}, \quad \cos \beta = \frac{F_2}{R}, \quad \cos \gamma = \frac{F_3}{R}. \tag{2.115}$$

NOTE. If three forces not in the same plane are not mutually at right angles to each other, the resultant force may be found by Equation (2.121).

Parallelopiped of Forces. The **resultant force** (R) of three forces F_1, F_2, and F_3, not lying in the same plane, is represented in magnitude and direction by the diagonal lying between those three sides of a parallelopiped which represent F_1, F_2, and F_3 in magnitude and direction.

The **resultant force** (R) of several forces lying in the same plane, if ΣF_x and ΣF are the algebraic sums of the components of the forces parallel to two rectangular axes $X'X$ and $Y'Y$, the angle between the resultant force and the axis $X'X$ being α:

$$R = \sqrt{(\Sigma F_x)^2 + (\Sigma F_y)^2}. \tag{2.116}$$

$$\tan \alpha = \frac{\Sigma F_y}{\Sigma F_x}, \quad \sin \alpha = \frac{\Sigma F_y}{R}, \quad \cos \alpha = \frac{\Sigma F_x}{R}. \tag{2.117}$$

The **perpendicular distance** (d) from a given point to the resultant force (R) of several forces lying in the same plane, if ΣM is the algebraic sum of the moments, about that point, of the separate forces:

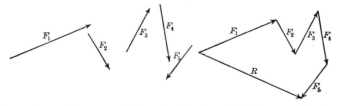

Fig. 2.23 Force polygon. The arrows indicate the directions of the forces, and for the given forces, they must point in the same direction around the polygon; but for the resultant force, they must point in the opposite direction or leading from the starting point of the first force to the end point of the last force.

$$d = \frac{\Sigma M}{R}.$$ (2.118)

Note. The resultant of several parallel forces is the algebraic sum of the forces (ΣF). If $\Sigma F = 0$, the resultant is a couple whose moment is ΣM.

Force Polygon.   The **resultant force** (R) of several forces $F_1, F_2, \ldots, F_n$, lying in the same plane, is represented in magnitude and direction by the closing side of a polygon in which the remaining sides represent the forces $F_1, F_2, \ldots, F_n$ in magnitude and direction (Fig. 2.23).

The **moment** (M) of a force F, about a line, is the product of the rectangular component of the force perpendicular to the line (the other component being parallel to the line) into the perpendicular distance between the line and this rectangular component; or the force F may be resolved into three rectangular components, one parallel and the other two perpendicular to the line, as in Fig. 2.24. The moment of the force about each axis is then obtained as follows:

$$M_x = yF \cos \gamma - zF \cos \beta.$$
$$M_y = zF \cos \alpha - xF \cos \gamma.$$ (2.119)
$$M_z = xF \cos \beta - yF \cos \alpha.$$

The **resultant force** (R) of several parallel forces, not lying in the same plane, is the algebraic sum (ΣF) of the forces.

Note.   If $\Sigma F = 0$, the resultant is a couple whose moments are ΣM_x, ΣM_y, and so on.

The **perpendicular distances** d_x and d_y from each of two axes $X'X$ and $Y'Y$ to the resultant force (R) of several parallel forces not lying in the same plane, if ΣM_x and ΣM_y are the algebraic sums of the moments of the separate forces about the axes $X'X$ and $Y'Y$, respectively:

$$d_x = \frac{\Sigma M_x}{R}, \qquad d_y = \frac{\Sigma M_y}{R}.$$ (2.120)

The **resultant force** (R) and direction (α, β, γ) of the resultant force of several forces not lying in the same plane, if ΣF_x, ΣF_y, and ΣF_z are the algebraic sums of the components parallel to three rectangular axes $X'X$, $Y'Y$, and $Z'Z$, and α, β, and

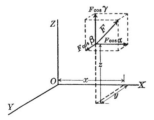

Fig. 2.24   Moment M of a force F.

γ are the angles which the resultant force makes with the axes $X'X$, $Y'Y$, and $Z'Z$, respectively:

$$R = \sqrt{(\Sigma F_x)^2 + (\Sigma F_y)^2 + (\Sigma F_z)^2}. \tag{2.121}$$

$$\cos \alpha = \frac{\Sigma F_x}{R}, \quad \cos \beta = \frac{\Sigma F_y}{R}, \quad \cos \gamma = \frac{\Sigma F_z}{R}. \tag{2.122}$$

The **resultant couple** (M) and direction $(\alpha_m, \beta_m, \gamma_m)$ of the axis of the resultant couple of several forces not acting in the same plane, if ΣM_x, ΣM_y, and ΣM_z are the algebraic sums of the moments about three rectangular axes $X'X$, $Y'Y$, and $Z'Z$, and α_m, β_m, and γ_m are the angles which the moment axis of the resultant couple makes with the axes $X'X$, $Y'Y$, and $Z'Z$, respectively:

$$M = \sqrt{(\Sigma M_x)^2 + (\Sigma M_y)^2 + (\Sigma M_z)^2}. \tag{2.123}$$

$$\cos \alpha_m = \frac{\Sigma M_x}{\Sigma M}, \quad \cos \beta_m = \frac{\Sigma M_y}{\Sigma M}, \quad \cos \gamma_m = \frac{\Sigma M_z}{\Sigma M}. \tag{2.124}$$

NOTE. In general, the resultant of several nonparallel forces not in the same plane is not a single force, but by the use of the principles above, the system may be reduced to a single force and a couple.

The **conditions of equilibrium** of several forces lying in the same plane, if ΣF_x and ΣF_y are the algebraic sums of the components parallel to two axes $X'X$ and $Y'Y$, and ΣM is the algebraic sum of the moments of the forces about any point:

$$\Sigma F_x = 0, \quad \Sigma F_y = 0, \quad \Sigma M = 0. \tag{2.125}$$

The **conditions of equilibrium** of several forces not lying in the same plane, if ΣF_x, ΣF_y, and ΣF_z are the algebraic sums of the components parallel to three axes $X'X$, $Y'Y$, and $Z'Z$ which intersect at a common point but do not lie in the same plane, and ΣM_x, ΣM_y, and ΣM_z are the algebraic sums of the moments of the forces about these three axes:

$$\Sigma F_x = 0, \quad \Sigma F_y = 0, \quad \Sigma F_z = 0. \tag{2.126}$$

$$\Sigma M_x = 0, \quad \Sigma M_y = 0, \quad \Sigma M_z = 0. \tag{2.127}$$

2.4.2 Stresses in Framed Structures*

Pratt Truss. Two live loads of 10 tons each as shown in Fig. 2.25.
(a) *Reactions* [use conditions of equilibrium, Equation (2.126)].

- By $\Sigma M = 0$, $\Sigma M_A = 0 = 10 \times 15 + 10 \times 30 - V_B \times 90$, $V_B = 5$ tons.
- By $\Sigma F_y = 0$, $20 - V_B = V_A$, $V_A = 15$ tons.

---

*Due to live loads only. The weight of the structure is neglected.

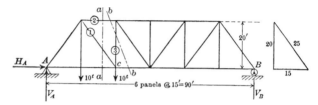

Fig. 2.25   Pratt truss.

- By $\Sigma F_x = 0$, $H_A = 0$. (Note that a roller is used at B, fixing the reaction there in a vertical direction.)

(b) *Stresses in bars.* To find the stress in a bar consider a plane cutting the bar to divide the truss into two parts; remove one part and replace that portion of the bars that are removed by their stresses which may now be treated as outer forces. These stresses are found by applying the equations of equilibrium. It is essential that only three of the bars that are cut shall have unknown stresses.

NOTE.   If tension is called positive and all unknown stresses are assumed to be tension stresses, a positive sign for the result indicates tension and a negative sign compression.

- Bar ①. Truss cut by plane *aa*. Consider the left portion. Let $V_①$ = the vertical component of $S_①$, the stress in bar ①. By $\Sigma F_y = 0$, $-V_A + 10 + V_① = 0$, $V_① = 5$, $S_② = (25/20) \times 5 = 6.25$ tons tension.
- Bar ②. Truss cut by plane *aa*. Take the moments about joint *c*. By $\Sigma M = 0$, $\Sigma M_c = 0 = V_A \times 30 - 10 \times 15 + S_② \times 20$.

$$S_② = \frac{-450 + 150}{20} = -15 = 15 \text{ tons compression.}$$

- Bar ③. Truss cut by plane *bb*. By $\Sigma F_y = 0$, $-V_A + 20 + S_③ = 0$, $S_③ = -5 = 5$ tons compression.

Roof Truss.   Two live loads of 3 tons each, as shown in Fig. 2.26.
(a) *Reactions* [use conditions of equilibrium, Equation (2.126)].

- By $\Sigma M = 0$, $\Sigma M_A = 0 = 3 \times 13.4 + 3 \times 26.8 - V_B \times 72$, $V_B = 1.67$ tons.

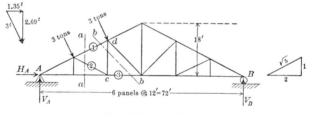

Fig. 2.26   Roof truss.

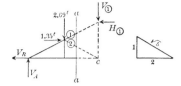

Fig. 2.27   Roof truss (left-hand portion).

- By $\Sigma F_y = 0$, $2 \times 2.69 - V_B - V_A = 0$, $V_A = 3.71$ tons.
- By $\Sigma F_x = 0$, $2.70 + H_A = 0$, $H_A = -2.70$ tons (i.e., acting to the left). (Note that a roller is used at B, fixing the reaction in a vertical direction.)
     (b) *Stresses in bars.* [See (b) under "Pratt Truss."]
- Bar ①. Truss cut by plane aa. Consider the left portion. Take moments about joint c. Let $H_①$ = horizontal component of $S_①$ (Fig. 2.27). By $\Sigma M = 0$, $\Sigma M_c = 0 = V_A \times 24 - 2.69 \times 12 + 1.35 \times 6 + H_① \times 12$.

$$H_① = -5.34, \quad S_① = \frac{\sqrt{5}}{2} \times 5.34 = 5.96 \text{ tons compression.}$$

- Bar ②. Truss cut by plane aa. Take moments about A. Let $V_②$ = vertical component of $S_②$. By $\Sigma M = 0$, $\Sigma M_A = 0 = 3 \times 13.4 + V_② \times 24$, $V_② = -1.67$ tons. $S_② = \sqrt{5} \times 1.67 = 3.73$ tons compression.
- Bar ③. Truss cut by plane bb. Consider the right portion, as fewer loads lie to the right of the cutting plane. Take moments about joint d. By $\Sigma M = 0$, $\Sigma M_d = 0 = -V_B \times 48 + S_③ \times 12$, $S_③ = 6.68$ tons tension.

2.5   PROPERTIES OF MATERIALS

(See Table 2.6 and Chapter 12.)

The **intensity of stress** is the stress per unit area, usually expressed in pounds per square inch. The simple term **stress** is normally used to indicate intensity of stress.

The **ultimate stress** is the greatest stress that can be produced in a body before rupture occurs.

The **allowable stress** or **working stress** is the intensity of stress that the material of a structure or a machine is designed to resist.

The **factor of safety** is a factor by which the ultimate stress is divided to obtain the allowable stress.

The **elastic limit** is the maximum intensity of stress to which a material may be subjected and return to its original shape upon the removal of the stress.

NOTE.   For stresses below the elastic limit the deformations are directly proportional to the stresses producing them; that is, Hooke's law holds for stresses below the elastic limit.

The **yield point** is the intensity of stress beyond which the change in length increases rapidly with little if any increase in stress.

The **modulus of elasticity** is the ratio of stress to strain, for stresses below the elastic limit.

Table 2.6   Mechanical Properties of Some Engineering Materials

Material	Equivalent	Ultimate strength, psi			Yield point, tension, psi	Modulus of elasticity, tension or compression, psi	Modulus of elasticity, shear, psi	Weight per cu in., lb
		Tension	Compression *	Shear				
Steel, forged-rolled:								
C, 0.10–0.20	SAE 1015	60,000	39,000	48,000	39,000	30,000,000	12,000,000	0.28
C, 0.20–0.30	SAE 1025	67,000	43,000	53,000	43,000	30,000,000	12,000,000	0.28
C, 0.30–0.40	SAE 1035	70,000	46,000	56,000	46,000	30,000,000	12,000,000	0.28
C, 0.60–0.80	....	125,000	65,000	75,000	65,000	30,000,000	12,000,000	0.28
Nickel	SAE 2330	115,000	....	92,000	......	30,000,000	12,000,000	0.28
Cast iron:								
Gray	ASTM 20	20,000	80,000	27,000	......	15,000,000	6,000,000	0.26
Gray	ASTM 35	35,000	125,000	44,000				0.26
Gray	ASTM 60	60,000	145,000	70,000		20,000,000	8,000,000	0.26
Malleable	SAE 32510	50,000	120,000	48,000		23,000,000	9,200,000	0.26
Wrought iron	....	48,000	25,000	38,000	25,000	27,000,000	....	0.28
Steel, cast:								
Low C		60,000	....	......	....	....	....	0.28
Medium C		70,000	....	......	....			0.28
High C		80,000	45,000	......	45,000	....	....	0.28
Aluminum alloy:								
Structural, No. 350		16,000	5,000	11,000	5,000	10,000,000	3,750,000	0.10
Structural, No. 17ST		58,000	35,000	35,000	35,000	10,000,000	3,750,000	0.10
Brass:								
Cast		40,000	....	......	....	....	....	0.30
Annealed		54,000	18,000	......	18,000			0.30
Cold-drawn		96,700	49,000	......	49,000	15,500,000	6,200,000	0.30
Bronze:								
Cast		22,000	....	......	....			0.31
Cold-drawn		85,000	....	......	....	15,000,000	6,000,000	0.31
Brick, clay:								
Grade SW	ASTM	......	3,000 (min) †	......				0.072
Grade MW	ASTM	......	2,500 (min)					
Grade NW	ASTM	......	1,500 (min)					
Concrete, 1:2:4 (28 days)	....	......	2,000	......		3,000,000		0.087
Stone	....	......	8,000					0.092
White oak:								
Parallel to grain		......	7,440	2,000	4,760 ‡	1,780,000		0.028
Across grain		800	....		1,320 ‡			
White pine:								
Parallel to grain		......	4,840	860	3,680 ‡	1,280,000		0.015
Across grain		300	....	......	550 ‡			
Southern longleaf pine:								
Parallel to grain		......	8,440	1,500	6,150 ‡	1,999,000		0.024
Across grain		470	....	......	1,199 ‡			

* The ultimate strength in compression for ductile materials is usually taken as the yield point. The bearing value for pins and rivets may be much higher, and for structural steel is taken as 90,000 psi.
† Average of five bricks.
‡ Proportional limit in compression.
Source: Urquhart, *Civil Engineering Handbook,* 4th Edition, 1959. Used with permission of McGraw-Hill Book Co.

NOTE.   Modulus of elasticity may also be defined as the stress that would produce a change of length of a bar equal to the original length of the bar, assuming the material to retain its elastic properties up to that point.

Poisson's ratio is the ratio of the relative change of diameter of a bar to its unit change of length under an axial load that does not stress it beyond the elastic limit.

NOTE.   Poisson's ratio is usually denoted by $1/m$. It varies for different materials but is usually about ¼.

The **intensity of stress**(s) due to a force of P pounds producing tension, compression, or shear on an area of A square inches, over which it is uniformly distributed:

$$s = \frac{P}{A} \quad \text{pounds/inch}^2. \tag{2.128}$$

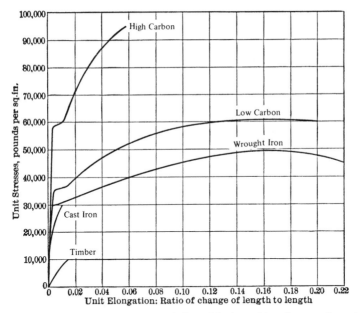

Fig. 2.28 Stress–strain diagram showing the relation of the intensities of stress of a material to the corresponding strains or deformations.

The **modulus of elasticity** (E) of a bar of A square inches cross-sectional area and l inches length, which undergoes a change of length of d inches under an axial load of P pounds:

$$E = \frac{Pl}{Ad} \quad \text{pounds/inch}^2. \tag{2.129}$$

NOTE.   The load must be such as to produce an intensity of stress below the elastic limit. If s is the intensity of stress produced and e the ratio of change of length to total length, $E = s/e$ and $e = s/E$.

The **change of length** (d) of a bar of A square inches cross-sectional area, l inches length, and E pounds per square inch modulus of elasticity of material, due to an axial load of P pounds:

$$d = \frac{Pl}{AE} \quad \text{inches.} \tag{2.130}$$

2.6 STRUCTURAL MATERIALS

Concrete, structural steel, and timber are most widely used as **structural materials;** they are described in Chapter 3. Several other common materials are described in the following sections.

2.6.1   Composite Materials[1-]

Composite materials are usually defined as those in which groupings of reinforcements are distributed in a weaker **matrix,** the **reinforcements** providing the principal element of strength and stiffness, while the matrix provides the geometric stability and alignment to the reinforcements. (**Reinforced concrete** using portland cement and steel reinforcing is a specialized form of composite material dealt with in Section 3.4.)

Composites require more care in their design than metallic materials and are often more costly. Their major advantage in stiffness, lightness, and generally greater fatigue life, together with often specific additional advantages, such as chemical resistance or electronic transparency, for many applications offset the initially higher cost. While their impact resistance is generally lower than that of metals, suitable design can largely overcome this characteristic. In some applications and for some composites, exposure to moisture or moderately elevated temperatures (e.g., in the range of 250°F) creates a reduction in strength or stiffness. The number of types of reinforcements is very broad, and theoretically, any material with significantly greater tensile strength than the matrix can be used. Figure 2.29 compares stiffness and toughness of commonly used composite materials to those of steel and aluminum.

The thickness (z axis) of the structure of composite materials is generally small compared to the x and y dimensions. Typical design values calculated from standard

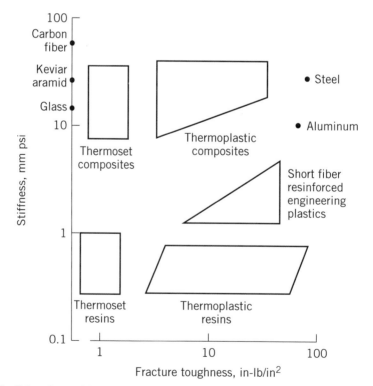

Fig. 2.29   Selected materials characteristics. (Redrawn with permission from *Mechanical Engineering,* April 1988. © Copyright 1988, ASME International.)

formulas may be found in such sources as the books by Lubin, *Handbook of Composites,*[6] Reinhart, *Engineered Materials Handbook.*[7] While useful for first approximations, these should not be used for the final design, as the materials may have a significant variation of properties from the theoretical.

These materials by their nature are **anisotropic,** and computer-aided design (normally, the **finite element method**) is used to establish the design of the composite. Both the tensile properties of reinforcements and the physical properties of matrices are highly predictable. However, the interaction between them, particularly when coupled with the method of lay of the reinforcement, its effective angularity, and so on, make accurate prediction of performance difficult. As a result, empirical and test data from trial configurations are often used to determine a final design.

*Because of the large number of nonmachine-controlled steps in manufacturing composites, extensive **testing** of the structure is usually performed. This establishes both areas of underdesign, where additional reinforcement is needed, as well as areas of overdesign, where material can be reduced. With the relatively high cost of composites, the design savings can be significant.

It is important to obtain specific design values from manufacturers where final matrix material properties are usually determined more exactly from experiments. Composite suppliers typically provide acceptable values on data sheets for their particular material. In the case of new composite designs, experimental batches of the materials are often prepared to permit testing of "as produced" properties. Table 2.7 indicates the general properties of some selected composites. As indicated above, these values are useful only for approximations and should not be used for final design.

The number of types of **reinforcements** is very broad, and theoretically, any material with significantly greater tensile strength than the matrix can be used. The actual number in use is more limited, due primarily to desirable physical characteristics and availability. The general family of these reinforcements uses filaments of relatively high strength, such as **fiberglass (S glass** and **E glass), carbon fiber,** and **Kevlar.**

Reinforcements are laid up in forms as **tows** (parallel fibers), tapes, yarns, woven fabric, chopped fiber, or molding compounds. The choice of fiber lay has a major influence on the strength and stiffness of the composite. For high production rates, injection molding is sometime used to produce the part. Injection molding generally

Table 2.7 Selected Composite Material Properties, 60% Fiber Volume

Properties	Epoxy 350°F Graphite	Glass	Polyimide (J-2) Kevlar Aramid	Polyether ketone Graphite	Polyimide K-III Graphite	Polyester (PET) Glass
Flex modulus, Msi	18	6.5	10	18	18	6.0
Flex strength, Ksi	240	188	103	217	220	150
G_{IC}, kJ/m^2	0.26	—	—	1.4	1.7	—
Use temperature, °C (dry and wet)	175	100	100	175	200	80
Projected relative price	18	1.5	7	18	50	1

Source: Reprinted with permission from *Mechanical Engineering,* April 1988. © Copyright 1988, ASME International.

Source: J. C. Leslie, in J. M. Margolis, ed., *Advanced Thermoset Composites,* Van Nostrand Reinhold, New York, 1986.

is limited to 30%, compared to a typical 60% fiber volume. For **filament-wound** structures, computer-controlled equipment assuring alignment and tensioning is essential.

Glass-Reinforced Composites

Fiberglass is the reinforcement most often used in fabricating structural composite materials. Three grades of glass are used: E, S, and **structural quartz.** E glass ("electrical glass"), the **borosilicate glass** most often used for glass fibers in conventional reinforced polymers, has a typical tensile strength of 500 kpsi. S glass is a **magnesia–alumina silicate** glass reinforcement designed to provide very high tensile strength, in the range of 665 kpsi.

Structural quartz has particular uses in cryogenic, electrical, and high-temperature application (its higher cost, on the order of 40 to 100 times more than E glass, limits its use to special applications). Structural quartz shares with Kevlar unique structural and transmission capability for radar and electronic packaging. It is in these fields that its most frequent applications are found.

Reinforcements

Regardless of the reinforcement, fiber size is small. Glass filament diameters are shown in Table 2.8. The plies of fiber can be as little as 0.005 to 0.01 in. thick, with the layup consisting of many plies laid up to provide directional strength where necessary. Fibers constitute 50 to 65% of the volume of the composite. The range to 50% is considered currently available commercial material. Volumes above 65% are very difficult to achieve.

Fiberglass composites have widely varying properties, depending on reinforcement type and fabrication method. Table 2.9 indicates the properties of some selected composites, depending on their method of fabrication.

Kevlar is the registered trademark for one group of E.I. du Pont's family of manufactured aromatic polyamide fibers. These are widely available in three grades: 29, 49, and 149. Kevlar demonstrates low thermal expansion, ranging from a negative coefficient of -1.1×10^{-6} in./in./°F (0 to 100°C) to -2×10^{-6} (200 to 260°C). It also has good chemical and thermal stability. Kevlar fabric has a strong affinity for moisture. It is necessary to dry Kevlar and control humidity exposure prior to and during impregnation with moisture-sensitive resin systems. In addition, it suffers a minor loss of strength from exposure to ultraviolet light. This would be a concern only with either very thin structures or those that are unpainted.

The term **carbon–graphite fibers** covers various fibers described as either carbon or graphite. These fibers can be classified into four groupings depending on their modulus levels: *standard* fibers at $33/35 \times 10^6$, *high strength* or *high strain* at $40/45 \times 10^6$, *high modulus* at $50/55 \times 10^6$, and *ultrahigh modulus* at $70/120 \times 10^6$. The high and ultrahigh types are not widely used, due to their very high cost.

Carbon–graphite fibers exhibit increasing negative coefficient of thermal expansion with increasing modulus. This characteristic is one of the major factors in selecting ultrahigh-modulus fibers. This makes it easier to design structures even for low-modulus fibers in which the overall **coefficient of expansion** approaches zero. These

Table 2.8   Designations for Glass Strand Filament Diameters

| Filament Designation | | Range for Filament Diameter Average | |
U.S. Units (letter)	SI Units μm^a	Inches	Micrometers[b]
B	3.5	0.00013–0.000159	3.30–4.05
C	4.5	0.00016–0.000189	4.06–4.82
D	5	0.00019–0.000229	4.83–5.83
DE	6	0.00023–0.000269	5.84–6.85
E	7	0.00025–0.000299	6.35–7.61
F	8	0.00030–0.000345	7.62–8.88
G	9	0.00035–0.000399	8.89–10.15
H	11	0.00040–0.000449	10.16–11.42
J	12	0.00045–0.000499	11.43–12.69
K	13	0.00050–0.000549	12.70–13.96
L	14	0.00055–0.000599	13.97–15.23
M	16	0.00060–0.000649	15.24–16.50
N	17	0.00065–0.000699	16.51–17.77
P	18	0.00070–0.000749	17.78–19.04
Q	20	0.00075–0.000799	19.05–20.31
R	21	0.00080–0.000849	20.32–21.58
S	22	0.00085–0.000899	21.59–22.85
T	23	0.00090–0.000949	22.86–24.12
U	24	0.00095–0.000999	24.13–25.40

Source: Reprinted with permission from Owens-Corning Fiberglass Corporation.
[a] 1 micrometer = 1 micron.
[b] The low values stated for each micrometer range are exact equivalents to inches, rounded to the nearest hundredth micrometer. The high values stated for each micrometer range are slightly higher than exact equivalents to inches to provide continuation between ranges. They are consistent for inch-pound and SI filament size descriptions commonly used in the industry. In some publications, the SI designation for H filament size has been shown as 10.

laminates also offer superior fatigue properties, far exceeding steel or aluminum in fatigue life, as well as superior **vibration damping** characteristics.

Specialized Fibers

Other more specialized fibers are used less often, primarily because of high cost. It is likely that in the near term their use will increase and their cost decline.

Boron fibers provide excellent compressive and high-modulus properties. **Silicon carbide fibers** offer superior compressive strengths in epoxy matrices. The fibers have excellent wettability for metal matrices and are useful up to 1200°C. **Ceramic fibers** have high melting temperatures, high modulus, overall high strength, and good resistance to chemical attack. These are potential reinforcements for metal or ceramic matrix composites. In organic matrices, they offer rigidity and abrasion and impact resistance. **Oxide fibers** have high electrical resistance and low moisture absorption and are suitable for electrical insulation applications. Oxide ceramic fibers may be the only workable choice for high-temperature applications where chemical reactivity, especially oxidation, quickly degrades nonoxides.

Alumina fibers (typically 99% or more pure) offer high modulus of elasticity and exceptional resistance to corrosive environments. Combining alumina with approxi-

Table 2.9 Typical Properties of Selected Fiberglass Composites

Polyester (Styrene Monomer)	ASTM Method	Open (Contact) Molding		Matched Die Compression Molding			Injection Molding (RTP)	Filament Winding	Pultrusion (Rod Stock)
		Layup	Spray-up	Bulk MLDG CPD	Sheet MLDG CPD	XMC			
Mechanical Properties									
Tensile strength,[a,b] psi × 10³	D638	9–50	5–18	3–10	8–20	70–90	6–30	80–200[e]	70–180
Tensile modulus, psi × 10³	D638	0.6–4.5	0.8–1.8	1–2.5	1–2.5	6.0–6.5	0.5–1.8	4–8	4–6
Compressive strength, psi × 10³	D695	18–50	15–30	15–30	15–30	24= / 72⊥	6–26	50–80	40–78
Shear strength	c	4–6				8.9		7–10	8–10
Impact strength, Izod, ft-lb/in. of notch	D256	5–30	5–15	1.5–1.0	8–22	65	1–5	40–60	
Physical Properties									
Specific gravity	D792	1.4–2.1	1.4–1.6	1.6–2.3	1.6–2.8	1.8–1.9	1.1–1.7	1.7–2.2	1.7–2.1
Density, lb/in³		0.051–0.076	0.051–0.058	0.058–0.083	0.058–0.094	0.065–0.069	0.040–0.061	0.061–0.079	0.061–0.076
Coefficient of thermal expansion, in./in./°F × 10⁻⁶	D696	1.0–1.8	1.2–2.0	0.8–1.2	0.8–1.2	0.4–0.5	1.0–3.5	0.4–0.6	0.4–0.6
Electrical Properties									
Dielectric constant 60 Hz	D150	3.8–6.0	3.7–6.0	5.3–7.3	4.4–6.3		2.4–4.2	4.2–5.3	4.0–6.0
10³ Hz		4.0–6.0	4.0–6.0	4.0–6.8	4.4–6.1		2.4–4.0		4.0–6.0
10⁸ Hz		3.5–5.5	3.6–6.0	5.2–6.4	4.2–8.0		2.4–3.9	4.0–5.2	4.0–6.0
Dissipation factor 60 Hz	D150	0.01–0.05	0.01–0.04	0.01–0.2	0.007–0.2		0.001–0.03		
10³ Hz		0.01–0.06	0.01–0.05	0.01–0.2	0.007–0.2		0.001–0.025	0.018–0.05	
10³ Hz		0.01–0.03	0.01–0.03	0.01–0.02	0.01–0.02		0.0015–0.026		
Dielectric strength (short time), V/mil	D149	300–600	200–450	300–450	300–450		400–600	200–400	300–350[f]

Source: Courtesy of PPG Industries, Inc.

[a]For XMC, values are parallel (=) to continuous reinforcement, XMC-3 provides 10–15 psi × 10³ transverse (⊥) tensile strength. For filament-wound structures values are by ASTM D2290 and D2343 for filament and strand tensile strengths, respectively.

[b]Ultimate tensile elongation ranges from 0.3 to 2.7% for FGRP thermoset polyesters, and from 1 to 5% for FGRTP thermoplastics.

[c]PPG in-plane, twin-notch compression test, except values for filament winding, which are apparent horizontal shear, ASTM D2344.

[d]Charpy values for HMC and XMC composites are 16 and 41 ft-lb/in. of notch, respectively.

[e]May be calculated as follows: Laminate sp. gr. equals the reciprocal of the sums of the ratios of weight fraction to sp. gr. for each component in the laminate.

[f]Range applies to sheet. Parallel dielectric strength for rods is 50 kV/in.

mately 20% by weight of zirconia provides about a 50% increase in tensile strength over plain alumina fiber. These **alumina–zirconia fibers** retain their mechanical properties after exposure to high temperatures, which makes them suitable as reinforcement of glass and ceramic matrix composites, where high fabrication temperatures are required.

Oxide–ceramic fibers are often not candidates for use in high-temperature composites, due to their loss of mechanical properties, which results from grain growth and creep at high temperatures. **Alumina–boria–silica fibers** possess superior handling qualities over other continuous polycrystalline fibers and can be used to produce textiles without the aid of fiber or metal additives.

Matrices

Composites typically use matrices of organic, metal, or ceramic material. **Organic** (often called *plastic*) **matrices** use thermoplastic or thermosetting materials. In general, thermoplastics are less costly. The general properties of thermosetting and thermoplastics are compared in Table 2.10.

The thermal effects of the curing process can be extremely important. The thermal coefficient of expansion for epoxy is 32×10^{-6} in./in./°F, while that of graphite fiber is -1.5 to 1.5×10^{-6} in./in./°F. Typically, thermoset epoxys are cured at 250 to 350°F and thermoplastic at 700°F. As a result, when cooled to room temperature, significant residual thermal stresses of as much as 40 to 60% of the ultimate strength of the laminate may remain with certain combinations of reinforcements and matrices. Some laminates have even failed when cooling down.

Table 2.10   Thermoset and Thermoplastic Trade-offs for Commercial Aircraft

Property	Thermosets	Thermoplastics
Resin cost	Low	Low to high
Prepregability	Excellent	Poor (new methods such as emulsion could change)
Prepreg tack/drape	Excellent	None (revised layup techniques required)
Volatile-free prepreg	Good	Good to excellent
Prepreg shelf life and out-time	Poor	Excellent
Prepreg quality assurance	Fair	Excellent
Prepreg cost	Good	High (new methods needed)
Composite processing	Slow	Slow (unless automated processes are developed)
Shrinkage	Moderate	Low
Composite mechanical properties	Good[a]	Good
Interlamiar fracture toughness	Low	High
Resistance to fluids and solvents	Good	Poor to good
Resistance to creep	Good	Currently not known
Crystallinity problems	None	Yes

Source: J. C. Leslie, in J. M. Margolis, ed., *Advanced Thermoset Composites,* Van Nostrand Reinhold, New York, 1986.
[a]Room for improvement in damage tolerance.

Design Considerations

When designing composite structures, particular care needs to be taken to assure that to the greatest extent possible, loads are resisted by fibers in tension. Where loads are multidimensional, multiaxis reinforcement is necessary. Thus some care needs to be taken to assure that the load conditions and direction(s) are reflected in the reinforcement orientation. Figure 2.30 indicates the way that load resistance changes with angularity of load for a typical material wound in only one direction.

Joining of composite structures favors the use of adhesive joining methods. Where holes are needed for bolting, for example, it is important that they be located precisely, as the composites have little tendency to yield and redistribute stress. Thus it is possible to have overstress, with fracture of the fiber structure at the point of load. This can be avoided in part by running plies of fibers at 45° around a hole to redis-

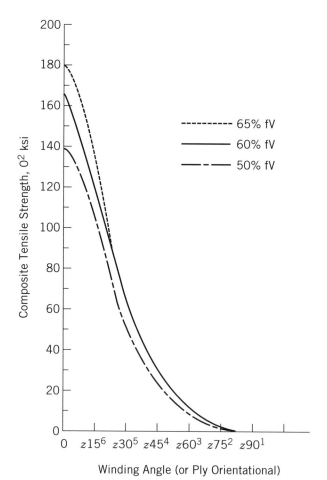

Fig. 2.30 Composite tensile E glass/epoxy, varies with winding angle and % fiber volume (fv). (In-house, first approximation data, Advanced Composite Products and Technology, Inc.). (Redrawn from J. C. Leslie, in J. M. Margolis, ed., *Advanced Thermoset Composites,* Van Nostrand Reinhold, New York, 1986.)

tribute stresses. Wherever possible it is preferable to drill holes rather than mold or lay them up, as drilling assures that fiber directions are not distorted around the hole, causing stress concentration. Many applications now use a combination of joints that are both adhesively bonded and mechanically fastened. This has the advantage of allowing some load distribution while providing a fastener to carry the principal portion of the load.

In fabricating custom components, the reinforcement is laid down in the proper orientation(s), and resins (matrix material) are either present with the reinforcement material or are introduced as a liquid, the composite is bagged, and the material is heated (generally in an autoclave) to temperatures in the range of 250 to 350°F to cure it and drive off excess resin. Following that, the composite is trimmed to final shape and assembled into a structure as appropriate. For glass-reinforced structures, curing often occurs at room temperature, the principal concern being ventilation provision for removal of the vapors from the curing resin.

2.6.2 Plastics (Polymers)*

Plastics can be classified several ways. From a performance standpoint they can be either commodity or engineering plastics. During fabrication a polymer will behave as a thermoplastic or a thermoset. Thermoplastics can be repeatedly softened by heat and shaped. By contrast, a thermoset can be shaped and cured by heat only once. Other classification schemes include crystalline and amorphous or synthetic and natural. These classifications will be useful in differentiating plastics.

Polyethylene. Polyethylenes (PEs) are crystalline thermoplastics that exhibit toughness, near-zero moisture absorption, excellent chemical resistance, excellent electrical insulating properties, low coefficient of friction, and ease of processing. Their heat deflection temperatures are reasonable but not high. High-density polyethylene (HDPE) exhibits greater stiffness, rigidity, improved heat resistance, and increased resistance to permeability than low-density polyethylene (LDPE). Some typical properties of PEs are listed in Table 2.11. The major use of HDPEs is in blow-molded bottles, automotive gas tanks, drums, and carboys; injection-molded material-handling pallets, trash and garbage cans, and household and automotive parts; and extruded pipe. LDPEs find major applications in film form for food packaging, as a vapor-barrier film; for extruded wire and cable insulation; and for bottles, closures, and toys.

Polypropylene. Polypropylene (PP) exhibits low density, rigidity, excellent chemical resistance, negligible water absorption, and excellent electrical properties. Its properties appear in Table 2.12. End uses for PP are in blow-molded bottles, closures, automotive parts, appliances, housewares, and toys. PP can be extruded into fibers and filaments for use in carpets, rugs, and cordage.

---

*Adapted from E. N. Peters, Chapter 17 in *Mechanical Engineers Handbook,* M. Kutz, ed., John Wiley & Sons, Inc., New York, 1986.

Table 2.11 Typical Property Values of Polyethylenes

	HDPE	LDPE
Density, Mg/m^3	0.96–0.97	0.91–0.93
Tensile modulus, GPa	0.76–1.0	—
Tensile strength, MPa	25–32	4–20
Elongation at break, %	500–700	275–600
Flexural modulus, GPa	0.8–1.0	0.21
Vicat soft point, °C	120–129	80–96
Brittle temperature, °C	−100 to −70	−85 to −35
Hardness, Shore	D60–D69	D45–D52
Dielectric constant, 10^6 Hz	—	2.3
Dielectric strength, MV/m	—	9–21
Dissipation factor, 10^6 Hz	—	0.0002
Linear mold shrinkage, in./in.	0.007–0.009	0.015–0.035

Source: Reprinted from E. N. Peters, Chapter 17 in *Mechanical Engineers Handbook,* M. Kutz, ed. Copyright © 1986 by John Wiley & sons, Inc. Reprinted with permission.

Polystyrene.   Polystyrene (PS) has excellent electrical insulating properties; however, it is brittle under impact and exhibits very poor resistance to surfactants and solvents. Its properties appear in Table 2.13. Ease of processing, rigidity, clarity, and low cost combine to support applications in toys, displays, and housewares. PS foams can readily be prepared and are characterized by excellent low thermal conductivity, high strength-to-weight ratio, low water absorption, and excellent energy absorption. These attributes have made PS foam of special interest as insulation boards for construction, protective packaging materials, insulated drinking cups, and flotation devices.

ABS.   ABS exhibits good impact strength, improved chemical resistance, and similar heat deflection temperature and rigidity. ABS is also opaque. It is suitable for tough consumer products; automotive parts; business machine housings; telephones; luggage; and pipe, fittings, and conduit.

Poly(vinyl chloride).   Poly(vinyl chloride) (PVC) has a low degree of crystallinity and good transparency, advantages in flame resistance, fair heat deflection temperature, good electrical resistance, and chemical resistance. However, PVC is

Table 2.12 Typical Property Values of Polypropylene

Density, Mg/m^3	0.90–0.91
Tensile modulus, GPa	1.8
Tensile strength, MPa	37
Elongation at break, %	10–50
Heat deflection temperature at 0.45 MPa, °C	100–105
Heat deflection temperature at 1.81 MPa, °C	60–65
Vicat soft point, °C	130–148
Linear thermal expansion, mm/mm · K	3.8×10^{-5}
Hardness, Shore	D76
Volume resistivity, Ω cm	1.0×10^{17}
Linear mold shrinkage, in./in.	0.01–0.02

Table 2.13 Typical Properties of Styrene Thermoplastics

Property	PS	SAN	IPS/HIPS	ABS
Density, Mg/m^3	1.050	1.080	1.020–1.040	1.050–1.070
Tensile modulus, GPa	2.76–3.10	3.4–3.9	2.0–2.4	2.5–2.7
Tensile strength, MPa	41.4–51.7	65–76	26–40	36–39
Yield elongation, %	1.5–2.5	—	—	2.2–2.6
Heat deflection temperature at 264 psi, °C	82–93	101–103	81	80–95
Vicat soft point, °C	98–107	110	88–101	90–99
Notched Izod, kJ/m	0.02	0.02	0.10–0.32	0.09–0.48
Linear thermal expansion, 10^{-5} mm/mm · K	5–7	6.4–6.7	7.0–7.5	7.5–9.5
Hardness, Rockwell	M60–M75	M80–M83	M45, L55	R69–R115
Linear mold shrinkage, in./in.	0.007	0.003–0.004	0.007	0.0055

difficult to process. The chlorine atoms also have a tendency to split out under the influence of heat during processing and heat and light during end use in finished products, producing discoloration and embrittlement. Therefore, special stabilizer systems are often used with PVC to retard degradation. There are two major subclassifications of PVC, rigid and flexible. Properties appear in Table 2.14.

Rigid PVC. PVC alone is a fairly good rigid polymer but is difficult to process and has low impact strength. Both of these properties are improved by the addition of elastomers or impact-modified graft copolymers such as ABS and impact acrylic polymers. These improve the melt flow during processing and improve the impact strength without seriously lowering the rigidity or the heat deflection temperature. Rigid PVCs are used in such applications as door and window frames; pipe, fittings, and conduit; building panels and siding; rainwater gutters and downspouts; credit cards; and flooring.

Table 2.14 Typical Property Values for Polyvinyl Chloride Materials

Property	General Purpose	Rigid	Rigid Foam	Plasticized	Copolymer
Linear mold shrinkage, in./in.	0.003	—	—	—	—
Density, Mg/m^3	1.400	1.340–1.390	0.750	1.290–1.340	1.370
Tensile modulus, GPa	3.45	2.41–2.45	—	—	3.15
Tensile strength, MPa	8.7	37.2–42.4	>13.8	14–26	52–55
Elongation at break, %	113	—	>40	250–400	—
Notched Izod, kJ/m	0.53	0.74–1.12	>0.06	—	0.02
Heat deflection temperature at 1.81 MPa, °C	77	73–77	65	—	65
Brittle temperature, °C	—	—	—	−60 to −30	—
Hardness	D85 (Shore)	R107–R112 (Rockwell)	D55 (Shore)	A71–A96 (Shore)	—
Linear thermal expansion, 10^{-5} mm/mm · K	7.00	5.94	5.58	—	—

Table 2.15   Typical Properties of Poly(methyl methacrylate)

Density, Mg/m³	1.180–1.190
Tensile modulus, GPa	3.10
Tensile strength, MPa	72
Elongation at break, %	5
Notched Izod, kJ/m	0.4
Heat deflection temperature at 1.81 MPa, °C	96
Hardness, Rockwell	M90–M100
Linear thermal expansion, 10^{-5} mm/mm · K	6.3
Continuous-service temperature, °C	88
Linear mold shrinkage, in./in.	0.002–0.008

Plasticized PVC.   Plasticized PVC is used for wire and cable insulation, outdoor apparel, rainwear, flooring, interior wall coverings, upholstery, automotive seat coverings, garden hose, toys, clear tubing, shoes, tablecloths and shower curtains.

PVC is also available in liquid formulations known as plastisols or organosols. These materials are used in coating fabrics, paper, and metal and are rotationally cast into dolls, balls, and so on.

Foamed PVC.   Rigid PVC can be foamed to a low-density cellular material that is used for decorative moldings and trim. Foamed plastisols add greatly to the softness and energy absorption already inherent in plasticized PVC, giving richness and warmth to leatherlike upholstery, clothing, shoe fabrics, handbags, luggage, and auto door panels and energy absorption for quiet and comfort in flooring, carpet backing, auto headliners, and so on.

PVC Copolymer.   Copolymerization of vinyl chloride with 10 to 15% vinyl acetate gives a vinyl copolymer with improved flexibility and less crystallinity than PVC, making such copolymers easier to process without detracting seriously from the rigidity and heat deflection temperature. These copolymers find primary applications in phonograph records, flooring, and solution coatings.

Poly(methyl methacrylate).   PMMA has excellent resistance to weathering, low water absorption, and good electrical resistivity. PMMA properties appear in Table 2.15. PMMA is used for glazing, lighting diffusers, skylights, outdoor signs, and automobile tail lights.

Engineering Thermoplastics

Engineering polymers comprise a special high-performance segment of synthetic plastic materials that offer premium properties. When properly formulated, they may be shaped into mechanically functional, semiprecision parts or structural components. Mechanically functional implies that the parts may be subject to mechanical stress, impact, flexure, vibration, sliding friction, temperature extremes, hostile environments, and so on, and continue to function.

As substitutes for metal in the construction of mechanical apparatus, engineering plastics offer advantages such as transparency, light weight, self-lubrication, and

Table 2.16 Typical Properties of Poly(butylene terephthalate)

Property	PBT	PBT + 40% Glass Fiber
Density, Mg/m^3	1.300	1.600
Flexural modulus, GPa	2.4	9.0
Flexural strength, MPa	88	207
Elongation at break, %	300	3
Notched Izod, kJ/m	0.06	0.12
Heat deflection temperature at 1.81 MPa, °C	54	232
Heat deflection temperature at 0.45 MPa, °C	154	232
Hardness, Rockwell	R117	M86
Linear thermal expansion, 10^{-5} mm/mm·K	9.54	1.89
Linear mold shrinkage, in./in.	0.020	<0.007

economy in fabricating and decorating. Replacement of metals by plastic is favored as the physical properties and operating temperature ranges of plastics improve and the cost of metals and their fabrication increases.

Polyesters (Thermoplastic). Poly(butylene terephthalate) (PBT) has low moisture absorption, extremely good self-lubrication, fatigue resistance, solvent resistance, and good maintenance of mechanical properties at elevated temperatures. Properties appear in Table 2.16. Applications of PBT include gears, rollers, bearings, housings for pumps and appliances, impellers, pulleys, switch parts, automotive components, and electrical/electronic components.

Polyamides (Nylon). Their key features include a high degree of solvent resistance toughness, and fatigue resistance. Nylons do exhibit a tendency to creep under applied load. Their properties appear in Table 2.17. The largest application of nylons is in fibers. Molded applications include automotive components, related machine

Table 2.17 Typical Properties of Nylons

Property	Nylon 6	Nylon 6 + 40% Glass Fiber	Nylon 66	Nylon 66 + 40% Glass Fiber
Density, Mg/m^3	1.130	1.460	1.140	1.440
Flexural modulus, GPa	2.8	10.3	2.8	9.3
Flexural strength, MPa	113	248	—	219
Elongation at break, %	150	3	60	4
Notched Izod, kJ/m	0.06	0.16	0.05	0.14
Heat deflection temperature at 1.81 MPa, °C	64	216	90	250
Heat deflection temperature at 0.45 MPa, °C	170	218	235	260
Hardness, Rockwell	R119	M92	R121	R119
Linear thermal expansion, 10^{-5} mm/mm·K	8.28	2.16	8.10	3.42
Linear mold shrinkage, in./in.	0.013	0.003	0.0150	0.0025

Table 2.18 Typical Properties of Polyacetals

Property	Polyacetal	Polyacetal + 40% Glass Fiber
Density, Mg/m^3	1.420	1.740
Flexural modulus, GPa	2.7	11.0
Flexural strength, MPa	107	117
Elongation at break, %	75	1.5
Notched Izod, kJ/m	0.12	0.05
Heat deflection temperature at 1.18 MPa, °C	124	164
Heat deflection temperature at 0.45 MPa, °C	170	167
Hardness, Rockwell	M94	R118
Linear thermal expansion, 10^{-5} mm/mm · K	10.4	3.2
Linear mold shrinkage, in./in.	0.02	0.003

parts (gears, cams, pulleys, rollers, boat propellers, etc.), appliance parts, and electrical insulation.

Polyacetals. Polyacetals exhibit rigidity, high strength solvent resistance, fatigue resistance, toughness, self-lubricity, and cold-flow resistance. They also exhibit a tendency to thermally unzip and hence are difficult to flame retard. Properties appear in Table 2.18. Applications of polyacetals include moving parts in appliances and machines (gears, bearings, bushings, etc.), in automobiles (e.g., door handles), and in plumbing (valves, pumps, faucets, etc.).

Polyphenylene Sulfide. It is characterized by high heat resistance, rigidity, excellent chemical resistance, low friction coefficient, good abrasion resistance, and electrical properties. Polyphenylene sulfides are somewhat difficult to process due to their very high melting temperature, relatively poor flow characteristics, and some tendency for slight cross-linking during processing. Properties appear in Table 2.19. The unreinforced resin is used only in coatings. The reinforced materials are used in aerospace applications, pump components, electrical/electronic components, appliance parts, and automotive applications.

Table 2.19 Typical Properties of Polyphenylene Sulfide

Property	Polyphenylene Sulfide + 40% Glass Fiber
Density, Mg/m^3	1.640
Tensile modulus, GPa	7.7
Tensile strength, MPa	135
Flexural modulus, GPa	11.7
Flexural strength, MPa	200
Elongation at break, %	1.3
Notched Izod, kJ/m	0.08
Heat deflection temperature at 1.81 MPa, °C	>260
Hardness, Rockwell	R123
Linear thermal expansion, 10^{-5} mm/mm · K	4.0
Linear mold shrinkage, in./in.	0.004
Constant-service temperature, °C	232

Table 2.20 Typical Properties of Polycarbonates

Property	Polycarbonate	Polycarbonate + 40% Glass Fiber
Density, Mg/m^3	1.200	1.520
Tensile modulus, GPa	2.4	11.6
Tensile strength, MPa	65	158
Flexural modulus, GPa	2.3	9.7
Flexural strength, MPa	93	186
Elongation at break, %	110	4
Notched Izod, kJ/m	0.86	0.13
Heat deflection temperature at 1.81 MPa, °C	132	146
Heat deflection temperature at 0.45 MPa, °C	138	154
Hardness, Rockwell	M70	M93
Linear thermal expansion, 10^{-5} mm/mm·K	6.75	1.67
Linear mold shrinkage, in./in.	0.006	0.0015
Constant-service temperature, °C	121	135

Polycarbonates.　Polycarbonates are among the stronger, tougher, and more rigid thermoplastics. Polycarbonates also show resistance to creep and excellent electrical-insulating characteristics. Polycarbonate properties appear in Table 2.20. Applications of polycarbonates include safety glazing, safety shields, nonbreakable windows, automobile tail lights, electrical relay covers, various appliance parts and housings, power tool housings, automotive fender extensions, and blow-molded bottles.

Polysulfone.　Polysulfone is characterized by excellent thermooxidative resistance, hydrolytic stability, and creep resistance. Polysulfone properties appear in Table 2.21. Typical applications of polysulfones include microwave cookware, medical equipment where sterilization by steam is required, coffee makers, and electrical-electronic components.

Modified Polyphenylene Ether.　Blends of poly(2,6-dimethyl phenylene ether) with styrenics (HIPS, ABS, etc.) form a family of modified polyphenylene-ether-based resins. Depending on the blend, these materials have a broad temperature use range. They are characterized by outstanding dimensional stability at elevated temperatures, outstanding hydrolytic stability, long-term stability under load, and excel-

Table 2.21 Typical Properties of Polysulfone

Density, Mg/m^3	1.240
Tensile modulus, GPa	2.48
Tensile strength, MPa	70
Flexural modulus, GPa	2.69
Flexural strength, MPa	106
Elongation at break, %	75
Notched Izod, kJ/m	0.07
Heat deflection temperature at 1.81 MPa, °C	174
Hardness, Rockwell	M69
Linear thermal expansion, 10^{-5} mm/mm·K	5.6
Linear mold shrinkage, in./in.	0.007
Constant-service temperature, °C	150

lent dielectric properties over a wide range of frequencies and temperatures. Their properties appear in Table 2.22. Modified polyphenylene ether applications include automotive applications (dashboards, trim, etc.), TV cabinets, electrical connectors, pumps, plumbing fixtures, and small appliance and business machine housings.

Polyimides. Thermoplastic and thermoset grades of polyimides are available. The thermoset polyimides are among the most heat resistant polymers; for example, they can withstand temperatures up to 250°C. Thermoplastic polyimides, which can be processed by standard techniques, fall into two main categories: polyetherimides and polyamideimides.

In general, polyimides have very good electrical properties, very good wear resistance, superior dimensional stability, outstanding flame resistance and high strength and rigidity. Polyimide properties appear in Table 2.23. Polyimide applications include gears, bushings, bearings, seals, insulators, electrical/electronic components (printed wiring boards, connectors, etc.), microwave oven components, and structural components.

Fluorinated Thermoplastics

In general, fluoropolymers, include inertness to most chemicals, resistance to high temperature, extremely low coefficient of friction, and excellent dielectric properties. Properties appear in Table 2.24. Mechanical properties are normally low but can be improved when reinforced with glass or carbon fibers or molybdenum disulfide fillers.

The difficulty of fluorochemical synthesis makes the price of fluoropolymers relatively high, and their uses are largely restricted to critical specialty applications.

Poly(tetrafluoroethylene). Prepared from tetrafluoroethylene, poly(tetrafluoroethylene) (PTFE) is a crystalline, very heat-resistant (up to 500°F), and outstanding chemical-resistant polymer, and it has the lowest coefficient of friction of any polymer. PTFE does not soften like other thermoplastics and has to be processed by nonconventional techniques (PTFE powder is compacted to the desired shape and sintered). PFTE applications include nonstick coatings on cookware; nonlubricated

Table 2.22 Typical Properties of Modified Polyphenylene Ethers

Property	190 Grade	225 Grade	300 Grade
Density, Mg/m³	1.080	1.090	1.060
Tensile modulus, GPa	2.5	2.4	—
Tensile strength, MPa	48	55	76
Flexural modulus, GPa	2.2	2.4	2.4
Flexural strength, MPa	56.5	76	104
Elongation at break, %	35	—	—
Notched Izod, kJ/m	0.37	0.32	0.53
Heat deflection temperature at 1.81 MPa, °C	88	107	149
Heat deflection temperature at 0.45 MPa, °C	96	118	157
Hardness, Rockwell	R115	R116	R119
Linear thermal expansion, 10^{-5} mm/mm·K	—	—	5.9
Linear mold shrinkage, in./in.	0.006	0.006	0.006
Constant-service temperature, °C	—	95	—

Table 2.23 Typical Properties of Polyimides

Property	Polyimide	Polyetherimide		Polyamideimide	
		Unfilled	30% Glass Reinforced	Unfilled	30% Glass Reinforced
Density, Mg/m³	—	1.27	1.51	1.38	1.57
Tensile modulus, GPa	2.65	0.30	0.90	—	1.15
Tensile strength, MPa	196.2	104.8	168.9	117.2	195.2
Elongation at break, %	90	60	3	10	5
Notched Izod, kJ/m	—	0.6	0.11	0.13	0.11
Heat deflection temperature at 1.81 MPa, °C	—	392	410	260	274
Heat deflection temperature at 0.45 MPa, °C	—	410	414	—	—
Hardness, Rockwell	—	R109	M125	E78	E94
Linear thermal expansion, 10^{-5} mm/mm·K	—	5.6	2.0	3.60	1.80
Linear mold shrinkage, in./in.	—	0.5	0.2	—	0.25

bearings; chemical-resistant pipe fittings, valves, and pump parts; high-temperature electrical parts; and gaskets, seals, and packings.

Poly(chlorotrifluoroethylene). Poly(chlorotrifluoroethylene) (CTFE) is less crystalline and exhibits higher rigidity and strength than PTFE; it is chemical resistant and has heat resistance up to 390°F. Unlike PTFE, it can be molded and extruded by

Table 2.24 Typical Properties of Fluoropolymers

Property	PTFE	CTFE	FEP	ETFE	ECTFE
Density, Mg/m³	2.160	2.100	2.150	1.700	1.680
Tensile modulus, GPa	—	14.3	—	—	—
Tensile strength, MPa	27.6	39.4	20.7	44.8	48.3
Elongation at break, %	~275	~150	~300	100–300	200
Notched Izod, kJ/m	—	0.27	0.15	—	—
Heat deflection temperature at 1.81 MPa, °C	—	75	—	71	77
Heat deflection temperature at 0.45 MPa, °C	—	126	—	104	116
Hardness	D55–65 (Shore)	D75–80 (Shore)	D55 (Shore)	D75 (Shore)	R93 (Rockwell)
Linear thermal expansion, 10^{-5} mm/mm·K	9.9	4.8	9.3	13.68	—
Dielectric strength, MV/m	23.6	19.7	82.7	7.9	19.3
Dielectric constant at 10^2 Hz	2.1	3.0	2.1	2.6	2.5
Dielectric constant at 10^3 Hz	2.1	2.7	—	2.6	2.5
Constant-service temperature, °C	260	199	204	—	150–170
Linear mold shrinkage, in./in.	0.033–0.053	0.008	—	—	<0.025

conventional processing techniques. CTFE applications include electrical insulation, cable jacketing, electrical and electronic coil forms, pipe and pump parts, valve diaphragms, and coatings for corrosive process industries and other industrial parts.

Fluorinated Ethylene–Propylene.   (FEP) Copolymerization of tetrafluoroethylene with some hexafluoropropylene produces a polymer with less crystallinity, lower melting point, and improved impact strength than PTFE. This copolymer can be molded by thermoplastic molding techniques. Fluorinated ethylene–propylene applications include wire insulation and jacketing, high-frequency connectors, coils, gaskets, and tube sockets.

Polyvinylidene Fluoride.   Polyvinylidene fluoride has high tensile strength and better ability to be processed but less thermal and chemical resistance than the previous fluoropolymers. Polyvinylidene fluoride applications include insulation, seals and gaskets, diaphragms, and piping.

Poly(ethylene trifluoroethylene).   Copolymerization of ethylene with trifluoroethylene produces poly(ethylene trifluoroethylene) (ETFE) with good high-temperature and chemical resistance; ETFE can be processed by conventional techniques. ETFE applications include molded labware, valve liners, electrical connectors, and coil bobbins.

Poly(ethylene chlorotrifluoroethylene).   The copolymer of ethylene and chlorotrifluoroethylene, poly(ethylene chlorotrifluoroethylene) (ECTFE), is strong and chemical and impact resistant. It can be processed by conventional techniques. ECTFE applications include wire and cable coatings, chemical-resistant coatings and linings, molded labware, and medical packaging.

Poly(vinyl fluoride).   Poly(vinyl fluoride) films exhibit excellent outdoor durability. Poly(vinyl fluoride) uses include glazing, lighting, and coatings on presurfaced exterior building panels.

Thermosets

Thermosetting polymers are used in molded and laminated plastics. Thermosets are generally catalyzed and/or heated to finish the polymerization reaction, cross-linking them to almost infinite molecular weight. This step is often referred to as cure. Such cured polymers cannot be reprocessed or reshaped.

Phenolic.   Phenolic resins have high hardness, rigidity, strength, heat resistance, chemical resistance, and good electrical properties. Phenolic applications include automotive uses, distributor caps, rotors, and brake linings; appliance parts, pot handles, knobs, and bases; electrical/electronic components, connectors, circuit breakers, and switches; and adhesives in laminates (e.g., plywood).

Epoxy Resins.   Cured epoxy resins exhibit hardness, strength, heat resistance, electrical resistance, and broad chemical resistance. Epoxy applications include glass-reinforced, high-strength composites used in aerospace, pipes, tanks, pressure vessels;

encapsulation or casting of various electrical and electronic components; adhesives; protective coatings in appliances, flooring, and industrial equipment; and sealants.

Unsaturated Polyesters.   Properly formulated glass-reinforced unsaturated polyesters are commonly referred to as sheet molding compound (SMC) or reinforced plastics. Unsaturated polyesters are thermosets and are quite distinct from thermoplastic polyesters.

In combination with reinforcing materials such as glass fibers, cured resins offer outstanding strength, high rigidity, high strength-to-weight ratio, impact resistance, and chemical resistance. The prime use of unsaturated polyesters is in combination with glass fibers in high-strength composites; these include transportation markets (body parts and components for automobiles, trucks, trailers, buses, and aircraft), marine uses (small- to medium-size boat hulls and associated marine equipment), building panels, housings, bathroom components (bathtubs and shower stalls), appliances, and electronic/electrical components.

Alkyd Resins.   Alkyds have excellent heat resistance; are dimensionally stable at high temperatures; and have excellent dielectric strength (>14 MV/m), high resistance to electrical leakage, and excellent arc resistance. Alkyd resin applications include drying oils in enamel paints; lacquers for automobiles and appliances; and molding compounds when formulated with reinforcements for electrical applications (circuit breaker insulation, encapsulation of capacitors and resistors, and coil forms).

Diallyl Phthalate.   Diallyl phthalate (DAP) has excellent dimension stability and high insulation resistance. In addition, DAP has high dielectric strength, excellent arc resistance, and chemical resistance.

Amino Resins.   The two main members of the amino family of thermosets are the melamine and urea resins. In general, these materials exhibit extreme hardness, scratch resistance, electrical resistance, and chemical resistance. DAP applications include electronic parts, electrical connectors, bases, and housings. DAP is also used as a coating and impregnating material. Melamine resins find use in colorful, rugged dinnerware; decorative laminates (countertops, tabletops, and furniture surfacing); electrical applications (switchboard panels, circuit breaker parts, arc barriers, and armature and slot wedges); and adhesives and coatings.

Urea resins are used in particleboard binders, decorative housings, closures, electrical parts, coatings, and paper and textile treatment.

2.6.3   Wire Rope[8,9]

Wire rope is available in a wide variety of materials and sizes up to 3 in. in diameter and breaking strengths over 600,000 lb. Figure 2.31 describes the common classifications of wire rope, and Table 2.25 lists the features of the more common types and sizes of wire rope. In selecting wire rope it is general practice to establish the safe working load at no more than one-fifth of the breaking strength. Fittings are designed to develop the full breaking strength of the rope and can be swaged, bolted, or adhesive bonded (generally, two-part epoxy systems). Increasingly, wire rope is specified in metric sizes as well as in FPS units. It is common practice to request factory

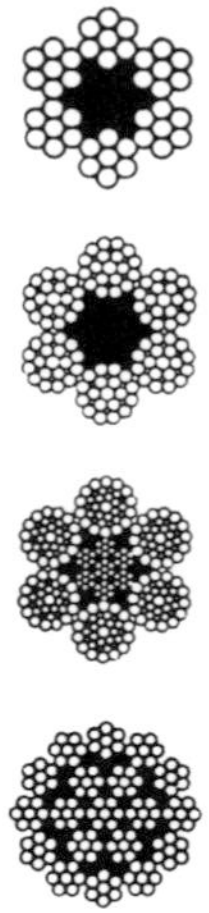

6 x 7 Class Wire Rope: 6 strands, 7 wires per strand

This construction is used where ropes are dragged over the ground or over rollers and resistance to wear abrasion are important factors. The wires are quite large and will stand a great deal of wear. The 6 x 7 is a stiff rope and needs sheaves and drums of large sizes. It will not stand bending stresses as well as ropes with a larger number of wires.

6 x 19 Class Wire Rope: 6 strands, nominally 19 main wires per strand

This class is most widely used and is found in its many variations throughout nearly all industries. With its combination of flexibility and wear resistance, rope in this class can be suited to the specific needs of diverse kinds of machinery and equipment. The designation of 6 x 19 is only nominal as the number of wires per strand ranges from 15 to 26.

6 x 37 Class Wire Rope: 6 strands, nominally 37 wires per strand

The 6 x 37 class of wire rope is characterized by the relatively large number of wires used in each strand. Ropes of this class are among the most flexible available, but their resistance to abrasion is less than the 6 x 19 class. The designation of 6 x 37 is again only nominal as in the 6 x 19 class.

19 x 7 Rotation Resistant Wire Rope

The 19 x 7 rotation resistant wire rope consists of an inner layer of 6 strands of 7 wires each, made left lang lay over a strand core, and an outer layer of 12 strands, each of 7 wires, made in right regular lay. It is this combination of opposing lays which enables the rope to resist the tendency to rotate when in service.

Fig. 2.31 Common classifications of wire rope. (Courtesy of Loos & Co. Reprinted with permission.)

Table 2.25   Wire Rope Features (6 × 9 and 6 × 37 Bright^a)

	Preformed
6 × 9 Classification	6 × 37 Classification

6 x 25 FILLER WIRE WITH FIBER CORE **6 x 19 SEALE WITH IWRC** **6 x 25 FILLER WIRE WITH IWRC** **6 x 26 WARRINGTON SEALE WITH IWRC**

6 x 36 WARRINGTON SEALE WITH FIBER CORE **6 x 36 WARRINGTON SEALE WITH IWRC**

| | Improved Plow Steel | | | | Extra Improved Plow Steel IWRC | |
| | Fiber Core | | IWRC | | | |
Diameter (in.)	Approx. Weight per Foot (lb)	Minimum Breaking Strength (net tons)	Approx. Weight per Foot (lb)	Minimum Breaking Strength (net tons)	Approx. Weight per Foot (lb)	Minimum Breaking Strength (net tons)
1.4	0.105	2.74	0.116	2.94	0.116	3.40
5/16	0.164	4.26	0.180	4.58	0.180	5.27
3/8	0.236	6.10	0.260	6.56	0.260	7.55
7/16	0.32	8.27	0.35	8.89	0.35	10.2
1/2	0.42	10.7	0.46	11.50	0.46	13.3
9/16	0.53	13.5	0.59	14.5	0.59	16.8
5/8	0.66	16.7	0.72	17.9	0.72	20.6
3/4	0.95	23.8	1.04	25.6	1.04	29.4
7/8	1.29	32.2	1.42	34.6	1.42	39.8
1	1.68	41.8	1.85	44.9	1.85	51.7

^{a}Same strength and weight data apply to 6 × 19 and 6 × 37 classifications.
Source: Courtesy of Loos & Co. Reprinted with permission.

test certificates, which document the physical properties of the wire rope. They are widely accepted by **OSHA** (Occupational Health and Safety Administration) and **ABS** (American Bureau of Shipping), which in many cases have jurisdiction for health and safety matters.

Sheave diameters less than 10 times the rope diameter substantially reduce its strength and fatigue life. Preferred sheave diameter ratios are in the range of 20 times the rope diameter, while for **aircraft service,** 35 is the preferred ratio. Wire rope is

Table 2.26   G Factor

Cable/Wire Rope	G Factor	Cable/Wire Rope	G Factor
1 × 7 302 S.S.	0.00000735	1 × 7 Galv.	0.00000661
1 × 19 302 S.S.	0.00000779	1 × 19 Galv.	0.00000698
7 × 7 302 S.S.	0.0000120	7 × 7 Galv.	0.0000107
7 × 19 302 S.S.	0.0000162	7 × 19 Galv.	0.0000140
6 × 19 IWRC 302 S.S.	0.0000157	6 × 19 IWRC Galv.	0.0000136
6 × 37 IWRC 302 S.S.	0.0000160	6 × 37 IWRC Galv.	0.0000144
19 × 7 302 S.S.	0.0000197	19 × 7 Galv.	0.0000178

Source: Courtesy of Loos & Co. Reprinted with permission.

typically available in either "bright" (ungalvanized steel), galvanized steel, 302 stainless steel, 316 stainless steel, 305 (nonmagnetic) stainless steel, and Monel. For the same size and construction, stainless steel wire rope will be about 5% weaker, and galvanized about 10% weaker than "bright" steel (although for most applications, this slight difference is overlooked and they are considered to have the same strength). Virtually all types are commercially available in plastic jacketed for corrosion resistance or underwater service. For resistance to abrasion, plastic impregnation is often specified. Wire rope is manufactured in a variety of wire and strand arrangements, the more common of which are shown in Table 2.25.

    Wire rope or cable stretch is of two types, **structural stretch,** slightly less than 1% of the cable length, which is the lengthening of the lay of the cable under load, and **elastic stretch,** which is the actual lengthening of the cable under its working load. Cables can be prestretched to remove the structural stretch. Elastic stretch can be calculated to within about 2% accuracy by

$$E = \frac{W}{D^2} \times G \qquad (2.131)$$

where E is the elastic stretch (% of length), W the load in pounds, D the diameter in inches, and G values are as listed in Table 2.26.

    Because the modulus of elasticity of the wires making up the rope change slightly with load, there will be some slight variation in the elastic stretch calculated above. The wire rope manufacturer should be consulted for cases where more exact calculations are necessary. All of the above assumes that the rope ends are prevented from rotating. Where rotation is allowed to occur, rope lengthening, caused by unwinding the lay of the rope introduced during manufacture, will occur.

2.7   FINITE ELEMENT ANALYSIS*

Of the many methodologies available for design analysis, perhaps the most widely used is **finite element analysis.**[10–16] In addition to the analytical information pre-

---

*Courtesy of MacNeal-Schwendler Corporation, Los Angeles.

sented here, testing methods to determine the physical properties of materials are discussed in Section 8.3.6. The goal of analysis is to synthesize or verify a design by modeling its behavior prior to manufacture.

Finite element analysis (FEA) has long been used to analyze aircraft, automobiles, buildings, consumer electronics, jet engines, heavy machinery, launch vehicles, medical instrumentation, piping, and virtually every other type of manufactured product. A common use is to simulate structural behavior, but other uses abound, such as acoustics, electromagnetics, thermal, and fluid flow. Finite element analysis is a procedure in which response to the environment is simulated. The overall process is comprised of:

- Creating the model (also called *finite element modeling*)
- Analyzing the model
- Assessing the results
- Optimizing the design, as necessary, to achieve better results

2.7.1 Creating the Model

In finite element modeling, a structure or component is analyzed by being meshed into an assemblage of small, discrete elements, their vertices being called *grid points* or *nodes*. The steps in creating the model are:

- Modeling the geometry
- Generating the elements
- Selecting the material
- Defining the environment

Creating an accurate geometric representation of the physical object is the first step in creating the model. There are several means of creating the finite element model, including:

- Using a CAD program
- Using a dedicated finite element modeling program
- Using both CAD and finite element modeling programs

Geometric modeling is the strength of a CAD program. **Finite element modeling** (meshing, creating element and material properties, applying loads, and editing) is the strength of a finite element modeling program. A combination of the two—using CAD for the geometric modeling, and then using a finite element modeler to add the remainder of the engineering input—provides the best overall solution as long as the CAD and finite element modeling programs can accurately and easily transmit information between them. CAD models often have a much finer level of detail than that required for analysis, and some type of **feature suppression,** manual or automatic, is usually done prior to analysis.

2.7.2 Generating the Elements

Accuracy of the computed response depends on the nature of the environment being simulated, adequacy of the finite element **mesh,** and use of the proper elements. These considerations are kept in mind when selecting and generating the elements. To simulate complex behavior, finite element programs contain a large library of element types (perhaps 30). Elements commonly used for structural applications include beams, plates, and solids. Other element types are also available, including axisymmetric, acoustic, aeroelastic, mass and damping, and heat transfer elements.

Two primary types of elements are available. **h elements** are the standard finite elements that have been available for the last 30 years. To create a refined model that accurately represents a complex stress gradient, a finer (denser) mesh is used with a greater numbers of elements. **p elements** are elements that can internally represent a high-order stress gradient. To create a refined model, the internal order of the p element is increased automatically. p elements have been available commercially for about 10 years.

Meshing is the process of generating elements. Meshing can be performed via automatic meshing, in which the mesh generating program determines the mesh, or via mapped meshing, in which the user explicitly defines the mesh density. Because no single method is best in every situation, finite element modeling programs contain both automatic and mapped meshing and provide ways to edit the mesh. Once meshed, the model may be comprised of several hundred to many thousands of elements, depending on the complexity of the geometry and the complexity of the internal stresses.

2.7.3 Specifying the Material

Structural behavior is a function of the stress–strain relationship of the material. There are many different material types to model complex material behavior, including:

- **Isotropic materials,** in which the material constants are the same in every direction
- **Anisotropic materials,** in which the material properties vary in each direction
- **Orthotropic materials,** in which the properties vary in two directions
- **Composite materials,** for analyzing laminated composites and sandwich structures
- **Nonlinear-elastic materials,** in which the stress–strain relationship is nonlinear and the material loads and unloads along the same path
- **Elastic–plastic materials,** in which plastic deformation occurs
- **Creep-dependent materials,** in which the quasistatic behavior of viscoelastic materials can be analyzed
- **Temperature-dependent materials,** whose constants are a function of temperature
- **Hyperelastic materials,** for large strain with **Poisson ratios** approaching 0.5

Because of the potentially large number of materials and material constants, a material properties database is useful for ensuring corporate-wide material property consistency.

2.7.4   Defining the Environment

Structures are subjected to stress, vibration, heat, shock, and noise, among other disturbances. These excitations comprise the operating environment that is simulated by applying loads and boundary conditions to the model. **Loads** define excitations that comprise the operating environment. They can be applied to geometry, to elements, and to grid points; they can be discrete (concentrated at a single location) or continuous (varying over a region, such as a pressure); and they can be static (time invariant) or functions of time, temperature, or displacement. **Boundary conditions** define how a structure is restrained, whether it be a building connected to the ground or a bracket connected to a wall. **Enforced motion** is a special kind of boundary condition in which there is a specified displacement of the structure's base.

2.7.5   Analyzing the Model

After the model is built, it is ready for analysis. Whereas the process of building the model is graphical and interactive, the solution process—which solves systems of equations—takes place in a background or batch mode. Several types of analyses are described below.

Linear Static Analysis

Linear static analysis represents the most commonly used finite element analysis capability for structural analysis. The term *linear* means that the material is linear-elastic and that the computed response—displacement or stress, for example—is linearly related to the applied force or temperature. In addition, deformations are assumed small with respect to the overall structural dimensions, and the direction of applied load is assumed to be constant. *Static* means that the applied forces, and the responses, are assumed invariant with time. The objective of linear static analysis is to compute displacements, internal forces and stresses, and reaction forces.

Inertia Relief Analysis

Static analysis using the finite element method assumes that the model is restrained or tied to ground and that there are no internal mechanisms (internal hinges that allow part of the model to behave as a rigid body). **Inertia relief analysis** is a special form of static analysis in which an unrestrained structure—such as an airplane in flight—or a mechanism can be analyzed.

Elastic Buckling Analysis

In linear static analysis, a structure is assumed to be in a state of stable equilibrium. As the applied load is removed, the structure returns to its undeformed position. Under certain combinations of loadings, however, the structure continues to deform without an increase in the loading, and the structure has become unstable; it has buckled. For **elastic,** or **linear, buckling,** it is assumed that there is no material yielding and there is no change in applied force direction during loading. **Elastic,** or **linear, buckling analysis** solves for the critical buckling load, that is, the load that will produce the onset of buckling.

Heat Transfer Analysis

Heat transfer analysis is performed to determine the temperatures and heat flows due to conduction, convection, and radiation. **Conduction** is the flow of heat through a solid. Conduction heat flow is governed by **Fourier's law,** which says that the amount of heat flow is proportional to the temperature gradient. The constant of proportionality is the thermal conductivity. **Convection** deals with the flow of heat at the boundaries of structures relative to adjacent fluid media. This heat flow is proportional to the temperature difference between the surface and the fluid. The proportionality constant is the convection heat transfer coefficient or film coefficient. **Thermal radiation** involves energy transport in the absence of an intermediate medium, with energy exchanged between bodies by electromagnetic surface interaction.

Virtually all thermal analyses involve nonlinear processes with temperature-dependent material properties, temperature-dependent boundary condition coefficients, and governing nonlinear transport relationships. These nonlinear aspects enter into both the study of thermal system analysis and thermal detailed analysis. System analysis is concerned with overall process energy flows and balances, such as the total on-orbit energy survey for a satellite. Detailed analysis is more important when considering an individual component of a thermal system where a thermal analysis often leads to a detailed thermal stress analysis. Whether system level or detailed level, analyses typically are performed for both steady-state conditions as well as transient conditions (time-varying behavior).

Normal Modes Analysis

Normal modes analysis computes the natural frequencies and mode shapes. The **natural frequencies** are the frequencies at which a structure will vibrate if subjected to a disturbance. For example, the strings of a piano are each tuned to vibrate at a specific frequency. The deformed shape at a specific natural frequency is called the **mode shape.** The natural frequency is sometimes called the *resonant frequency* or **eigenvalue,** and the mode shape is sometimes called the **eigenvector.** Normal modes analysis is also called *real eigenvalue analysis.*

Normal modes analysis computes the dynamic characteristics of a system. Because there is no applied force, the mode shape is not a true physical deformation but rather, can be scaled by an arbitrary factor. Scaling is often done so that there is a unit modal mass in each mode, and it can be done such that the maximum displacement has a value of 1.0. Typically, only the lowest frequencies and modes are computed. For example, if a model has 100,000 degrees of freedom, only the lowest 50 modes may be of interest. Because no single eigenvalue extraction method is perfect for all models, most programs provide multiple methods, with **Lanczos, subspace iteration, inverse power,** and **Householder–Givens** being the most popular.

Frequency Response Analysis

Frequency response analysis computes the steady-state response to **oscillatory input.** Examples of oscillatory excitation include rotating machinery, unbalanced tires, and helicopter blades. In frequency response analysis the excitation is defined at each forcing frequency, and responses—typically displacements, accelerations, and stresses—are computed at those frequencies. The responses are complex numbers represented by real and imaginary components or magnitude and phase (with respect to the applied force).

Transient Response Analysis

Transient response analysis computes the response due to **time-varying excitation.** Examples of transient excitation include earthquakes, wind gusts, and impacts. The excitation is defined as a function of time, and responses are computed at discrete time intervals. Typical responses include displacements, accelerations, and stresses.

Random Vibration Analysis

Random vibration is vibration that can be described only in a statistical sense. The instantaneous magnitude is not known at any given time; rather, the magnitude is expressed in terms of its statistical properties, such as mean value, standard deviation, and probability of exceeding a certain value. Random excitations—such as ocean wave heights and frequencies, and jet engine noise—are usually described in terms of an input PSD (power spectral density) function that defines energy per frequency. An excitation PSD may be expressed in units of g^2/Hz for acceleration base motion and psi^2/Hz for applied pressure. Random response output consists of the response PSD (often in terms of $stress^2/Hz$) and RMS (root-mean-square) values of response.

Response Spectrum Analysis

Response spectrum analysis is an approximate method of computing the peak response of a structure or component to transient excitation. This method is used in civil engineering to predict the peak response of a component (e.g., equipment) in a building subjected to earthquake excitation. Response spectrum analysis is also called **shock spectrum analysis.**

There are two parts to response spectrum analysis: (1) generation of the spectrum, and (2) use of the spectrum for response analysis. Figure 2.32 depicts the steps in generating a response spectrum. Once a spectrum is computed, it can be used for stress analysis of the component by computing the stress in each of the component's

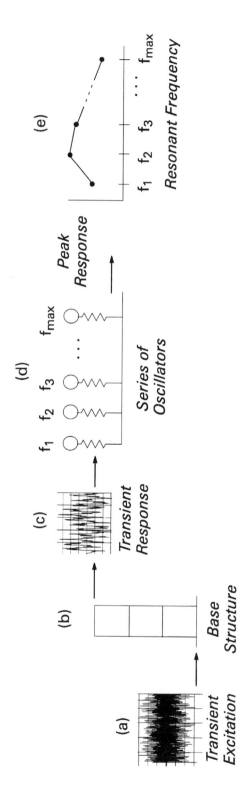

Transient excitation (a) is applied to a base structure (b), from which transient response (c) is computed for each floor. This response is applied to a series of oscillators (d), for which the peak response is plotted (e). Steps (d) and (e) are repeated for different damping values to form response spectra.

Fig. 2.32 Generation of a response spectrum. (Courtesy of MacNeal-Schwendler Corporation, Los Angeles.)

modes and then summing them across all modes of interest. Because the spectrum contains only the magnitude of response, various methods are used to compute the sum of the modal responses.

Acoustic Analysis

Acoustic analysis computes the response of a structure to acoustic excitation, such as that due to rocket or jet engine noise, or mechanically induced noise such as that found in the interior of an automobile. Acoustic analysis can be performed by finite element programs, and in the case of external acoustic excitation, in which the noise is radiated into space, other analytical methods, such as the **boundary element** method, are appropriate. Acoustic analysis can be performed in a deterministic manner when the noise source is well known and the analysis is done as a frequency or transient response analysis. Acoustic analysis can also be performed in a probabilistic manner when the noise source is understood only in statistical terms and the analysis is done as a random response analysis. Responses in acoustic analysis are typically the sound pressure levels at various locations.

Complex Eigenvalue Analysis

Complex eigenvalue analysis is used to compute the **damped modes** and is also used to assess the stability of systems such as servomechanisms. Complex eigenvalue analysis solves for the damped eigenvalues and eigenvectors (note that in normal modes analysis, the system damping is neglected).

Nonlinear Analysis

The response of many structures and components is not directly proportional to the applied loads or temperatures. This is because materials are not linearly elastic, the response may have large deformations or rotations, the direction of the applied force may change as the structure deforms, or parts of the structure come into contact with each other. **Nonlinear analysis** falls into three broad categories:

1. **Material nonlinear behavior,** including small strain plasticity, nonlinear elasticity, creep, thermoelasticity, hyperelasticity, and large strain plasticity.
2. **Geometric nonlinear behavior,** including large geometric deformations and rotations, snap-back and snap-through phenomena, and postyield buckling
3. **Contact,** including structure–structure contact and gaps opening and closing

Analyses are *incremental* (in which the load is applied in small steps) and *iterative* (in which multiple iterations are required to reach equilibrium). Because of this complexity, and because no single solution is best for all types of nonlinearities, finite element analysis programs that include nonlinear analysis provide multiple solution choices.

Fatigue Analysis

Fatigue analysis is the process of predicting the length of time a component or structure will last before breaking apart. Fatigue analysis also determines when and where a crack will initiate, as well as the prediction of **crack propagation.** These types of analyses generally require stress or strain information, cyclic material properties, and a definition of the variation of loading with time. Fatigue analyses rely on stress or strain results from other types of linear structural analyses, such as linear static, transient response, frequency response, or random vibration. Fatigue analysis is based on small amounts of local plasticity (not gross yielding) causing damage over a period of time. Elastic stress or strain results from linear structural analyses are corrected for plasticity within the fatigue analysis itself by the use of various empirical methods and cyclic material properties. From this the cumulative damage is determined, summed, and reported as a **life value.**

Electromagnetic Analysis*

Finite element analysis is often used to compute electromagnetic fields, and from the fields important parameters such as power loss and force are computed. Numerous solution types are possible. Static (dc) solutions include **electrostatics, magneto-statics,** and **current flow** solutions. **Frequency-domain** (ac) solutions are usually direct frequency solutions, but modal frequency solutions may be preferred for certain high-frequency problems such as **microwave** filters. Microwave devices are often analyzed using normal modes analysis to predict their resonant frequencies. Transient solutions are used when time-varying signals are nonsinusoidal, such as square voltage pulses used in computers. Because magnetic permeability is often extremely nonlinear, nonlinear static and nonlinear transient solutions are important. Coupled electromagnetic and structural nonlinear transient finite element analysis can be used to predict the motion of electromagnetic actuators such as fuel injectors.

Plastic Flow Analysis

Plastic flow analysis is used in the simulation of the **injection molding** process. The analysis provides studies to evaluate runners and gating schemes for the plastic flow in the mold cavity. It determines various filling patterns and identifies areas of high stresses, local overpacking of the plastic material, weld line positions, air traps, and flow/shrinkage defects. The benefits are to predict the moldability of the plastic part, that is, whether or not the mold cavity can be filled without building an actual prototype. Also, the plastic flow analysis helps identify appropriate processing conditions.

Kinematic Analysis**

Kinematic analysis computes the behavior of a mechanical system of both rigid and flexible elements interconnected by joints and forces to determine the range of mo-

*Courtesy of ANSOFT, Pittsburgh, Pa.
**Courtesy of Brant Ross, Mechanical Dynamics, Ann Arbor, Mich.

tion, accelerations, forces, lockup positions, work envelopes, and interference detections of an assembled system. Displacements are often large, on the order of the overall dimensions of the system. The benefits are in evaluating critical design positions of moving parts within a mechanical system. The results from kinematic analysis are often linked with flexible-body analysis (typical of most finite element programs). For example, a kinematic analysis of a truck going around a corner would compute the rigid-body inertial forces, which can then be transferred to a structural analysis program to compute distortions and stresses.

Crash Analysis

Crash analysis simulates high-speed crash studies for safety considerations. Crash analysis, also called *impact analysis,* is performed as a transient response solution with highly nonlinear materials. The widest use of crash analysis is in the automotive industry, where the safety of passengers is highly critical. The computer simulations provide better design of the overall structure and restraint systems in the automobile.

Aeroelastic Analysis

Aeroelastic analysis takes into account **aerodynamic,** inertial, and structural forces. It is important in the design of aircraft, launch vehicles, helicopters, suspension bridges, tall buildings, and power lines. Aeroelastic analysis can be performed as static analyses (in which the dynamic structural effects are ignored, such as in trimmed flight) or as dynamic analyses (when turbulence is present). Dynamic instabilities, such as **flutter** and **galloping,** can also be analyzed.

Substructuring Analysis

Finite element analyses can be performed on the entire model or on pieces at a time. When pieces are to be analyzed, and then the overall or coupled results computed, **substructuring** is used. Substructuring is useful when multiple companies are engaged in the design and analysis, such as when spacecraft are analyzed (one company models the launch vehicle, another the rocket motor, and still another the satellite—and then the models are combined into one). Substructure analysis types include linear statics and dynamics (normal modes, frequency response, and transient response). Substructuring is also called *superelement analysis.*

2.7.6 Assessing the Results

Once the analysis has been run, results are assessed by comparing them to design allowables. In years past this assessment was typically done by looking at printed computer output, but today most of the assessment is done in an interactive graphical manner using a finite element modeler (which usually also contains results or processing capabilities) or a dedicated plotting program. **Plotted output** is generally available in two forms:

1. **Structure plots,** which display the deformed shape and stress and temperature contours

2. **X-Y plots,** which display a single element(s) or grid point(s) results versus time or frequency

Results are compared to design allowables, such as the maximum permissible stress or deflection. These are obtained from published standards (such as MIL specifications), from materials properties (the yield stress, for example), or from industry or company standards. Margins of safety are usually included.

2.7.7 Optimizing the Design

The design needs to be modified when three conditions exist:

1. The computed results exceed the design allowables, in which case the design is changed by adding material or adding additional load-carrying members.

2. The computed results do not match available test data, in which case the model is refined to get a better match.

3. The computed results are well below the design allowables, in which case the design can be made lighter by removing material.

An optimal design is one that minimizes or maximizes an objective, subjective to constraints. In structural analysis, weight is often the objective, and it is to be minimized. Constraints take two forms: **response constraints** (such as a maximum allowable stress) and **design constraints** (such as a minimum allowable plate thickness).

Optimization can be performed in a brute-force manner by postulating design changes and rerunning the analysis. Optimization is best performed by using a finite element program that contains an optimization capability, in which case the objective, response constraints, and design constraints are specified and the program automatically iterates by using efficient optimization algorithms to obtain a better design.

2.8 POWER TRANSMISSION

2.8.1 Shafts

The **maximum intensity of shear** (s) in a shaft of r inches radius and J_0 inches[4] polar moment of inertia due to a torque (twisting moment) of M inch-pounds:

$$s = \frac{Mr}{J_0} \quad \text{pounds/inch}^2. \tag{3.132}$$

NOTE.   For a solid round shaft, $s = 2M/\pi r^3$.

The **angle** (θ) of twist in a solid circular shaft, or r inches radius, l inches in length, and with E_s pounds/inch^2 modulus of elasticity in shear, due to a torque of M inch-pounds:

$$\theta = \frac{2Ml}{\pi r^4 E_s} \quad \text{radians.} \tag{2.133}$$

NOTE. E_s for steel is commonly taken as 12,000,000.

The **horsepower** (P) transmitted by a shaft making n revolutions/minute under a torque of M inch-pounds:

$$P = \frac{2\pi n M}{33,000 \times 12} \quad \text{horsepower.} \tag{2.134}$$

The **diameter** (d) of a solid circular shaft to transmit P horsepower at n revolutions/minute with a fiber stress in shear of s pounds/inch^2:

$$d = \sqrt[3]{\frac{321,000P}{ns}} \quad \text{inches.} \tag{2.135}$$

The **maximum intensity of shearing stress** (s') and of **tensile or compression stress** (f') due to combined twisting and bending in a shaft where s is the maximum intensity of shear due to the torque and f is the maximum intensity of tension or compression due to the bending:

$$s' = \tfrac{1}{2}\sqrt{4s^2 + f^2} \quad \text{pounds/inch}^2. \tag{2.136}$$

$$f' = \tfrac{1}{2}f + \tfrac{1}{2}\sqrt{4s^2 + f^2} \quad \text{pounds/inch}^2. \tag{2.137}$$

2.8.2   V-Belt Drives[17]

Belt-type **power transmission** almost exclusively utilizes **V-belt** rather than flat-belt systems. V-belts are available in **classical** (ANSI/RMA* IP-20-1988) and **narrow** (IP-22-1991) types and are interchangeable between English and metric sizes. (Each standard contains both English and metric detail.) Narrow to classical belts are not interchangeable. Belt sizes are standardized in the English system in cross sections identified as A, B, C, and D (classical) and 3V, 5V, and 8V (narrow) in specific lengths from 26 to 660 in. and are utilized as single or multiple belts. They are available in matched sets of up to 12 or more belts, the limitation on multiple-belt matched set drives being the availability of sheaves having the required number of grooves. Previous systems of power ratings of belts have been discontinued in favor of the use of more detailed specific formulas to establish exact requirements. The formulas for determining power transmission capability of the various classical and narrow belts in both standard and higher-capacity constructions (designated by an "x" suffix) are provided in the IP 20 and 22 standards.

   Standard V-belts are fully sealed (although the highest-capacity belts have no covering), oil resistant, and static conductive to avoid electrical buildup. Advanced versions of the standard V-belts are available which have higher-modulus tensile

*RMA, Rubber Manufacturers Association.

members [fiberglass and polyamide (Kevlar)] for high loads, link type for ease of installation, toothed or cogged for synchronous (zero-slip) applications, and so on. Although intended to be used with mating sheaves, V-belts can operate over flat-faced pulleys for some applications. Both belts and sheaves are available in pre-engineered sizes and matched sets and are widely stocked.

2.8.3   Variable-Speed Drives

Many **variable-speed drives** utilizing ingenious mechanisms have been developed. One type in extremely wide use for industrial application is the Reeves drive (Fig. 2.33). This utilizes a relatively wide and fairly stiff V-belt running over sheaves whose faces can be moved closer or farther apart, resulting in a changed effective pitch diameter for the sheave, thus yielding different speeds. These drives have proven to be relatively simple, maintenance free, and have a high degree of reliability. The speed can only be changed when the drive is operating, however, not at rest.

Variable-speed drives are also available which utilize a **hydraulic coupling** system between the prime mover and the driven equipment. These systems typically utilize a coupling in which the amount of fluid between two impellers can be varied, thus permitting varying amounts of slippage, and providing for variations of speed. Because of the slippage and heating that occurs within the driven fluid (typically, a synthetic oil), a heat exchange and filtration system is necessary to maintain proper viscosity of the coupling fluid.

Variable-speed drives utilizing **electrical couplings** operate on a principle similar to that of hydraulic couplings. Speed differences (slippage) result in eddy current heating, which must be dissipated. Usually, this is handled by forced airflow, although in some cases fluid coolers can be used to remove this heat. There are several types of **variable-speed ac motors,** the choice of which is determined by the size, application, cost, and efficiency. The **wound-rotor induction motor** can be controlled from zero speed to full speed, but because of the high losses in the external rotor

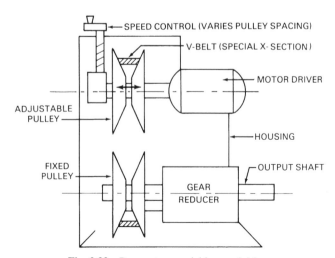

Fig. 2.33   Reeves-type variable-speed drive.

resistances, it is not often used in the larger ratings where operation at less than full speed is required for other than short periods. A more efficient variation of the wound-rotor application is termed the **wound-rotor slip recovery drive.** Here the rotor is connected to an ac/dc converter, which drives a dc/ac inverter. The inverter ac output is three-phase, 60 Hz, and may be coupled into the source supplying the motor by a transformer. Another system with high efficiency is the **variable-speed synchronous motor.** The source at 60 Hz is converted to dc, which supplies an inverter, the output of which can be controlled to provide a variable frequency and voltage to supply the synchronous motor. Both of the latter two systems are quite efficient over their entire speed range, which, in many applications, will offset their higher initial cost.

Dc motors of the series type are also used to provide speed variation, but can accelerate to infinite speed on loss of the driven load. Thus, for application, where loads may change suddenly, such as a belt drive which breaks, these motors may increase their speed very rapidly and the motor(s) may self-destruct due to centrifugal force.

REFERENCES

1. *Composites Fabrication,* Sept. 1995.
2. "Design Guide for Advanced Composites Applications," *Advanced Composites Magazine,* 1993.
3. J. C. Leslie, *Properties of Advanced Composites,* Advanced Composite Products and Technology, Inc., Huntington Beach, CA, 1984, Rev. 1994.
4. *Mechanical Engineering,* Apr. 1988.
5. *New Structural Materials Technology,* Technical Memorandum, U.S. Congress, Office of Technology Assessment, Washington, D.C., 1986.
6. G. Lubin, *Handbook of Composites,* Van Nostrand Reinhold, New York, 1982.
7. J. Reinhart, *Engineered Materials Handbook,* Vol. 1, *Composites,* ASM International, Metals Park, Ohio, 1987.
8. *Wire Rope, Cable, etc.,* Loos & Co., Pomfret, Conn., Nov. 1, 1993.
9. *Wire Rope Users Manual,* 3rd ed., Wire Rope Technical Board, Woodstock, MD, 1993.
10. K. Bathe and E. L. Wilson, *Numerical Methods in Finite Element Analysis,* Prentice Hall, Upper Saddle River, N.J., 1976.
11. R. D. Cook, D. S. Malkus, and M. E. Plesha, *Concepts and Applications of Finite Element Analysis,* 3rd ed., John Wiley & Sons, Inc., New York, 1989.
12. *Introduction to MSC/NASTRAN,* MacNeal-Schwendler Corporation, Los Angeles, 1994.
13. R. MacNeal, *Finite Elements: Their Design and Performance,* Marcel Dekker, Inc., New York, 1994.
14. J. S. Przemieniecki, *Theory of Matrix Structural Analysis,* McGraw-Hill Book Company, New York, 1968.
15. B. Ross, "Kinematic Analysis," Mechanical Dynamics, Inc., Ann Arbor, MI, private communication, 1996.
16. O. C. Zienkiewicz, *The Finite Element Method,* 3rd ed., McGraw-Hill Book Company, New York, 1977.
17. *Specifications for Drives,* Engineering Standard[s], ANSI/RMA IP-20-1988 and IP-22-1991.

CHAPTER 3
STRUCTURES

3.1 LOADS AND CODE REQUIREMENTS

Loads are established for many designs from **code** requirements which prescribe minimum live, dead, wind, seismic, and other loadings. Where designs are more complex, it is often necessary to establish specific loadings for the structure or portions of it, depending on its function. Typically, structures are loaded in several different ways, often at the same time; as a result, considerable care must be taken to identify the various types of loads and their combinations. In some cases, loads occurring during the erection of the structure are larger than the working loads during the life of the structure, and this possibility needs to be kept in mind.

3.1.1 Ultimate Strength Design Versus Working Stress Design

Working stress design is currently being replaced by **ultimate strength design** for most materials. Concrete design codes have been devoted exclusively to the strength design method since 1977. Steel design codes are still based on both methods; however, college courses are currently (1997) taught using the strength design method, also known as load resistance factor design (LRFD) for steel. Timber is the only material that is still based on the working stress design.

Working stress design is based on the comparison of stresses resulting from design loads to allowable stresses dictated by the codes. The factor of safety is inherent in the allowable stresses. Designs are intended to remain within the elastic range. Ultimate strength design is based on the requirement that the computed nominal strengths reduced by specific reduction factors equal or exceed the design load effects multiplied by specific load factors.

3.1.2 Codes, References, and Standards

The principal codes governing the development of structural designs are the Uniform Building Code, the BOCA National Building Code, and the Standard Building Code. These codes are mandated for different parts of the United States, and while their content is similar, care should be taken to assure that the correct code is utilized for the specific jurisdiction in which the work is to be performed. For the use of major structural materials, the *Building Code Requirements for Reinforced Concrete* (American Concrete Institute), the *Manual of Steel Construction* (American Institute of Steel Construction), and the *Timber Construction Manual* (American Institute of Timber

Constitution) are considered the authoritative sources for design requirements. Regardless of the jurisdiction, local licensing and permitting bodies should be consulted to assure code jurisdiction, local ordinance requirements, and such.

3.2 BEAMS

The **vertical shear** at any section of a beam is equal to the algebraic sum of all the vertical forces on one side of the section. The shear is positive when the part of the beam to the left of the section tends to move upward under the action of the resultant of the vertical forces.

NOTE. In the study of beams, the reactions must be treated as applied loads and included in shear and moment. A section is always taken as cut by a plane normal to the axis of the beam. In all cases vertical means normal to the axis.

The **bending moment** at any section of a beam is equal to the algebraic sum of the moments, about the center of gravity of the section, of all the forces on one side of the section. Moment that causes compression in the upper fibers of a beam is positive.

NOTE. The maximum moment occurs at a section where the shear is zero. A diagram of shears or of moments is a curve the ordinate to which at any section shows the value of the shear or moment at that section.

The **neutral axis** of a beam is the plane that undergoes no change in length due to the bending and along which the direct stress is zero. The fibers on one side of the neutral axis are stressed in tension and on the other side in compression and the intensities of these stresses in homogeneous beams are directly proportional to the distances of the fibers from the neutral axis.

NOTE. The neutral axis at any section in a beam subject to bending only passes through the center of gravity of that section.

The **elastic curve** of a beam is the curve formed by the neutral plane when the beam deflects due to bending.

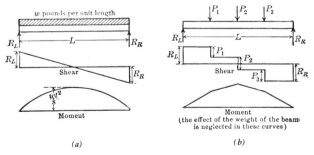

(a)   *(b)*

Fig. 3.1 Moment and shear diagram for a simple beam: (*a*) with a uniformly distributed load; (*b*) with concentrated loads.

The **equation of the elastic curve** of a beam of I inches^4 moment of inertia and a modulus of elasticity of the material of E pounds per square inch, if x and y in inches are the abscissa and ordinate, respectively, of a point on the neutral axis referred to rectangular coordinates through the points of support, and M is the moment in inch pounds at that point:

$$M = EI \frac{d^2y}{dx^2}. \tag{3.1}$$

NOTE. The equation of the elastic curve is used to find the slope and deflection of a beam under loading. A single integration gives the slope, integrating twice gives the deflection; in each case, however, the proper value of the constant of integration must be determined.

The **three-moment equation** gives the ratio between the moments M_a, M_b, and M_c at three consecutive points of support (a, b, and c) on a beam continuous over three or more supports.

CASE I. Concentrated loads (see Fig. 3.2)

$$M_a l_1 + 2M_b(l_1 + l_2) + M_c l_2 = P_1 l_1^2(k_1^3 - k_1) + P_2 l_2^2(3k_2^2 - k_2^3 - 2k_2). \tag{3.2}$$

CASE II. Uniformly distributed load (see Fig. 3.3)

$$M_a l_1 + 2M_b(l_1 + l_2) + M_c l_2 = -\tfrac{1}{4} w_1 l_1^3 - \tfrac{1}{4} w_2 l_2^3. \tag{3.3}$$

The **intensity of stress** (s) in tension or compression on a fiber y inches distant from the center of gravity of a section of a beam with I inches^4 moment of inertia, due to a bending moment of M pound-inches:

$$s = \frac{My}{I} \text{ pounds/inch}^2. \tag{3.4}$$

The **intensity of stress** (s) on the outer fiber of a solid rectangular beam h inches in depth and b inches in width due to a bending moment of M pound-inches:

$$s = \frac{6M}{bh^2} \text{ pounds/inch}^2. \tag{3.5}$$

The **intensity of stress** (s) in a fiber y inches distant from the center of gravity of a section of a beam of A inches^2 area and I inches^4 moment of inertia, due to an axial load (parallel to axis of beam) of P pounds and a bending moment of M pound-inches:

$$s = \frac{P}{A} \pm \frac{My}{I} \text{ pounds/inch}^2. \tag{3.6}$$

(text continues on page 166)

Table 3.1 Beams Under Various Loadings

Beam, Loading and Moment Diagram	Reaction	Bending Moment	Deflection
	$R_L = R_R = \dfrac{wl}{2}$	$M_x = \dfrac{wlx}{2} - \dfrac{wx^2}{2}$ $M_{\max} = \dfrac{wl^2}{8}$	$d_{\max} = \dfrac{5wl^4}{384EI}$
	$R_L = R_R = \dfrac{P}{2}$	$M_x = \dfrac{Px}{2}$ $M_{\max} = \dfrac{Pl}{4}$	$d_{\max} = \dfrac{Pl^3}{48EI}$
	$R_L = \dfrac{Pb}{1}$ $R_R = \dfrac{Pa}{1}$	$M_{x_1} = \dfrac{Pbx_1}{1}$ $M_{x_2} = \dfrac{Pax_2}{1}$ $M_{\max} = \dfrac{Pab}{1}$	$d_{\max} = \dfrac{Pab(2a + b)\sqrt{3b(2a + b)}}{27EI}$
	$R_L = R_R = P$	$M_x = Px$ $M_{\max} = Pa$	$d_{\max} = \dfrac{Pa}{6EI}\left(\dfrac{3}{4}l^2 - a^2\right)$

$$R_L = \frac{wb(2c + b)}{2l}$$

$$R_R = \frac{wb(2a + b)}{2l}$$

$$M_x = R_L x - \frac{w(x - a)^2}{2}$$

$$M_{\max} = R_L\left(a + \frac{R_L}{2w}\right)$$

$$d_{\max} = \frac{0.013044 Wl^3}{EI}$$

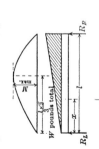

$$R_L = \frac{1}{3}W$$

$$R_R = \frac{2}{3}W$$

$$M_x = \frac{Wx}{3}\left(1 - \frac{x^2}{l^2}\right)$$

$$M_{\max} = \frac{2Wl}{9\sqrt{3}}$$

$$d_{\max} = \frac{Wl^3}{60EI}$$

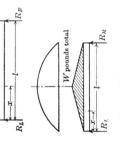

$$R_L = R_R = \frac{W}{2}$$

$$M_x = W_x\left(\frac{1}{2} - \frac{2x^2}{3l^2}\right)$$

$$M_{\max} = \frac{Wl}{6}$$
(at center)

$$R_L = P$$

$$M_x = Px$$
$$M_{\max} = Pl$$

$$d_{\max} = \frac{Pl^3}{3EI}$$

$$R_L = wl$$

$$M_x = \frac{wx^2}{2}$$
$$M_{\max} = \frac{wl^2}{2}$$

$$d_{\max} = \frac{wl^4}{8EI}$$

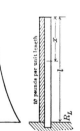

Table 3.1 (*Continued*)

Beam, Loading and Moment Diagram	Reaction	Bending Moment	Deflection
W pounds total	$R_L = W$	$Mx = \dfrac{Wx^3}{3l^2}$ $M_{\max} = \dfrac{Wl}{3}$	$d_{\max} = \dfrac{Wl^3}{15EI}$
	$R_L = R_R = P$	$M_x = Px$ $M_{\max} = Pa$	$d_{\text{end}} = \dfrac{Pa^2(2a + 3l)}{6\,EI}$ $d_{\text{center}} = -\dfrac{Pal^2}{8\,EI}$
w pounds per unit length	$R_L = R_R = \dfrac{w(1 + 2a)}{2}$	M at R_L and $R_R = \dfrac{wa^2}{2}$ $M_{\text{center}} = \dfrac{w(l^2 - 4a^2)}{8}$	
	$R_L = \dfrac{11}{16}P$ $R_R = \dfrac{5}{16}P$	$M_x = \dfrac{5}{16}Px$ $M_{\max} = \dfrac{3}{16}Pl$	$d_{\max} = \sqrt{\dfrac{1}{5}}\dfrac{Pl^3}{48EI}$ at $x = l\sqrt{\dfrac{1}{5}}$

$$R_L = \frac{5}{8} wl$$

$$R_R = \frac{3}{8} wl$$

$$M_x = \frac{wx^2}{2}\left(\frac{3}{4} - \frac{x}{l}\right)$$

$$M_{max} = \frac{wl^2}{8} \ (\text{at } R_L)$$

$$d_{center} = \frac{wl^4}{192EI}$$

$$d_{max} = \frac{wl^4}{185EI}$$

$$\text{at } x = 0.4215l$$

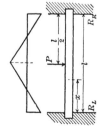

$$R_L = R_R = \frac{P}{2}$$

$$M_x = \frac{Pl}{2}\left(\frac{x}{l} - \frac{1}{4}\right)$$

$$\left\{ \begin{array}{l} M_{max} = \dfrac{Pl}{8} \\ (\text{at supports}) \\ M_{max} = \dfrac{Pl}{8} \\ (\text{at center}) \end{array} \right.$$

$$d_{max} = \frac{Pl^3}{192EI}$$

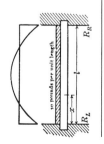

$$R_L = R_R = \frac{wl}{2}$$

$$M_x = \frac{wl^2}{2}\left(\frac{1}{6} - \frac{x}{l} + \frac{x^2}{l^2}\right)$$

$$M_{max} = \frac{wl^2}{12}$$

$$(\text{at supports})$$

$$d_{max} - \frac{wl^4}{384 \, EI}$$

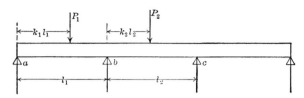

Fig. 3.2   Three-moment equation, concentrated loads.

The **maximum moment** (*M*) that can be carried by a beam with *I* inches^4 moment of inertia and *y* inches greatest distance from center of gravity to outer fiber without exceeding an intensity of stress of *s* pounds/inch^2 in the outer fiber:

$$M = \frac{sw}{y} \quad \text{pound-inches.} \tag{3.7}$$

The **section modulus** (*S*) of a section of a beam with *I* inches^4 moment of inertia and *y* inches distance from the center of gravity to the outer fiber:

$$S = \frac{I}{y} \quad \text{inches}^3. \tag{3.8}$$

The **intensity of stress** (*s*) on the outer fiber of a beam of section modulus of *S* inches^3, due to a bending moment of *M* pound-inches:

$$s = \frac{m}{S} \quad \text{pounds/inch}^2. \tag{3.9}$$

The **intensity of longitudinal shear** (*s*) along a plane *XX* at the section of a beam where the total vertical shear is *V* pounds if *I* inches^4 is the moment of inertia of the total section about its center of gravity axis, *b* the width of the beam at lane *XX*, and *Q* inches^3 the statical moment, taken about the center of gravity axis, of that portion of the section that lies outside the axis *XX*:

$$s = \frac{VQ}{bI} \quad \text{pounds/inch}^2. \tag{3.10}$$

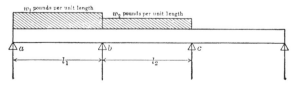

Fig. 3.3   Three-moment equation, uniformly distributed loads.

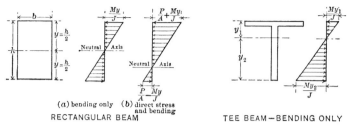

(a) bending only (b) direct stress
                   and bending
RECTANGULAR BEAM

TEE BEAM—BENDING ONLY

Fig. 3.4 Stress distribution in beam.

NOTE. The maximum intensity of shear always occurs at the center of gravity of the section of a beam.

The **maximum intensity of shear** (s) is a rectangular beam A inches^2 in area at a section where the total vertical shear is V pounds:

$$s = \frac{3}{2}\frac{V}{A} \quad \text{pounds/inch}^2. \tag{3.11}$$

NOTE. The intensity of vertical shear is equal to that of the longitudinal shear acting at right angles to it. The intensity of vertical shear is obtained by the formula $s = VQ/bI$.

3.3 COLUMNS

Columns are classified as long or short, depending on their failure mode. If a column is sufficiently slender and failure is by buckling or elastic instability, it is considered a **long column.** A **short column** will fail when the maximum fiber stress due to direct compression and bending reaches the yield stress of the material. For design purposes columns having l/r ratios of 50 or less normally fail as short columns and are analyzed as such. Those with l/r ratios of 60 to 120 are often analyzed as long columns. Columns with an l/r ratio greater that 120 should always be considered as long columns. Long columns are often braced to reduce their effective length (l) and care needs to be taken to assure that the bracing is adequate in both lateral axes.

End fixity will effectively reduce the free length l of the column and increase its strength. This is reflected in corrections found in both the Euler and Gordon formulas. In other cases the column stress levels are computed using a term kl/r, where the term k reflects the degree of end fixity. Where some degree of fixity occurs consideration of end conditions will often permit use of a lighter member. Because timber

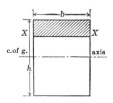

Fig. 3.5 Intensity of longitudinal shear.

columns respond slightly differently to loads, they are discussed separately (see Section 3.6.5).

Euler's formula for the ultimate average intensity of stress (f) on a column l inches in length, with a least radius of gyration of r inches and of material of E pounds/inch^2 modulus of elasticity. f should not exceed the elastic limit.

$$\text{Column with ends pinned,} \quad f = \pi^2 E \left(\frac{r}{l}\right)^2 \quad \text{pounds/inch}^2. \tag{3.12}$$

$$\text{Column with ends fixed,} \quad f = 4\pi^2 E \left(\frac{r}{l}\right)^2 \quad \text{pounds/inch}^2. \tag{3.13}$$

$$\text{Column with one end fixed} \quad f = \frac{9}{4}\pi^2 E \left(\frac{r}{l}\right)^2 \quad \text{pounds/inch}^2. \tag{3.14}$$
and one end pinned.

The **Gordon formula** for allowable average intensity of stress (f) on a column l inches in length, with a least radius of gyration of r inches and a maximum allowable compression stress of f_c pounds/inch^2 on the material:

$$f = \frac{f_c}{1 + (1/c)(l/r)^2} \quad \text{pounds/inch}^2. \tag{3.15}$$

NOTE: The following values of c are commonly used for steel columns:

Column with ends pinned	9,000
Column with ends fixed	20,000
Column with one end fixed and one end pinned	36,000

The **straight-line formula** for the allowable average intensity of stress (f) in a column l inches in length, with a least radius of gyration of r inches and a maximum allowable compression stress of f_c pounds/inch^2 on the material:

$$f = f_c - c\left(\frac{l}{r}\right) \quad \text{pounds/inch}^2. \tag{3.16}$$

The **maximum intensity of stress** (s) in a column of A square inches area of cross section, l inches length, I inches^4 moment of inertia about the axis about which bending occurs and y inches distance from that axis to the most stressed fiber, due to an axial load of P pounds and a bending moment of M inch-pounds:

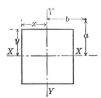

Fig. 3.6   Maximum intensity of stress.

$$s = \frac{P}{A} + \frac{My}{I - Pl^2/cE} \quad \text{pounds/inch}^2 \text{ (approx.).} \tag{3.17}$$

NOTE. The constant c for the common case of pin-ended columns subject to bending due to a uniformly distributed load may be taken as 10.

The **maximum intensity of stress** (s) in a short column of A square inches area of cross section, due to a load of P pounds applied a inches distant from the X axis of symmetry and b inches distant from the Y axis of symmetry, if I_x inches^4 is the moment of inertia about the X axis, y inches the distance from the X axis to the most stressed fiber, I_y inches^4 the moment of inertia about the Y axis, and x inches the distance from Y axis to the most stressed fiber:

$$s = \frac{P}{A} + \frac{Pay}{I_x} + \frac{Pbx}{I_y} \quad \text{pounds/inch}^2. \tag{3.18}$$

3.4   REINFORCED CONCRETE[1,2]

3.4.1   Codes

The American Concrete Institute (ACI) publishes the Building Code Requirements for Reinforced Concrete, ACI 318. This code has been essentially adopted into the governing building codes of the various regions of the United States. The ACI also publishes the Manual of Concrete Practice, which is a five-part compilation of current ACI standards and committee reports and is broadly accepted as the standard for concrete work of all types. The five parts of the standards and committee reports are:

Part 1	Materials and General Properties of Concrete
Part 2	Construction Practices and Inspection, Pavements
Part 3	Use of Concrete in Buildings: Design, Specifications, and Related Topics
Part 4	Bridges, Substructures, Sanitary, and Other Special Structures: Structural Properties
Part 5	Masonry, Precast Concrete, Special Processes

3.4.2   Cement Types

Portland cement is produced in five types as defined by ASTM Specification C150, Standard Specification for Portland Cement.

- **Type I—Normal Portland Cement:** used when the special properties specified for any other type are not required.
- **Type II—Modified Portland Cement:** for general use, especially when moderate sulfate resistance or moderate heat of hydration is required.
- **Type III—High Early Strength Portland Cement:** for use when high early strength, shorter curing, or early form stripping is required
- **Type IV—Low Heat Portland Cement:** for use when low heat of hydration is desired. Of particular use for mass concrete work such as dams. Fairly slow in gaining strength and not widely used.
- **Type V—High Sulfate Resistance Portland Cement:** for use when high sulfate resistance is desired. Of particular use for concrete subject to a corrosive environment.

Air-entraining portland cements are produced as types IA, IIA, and IIIA and are used where resistance to thawing and freezing are a significant concern. Types I and II are most commonly used. The use of specialty cements is not as common because of readily available concrete admixtures that accomplish the desired effect. Examples of adixtures are accelerators, air-entraining agents, and corrosion inhibitors.

3.4.3 Concrete Properties

Concrete is comprised of a combination of cement, sand, **aggregate,** and water. Sand and aggregate conform to the requirements of ASTM C33. The maximum size of aggregate is normally specified for a concrete mix and is typically on the order of $\frac{3}{4}$ to $1\frac{1}{2}$ in. The aggregates occupy about 70 to 75% of the total volume of the hardened concrete.

Concrete has a very high compressive strength but is relatively weak in tension. Concretes typically have compressive strengths, f'_c that range from 3000 to 5000 psi (20 to 50 MPa). Both the tensile and shear strength are a function of the square root of f'_c and are generally between 10 to 15% and 20%, respectively, of the compressive strength. The unit weight of concrete is typically taken as 150 lb/ft^3. For normal-weight concrete,

$$E_c = 57,000 \sqrt{f'_c} \tag{3.19}$$

where E_c is the modulus of concrete (psi) and f'_c is the specified compressive strength at 28 days (psi).

3.4.4 Concrete Mix Design

Mix designs are proportioned so that the resulting concrete has adequate strength, proper workability for placing, and is economical. The latter dictates the use of the minimum amount of cement (which is the most costly component) to achieve the desired strength. A well-graded aggregate will minimize the volume of voids and require less cement paste to fill these voids.

Table 3.2 Variation of Compressive Strength with Age[a]

Water–Cement Ratio by Volume, (gal per bag of cement)	3 days	7 days	28 days	3 months	1 year
5	40	75	100	125	145
7	30	65	100	135	155
9	25	50	100	145	165

Source: T. Baumeister, E. A. Avallone, and T. Baumeister III, *Mark's Standard Handbook for Mechanical Engineers,* 8th ed. © 1978 McGraw-Hill Book Company. Used with permission.
[a](Strength at 28 days taken as 100).

The amount of water used is dictated by the **water/cement ratio.** The strength of the concrete depends primarily on this ratio and decreases directly with increasing water/cement ratio. There is a minimum amount of water required for the cement hydration (curing) process. A water/cement ratio of 0.35 to 0.40 by weight corresponds to 4 to 4.5 gal of water per sack of cement and is often considered the minimum practical. For concretes using well-proportioned mixes, the characteristics in Tables 3.2 and 3.3 are typical. The proportion of some typical concrete mixes is given in Table 3.4.

3.4.5 Reinforcing Steel[3]

Reinforcing steel is a high-strength material compared to concrete. It is used in concrete primarily to resist tensile forces, although it is sometimes used to resist compressive stresses in order to reduce concrete cross sections. Deformed bars conforming to ASTM A615, Grades 40 and 60, are commonly used, with yield strengths of 40,000 and 60,000 psi (276 and 414 MPa), respectively. Bar sizes are designated by numbers, with the number designation equivalents to $\frac{1}{8}$ in. of the bar diameter. Table 3.5 shows a listing of areas and perimeters of standard deformed bars. Grade 60 reinforcing is typically used for No. 5 bars and larger, Grade 40 for Nos. 3 and 4 bars.

Reinforcing bars can be spliced by lapping a required length in accordance with ACI code requirements, typically 20 bar diameters. Other methods of splicing include mechanical coupling, and welding. When welding of reinforcing is required, ASTM A706 bars are specified, due to their improved chemical composition.

The relationship between steel and concrete stress is given by the modular ratio

Table 3.3 Strength of Plain Concrete at 28 Days

Max water content, gal per bag of cement	5	5.5	6	6.5	7.0
Compressive strength, lb/in²	4000	3700	3350	3000	2650
Modulus of rupture, lb/in²	650	625	600	550	500
Tensile strength (split cylinder method), lb/in²	350	325	300	275	250

Source: T. Baumeister, E. A. Avallone, and T. Baumeister III, *Mark's Standard Handbook for Mechanical Engineers,* 8th ed. © 1978 McGraw-Hill Book Company. Used with permission.

Table 3.4   Comparison of Quantities and Properties of Some Typical Concrete Mixes[a,b]

	Mixture		
	Rich	Medium	Lean
Cement/water ratio (by solid volume)	0.9	0.6	0.4
Predicted 28-day strength, psi (w/c basis)	6,000	4,000	2,000
Workability, slump (in.)	3.0	3.0	3.0
Workability, texture	Plastic	Plastic	Plastic
Quantity in 1 unit volume concrete by solid (absolute) volume: cement (c)	0.153	0.102	0.068
F.A. (a)	0.247	0.298	0.332
C.A. (b)	0.430	0.430	0.430
Water (w)	0.170	0.170	0.170
Air (assumed 0)	0	0	0
Total	1.000	1.000	1.000
Water/cement ratio: gal per bag cement	4.0	6.0	9.0
Weight	0.34	0.52	0.80
Bulk volume	0.54	0.80	1.20
Solid volume	1.11	1.67	2.50
Voids/cement ratio: solid volume	1.11	1.67	2.50
Proportions: by weight	1:1.4:2.4	1:2.5:3.6	1:4.1:5.3
Bulk volume	1:1.2:2.3	1:2.2:3.4	1:3.7:5.1
Solid volume	1:1.6:2.8	1:2.9:4.2	1:4.9:6.3
Ratio $b \div b_0$	0.70	0.70	0.70
Ratio C.A. to F.A. (i.e., $b \div a$)	1.74	1.44	1.30
Quantity in 1 yd³: cement, bags	8.64	5.76	3.84
F.A., lb	1,103	1,331	1,487
C.A., lb	1,920	1,920	1,920
Water, gal	34.3	34.3	34.3
Yield, ft³ concrete per bag cement	3.1	4.7	7.0
Strength-economy index (psi per bag per yd³)	695	695	469
Weight fresh concrete			
Lb/ft³	153	151	150
Lb/yd³	4,122	4,079	4,055

[a]Materials: F. A., washed, sanded, graded, 0 to 4 (Standard Ottawa to Fine Sands); sp. gr., 2.65; bulk weight, 103.5 lb/ft³. C. A., river gravel, graded, ⅜ to 1½ in.; sp. gr., 2.65; bulk weight, 98.3 lb/ft³. Cement, sp. gr., 3.15.

[b]Although quantities and relative proportions are representative of those for rich, medium, and lean mixtures, in general they are not identical with what would be obtained by some other (less direct) basis of adjustment, such as varying total aggregate at a constant ratio of coarse to fine, along with appropriate alterations in water, and/or cement.

Source: L. C. Urquhart, *Civil Engineering Handbook,* 4th ed. © 1959 McGraw-Hill Book Company. Used with permission.

$$n = \frac{E_s}{E_c} \qquad (3.20)$$

where E_s = modulus of elasticity of steel, usually taken as 29,000,000 psi (200 × 10³ MPa), and E_c = modulus of elasticity of concrete, usually taken as 3,000,000 psi (21 × 10³ MPa). Table 3.6 gives the normal range of values of n. These data for various strengths of concrete are shown in Table 3.3. The value of n used is the nearest whole number and normally ranges between 7 and 10.

Table 3.5   Areas and Perimeters of Standard Deformed Bars

Bar Designation No.	Nominal Diameter (in.)	Area (in²)	Perimeter (in.)
3	0.375	Area	0.11
		Perimeter	1.178
4	0.500	Area	0.20
		Perimeter	1.571
5	0.625	Area	0.31
		Perimeter	1.963
6	0.750	Area	0.44
		Perimeter	2.356
7	0.875	Area	0.60
		Perimeter	2.749
8	1.000	Area	0.79
		Perimeter	3.142
9	1.128	Area	1.00
		Perimeter	3.544
10	1.270	Area	1.27
		Perimeter	3.990
11	1.410	Area	1.56
		Perimeter	4.430
14	1.693	Area	2.25
		Perimeter	5.32
18	2.257	fArea	4.00
		Perimeter	7.09

Source: H. Parker and H. D. Hauf, *Simplified Design of Reinforced Concrete,* 4th ed., John Wiley & Sons, Inc., New York, 1976.

Minimum **cover** for rebar is prescribed in the ACI Code. The cover is dependent on the aggregate size and whether the concrete is in contact with earth. For nominal ¾ in. minus aggregate mix, the cover is normally 2 in. In some circumstances this can be reduced to 1½ in. For reinforcing in foundations, this cover can be up to 3 in.

Welded wire fabric is typically used for reinforcing slabs, although it is seeing increased use for walls and footings and for ties and stirrups for column, beam, and joist cage reinforcement. It is usually manufactured in 5- to 8-ft wide sheets and rolls. Sheets up to 12 ft wide and 40 ft long are produced, primarily for highway

Table 3.6   Modulus of Elasticity of Normal-Weight Concrete

Ultimate Compressive Strength at 28-Day Period, f_c' (psi)	Modulus of Elasticity of Concrete E_c (psi)	$n = \dfrac{E_s}{E_c}$
2500	2,880,000	10
3000	3,150,000	9
4000	3,640,000	8
5000	4,070,000	7

Source: H. Parker and H. D. Hauf, *Simplified Design of Reinforced Concrete,* 4th ed., John Wiley & Sons, Inc., New York, 1976.

paving and precast components. Wire is designated as either "W" or "D", for plain or deformed wire. For a style designation of 12×12—D $10 \times$ D 10, the first set of numbers is the wire spacing in inches for both transverse and longitudinal directions, and the second set of numbers, the cross-sectional areas of the respective wires in square inches multiplied by 100 (e.g., 0.10 in$^2 \times 100 = 10$).

For slabs on grade less than 5 in., a single layer of welded wire is placed in the middle of the slab, while for slabs 6 in. and greater, the top cover is one-third of the slab depth. Tables 3.7 and 3.8 give common sizes and steel areas for welded wire reinforcement.

Reinforcement Locations

Although theoretically all compression loads can be carried by the concrete, in fact many designs include reinforcement for compression as well as for tension and shear. For designs with rebar not handling compression, both positive and negative moments must be considered yielding designs with rebar both above and below the neutral axis, and in some cases added rebar at points of high shear loads (e.g., supports). Stirrups are frequently found in major beams to avoid the development of shear cracking at loading points. Economy of design, with the areas of concrete and rebar proportional to their load-carrying ability, including rebar cover and spacing requirements, frequently results in a T-configuration for beams with the bulk of the rebar found below the neutral axis. Figure 3.7 indicates typical reinforcement locations.

3.4.6 Load Factors and Strength Reduction Factors

Current concrete codes utilize the strength design method, which requires the design strength of a member at any section to be equal to or greater than the strength required as calculated by the code-specified factored load combinations. That is,

$$\text{design strength} = \text{strength reduction factor } (\Phi) \times \text{nominal strength} \quad (3.21)$$

The strength reduction factor (Φ) accounts for such things as the probability of understrength of a member, the degree of ductility and required reliability, and the importance of the member in the structure. The value of Φ equals 0.90 for bending and tension, 0.70 to 0.75 for compression, and 0.85 for shear and torsion.

The nominal strength of a member is calculated from the code provisions. That is,

$$\text{required strength} = \text{load factor} \times \text{service load effects} \quad (3.22)$$

The load factor is an overload factor accounting for the probable variation in service loads. Service loads are loads specified in the general building code. The load factors are generally 1.4 for dead loads and 1.7 for live loads. The loads are combined in accordance with the load combinations given in the concrete code to compute the strength required.

Table 3.7 Common Styles of Metric Welded Wire Reinforcement (WWR) with Equivalent U.S. Customary Units[3]

Common Styles of Metric Welded Wire Reinforcement (WWR) With Equivalent US Customary Units[3]

	A_s (mm²/m)	Metric Styles (MW = Plain wire)[2]	Wt. (kg/m²)	Equivalent Inch-Pound Styles (W = Plain Wire)[2]	A_s (in²/ft)	Wt. (lbs/CSF)
A[1 & 4]	88.9	102x102 - MW9xMW9	1.51	4x4 - W1.4xW1.4	.042	31
	127.0	102x102 - MW13xMW13	2.15	4x4 - W2.0xW2.0	.060	44
	184.2	102x102 - MW19xMW19	3.03	4x4 - W2.9xW2.9	.087	62
	254.0	102x102 - MW26xMW26	4.30	4x4 - W4.0xW4.0	.120	88
	59.3	152x152 - MW9xMW9	1.03	6x6 - W1.4xW1.4	.028	21
	84.7	152x152 - MW13xMW13	1.46	6x6 - W2.0xW2.0	.040	30
	122.8	152x152 - MW19xMW19	2.05	6x6 - W2.9xW2.9	.058	42
	169.4	152x152 - MW26xMW26	2.83	6x6 - W4.0xW4.0	.080	58
B[1]	196.9	102x102 - MW20xMW20	3.17	4x4 - W3.1xW3.1	.093	65
	199.0	152x152 - MW30xMW30	3.32	6x6 - W4.7xW4.7	.094	68
	199.0	305x305 - MW61xMW61	3.47	12x12 - W9.4xW9.4	.094	71
	362.0	305x305 - MW110xMW110	6.25	12x12 - W17.1xW17.1	.171	128
C[1]	342.9	152x152 - MW52xMW52	5.66	6x6 - W8.1xW8.1	.162	116
	351.4	152x152 - MW54xMW54	5.81	6x6 - W8.3xW8.3	.166	119
	192.6	305x305 - MW59xMW59	8.25	12x12 - W9.1xW9.1	.091	69
	351.4	305x305 - MW107xMW107	9.72	12x12 - W16.6xW16.6	.166	125
D[1]	186.3	152x152 - MW28xMW28	3.22	6x6 - W4.4xW4.4	.088	63
	338.7	152x152 - MW52xMW52	5.61	6x6 - W8xW8	.160	115
	186.3	305x305 - MW57xMW57	3.22	12x12 - W8.8xW8.8	.088	66
	338.7	305x305 - MW103xMW103	5.61	12x12 - W16xW16	.160	120
E[1]	177.8	152x152 - MW27xMW27	3.08	6x6 - W4.2xW4.2	.084	60
	317.5	152x152 - MW48xMW48	5.52	6x6 - W7.5xW7.5	.150	108
	175.7	305x305 - MW54xMW54	3.08	12x12 - W8.3xW8.3	.083	63
	317.5	305x305 - MW97xMW97	5.52	12x12 - W15xW15	.150	113

[1] Group A - Compares areas of WWR at f_y = 60,000 psi with other reinforcing at f_y = 60,000 psi
Group B - Compares areas of WWR at f_y = 70,000 psi with other reinforcing at f_y = 60,000 psi
Group C - Compares areas of WWR at f_y = 72,500 psi with other reinforcing at f_y = 60,000 psi
Group D - Compares areas of WWR at f_y = 75,000 psi with other reinforcing at f_y = 60,000 psi
Group E - Compares areas of WWR at f_y = 80,000 psi with other reinforcing at f_y = 60,000 psi

[2] Wires may also be deformed, use prefix MD or D, except where only MW or W is required by building codes (usually less than a MW26 or W4). Also wire sizes can be specified in 1mm² (metric) or .001 in² (inch-pound) increments.
[3] For other available styles or wire sizes, consult other WRI publications or discuss with WWR manufacturers.
[4] Styles may be obtained in roll form. Note: It is recommended that rolls be straightened and cut to size before placement.

Source: Reprinted with permission from the Wire Reinforcement Institute.

Table 3.8   Metric Wire Area, Diameters, and Mass with Equivalent Inch-Pound Units[a]

Metric Units[b]				Inch-Pound Units[c]				
Size[d] (MW = Plain) (mm²)	Area (mm²)	Diameter (mm)	Mass (kg/m)	Size[d] (W = Plain) (in² × 100)	Area (in²)	Diameter (in.)	Weight (lb/ft)	Gage Guide
MW290	290	19.22	2.27	W45	0.450	0.757	1.530	
MW200	200	15.95	1.57	W31	0.310	0.628	1.054	
MW130	130	12.9	1.02	W20.2	0.202	0.507	0.687	
								7/0
MW120	120	12.4	0.941	W18.6	0.186	0.487	0.632	
								6/0
MW100	100	11.3	0.784	W15.5	0.155	0.444	0.527	
								5/0
MW90	90	10.7	0.706	W14.0	0.140	0.422	0.476	
MW80	80	10.1	0.627	W12.4	0.124	0.397	0.422	
								4/0
MW70	70	9.4	0.549	W10.9	0.109	0.373	0.371	
								3/0
MW65	65	9.1	0.510	W10.1	0.101	0.359	0.343	
MW60	60	8.7	0.470	W9.3	0.093	0.344	0.316	
								2/0
MW55	55	8.4	0.431	W8.5	0.085	0.329	0.289	
MW50	50	8.0	0.392	W7.8	0.078	0.314	0.263	
								1/0
MW45	45	7.6	0.353	W7.0	0.070	0.298	0.238	
								1
MW40	40	7.1	0.314	W6.2	0.062	0.283	0.214	
MW35	35	6.7	0.274	W5.4	0.054	0.262	0.184	2
MW30	30	6.2	0.235	W4.7	0.047	0.245	0.160	3
MW26	26	5.7	0.204	W4.0	0.040	0.226	0.136	4
MW25	25	5.6	0.196	W3.9	0.039	0.223	0.133	
MW20	20	5.0	0.157	W3.1	0.031	0.199	0.105	
MW19	19	4.9	0.149	W2.9	0.029	0.192	0.098	6
MW15	15	4.4	0.118	W2.3	0.023	0.171	0.078	
								8
MW13	13	4.1	0.102	W2.0	0.020	0.160	0.068	
MW10	10	3.6	0.078	W1.6	0.016	0.143	0.054	
MW9	9	3.4	0.071	W1.4	0.014	0.135	0.048	10

Source: Reprinted with permission from the Wire Reinforcement Institute.

[a]For other available wire sizes, consult other WRI publications or discuss with WWF manufacturers.
[b]Metric wire sizes can be specified in 1-mm² increments.
[c]Inch-pound sizes can be specified in 0.001-in.² increments.
[d]Wires may be deformed; use prefix MD or D, except where only MW or W is required by building codes (usually less than MW26 or W4).

3.4.7   Foundations

Foundations must be designed for the full range and variations of loadings, including settlement that may occur. Care should be exercised to determine the requirements of applicable codes, which often establish specific design criteria. In addition to simple bearing loads, foundation loads may include moments, eccentric loadings, and uplift; their design must consider each of these as appropriate.

For simple foundations in the absence of seismic, vibratory, impact, or cyclical loadings (as occurs with machinery unbalance), the dead load is often increased

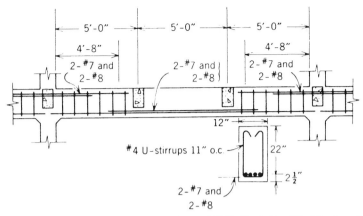

Fig. 3.7   Reinforcement locations. (From H. Parker and H. D. Hauf, *Simplified Design of Reinforced Concrete*, 4th ed., John Wiley & Sons, Inc., New York, 1976.)

arbitrarily by one-half to provide an allowance for live loads. For other more severe service, specific loads should be established based on occupancy, or from codes or equipment suppliers.

Spread foundations (or footings) of the pier, grade beam, or mat variety are the most common type, with reinforced concrete used almost universally. Although their section and proportions vary, care must be taken to assure that loads are transmitted to the centroid of the section and are not eccentric. Designers should avoid a condition where a continuous foundation rests on different materials, such as rock and engineered fill. The **differential settlement** that occurs in such a case is virtually certain to create problems, including possible foundation failure. In cases where it is necessary to found the structure on different materials, special measures, such as load-spanning grade beams, must be taken to avoid this problem. Reinforcement of all foundations is essential and minimally consists of relatively small-size rebar placed below the neutral axis to resist tensile stresses.

The **bearing value** of the material beneath the footing usually determines its base (largest dimension). The size of the foundation can be no smaller than the bearing value of the foundation material. For all but the most simple structures and particularly those where settlement can be a major problem, such as tall chimneys, actual bearing tests should be run. In the absence of specific code requirements, a conservative design for simple structures can utilize the bearing values in Table 3.9.

Pile foundations utilize either **bearing** or **friction** piles, the piles being of precast or cast in place concrete, steel, or wood. Typical bearing values for wood piles is 15 to 20 tons each with concrete or steel piles designed for loadings to 30 tons or more. Friction pile capacity is dependent on the pile shape, material, and the adhesion coefficient, which varies extremely widely depending on the type and consistency of the soil. Test piles, to assure load capacity, are essential where friction piles are contemplated.

Minimum pile spacing is usually 3 ft on center, and loads are transferred into and out of the piles by concrete pile caps often cast in place. Wood piles are subject to rapid deterioration at a wetted zone, where alternate wetting and drying occurs, as

Table 3.9   Typical Foundation Bearing Values

Material	Maximum Allowable Bearing Value, ksf
Sound, solid bedrock, without fissures	150–200
Laminated rock in sound condition, minor cracking only	50–70
Sound shale or hardpan	16–20
Compacted gravel or compacted sand/gravel	8–10
Loose gravel, coarse sand, compacted fine sand	4–8
Clays (vary widely)	2–10[a]

[a] Testing should be performed.

in pier construction. For this use, wood piles are normally cut off below the wet zone (or water line) and the pile cap cast in the wet zone.

Piling foundations consist of a series of individual elements (the piles), which depending on their geometric arrangement, usually see different loads. As a result, the pattern of piles that make up the foundation must be carefully engineered to assure adequacy under the variety of loading conditions that occur. Under some loading conditions, undesirable uplift forces may be placed on piles. Some effort should be taken to avoid this condition if possible.

Wherever possible, test pile loadings should be conducted. Typically, a test pile is loaded to twice its planned working load and passes the test if there is no settlement after 24 hours and settlement (after unloading) of not more than 0.01 in. per ton of test load. Although this does not establish the actual strength of the pile, it assures a minimum factor of safety of 2.

Preservation of wood piles is usually performed and consists of pressure impregnation of creosote, copper-bearing chemicals, or other materials. Steel piles are often coated with coal tar derivatives.

3.4.8   Slabs on Grade

Although any reasonable concrete mix can be used, ACI 302 classifies floors (and slabs) as shown in Tables 3.10 and 3.11. For exposure to freezing conditions while in the wet condition, air-entrained concrete is widely used. Upon placing, upward water migration will tend to reduce the strength and durability of the surface; thus some care to assure that the water/cement ratio is controlled properly is important. In general, for slabs carrying commercial or industrial traffic, surfaces are hard-steel trowled for flatness and/or smoothness. Where necessary, hardeners can either be bedded into the slab or trowled into its upper surface.

Reinforcing should include at least minimum amounts of steel to avoid excessive temperature-induced shrinkage cracking. The minimum ratio of reinforcement area to the gross concrete area is on the order of 0.0018 to 0.0020. For a 6-in. slab this equates to No. 4 rebar at 18 in. each way. Reinforcement should have at least 2 in. of cover top (and bottom) of slab and for lower slab reinforcement be supported on precast concrete blocks about 4 in. square (minimum) or their equivalent.

Joints should be placed in slabs to provide either isolation, shrinkage (cracking) control, or construction **joints** at predetermined locations or at the ends of pours.

Table 3.10 Floor Classifications

Class	Anticipated Type of Traffic	Use	Special Considerations	Final Finish
1	Light foot	Residential surfaces; mainly with floor coverings	Grade for drainage; level slabs suitable for applied coverings; curing	Single troweling
2	Foot	Offices and churches, usually with floor covering; decorative	Surface tolerance (including elevated slabs), nonslip aggregate in specific areas	Single troweling; nonslip finish where required As required
			Colored mineral aggregate, hardener or exposed aggregate, artistic joint layout	
3	Foot and pneumatic wheels	Exterior walls, driveways, garage floors, sidewalks	Grade for drainage; proper air content; curing; see Chapter 5 of the standard for specific durability requirements	Float, trowel, or broom finish
4	Foot and light vehicular traffic	Institutional and commercial	Level slab suitable for applied coverings, nonslip aggregate for specific areas and curing	Normal steel trowel finish
5	Industrial vehicular traffic with pneumatic wheels	Light-duty industrial floors for manufacturing, processing, and warehousing	Good uniform subgrade, surface tolerance, joint layout, abrasion resistance, curing	Hard steel trowel finish
6	Industrial vehicular traffic with hard wheels	Industrial floors subject to heavy traffic; may be subject to impact loads	Good uniform subgrade, surface tolerance, joint layout, load transfer, abrasion resistance, curing	Special metallic or mineral aggregate, repeated hard steel troweling
7	Industrial vehicular traffic with hard wheels	Bonded two-course floors subject to heavy traffic and impact	*Base slab:* good uniform subgrade, reinforcement, joint layout, level surface, curing	Clean-textured surface suitable for subsequent bonded topping
			Topping: composed of well-graded all-mineral or all-metallic aggregate, mineral or metallic aggregate applied to high-strength plain topping to toughen, surface tolerance, curing	Special power floats with repeated steel trowelings

Table 3.10 (*Continued*)

Class	Anticipated Type of Traffic	Use	Special Considerations	Final Finish
8	As in Class 4, 5, or 6	Unbonded toppings: freezer floors on insulation, on old floors, or where construction schedule dictates	Bond breaker on old surface, mesh reinforcement, minimum thickness 3 in. (nominal 75 mm), abrasion resistance and curing	Hard steel trowel finish
9	Superflat or critical surface tolerance required; special materials-handling vehicles or robotics requiring specific tolerances	Narrow-aisle, high-bay warehouses; television studios	Varying concrete quality requirements, shake-on hardeners cannot be used unless special application and great care are employed, proper joint arrangement, F_F 35 to F_F 125 (F_F 100 is "superflat" floor)	Strictly follow finishing techniques as indicated in Section 7.15 of the standard

Source: ACI Standard 302.1R; Reprinted with permission of the American Concrete Institute, Farmington Hills, Mich.

Table 3.11   Recommended Slump at Point of Placement and Strength for Each Class of Concrete Floor

Floor Class[a,b]	28-Day Compressive Strength[c]		Slump[d,e]	
	psi	MPa	in.	mm
1	3000	21	5	125
2	3500	24	5	125
3	3500	24	5	125
4	4000	28	5	125
5	4000	28	4	100
6	4500	31	4	100
7 Base	3500	24	4	100
8 Toppings[f,g]	5000–8000	35–55	2	50
9 Superflat	4000 or higher	28	5	125

Source: ACI Standard 302.1R; reprinted with permission of the American Concrete Institute, Farmington Hill, Mich.

[a] For concrete made with normal weight aggregate and exposed to freezing and thawing, air content should conform to the limits given in Table 5.2.7.a of the standard and have a maximum water-cement ratio of 0.50 [roughly equivalent to 4000 psi (28 MPa) or more]. Nonreinforced concrete subjected to deicing salts should have a water-cement ratio no greater than 0.45 [roughly 4500 psi (31 MPa) or more]. Reinforced concrete exposed to brackish water, seawater, or deicing salts should have a maximum water-cement ratio of 0.40 [roughly 5000 psi (34 MPa) or more]. Structural lightweight aggregate concretes should have the air contents given in Table 5.2.7.b of the standard. For lightweight concretes subject to freezing and thawing, recommendations for air and cement contents should be secured from the lightweight aggregate manufacturer.

[b] Refer to Table 1.1 of the standard (Table 3.10 here) for floor usage and requirements.

[c] On Class 2 through 9 floors, compressive strength of concrete slab before allowing construction traffic should be at least 1800 psi (12 MPa). "Strength" refers to compressive strength of cylinders that have been continuously moist cured and tested according to applicable ASTM standards at 7 and 28 days. For each strength requirement at a designated age, the strength level of the concrete will be considered satisfactory if the average of all sets of three consecutive strength test results equal or exceed the recommended values above, and no individual strength (average of two cylinders) falls below the strength specified by more than 500 psi (3.5 MPa).

[d] Maximum slump measured at point of placement, e.g., at the discharge end of the pump line if concrete is pumped. Refer to comments on workability and placement in Section 5.2.5 of the standard. The slump of concrete as used should be such as not to cause objectionable aspects of higher slump concretes as discussed in Section 5.2.5 of the standard. A high-range water-reducing admixture (superplasticizer) meeting the requirements of ACTM C494 (Type F or G) or a combination of admixtures meeting the requirements of ASTM C494 (Type A, C, D, E, F or G) may be used to increase this slump level, providing the resulting mix conforms to all other requirements.

[e] ACI 223 gives slump requirements for shrinkage-compensating concrete.

[f] The strength required will depend on severity of usage. The ranges shown will cover most situations.

[g] Maximum aggregate size not greater than one-fourth the thickness of unbonded topping.

Isolation joints typically utilize special filler material, while contraction joints can be hand tooled, machine cut, or may use prefabricated materials; these joints should have a depth at least one-fourth the thickness of the slab. Joints are normally located to divide the slab into about 100-ft^2 sections having roughly equal sides. Construction joints normally extend through the entire slab thickness and utilize some form of keying to transmit loads. Often, the keying uses rebar dowels epoxied or cast into the previous pour.

Where water or water vapor migration is likely, **vapor barriers** are installed beneath the slab. Typically, $\frac{3}{4}$ in. minus compacted gravel fill is placed over 4-mil (0.004 in.) or thicker polyethylene with its laps sealed. Some care needs to be taken during fill placement to avoid puncturing the barrier and ruining the watertightness. The fill is normally wetted prior to placing of the slab concrete to avoid pulling water out of the lower portion of the slab. Where severe water conditions exist, the fill is deepened and may be placed in two or more graded layers to assure underslab drainage. Walls and footings are typically isolated from the slab by asphaltic impregnated materials or in some cases $\frac{3}{4}$-in. polystyrene sheets.

3.4.9   Precast Components

Precast concrete is widely used in both buildings and bridges. The precast components, such as beams, columns, and walls, are erected and "tied" together during construction to form the entire structure. These components are typically cast in a precast plant under rigid quality control conditions. Precast concrete lends itself to fast-track construction methods due to the continuous and uninterrupted erection of the components. Special care must be taken in the design and construction of the connections to ensure adequate load transfer.

Precast beams and columns are either conventionally reinforced or reinforced with prestressing strands, or both. In prestressed precast concrete high-strength strands on the order of 270 ksi minimum tensile strength are typically pretensioned prior to placing the concrete. This precompresses the concrete, thereby increasing the compression zone, resulting in a more efficient member.

3.5   STRUCTURAL SYSTEMS[4]

3.5.1   Framing Systems and Connections

Structural framing systems can be classified as **nonrigid, rigid,** and **composite. Nonrigid systems** utilize the structural members acting as essentially independent elements transmitting all ordinary forces but not transmitting moments to each other or to a foundation system; their analysis is relatively straightforward. **Rigid systems** have the ability to transmit all normal loads, including moment, and this feature makes their analysis more difficult. Finally, **composite systems** utilize the interaction between concrete elements such as slabs, beams, and panels and the steel members, often by embedding the steel members within the concrete. This may permit a reduction in the quantity of steel and concrete, strengthens the structure, and provides an improved load-carrying ability.

Nonrigid framing systems are often divided into bearing wall, skeleton, long span, or curtain wall types. The names are self-explanatory except for skeleton framing, which utilizes columns, beams, and stringers (or purlins) to carry floor, wall, or roof loads. These loads are carried to individual intermediate foundations, as in the case of interior columns in a warehouse, or to exterior wall footings, or bearing walls. Since the connections between the members are relatively modest, the design of these systems does not take credit for any moment transmission, but rather are based on, shear and thrust. Bending moments are normally resisted by stresses within an individual member. Analysis is straightforward and the structure is **determinate.**

Rigid framing systems transmit moment and in simple form are often analyzed as a two- or three-hinged frame, depending on the type of connection at midspan, as in Fig. 3.8. To resist moment, knees are normally deepened, and stiffening of the knee flange or web to resist buckling is common. Since the base connection is for analysis, considered hinged, and cannot resist moment, it is common on spans of 40 feet or more, or in some cases shorter if loads are high, to resist the lateral reaction by tieing the foundations together, often with adjustable tie rods.

Analysis of a three-hinged frame is determinate. Analysis of a two-hinged frame or a frame with rigid supports is **indeterminate,** and analysis proceeds by assuming moments and member size and solving for the stresses resulting. This solution is

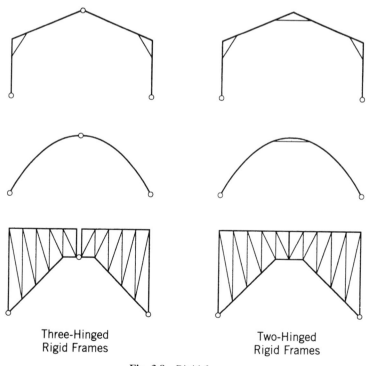

Three-Hinged   Two-Hinged
Rigid Frames   Rigid Frames

Fig. 3.8 Rigid frames.

refined by iteration until a satisfactory solution results. In more recent years advanced computer programs perform all of this analysis. Numerous manufacturers produce prefabricated rigid frames in shapes and spans for a wide variety of loadings and can provide precalculated values for all members in a typical framing system.

Composite systems are widely found and utilize primarily the compressive strength of concrete and tie the concrete slabs, beams, and panels into the structural steel frame of the building. This permits a reduction in the amount of steel required for the frame and provides a stiffer frame with an improved ability to dampen vibrations. Because of the interaction between the concrete and steel, and the load transfer between them, the design of these structures is complex and beyond the scope of this book.

Connections in steel-framed structures use welds, high-strength bolts, commercial bolts, and rivets. Welding has the widest application because of its high strength, lightness, and good appearance. It also has the advantage of being able to provide rigid moment connections easily. High-strength bolting is widely used for field connections because of its ease of fit-up and speed, but because of the high cost of the bolts themselves is not widely used for shop fabrication. Commercial bolting is widely used for secondary structural members and light structures. Riveting is only rarely used today (1998).

Most structures utilize simple connections that allow some rotation and are relatively flexible. Although some moment transfer occurs, for design purposes it is often disregarded, and the connections are assumed to transfer shear or thrust only. Rigid connections transmit moment and are usually assumed to allow no joint rotation. For these connections, welding is almost universally used.

The *Manual of Steel Construction*[5] published by the American Institute of Steel Construction provides precalculated load values of both bolted and welded connections for normal loadings as well as methodology for calculating these connection capacities for eccentric loadings. The manual also includes suggested details for beam framing, column bases, column splices, and miscellaneous details.

Analysis of structural systems utilizes the **elastic theory** for most work. In this system it is assumed that materials are only stressed within their elastic range and that yield does not occur. **Plastic analysis** is used in cases where a more refined analysis is required and loads imposed are severe and of short duration, such as seismic or impact loads. This methodology produces higher levels of stress within the materials and can only be justified for limited cases. The analysis assumes some yielding at load points, causing the development of plastic hinges with load redistribution to lower stressed portions of the member. Considerable care must be taken to perform the analysis under all conceivable loading conditions, as design margins are considerably reduced.

3.5.2 Selection of the Appropriate Structural Steel

ASTM A36 is the all-purpose grade steel widely used in building and bridge construction. ASTM A529 structural carbon steel, ASTM A441 and A572 high-strength low-alloy structural steels, ASTM A242 and A588 atmospheric corrosion-resistant high-strength low-alloy structural steels, and ASTM A514 quenched and tempered alloy structural steel plate may each have certain advantages over ASTM A36, depending on the application. These **high-strength steels** have proven to be economical choices where lighter members, resulting from use of higher allowable stresses, are not penalized because of instability, local buckling, deflection, or other similar reasons. They are frequently used in tension members, beams in continuous and composite construction where deflections can be minimized, and columns having low slenderness ratios. The reduction of dead load, and associated savings in shipping costs, can be significant factors. However, higher-strength steels are not to be used indiscriminately. Effective use of all steels depends on thorough cost and engineering analysis.

With suitable procedures and precautions, all steels listed in the AISC Specification are suitable for welded fabrication.

ASTM A242 and A588 **atmospheric corrosion-resistant** high-strength low-alloy steels are more expensive than the high-strength low-alloy steels. They are suitable for use in the bare (uncoated) condition, where exposure to normal atmosphere causes a tightly adherent oxide to form on the surface, protecting the steel from further oxidation. The reduction of maintenance resulting from use of these steels often offsets their higher initial cost. Designers should consult the steel producers on the corrosion-resistant properties and limitations of these steels prior to use in the bare (uncoated) condition.

When either A242 or A588 steel is exposed to a more corrosive atmospheric environment, its use in the coated condition provides longer coating life than with other structural steels. It should be noted that A588 steel, in addition to its ability to resist atmospheric corrosion, offers the advantage of being the only listed "as-rolled" structural steel available at 50 ksi minimum specified yield stress in thickness up to 4 in. (inclusive).

Table 3.12 indicates the tensile properties of three of the most commonly used materials. The AISC *Manual of Steel Construction* should be referred to for the stress ratings and availability of other materials.

3.5.3 Structural Shapes: Designations, Dimensions, and Properties

The letter M designates shapes that cannot be classified as W, HP, or S shapes. Similarly, MC designates channels that cannot be classified as C shapes. Because

Table 3.12 Tensile Properties of Several Structural Materials: Shapes, Plates, and Bars

Steel Type	ASTM Designation	Minimum Yield Stress, F_y (ksi)	Tensile Stress, F_u (ksi)[a]
Carbon	A36	36	58–80
High-strength	A572 Grade 42	42	60
Low-alloy	A572 Grade 50	50	65[b]
Corrosion-resistant high-strength, low-alloy	A588	50	70

[a]Minimum unless range is given.
[b]Some size limitation for plates and bars.

many of the M and MC shapes are available only from a limited number of producers, or are infrequently rolled, their availability should be checked prior to specifying these shapes. They have various slopes on their inner flange surfaces, for which dimensions may be obtained from the respective producing mills. The flange thickness given in Table 3.13 is the average flange thickness.

In calculating the theoretical weights, properties, and dimensions of the rolled shapes, fillets, and roundings have been included for all shapes except angles. The properties of these rolled shapes are based on the smallest theoretical size fillets produced; dimensions for detailing are based on the largest theoretical size fillets produced. These properties and dimensions are either exact or slightly conservative for all producers who offer them. Maximum lengths available vary widely with producers but a conservative range is from 60 to 75 ft.

3.5.4 Tables of Shapes

For convenience, larger and less commonly available shapes and tees have been omitted from Table 3.13. For complete listings, refer to the AISC *Manual of Steel Construction*.

3.6 STRUCTURAL TIMBER

3.6.1 Codes

Several industrial associations provide information and standards to the timber industry. Among those best known are the American Institute of Timber Construction (AITC) and the American Forest & Paper Association (AFPA). The AITC has developed extensive design data in their *Timber Construction Manual.*[6] In addition, they have developed additional standards, some of which are listed below.

AITC 104-84	Typical Construction Details
AITC 108-93	Standard for Heavy Timber Construction
AITC 109-90	Standard for Preservative Treatment of Structural Glued Laminated Timber
AITC 110-84	Standard Appearance Grades for Structural Glued Laminated Timber
AITC 111-79	Recommended Practice for Protection of Structural Glued Laminated Timber During Transit, Storage and Erection
AITC-112-93	Standard for Tongue-and-Groove Heavy Timber Roof Decking
AITC-113-93	Standard for Dimensions of Structural Glued Laminated Timber

Table 3.13 Selected Structural Shapes: Dimensions and Properties

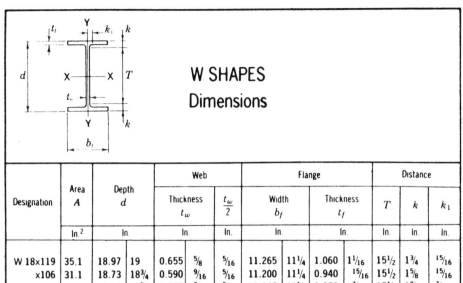

W SHAPES
Dimensions

Designation	Area A (In.²)	Depth d (In.)		Web Thickness t_w (In.)		$\dfrac{t_w}{2}$ (In.)	Flange Width b_f (In.)		Flange Thickness t_f (In.)		T (In.)	k (In.)	k_1 (In.)
W 18×119	35.1	18.97	19	0.655	5/8	5/16	11.265	11¼	1.060	1¹/₁₆	15½	1¾	15/16
×106	31.1	18.73	18¾	0.590	9/16	5/16	11.200	11¼	0.940	15/16	15½	1⅝	15/16
× 97	28.5	18.59	18⅝	0.535	9/16	5/16	11.145	11⅛	0.870	7/8	15½	1⁹/₁₆	7/8
× 86	25.3	18.39	18⅜	0.480	½	¼	11.090	11⅛	0.770	¾	15½	1⁷/₁₆	7/8
× 76	22.3	18.21	18¼	0.425	7/16	¼	11.035	11	0.680	11/16	15½	1⅜	13/16
W 18× 71	20.8	18.47	18½	0.495	½	¼	7.635	7⅝	0.810	13/16	15½	1½	7/8
× 65	19.1	18.35	18⅜	0.450	7/16	¼	7.590	7⅝	0.750	¾	15½	1⁷/₁₆	7/8
× 60	17.6	18.24	18¼	0.415	7/16	¼	7.555	7½	0.695	11/16	15½	1⅜	13/16
× 55	16.2	18.11	18⅛	0.390	⅜	3/16	7.530	7½	0.630	⅝	15½	1⁵/₁₆	13/16
× 50	14.7	17.99	18	0.355	⅜	3/16	7.495	7½	0.570	9/16	15½	1¼	13/16
W 16×100	29.4	16.97	17	0.585	9/16	5/16	10.425	10⅜	0.985	1	13⅝	1¹¹/₁₆	15/16
× 89	26.2	16.75	16¾	0.525	½	¼	10.365	10⅜	0.875	7/8	13⅝	1⁹/₁₆	7/8
× 77	22.6	16.52	16½	0.455	7/16	¼	10.295	10¼	0.760	¾	13⅝	1⁷/₁₆	7/8
× 67	19.7	16.33	16⅜	0.395	⅜	3/16	10.235	10¼	0.665	11/16	13⅝	1⅜	13/16
W 16× 57	16.8	16.43	16⅜	0.430	7/16	¼	7.120	7⅛	0.715	11/16	13⅝	1⅜	7/8
× 50	14.7	16.26	16¼	0.380	⅜	3/16	7.070	7⅛	0.630	⅝	13⅝	1⁵/₁₆	13/16
× 45	13.3	16.13	16⅛	0.345	⅜	3/16	7.035	7	0.565	9/16	13⅝	1¼	13/16
× 40	11.8	16.01	16	0.305	5/16	3/16	6.995	7	0.505	½	13⅝	1³/₁₆	13/16
× 36	10.6	15.86	15⅞	0.295	5/16	3/16	6.985	7	0.430	7/16	13⅝	1⅛	¾
W 14×426	125.0	18.67	18⅝	1.875	1⅞	15/16	16.695	16¾	3.035	3¹/₁₆	11¼	3¹¹/₁₆	1⁹/₁₆
×398	117.0	18.29	18¼	1.770	1¾	7/8	16.590	16⅝	2.845	2⅞	11¼	3½	1½
×370	109.0	17.92	17⅞	1.655	1⅝	13/16	16.475	16½	2.660	2¹¹/₁₆	11¼	3⁵/₁₆	1⁷/₁₆
×342	101.0	17.54	17½	1.540	1⁹/₁₆	13/16	16.360	16⅜	2.470	2½	11¼	3⅛	1⅜
×311	91.4	17.12	17⅛	1.410	1⁷/₁₆	¾	16.230	16¼	2.260	2¼	11¼	2¹⁵/₁₆	1⁵/₁₆
×283	83.3	16.74	16¾	1.290	1⁵/₁₆	11/16	16.110	16⅛	2.070	2¹/₁₆	11¼	2¾	1¼
×257	75.6	16.38	16⅜	1.175	1³/₁₆	⅝	15.995	16	1.890	1⅞	11¼	2⁹/₁₆	1³/₁₆
×233	68.5	16.04	16	1.070	1¹/₁₆	9/16	15.890	15⅞	1.720	1¾	11¼	2⅜	1³/₁₆
×211	62.0	15.72	15¾	0.980	1	½	15.800	15¾	1.560	1⁹/₁₆	11¼	2¼	1⅛
×193	56.8	15.48	15½	0.890	7/8	7/16	15.710	15¾	1.440	1⁷/₁₆	11¼	2⅛	1¹/₁₆
×176	51.8	15.22	15¼	0.830	13/16	7/16	15.650	15⅝	1.310	1⁵/₁₆	11¼	2	1¹/₁₆
×159	46.7	14.98	15	0.745	¾	⅜	15.565	15⅝	1.190	1³/₁₆	11¼	1⅞	1
×145	42.7	14.78	14¾	0.680	11/16	⅜	15.500	15½	1.090	1¹/₁₆	11¼	1¾	1

Source: American Institute of Steel Construction.

Table 3.13 *(Continued)*

W SHAPES
Properties

Nominal Wt. per Ft	Compact Section Criteria				r_T	$\dfrac{d}{A_f}$	Elastic Properties						Torsional constant	Plastic Modulus	
	$\dfrac{b_f}{2t_f}$	F_y'	$\dfrac{d}{t_w}$	F_y'''			Axis X-X			Axis Y-Y			J	Z_x	Z_y
							I	S	r	I	S	r			
Lb.		Ksi		Ksi	In.		In.⁴	In.³	In.	In.⁴	In.³	In.	In.⁴	In.³	In.³
119	5.3	—	29.0	—	3.02	1.59	2190	231	7.90	253	44.9	2.69	10.6	261	69.1
106	6.0	—	31.7	—	3.00	1.78	1910	204	7.84	220	39.4	2.66	7.48	230	60.5
97	6.4	—	34.7	54.7	2.99	1.92	1750	188	7.82	201	36.1	2.65	5.86	211	55.3
86	7.2	—	38.3	45.0	2.97	2.15	1530	166	7.77	175	31.6	2.63	4.10	186	48.4
76	8.1	64.2	42.8	36.0	2.95	2.43	1330	146	7.73	152	27.6	2.61	2.83	163	42.2
71	4.7	—	37.3	47.4	1.98	2.99	1170	127	7.50	60.3	15.8	1.70	3.48	145	24.7
65	5.1	—	40.8	39.7	1.97	3.22	1070	117	7.49	54.8	14.4	1.69	2.73	133	22.5
60	5.4	—	44.0	34.2	1.96	3.47	984	108	7.47	50.1	13.3	1.69	2.17	123	20.6
55	6.0	—	46.4	30.6	1.95	3.82	890	98.3	7.41	44.9	11.9	1.67	1.66	112	18.5
50	6.6	—	50.7	25.7	1.94	4.21	800	88.9	7.38	40.1	10.7	1.65	1.24	101	16.6
100	5.3	—	29.0	—	2.81	1.65	1490	175	7.10	186	35.7	2.51	7.73	198	54.9
89	5.9	—	31.9	64.9	2.79	1.85	1300	155	7.05	163	31.4	2.49	5.45	175	48.1
77	6.8	—	36.3	50.1	2.77	2.11	1110	134	7.00	138	26.9	2.47	3.57	150	41.1
67	7.7	—	41.3	38.6	2.75	2.40	954	117	6.96	119	23.2	2.46	2.39	130	35.5
57	5.0	—	38.2	45.2	1.86	3.23	758	92.2	6.72	43.1	12.1	1.60	2.22	105	18.9
50	5.6	—	42.8	36.1	1.84	3.65	659	81.0	6.68	37.2	10.5	1.59	1.52	92.0	16.3
45	6.2	—	46.8	30.2	1.83	4.06	586	72.7	6.65	32.8	9.34	1.57	1.11	82.3	14.5
40	6.9	—	52.5	24.0	1.82	4.53	518	64.7	6.63	28.9	8.25	1.57	0.79	72.9	12.7
36	8.1	64.0	53.8	22.9	1.79	5.28	448	56.5	6.51	24.5	7.00	1.52	0.54	64.0	10.8
426	2.8	—	10.0	—	4.64	0.37	6600	707	7.26	2360	283	4.34	331	869	434
398	2.9	—	10.3	—	4.61	0.39	6000	656	7.16	2170	262	4.31	273	801	402
370	3.1	—	10.8	—	4.57	0.41	5440	607	7.07	1990	241	4.27	222	736	370
342	3.3	—	11.4	—	4.54	0.43	4900	559	6.98	1810	221	4.24	178	672	338
311	3.6	—	12.1	—	4.50	0.47	4330	506	6.88	1610	199	4.20	136	603	304
283	3.9	—	13.0	—	4.46	0.50	3840	459	6.79	1440	179	4.17	104	542	274
257	4.2	—	13.9	—	4.43	0.54	3400	415	6.71	1290	161	4.13	79.1	487	246
233	4.6	—	15.0	—	4.40	0.59	3010	375	6.63	1150	145	4.10	59.5	436	221
211	5.1	—	16.0	—	4.37	0.64	2660	338	6.55	1030	130	4.07	44.6	390	198
193	5.5	—	17.4	—	4.35	0.68	2400	310	6.50	931	119	4.05	34.8	355	180
176	6.0	—	18.3	—	4.32	0.74	2140	281	6.43	838	107	4.02	26.5	320	163
159	6.5	—	20.1	—	4.30	0.81	1900	254	6.38	748	96.2	4.00	19.8	287	146
145	7.1	—	21.7	—	4.28	0.88	1710	232	6.33	677	87.3	3.98	15.2	260	133

Table 3.13 *(Continued)*

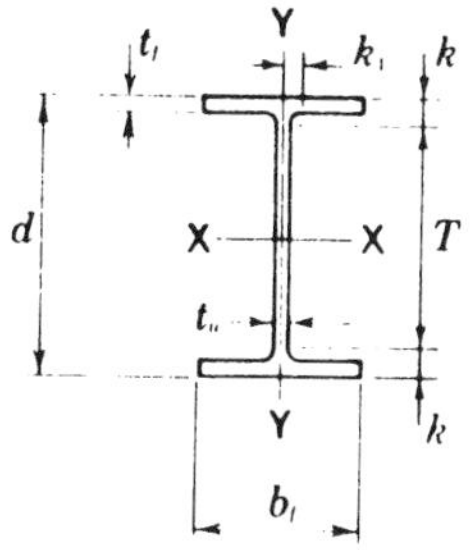

W SHAPES
Dimensions

Designation	Area A	Depth d		Web Thickness t_w		Web $\frac{t_w}{2}$	Flange Width b_f		Flange Thickness t_f		Distance T	Distance k	Distance k_1
	In.^{2}	In.		In.		In.	In.		In.		In.	In.	In.
W 14x132	38.8	14.66	14 5/8	0.645	5/8	5/16	14.725	14 3/4	1.030	1	11 1/4	1 11/16	15/16
x120	35.3	14.48	14 1/2	0.590	9/16	5/16	14.670	14 5/8	0.940	15/16	11 1/4	1 5/8	15/16
x109	32.0	14.32	14 3/8	0.525	1/2	1/4	14.605	14 5/8	0.860	7/8	11 1/4	1 9/16	7/8
x 99	29.1	14.16	14 1/8	0.485	1/2	1/4	14.565	14 5/8	0.780	3/4	11 1/4	1 7/16	7/8
x 90	26.5	14.02	14	0.440	7/16	1/4	14.520	14 1/2	0.710	11/16	11 1/4	1 3/8	7/8
W 14x 82	24.1	14.31	14 1/4	0.510	1/2	1/4	10.130	10 1/8	0.855	7/8	11	1 5/8	1
x 74	21.8	14.17	14 1/8	0.450	7/16	1/4	10.070	10 1/8	0.785	13/16	11	1 9/16	15/16
x 68	20.0	14.04	14	0.415	7/16	1/4	10.035	10	0.720	3/4	11	1 1/2	15/16
x 61	17.9	13.89	13 7/8	0.375	3/8	3/16	9.995	10	0.645	5/8	11	1 7/16	15/16
W 12x190	55.8	14.38	14 3/8	1.060	1 1/16	9/16	12.670	12 5/8	1.735	1 3/4	9 1/2	2 7/16	1 3/16
x170	50.0	14.03	14	0.960	15/16	1/2	12.570	12 5/8	1.560	1 9/16	9 1/2	2 1/4	1 1/8
x152	44.7	13.71	13 3/4	0.870	7/8	7/16	12.480	12 1/2	1.400	1 3/8	9 1/2	2 1/8	1 1/16
x136	39.9	13.41	13 3/8	0.790	13/16	7/16	12.400	12 3/8	1.250	1 1/4	9 1/2	1 15/16	1
x120	35.3	13.12	13 1/8	0.710	11/16	3/8	12.320	12 3/8	1.105	1 1/8	9 1/2	1 13/16	1
x106	31.2	12.89	12 7/8	0.610	5/8	5/16	12.220	12 1/4	0.990	1	9 1/2	1 11/16	15/16
x 96	28.2	12.71	12 3/4	0.550	9/16	5/16	12.160	12 1/8	0.900	7/8	9 1/2	1 5/8	7/8
x 87	25.6	12.53	12 1/2	0.515	1/2	1/4	12.125	12 1/8	0.810	13/16	9 1/2	1 1/2	7/8
x 79	23.2	12.38	12 3/8	0.470	1/2	1/4	12.080	12 1/8	0.735	3/4	9 1/2	1 7/16	7/8
x 72	21.1	12.25	12 1/4	0.430	7/16	1/4	12.040	12	0.670	11/16	9 1/2	1 3/8	7/8
x 65	19.1	12.12	12 1/8	0.390	3/8	3/16	12.000	12	0.605	5/8	9 1/2	1 5/16	13/16
W 12x 58	17.0	12.19	12 1/4	0.360	3/8	3/16	10.010	10	0.640	5/8	9 1/2	1 3/8	13/16
x 53	15.6	12.06	12	0.345	3/8	3/16	9.995	10	0.575	9/16	9 1/2	1 1/4	13/16
W 12x 50	14.7	12.19	12 1/4	0.370	3/8	3/16	8.080	8 1/8	0.640	5/8	9 1/2	1 3/8	13/16
x 45	13.2	12.06	12	0.335	5/16	3/16	8.045	8	0.575	9/16	9 1/2	1 1/4	13/16
x 40	11.8	11.94	12	0.295	5/16	3/16	8.005	8	0.515	1/2	9 1/2	1 1/4	3/4
W 12x 35	10.3	12.50	12 1/2	0.300	5/16	3/16	6.560	6 1/2	0.520	1/2	10 1/2	1	9/16
x 30	8.79	12.34	12 3/8	0.260	1/4	1/8	6.520	6 1/2	0.440	7/16	10 1/2	15/16	1/2
x 26	7.65	12.22	12 1/4	0.230	1/4	1/8	6.490	6 1/2	0.380	3/8	10 1/2	7/8	1/2

AMERICAN INSTITUTE OF STEEL CONSTRUCTION

Table 3.13 (*Continued*)

W SHAPES
Properties

Nominal Wt. per Ft.	Compact Section Criteria				r_T	$\dfrac{d}{A_f}$	Elastic Properties						Torsional constant	Plastic Modulus	
	$\dfrac{b_f}{2t_f}$	F_y'	$\dfrac{d}{t_w}$	F_y'''			Axis X-X			Axis Y-Y				Z_x	Z_y
							I	S	r	I	S	r	J		
Lb.		Ksi		Ksi	In.		In.⁴	In.³	In.	In.⁴	In.³	In.	In.⁴	In.³	In.³
132	7.1	—	22.7	—	4.05	0.97	1530	209	6.28	548	74.5	3.76	12.3	234	113
120	7.8	—	24.5	—	4.04	1.05	1380	190	6.24	495	67.5	3.74	9.37	212	102
109	8.5	58.6	27.3	—	4.02	1.14	1240	173	6.22	447	61.2	3.73	7.12	192	92.7
99	9.3	48.5	29.2	—	4.00	1.25	1110	157	6.17	402	55.2	3.71	5.37	173	83.6
90	10.2	40.4	31.9	—	3.99	1.36	999	143'	6.14	362	49.9	3.70	4.06	157	75.6
82	5.9	—	28.1	—	2.74	1.65	882	123	6.05	148	29.3	2.48	5.08	139	44.8
74	6.4	—	31.5	—	2.72	1.79	796	112	6.04	134	26.6	2.48	3.88	126	40.6
68	7.0	—	33.8	57.7	2.71	1.94	723	103	6.01	121	24.2	2.46	3.02	115	36.9
61	7.7	—	37.0	48.1	2.70	2.15	640	92.2	5.98	107	21.5	2.45	2.20	102	32.8
190	3.7	—	13.6	—	3.50	0.65	1890	263	5.82	589	93.0	3.25	48.8	311	143
170	4.0	—	14.6	—	3.47	0.72	1650	235	5.74	517	82.3	3.22	35.6	275	126
152	4.5	—	15.8	—	3.44	0.79	1430	209	5.66	454	72.8	3.19	25.8	243	111
136	5.0	—	17.0	—	3.41	0.87	1240	186	5.58	398	64.2	3.16	18.5	214	98.0
120	5.6	—	18.5	—	3.38	0.96	1070	163	5.51	345	56.0	3.13	12.9	186	85.4
106	6.2	—	21.1	—	3.36	1.07	933	145	5.47	301	49.3	3.11	9.13	164	75.1
96	6.8	—	23.1	—	3.34	1.16	833	131	5.44	270	44.4	3.09	6.86	147	67.5
87	7.5	—	24.3	—	3.32	1.28	740	118	5.38	241	39.7	3.07	5.10	132	60.4
79	8.2	62.6	26.3	—	3.31	1.39	662	107	5.34	216	35.8	3.05	3.84	119	54.3
72	9.0	52.3	28.5	—	3.29	1.52	597	97.4	5.31	195	32.4	3.04	2.93	108	49.2
65	9.9	43.0	31.1	—	3.28	1.67	533	87.9	5.28	174	29.1	3.02	2.18	96.8	44.1
58	7.8	—	33.9	57.6	2.72	1.90	475	78.0	5.28	107	21.4	2.51	2.10	86.4	32.5
53	8.7	55.9	35.0	54.1	2.71	2.10	425	70.6	5.23	95.8	19.2	2.48	1.58	77.9	29.1
50	6.3	—	32.9	60.9	2.17	2.36	394	64.7	5.18	56.3	13.9	1.96	1.78	72.4	21.4
45	7.0	—	36.0	51.0	2.15	2.61	350	58.1	5.15	50.0	12.4	1.94	1.31	64.7	19.0
40	7.8	—	40.5	40.3	2.14	2.90	310	51.9	5.13	44.1	11.0	1.93	0.95	57.5	16.8
35	6.3	—	41.7	38.0	1.74	3.66	285	45.6	5.25	24.5	7.47	1.54	0.74	51.2	11.5
30	7.4	—	47.5	29.3	1.73	4.30	238	38.6	5.21	20.3	6.24	1.52	0.46	43.1	9.56
26	8.5	57.9	53.1	23.4	1.72	4.95	204	33.4	5.17	17.3	5.34	1.51	0.30	37.2	8.17

Table 3.13 *(Continued)*

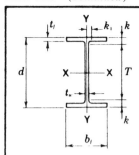

W SHAPES
Dimensions

Designation	Area A	Depth d		Web Thickness t_w		$\dfrac{t_w}{2}$	Flange Width b_f		Flange Thickness t_f		T	k	k_1
	In.²	In.		In.		In.	In.		In.		In.	In.	In.
W 10x112	32.9	11.36	11 3/8	0.755	3/4	3/8	10.415	10 3/8	1.250	1 1/4	7 5/8	1 7/8	15/16
x100	29.4	11.10	11 1/8	0.680	11/16	3/8	10.340	10 3/8	1.120	1 1/8	7 5/8	1 3/4	7/8
x 88	25.9	10.84	10 7/8	0.605	5/8	5/16	10.265	10 1/4	0.990	1	7 5/8	1 5/8	13/16
x 77	22.6	10.60	10 5/8	0.530	1/2	1/4	10.190	10 1/4	0.870	7/8	7 5/8	1 1/2	13/16
x 68	20.0	10.40	10 3/8	0.470	1/2	1/4	10.130	10 1/8	0.770	3/4	7 5/8	1 3/8	3/4
x 60	17.6	10.22	10 1/4	0.420	7/16	1/4	10.080	10 1/8	0.680	11/16	7 5/8	1 5/16	3/4
x 54	15.8	10.09	10 1/8	0.370	3/8	3/16	10.030	10	0.615	5/8	7 5/8	1 1/4	11/16
x 49	14.4	9.98	10	0.340	5/16	3/16	10.000	10	0.560	9/16	7 5/8	1 3/16	11/16
W 10x 45	13.3	10.10	10 1/8	0.350	3/8	3/16	8.020	8	0.620	5/8	7 5/8	1 1/4	11/16
x 39	11.5	9.92	9 7/8	0.315	5/16	3/16	7.985	8	0.530	1/2	7 5/8	1 1/8	11/16
x 33	9.71	9.73	9 3/4	0.290	5/16	3/16	7.960	8	0.435	7/16	7 5/8	1 1/16	11/16
W 10x 30	8.84	10.47	10 1/2	0.300	5/16	3/16	5.810	5 3/4	0.510	1/2	8 5/8	15/16	1/2
x 26	7.61	10.33	10 3/8	0.260	1/4	1/8	5.770	5 3/4	0.440	7/16	8 5/8	7/8	1/2
x 22	6.49	10.17	10 1/8	0.240	1/4	1/8	5.750	5 3/4	0.360	3/8	8 5/8	3/4	1/2
W 8x67	19.7	9.00	9	0.570	9/16	5/16	8.280	8 1/4	0.935	15/16	6 1/8	1 7/16	11/16
x58	17.1	8.75	8 3/4	0.510	1/2	1/4	8.220	8 1/4	0.810	13/16	6 1/8	1 5/16	11/16
x48	14.1	8.50	8 1/2	0.400	3/8	3/16	8.110	8 1/8	0.685	11/16	6 1/8	1 3/16	5/8
x40	11.7	8.25	8 1/4	0.360	3/8	3/16	8.070	8 1/8	0.560	9/16	6 1/8	1 1/16	5/8
x35	10.3	8.12	8 1/8	0.310	5/16	3/16	8.020	8	0.495	1/2	6 1/8	1	9/16
x31	9.13	8.00	8	0.285	5/16	3/16	7.995	8	0.435	7/16	6 1/8	15/16	9/16
W 8x28	8.25	8.06	8	0.285	5/16	3/16	6.535	6 1/2	0.465	7/16	6 1/8	15/16	9/16
x24	7.08	7.93	7 7/8	0.245	1/4	1/8	6.495	6 1/2	0.400	3/8	6 1/8	7/8	9/16
W 8x21	6.16	8.28	8 1/4	0.250	1/4	1/8	5.270	5 1/4	0.400	3/8	6 5/8	13/16	1/2
x18	5.26	8.14	8 1/8	0.230	1/4	1/8	5.250	5 1/4	0.330	5/16	6 5/8	3/4	7/16
W 8x15	4.44	8.11	8 1/8	0.245	1/4	1/8	4.015	4	0.315	5/16	6 5/8	3/4	1/2
x13	3.84	7.99	8	0.230	1/4	1/8	4.000	4	0.255	1/4	6 5/8	11/16	7/16
x10	2.96	7.89	7 7/8	0.170	3/16	1/8	3.940	4	0.205	3/16	6 5/8	5/8	7/16
W 6x25	7.34	6.38	6 3/8	0.320	5/16	3/16	6.080	6 1/8	0.455	7/16	4 3/4	13/16	7/16
x20	5.87	6.20	6 1/4	0.260	1/4	1/8	6.020	6	0.365	3/8	4 3/4	3/4	7/16
x15	4.43	5.99	6	0.230	1/4	1/8	5.990	6	0.260	1/4	4 3/4	5/8	3/8
W 6x16	4.74	6.28	6 1/4	0.260	1/4	1/8	4.030	4	0.405	3/8	4 3/4	3/4	7/16
x12	3.55	6.03	6	0.230	1/4	1/8	4.000	4	0.280	1/4	4 3/4	5/8	3/8
x 9	2.68	5.90	5 7/8	0.170	3/16	1/8	3.940	4	0.215	3/16	4 3/4	9/16	3/8

Table 3.13 (*Continued*)

W SHAPES
Properties

Nominal Wt. per Ft.	Compact Section Criteria				r_T	$\frac{d}{A_f}$	Elastic Properties						Torsional constant J	Plastic Modulus	
	$\frac{b_f}{2t_f}$	F_y'	$\frac{d}{t_w}$	F_y'''			Axis X-X			Axis Y-Y				Z_x	Z_y
							I	S	r	I	S	r			
Lb.		Ksi		Ksi	In.		In.⁴	In.³	In.	In.⁴	In.³	In.	In.⁴	In.³	In.³
112	4.2	—	15.0	—	2.88	0.87	716	126	4.66	236	45.3	2.68	15.1	147	69.2
100	4.6	—	16.3	—	2.85	0.96	623	112	4.60	207	40.0	2.65	10.9	130	61.0
88	5.2	—	17.9	—	2.83	1.07	534	98.5	4.54	179	34.8	2.63	7.53	113	53.1
77	5.9	—	20.0	—	2.80	1.20	455	85.9	4.49	154	30.1	2.60	5.11	97.6	45.9
68	6.6	—	22.1	—	2.79	1.33	394	75.7	4.44	134	26.4	2.59	3.56	85.3	40.1
60	7.4	—	24.3	—	2.77	1.49	341	66.7	4.39	116	23.0	2.57	2.48	74.6	35.0
54	8.2	63.5	27.3	—	2.75	1.64	303	60.0	4.37	103	20.6	2.56	1.82	66.6	31.3
49	8.9	53.0	29.4	—	2.74	1.78	272	54.6	4.35	93.4	18.7	2.54	1.39	60.4	28.3
45	6.5	—	28.9	—	2.18	2.03	248	49.1	4.32	53.4	13.3	2.01	1.51	54.9	20.3
39	7.5	—	31.5	—	2.16	2.34	209	42.1	4.27	45.0	11.3	1.98	0.98	46.8	17.2
33	9.1	50.5	33.6	58.7	2.14	2.81	170	35.0	4.19	36.6	9.20	1.94	0.58	38.8	14.0
30	5.7	—	34.9	54.2	1.55	3.53	170	32.4	4.38	16.7	5.75	1.37	0.62	36.6	8.84
26	6.6	—	39.7	41.8	1.54	4.07	144	27.9	4.35	14.1	4.89	1.36	0.40	31.3	7.50
22	8.0	—	42.4	36.8	1.51	4.91	118	23.2	4.27	11.4	3.97	1.33	0.24	26.0	6.10
67	4.4	—	15.8	—	2.28	1.16	272	60.4	3.72	88.6	21.4	2.12	5.06	70.2	32.7
58	5.1	—	17.2	—	2.26	1.31	228	52.0	3.65	75.1	18.3	2.10	3.34	59.8	27.9
48	5.9	—	21.3	—	2.23	1.53	184	43.3	3.61	60.9	15.0	2.08	1.96	49.0	22.9
40	7.2	—	22.9	—	2.21	1.83	146	35.5	3.53	49.1	12.2	2.04	1.12	39.8	18.5
35	8.1	64.4	26.2	—	2.20	2.05	127	31.2	3.51	42.6	10.6	2.03	0.77	34.7	16.1
31	9.2	50.0	28.1	—	2.18	2.30	110	27.5	3.47	37.1	9.27	2.02	0.54	30.4	14.1
28	7.0	—	28.3	—	1.77	2.65	98.0	24.3	3.45	21.7	6.63	1.62	0.54	27.2	10.1
24	8.1	64.1	32.4	63.0	1.76	3.05	82.8	20.9	3.42	18.3	5.63	1.61	0.35	23.2	8.57
21	6.6	—	33.1	60.2	1.41	3.93	75.3	18.2	3.49	9.77	3.71	1.26	0.28	20.4	5.69
18	8.0	—	35.4	52.7	1.39	4.70	61.9	15.2	3.43	7.97	3.04	1.23	0.17	17.0	4.66
15	6.4	—	33.1	60.3	1.03	6.41	48.0	11.8	3.29	3.41	1.70	0.876	0.14	13.6	2.67
13	7.8	—	34.7	54.7	1.01	7.83	39.6	9.91	3.21	2.73	1.37	0.843	0.09	11.4	2.15
10	9.6	45.8	46.4	30.7	0.99	9.77	30.8	7.81	3.22	2.09	1.06	0.841	0.04	8.87	1.66
25	6.7	—	19.9	—	1.66	2.31	53.4	16.7	2.70	17.1	5.61	1.52	0.46	18.9	8.56
20	8.2	62.1	23.8	—	1.64	2.82	41.4	13.4	2.66	13.3	4.41	1.50	0.24	14.9	6.72
15	11.5	31.8	26.0	—	1.61	3.85	29.1	9.72	2.56	9.32	3.11	1.46	0.10	10.8	4.75
16	5.0	—	24.2	—	1.08	3.85	32.1	10.2	2.60	4.43	2.20	0.966	0.22	11.7	3.39
12	7.1	—	26.2	—	1.05	5.38	22.1	7.31	2.49	2.99	1.50	0.918	0.09	8.30	2.32
9	9.2	50.3	34.7	54.8	1.03	6.96	16.4	5.56	2.47	2.19	1.11	0.905	0.04	6.23	1.72

Table 3.13 *(Continued)*

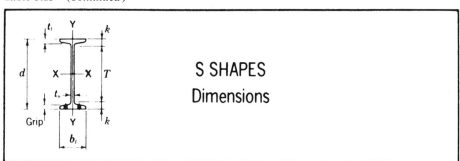

S SHAPES
Dimensions

Designation	Area A	Depth d	Web		Flange		Distance		Grip	Max. Flge. Fastener
			Thickness t_w	$\frac{t_w}{2}$	Width b_f	Thickness t_f	T	k		
	In.^{2}	In.	In.	In.	In.	In.	In.	In.	In.	In.
S 18×70	20.6	18.00	18	0.711	$^{11}/_{16}$	$^3/_8$	6.251	$6^1/_4$	0.691	$^{11}/_{16}$	15	$1^1/_2$	$^{11}/_{16}$	$^7/_8$
×54.7	16.1	18.00	18	0.461	$^7/_{16}$	$^1/_4$	6.001	6	0.691	$^{11}/_{16}$	15	$1^1/_2$	$^{11}/_{16}$	$^7/_8$
S 15×50	14.7	15.00	15	0.550	$^9/_{16}$	$^5/_{16}$	5.640	$5^5/_8$	0.622	$^5/_8$	$12^1/_4$	$1^3/_8$	$^9/_{16}$	$^3/_4$
×42.9	12.6	15.00	15	0.411	$^7/_{16}$	$^1/_4$	5.501	$5^1/_2$	0.622	$^5/_8$	$12^1/_4$	$1^3/_8$	$^9/_{16}$	$^3/_4$
S 12×50	14.7	12.00	12	0.687	$^{11}/_{16}$	$^3/_8$	5.477	$5^1/_2$	0.659	$^{11}/_{16}$	$9^1/_8$	$1^7/_{16}$	$^{11}/_{16}$	$^3/_4$
×40.8	12.0	12.00	12	0.462	$^7/_{16}$	$^1/_4$	5.252	$5^1/_4$	0.659	$^{11}/_{16}$	$9^1/_8$	$1^7/_{16}$	$^5/_8$	$^3/_4$
S 12×35	10.3	12.00	12	0.428	$^7/_{16}$	$^1/_4$	5.078	$5^1/_8$	0.544	$^9/_{16}$	$9^5/_8$	$1^3/_{16}$	$^1/_2$	$^3/_4$
×31.8	9.35	12.00	12	0.350	$^3/_8$	$^3/_{16}$	5.000	5	0.544	$^9/_{16}$	$9^5/_8$	$1^3/_{16}$	$^1/_2$	$^3/_4$
S 10×35	10.3	10.00	10	0.594	$^5/_8$	$^5/_{16}$	4.944	5	0.491	$^1/_2$	$7^3/_4$	$1^1/_8$	$^1/_2$	$^3/_4$
×25.4	7.46	10.00	10	0.311	$^5/_{16}$	$^3/_{16}$	4.661	$4^5/_8$	0.491	$^1/_2$	$7^3/_4$	$1^1/_8$	$^1/_2$	$^3/_4$
S 8×23	6.77	8.00	8	0.441	$^7/_{16}$	$^1/_4$	4.171	$4^1/_8$	0.426	$^7/_{16}$	6	1	$^7/_{16}$	$^3/_4$
×18.4	5.41	8.00	8	0.271	$^1/_4$	$^1/_8$	4.001	4	0.426	$^7/_{16}$	6	1	$^7/_{16}$	$^3/_4$
S 7×15.3	4.50	7.00	7	0.252	$^1/_4$	$^1/_8$	3.662	$3^5/_8$	0.392	$^3/_8$	$5^1/_8$	$^{15}/_{16}$	$^3/_8$	$^5/_8$
S 6×17.25	5.07	6.00	6	0.465	$^7/_{16}$	$^1/_4$	3.565	$3^5/_8$	0.359	$^3/_8$	$4^1/_4$	$^7/_8$	$^3/_8$	$^5/_8$
×12.5	3.67	6.00	6	0.232	$^1/_4$	$^1/_8$	3.332	$3^3/_8$	0.359	$^3/_8$	$4^1/_4$	$^7/_8$	$^3/_8$	—

Table 3.13 *(Continued)*

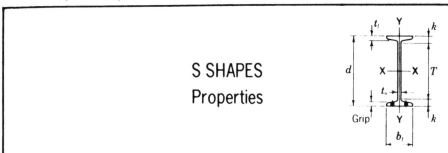

S SHAPES
Properties

Nominal Wt. per Ft	Compact Section Criteria				r_T	$\dfrac{d}{A_f}$	Elastic Properties						Torsional constant	Plastic Modulus	
	$\dfrac{b_f}{2t_f}$	F_y'	$\dfrac{d}{t_w}$	F_y'''			Axis X-X			Axis Y-Y			J	Z_x	Z_y
							I	S	r	I	S	r			
Lb.		Ksi		Ksi	In.		In.⁴	In.³	In.	In.⁴	In.³	In.	In.⁴	In.³	In.³
70	4.5	—	25.3	—	1.36	4.17	926	103	6.71	24.1	7.72	1.08	4.15	125	14.4
54.7	4.3	—	39.0	43.3	1.37	4.34	804	89.4	7.07	20.8	6.94	1.14	2.37	105	12.1
50	4.5	—	27.3	—	1.26	4.28	486	64.8	5.75	15.7	5.57	1.03	2.12	77.1	9.97
42.9	4.4	—	36.5	49.6	1.26	4.38	447	59.6	5.95	14.4	5.23	1.07	1.54	69.3	9.02
50	4.2	—	17.5	—	1.25	3.32	305	50.8	4.55	15.7	5.74	1.03	2.82	61.2	10.3
40.8	4.0	—	26.0	—	1.24	3.46	272	45.4	4.77	13.6	5.16	1.06	1.76	53.1	8.85
35	4.7	—	28.0	—	1.16	4.34	229	38.2	4.72	9.87	3.89	0.980	1.08	44.8	6.79
31.8	4.6	—	34.3	56.2	1.16	4.41	218	36.4	4.83	9.36	3.74	1.00	0.90	42.0	6.40
35	5.0	—	16.8	—	1.10	4.12	147	29.4	3.78	8.36	3.38	0.901	1.29	35.4	6.22
25.4	4.7	—	32.2	63.9	1.09	4.37	124	24.7	4.07	6.79	2.91	0.954	0.60	28.4	4.96
23	4.9	—	18.1	—	0.95	4.51	64.9	16.2	3.10	4.31	2.07	0.798	0.55	19.3	3.68
18.4	4.7	—	29.5	—	0.94	4.70	57.6	14.4	3.26	3.73	1.86	0.831	0.34	16.5	3.16
15.3	4.7	—	27.8	—	0.87	4.88	36.7	10.5	2.86	2.64	1.44	0.766	0.24	12.1	2.44
17.25	5.0	—	12.9	—	0.81	4.69	26.3	8.77	2.28	2.31	1.30	0.675	0.37	10.6	2.36
12.5	4.6	—	25.9	—	0.79	5.02	22.1	7.37	2.45	1.82	1.09	0.705	0.17	8.47	1.85

Table 3.13 *(Continued)*

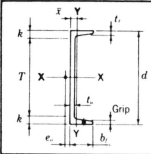

CHANNELS

AMERICAN STANDARD

Dimensions

Designation	Area A	Depth d	Web			Flange				Distance		Grip	Max. Flge. Fastener
			Thickness t_w		$\dfrac{t_w}{2}$	Width b_f		Average thickness t_f		T	k		
	In.^{2}	In.	In.		In.	In.		In.		In.	In.	In.	In.
C 12×30	8.82	12.00	0.510	½	¼	3.170	3⅛	0.501	½	9¾	1⅛	½	⅞
×25	7.35	12.00	0.387	⅜	³⁄₁₆	3.047	3	0.501	½	9¾	1⅛	½	⅞
×20.7	6.09	12.00	0.282	⁵⁄₁₆	⅛	2.942	3	0.501	½	9¾	1⅛	½	⅞
C 10×30	8.82	10.00	0.673	¹¹⁄₁₆	⁵⁄₁₆	3.033	3	0.436	⁷⁄₁₆	8	1	⁷⁄₁₆	¾
×25	7.35	10.00	0.526	½	¼	2.886	2⅞	0.436	⁷⁄₁₆	8	1	⁷⁄₁₆	¾
×20	5.88	10.00	0.379	⅜	³⁄₁₆	2.739	2¾	0.436	⁷⁄₁₆	8	1	⁷⁄₁₆	¾
×15.3	4.49	10.00	0.240	¼	⅛	2.600	2⅝	0.436	⁷⁄₁₆	8	1	⁷⁄₁₆	¾
C 9×20	5.88	9.00	0.448	⁷⁄₁₆	¼	2.648	2⅝	0.413	⁷⁄₁₆	7⅛	¹⁵⁄₁₆	⁷⁄₁₆	¾
×15	4.41	9.00	0.285	⁵⁄₁₆	⅛	2.485	2½	0.413	⁷⁄₁₆	7⅛	¹⁵⁄₁₆	⁷⁄₁₆	¾
×13.4	3.94	9.00	0.233	¼	⅛	2.433	2⅜	0.413	⁷⁄₁₆	7⅛	¹⁵⁄₁₆	⁷⁄₁₆	¾
C 8×18.75	5.51	8.00	0.487	½	¼	2.527	2½	0.390	⅜	6⅛	¹⁵⁄₁₆	⅜	¾
×13.75	4.04	8.00	0.303	⁵⁄₁₆	⅛	2.343	2⅜	0.390	⅜	6⅛	¹⁵⁄₁₆	⅜	¾
×11.5	3.38	8.00	0.220	¼	⅛	2.260	2¼	0.390	⅜	6⅛	¹⁵⁄₁₆	⅜	¾
C 6×13	3.83	6.00	0.437	⁷⁄₁₆	³⁄₁₆	2.157	2⅛	0.343	⁵⁄₁₆	4⅜	¹³⁄₁₆	⁵⁄₁₆	⅝
×10.5	3.09	6.00	0.314	⁵⁄₁₆	³⁄₁₆	2.034	2	0.343	⁵⁄₁₆	4⅜	¹³⁄₁₆	⅜	⅝
× 8.2	2.40	6.00	0.200	³⁄₁₆	⅛	1.920	1⅞	0.343	⁵⁄₁₆	4⅜	¹³⁄₁₆	⁵⁄₁₆	⅝
C 4× 7.25	2.13	4.00	0.321	⁵⁄₁₆	³⁄₁₆	1.721	1¾	0.296	⁵⁄₁₆	2⅝	¹¹⁄₁₆	⁵⁄₁₆	⅝
× 5.4	1.59	4.00	0.184	³⁄₁₆	¹⁄₁₆	1.584	1⅝	0.296	⁵⁄₁₆	2⅝	¹¹⁄₁₆	—	—
C 3× 6	1.76	3.00	0.356	⅜	³⁄₁₆	1.596	1⅝	0.273	¼	1⅝	¹¹⁄₁₆	—	—
× 5	1.47	3.00	0.258	¼	⅛	1.498	1½	0.273	¼	1⅝	¹¹⁄₁₆	—	—
× 4.1	1.21	3.00	0.170	³⁄₁₆	¹⁄₁₆	1.410	1⅜	0.273	¼	1⅝	¹¹⁄₁₆	—	—

Table 3.13   (*Continued*)

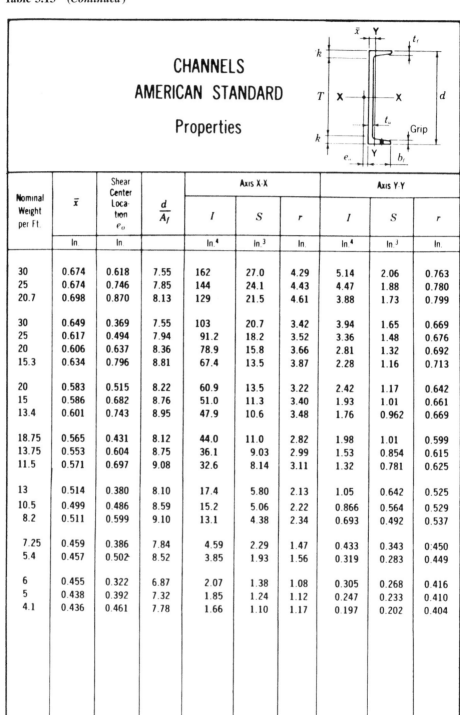

Nominal Weight per Ft.	$\bar{x}$	Shear Center Location e_o	$\dfrac{d}{A_f}$	Axis X-X			Axis Y-Y		
				I	S	r	I	S	r
	In.	In.		In.⁴	In.³	In.	In.⁴	In.³	In.
30	0.674	0.618	7.55	162	27.0	4.29	5.14	2.06	0.763
25	0.674	0.746	7.85	144	24.1	4.43	4.47	1.88	0.780
20.7	0.698	0.870	8.13	129	21.5	4.61	3.88	1.73	0.799
30	0.649	0.369	7.55	103	20.7	3.42	3.94	1.65	0.669
25	0.617	0.494	7.94	91.2	18.2	3.52	3.36	1.48	0.676
20	0.606	0.637	8.36	78.9	15.8	3.66	2.81	1.32	0.692
15.3	0.634	0.796	8.81	67.4	13.5	3.87	2.28	1.16	0.713
20	0.583	0.515	8.22	60.9	13.5	3.22	2.42	1.17	0.642
15	0.586	0.682	8.76	51.0	11.3	3.40	1.93	1.01	0.661
13.4	0.601	0.743	8.95	47.9	10.6	3.48	1.76	0.962	0.669
18.75	0.565	0.431	8.12	44.0	11.0	2.82	1.98	1.01	0.599
13.75	0.553	0.604	8.75	36.1	9.03	2.99	1.53	0.854	0.615
11.5	0.571	0.697	9.08	32.6	8.14	3.11	1.32	0.781	0.625
13	0.514	0.380	8.10	17.4	5.80	2.13	1.05	0.642	0.525
10.5	0.499	0.486	8.59	15.2	5.06	2.22	0.866	0.564	0.529
8.2	0.511	0.599	9.10	13.1	4.38	2.34	0.693	0.492	0.537
7.25	0.459	0.386	7.84	4.59	2.29	1.47	0.433	0.343	0:450
5.4	0.457	0.502	8.52	3.85	1.93	1.56	0.319	0.283	0.449
6	0.455	0.322	6.87	2.07	1.38	1.08	0.305	0.268	0.416
5	0.438	0.392	7.32	1.85	1.24	1.12	0.247	0.233	0.410
4.1	0.436	0.461	7.78	1.66	1.10	1.17	0.197	0.202	0.404

CHANNELS
AMERICAN STANDARD
Properties

Table 3.13   (*Continued*)

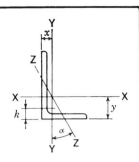

ANGLES
Equal legs and unequal legs
Properties for designing

Size and Thickness	k	Weight per Foot	Area	AXIS X-X				AXIS Y-Y				AXIS Z-Z	
				I	S	r	y	I	S	r	x	r	Tan
In.	In.	Lb.	In.^{2}	In.^{4}	In.^{3}	In.	In.	In.^{4}	In.^{3}	In.	In.	In.	α
L 6 x6 x1	1½	37.4	11.0	35.5	8.57	1.80	1.86	35.5	8.57	1.80	1.86	1.17	1.000
⅞	1⅜	33.1	9.73	31.9	7.63	1.81	1.82	31.9	7.63	1.81	1.82	1.17	1.000
¾	1¼	28.7	8.44	28.2	6.66	1.83	1.78	28.2	6.66	1.83	1.78	1.17	1.000
⅝	1⅛	24.2	7.11	24.2	5.66	1.84	1.73	24.2	5.66	1.84	1.73	1.18	1.000
½	1	19.6	5.75	19.9	4.61	1.86	1.68	19.9	4.61	1.86	1.68	1.18	1.000
⅜	⅞	14.9	4.36	15.4	3.53	1.88	1.64	15.4	3.53	1.88	1.64	1.19	1.000
L 6 x4 x 1¾	1¼	23.6	6.94	24.5	6.25	1.88	2.08	8.68	2.97	1.12	1.08	0.860	0.428
⅝	1⅛	20.0	5.86	21.1	5.31	1.90	2.03	7.52	2.54	1.13	1.03	0.864	0.435
½	1	16.2	4.75	17.4	4.33	1.91	1.99	6.27	2.08	1.15	0.987	0.870	0.440
⅜	⅞	12.3	3.61	13.5	3.32	1.93	1.94	4.90	1.60	1.17	0.941	0.877	0.446
L 6 x3½x 1 ⅜	⅞	11.7	3.42	12.9	3.24	1.94	2.04	3.34	1.23	0.988	0.787	0.767	0.350
5⁄16	13⁄16	9.8	2.87	10.9	2.73	1.95	2.01	2.85	1.04	0.996	0.763	0.772	0.352
L 4 x4 x ¾	1⅛	18.5	5.44	7.67	2.81	1.19	1.27	7.67	2.81	1.19	1.27	0.778	1.000
⅝	1	15.7	4.61	6.66	2.40	1.20	1.23	6.66	2.40	1.20	1.23	0.779	1.000
½	⅞	12.8	3.75	5.56	1.97	1.22	1.18	5.56	1.97	1.22	1.18	0.782	1.000
⅜	¾	9.8	2.86	4.36	1.52	1.23	1.14	4.36	1.52	1.23	1.14	0.788	1.000
5⁄16	11⁄16	8.2	2.40	3.71	1.29	1.24	1.12	3.71	1.29	1.24	1.12	0.791	1.000
¼	⅝	6.6	1.94	3.04	1.05	1.25	1.09	3.04	1.05	1.25	1.09	0.795	1.000
L 4 x3½x ½	15⁄16	11.9	3.50	5.32	1.94	1.23	1.25	3.79	1.52	1.04	1.00	0.722	0.750
⅜	13⁄16	9.1	2.67	4.18	1.49	1.25	1.21	2.95	1.17	1.06	0.955	0.727	0.755
5⁄16	¾	7.7	2.25	3.56	1.26	1.26	1.18	2.55	0.994	1.07	0.932	0.730	0.757
¼	11⁄16	6.2	1.81	2.91	1.03	1.27	1.16	2.09	0.808	1.07	0.909	0.734	0.759
L 4 x3 x ⅜	13⁄16	8.5	2.48	3.96	1.46	1.26	1.28	1.92	0.866	0.879	0.782	0.644	0.551
5⁄16	¾	7.2	2.09	3.38	1.23	1.27	1.26	1.65	0.734	0.887	0.759	0.647	0.554
¼	11⁄16	5.8	1.69	2.77	1.00	1.28	1.24	1.36	0.599	0.896	0.736	0.651	0.558

Angles in shaded rows may not be readily available.   Availability is subject to rolling accumulation and geographical location, and should be checked with material suppliers.

Table 3.13 *(Continued)*

ANGLES
Equal legs and unequal legs
Properties for designing

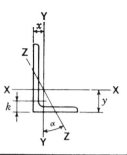

Size and Thickness	k	Weight per Foot	Area	AXIS X-X				AXIS Y-Y				AXIS Z-Z	
				I	S	r	y	I	S	r	x	r	Tan
In.	In.	Lb.	In.^{2}	In.^{4}	In.^{3}	In.	In.	In.^{4}	In.^{3}	In.	In.	In.	α
L 3 x3 x ½	13/16	9.4	2.75	2.22	1.07	0.898	0.932	2.22	1.07	0.898	0.932	0.584	1.000
3/8	11/16	7.2	2.11	1.76	0.833	0.913	0.888	1.76	0.833	0.913	0.888	0.587	1.000
5/16	5/8	6.1	1.78	1.51	0.707	0.922	0.865	1.51	0.707	0.922	0.865	0.589	1.000
1/4	9/16	4.9	1.44	1.24	0.577	0.930	0.842	1.24	0.577	0.930	0.842	0.592	1.000
3/16	1/2	3.71	1.09	0.962	0.441	0.939	0.820	0.962	0.441	0.939	0.820	0.596	1.000
L 3 x2½x 3/8	3/4	6.6	1.92	1.66	0.810	0.928	0.956	1.04	0.581	0.736	0.706	0.522	0.676
1/4	5/8	4.5	1.31	1.17	0.561	0.945	0.911	0.743	0.404	0.753	0.661	0.528	0.684
3/16	9/16	3.39	0.996	0.907	0.430	0.954	0.888	0.577	0.310	0.761	0.638	0.533	0.688
L 3 x2 x 3/8	11/16	5.9	1.73	1.53	0.781	0.940	1.04	0.543	0.371	0.559	0.539	0.430	0.428
5/16	5/8	5.0	1.46	1.32	0.664	0.948	1.02	0.470	0.317	0.567	0.516	0.432	0.435
1/4	9/16	4.1	1.19	1.09	0.542	0.957	0.993	0.392	0.260	0.574	0.493	0.435	0.440
3/16	1/2	3.07	0.902	0.842	0.415	0.966	0.970	0.307	0.200	0.583	0.470	0.439	0.446
L 2½x2½x 3/8	11/16	5.9	1.73	0.984	0.566	0.753	0.762	0.984	0.566	0.753	0.762	0.487	1.000
5/16	5/8	5.0	1.46	0.849	0.482	0.761	0.740	0.849	0.482	0.761	0.740	0.489	1.000
1/4	9/16	4.1	1.19	0.703	0.394	0.769	0.717	0.703	0.394	0.769	0.717	0.491	1.000
3/16	1/2	3.07	0.902	0.547	0.303	0.778	0.694	0.547	0.303	0.778	0.694	0.495	1.000
L 2½x2 x 3/8	11/16	5.3	1.55	0.912	0.547	0.768	0.831	0.514	0.363	0.577	0.581	0.420	0.614
5/16	5/8	4.5	1.31	0.788	0.466	0.776	0.809	0.446	0.310	0.584	0.559	0.422	0.620
1/4	9/16	3.62	1.06	0.654	0.381	0.784	0.787	0.372	0.254	0.592	0.537	0.424	0.626
3/16	1/2	2.75	0.809	0.509	0.293	0.793	0.764	0.291	0.196	0.600	0.514	0.427	0.631
L 2 x2 x 3/8	11/16	4.7	1.36	0.479	0.351	0.594	0.636	0.479	0.351	0.594	0.636	0.389	1.000
5/16	5/8	3.92	1.15	0.416	0.300	0.601	0.614	0.416	0.300	0.601	0.614	0.390	1.000
1/4	9/16	3.19	0.938	0.348	0.247	0.609	0.592	0.348	0.247	0.609	0.592	0.391	1.000
3/16	1/2	2.44	0.715	0.272	0.190	0.617	0.569	0.272	0.190	0.617	0.569	0.394	1.000
1/8	7/16	1.65	0.484	0.190	0.131	0.626	0.546	0.190	0.131	0.626	0.546	0.398	1.000

Angles in shaded rows may not be readily available. Availability is subject to rolling accumulation and geographical location, and should be checked with material suppliers.

AMERICAN INSTITUTE OF STEEL CONSTRUCTION

Source: Ref. 5. Courtesy of the American Institute of Steel Construction, Chicago.

AITC 117-93   Design Standard Specifications for
Structural Glued Laminated Timber
of Softwood Species

Being a natural product, wood will vary widely both between and within a species. Further, its structure is nonuniform, and hence loading direction, duration, and type are of major importance. Its properties will also change depending on its moisture content and similar factors. The industry standards provide for corrections to design values to account for these factors.

Wood members change less in dimension with variations in temperature than do other materials. Usually, the effects of thermal expansion are negligible compared to dimensional changes (shrinking or swelling) due to changes in moisture content, and the two changes tend to offset each other. Longitudinal dimensional changes can be neglected for most structural designs.

3.6.2 Standard Sizes

Standard sizes for finished lumber are shown in Table 3.14 and 3.15.

3.6.3 Stress-Graded Design Values

Design values for stress-graded lumber are given in Table 3.16. The allowable stresses to be used in design must, of course, conform to the requirements of the local building code. Many municipal codes are revised only infrequently and con-

Table 3.14 Standard Sizes for Finished Dry[a] Lumber

Nominal Thickness (in.)	Surfaced Thickness (in.)	Nominal Width (in.)	Surfaced Width (in.)
$\frac{3}{8}$	$\frac{5}{16}$	2	$1\frac{1}{2}$
$\frac{1}{2}$	$\frac{7}{16}$	3	$2\frac{1}{2}$
$\frac{5}{8}$	$\frac{9}{16}$	4	$3\frac{1}{2}$
$\frac{3}{4}$	$\frac{5}{8}$	5	$4\frac{1}{2}$
1	$\frac{3}{4}$	6	$5\frac{1}{2}$
$1\frac{1}{4}$	1	7	$6\frac{1}{2}$
$1\frac{1}{2}$	$1\frac{1}{4}$	8 and wider $\frac{3}{4}$ off nominal	
$1\frac{3}{4}$	$1\frac{3}{8}$		
2	$1\frac{1}{2}$		
$2\frac{1}{2}$	2		
3	$2\frac{1}{2}$		
$3\frac{1}{2}$	3		
4	$3\frac{1}{2}$		

Standard lengths of finish are 3 ft and longer in multiples of 1 ft. In the Superior grade, 3% of 3 and 4 ft and 7% of 3 to 6 ft are permitted. In the Prime grade, 20% of 3 to 6 ft is permitted.

Source: H. Parker and H. D. Hauf, *Simplified Design of Structural Wood,* 4th ed., John Wiley & Sons, Inc., New York, 1979.

[a]Dry is defined as having a moisture content (by weight) of 19% or less.

Table 3.15 Properties of Structural Lumber: Standard Drressed Sizes[a]

Sizes

Nominal size (in.) b d	Dressed size (in.) b d	Area of section (sq in.) A	Moment of inertia (in.^{4}) I	Section modulus (in.^{3}) S	Weight per linear foot†
2 × 4	$1\frac{1}{2} \times 3\frac{1}{2}$	5.250	5.359	3.063	1.458
2 × 6	$1\frac{1}{2} \times 5\frac{1}{2}$	8.250	20.797	7.563	2.292
2 × 8	$1\frac{1}{2} \times 7\frac{1}{4}$	10.875	47.635	13.141	3.021
2 × 10	$1\frac{1}{2} \times 9\frac{1}{4}$	13.875	98.932	21.391	3.854
2 × 12	$1\frac{1}{2} \times 11\frac{1}{4}$	16.875	177.979	31.641	4.688
2 × 14	$1\frac{1}{2} \times 13\frac{1}{4}$	19.875	290.775	43.891	5.521
3 × 2	$2\frac{1}{2} \times 1\frac{1}{2}$	3.750	0.703	0.938	1.042
3 × 4	$2\frac{1}{2} \times 3\frac{1}{2}$	8.750	8.932	5.104	2.431
3 × 6	$2\frac{1}{2} \times 5\frac{1}{2}$	13.750	34.661	12.604	3.819
3 × 8	$2\frac{1}{2} \times 7\frac{1}{4}$	18.125	79.391	21.901	5.035
3 × 10	$2\frac{1}{2} \times 9\frac{1}{4}$	23.125	164.886	35.651	6.424
3 × 12	$2\frac{1}{2} \times 11\frac{1}{4}$	28.125	296.631	52.734	7.813
3 × 14	$2\frac{1}{2} \times 13\frac{1}{4}$	33.125	484.625	73.151	9.201
3 × 16	$2\frac{1}{2} \times 15\frac{1}{4}$	38.125	738.870	96.901	10.590
4 × 2	$3\frac{1}{2} \times 1\frac{1}{2}$	5.250	0.984	1.313	1.458
4 × 3	$3\frac{1}{2} \times 2\frac{1}{2}$	8.750	4.557	3.646	2.431
4 × 4	$3\frac{1}{2} \times 3\frac{1}{2}$	12.250	12.505	7.146	3.403
4 × 6	$3\frac{1}{2} \times 5\frac{1}{2}$	19.250	48.526	17.646	5.347
4 × 8	$3\frac{1}{2} \times 7\frac{1}{4}$	25.375	111.148	30.661	7.049
4 × 10	$3\frac{1}{2} \times 9\frac{1}{4}$	32.375	230.840	49.911	8.933
4 × 12	$3\frac{1}{2} \times 11\frac{1}{4}$	39.375	415.283	73.828	10.938
4 × 14	$3\frac{1}{2} \times 13\frac{1}{4}$	46.375	678.475	102.411	12.877
4 × 16	$3\frac{1}{2} \times 15\frac{1}{4}$	53.375	1,034.418	135.66	14.828
6 × 2	$5\frac{1}{2} \times 1\frac{1}{2}$	8.250	1.547	2.063	2.292
6 × 3	$5\frac{1}{2} \times 2\frac{1}{2}$	13.750	7.161	5.729	3.819
6 × 4	$5\frac{1}{2} \times 3\frac{1}{2}$	19.250	19.651	11.229	5.347
6 × 6	$5\frac{1}{2} \times 5\frac{1}{2}$	30.250	76.255	27.729	8.403
6 × 8	$5\frac{1}{2} \times 7\frac{1}{2}$	41.250	193.359	51.563	11.458
6 × 10	$5\frac{1}{2} \times 9\frac{1}{2}$	52.250	392.963	82.729	14.514
6 × 12	$5\frac{1}{2} \times 11\frac{1}{2}$	63.250	697.068	121.229	17.569
6 × 14	$5\frac{1}{2} \times 13\frac{1}{2}$	74.250	1,127.672	167.063	20.625
6 × 16	$5\frac{1}{2} \times 15\frac{1}{2}$	85.250	1,706.776	220.229	23.681

sequently may not be in agreement with current industry-recommended stress levels. The design values shown are those given in the *National Design Specification for Wood Construction*[7] and recommended by the American Forest & Paper Association.

Machine-stress-rated (MSR) **lumber** is also available. In the process, each piece of lumber is machine tested (nondestructively) to establish its modulus of elasticity. This lumber is also required to meet certain visual grading requirements.

Table 3.15 (*Continued*)

Sizes (*continued*)

Nominal size (in.) b d	Dressed size (in.) b d	Area of section (sq in.) A	Moment of inertia (in.^{4}) I	Section modulus (in.^{3}) S	Weight per linear foot[a]
8 × 2	$7\frac{1}{4} \times 1\frac{1}{2}$	10.875	2.039	2.719	3.021
8 × 3	$7\frac{1}{4} \times 2\frac{1}{2}$	18.125	9.440	7.552	5.035
8 × 4	$7\frac{1}{4} \times 3\frac{1}{2}$	25.375	25.904	14.802	7.049
8 × 6	$7\frac{1}{2} \times 5\frac{1}{2}$	41.250	103.984	37.813	11.458
8 × 8	$7\frac{1}{2} \times 7\frac{1}{2}$	56.250	263.672	70.313	15.625
8 × 10	$7\frac{1}{2} \times 9\frac{1}{2}$	71.250	535.859	112.813	19.792
8 × 12	$7\frac{1}{2} \times 11\frac{1}{2}$	86.250	950.547	165.313	23.958
8 × 14	$7\frac{1}{2} \times 13\frac{1}{2}$	101.250	1,537.734	227.813	28.125
8 × 16	$7\frac{1}{2} \times 15\frac{1}{2}$	116.250	2,327.422	300.313	32.292
10 × 2	$9\frac{1}{4} \times 1\frac{1}{2}$	13.875	2.602	3.469	3.854
10 × 3	$9\frac{1}{4} \times 2\frac{1}{2}$	23.125	12.044	9.635	6.424
10 × 4	$9\frac{1}{4} \times 3\frac{1}{2}$	32.375	33.049	18.885	8.993
10 × 6	$9\frac{1}{2} \times 5\frac{1}{2}$	52.250	131.714	47.896	14.514
10 × 8	$9\frac{1}{2} \times 7\frac{1}{2}$	71.250	333.984	89.063	19.792
10 × 10	$9\frac{1}{2} \times 9\frac{1}{2}$	90.250	678.755	142.896	25.069
10 × 12	$9\frac{1}{2} \times 11\frac{1}{2}$	109.250	1,204.026	209.396	30.347
10 × 14	$9\frac{1}{2} \times 13\frac{1}{2}$	128.250	1,947.797	288.563	35.625
10 × 16	$9\frac{1}{2} \times 15\frac{1}{2}$	147.250	2,948.068	380.396	40.903
10 × 18	$9\frac{1}{2} \times 17\frac{1}{2}$	166.250	4,242.836	484.896	46.181
12 × 2	$11\frac{1}{4} \times 1\frac{1}{2}$	16.875	3.164	4.219	4.688
12 × 3	$11\frac{1}{4} \times 2\frac{1}{2}$	28.125	14.648	11.719	7.813
12 × 4	$11\frac{1}{4} \times 3\frac{1}{2}$	39.375	40.195	22.969	10.938
12 × 6	$11\frac{1}{2} \times 5\frac{1}{2}$	63.250	159.443	57.979	17.569
12 × 8	$11\frac{1}{2} \times 7\frac{1}{2}$	86.250	404.297	107.813	23.958
12 × 10	$11\frac{1}{2} \times 9\frac{1}{2}$	109.250	821.651	172.979	30.347
12 × 12	$11\frac{1}{2} \times 11\frac{1}{2}$	132.250	1,457.505	253.479	36.736
12 × 14	$11\frac{1}{2} \times 13\frac{1}{2}$	155.250	2,357.859	349.313	43.125
12 × 16	$11\frac{1}{2} \times 15\frac{1}{2}$	178.250	3,568.713	460.479	49.514
14 × 16	$13\frac{1}{2} \times 15\frac{1}{2}$	209.250	4,189.359	540.563	58.125
14 × 18	$13\frac{1}{2} \times 17\frac{1}{2}$	236.250	6,029.297	689.063	65.625
14 × 20	$13\frac{1}{2} \times 19\frac{1}{2}$	263.250	8,341.734	855.563	73.125
14 × 22	$13\frac{1}{2} \times 21\frac{1}{2}$	290.250	11,180.672	1,040.063	80.625

Source: Compiled from data in the 1982 edition of the *National Design Specification for Wood Construction*. Courtesy of the National Forest Products Association.

[a] Based on an assumed average weight of 40 lb/ft^3.

Table 3.16 Selected Stress Values for Timber

Commercial Grade	Size Classification	Bending F_b	Tension Parallel to Grain F_t	Shear Parallel to Grain F_v	Compression Perpendicular to Grain $F_{c\perp}$	Compression Parallel to Grain F_c	Modulus of Elasticity E	Grading Rules Agency
Douglas Fir-Larch								
Select Structural	2–4 in. thick	1450	1000	95	625	1700	1,900,000	
No. 1 and better		1150	775	95	625	1500	1,800,000	
No. 1		1000	675	95	625	1450	1,700,000	WCLIB
No. 2	2 in. and wider	875	575	95	625	1300	1,600,000	WWPA
No. 3		500	325	95	625	750	1,400,000	
Stud		675	450	95	625	825	1,400,000	
Construction	2–4 in. thick	1000	650	95	625	1600	1,500,000	
Standard		550	375	95	625	1350	1,400,000	
Utility	2–4 in. wide	275	175	95	625	875	1,300,000	
Redwood								
Clear Structural	2–4 in. thick	1750	1000	145	650	1850	1,400,000	
Select Structural		1350	800	80	650	1500	1,400,000	
Select Structural, open grain		1100	625	80	425	1100	1,100,000	
No. 1	2–4 in. thick	975	575	80	650	1200	1,300,000	
No. 1, open grain		775	450	80	425	900	1,100,000	RIS
No. 2	2 in. and wider	925	525	80	650	950	1,200,000	
No. 2, open grain		725	425	80	425	700	1,000,000	
No. 3		525	300	80	650	550	1,100,000	
No. 3, open grain		425	250	80	425	400	900,000	
Stud		575	325	80	425	450	900,000	
Construction	2–4 in. thick	825	475	80	425	925	900,000	

Mixed Southern Pine

Grade	Size						SPIB
Select Structural		2050	1200	100	565	1800	1,600,000
No. 1	2–4 in. thick	1450	875	100	565	1650	1,500,000
No. 2		1300	775	90	565	1650	1,400,000
No. 3	2–4 in. thick	750	450	90	565	950	1,200,000
Stud		775	450	90	565	950	1,200,000
Construction	2–4 in. thick	1000	600	100	565	1700	1,300,000
Standard		550	325	90	565	1450	1,200,000
Utility	4 in. wide	275	150	90	565	950	1,100,000

Source: Courtesy of the American Forest & Paper Association, Washington, D.C.

Of great importance is the *National Design Specification* (NDS) for *Wood Construction* promulgated by the American Forest & Paper Association (formerly, National Forest Products Association). This standard provides detailed data on design values of stress for various timber species in various grades and types of loading. In addition, it provides a series of correction factors to be used depending on such things as size, flat use, and wet service. It should be used in conjunction with the AITC *Manual*. Selected portions of the NDS specification are shown in Table 3.16.

Where more general designs are being developed and less precision in stress values is necessary, the *Wood Handbook* published by the U.S. Department of Agriculture, Forest Products Laboratory, Madison, Wisconsin, provides convenient design information.

3.6.4   Glulams[8]

Wooden structural members in large sizes have become increasingly costly and difficult to locate. As a result, structural glued laminated timber members (**glulams**) have been developed and have made possible the production of structural timbers in a wide variety of sizes and shapes. Glulams are used as load-carrying structural framing for roofs and other structural portions of buildings, and for other construction, such as bridges, towers, and marine installations. The term *structural glued laminated timber* refers to an engineered, stress-rated product of a timber laminating plant, comprising assemblies of suitably selected and prepared wood laminations securely bonded together with adhesives. The grain of all laminations is approximately parallel longitudinally. The individual laminations do not exceed 2 in. (50 mm) in net thickness. Laminations may be comprised of pieces end-joined to form any lengths, pieces placed or glued edge to edge to make wider ones, or pieces bent to a curved form during gluing.

Adhesives for the manufacture of glulam must comply with Voluntary Product Standard PS 56-73. Wet-use adhesives must be used if the members are subject to occasional or continuous wetting, or for applications, either exterior or interior, where the moisture content of the wood will exceed 16%.

Timber construction has historically been recognized as an economical type of construction. Laminated wood does not require the added expense of false ceilings to cover or disguise the structural framework. Glulam members can be used to provide long clear spans eliminating interior walls and supports.

Since wood is relatively inert chemically, under normal conditions it is not subject to chemical change or deterioration. It is resistant to most acids, rust, and other corrosive agents. Thus, it is often used where chemical deterioration eliminates use of other structural materials.

Heavy timber sizes used in glulam construction are difficult to ignite. Glulam burns slowly and resists heat penetration by the formation of self-insulating char. In a large member subjected to fire, the uncharged inner portion maintains its strength. Glulams are available in a variety of appearance grades and finishes. Stress grades can be specified and range from 1600 to 2400 psi (11.2 to 16.6 MPa), extreme fiber tension when loaded in bending, with compression values, parallel to grain, from 1050 to 1700 psi (7.2 to 11.7 MPa). Sizes of glulams most readily available range from 3⅛ to 10¾ in. (79 to 273 mm) in width with depths from 4½ to 60 in. (114 mm to 1.5 m). Precalculated capacity tables and additional data are available from the American Institute of Timber Construction.

3.6.5 Timber Columns[9]

The **maximum allowable unit compressive stress** (F_c') in axially loaded, square or rectangular, simple solid columns can be computed based on column length classification:

$$\text{Short columns:} \qquad F_c' = F_c, \qquad (3.23)$$

$$\text{Intermediate columns:} \quad F_c' = F_c\left[1 - \frac{1}{3}\left(\frac{l/d}{K}\right)^4\right], \qquad (3.24)$$

$$\text{Long columns:} \qquad F_c' = \frac{0.30E}{(l/d)^2}, \qquad (3.25)$$

where l/d = slenderness ratio (unbraced height/least side, both in inches), E = modulus of elasticity, F_c' = allowable unit stress in compression parallel to grain, adjusted for l/d ratio (above), with the limiting magnitude of F_c' equal to F_c, the design value for compression parallel to grain for the species and grade of lumber used.

Short columns have an l/d ratio of 11 or less. Intermediate columns have an l/d ratio greater than 11 but less than K, where $K = 0.671(E/F_c)^{1/2}$. Long columns have an l/d ratio of K or greater.

Values of K for Selected Values of E (psi) and F_c (psi)

| E = 1,100,000 | | E = 1,300,000 | | E = 1,400,000 | |
F_c	K	F_c	K	F_c	K
625	28.15	1050	23.61	625	31.76
700	26.60	1200	22.09	725	29.49
900	23.46	1650	18.83		

| E = 1,500,000 | | E = 1,600,000 | | E = 1,700,000 | |
F_c	K	F_c	K	F_c	K
775	29.52	925	27.91	1200	25.26
		1000	26.84	1350	23.81
		1100	25.59		
		1150	25.03		
		1300	23.54		

Source: H. Parker and H. D. Hauf, *Simplified Design of Structural Wood,* 4th ed., John Wiley & Sons, Inc., New York, 1979.

These formulas apply to square-end, simple solid columns as well as to the pin-end condition from which they were derived. They are appropriate for wood columns subjected to normal loading and used in dry locations.

3.6.6   Connections

Timber connections are achieved by either mechanical fasteners or gluing. Mechanical fasteners are normally used for both factory-fabricated and field connections, while glued systems normally are used for factory or prefabricated joints. Glued systems depend on a series of particular characteristics to achieve a satisfactory joint, often including proprietary compounding of the adhesive material and the jointing method. For glued applications, the engineer should consult adhesive manufacturers.

For mechanical fasteners a wide body of knowledge has been developed, often reflected in codes and standards. In general, mechanical fasteners include nails, staples, screws, metal connector plates, bolts or dowels, lag screws, and split-ring or shear plate connectors. Connections are rated by load direction (lateral or withdrawal), wood species, wood moisture content, geometry of the joint, number and pattern of multiple fasteners, and similar factors. For commonly used connections, many building codes include standard nailing requirements, usually in tabular form. Even with lightly loaded joints care should be taken to avoid introducing eccentric loading or moments which may split the wood or open the joint. For more complex or heavily loaded joints, including such subjects as timber bridges and timber piling, the AITC *Timber Construction Manual*[6] contains extensive design information and methodology.

3.6.7   Preservation

Since wood is relatively chemically inert, under conditions where its internal moisture content is less than 20%, it has a virtually indefinite life and is not subject to chemical change or rapid deterioration. It is resistant to most acids and other corrosive agents. Preservatives are available that extend the life of timber when used in buried, intermittently wet, or submerged environments. The most frequently used preservatives are the following:

- Creosote and creosote/coal tar compounds, usually pressure impregnated, are suitable for the most severe conditions but have a dark oily appearance. They stain other materials, such as painted surfaces, plaster, wallboard, and so on. To avoid this, roofing felt is often used to cover the creosoted surfaces.
- Oil preservatives are classified into three commonly used types, depending on the hydrocarbon carrier, A, C, or D. The active preservation ingredient is usually pentachlorophenol or, less frequently, copper napthenate. Type A treatment uses a petroleum distillate or a blend of them with cosolvents, type C uses light hydrocarbons, and type D uses cholorinated hydrocarbons with methylene chloride. Of these, type C has the least effect on the surface appearance and effect, and later painting and staining can readily be accomplished. Types A and D affect the surface appearance, and later painting or staining is difficult (see AWPA Standard P9).
- Water-based preservatives are also widely used. (AWPA Standard P5 describes these in more detail.) Generally, these are copper based, which leaves a greenish cast on the treated material. Other types leave either a gray-green or brown color, depending on the chemical used.

3.7 FABRICATION

3.7.1 Welding Processes

As a method of joining metals, **welding** offers the opportunity to achieve a more efficient use of the materials and faster fabrication and erection. Welding also permits the designer to develop and use new and aesthetically appealing designs and saves weight because connecting plates are not needed and allowances need not be made for reduced load-carrying ability due to holes for bolts, rivets, and so on.

The welding process joins two pieces of metal together by establishing a metallurgical bond between them. There are many different types of processes, most of which use a fusion technique. The two most widely used are **arc welding** and **gas welding.**

Arc Welding

The arc welding process obtains the intense heat needed to liquefy the metals to be joined from an electric arc that is developed between the workpiece to be welded and the electrode. The electrode may be either consumable or nonconsumable. An arc temperature of approximately 6500°F (3600°C) is created as a result of a continual electrical discharge between the workpiece and the electrode, which in the case of a consumable electrode, melts along with the surface of the workpiece. A weld bead is formed by the deposition of metal from the electrode by slowly moving the electrode along the workpiece, generally with a weaving motion. This process is repeated in layers, as necessary, with intermediate removal of surface slag after each welding pass, to create the necessary size weld (or thickness). The slag formed results from melting of the specially designed electrode coating or flux which **scavenges weld impurities** and produces an **inert gas cover** over the liquid weld metal to avoid its contamination and weakening by contacting the air. The arc welding process requires a continuous supply of electrical current having adequate voltage and amperage to maintain the arc, and may be either ac or dc. Typical voltage ranges are from 20 to 80 with amperage ranging from 50 to 500.

Principal Arc Welding Processes

Shielded Metal Arc. This process uses hand-held, "stick"-type, coated electrodes. Advantages of this process are minimal cost, flexibility, and ease of use. The electrode coating is the flux. Weld quality is largely dependent on the skill of the operator.

Submerged Arc. A machine-controlled electrode (wire feed and travel speed) is submerged in a blanket of granular flux. High deposition rates and deep penetration are characteristic with this process, making it economical for large welds. Joint must be flat or horizontal, and thus welding positioners may be necessary. The final weld surface is generally very smooth and blends well with the base metal.

Flux-Cored Arc.   This process typically uses a hand-held wire gun with a continuous feed wire containing a flux internal to the wire (a hollow wire filled with a flux material). Higher deposition rates possible with the flux-cored arc process permit increased tolerance for poor joint fit. Wind shelters are not generally required, which permits increased ventilation around operators and the work.

Gas-Shielded Arc.   This process is similar to the self-shielded flux-cored process except that shielding is accomplished by an external (annular) gas stream. Typical shielding gases are **CO_2, helium,** or **argon. Gas metal arc** (MIG), or **gas tungsten arc** (TIG) welding are two popular processes using this technique. MIG filler metal is fed through the center of a welding gun wire guide. The TIG process creates the arc using a tungsten electrode that is not consumed with the desired bare filler metal rod separately added to the molten weld metal pool under the inert gas blanket. The TIG process may also be used where additional filler metal is not required. These processes are very flexible and are suitable for alloy or thin-section work.

Other Frequently Used Arc Welding Processes

Electroslag Welding.   Used for welding very heavy and thick metallurgical sections.

Plasma Arc Welding.   Sometimes called *plasma jet welding*. Frequently used for metal overlaying and may also be used for cutting.

Electron Beam Welding.   Used for deep penetration welds requiring minimal distortion and metallurgical structure damage, usually accomplished in a vacuum chamber.

Resistance Welding.   Spot, seam, and roll-spot welds created by passing a localized current through parts to be joined, with the parts held together under pressure by electrodes. Weld results from resistance heating of a localized area.

Stud Welding.   Welding of a metal stud to another part by creating an arc between the two which produces a molten puddle. The stud is forced into the molten metal which is then permitted to solidify.

Gas Welding.   Gas welding is a process whereby metal surfaces to be joined are melted, using a fuel gas–oxygen flame, and caused to flow together without the application of pressure to the parts being joined. A filler metal may or may not be used. The most widely used source of heat is a combination of **oxygen** and **acetylene gases,** which are mixed in an oxyacetylene welding torch. The flame temperature of this mixture reaches approximately 5600°F (3100°C). Other fuel gases such as **propane, butane,** and **natural gas** can be used for welding and soldering of nonferrous metals, but these fuels do not provide adequate heat or proper inert atmosphere for welding of ferrous metals.

The gases used are stored in separate heavy metal cylinders, withdrawal being regulated by a pressure regulator on each cylinder. The gas passes through flexible rubber hoses to a welding torch where the combination of gases can be adjusted to obtain the desired characteristics.

The welder has considerable control over the temperature of the metal in the weld zone when using the gas welding technique. Deposition rates of weld filler metal are easily controlled because the source of heat and source of filler metal are separate. Heat can be applied to either the base metal or the filler metal while both are still within the flame envelope. These characteristics make gas welding well suited for joining thin metal sections and where fit-up is poor. Gas welding is not as economic a method of joining heavy sections as arc welding. Equipment required to perform gas welding is relatively inexpensive and portable. It can be used to preheat, postheat, weld, braze, and may be converted for oxygen cutting. The gas welding process is well adapted to short production runs and field repairs and alterations.

Distortion

Severe physical distortion can result from the numerous variables of the welding process and procedures used and from the physical design, sequence of welding operations, inadequate fixtures, improper welding parameters, and so on. Control of distortion as a result of shrinkage becomes extremely important. This is particularly true where welded structures consist of material having widely varying thicknesses because more massive weldments generally have a greater shrinkage than smaller weldments. The susceptibility to distortion increases, with welds that are not symmetrical, about the neutral axis of the section, where materials of greatly different thickness are welded together or where welds lie in different planes. Further, weld distortion may be increased where weld-deposited cross sections vary widely or where the rate of deposition of weld metal varies widely (e.g., where material is deposited at very high rates). The actual design of welded joints to reduce distortion is beyond the scope of this book, but references at the end of this chapter provide more specific guidance. If the sequence of welding is not carefully chosen, there may be cracking induced in initially deposited welds from subsequently deposited heavier welds. Further, if the welding sequence is not considered during the design of the weldment and not planned for in the course of fabrication, distortions in the positioning, angularity, and so on, of the various members making up the weldment will likely occur. For this reason, large carbon steel weldments are frequently **stress relieved** after welding using temperatures of 1100 to 1200°F (590 to 650°C) for a period of time approaching 1 hour per inch (25 mm) of thickness of the heaviest section, up to a maximum of 8 hours. This stress relief substantially reduces the amount of residual stress remaining in the welds and provides a way in which the welded structure can relax the thermally (weld-) induced stresses and provides a stable structure suitable for final machining.

3.7.2 Weld Strength

The strength of a welded structure is dependent on many factors, including: the load-carrying ability of the base metal, design of the weld joint and amount and configuration of deposited weld metal, type of weld metal, the variables of the individual welding procedures, and residual—after-welding—stresses. The load-carrying ability

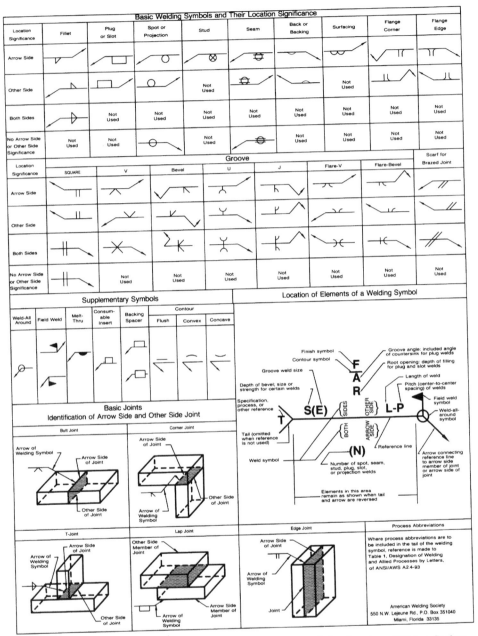

Fig. 3.9 Standard welding symbols. (From Ref. 11. Courtesy of the American Welding Society, Miami, Fla.)

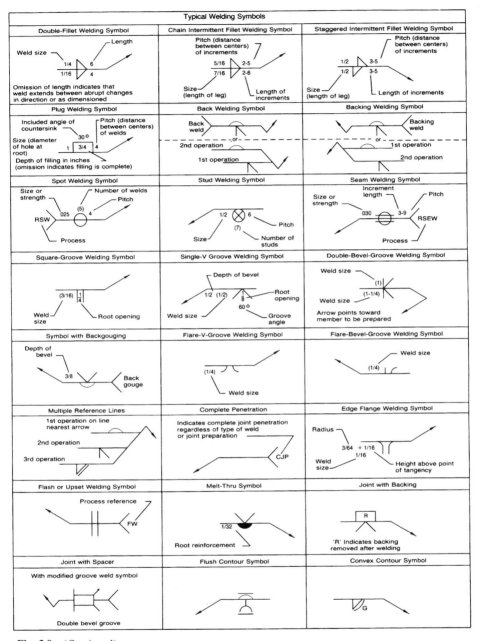

Fig. 3.9 (*Continued*)

can be related to the weld strength by knowing the type of stresses involved and the weld metal cross-sectional area that is carrying the load. Assurance that welds perform as desired is obtained by establishing welding procedures that define all essential variables, performing procedure qualification tests including destructive testing of the weld sample joints, assuring that welding of critical structures and components is accomplished by qualified welders, and, when welding is to be performed in accordance with code (e.g., ASME, AWS) requirements, verify that a quality control program is in effect to assure that code requirements are met.

The weld metal is as strong as and generally stronger than the adjacent base metal being welded. Thus, strength calculations of full penetration butt welds generally are not necessary, the critical requirements being selection of a weld filler material compatible with and as strong as the base metal, and selection of the proper weld procedure. In application, fillet welds required may vary from 80% to less than 50%, the size of the thickness of the plate depending on whether the design is based on rigidity or actual weld strength.

Allowable Shear and Unit Forces

Table 3.17 presents the allowable shear values for various weld-metal strength levels and the more common fillet weld sizes. These values are for equal-leg fillet welds where the effective throat (t_c) equals $0.707 \times$ leg size (ω):

$$F = 0.707\omega \times \tau, \tag{3.26}$$

where $\tau = 0.30 \, (EXX)$.

Table 3.17   Allowable Load for Various Sizes of Fillet Welds

	Strength Level of Weld Metal (EXX)					
	60	70	80	90	100	110
	Allowable Shear Stress on Throat of Fillet Weld or Partial-Penetration Groove Weld (1000 psi)					
$\tau =$	18.0	21.0	24.0	27.0	30.0	33.0
	Allowable Unit Force on Fillet Weld (1000 psi/linear in.)					
$f =$	12.73ω	14.85ω	16.97ω	19.09ω	21.21ω	23.33ω
Leg Size ω (in.)	Allowable Unit Force for Various Sizes of Fillet Welds (1000 lbs/linear in.)					
1	12.73	14.85	16.97	19.09	21.21	23.33
7/8	11.14	12.99	14.85	16.70	18.57	20.41
3/4	9.55	11.14	12.73	14.32	15.92	17.50
5/8	7.96	9.28	10.61	11.93	13.27	14.58
1/2	6.37	7.42	8.48	9.54	10.61	11.67
7/16	5.57	6.50	7.42	8.35	9.28	10.21
3/8	4.77	5.57	6.36	7.16	7.95	8.75
5/16	3.98	4.64	5.30	5.97	6.63	7.29
1/4	3.18	3.71	4.24	4.77	5.30	5.83
3/16	2.39	2.78	3.18	3.58	3.98	4.38
1/8	1.59	1.86	2.12	2.39	2.65	2.92
1/16	.795	.930	1.06	1.19	1.33	1.46

Source: Ref. 10. Courtesy of the Lincoln Electric Company.

3.7.3 Riveted Joints

The **shearing strength** (r_s) of a rivet d inches in diameter, with an allowable stress in shear of f_s pounds/inch²:

$$r_s = \frac{\pi d^2}{4} f_s \quad \text{pounds.} \tag{3.27}$$

The **bearing strength** (r_b) of a rivet d inches in diameter, with an allowable stress in bearing of f_b pounds/inch², against a plate t inches in thickness:

$$r_b = dt f_b \quad \text{pounds.} \tag{3.28}$$

The **total stress** (r) on each of n rivets resisting a pull or thrust of P pounds:

$$r = \frac{P}{N} \quad \text{pounds.} \tag{3.29}$$

The **total stress** (r_m) on the most stressed rivet of a group of rivets resisting the action of a couple of M inch-pounds, if y is the distance in inches from the center of gravity of the group of rivets to the outermost rivet and Σy^2 is the sum of the squares of the distances from the center of gravity of the group to each of the rivets (Fig. 3.10):

$$r_m = \frac{My}{\Sigma y^2} \quad \text{pounds.} \tag{3.30}$$

The **resistance to moment** (M) of a group of rivets, if the distance of the outermost rivet from the center of gravity of the group is y inches and the sum of the squares of the distance from the center of gravity of the group to each of the rivets is Σy^2 and r is the total allowable stress on a rivet:

$$M = \frac{r \Sigma y^2}{y} \quad \text{inch-pounds.} \tag{3.31}$$

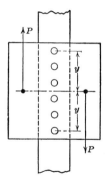

Fig. 3.10   Total stress (couple).

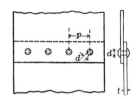

Fig. 3.11   Resistance to tearing.

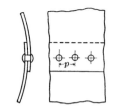

Fig. 3.12   Single-riveted lap joint.

The **resistance to tearing** (T) between rivets, of a plate t inches in thickness in which rivets of d inches diameter are placed with p inches pitch, if the allowable intensity of stress of the plate in tension is f_1 pounds/inch^2:

$$T = t(p - d)f, \quad \text{pounds.} \tag{3.32}$$

The **efficiency of a riveted joint** is the ratio of the least strength of the joint to the tensile strength of the solid plate.

Single-Riveted Lap Joint

$$\text{Shearing one rivet} = \frac{\pi d^2}{4} f_s.$$

$$\text{Tearing plate between rivets} = (p - d)tf_t. \tag{3.33}$$

$$\text{Crushing of rivet or plate} = dtf_b.$$

where f_s is the allowable shearing stress in pounds/inch^2, f_b is the allowable bearing stress in pounds/inch^2, f_t is the allowable tension stress in pounds/inch^2, d is the diameter of rivet in inches, and t is the thickness of plate in inches.

Double-Riveted Lap Joint

$$\text{Shearing two rivets} = \frac{2\pi d^2}{4} f_s.$$

$$\text{Tearing between two rivets} = (p - d)tf_t. \tag{3.34}$$

$$\text{Crushing in front of rivets} = 2dtf_b.$$

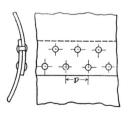

Fig. 3.13   Double-riveted lap joint.

REFERENCES

1. *Notes on ACI 318-95 Building Code Requirements for Structural Concrete,* Portland Cement Association, Skokie, Il., 1996.

2. R. Park and T. Paulay, *Reinforced Concrete Structures,* John Wiley & Sons, Inc., New York, 1975.

3. Wire Reinforcement Institute, Findley, Ohio, private communication, 1997.

4. J. C. McCormac, *Structural Steel Design,* 3rd ed., Harper & Row, Publishers, Inc., New York, 1981.

5. *Manual of Steel Construction,* 9th ed. American Institute of Steel Construction, Chicago, 1989.

6. *Timber Construction Manual,* 4th ed., American Institute of Timber Construction, Englewood, Colo., 1994.

7. *National Design Specification for Wood Construction,* American Forest & Paper Association, Washington, D.C., 1991.

8. *Glulam Systems,* American Institute of Timber Construction, Englewood, Colo., 1980.

9. H. Parker and H. D. Hauf, *Simplified Design of Structural Wood,* 4th ed., John Wiley & Sons, Inc., New York, 1979.

10. *Procedure Handbook of Arc Welding,* 12th ed., Lincoln Electric Company, Cleveland, Ohio, 1973.

11. *American Welding Society Standard Welding Symbols,* American Welding Society, Miami, Fla., 1996.

CHAPTER 4
FLUID MECHANICS

4.1 HYDROSTATICS

The **pressure** (p) due to a head of h feet in a liquid weighing w pounds/foot³:

$$p = \frac{wh}{144} \quad \text{pounds/inch}^2. \tag{4.1}$$

NOTE. In water, the pressure corresponding to a head of h feet is $0.434h$ lb/in² (3 kPa).

The **head** (h) corresponding to a pressure of p pounds/foot² in a liquid weighing w pounds/foot³:

$$h = \frac{p}{w} \quad \text{feet}. \tag{4.2}$$

NOTE. In water, the head corresponding to pressure of p pounds/inch² is $2.3p$ feet.

The **total normal pressure** (P) on a plane or curved surface A square feet in area immersed in a liquid weighing w pounds/foot³ with a head of h_0 feet on its center of gravity:

$$P = wAh_0 \quad \text{pounds}. \tag{4.3}$$

NOTE. The total pressure on a plane surface may be represented by a resultant force of P pounds acting normally to the area at its center of pressure.

The **component of normal pressure** (P_c) on a plane area of A square feet with h_0 feet head on its center of gravity and a projection of A_c square feet on a plane perpendicular to the component of pressure:

$$P_c = wA_c h_0 \quad \text{pounds}. \tag{4.4}$$

The **vertical component of pressure** (P_v) on a plane area of A square feet with h_0 feet head on its center of gravity and A_h square feet horizontal projection of area:

$$P_v = wA_h h_0 \quad \text{pounds}. \tag{4.5}$$

The **horizontal component of pressure** (P_h) on any area of A square feet with

Table 4.1   Water Equivalents

1 U.S. gallon water =	0.1337	cubic foot
=	231.	cubic inches
=	0.833	British Imperial gallon
=	3.785	liters
=	3,785.	cubic centimeters (milliliters)
=	8.33	pounds
1 cubic foot water =	7.48	U.S. gallons
=	62.43	pounds (at greatest density, 39.2°F)
1 acre-inch	= 27,154.	U.S. gallons
1 acre-foot	= 325,851.	U.S. gallons
1 second-foot =	1.	cubic feet per second = 60. cubic feet per minute
=	7.48	U.S. gallons per second = 448.8 U.S. gallons per minute
1 miner's inch	= 1.2 to 1.76	cubic feet per minute (varies in different states)
1 cubic meter =	1,000.	liters
=	264.2	U.S. gallons
=	220.	British imperial gallons
=	35.31	cubic feet
1 boiler hp-hr =	4.	gallons water evaporated per hour

Source: Reprinted from *Water and Waste Treatment Data Book.* © 1981 The Permutit Co., Inc.

A_v square feet vertical projection of area and h_0 feet head on the center of gravity of the projected area:

$$P_h = wA_v h_0 \quad \text{pounds.} \tag{4.6}$$

The **resultant pressure** (P_{bc}) on an area bc of A_{bc} square feet with a head above its base of h_1 feet on one side and h_2 feet on the other side, or a difference of head of h feet:

$$P_{bc} = wA_{bc}(h_1 - h_2) = wA_{bc}h \quad \text{pounds.} \tag{4.7}$$

For an allowable design stress, usually established by the governing code (e.g., ASME, API, AWWA) the thickness t required for a pipe of d inches internal diameter to withstand a pressure of p pounds/inch^2 with a design (tensile) stress of f pounds/inch^2:

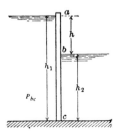

Fig. 4.1   Resultant pressure.

$$t = \frac{pd}{2f} \quad \text{inches.} \tag{4.8}$$

The **difference in water pressure** $(p_1 - p_2)$ in two pipes as indicated by a differential gauge with an oil of specific gravity s, when the difference in level of the surfaces of separation of the oil and water is z feet and the difference in level of the two pipes is h feet.

CASE I.   When the oil has a specific gravity less than 1 (see Fig. 4.2)

$$p_1 - p_2 = 0.434[z(1 - s) - h] \quad \text{pounds/inch}^2. \tag{4.9}$$

CASE II.   When the oil has a specific gravity greater than 1 (see Fig. 4.3)

$$p_1 - p_2 = 0.434[z(s - 1) - h] \quad \text{pounds/inch}^2. \tag{4.10}$$

4.2 HYDRODYNAMICS

Conservation of Energy.   The law of conservation of energy states that with steady flow the total energy at any section is equal to the total energy at any further section in the direction of flow, plus the loss of energy due to friction in the distance between the two sections. Thus the various forms of energy are interchangeable and their sum is constant (less frictional losses).

The **pressure energy** (W_{pr}) per pound of water weighing w pounds/foot^3 due to a pressure of p pounds/foot^2:

$$W_{pr} = \frac{p}{w} = 0.016p \quad \text{foot-pounds.} \tag{4.11}$$

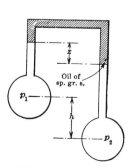

Fig. 4.2 Differences in water pressure, specific gravity under 1.

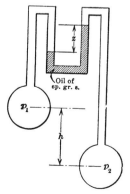

Fig. 4.3 Differences in water pressure, specific gravity over 1.

The **potential energy** (W_p) per pound of water due to a height of z feet of the center of gravity of the section above the datum level:

$$W_p = z \quad \text{foot-pounds.} \tag{4.12}$$

The **kinetic energy** (W) per pound of water due to a velocity of v feet/second, the acceleration due to gravity being g feet/second²:

$$W = \frac{v^2}{2g} \quad \text{foot-pounds.} \tag{4.13}$$

Bernoulli's Theorem.   In steady flow the total head (pressure head plus potential head plus velocity head) at any section is equal to the total head at any further section in the direction of flow, plus the lost head due to friction between these two sections.

$$\frac{p_1}{w} + z_1 + \frac{v_1^2}{2g} = \frac{p_2}{w} + z_2 + \frac{v_2^2}{2g} + \text{lost head.} \tag{4.14}$$

NOTE.   This is also known as the conservation of energy equation.

The **power** (P) available at a section of A square feet area in a moving stream of water, due to a pressure of p pounds/foot², a velocity of v feet/second, and a height of z feet above the datum level:

$$P = wvA \left(\frac{p}{w} + z + \frac{v^2}{2g} \right) \quad \text{foot-pounds/second.} \tag{4.15}$$

The **horsepower** (hp) available at any section of a stream:

$$\text{hp} = \frac{wvA(p/w + z + v^2/2g)}{550} \quad \text{horsepower.} \tag{4.16}$$

The **power** (P) available in a jet A square feet in area discharging with a velocity of v feet/second:

$$P = \frac{wv^3A}{2g} \quad \text{foot-pounds/second.} \tag{4.17}$$

4.2.1   Orifices

The **theoretical velocity of discharge** (v) through an orifice due to a head of h feet over the center of gravity of the orifice:

$$v = \sqrt{2gh} \quad \text{feet/second.} \tag{4.18}$$

The **actual velocity of discharge** (v) if the coefficient of velocity for the orifice is c_v:

$$v = c_v\sqrt{2gh} \quad \text{feet/second.} \tag{4.19}$$

The **quantity of discharge** (Q) through an orifice A square feet in area due to a head of h feet over the center of gravity of the orifice if the coefficient of discharge is c:

$$Q = cA\sqrt{2gh} \quad \text{feet}^3/\text{second.} \tag{4.20}$$

NOTE.   Orifice coefficients are given in Chapter 12.

The **quantity of discharge** (Q) through a submerged orifice A square feet in area due to a head of h_1 feet on one side of the orifice and h_2 feet on the other side, the coefficient of discharge being c:

$$Q = cA\sqrt{2g(h_1 - h_2)} \quad \text{feet}^3/\text{second.} \tag{4.21}$$

NOTE.   If $h = h_1 - h_2$, $Q = cA(2gh)^{1/2}$ feet^3/second.

The **quantity of discharge** (Q) through a large rectangular orifice b feet in width with a small head of h_1 feet above the top of the orifice and a head of h_2 feet above the bottom of the orifice, the coefficient of discharge being c:

$$Q = \tfrac{2}{3}cb\sqrt{2g}(h_2^{3/2} - h_1^{3/2}) \quad \text{feet}^3/\text{second.} \tag{4.22}$$

The **velocity of discharge** (v) and **quantity of discharge** (Q) through an orifice A_1 square feet in area, considering the velocity of approach in the approach channel of A_2 square feet area, due to a pressure head of h feet, if the coefficient of discharge is c and the coefficient of velocity is c_v:

$$v = c_v\sqrt{\frac{2gh}{1 - (A_1c/A_2)^2}} \quad \text{feet/second.} \tag{4.23}$$

The **time** (t) to lower the water in a vessel of A_1 square feet constant cross section through an orifice A_2 square feet in area, from an original head of h_1 feet over the orifice to a final head of h_2 feet:

$$t = \frac{2A_1}{cA_2\sqrt{2g}}(\sqrt{h_1} - \sqrt{h_2}) \quad \text{seconds.} \tag{4.24}$$

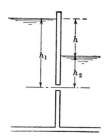

Fig. 4.4   Quantity of discharge, submerged orifice.

Note.   In general, problems involving the time required to lower the water in a reservoir of any cross section may be solved thus: Let A = cross-sectional area of the reservoir (this may be a variable in terms of h), Q = the rate of discharge through an orifice (or weir) as given by the ordinary formula, and h_1 and h_2 the initial and final heads.

$$t = \int_{h_2}^{h_1} \frac{Adh}{A} \quad \text{seconds.}$$

For a suppressed weir this would be

$$t = \int_{h_2}^{h_1} \frac{Adh}{3.33bh^{3/2}} \quad \text{seconds.}$$

The **mean velocity of discharge** (v_m) in lowering water in a vessel of constant cross section if the initial velocity of discharge is v_1 feet/second and the final velocity is v_2 feet/second:

$$v_m = \frac{v_1 + v_2}{2} \quad \text{feet/second.} \tag{4.25}$$

The **constant head** (h_m) that will produce the same mean velocity of discharge as is produced in lowering the water in a vessel of constant cross section from an initial head of h_1 feet over the orifice to a final head of h_2 feet:

$$h_m = \left(\frac{\sqrt{h_1} + \sqrt{h_2}}{2} \right)^2 \quad \text{feet.} \tag{4.26}$$

4.2.2   Weirs

The **theoretical discharge** (Q) over a rectangular weir b feet in width due to a head of H feet over the crest:

$$Q = \tfrac{2}{3}b\sqrt{2g}H^{3/2} \quad \text{feet}^3/\text{second.} \tag{4.27}$$

Note.   If the velocity head due to the velocity of approach v feet/second in the channel back of the weir is h feet: $Q = \tfrac{2}{3}b(2g)^{1/2}[(H + h)^{3/2} - h^{3/2}]$ feet^3/second. The actual discharge may be obtained by multiplying the theoretical discharge by a coefficient c which varies from 0.60 to 0.63 for contracted weirs (approach channel width > weir width) and from 0.62 to 0.65 for suppressed weirs (approach channel width = weir width).

The **Francis formula for discharge** (Q) over a rectangular weir b feet in width due to a head of H feet over the crest for a contracted weir:

$$Q = 3.33(b - 0.2H)H^{3/2} \quad \text{feet}^3/\text{second.} \tag{4.28}$$

For a contracted weir considering the velocity head h due to the velocity of approach.

Fig. 4.5 Discharge over a triangle weir.

$$Q = 3.33(b - 0.2H)[(H + h)^{3/2} - h^{3/2}] \quad \text{feet}^3/\text{second}. \tag{4.29}$$

NOTE. In case contraction occurs on only one side of the weir, the term for width becomes $b - 0.1H$.

The **Bazin formula for discharge** (Q) over a rectangular suppressed weir b feet in width due to a head of H feet over the crest and a height p feet of the crest above the bottom of the channel:

$$Q = \left(0.405 + \frac{0.00984}{H}\right)\left[1 + 0.55\left(\frac{H}{p + H}\right)^2\right]b\sqrt{2g}H^{3/2} \quad \text{feet}^3/\text{second}. \tag{4.30}$$

Fteley and Stearns' formula for discharge (Q) over a suppressed weir b feet in width due to a head of H feet over the crest:

$$Q = 3.31bH^{3/2} + 0.007b \quad \text{feet}^3/\text{second}. \tag{4.31}$$

NOTE. Considering the velocity head h due to the velocity of approach. $Q = 3.31b(H + 1.5h)^{3/2} + 0.007b$ feet^3/second.

The **discharge** (Q) **over a triangular weir,** with the sides making an angle of α degrees with the vertical, due to a head of H feet over the crest (Fig. 4.5):

$$Q = c\,{}^{8}\!/_{15}\tan\alpha\sqrt{2g}H^{5/2} \quad \text{feet}^3/\text{second}. \tag{4.32}$$

NOTE. If $\alpha = 45°$ (90° notch), $Q = 2.53H^{5/2}$ feet^3/second.

The **discharge** (Q) **over a trapezoidal weir** is computed by adding the discharge over a suppressed weir b feet in width to that over the triangular weir formed by the sloping sides.

The **Bazin formula**[1] **for discharge** (Q) over a submerged sharp-crested weir b feet in width due to a head of H feet over the crest on the upstream side with a depth of submergence of D, a crest height of P above channel bottom, and a difference in

Fig. 4.6 Discharge over a trapezoidal weir.

Fig. 4.7 Discharge over a trapezoidal weir.

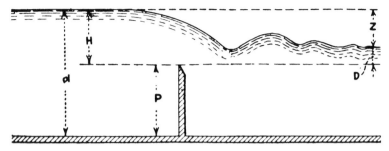

Fig. 4.8 Submerged weir.

water surface elevation above and below the weir of Z with $d = A/b$ A = cross-sectional area of the approach channel and b = effective width of the weir including the effect of end contractions = total width − 0.1 × number of contractions × H (or width − 0.2H, for contractions at both ends):

$$Q = bH^{3/2} \left(3.248 + \frac{0.079}{H} \right) \left(1 + 0.55 \frac{H^2}{d^2} \right) \left(1.05 + 0.21 \frac{D}{P} \right) \sqrt[3]{\frac{z}{h}}. \quad (4.33)$$

4.2.3   Flow in Open Channels

The **Chezy formula for quantity** (Q) and **velocity** (v) of flow in an open channel A square feet in sectional area with p feet wetted perimeter, r feet hydraulic radius, h feet drop of water surface in distance l feet, and slope s of water surface $(s = h/l)$:

$$r = \frac{A}{p} \quad \text{feet} \qquad v = c\sqrt{rs} \quad \text{feet/second.} \qquad (4.34)$$

$$Q = Av \quad \text{feet}^3/\text{second.} \qquad (4.35)$$

NOTE. c is the coefficient and is usually found either by the Kutter formula or by the Bazin formula.

Kutter Formula

$$v = c\sqrt{rs} \quad \text{feet/second,} \qquad (4.36)$$

where

$$c = \frac{41.6 + 1.811/n + 0.00281/s}{1 + (41.6 + 0.00281/s)n/\sqrt{r}} .$$

NOTE.   Specific values of c are given in Chapter 12. n is the coefficient of roughness and has the following values:

Channel Lining	n
Smooth wooden flume	0.009
Neat cement and glazed pipe	0.010
Unplaned timber	0.012
Ashlar and brickwork	0.013
Rubble masonry	0.017
Very firm gravel	0.020
Earth free from stone and weeds	0.025
Earth with stone and weeds	0.030
Earth in bad condition	0.035

Bazin Formula

$$v = c\sqrt{rs} \quad \text{feet/second,} \tag{4.37}$$

where

$$c = \frac{87}{0.552 + m/\sqrt{r}}.$$

NOTE.   Specific values of c are given in Chapter 12. m is the coefficient of roughness and has the following values:

Channel Lining	m
Smooth cement or matched boards	0.06
Planks and bricks	0.16
Masonry	0.46
Regular earth beds	0.85
Canals in good order	1.30
Canals in bad order	1.75

4.2.4   Jet Dynamics

The **reaction of a jet** (P) A square feet in area, the head on the orifice being h feet and the weight of the liquid w pounds/foot^3:

$$P = 2Awh \quad \text{pounds (theoretical).} \tag{4.38}$$

NOTE. P equals about $1.2AwH$ pounds (actual).

The **energy of a jet** (W) discharging with a velocity of v feet/second:

$$W = \frac{wv^3A}{2g} \quad \text{foot-pounds.} \tag{4.39}$$

NOTE.   If h_v is the velocity head and Q $(= A_v)$ is the quantity of flow in feet³/second, $W = wQh_v$ foot-pounds.

The **force** (F) exerted on a fixed curve vane by a jet A square feet in area and v feet/second velocity:

$$F = \frac{Awv^2}{g}\sqrt{2(1 - \cos \alpha)} \quad \text{pounds.} \tag{4.40}$$

The **vertical component of force** (F_v) exerted by a jet on a fixed curved vane:

$$F_v = \frac{Awv^2}{g}\sin \alpha \quad \text{pounds.} \tag{4.41}$$

The **horizontal component of force** (F_h) exerted by a jet on a fixed curved vane:

$$F_h = \frac{Awv^2}{g}(1 - \cos \alpha) \quad \text{pounds.} \tag{4.42}$$

The **force** (F) exerted by a jet on a flat fixed plate perpendicular to the jet:

$$F = \frac{Awv^2}{g} \quad \text{pounds.} \tag{4.43}$$

The **force** (F) exerted on a moving curved vane by a jet A square feet in area with a velocity of v feet/second, the vane moving in the direction of the flow of the jet with a velocity of v_0 feet/second:

$$F = \frac{wA(v - v_0)^2}{g}\sqrt{2(1 - \cos \alpha)} \quad \text{pounds.} \tag{4.44}$$

The **vertical component of force** (F_v) exerted by a jet on a moving curved vane:

$$F_v = \frac{wA(v - v_0)^2}{g}\sin \alpha \quad \text{pounds.} \tag{4.45}$$

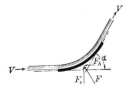

Fig. 4.9   Force exerted on fixed curved vane.   **Fig. 4.10**   Force exerted on moving curved vane.

The **horizontal component of force** (F_h) exerted by a jet on a moving curved vane:

$$F_h = \frac{wA(v - v_0)^2}{g}(1 - \cos \alpha) \quad \text{pounds.} \tag{4.46}$$

NOTE.   If there is a series of vanes, $F_h = (wAv/g)(v - v_0)(1 - \cos \alpha)$ pounds. $F_v = (wAv/g)(v - v_0) \sin \alpha$ pounds.

The **power** (P) exerted on a (moving) vane:

$$P_h = F_h v_0 \quad \text{foot-pounds/second.} \tag{4.47}$$

NOTE.   Maximum efficiency for a series of vanes occurs where $v_0 = v/2$ if there is no friction loss; then, $P = [wAv^3/(4g)](1 - \cos \alpha)$ foot-pounds/second.

4.2.5   Venturi Meter

The **quantity of water** (Q) flowing through a Venturi meter with an area of A_1 square feet in the main pipe and an area A_2 square feet in the throat and a pressure head of h_1 feet in the main pipe and of h_2 feet in the throat, if the coefficient of the meter is c:

$$Q = c \frac{A_1 A_2}{\sqrt{A_1^2 - A_2^2}} \sqrt{2g(h_1 - h_2)} \quad \text{feet}^3/\text{second.} \tag{4.48}$$

4.3   FLOW THROUGH PIPES

Solution by Bernoulli's Theorem

If the total head at any point in the pipe system (preferably at the source) is known, the velocity of discharge at the end can be computed by applying Bernoulli's theorem between these two points, provided that the losses of head can be determined. Following are expressions for the important losses of head that may occur.*

The **Darcy–Weisbach formula for friction loss** (h_f) in a pipe of d feet internal diameter and l feet length with a velocity of v feet/second and a friction factor f:

$$h_f = f \frac{l}{d} \frac{v^2}{2g} \quad \text{feet.} \tag{4.49}$$

The **Reynolds number** (R_e) is a dimensionless value that relates fluid viscosity and inertial factors. It is of particular use for predicting performance of full-size

---

*These formulas apply to pipes flowing full under pressure; otherwise, the pipe should be treated as an open channel.

components from model tests and for establishing flow regimes in piping systems to predict fluid friction (head loss):

$$R_e = \frac{\rho v d}{\mu} \tag{4.50}$$

where ρ is the specific mass or density (lb/ft³), d the length dimension (i.e., inside diameter for a round pipe ft), μ the absolute viscosity (lb-sec/ft² or slugs/ft-sec), and v the velocity (ft/sec).

There are three general cases of flow defined by R_e:

0–3000	**Laminar flow**	viscous regime
2000–3000	**Transition flow**	viscous regime
>3000	**Turbulent flow**	inertia regime

Piping velocities for system designs that balance piping cost and friction loss are normally in the Turbulent Flow Regime (unless velocities are quite low or viscosities high); thus, where precalculated head loss tables for piping flow are utilized, they are computed on the turbulent flow basis.

The **friction factor** (F) can also be determined:

$$h_f = \frac{64}{R_e} \frac{l}{d} \frac{v^2}{2g} \quad \text{feet.} \tag{4.51}$$

Thus, $F = 64/R_e$ (laminar flow).

NOTE.   A mean value for the friction factor for clean-cast iron pipe is 0.02. Chapter 12 gives values for various sizes of pipes and different velocities. In long pipelines, 1000 ft (300 m) or more, it is accurate enough to consider that the total head H is used up in overcoming friction in the pipe. Then

$$H = f \frac{l}{d} \frac{v^2}{2g} \quad \text{feet,} \tag{4.52}$$

and $Q = Av$ feet³/second.

The **loss at entrance to a pipe** (h_e) if the velocity of flow in the pipe is v feet/second and the entrance is sharp cornered:

$$h_e = 0.5 \frac{v^2}{2g} \quad \text{feet.} \tag{4.53}$$

The **loss due to sudden expansion** (h_x) where one pipe is abruptly followed by a second pipe of larger diameter, if the velocity in the smaller pipe is v_1 feet/second and that in the larger pipe is v_2 feet/second:

$$h_x = \frac{(v_1 - v_2)^2}{2g} \quad \text{feet.} \tag{4.54}$$

The **loss due to sudden contraction** (h_c) where one pipe is abruptly followed by a second pipe of smaller diameter, if the velocity in the smaller pipe is v feet/second and c_c is a coefficient:

$$h_c = c_c \frac{v^2}{2g} \quad \text{feet.} \tag{4.55}$$

NOTE. Values of c_c:

Ratio of areas	0.1	0.2	0.3	0.4	0.5	0.8	1.00
c_c	0.362	0.338	0.308	0.267	0.221	0.053	0.00

The **loss due to bends** (h_b):

$$h_b = c_b \frac{v^2}{2g} \quad \text{feet.} \tag{4.56}$$

NOTE. Values of c_b (d is the diameter of the pipe in feet and r is the radius of the bend in feet):

d/r	0.2	0.4	0.6	0.8	1.00
c_b	0.131	1.138	0.158	0.206	0.294

The **nozzle loss** (h_n) if the velocity of discharge is v feet/second and the velocity coefficient of the nozzle is c_v:

$$h_n = \left(\frac{1}{c_v^2} - 1\right) \frac{v^2}{2g} \quad \text{feet.} \tag{4.57}$$

The **quantity of discharge** (Q) in a pipe A square feet in area where the velocity is v feet/second:

$$Q = Av \quad \text{feet}^3/\text{second.} \tag{4.58}$$

The **diameter of pipe** (d) required to deliver Q feet^3 of water/second under a head of h feet if the friction factor is f:

$$d = \sqrt[5]{\frac{fl}{2gh} \left(\frac{4Q}{\pi}\right)^2} \quad \text{feet.} \tag{4.59}$$

The **hydraulic gradient** is a line the ordinates to which show the pressure heads at the different points in the pipe system. It may also be defined as the line to which water would rise in piezometer tubes placed at intervals along the pipe.

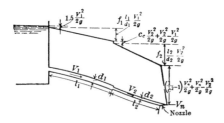

Fig. 4.11 Hydraulic gradient.

Solution by Chezy Formula

The **quantity** (Q) and **velocity** (v) **of flow through a pipe** when the hydraulic radius is r feet, and the slope of the hydraulic gradient is s, and the coefficient for the Chezy formula is c:

$$v = c\sqrt{rs} \quad \text{feet/second.} \tag{4.60}$$
$$Q = Av \quad \text{feet}^3/\text{second.} \tag{4.61}$$

NOTE. r equals the area in square feet divided by the wetted perimeter in feet, and s equals the head in feet divided by the length of the pipe in feet, or the slope of the hydraulic gradient.

4.4 FRICTION LOSS CALCULATIONS

An alternative and widely used method of calculating friction loss is to express all valves and fittings in the system in terms of **equivalent feet** of piping, adding this to the actual length of piping and establishing an equivalent length of piping for the entire system. This is then multiplied by values of head loss precalculated for that flow rate [usually expressed per 100 ft (30 m) of pipe length] to determine the piping system head loss. To this must be added head loss (or gain) for elevation differences as well as pressure drop (expressed in feet of head) for equipment such as strainers, heat exchangers, and vessels in the system, to determine the total system head loss. The method has been proven to be of suitable accuracy for the great majority of applications and has the added advantage of rapidity.

4.4.1 Friction Loss Graphs

Figure 4.12 and Table 4.2 express the resistance of valves and fittings in terms of length or pipe diameters. The **pipe diameter equivalent** method (Table 4.2) (which must be converted to equivalent feet) tends to be more accurate.

4.4.2 Friction Loss Tables[2]

Friction Losses in Pipes Carrying Water. Among the many empirical formulas for friction losses that have been proposed, that of Williams and Hazen has been most widely used. In a convenient form it reads

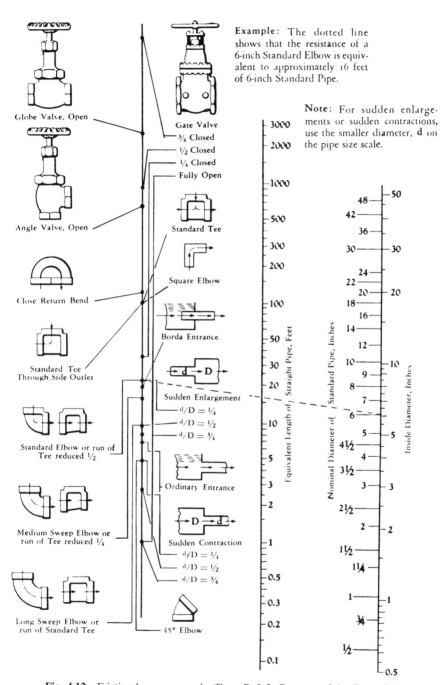

Example: The dotted line shows that the resistance of a 6-inch Standard Elbow is equivalent to approximately 16 feet of 6-inch Standard Pipe.

Note: For sudden enlargements or sudden contractions, use the smaller diameter, **d** on the pipe size scale.

Fig. 4.12 Friction loss nomograph. (From Ref. 2. Courtesy of the Crane Co.)

Table 4.2   Representative Equivalent Length in Pipe Diameters (L/D) of Various Valves and Fittings

	Description of Product			Equivalent Length In Pipe Diameters (L/D)
Globe Valves	Conventional	With no obstruction in flat, bevel, or plug type seat	Fully open	340
		With wing or pin guided disc	Fully open	450
	Y-Pattern	(No obstruction in flat, bevel, or plug type seat)		
		− With stem 60 degrees from run of pipe line	Fully open	175
		− With stem 45 degrees from run of pipe line	Fully open	145
Angle Valves	Conventional	With no obstruction in flat, bevel, or plug type seat	Fully open	145
		With wing or pin guided disc	Fully open	200
Gate Valves	Conventional Wedge Disc, Double Disc, or Plug Disc		Fully open	13
			Three-quarters open	35
			One-half open	160
			One-quarter open	900
	Pulp Stock		Fully open	17
			Three-quarters open	50
			One-half open	260
			One-quarter open	1200
	Conduit Pipe Line		Fully open	3**
Check Valves	Conventional Swing		0.5†...Fully open	135
	Clearway Swing		0.5†...Fully open	50
	Globe Lift or Stop		2.0†...Fully open	Same as Globe
	Angle Lift or Stop		2.0†...Fully open	Same as Angle
	In-Line Ball	2.5 vertical and 0.25 horizontal†...Fully open		150
Foot Valves with Strainer	With poppet lift-type disc		0.3†...Fully open	420
	With leather-hinged disc		0.4†...Fully open	75
Butterfly Valves (6-inch and larger)			Fully open	20
Cocks	Straight-Through	Rectangular plug port area equal to 100% of pipe area	Fully open	18
	Three-Way	Rectangular plug port area equal to 80% of pipe area (fully open)	Flow straight through	44
			Flow through branch	140
Fittings	90 Degree Standard Elbow			30
	45 Degree Standard Elbow			16
	90 Degree Long Radius Elbow			20
	90 Degree Street Elbow			50
	45 Degree Street Elbow			26
	Square Corner Elbow			57
	Standard Tee	With flow through run		20
		With flow through branch		60
	Close Pattern Return Bend			50

**Exact equivalent length is equal to the length between flange faces or welding ends.
†Minimum calculated pressure drop (psi) across valve to provide sufficient flow to life disc fully.
Source: Ref. 2. Courtesy of the Crane Co.

$$f = 0.2083 \left(\frac{100}{C}\right)^{1.85} \times \left(\frac{q^{1.84}}{d^{4.87}}\right), \qquad (4.62)$$

where f is the friction head in feet of liquid per 100 ft of pipe (if desired in lb/in.^{2}, multiply $f \times 0.433 \times$ sp. gr.), d the inside diameter of pipe (in.), q the flow (gal/min), and C is a constant accounting for surface roughness. This formula gives accurate values when the kinematic viscosity of the liquid is about 1.1 centistokes or 31.5 SSU, which is the case with water at about 60°F (15.6°C). The viscosity of water varies with the temperature from 1.8 at 32°F (0°C) to 0.29 centistokes at 212°F (100°C). Tables 4.3 and 4.4 are therefore subject to this error, which may increase the friction loss as much as 20% at 32°F (0°C) and decrease it as much as 20% at

Table 4.3   Pipe *C* Values

TYPE OF PIPE	VALUES OF C		
	Range — High = best, smooth, well laid — Low = poor or corroded	Average value for good, clean, new pipe	Commonly used value for design purposes
Cement—Asbestos..	160–140	150	140
Fibre..	—	150	140
Bitumastic-enamel-lined iron or steel centrifugally applied..	160–130	148	140
Cement-lined iron or steel centrifugally applied............	—	150	140
Copper, brass, lead, tin or glass pipe and tubing...........	150–120	140	130
Wood-stave..	145–110	120	110
Welded and seamless steel................................	150–80	140	100
Continuous-interior riveted steel (no projecting rivets or joints..	—	139	100
Wrought-iron..	150–80	130	100
Cast-iron..	150–80	130	100
Tar-coated cast-iron.......................................	145–80	130	100
Girth-riveted steel (projecting rivets in girth seams only)...	—	130	100
Concrete...	152–85	120	100
Full-riveted steel (projecting rivets in girth and horizontal seams)...	—	115	100
Vitrified..	—	110	100
Spiral-riveted steel (flow with lap)........................	—	110	100
Spiral-riveted steel (flow against lap)......................	—	100	90
Corrugated steel...	—	60	60

Value of C.........................	150	140	130	120	110	100	90	80	70	60
Multiplier to correct tables........	.47	.54	.62	.71	.84	1.0	1.22	1.50	1.93	2.57

Source: Ref. 3. Used with permission of Ingersoll-Rand Company, © 1965.

212°F (100°C). Note that the tables may be used for any liquids having a viscosity of the same order as just indicated.

Values of *C* for various types of pipe are given in Table 4.3 together with the corresponding multiplier that should be applied to the tabulated values of the head loss in Table 4.4.

4.5   PUMPS

Pumping is a major requirement in many processes and can be carried out by a variety of devices. The most widely used pumps are of the centrifugal type.

4.5.1   Centrifugal Pumps

For centrifugal pumps pumping water:

$$\text{brake hp} = \frac{\text{gpm} \times H \times s}{3960 \times \text{pump eff. (overall)}}, \qquad (4.63)$$

where H is the total head (in feet) of liquid pumped and s is the specific gravity [water at 60°F (15.6°C) = 1.0].

Table 4.4 Friction Losses in Pipe: $C = 100$ (for Old Pipe)

½ Inch

FLOW US gal per min	Standard Wt Steel .622" inside dia			Extra Strong Steel .546" inside dia		
	Velocity ft per sec	Velocity head ft	Head loss ft per 100 ft	Velocity ft per sec	Velocity head ft	Head loss ft per 100 ft
0.5	.528	.00	.582	.686	.01	1.10
1.0	1.06	.02	2.10	1.37	.03	3.96
1.5	1.58	.04	4.44	2.06	.07	8.38
2.0	2.11	.07	7.57	2.74	.12	14.3
2.5	2.64	.11	11.4	3.43	.18	21.6
3.0	3.17	.16	16.0	4.11	.26	30.2
3.5	3.70	.21	21.3	4.80	.36	40.2
4.0	4.23	.28	27.3	5.48	.47	51.4
4.5	4.75	.35	33.9	6.17	.59	64.0
5.0	5.28	.43	41.2	6.86	.73	77.7
5.5	5.81	.52	49.2	7.54	.88	92.7
6.0	6.34	.62	57.8	8.23	1.05	109
6.5	6.87	.73	67.0	8.91	1.23	126
7.0	7.39	.85	76.8	9.60	1.43	145
7.5	7.92	.97	87.3	10.3	1.6	165
8.0	8.45	1.11	98.3	11.0	1.9	185
8.5	8.98	1.25	110	11.6	2.1	207
9.0	9.51	1.4	122	12.3	2.4	231
9.5	10.0	1.6	135	13.0	2.6	255
10	10.6	1.7	149	13.7	2.9	280

¾ Inch

FLOW US gal per min	Standard Wt Steel .824" inside dia			Extra Strong Steel .742" inside dia		
	Velocity ft per sec	Velocity head ft	Head loss ft per 100 ft	Velocity ft per sec	Velocity head ft	Head loss ft per 100 ft
1.5	.903	.01	1.13	1.11	.02	1.
2.0	1.20	.02	1.93	1.48	.03	3.
2.5	1.51	.04	2.91	1.86	.05	4.
3.0	1.81	.05	4.08	2.23	.08	6.
3.5	2.11	.07	5.42	2.60	.11	9.
4.0	2.41	.09	6.94	2.97	.14	11.
4.5	2.71	.11	8.63	3.34	.17	14.
5	3.01	.14	10.5	3.71	.21	17.
6	3.61	.20	14.7	4.45	.31	24.
7	4.21	.28	19.6	5.20	.42	32.
8	4.82	.36	25.0	5.94	.55	41.
9	5.42	.46	31.1	6.68	.69	51.
10	6.02	.56	37.8	7.42	.86	63.
11	6.62	.68	45.1	8.17	1.04	75.
12	7.22	.81	53.0	8.91	1.23	88.
13	7.82	.95	61.5	9.63	1.44	102
14	8.43	1.10	70.5	10.4	1.7	117
16	9.63	1.44	90.2	11.9	2.2	150
18	10.8	1.8	112	13.4	2.8	187
20	12.0	2.2	136	14.8	3.4	227

1 Inch

FLOW US gal per min	Standard Wt Steel 1.049" inside dia			Extra Strong Steel .957" inside dia		
	Velocity ft per sec	Velocity head ft	Head loss ft per 100 ft	Velocity ft per sec	Velocity head ft	Head loss ft per 100 ft
2	.742	.01	.595	.892	.01	.930
3	1.11	.02	1.26	1.34	.03	1.97
4	1.49	.03	2.14	1.79	.05	3.28
5	1.86	.05	3.24	2.23	.08	5.07
6	2.23	.08	4.54	2.68	.11	7.10
8	2.97	.14	7.73	3.57	.20	12.1
10	3.71	.21	11.7	4.45	.31	18.3
12	4.46	.31	16.4	5.36	.45	25.6
14	5.20	.42	21.8	6.25	.61	34.0
16	5.94	.55	27.9	7.14	.79	43.6
18	6.68	.69	34.7	8.03	1.00	54.2
20	7.43	.86	42.1	8.92	1.24	65.8
22	8.17	1.04	50.2	9.82	1.50	78.5
24	8.91	1.23	59.0	10.7	1.8	94.4
26	9.66	1.45	68.4	11.6	2.1	107

1½ Inch

FLOW US gal per min	Standard Wt Steel 1.610" inside dia			Extra Strong Steel 1.500" inside dia		
	Velocity ft per sec	Velocity head ft	Head loss ft per 100 ft	Velocity ft per sec	Velocity head ft	Head loss ft per 100 ft
4	.63	.01	.267	.73	.01	.
5	.79	.01	.403	.91	.01	.
6	.95	.01	.565	1.09	.02	.
7	1.10	.02	.751	1.27	.03	1.
8	1.26	.02	.962	1.45	.03	1.
9	1.42	.03	1.20	1.63	.04	1.
10	1.58	.04	1.45	1.82	.05	2.
12	1.89	.06	2.04	2.18	.07	2.
14	2.21	.08	2.71	2.54	.10	3.
16	2.52	.10	3.47	2.90	.13	4.
18	2.84	.13	4.31	3.27	.17	6.
20	3.15	.15	5.24	3.63	.20	7.
22	3.47	.19	6.25	3.99	.25	8.
24	3.78	.22	7.34	4.36	.30	10.
26	4.10	.26	8.51	4.72	.35	12.
28	4.41	.30	9.76	5.08	.40	13.
30	4.73	.35	11.1	5.45	.46	15.
32	5.04	.39	12.5	5.81	.52	17.
34	5.36	.45	14.0	6.17	.59	19.
36	5.67	.50	15.5	6.54	.66	21.
38	5.99	.56	17.2	6.90	.74	24.
40	6.30	.62	18.9	7.26	.82	26.
42	6.62	.68	20.7	7.63	.90	29.
44	6.93	.75	22.5	7.99	.99	31.
46	7.25	.82	24.5	8.35	1.08	34.
48	7.57	.89	27.1	8.72	1.18	37.
50	7.88	.97	28.5	9.08	1.28	40.
55	8.67	1.17	34.0	9.99	1.55	49.
60	9.46	1.39	40.0	10.9	1.8	56.
65	10.2	1.6	46.4	11.8	2.2	65.

Source: Ref. 3. Used with permission of Ingersoll-Rand Company, © 1965.

Table 4.4 (*Continued*)

2 Inch

FLOW US gal per min	Standard Wt Steel 2.067" inside dia			Extra Strong Steel 1.939" inside dia		
	Velocity ft per sec	Velocity head ft	Head loss ft per 100 ft	Velocity ft per sec	Velocity head ft	Head loss ft per 100 ft
5	.48	.00	.120	.54	.00	.163
6	.57	.01	.167	.65	.01	.229
7	.67	.01	.223	.76	.01	.304
8	.77	.01	.285	.87	.01	.389
9	.86	.01	.355	.98	.01	.484
10	.96	.01	.431	1.09	.02	.588
12	1.15	.02	.604	1.30	.03	.824
14	1.34	.03	.803	1.52	.04	1.10
16	1.63	.04	1.03	1.74	.05	1.40
18	1.72	.05	1.28	1.96	.06	1.74
20	1.91	.06	1.55	2.17	.07	2.12
22	2.10	.07	1.85	2.39	.09	2.53
24	2.29	.08	2.18	2.61	.11	2.97
26	2.49	.10	2.52	2.83	.12	3.44
28	2.68	.11	2.89	3.04	.14	3.95
30	2.87	.13	3.29	3.26	.17	4.49
35	3.35	.17	4.37	3.80	.22	5.97
40	3.82	.23	5.60	4.35	.29	7.64
45	4.30	.29	6.96	4.89	.37	9.50
50	4.78	.36	8.46	5.43	.46	11.5
55	5.26	.43	10.1	5.98	.56	13.7
60	5.74	.51	11.9	6.52	.66	16.2
65	6.21	.60	13.7	7.06	.77	18.8
70	6.69	.70	15.8	7.61	.90	21.5
75	7.17	.80	17.9	8.15	1.03	24.5
80	7.65	.91	20.2	8.69	1.17	27.6
85	8.13	1.03	22.6	9.03	1.27	30.8
90	8.61	1.15	25.1	9.78	1.49	34.3
95	9.08	1.28	27.7	10.3	1.6	37.9
100	9.56	1.42	30.5	10.9	1.8	41.6
110	10.5	1.7	36.4	12.0	2.2	49.7
120	11.5	2.1	42.7	13.0	2.6	58.3
130	12.4	2.4	49.6	14.1	3.1	67.7
140	13.4	2.8	56.9	15.2	3.6	77.6
150	14.3	3.2	64.7	16.3	4.1	88.4

3 Inch

FLOW US gal per min	Cast Iron 3.0" inside dia			Std Wt Steel 3.068" inside dia			Extra Strong Steel 2.900" inside dia		
	Velocity ft per sec	Velocity head ft	Head loss ft per 100 ft	Velocity ft per sec	Velocity head ft	Head loss ft per 100 ft	Velocity ft per sec	Velocity head ft	Head loss ft per 100 ft
10	.45	.00	.070	.43	.00	.063	.49	.00	.083
15	.68	.01	.149	.65	.01	.134	.73	.01	.176
20	.91	.01	.254	.87	.01	.227	.97	.02	.299
25	1.13	.02	.383	1.09	.02	.344	1.21	.02	.452
30	1.36	.03	.537	1.30	.03	.481	1.45	.03	.633
35	1.59	.04	.714	1.52	.04	.640	1.70	.04	.842
40	1.82	.05	.914	1.74	.05	.820	1.94	.06	1.08
45	2.04	.06	1.14	1.95	.06	1.02	2.18	.07	1.34
50	2.27	.08	1.38	2.17	.07	1.24	2.43	.09	1.63
55	2.50	.10	1.64	2.39	.09	1.47	2.67	.11	1.94
60	2.72	.12	1.94	2.60	.11	1.74	2.91	.13	2.28
65	2.95	.14	2.24	2.82	.12	2.01	3.16	.15	2.65
70	3.18	.16	2.57	3.04	.14	2.31	3.40	.18	3.04
75	3.40	.18	2.92	3.25	.16	2.62	3.64	.21	3.45
80	3.63	.20	3.30	3.47	.19	2.96	3.88	.23	3.89
85	3.86	.23	3.69	3.69	.21	3.31	4.12	.26	4.35
90	4.09	.26	4.10	3.91	.24	3.67	4.37	.29	4.83
95	4.31	.29	4.53	4.12	.26	4.06	4.61	.33	5.34
100	4.54	.32	4.98	4.34	.29	4.47	4.85	.36	5.87
110	4.99	.39	5.94	4.77	.35	5.33	5.33	.44	7.01
120	5.45	.46	6.98	5.21	.42	6.26	5.81	.52	8.23
130	5.90	.54	8.09	5.64	.49	7.26	6.30	.62	9.54
140	6.35	.63	9.28	6.08	.57	8.32	6.79	.71	10.9
150	6.81	.72	10.6	6.51	.66	9.48	7.28	.82	12.5
160	7.26	.82	11.9	6.94	.75	10.7	7.76	.93	14.0
180	8.16	1.03	14.8	7.81	.95	13.2	8.72	1.01	17.4
200	9.08	1.28	18.0	8.68	1.17	16.1	9.70	1.46	21.2
220	9.99	1.55	21.4	9.55	1.42	19.2	10.7	1.78	25.3
240	10.9	1.8	25.2	10.4	1.7	22.6	11.6	2.07	29.7
260	11.8	2.2	29.2	11.3	2.0	26.2	12.6	2.46	34.4
280	12.7	2.5	33.5	12.2	2.3	30.4	13.6	2.88	39.5
300	13.6	2.9	38.0	13.0	2.6	34.1	14.5	3.26	44.8
320	14.5	3.3	42.8	13.9	3.0	38.4	15.5	3.77	50.5
340	15.4	3.7	47.9	14.8	3.4	43.0	16.5	4.22	56.5
360	16.3	4.1	53.3	15.6	3.8	47.8	17.5	4.73	62.8

4 Inch

FLOW US gal per min	Cast Iron 4.0" inside dia			Std Wt Steel 4.026" inside dia			Extra Strong Steel 3.826" inside dia		
	Velocity ft per sec	Velocity head ft	Head loss ft per 100 ft	Velocity ft per sec	Velocity head ft	Head loss ft per 100 ft	Velocity ft per sec	Velocity head ft	Head loss ft per 100 ft
70	1.79	.05	.635	1.76	.05	.615	1.95	.06	.789
80	2.04	.06	.813	2.02	.06	.788	2.23	.08	1.01
90	2.30	.08	1.01	2.27	.08	.980	2.51	.10	1.26
100	2.55	.10	1.23	2.52	.10	1.19	2.79	.12	1.53
110	2.81	.12	1.47	2.77	.12	1.42	3.07	.15	1.82
120	3.06	.15	1.72	3.02	.14	1.67	3.35	.17	2.14
130	3.32	.17	2.00	3.28	.17	1.93	3.63	.20	2.48
140	3.57	.20	2.29	3.53	.19	2.22	3.91	.24	2.84
150	3.83	.23	2.61	3.78	.22	2.53	4.19	.27	3.24
160	4.08	.26	2.93	4.03	.25	2.84	4.47	.31	3.64
170	4.34	.29	3.28	4.29	.29	3.18	4.75	.35	4.07
180	4.60	.33	3.64	4.54	.32	3.53	5.02	.39	4.52
190	4.86	.37	4.03	4.79	.36	3.90	5.30	.44	5.00
200	5.11	.41	4.43	5.05	.40	4.29	5.58	.48	5.50
220	5.62	.49	5.28	5.55	.48	5.12	6.14	.59	6.56
240	6.13	.58	6.21	6.05	.57	6.01	6.70	.70	8.93
260	6.64	.69	7.20	6.55	.67	6.97	7.26	.82	10.2
280	7.15	.79	8.25	7.06	.77	8.00	7.82	.95	11.6
300	7.66	.91	9.38	7.57	.89	9.09	8.38	1.09	13.1
320	8.17	1.04	10.6	8.07	1.01	10.2	8.94	1.24	13.1
340	8.68	1.17	11.8	8.58	1.14	11.5	9.50	1.40	14.7
360	9.19	1.31	13.1	9.08	1.28	12.7	10.0	1.6	16.3
380	9.70	1.46	14.5	9.59	1.43	14.1	10.6	1.7	18.0
400	10.2	1.6	16.0	10.1	1.6	15.5	11.2	1.9	19.8
420	10.7	1.8	17.5	10.6	1.7	16.9	11.7	2.1	21.7
440	11.2	1.9	19.0	11.1	1.9	18.5	12.3	2.3	23.6
460	11.7	2.1	20.7	11.6	2.1	20.7	12.8	2.5	25.7
480	12.3	2.3	22.4	12.1	2.3	21.7	13.4	2.8	27.8
500	12.8	2.5	24.1	12.6	2.5	23.4	14.0	3.0	30.0
550	14.0	3.0	28.8	13.9	3.0	27.9	15.3	3.6	35.7

6 Inch

FLOW US gal per min	Cast Iron 6.0" inside dia			Std Wt Steel 6.065" inside dia			Extra Strong Steel 5.761" inside dia		
	Velocity ft per sec	Velocity head ft	Head loss ft per 100 ft	Velocity ft per sec	Velocity head ft	Head loss ft per 100 ft	Velocity ft per sec	Velocity head ft	Head loss ft per 100 ft
100	1.13	.02	.171	1.11	.02	.162	1.23	.02	.208
120	1.36	.03	.239	1.33	.03	.227	1.48	.03	.292
140	1.59	.04	.318	1.56	.04	.302	1.72	.05	.388
160	1.82	.05	.408	1.78	.05	.387	1.97	.06	.497
180	2.04	.06	.507	2.00	.06	.481	2.22	.08	.618
200	2.27	.08	.616	2.22	.08	.584	2.46	.09	.751
220	2.50	.10	.735	2.44	.09	.697	2.71	.11	.895
240	2.72	.12	.863	2.67	.11	.819	2.96	.14	1.03
260	2.95	.14	1.00	2.89	.13	.950	3.20	.16	1.22
280	3.18	.16	1.15	3.11	.15	1.09	3.45	.19	1.40
300	3.40	.18	1.31	3.33	.17	1.24	3.69	.21	1.59
320	3.64	.21	1.47	3.56	.20	1.39	3.94	.24	1.79
340	3.86	.23	1.64	3.78	.22	1.56	4.19	.27	2.00
360	4.08	.26	1.83	4.00	.25	1.73	4.43	.31	2.23
380	4.31	.29	2.02	4.22	.28	1.92	4.68	.34	2.46
400	4.55	.32	2.22	4.44	.31	2.11	4.93	.38	2.71
450	5.11	.41	2.76	5.00	.39	2.62	5.54	.48	3.36
500	5.68	.50	3.36	5.56	.48	3.19	6.16	.59	4.09
550	6.25	.61	4.00	6.11	.58	3.80	6.77	.71	4.88
600	6.81	.72	4.70	6.66	.69	4.46	7.39	.85	5.73
650	7.38	.85	5.45	7.22	.81	5.17	8.00	.99	6.64
700	7.95	.98	6.25	7.78	.94	5.93	8.63	1.16	7.62
750	8.52	1.13	7.10	8.34	1.08	6.74	9.24	1.33	8.66
800	9.08	1.28	8.00	8.90	1.23	7.60	9.85	1.51	9.75
850	9.65	1.45	8.95	9.45	1.39	8.50	10.5	1.7	10.9
900	10.2	1.6	9.95	10.0	1.6	9.44	11.1	1.9	12.1
950	10.8	1.8	11.0	10.5	1.7	10.2	11.7	2.1	13.4
1000	11.4	2.0	12.1	11.1	1.9	11.5	12.3	2.4	14.7
1100	12.5	2.4	14.4	12.2	2.3	13.7	13.5	2.8	17.6
1200	13.6	2.9	16.9	13.3	2.7	16.1	14.8	3.4	20.7

Table 4.4   (Continued)

8 Inch

FLOW US gal per min	Cast Iron 8.0″ inside dia			Std Wt Steel 7.981″ inside dia			Extra Strong Steel 7.625″ inside dia		
	Velocity ft per sec	Velocity head ft	Head loss ft per 100 ft	Velocity ft per sec	Velocity head ft	Head loss ft per 100 ft	Velocity ft per sec	Velocity head ft	Head loss ft per 100 ft
180	1.15	.02	.125	1.15	.02	.126	1.26	.02	.158
190	1.21	.02	.138	1.22	.02	.140	1.33	.03	.175
200	1.28	.03	.152	1.28	.03	.154	1.41	.03	.192
220	1.40	.03	.181	1.41	.03	.183	1.55	.04	.229
240	1.53	.04	.213	1.54	.04	.215	1.69	.04	.269
260	1.66	.04	.247	1.67	.04	.250	1.83	.05	.312
280	1.79	.05	.283	1.80	.05	.286	1.97	.06	.358
300	1.91	.06	.322	1.92	.06	.325	2.11	.07	.406
350	2.24	.08	.428	2.24	.08	.433	2.46	.09	.540
400	2.56	.10	.548	2.57	.10	.554	2.81	.12	.692
450	2.87	.13	.681	2.88	.13	.689	3.16	.15	.860
500	3.19	.16	.828	3.20	.16	.838	3.51	.19	1.05
550	3.51	.19	.987	3.52	.19	.999	3.86	.23	1.25
600	3.83	.23	1.16	3.85	.23	1.17	4.22	.28	1.46
650	4.15	.27	1.34	4.17	.27	1.36	4.57	.32	1.70
700	4.47	.31	1.54	4.49	.31	1.56	4.92	.38	1.95
750	4.79	.36	1.75	4.81	.36	1.77	5.27	.43	2.21
800	5.11	.41	1.97	5.13	.41	1.99	5.62	.49	2.49
850	5.43	.46	2.21	5.45	.46	2.23	5.97	.55	2.79
900	5.75	.51	2.46	5.77	.52	2.48	6.32	.62	3.10
950	6.06	.57	2.71	6.09	.58	2.74	6.67	.69	3.43
1000	6.38	.63	2.98	6.41	.64	3.02	7.03	.77	3.77
1100	7.03	.77	3.56	7.05	.77	3.60	7.83	.95	4.49
1200	7.66	.91	4.18	7.69	.92	4.23	8.43	1.10	5.28
1300	8.30	1.07	4.85	8.33	1.08	4.90	9.13	1.30	6.12
1400	8.95	1.24	5.56	8.97	1.25	5.62	9.83	1.50	7.02
1500	9.58	1.43	6.32	9.61	1.44	6.39	10.5	1.7	7.98
1600	10.2	1.6	7.12	10.3	1.7	7.20	11.2	2.0	8.99
1800	11.5	2.1	8.85	11.5	2.1	8.95	12.6	2.5	11.2
2000	12.8	2.6	10.8	12.8	2.5	10.9	14.1	3.1	13.6

10 Inch

FLOW US gal per min	Cast Iron 10.0″ inside dia			Standard Wt Steel 10.02″ inside dia		
	Velocity ft per sec	Velocity head ft	Head loss ft per 100 ft	Velocity ft per sec	Velocity head ft	Head loss ft per 100 ft
500	2.04	.06	.280	2.04	.06	
550	2.24	.08	.333	2.24	.08	
600	2.45	.09	.392	2.44	.09	
650	2.65	.11	.454	2.64	.11	
700	2.86	.13	.521	2.85	.13	
800	3.26	.17	.667	3.25	.16	
900	3.67	.21	.829	3.66	.21	
1000	4.08	.26	1.01	4.07	.26	
1100	4.49	.31	1.20	4.48	.31	
1200	4.90	.37	1.41	4.89	.37	
1300	5.31	.44	1.64	5.30	.44	
1400	5.71	.51	1.88	5.70	.50	
1500	6.12	.58	2.13	6.10	.58	
1600	6.53	.66	2.40	6.51	.66	
1700	6.94	.75	2.69	6.92	.74	
1800	7.35	.84	2.99	7.32	.83	
1900	7.76	.94	3.30	7.73	.93	
2000	8.16	1.03	3.63	8.14	1.03	
2200	8.98	1.25	4.33	8.95	1.24	
2400	9.80	1.49	5.09	9.76	1.48	
2600	10.6	1.7	5.90	10.6	1.7	
2800	11.4	2.0	6.77	11.4	2.0	
3000	12.2	2.3	7.69	12.2	2.3	
3200	13.1	2.7	8.66	13.0	2.7	
3400	13.9	3.0	9.69	13.8	3.0	
3600	14.7	3.4	10.8	14.6	3.3	
3800	15.5	3.7	11.9	15.5	3.7	
4000	16.3	4.1	13.1	16.3	4.1	
4500	18.4	5.3	16.3	18.3	5.2	
5000	20.4	6.5	19.8	20.3	6.4	

12 Inch

FLOW US gal per min	Cast Iron 12.0″ inside dia			Standard Wt Steel 12.000″ inside dia		
	Velocity ft per sec	Velocity head ft	Head loss ft per 100 ft	Velocity ft per sec	Velocity head ft	Head loss ft per 100 ft
800	2.27	.08	.275	2.27	.08	.275
900	2.56	.10	.341	2.56	.10	.341
1000	2.84	.13	.415	2.84	.13	.415
1100	3.12	.15	.495	3.12	.15	.495
1200	3.41	.18	.581	3.41	.18	.581
1300	3.69	.21	.674	3.69	.21	.674
1400	3.98	.25	.773	3.98	.25	.773
1500	4.26	.28	.878	4.26	.28	.878
1600	4.55	.32	.990	4.55	.32	.990
1800	5.11	.41	1.23	5.11	.41	1.23
2000	5.68	.50	1.50	5.68	.50	1.50
2200	6.25	.61	1.78	6.25	.61	1.78
2400	6.81	.72	2.10	6.81	.72	2.10
2600	7.38	.85	2.43	7.38	.85	2.43
2800	7.95	.98	2.78	7.95	.98	2.78
3000	8.52	1.13	3.17	8.52	1.13	3.17
3500	9.95	1.54	4.21	9.95	1.54	4.21
4000	11.4	2.0	5.39	11.4	2.0	5.39
4500	12.8	2.5	6.70	12.8	2.5	6.70
5000	14.2	3.1	8.15	14.2	3.1	8.15
5500	15.6	3.8	9.72	15.6	3.8	9.72
6000	17.0	4.5	11.4	17.0	4.5	11.4
6500	18.4	5.3	13.2	18.4	5.3	13.2
7000	19.9	6.2	15.2	19.9	6.2	15.2
7500	21.3	7.1	17.3	21.3	7.1	17.3
8000	22.7	8.0	19.4	22.7	8.0	19.4
8500	24.2	9.1	21.7	24.2	9.1	21.7
9000	25.6	10.2	24.2	25.6	10.2	24.2
9500	27.0	11.3	26.7	27.0	11.3	26.7
10000	28.4	12.5	29.4	28.4	12.5	29.4

18 Inch

FLOW US gal per min	Cast Iron 18.0″ inside dia			Steel 17 18″ inside dia		
	Velocity ft per sec	Velocity head ft	Head loss ft per 100 ft	Velocity ft per sec	Velocity head ft	Head loss ft per 100 ft
1000	1.26	.02	.058	1.38	.03	.072
1200	1.53	.04	.081	1.66	.04	.101
1400	1.78	.05	.108	1.94	.06	.135
1600	2.03	.06	.138	2.21	.08	.173
1800	2.27	.08	.171	2.49	.10	.215
2000	2.52	.10	.208	2.77	.12	.261
2500	3.15	.15	.314	3.46	.19	.394
3000	3.78	.22	.440	4.15	.27	.553
3500	4.41	.30	.586	4.85	.37	.735
4000	5.04	.39	.750	5.54	.48	.941
4500	5.67	.50	.932	6.23	.60	1.17
5000	6.30	.62	1.13	6.92	.74	1.42
6000	7.56	.89	1.59	8.31	1.1	1.99
7000	8.83	1.2	2.11	9.70	1.5	2.65
8000	10.1	1.6	2.70	11.1	1.9	3.39
9000	11.3	2.0	3.36	12.5	2.4	4.22
10000	12.6	2.5	4.08	13.8	3.0	5.12
12000	15.3	3.6	5.72	16.6	4.3	7.18
14000	17.8	4.9	7.61	19.4	5.8	9.55
16000	20.3	6.4	9.74	22.1	7.6	12.2
18000	22.7	8.0	12.1	24.9	9.6	15.2
20000	25.2	9.9	14.7	27.7	11.9	18.5
22000	27.7	11.9	17.6	30.3	14.3	22.0
24000	30.6	14.6	20.6	33.2	17.1	25.9
26000	32.8	16.7	23.9	36.0	20.1	30.0
28000	35.5	19.6	27.4	38.8	23.4	34.4
30000	37.8	22.2	31.2	41.5	26.8	39.1
32000	40.6	25.6	35.1	44.3	30.5	44.1
34000	42.8	28.5	39.4	47.1	34.5	49.4
36000	45.4	32.0	43.7	49.9	38.7	54.8

Pump efficiencies typically vary from 50 to 90%, depending on size, specific speed, and rotative speed. Close to the rating point a properly chosen pump and driver will have a combined efficiency of 80 to 90%.

Other useful relationships:

$$\text{At constant rpm:} \quad \frac{\text{capacity } A}{\text{capacity } B} = \frac{\text{impeller dia. } A}{\text{impeller dia. } B} = \frac{H_A}{H_B}. \tag{4.64}$$

$$\text{At constant impeller dia.:} \quad \frac{\text{capacity } A}{\text{capacity } B} = \frac{N_A}{N_B} = \frac{H_A}{H_B} \tag{4.65}$$

and

$$\frac{\text{brake hp}_A}{\text{brake hp}_B} = \frac{N_A^3}{N_B^3}, \tag{4.66}$$

where $N_{A,B}$ is the rotative speed (rpm). These relationships should only be applied to relatively small changes, that is, 10 to 15% maximum.

The **specific speed** is defined as

$$N_s = \frac{NQ}{H^{3/4}} \tag{4.67}$$

where N_s is the specific speed (dimensionless).

Centrifugal pumps are broadly classified into types based on specific speed (Fig. 4.13). Centrifugal pumps have a head/flow relationship that typically has a rising characteristic. If the system head/flow curve is plotted with the pump curve, their intersection will be the operating point for the pump/pumping system (see Fig. 4.14).

While less of a problem today than formerly, care should still be taken to be certain the pump head curve is *continuously* rising to avoid a condition of instability and "hunting." In Figure 4.15, two flows, a and b, correspond to h'. The pump will "hunt" (oscillate) between them and its operation will be unstable.

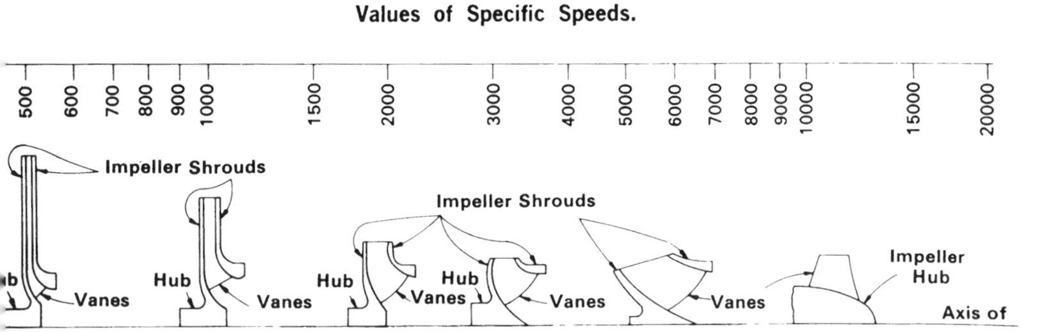

Values of Specific Speeds.

Radial-Vane Area   Francis-Vane Area   Mixed-Flow Area   Axial-Flow Area   Rotation

Fig. 4.13 Comparison of pump profiles. (From Ref. 4. Courtesy of the Hydraulic Institute.)

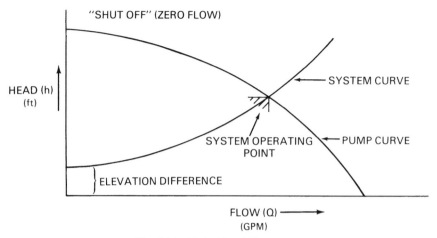

Fig. 4.14 Typical head/flow curve.

4.5.2 Other Pumps

Other types of pumps that are widely used are of the positive displacement type:

- Screw
- Gear
- Piston (reciprocating)
- Diaphragm
- Vane

4.5.3 Pump Application

Capacity. The theoretical cubic feet per minute displacement (V) of a pump that makes N pumping strokes of L feet forward per minute and has a piston of A square inches effective area is

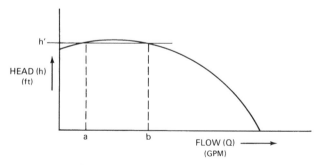

Fig. 4.15 Unstable head/flow curve.

$$V = \frac{ALN}{144} \quad \text{feet}^3/\text{minute}. \tag{4.68}$$

NOTE. If the pump is double-acting, the total displacement is the sum of the displacements on the forward and return strokes, the effective area varies for the two sides of the piston. Due to clearance, slip, imperfect valve action, and so on, the actual displacement is reduced as much as 50% in some cases.

Water Horsepower. For a pump that discharges G pounds of water per minute through a total head H feet, the water horsepower (whp) is

$$\text{whp} = \frac{GH}{33,000} \quad \text{horsepower.} \tag{4.69}$$

NOTE. The total head must include the suction lift, the discharge lift, friction and velocity heads.

Overall Thermal Efficiency. For steam-driven pumping units it is customary to express the ratio of the heat actually converted into work in lifting the water to the heat supplied as overall thermal efficiency (ϵ_t), hence

$$\epsilon_t = \frac{2545}{w_a(h_1 - h_{f2})}, \tag{4.70}$$

where w_a is the actual steam consumption in pounds per water horsepower-hour.

Duty. The term *duty* is applied to steam-driven pumping units to indicate the foot-pounds of work done for every million Btu supplied. For a pump that discharges G pounds of water per minute through a total head of H feet while using M pounds of steam per minute with Q Btu available per pound, the duty (D) is

$$\begin{aligned} D &= \frac{GH}{MQ} \times 10^6 \quad \text{foot-pounds/million Btu} \\ &= \frac{\text{w.h.p.}}{MQ} \times 33 \times 10^9 \quad \text{foot-pounds/million Btu} \\ &= \epsilon_t \times 778 \times 10^6 \quad \text{foot-pounds/million Btu.} \end{aligned} \tag{4.71}$$

Pump drivers include the following principal types:

- Electric motors
- Engines—internal combustion and other
- Compressed air, steam, and so forth

Engine drivers, particularly for positive development pump applications should include a clutch device to permit engine starting under a no-load condition.

The **design life of pumps** can range up to 40 years for carefully engineered and well-maintained installations; however, as with any machine, allowance for wear must be considered. Even pumping clear water without suspended material, pumps will suffer wear over an extended time.

Typically, pump requirements are stated for clear water service and often include at rated head, flow allowances of 7 to 10% for wear plus a 5% additional margin when fully worn. Centrifugal pumps are frequently furnished with replacement wear rings to permit restoration to original internal clearances over the pump life; other type pumps achieve this through replacement of sleeves, liners, and so on.

Pumps for abrasive service are available with a variety of types of replaceable liners (rubber, abrasion resistant, metallic alloys, etc.). Pumps for hazardous or sensitive materials can be obtained where the surfaces in contact with the fluid are either of compatible materials or isolated from the driver. Zero leakage ("canned") pumps have also been developed for particular services.

The **net positive suction head** (NPSH) defines the margin of absolute pressure required at the pump centerline to avoid flashing due to reductions in vapor pressure. Flashing (vaporization) can occur either in the suction or inlet piping or at the pump entrance, with loss of fluid flow into the pump or internally (usually at the impeller) causing cavitation which can be extremely destructive to the pump. Pump manufacturers normally provide data on required NPSH and recommended inlet piping to avoid the problem. NPSH requirements vary with pump sizes, types, and speeds.

4.6 PIPING SYSTEMS[2,5]

4.6.1 Design Considerations

The principal design considerations for fluid piping systems flowing full are: flow requirements; head loss (flow resistance); corrosion/erosion resistance; resistance to fouling; handling strength; temperature; operating, and test pressure requirements; and hanging and supports for the piping system.

Flow requirements are established by the process requirements and must consider maximum and in some cases minimum required flows as well as the normal flow for the system. **Economic sizing** of pressure piping systems would typically result in fluid flow velocities of 6 to 8 fps (1.8 to 2.4 mps) to 12-in. (300 mm) diameter lines and in larger lines above 12 in. (300 mm) of 10 to 15 fps (3 to 5 mps). For systems carrying steam the following values are typical: saturated, 100 to 175 fps (30 to 55 mps); superheated, 125 to 350 fps (40 to 110 mps). For complex or costly systems, detailed economic studies to evaluate pipe size versus system and pumping cost(s) are necessary. Caution must be exercised for high-velocity systems to avoid excessive erosion, noise, and vibration.

The **head loss** or **flow resistance** for the system is a function of both required flow(s) and the smoothness of the interior of the pipe. It is also a function of the head loss (pressure drop) through components such as heat exchangers, pressure vessels, valves, and fittings, as well as the elevation differences existing in the system. Excluding elevation differences that are constant, head losses will increase roughly as the square of the velocity (or flow).

Corrosion resistance can be achieved either by material selection or by the use of greater design allowances, that is, thicker wall piping. For conventional cold water service to 350°F (180°C), Carbon steel, ASTM A106, Gr. B or ASTM A53, Gr. A or B are frequently specified. For other fluids, particularly those which are corrosive, considerable care should be taken in materials selection. A wide range of materials having high degrees of resistance to chemical attack are available, including plastic, high-nickel steel alloys, cast iron, glazed tile, and tempered glass. Each has its advantages and a proper selection should include consideration of each. Particular thought must be given to temperature requirements which might, for example, eliminate a highly corrosion-resistant plastic from consideration.

Erosion resistance can be achieved by providing greater design allowances or by selecting a piping material that is either very hard (such as high-nickel alloy piping) or relatively elastic (such as polyethylene or rubber-lined steel pipe). The plastics are relatively cheap and easy to assemble, but transitions and valve installations are sometimes awkward; further they are not usually able to accept elevated temperatures or high pressure because of their relatively low allowable (piping wall) tensile stress.

Handling strength should be considered where piping systems require long spans or have risk of mechanical damage. The risk of damage may be present during construction and installation as well as during operation, and this may require additional supports, barriers, or in some cases, the use of heavier wall piping. For some critical services **concentric piping systems** are used, with the inner line carrying the fluid under pressure and the outer line acting as a guard pipe or return line. Concentric piping can be used for steam jacketing (heating) of fluid in the inner line. It is expensive to fabricate and install, however; heating (tracing) of piping by a smaller line in contact with the larger line or by electric resistance heaters attached to the piping outer surface may be more economic.

Pressure boundary integrity of piping systems, including valves, instruments, and so on, is assured through a **hydrostatic (hydro) test.** These hydro tests require venting and completely filling the system with water at a temperature above the NDT (nil ductility temperature—brittle failure temperature of the piping system material) and then pressurizing the system above maximum operating pressure. The hydro test pressures are usually a requirement of the applicable code (ANSI B31.1, ASME Boiler and Pressure Vessel Code, API Codes, etc.) and are typically on the order of 125 to 150% of the system maximum operating pressure. For systems not fully filled with water during operation, piping and equipment supports should be checked for adequacy during this flooded condition.

Pipe wall thickness is designated by scheduled numbers in which Schedule 40 is standard wall, with Schedules 20 and 10 being progressively thinner and Schedules 80, 120, and 160 being thicker; ANSI standards B36.10 and B36.19 provide further information. For characteristics of pipe of various schedules up to 12 in (305 mm) in diameter, see Chapter 12.

4.6.2   Pipe Supports

Hangers or **pipe supports** are of three general types: rigid, variable support, and constant support. **Rigid hangers,** as their name implies, provide a rigid connection to the support structure, and although they may provide for horizontal linear expansion (through rollers, etc.), they do not provide for vertical motion. Vertical motion of systems or components having variable weight or large temperature ranges (causing changes in length) is provided for by either variable support or constant support hangers. **Variable support hangers** generally use a spring system to compensate for the load, but permit some movement, dependent on the load and the spring constant. **Constant support hangers** generally have a linkage on a spring system to compensate for the load and maintain the supported element at a fixed location.

Hanger-type information and design guides are available from several major manufacturers. A basic requirement is to avoid excessive span between hangers and to arrange fluid lines to permit natural drainage. Provision for expansion and protection of insulation are also required.

4.6.3 Piping Materials

Material selection of piping systems requires consideration of the fluid carried, its properties, system life, maintenance requirements, and so on. Table 4.5 is a guide to some typical applications. Where fluids are extremely corrosive or abrasive, more specialized materials may be required.

4.6.4 Valve and Fitting Material Selection

Piping equipment commonly used in industry falls into four basic material groups: bronze, cast iron, malleable iron, and steel. There are also one or more variants in each of these groups and each variant has individual service characteristics.

Bronze should not be used for temperatures exceeding 550°F (290°C). One manufacturer produces two grades of bronze. Normal bronze, an alloy of copper, tin, lead, and zinc, is widely used in valves and fittings for temperatures up to 450°F (230°C). Special bronze is a high-grade alloy used in piping equipment for higher pressures and for temperatures up to 550°F (290°C).

Cast iron is regularly made in two grades: Cast iron and High-tensile iron. It should not be used for temperatures exceeding 450°F (230°C). Cast iron is commonly used for small valves and fittings having light metal sections. High-tensile cast iron is a high-strength alloy cast iron used principally for castings for large size valves.

Malleable iron is particularly suited for use in screwed fittings, unions, and so on, and also is used to some extent for valves and flanges. It is characterized by pressure tightness, stiffness, and toughness, and is especially valuable for piping systems subject to expansion and contraction stresses and shock.

Steel is recommended for high pressure and temperatures and for services where working conditions, either internal or external, may be too severe for bronze or iron. Its superior strength and toughness and its resistance to piping strains, vibration, shock, low temperature, and damage by fire afford reliable protection when safety and utility are desired. Many different types of steel are both necessary and available because of the widely diversified services steel valves and fittings perform.

4.6.5 Piping Fabrication

The **manner of fabrication of piping** is of increasing importance. More and more welded piping systems are being used where threaded or flanged systems have been

Table 4.5 Typical Piping Materials[a]

Material	Typical Temperature Limit	Material Type	Comments
Low-allow steel 2¼ Cr, 1 Mo	1015°F (546°C)	ASTM A335, Gr. P22 (seamless) or ASTM A369, Gr. FP22 (seamless)	Steam to 24 in. (610 mm) To 2650 psig (18,300 kPa)
Low-alloy steel, 2¼ Cr, 1 Mo	975°F (524°C)	ASTM A155, Gr. 2¼ Cr Class 1 (seam-welded)	Steam/water to 24 in. (610 mm) To 900 psig (6200 kPa)
Low-alloy steel, 1¼ Cr	850°F (454°C)	ASTM A335, Gr. P-11 or seam-welded ASTM A-155, Class 2 Gr., 1¼ Cr	General-purpose low-alloy steel
Stainless steel	1100°F (590°C)	ASTM A376, Gr. TP 304 or A312 Gr. TP 304 or 304L seamless (seam-welded frequently used for lower temperature and pressures)	General-purpose stainless steel, typically 8 in. (200 mm) and smaller
Carbon steel	850°F (454°C) 350°F (177°C)	ASTM SA106, Gr. B (seamless) ASTM A53, Gr. A or B seamless (or ASTM A120 for gravity drains)	Steam/water To 24 in. (610 mm) General-purpose carbon steel; water, air, steam To 24 in. (610 mm)
Red brass	250°F (120°C)	ASTM B43	Instrument air to 4 in. (100 mm)
Copper tubing	400°F[b] (204°C)	ASTM B88 and ANSI H23.1	Water to 4 in. (100 mm)
Cast iron	150°F (67°C)	ANSI A21.11 and ANSI A21.6	Water, sewage underground service mechanical or push-on joints—to 24 in (610 mm)
Ductile cast iron	150°F (67°C)	ANSI A21.51	Water, sewage
Concrete	150°F (67°C)	ASTM C-76, Class III Wall A or Class IV Wall B	Wastewater, nonpressure drainage
Vitrified clay	150°F (67°C)	Extra strength ASTM C-700	Nonpressure sewer
Polyvinyl chloride	140°F (60°C)	ASTM D1784, Class 12454-B	Chemical drains To 6 in. (150 mm)

[a]Pressure requirements will establish pipe wall thickness; see applicable design code.
[b]May depend on joint type (i.e., mechanical, solder, etc.).

Fig. 4.16 Butt-welded pipe joint. (From Ref. 5. Courtesy of the Crane Co.)

Fig. 4.17 Butt-welded tee. (From Ref. 5. Courtesy of the Crane Co.)

used in the past. The practice of welding pipe joints is now used for all sizes of high-pressure, high-temperature lines and for general service installations.

Butt-welded joints are designed to combine serviceability with ease of installation. Butt-welding consists of beveling the two ends, lining up the two openings, and then making the circumferential butt-welds.

Socket-welded joints simplify the welder's tasks. Socket-welding fittings have deep sockets with ample "come and go," hence pipe need not be cut to precise lengths.

The socket-type weld has advantages over the butt-weld that recommend it strongly for smaller size piping. Pipe doesn't have to be cut accurately unless it must be butted against the fitting shoulder. Since the pipe end slips into, and is supported by, the socket, the joint is self-aligning. Tack welding, special clamps to line up and hold the joint, and backing rings are unnecessary. Pipe used in socket-welding does not require beveling. "Icicles" and weld spatter cannot enter the pipe.

Welded flanged joints are also used in joining flanges to piping. These welded joints are superior to screwed flanged joints because the possibility of leakage through the thread is eliminated. The full thickness of the pipe wall is maintained, and the welded flange becomes an integral part of the pipe. Two typical methods for making such joints are shown in Fig. 4.19.

Flanges are available in flat face, raised face, or various standard ring joint types to suit different pressure and gasket requirements.

Valves and in-line instruments are available in either threaded, butt-weld, or flanged-end types. Fittings are available in threaded, butt-weld, flanged, and socket-weld types. Flanges are available as weld neck (weld end) or slip-on types.

Valves and flanges are available in standardized ratings for carbon or alloy steel materials and can be applied with confidence to low-temperature systems, under 600°F (320°C), for the stated pressure. Higher temperatures or nonsteel valves require temperature derating. For large-diameter piping systems such as those 48 inches

Fig. 4.18 Socket-welded ell. (From Ref. 5. Courtesy of the Crane Co.)

welding neck   slip-on   Cranelap

Fig. 4.19 Pipe flanges. (From Ref. 5. Courtesy of the Crane Co.)

(1200 mm) and larger, valves and flanges carrying other pressure ratings such as 25, 50, 75, and 125 pounds (11, 23, 34, and 57 kg) are also available, usually on special order. Manufacturers should be contacted for further information.

4.6.6 Valve Types

The **selection of valves** for hydraulic systems is of fundamental importance to proper system operation. The following must be considered:

1. Is the installation high- or low-pressure/temperature?
2. What fluids will be sent through them?
3. Will operating conditions be moderate or severe (i.e., frequency of operation, high pressure drop, hostile environment, etc.)?
4. Is throttling control or on–off control required?
5. How much headroom must be allowed for valve stems?
6. What size is the line?
7. Will the valves have to be dismantled frequently for inspection and servicing?
8. Is the installation relatively permanent, or must the piping be broken into frequently?

The principal types of valves are **gate, globe, angle, check, butterfly,** and **ball.**

Gate Valves

Fluids flow through gate valves in a straight line. This construction offers little resistance to flow and reduces pressure drop to a minimum. A gatelike disc, actuated

Table 4.6 Standardized Valve and Flange Ratings

ANSI Class	PN (Metric) Designation
150	20
300	50
400	68
600	100
900	150
1500	250
2500	420

by a stem screw and handwheel, moves up and down at right angles to the path of flow, and seats against two seat faces to shut off flow.

Gate valves are best for services that require infrequent valve operation, and where the disc is kept either fully opened or closed. They are not practical for throttling. With the usual type of gate valve, close regulation is impossible. Velocity of flow against a partly opened disc may cause vibration and chattering, and result in damage to the seating surfaces. Also, when throttled, the disc is subjected to severe wire-drawing erosive effects.

Globe and Angle Valves

Fluids change direction when flowing through a globe valve. The seat construction increases resistance to and permits close regulation of fluid flow, i.e. throttling. Disc and seat can be quickly and conveniently reseated or replaced. This feature makes them ideal for services that require frequent valve maintenance. Shorter disc travel saves operators' time when valves must be operated frequently.

Angle valves have the same operating characteristics as globe valves. Used when making a 90° turn in a line, an angle valve reduces the number of joints and saves make-up time. It also gives less restriction to flow than the elbow and globe valve it displaces.

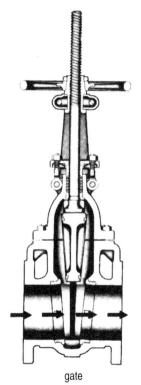

gate

Fig. 4.20 Gate valve. (From Ref. 5. Courtesy of the Crane Co.)

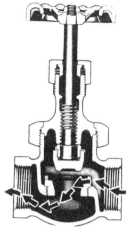

Fig. 4.21 Globe valve. (From Ref. 5. Courtesy of the Crane Co.)

Check Valves

Check valves are used to prevent back flow in lines. Check valves conform to one of the two basic patterns. Flow moves through swing check valves (Fig. 4.23) in approximately a straight line comparable to that in gate valves. In lift check valves (Fig. 4.24), flow moves through the body in a changing course as in globe and angle valves. In both swing and lift types, flow keeps the valve open while gravity and reversal of flow close it automatically.

Butterfly Valves

Butterfly valves are of the "quarter-turn" family and are so designated because a 90° turn of their operator fully opens or closes the valve. The valves utilize elastomer

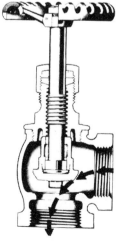

Fig. 4.22 Angle valve. (From Ref. 5. Courtesy of the Crane Co.)

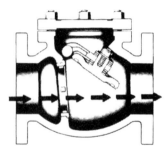

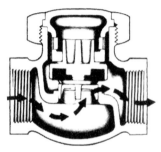

Fig. 4.23 Swing check valve. (From Ref. 5.   **Fig. 4.24** Lift check valve. (From Ref. 5.
Courtesy of the Crane Co.)   Courtesy of the Crane Co.)

seals and their popularity can be attributed to the improvements made in elastomer
materials. They are well suited to wide-open or fully closed position and in some
cases may be used for noncritical throttling applications. They are generally lighter
in weight than conventional valves. The position of the lever indicates whether they
are wide open, partially open, or fully closed. They are easily adapted to lever,
manual, gear, cylinder, or motor operation.

Ball Valves

The advantages of quarter-turn ball valves include straight-through flow, minimum
turbulence, low torque, tight closure, and compactness. Reliable operation, easy main-

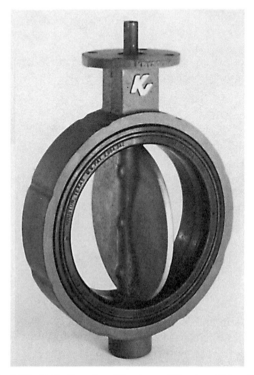

Fig. 4.25 Butterfly valve. (Courtesy of Keystone Valve Co.)

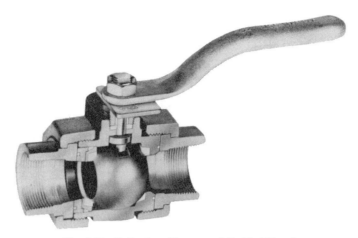

Fig. 4.26 Ball valve. (Courtesy of Pacific Valves.)

tenance, and long-life economy justify their extensive application. Industrial, chemical, petrochemical, refinery, pulp and paper, gas transmission, water works and sewage, and power plants are utilizing ball valves where other types of valves have proven inadequate.

Relief Devices

Relief valves are intended to open when the system pressure reaches a desired value. They are frequently of the angle body design with a spring whose compression can be varied, thus establishing the system discharge pressure. Normally they are provided with a lever that permits periodic exercising (testing) to assure operation when needed. Some larger relief valves utilize an electric solenoid "pilot" which initiates operation of the main disc. For certain critical safety applications, valve performance testing may be required. Principal codes such as that of the American Petroleum Institute and the ASME Boiler and Pressure Vessel Code require their use.

Rupture diaphragms are used in lieu of relief valves for applications where usage is rare and reseating is not necessary. Both relief valves and rupture diaphragms can be installed to protect against either **excess pressure** or potential collapse due to **vacuum.**

4.6.7 Seating Materials

The **seat** and **disc** constitute the heart of a valve and do most of its work. Valve manufacturers provide a wide choice of seating materials for increasing pressure or temperatures or for more rigorous service. For relatively low pressures and temperatures, say 500 psi (3450 kPa) or 350°F (180°C) and for ordinary fluids, seating materials are not a particularly difficult problem. Bronze and iron valves usually have bronze or bronze-faced seating surfaces, or iron valves may be all iron. Nonmetallic "composition" discs are available for tight seating on hard-to-hold fluids such as air or gasoline. Valves are available with **backseating capability** to permit use in vacuum service or for maintenance of packing while under pressure.

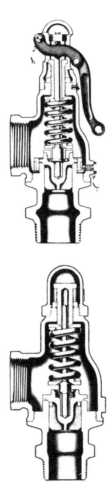

Fig. 4.27   Relief valve. (From Ref. 5. Courtesy of the Crane Co.)

As pressures and temperatures increase or as the service becomes more severe, careful consideration must be given to many factors, no one of which can be over-emphasized to the detriment of others. Long, trouble-free life requires the proper combination of hardness, wear-resistance, and resistance to corrosion, erosion, galling, seizing, and temperature. A successful combination in one instance may not serve equally well in all others.

The selection of seating materials for corrosive fluids, regardless of pressure or temperature, is almost endless. Included are many types of alloys as well as linings or coatings of many kinds. The safest policy in specifying seating materials is close adherence to valve manufacturer's recommendations, usually found in catalogs, or supplied on request.

4.6.8   Control Devices

Manual actuators are the most commonly found actuators in control processes. Typical of them are handwheels on valves, adjustable weirs, and adjustable orifices.

Major advantages to the manual devices are low cost, dependability, and relative freedom from wear and operational problems. A general disadvantage is lack of control response; they do not have automatic feedback from the process to provide any form of control action, and they have a slow response time because the intervention of an operator is usually necessary for adjustments.

Air actuators are widely used in industry today and take two general forms: **diaphragm actuators,** typically used with control (modulating) valves, and **air cylinders,** usually found on on–off service. Diaphragm actuators are sometimes of considerable diameter [e.g., 24 in. (0.6 m)] to provide sufficient force to overcome unbalanced flow forces across the valve seat as well as internal friction from stem packing, guides, and so on. They operate from 3 to 15 psig (48 to 103 kPa) as well as 6 to 30 psig (41 to 207 kPa) of air or in some air cylinders utilizing 125-psig (870-kPa) service air or higher pressures where there is a desire to minimize cylinder diameter. They provide modulating valve stem movement and can operate in any position, although a diaphragm or cylinder arrangement above the valve body is preferred in order to avoid moisture accumulation within the diaphragm or cylinder. Major advantages of air actuators are low cost and flexibility in application, since they can be used with a wide range of air pressures and in some cases can operate directly from a low-pressure control transmitter. Response time may be delayed due to compressibility of the actuating fluid.

For systems utilizing **instrument air systems,** air is normally provided by **special compressors** which are of the **nonlubricated** variety, typically carbon or Teflon ring. Oil contamination must be avoided in these systems as it may cause plugging of the fine orifices and flapper valves found within instruments. To avoid dew point change condensation from expansion of air both at the compressor and at point of use, **air dryers** are normally installed immediately downstream of the instrument air compressors. The air dryers are usually of the regenerative type and provide air having a dew point of approximately −40°F (−40°C) at standard conditions. Air actuators have been widely used for many years, are well proven, and are available in standard designs and sizes for a variety of services and applications.

Diaphragm-type actuators provide linear motion, limited to the amount of diaphragm deflection permitted, frequently to 4 in. (100 mm). Wear over a period of years in some installations can become a problem because the diaphragms occasionally tend to harden and crack. Piston seal wear can be a problem with air cylinder actuators, and "boots" are sometimes installed to keep contaminants off the piston rod surface. A typical diaphragm actuated control valve is shown in Fig. 7.12.

Hydraulic actuators are used for systems that require rapid response or large forces. They can range from low-pressure systems to those that operate on 3000 to 5000 psig (21–35 × 10³ kPa) pressure and that generally require a local, dedicated source of high-pressure fluid such as a special pump unit. Usually the working fluid is **oil,** the control mode being exercised by **spool-type valves** operating through various drilled and ported blocks having very close clearances and usually O-ring seals to minimize bypass leakage and thus avoid valve drift.

Filtration of the fluid to the 1-μm range designed to remove particulate material, is an essential and usually integral part of the high-pressure hydraulic package. The hydraulic actuators have the advantage of very rapid response rates, they can be small and light in weight, and are fairly easy to install and tune up. The cost for providing a dedicated source of hydraulic fluid and drainage lines back to a sump are the principal disadvantages. Some types of hydraulic fluids also present a fire hazard and special precautions such as extinguishing systems may be necessary.

Stored energy actuators are used for on–off or single-stroke service, generally of either the weight (gravity), compressed air, compressed spring, or explosive type. These actuators are "tailored" (although not necessarily custom designed) for the specific application and are frequently specified to provide extremely rapid opening or closure [e.g., $\frac{1}{10}$ to $\frac{2}{10}$ sec for valves as large as 36 in. (0.9 m)]. The explosive type is generally limited to the smaller sizes (to 3 in.; 76 mm) and, unlike the other types, is designed for one cycle and is not reusable.

Solenoid valves are suitable for on–off service and utilize a solenoid to move the valve disc to either the fully open or fully closed position. They are usually of the globe type, utilizing a circular disc and seat and are normally of moderate size—2 in. (51 mm) and smaller. For larger sizes the solenoid may be used as a pilot to operate an air or hydraulic actuator. Response time is very rapid; $\frac{1}{10}$ sec is typical. In addition, modulating solenoid valves are available for special applications.

4.6.9 Motor Actuators[6]

Motor actuators (valve operators) provide a means for opening and closing valves of all sizes and types without manual effort. Typically, they are electric motor driven and mount directly on the valve topworks. They are particularly useful for high-pressure applications since their gearing (often worm) provides very large forces for valve stem movement. Suitable for remote or automatic control, they are ideal for valves mounted in an inaccessible or remote location or for integration with process control field networks. Some types are suitable for mounting in other than a vertical position, an advantage in cramped space. Actuators may be of substantial weight, and this must be considered in their application. They are available in several types suitable for use on both long-stroke valves such as gate valves or sluice gates, or for quarter-turn operation for use on ball or butterfly valves. Although they can be used on throttling service, they are not generally used for control valve applications. The most sophisticated actuators use some form of torque-limiting system to avoid over-stroking the valve and damaging the seat or disc. The torque-limiting systems now utilized take a series of readings electronically, including motor torque, motor speed, and actuator position, to achieve accurate sealing and protect valve components such as seats or discs. Modern diagnostics compare valve torque profiles to identify wear and other changes in valve characteristics over time.

Motor operator closure times range from several seconds to several minutes, depending on valve size, motor sizing, gearing selected, and so on. For gate valves of 2 to 6 in., typical stroke times are 10 to 30 seconds full open to full closed. For 12-in. valves, 60 seconds, and 24-in., 120 seconds. Figure 4.28 shows a typical motor operator.

4.6.10 Vents and Drains[7]

Venting and **drainage** capability should be included in all piping systems. Vents are needed to permit complete filling of the system and avoid stoppage of flow or flow surges and interruptions due to air entrainment. Drains are necessary to permit ready

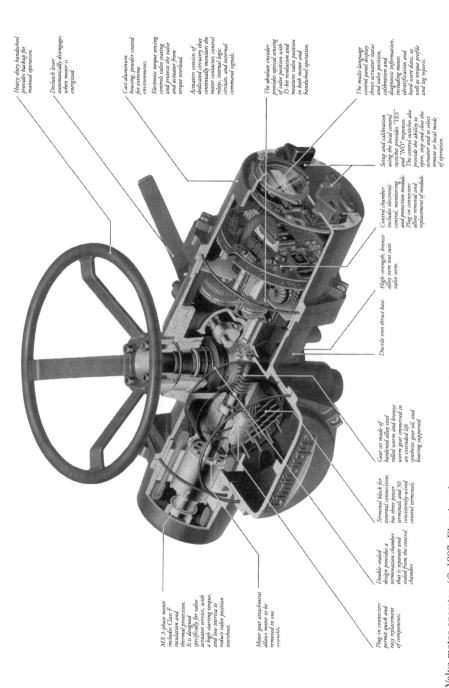

Heavy-duty handwheel provides backup for manual operation.

Declutch lever automatically disengages when motor is energized.

Cast aluminum housing, powder coated for extreme environments.

Electronic torque sensing controls valve seating and protects the valve and actuator from torque overload.

Actuation consists of dedicated circuitry that continually monitors the motor contactor, control relays, internal logic circuits, and external command signals.

The absolute encoder provides optical sensing of valve position with 15-bit resolution and measures valve position in both motor and handwheel operation.

The multi-language control panel display shows actuator status and valve position, calibration and diagnostic information, including motor identification and hard-ware data, as well as torque profile and log reports.

Setup and calibration using the local control switches provides "YES" and "NO" responses. The control switches also provide the ability to open, stop and close the actuator and to select remote or local mode of operation.

Control chamber includes electronic control, monitoring and protection module. Plug-in connectors allow removal and replacement of module.

High-strength, bronze alloy stem nut suits valve stem.

Ductile iron thrust base.

Gear set made of hardened alloy steel rolled worm and bronze worm gear immersed in an extended life, synthetic gear oil, and bearing supported.

Terminal block for external connections has three power terminals and 50 consistently-wired control terminals.

Double-sealed design provides a termination chamber that is separate and sealed from the control chamber.

Plug-in connectors permit quick and easy replacement of components.

MX 3-phase motor includes Class F insulation and thermal protection. It is designed specifically for valve actuator service, with a high starting torque, and low inertia to reduce valve position overshoot.

Motor gear attachment allows motor to be removed in one assembly.

Fig. 4.28 Valve motor operator. (© 1997, Electric valve actuator; courtesy of Limitorque Corporation, Lynchburg, Va.)

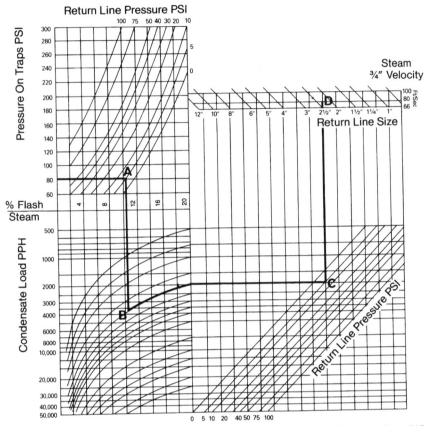

1. From pressure upstream of trap, move horizontally to pressure in return line - "A"
2. Drop vertically to condensate load in PPH - "B"
3. Follow curve to R.H. scale and across to same return line pressure - "C"
4. Move upward to return line flash velocity - say 80 ft/sec maximum - "D"
5. Read return line size

Fig. 4.29   Steam trap sizing. (© Spirax Sarco, Inc., 1992, *Design of Fluid Systems—Hook Ups.*)

removal of the working fluid by gravity. Where drainage is a process requirement, some slight inclination of the piping system, usually 1 in 12, is provided. For other systems, typically large diameter, no slope may be necessary if drain provisions are provided at low points.

Both venting and drainage can be achieved by automatic valves or, for less frequent use, by manual valves. Vents and drains are normally provided at the high and low points of a system, respectively, and a typical system may require several of each. For certain flammable fluids, **flame arrestors** are incorporated into the vents.

Automatic drainage of condensate from steam systems is normally accomplished by **steam traps.** The principal types include, float, bucket, thermostatic, bimetallic, and combinations of them. Correct sizing of the traps is important since the presence of condensate in the system will degrade its thermal performance. For systems with rotating or reciprocating machinery the presence of **undrained water** can present a significant hazard, leading to severe damage to the machinery and perhaps injury or loss of life to nearby personnel.

Sizing is usually based on the cold startup of the system, which provides in most cases several multiples of capacity needed for steady-state operation. Figure 4.29 permits sizing of the traps where system pressure is above 30 psig. Particular care needs to be taken to assure that the return line is of adequate size to handle the condensate flow. Typically, return lines are several pipe sizes (diameters) larger than the trap, although this varies depending on the pressure conditions in the system.

4.6.11 Water Hammer

Water hammer caused by rapid closure of valving or rapid change in flow rate can be extremely destructive and is a potential problem in any filled, noncompressible, fluid dynamic system. The hammer is caused by the conversion of the fluid momentum forces into shock waves due to the incompressibility of the fluid. The potential for water hammer increases with the momentum (either mass or velocity) of the fluid flowing in the system. To overcome this, **surge chambers** or **shock suppressors** are used where rapid valve closing may occur. (Surge chambers may also be required on the discharge of reciprocating pumps to dampen pulsations.) Where practical, hammer can be avoided by providing slower-closing valves, surge chambers, or **pressure relief devices.**

REFERENCES

1. L. C. Urquhart, *Civil Engineering Handbook,* 4th ed., McGraw-Hill Book Company, New York, 1959.
2. *Flow of Fluids Through Valves, Fittings and Pipe,* Technical Paper 410, Crane Co., New York, 1980.
3. *Cameron Hydraulic Data,* Ingersoll-Dresser Pumps, Liberty Corner, N.J., 1994.
4. *Hydraulic Institute Standards for Centrifugal, Rotary and Reciprocating Pumps,* Hydraulic Institute, Cleveland, Ohio, 1982.
5. *Piping Pointers—Application and Maintenance of Valves and Piping Equipment,* Publication VC-1013 A, Crane Co., New York, 1982.
6. Limitorque Corporation, Lynchburg, Va., private communication, 1996.
7. *Design of Fluid Systems—Hook-ups,* Spirax Sarco, Inc., Allentown, Pa., 1992.

CHAPTER 5
THERMODYNAMICS/HEAT TRANSFER

5.1 HEAT

In the following formulas, when specific units are not stated, any units may be used provided identical properties are expressed in the same units. Absolute pressure is indicated by p, total volume by V, specific volume by v, absolute temperature by T, and thermometer temperature by t. In all formulas containing indicated units, the temperature is measured in Fahrenheit degrees.

Measurement of Heat

Heat is the transient form of energy transmitted from one body to another when the two bodies are not at the same temperature. The ratio of the quantity of heat required to increase the temperature of a body in a specified state to that required to increase the temperature of an equal mass of water through the same temperature is called the specific heat of the body.

The unit of energy commonly used in the measurement of heat is the British thermal unit (Btu) which equals 2.930×10^{-4} kilowatt-hours or substantially $\frac{1}{180}$th of the quantity of heat required to raise one pound of water from 32 to 212°F at standard atmospheric pressure.

The **quantity of heat** (Q) added to M pounds of a substance having a constant specific heat (c) and causing the temperature to increase from t_1 to t_2 is

$$Q = Mkc(t_2 - t_1) \quad \text{units.} \tag{5.1}$$

NOTE. See Table 5.1 for definite values of c. The constant k depends on the units of measurement, as follows:

Q	M	$t_2 - t_1$	k
gram-calories	grams	°C	1
kilogram-calories	kilograms	°C	1
British thermal units	pounds	°C	1.8
British thermal units	pounds	°F	1
joules	grams	°C	4.18
joules	pounds	°F	1054
kilowatthours	kilograms	°C	1.16×10^{-3}
kilowatthours	pounds	°F	2.93×10^{-4}

Table 5.1 Mean Specific Heat of Gases at Constant Pressure (from 32°F to t°F in Btu/ lb per °F)[a]

Gas	Temperature (°F)					
	100	300	500	1000	1500	2000
Air	0.240	0.241	0.243	0.249	0.257	0.263
Oxygen, O_2	0.218	0.222	0.225	0.235	0.243	0.249
Nitrogen, N_2	0.248	0.249	0.251	0.256	0.262	0.270
Hydrogen, H_2	3.41	3.44	3.45	3.47	3.51	3.55
Water vapor, H_2O	0.444	0.449	0.454	0.472	0.493	0.516
Carbon monoxide, CO	0.248	0.249	0.251	0.258	0.265	0.273
Carbon dioxide, CO_2	0.200	0.213	0.224	0.246	0.262	0.274
Typical flue gas of average bituminous coal, based on 20% excess air and 5% H_2O	0.243	0.247	0.250	0.260	0.269	0.277

[a] The volume of 1 lb of a gas at any given temperature and pressure may be found from

$$V = \frac{t + 460}{W \times P \times 35.38}$$

where V is the volume (ft³), t is the temperature (°F), W is the weight of gas (lb/ft³) from Chapter 12, and P is the absolute pressure (psi).

If the specific heat varies with the temperature (as it usually does) according to the relation

$$c = a + bt + ft^2 \tag{5.2}$$

where a, b, and f are constants which are determined by experiment, then

$$Q = Mk\left[a(t_2 - t_1) + \frac{b}{2}(t_2^2 - t_1^2) + \frac{f}{3}(t_2^3 - t_1^3)\right] \quad \text{units.} \tag{5.3}$$

The quantity of heat added to a substance can also be determined by applying the First Law of Thermodynamics, which states that energy can be neither created nor destroyed. This principle may be expressed in an equation as follows:

$$Q = 778(W + \Delta E) \quad \text{Btu,} \tag{5.4}$$

where Q is the heat interchange in Btu, W is the external work done in foot-pounds, and ΔE is the change in the internal energy in foot-pounds.

NOTE.   Although the accepted value for the mechanical equivalent of heat is 778.26 foot-pounds, it is sufficient to use 778 foot-pounds in the solution of most engineering problems.

Influence of Heat on the Length of a Solid Body

If heat is applied to a solid body which has a length l_0 at a temperature of 0 degrees Celsius, the length l_t at a temperature of t degrees Celsius is

$$l_t = l_0(l + at).\tag{5.5}$$

NOTE. See Chapter 12 for definite values of a (the Celsius mean coefficient of linear expansion). The mean coefficient of cubical expansion equals $3a$, approximately. When the temperature is expressed in Fahrenheit degrees, Equation (5.5) becomes

$$l_t = l_{32}\{1 + a[(t - 32)/1.8]\}.$$

Measurement of External Work

External work is the result of a force acting through a distance to overcome external resistances. For mechanical processes this work may be expressed by

$$W = 144 \int_{V_1}^{V_2} p \, dV \quad \text{foot-pounds}\tag{5.6}$$

where p is the intensity of pressure in pounds absolute per square inch and V is the total volume expressed in cubic feet.

If the specific heat varies with the temperature according to the relation

$$c = a + bt + ft^2,\tag{5.7}$$

then

$$\Delta s = M \int_{T_1}^{T_2} \frac{c \, dT}{T} = M \left[(a - 460b + 211{,}600f) \ln \frac{T_2}{T_1} \right.$$

$$\left. + (b - 920f)(T_2 - T_1) + \frac{f}{2}(T_2^2 - T_1^2) \right] \quad \text{units of entropy.}\tag{5.8}$$

If the temperature remains constant, then

$$\Delta s = M \left(\frac{Q}{T} \right) \quad \text{units of entropy.}\tag{5.9}$$

If no heat is added to or rejected from the substance and the expansion or compression is frictionless (i.e., reversible), then

$$\Delta s = 0 \quad \text{units of entropy.}$$

Steady Flow

When the same quantity of working fluid progresses continuously and uniformly in one direction, the process is termed the steady flow condition. If the conservation of energy principle is applied to such a process between two sections, 1 and 2, the equation for each pound of working fluid, in its simplest form for engineering applications, is

$$E_1 + 144p_1v_1 + \frac{U_1^2}{64.4} = E_2 + 144p_2v_2 + \frac{U_2^2}{64.4} + W + 778Q_{loss} \quad \text{foot-pounds.}$$

(5.10)

where E is the internal energy in foot-pounds, $144pv$ is the flow work in foot-pounds, $U^2/64.4$ is the kinetic energy in foot-pounds, W is the external work done in foot-pounds, and Q_{loss} is the heat lost to (negative) or gained from (positive) the surroundings in Btu.

NOTE. In the expression $U^2/64.4$, U is the velocity of the fluid flowing in feet per second and 64.4 is equal to $2g$ [where g is the acceleration due to gravity and is assumed to equal 32.2 feet/second^2 (9.81 m/s^2)].

Enthalpy

The combination $E + 144pv$ occurs so often that the special term "enthalpy" has been universally adopted. This property is represented by the symbol h when the combination is expressed in Btu; that is, $(E + 144pv)/778$.

5.2 PERFECT GASES

Characteristic Equation

The relation between pressure, volume, and absolute temperature can be determined by combining the two experimental gas laws, those of Boyle and Charles, or Gay-Lussac. This relation for any two conditions 1 and 2 may be expressed by

$$\frac{p_1V_1}{T_1} = \frac{p_2V_2}{T_2} = MR.$$

(5.11)

Values of the gas constant (R) for various gases are given, as follows: air, 53.3; carbon dioxide, 35.1; carbon monoxide, 55.1; helium, 386; hydrogen, 767; nitrogen, 55.1; oxygen, 48.3.

Fundamental Equations

The dual relation between pressure and volume, temperature and pressure, or temperature and volume for many changes met in practice may be represented by exponential equations. These equations are

$$pV^n = \text{constant}, \quad Tp^{(1-n)/n} = \text{constant}, \quad TV^{n-1} = \text{constant}.$$

(5.12)

The **exponent** (n) may be determined from the following relations

$$n = \frac{\log p_1 - \log p_2}{\log V_2 - \log V_1},$$ (5.13)

or

$$n = \frac{c - c_p}{c - c_v} = \frac{c - kc_v}{c - c_v}$$ (5.14)

where c is the specific heat for a polytropic change of (pV^n = constant), c_p is the specific heat at constant pressure, and c_v is the specific heat at constant volume and k is c_p/c_v (Table 5.1).

NOTE. Another useful relation between c_p and c_v is $778(c_p - c_v) = R$.
   The **change in the internal energy** (ΔE) is independent of the path and may be determined by the following equations:

$$\Delta E = 778 M c_v (T_2 - T_1) \text{ foot-pounds}$$ (5.15)

$$= M(778 c_p - R)(T_2 - T_1) \quad \text{foot-pounds}$$ (5.16)

$$= \frac{MR(T_2 - T_1)}{k - 1} \quad \text{foot-pounds}$$ (5.17)

$$= \frac{144(p_2 V_2 - p_1 V_1)}{k - 1} \quad \text{foot-pounds.}$$ (5.18)

The **change of entropy** (Δs) is also independent of the path and may be determined by the following equations:

$$\Delta s = M \left[c_v \ln \frac{T_2}{T_1} + (c_p - c_v) \ln \frac{V_2}{V_1} \right] \quad \text{units of entropy}$$ (5.19)

$$= M \left[c_p \ln \frac{T_2}{T_1} - (c_p - c_v) \ln \frac{P_2}{P_1} \right] \quad \text{units of entropy}$$ (5.20)

$$= M \left[cv \ln \frac{P_2}{P_1} + c_p \ln \frac{V_2}{V_1} \right] \quad \text{units of entropy.}$$ (5.21)

The heat interchange and the external work done are dependent on the path or the character of the change that takes place between two conditions 1 and 2. Of the innumerable possible changes, the only ones of importance to the engineer are: constant pressure changes, during which the pressure remains constant; constant volume changes, during which the volume remains constant; isothermal changes, during which the temperature remains constant; adiabatic changes, during which no heat is received from or rejected to external bodies; polytropic changes, during which the heat supplied to or withdrawn from the gas by external bodies is directly proportional to the change in temperature. A summary of the convenient formulas for these paths is given in Table 5.2.

Table 5.2 Summary of Convenient Formulas for Perfect Gases Between Conditions 1 and 2

Path	pV-Relation	Heat Interchange (B.t.u.)	External Work Done (ft.-lbs.)	Change of Entropy (units of entropy)	Specific Heat
Constant Pressure	$p_1V_1^0 = p_2V_2^0$ $p = \text{const.}$	$Mc_p(T_2 - T_1)$ $\dfrac{MR}{778}\left(\dfrac{k}{k-1}\right)(T_2 - T_1)$	$144\,p(V_2 - V_1)$ $MR(T_2 - T_1)$	$Mc_p \ln\dfrac{T_2}{T_1}$ $Mc_p \ln\dfrac{V_2}{V_1}$	c_p
Constant Volume	$p_1V_1^\infty = p_2V_2^\infty$ $V = \text{const.}$	$Mc_v(T_2 - T_1)$ $\dfrac{MR}{778}\left(\dfrac{1}{k-1}\right)(T_2 - T_1)$	0	$Mc_v \ln\dfrac{T_2}{T_1}$ $Mc_v \ln\dfrac{p_2}{p_1}$	c_v
Isothermal or Isodynamic	$p_1V_1 = p_2V_2$ $pV = \text{const.}$	$MT(s_2 - s_1)$ $0.1851\,p_1V_1 \ln\dfrac{V_2}{V_1}$ $\dfrac{MRT}{778}\ln\dfrac{V_2}{V_1}$	$144\,p_1V_1 \ln\dfrac{V_2}{V_1}$ $MRT \ln\dfrac{V_2}{V_1}$	$\dfrac{Q}{T}$	∞
Reversible Adiabatic or Isentropic	$p_1V_1^k = p_2V_2^k$ $pV^k = \text{const.}$	0	$\dfrac{144(p_1V_1 - p_2V_2)}{k-1}$ $\dfrac{MR(T_1 - T_2)}{k-1}$	0	0
Polytropic	$p_1V_1^n = p_2V_2^n$ $pV^n = \text{const.}$	$Mc_v\left(\dfrac{n-k}{n-1}\right)(T_2 - T_1)$	$\dfrac{144(P_1V_1 - p_2V_2)}{n-1}$ $\dfrac{MR(T_1 - T_2)}{n-1}$	$Mc_v\left(\dfrac{n-k}{n-1}\right)\ln\dfrac{T_2}{T_1}$	$c_v\left(\dfrac{n-k}{n-1}\right)$

5.3 LIQUIDS AND VAPORS

Physical Conditions

A liquid and its vapor may exist in either of the following six conditions:

1. Compressed or subcooled liquid is a liquid at a temperature less than the saturation temperature corresponding to the pressure.
2. Saturated liquid is a liquid which under its given pressure will begin to vaporize when heat is added to it.
3. Saturated vapor is a vapor which under its given pressure will start changing to the liquid form when heat is removed.
4. Wet vapor is a physical mixture of saturated liquid and saturated vapor. In each pound of mixture, the fractional part by weight which is saturated vapor is designated by the symbol x.
5. Superheated vapor is a vapor the temperature of which is greater than the saturation temperature corresponding to the pressure imposed on it.
6. Supersaturated vapor is vapor the temperature and specific volume of which are less than those corresponding to the saturated condition for the pressure imposed on it. This condition occurs only during rapid expansion as in nozzles and is of special importance to turbine designers.

Properties of Liquids and Vapors

The various properties of the more important liquids and vapors are available in tabulated form. In general the properties given are pressure (p), temperature (t), specific volume (v), enthalpy (h), internal energy (E), and entropy (s).

Properties of Steam

The properties of saturated liquids, saturated vapors, and superheated vapors are given in Chapter 12. The properties of compressed liquids can be computed from tables in Keenan and Keyes.[1] The properties of wet vapor can be computed from the saturated properties as follows:

$$v = v_f + xv_{fg}, \; h = h_f + xh_{fg}, \; s = s_f + xs_{fg}. \qquad (5.22)$$
$$E = u_f + xu_{fg} = h - 0.1851pv.$$

Thermodynamic Processes

The heat interchange (Q), work done (W), change in the internal energy (ΔE), and change of entropy (Δs) for processes most frequently met in engineering practice are given in Table 5.3.

Table 5.3　Summary of Convenient Formulas for Steam Between States 1 and 2

Path	Heat Interchange (B.t.u.)	Work Done (B.t.u.)	Change in the Internal Energy (B.t.u.)	Change of Entropy (units of entropy)
Constant Pressure = const.	$M(h_2 - h_1)$	$0.1851\,p\,(V_2 - V_1)$	$M(E_2 - E_1)$	$M(s_2 - s_1)$
Constant Volume = const.	$M(E_2 - E_1)$	0	$M(E_2 - E_1)$	$M(s_2 - s_1)$
Reversible Adiabatic = const.	0	$M(E_1 - E_2)$	$M(E_2 - E_1)$	0
Isothermal = const.	$MT(s_2 - s_1)$	$M[T(s_2 - s_1) - (E_2 - E_1)]$	$M(E_2 - E_1)$	$M(s_2 - s_1)$
Isodynamic = const.	$\dfrac{M(T_1 + T_2)}{2}(s_2 - s_1)$	$\dfrac{M(T_1 + T_2)}{2}(s_2 - s_1)$	0	$M(s_2 - s_1)$
Exponential pv^n = const.	$W + \Delta E$	$\dfrac{0.1851\,(p_1V_1 - p_2V_2)}{n - 1}$	$M(E_2 - E_1)$	$M(s_2 - s_1)$

5.4   FLOW OF GASES AND VAPORS

Steady Flow Equation

Equation (5.10) for the steady flow of the working fluid when applied to nozzles and orifices, provided there is no loss or gain of heat, no friction, and no external work is done, reduces to

$$E_1 + 144p_1v_1 + \frac{U_1^2}{64.4} = E_2 + 144p_2v_2 + \frac{U_2^2}{64.4} \quad \text{foot-pounds.} \qquad (5.23)$$

Velocity

Since h may be substituted for $E + 144pv$ divided by 778, the change in kinetic energy is

$$\frac{U_2^2 - U_1^2}{64.4} = 778(h_1 - h_2) \quad \text{foot-pounds.} \qquad (5.24)$$

If the initial velocity U_1 is small, it may be neglected, giving

$$\frac{U_2^2}{64.4} = 778(h_1 - h_2) \quad \text{foot-pounds,} \qquad (5.25)$$

or (5.26)

$$U_2 = 223.8\sqrt{h_1 - h_2} \quad \text{feet/second.}$$

Weight

The weight of working fluid flowing must be constant throughout the process. If G represents this weight in pounds/second, a the area in square feet, U the velocity in feet/second, and v the specific volume in feet³/pound at any pressure, then

$$G = \frac{a_1 U_1}{v_1} = \frac{a_2 U_2}{v_2} \quad \text{pounds/second.} \qquad (5.27)$$

This weight is a maximum when the absolute pressure P_t at the throat is

$$p_t = p_1 \left(\frac{2}{n + 1}\right)^{n/(n-1)} \quad \text{pounds/inch}^2. \qquad (5.28)$$

For dry saturated steam $n = 1.135$, then

$$p_t = 0.58p_1 \quad \text{pounds/inch}^2. \tag{5.29}$$

For superheated steam $n = 1.30$, then

$$p_t = 0.55p_1 \quad \text{pounds/inch}^2. \tag{5.30}$$

For diatomic (two atoms per molecule) gases $n = 1.40$, then

$$p_t = 0.53p_1 \quad \text{pounds/inch}^2. \tag{5.31}$$

The pressure (p_t) that makes the weight of working fluid flowing a maximum is called **critical pressure.** When the final absolute pressure p_2 is less than the critical pressure, then the weight discharged remains constant and the term applied for such a condition is unretarded flow. When the final absolute pressure is greater than the critical pressure, then the weight discharged decreases as p_2 increases, and the flow is said to be retarded.

5.4.1 Flow of Gases

Velocity

For a perfect gas, $c_p T$ may be substituted for h, and k for n. Equation (5.26) may be modified to give

$$U_2 = 223.8\sqrt{c_p(T_1 - T_2)} \quad \text{feet/second} \tag{5.32}$$

or

$$U_2 = \sqrt{\frac{2gkp_1v_1}{k-1}\left[1 - \left(\frac{p_2}{p_1}\right)^{(k-1)/k}\right]} \quad \text{feet/second.} \tag{5.33}$$

Weight

Substituting Equation (5.33) in Equation (5.27) gives

$$G = a_2\sqrt{\frac{2gk}{k-1}\left(\frac{p_1}{v_1}\right)\left[\left(\frac{p_2}{p_1}\right)^{2/k} - \left(\frac{p_2}{p_1}\right)^{(k+1)/k}\right]} \quad \text{pounds/second.} \tag{5.34}$$

It is more convenient to use the throat area a_t. The equations used must be classified depending on the type of flow, retarded or unretarded. All units in feet.

For retarded flow of diatomic gases $(k = 1.40)$, Equation (5.34) reduces to

$$G = \frac{15.03 a_t p_1}{\sqrt{RT_1}} \sqrt{\left(\frac{p_2}{p_1}\right)^{1.43} - \left(\frac{p_2}{p_1}\right)^{1.71}} \quad \text{pounds/second,} \tag{5.35}$$

and for air ($R = 53.34$)

$$G = \frac{2.056 a_t p_1}{\sqrt{T_1}} \sqrt{\left(\frac{p_2}{p_1}\right)^{1.43} - \left(\frac{p_2}{p_1}\right)^{1.71}} \quad \text{pounds/second.} \tag{5.36}$$

Fliegner's empirical formula for the retarded flow of air (inch units) is

$$G = \frac{1.06 a_t}{\sqrt{T_1}} \sqrt{p_2(p_1 - p_2)} \quad \text{pounds/second.} \tag{5.37}$$

For the unretarded flow of diatomic gases ($k = 1.40$), Equation (5.34) reduces (inch or foot units) to

$$G = \frac{3.885 a_t p_1}{\sqrt{RT_1}} \quad \text{pounds/second,} \tag{5.38}$$

and for air ($R = 53.34$)

$$G = \frac{0.532 a_t p_1}{\sqrt{T_1}} \quad \text{pounds/second.} \tag{5.39}$$

Fliegner's empirical formula for the unretarded flow of air is

$$G = \frac{0.53 a_t p_1}{\sqrt{T_1}} \quad \text{pounds/second.} \tag{5.40}$$

5.4.2   Flow of Steam

Velocity

For steam, since the enthalpy (h) may be determined from the tables in Chapter 12, the velocity is

$$U_2 = 223.8 \sqrt{h_1 - h_2} \quad \text{feet/second.} \tag{5.41}$$

Weight

Although it is possible to deduce equations involving the exponent n, the most convenient form is

$$G = \frac{a_2 U_2}{v_2} \quad \text{pounds/second.} \tag{5.42}$$

If the throat area a_t is used,

$$G = \frac{a_t U_t}{v_t} \quad \text{pounds/second.} \tag{5.43}$$

The throat pressure equals p_2 for retarded flow and equals the critical pressure ($0.58p_1$ for wet or dry saturated vapor and $0.55p_1$ for superheated vapor) for unretarded flow.

Rankine's empirical formula for the retarded flow of dry saturated steam

$$G = 0.0292a_t\sqrt{p_2(p_1 - p_2)} \quad \text{pounds/second.} \tag{5.44}$$

Rankine's empirical formula for the unretarded flow of dry saturated vapor is

$$G = \frac{a_t p_1}{70} \quad \text{pounds/second.} \tag{5.45}$$

Grashof's empirical formula for the unretarded flow of dry saturated vapor (inch units) is

$$G = \frac{a_t p_1^{0.97}}{60} \quad \text{pounds/second.} \tag{5.46}$$

Equations (5.44), (5.45), and (5.46) should be divided by $(x_1)^{1/2}$ for wet steam and by $1 + 0.00065\,\Delta t$ for superheated steam.

NOTE. Δt is the number of Fahrenheit degrees of superheat.

5.4.3 Steam Calorimeters

Throttling Calorimeter

Equation (5.23) for the steady flow of the working fluid when applied to a throttling calorimeter, provided there is no loss or gain of heat, no external work is done, and the initial and final velocities are equal or negligible, reduces to

$$h_1 = h_2 \quad \text{Btu.} \tag{5.47}$$

This calorimeter is limited in its use because the steam in the calorimeter must be superheated. For reliable results, at least 10° of superheat must be available. The quality (x) can be determined from

$$x = \frac{h_2 - h_{f_1}}{h_{fg_1}}.$$ (5.48)

NOTE.   The percentage priming equals $(1 - x)100$.

Separating Calorimeters

A separating calorimeter is designed to separate the moisture from the steam. The drip (G_m) is collected and weighed. The saturated steam (G_s), for the same time interval, may be condensed and weighed or discharged through an orifice. The priming $(1 - x)$ can be determined from

$$1 - x = \frac{G_m}{G_s + G_m}.$$ (5.49)

5.5   THERMAL EQUIPMENT

5.5.1   Steam Engines

Mean Effective Pressure

The indicator card shows the pressure distribution in the cylinder of a steam engine at every point of the working and exhaust strokes. The mean effective pressure (M.E.P. or P) in pounds per square inch is

$$\text{M.E.P.} = \frac{aS}{l} \quad \text{pounds per square inch,}$$ (5.50)

where a is the area of the card in square inches, l is the length of the card in inches, and S is the scale of the indicator spring in pounds per square inch per inch of height, or

$$\text{M.E.P.} = \left[p_1 \times C + p_1(C + Cl) \ln \frac{R + Cl}{C + Cl} - p_2(1 - K) \right.$$
$$\left. - p_2(K + Cl) \ln \frac{K + Cl}{Cl} \right] \times \text{D.F.} \quad \text{pounds/inch}^2,$$ (5.51)

where p_1 is the admission pressure in pounds absolute per square inch, p_2 is the exhaust pressure in pounds absolute per square inch, C is the percent cut-off, R is the percent release, K is the percent compression, and Cl is the percent clearance. The diagram factor (D.F.) is usually between 0.85 and 0.95.

NOTE.   The percent events are expressed as decimal fractions.

In compound engines it is customary to neglect the clearance in the conventional card. If E represents the expansion ratio,

$$\text{M.E.P.} = \left(p_1 \times \frac{1}{E} + p_1 \times \frac{1}{E} \ln E - p_2\right) \times \text{D.F.} \quad \text{pounds/inch}^2. \quad (5.52)$$

Rankine Efficiency

The Rankine cycle for a steam engine consists of the reception of heat energy at constant pressure, a frictionless adiabatic expansion to the exhaust pressure and the rejection of heat energy at constant pressure. The approximate efficiency for such a cycle (ϵ_R) is

$$\epsilon_R = \frac{Q_1 - Q_2}{Q_1} = \frac{h_1 - h_2}{h_1 - h_{f_2}} = \frac{2545}{w_R(h_1 - h_{f_2})}, \quad (5.53)$$

where h_1 is the enthalpy at the initial conditions, h_2 is the enthalpy after isentropic expansion to the exhaust pressure, h_{f_2} is the enthalpy of the saturated liquid at the exhaust pressure, and w_R is the ideal steam consumption in pounds per horsepower-hour.

Thermal Efficiency

The actual cycle for a steam engine takes into account the fact that heat losses occur in the actual steam engine. The efficiency for such a cycle (ϵ_T) is

$$\epsilon_T = \frac{Q_1 - Q_2 - Q_{\text{losses}}}{Q_1} = \frac{2545}{w_A(h_1 - h_{f_2})}, \quad (5.54)$$

where w_A is the actual steam consumption in pounds per horsepower-hour.

Engine Efficiency or Rankine Cycle Ratio

The ratio of the ideal steam consumption per horsepower-hour (w_R) to the actual steam consumption in pounds per horsepower-hour (w_A) is termed engine efficiency (ϵ_E), cylinder efficiency, or Rankine cycle ratio, hence

$$E = \frac{w_R}{w_A} = \frac{2545}{w_A(h_1 - h_2)} = \frac{\epsilon_T}{\epsilon_R}. \quad (5.55)$$

Brake Horsepower

The output of a steam engine may be determined by means of a friction brake. If F represents the net force in pounds acting at a distance R feet from the center of the shaft rotating at N revolutions per minute, then the brake horsepower (bhp) is

$$\text{bhp} = \frac{2\pi RNF}{33,000} \quad \text{horsepower}. \quad (5.56)$$

Mechanical Efficiency

The ratio of the brake horsepower (bhp) to the indicated horsepower (ihp) is termed mechanical efficiency (ϵ_M), hence

$$\epsilon_M = \frac{\text{bhp}}{\text{ihp}}. \tag{5.47}$$

Carnot Efficiency

The Carnot cycle consists of the reception and rejection of heat energy, each at constant temperature together with frictionless adiabatic expansion and compression. The efficiency for such a cycle (ϵ_c) is

$$\epsilon_c = \frac{Q_1 - Q_2}{Q_1} = \frac{T_1 - T_2}{T_1}. \tag{5.58}$$

(This formula is extremely widely used and is applicable to any heat cycle.)

5.5.2   Steam Turbines

The **steam flow required** (F) can be determined:

$$F = \frac{R_T}{E_E} \times h_p \tag{5.59}$$

where F is the flow required (lb/hr), R_T is the steam rate—theoretical (lb/hr), h_p is the power output (hp or kW), and E_E is the engine efficiency (%). **Theoretical steam rates** and **efficiencies** vary widely depending on machine size and application, but can be obtained from the manufacturers. As a guide some approximate rates and efficiencies are shown in Tables 5.4 and 5.5. For large-capacity turbine applications **regenerative feedwater heating** is used to reduce the overall heat rate for the turbine/feedwater cycle. As the number of stages of heating increase, the incremental increase in efficiency declines (Table 5.6).

Table 5.4   Typical Theoretical Steam Rates (lb/kWh)

Exhaust Pressure	Throttle Conditions (psi/°F)				
	150/Sat.	400/650	600/825	1200/825	1800/1000 Reheat
2.5 in. Hg abs.	10.9	8.04	6.92	6.58	5.77
4.0 in. Hg abs.	11.8	8.52	7.28	6.90	6.01
5 psig	21.7	13.0	10.4	9.40	7.87
50 psig	46.0	19.4	14.3	12.2	9.77
100 psig	—	26.5	18.1	14.4	11.2
200 psig	—	48.2	27.0	18.5	13.6

Table 5.5 Typical Turbine Efficiencies (%)

Turbine Size (kW)	150/Sat	Throttle Conditions (psi/°F) 600/825	1800/1000 Reheat
1,000	64	64	—
10,000	78	77	72
100.000	—	82	81

5.5.3 Steam Boilers

Maximum Allowable Working Pressure. For a steam boiler drum with a shell of r inches radius, t inches thick, an ultimate tensile strength of f pounds/inch^2, factor of safety (FS), and an efficiency of the longitudinal joint ϵ percent, the maximum allowable working pressure (p) in pounds/inch^2 gauge is

$$p = \frac{ft\epsilon}{\text{FS} \cdot r} \quad \text{pounds/inch}^2. \tag{5.60}$$

Thickness of Bumped Head. For a bumped head of bumped radius r inches, working pressure of p pounds/inch^2 gauge, an ultimate tensile strength of f pounds/inch^2, and a factor of safety F.S., the thickness (t) in inches is

$$t = \frac{\text{FS} \cdot rp}{Kf} \quad \text{inches.} \tag{5.61}$$

NOTE. $K = 1$ for convex heads and $K = 0.6$ for concave heads. The factor of safety (FS) is usually taken as 5.

NOTE. Equations (5.60) and (5.61) may be superceded by the requirements of a local code or the ASME Boiler and Pressure Vessel Code. The local licensing authority should be consulted.

Boiler Horsepower. One boiler horsepower is the evaporation of 34.5 lb (15.7 kg) of water/hour at 212°F (100°C) and atmospheric pressure. If G_a pounds of water/hour enters the boiler with an enthalpy h_{f_2} and leaves as steam with an enthalpy h_1, the boiler horsepower (P_B) is

Table 5.6 Approximate Reduction in Heat Rate (%), Cumulative

Number of Stages of Feedwater Heaters	Throttle Conditions (psi/°F) 400/600	1800/1000 Reheat
1	5.0	7.0
2	7.5	9.5
4	9.0	11.0
6	10.0	12.0

$$P_B = \frac{G_a(h_1 - h_{f_2})}{33,475} \quad \text{horsepower.} \tag{5.62}$$

Equivalent Evaporation. The numerator of Equation (5.62) represents the actual heat absorbed per hour. Because 970.3 Btu (245 kcal) is required to evaporate 1 lb (0.454 kg) of water at 212°F (100°C), the equivalent evaporation (G_e) in pounds per hour is

$$G_e = \frac{G_a(h_1 - h_{f_2})}{970.3} \quad \text{pounds/hour.} \tag{5.63}$$

Factor of Evaporation. To determine the equivalent evaporation per hour (G_e) it is necessary to multiply the actual evaporation per hour (G_a) by a factor. This factor is termed factor of evaporation (F) and is

$$F = \frac{h_1 - h_{f_2}}{970.3}. \tag{5.64}$$

Boiler Efficiency. The ratio of the heat absorbed in the boiler to the heat supplied by the fuel is termed boiler efficiency ϵ_B and is

$$\epsilon_B = \frac{G_a(h_1 - h_{f_2})}{G_f(H)} \tag{5.65}$$

where G_f is the weight of fuel in pounds per hour and H is the calorific heating value in Btu per pound of fuel.

5.5.4 Chimneys and Drafts

Intensity of Draft. For a chimney H feet high whose gases have an absolute temperature of T_1 degrees and an outside absolute temperature of T_2 degrees, the intensity of the draft (D) in inches of water is

$$D = 7.64H \left(\frac{1}{T_2} - \frac{1}{T_1} \right) \quad \text{inches of water.} \tag{5.66}$$

NOTE. This formula neglects the effect of friction. For a chimney with a friction factor f, H feet high, C feet in circumference, A square feet in passage area, and discharging G pounds of gases per second, the draft loss (d) in inches of water is

$$d = \frac{f G^2 C H}{A^3} \quad \text{inches of water,} \tag{5.67}$$

where $f = 0.0015$ for steel stacks with gases at 600°F (316°C), 0.0011 at 810°F (432°C); and 0.0020 for brick or brick-lined stacks with gases at 650°F (343°C), 0.0015 at 810°F (432°C).

Effective Area of Chimney. The retardation of ascending gases by friction within the stack has the effect of decreasing the inside cross-sectional area or of lining the chimney with a layer of gas with no velocity. If the thickness of this lining is assumed to be 2 in. (5 cm) for all chimneys, the effective area (E) for square or round chimneys with A square feet of passage area is approximately

$$E = A - 0.6\sqrt{A} \quad \text{feet}^2. \tag{5.68}$$

Boiler Horsepower. For a chimney H feet high with A square feet of passage area, **Kent's empirical formula** for the boiler horsepower (P_B) is

$$P_B = 3.33(A - 0.6\sqrt{A})\sqrt{H} \quad \text{horsepower.} \tag{5.69}$$

NOTE. This formula is based on the assumptions that the boiler horsepower capacity varies as the effective area (E) and the available draft is sufficient to effect combustion of 5 lb (2.27 kg) of coal per hour per rated horsepower (the water-heating surface divided by 10).

5.5.5 Internal Combustion Engines

Compression Ratio. The ratio of the total volume at the beginning of compression (V_1) and the volume at the end of compression or clearance volume (V_c) is a significant ratio for the various internal combustion engine cycles. This ratio is known as the compression ratio (r_k) and may also be expressed in terms of the piston displacement (P.D.), hence

$$r_k = \frac{V_1}{V_c} = \frac{\text{P.D.} + V_c}{V_c}. \tag{5.70}$$

Otto Efficiency. The Otto cycle consists of a frictionless adiabatic compression, constant-volume burning, frictionless adiabatic expansion, and rejection of heat energy at constant volume. If 1 and 2 are used to designate the states at the beginning and end of compression, respectively, the efficiency for such a cycle (ϵ_0) is

$$\epsilon_0 = 1 - \frac{T_1}{T_2} = 1 - \left(\frac{p_1}{p_2}\right)^{(k-1)/k} = 1 - \left(\frac{V_2}{V_1}\right)^{k-1}$$

$$= 1 - \left(\frac{V_c}{\text{P.D.} + V_c}\right)^{k-1} = 1 - \left(\frac{1}{r_k}\right)^{k-1}. \tag{5.71}$$

NOTE. For explanation of the exponent k, see Section 5.2.

Joule or Brayton Efficiency. The Joule cycle consists of a frictionless adiabatic compression, constant-pressure burning, frictionless adiabatic expansion, and rejection of heat energy at constant pressure. If 1 and 2 are used to designate the states at the beginning and end of compression, respectively, the efficiency for such a cycle (ϵ_J) is

$$\epsilon_J = 1 - \frac{T_1}{T_2} = 1 - \left(\frac{p_1}{p_2}\right)^{(k-1)/k} = 1 - \left(\frac{V_2}{V_1}\right)^{k-1} = 1 - \left(\frac{1}{r_k}\right)^{k-1}. \quad (5.72)$$

Diesel Efficiency. The Diesel cycle consists of a frictionless adiabatic compression, constant-pressure burning, frictionless adiabatic expansion, and the rejection of heat energy at constant volume. If 1 and 2 are used to designate the states at the beginning and end of compression, respectively, and 3 and 4 are used to designate the states at the beginning and end of expansion, respectively, then the efficiency for such a cycle (ϵ_D) is

$$\epsilon_D = 1 - \frac{1}{k}\left(\frac{T_4 - T_1}{T_3 - T_2}\right) = 1 - \left(\frac{1}{r_k}\right)^{k-1}\left[\frac{r_c^k - 1}{k(r_c - 1)}\right], \quad (5.73)$$

where r_c represents the ratio of the total volume at the end of burning (V_3) to the volume at the start of burning (V_2). This ratio is termed cut-off ratio.

Thermal Efficiency. The actual cycle for an internal combustion engine takes into account the fact that heat losses occur in the actual engine. The efficiency for such a cycle (ϵ_T) is

$$\epsilon_T = \frac{Q_1 - Q_2 - Q_{\text{losses}}}{Q_1} = \frac{2545}{w_A(H_1)}, \quad (5.74)$$

where w_A is the actual fuel consumption in pounds per horsepower-hour and H_1 is the calorific heating value of the fuel per pound.

Engine Efficiency. The ratio of the ideal fuel consumption per horsepower-hour (w_I) to the actual fuel consumption per horsepower-hour (w_A) is termed engine efficiency (ϵ_E), hence

$$\epsilon_E = \frac{w_I}{w_A} = \frac{\epsilon_T}{\epsilon_I}. \quad (5.75)$$

NOTE. The ideal efficiency (ϵ_I) can be either the Otto, Joule, or Diesel cycle, depending on which one of these cycles the actual engine is operating.

Brake Horsepower. The output of an internal combustion engine may be determined by means of a friction brake. Equation (5.56) is applicable in this case.

The empirical equation for determining the brake horsepower (bhp) for an engine with n cylinders of d inches diameter and L inches stroke at N revolutions/minute, the clearance being m percent of the stroke, is

$$\text{bhp} = \frac{d^2 LnN}{14,000}\left(0.48 - \frac{1}{10m}\right) \text{ horsepower.} \quad (5.76)$$

Diameter. If an engine cylinder is designed for maximum obtainable indicated horsepower (ihp) with a mean effective pressure (M.E.P.) pounds/inch^2, the number of explosions per minute at full load being y and the stroke L in feet being x times the diameter (d) in feet,

$$d = \sqrt[3]{\frac{300(\text{ihp})}{(\text{M.E.P.})xy}} \quad \text{feet.} \tag{5.77}$$

5.5.6 Vacuum Pumps

Vacuum pumps are frequently used for processes where the presence of air is detrimental, where simulation of conditions in outer space is necessary, and so on. Vacuum pumps are typically integrated into a system in which the pump is one component in a process train that may consist of the system or test chamber, the pump **"backing" system,** instrumentation, and control components. Some selected data on vacuum pumps are listed in Table 5.7.

Lobe-type pumps are typically high capacity and generally used as the first stage on "rough" (i.e., low-vacuum) systems; they would normally discharge into a **Stokes pump** (positive displacement), which in turn discharges to the atmosphere.

Oil diffusion pumps are of moderate capacity and need to discharge into a "backing train" typically consisting of one or more lobe-type pumps, discharging into a Stokes-type pump; but care in the design of the installation is required to assure matching of capacities. Also **cold traps** and **baffles** must be provided to avoid oil **backstreaming** out of the pump and into the vacuum chamber/system with a potential for contamination of the system.

Table 5.7 Typical Vacuum Pump Data

Type	Typical Vacuum Limit	Capacity	Remarks
Ejector (steam jets)	½ in. Hg	Moderate	Efficient, low cost, no moving parts
Rotary			
Nash	3 in. Hg 0.7 in. Hg (multistage)	Can be high	Can have high water requirements
Stokes	to 10^{-2} torr[a]	Limited	
Lobe type	to 10^{-1} torr[a]	High	
Oil diffusion	to 10^{-9} torr[a]	Moderate	Backing train required
Cryo pumps	to 10^{-12} torr[a]	Very high	Depends on Cryo fluid temperature (generally ± 20 K) and pumping area available

[a]Typical values with care can be one to two orders of magnitude lower. 1 torr = ¹⁄₇₆₀ of 1 atm (approx. 1 mmHg).

Cryopumping provides very high capacity but because of limited heat absorption capability, shielding from radiant heat sources usually by liquid nitrogen-cooled panels at $-320°F$ $(-196°C)$ is required.

5.5.7 Air Compressors

Capacity. The capacity of an air compressor is usually measured in terms of the cubic feet of "free air" handled per minute (V_a), which is air at atmospheric pressure (p_a) and atmospheric temperature (t_a). If P.D. represents the piston displacement in cubic feet per minute of a compressor operating with suction pressure p_a pounds absolute per square inch, a discharge pressure p_2 pounds absolute per square inch, compression according to the law pV^n = a constant, and m percent clearance (expressed as a decimal), then

$$V_a = \text{P.D.} \left[1 + m - m \left(\frac{p_2}{p_a} \right)^{1/n} \right] \quad \text{feet}^3/\text{minute}. \tag{5.78}$$

NOTE. For general expressions relating pressure, volume, and temperature, see Section 5.2. For air compressors, n = 1.20 to 1.35.

Volumetric Efficiency. The ratio of the volume of free air (V_a) to the piston displacement (P.D.) is termed displacement or volumetric efficiency (ϵ_v). For single-stage compression

$$\epsilon_v = \frac{V_a}{\text{P.D.}} = 1 + m - m \left(\frac{p_2}{p_a} \right)^{1/n}. \tag{5.79}$$

For multistage air compressors

$$\epsilon_v = \frac{V_a}{\text{P.D.}} = 1 + m - m \left(\frac{p_x}{p_a} \right)^{1/n}, \tag{5.80}$$

where p_x is the discharge pressure in pounds absolute per square inch leaving the first-stage cylinder. For two-stage compression

$$p_x = \sqrt{p_a p_2} \quad \text{pounds/inch}^2. \tag{5.81}$$

For three-stage compression

$$p_x = \sqrt[3]{p_a^2 p_2} \quad \text{and} \quad p_y = \sqrt[3]{p_a p_2^2} \quad \text{pounds/inch}^2, \tag{5.82}$$

where p_y is the discharge pressure in pounds absolute per square inch leaving the second-stage cylinder.

Power. The power (P) required in foot-pounds per minute to compress V_a cubic feet of free air per minute polytropically according to the law pV^n = a constant, from atmospheric pressure p_a to a discharge pressure p_2 for single-stage compression is

$$P = 144 p_a V_a \frac{n}{n-1} \left[\left(\frac{p_2}{p_a} \right)^{(n-1)/n} - 1 \right] \quad \text{foot-pounds/minute.} \qquad (5.83)$$

For two-stage compression

$$P = 288 p_a V_a \frac{n}{n-1} \left[\left(\frac{p_2}{p_a} \right)^{(n-1)/(2n)} - 1 \right] \quad \text{foot-pounds/minute.} \qquad (5.84)$$

For three-stage compression

$$P = 432 p_a V_a \frac{n}{n-1} \left[\left(\frac{p_2}{p_a} \right)^{(n-1)/(3n)} - 1 \right] \quad \text{foot-pounds/minute.} \qquad (5.85)$$

The power (P) required in foot-pounds per minute to compress V_a cubic feet of free air per minute isothermally according to the law $pV^n = $ a constant, from atmospheric pressure (p_a) to a discharge pressure (p_2) is

$$P = 144 p_a V_a \ln \frac{p_2}{p_a} \quad \text{foot-pounds/minute.} \qquad (5.86)$$

Efficiency of Compression. The ratio of the isothermal power to the polytropic power is termed the efficiency of compression (ϵ_c). For single-stage compression

$$\epsilon_c = \frac{\ln (p_2/p_a)}{\frac{n}{n-1} \left[\left(\frac{p_2}{p_a} \right)^{(n-1)/n} - 1 \right]}. \qquad (5.87)$$

For two-stage compression

$$\epsilon_c = \frac{\ln (p_2/p_a)}{\frac{2n}{n-1} \left[\left(\frac{p_2}{p_a} \right)^{(n-1)/(2n)} - 1 \right]}. \qquad (5.88)$$

For three-stage compression

$$\epsilon_c = \frac{\ln (p_2/p_a)}{\frac{3n}{n-1} \left[\left(\frac{p_2}{p_a} \right)^{(n-1)/(3n)} - 1 \right]}. \qquad (5.89)$$

5.6 REFRIGERATION

Ideal Compression Refrigeration Cycle. In order to simplify the references to conditions in the ideal compression refrigeration cycle, a complete description of the cycle is given. The compressor draws the vapor (usually saturated or slightly superheated) from the evaporator at condition 1, compresses it adiabatically and without friction to condition 2 in the superheated region, and then discharges the vapor to the condenser. The cooling water condenses the vapor to a saturated liquid at condition 3. The liquid is drawn off and then passes through an expansion valve to condition 4. This partially vaporized liquid now enters the evaporator where further evaporation takes place before entering the compressor.

Refrigerating Effect.   The refrigerant enters the evaporator with an enthalpy of h_{f_3} Btu/pound and leaves with an enthalpy of h_1 Btu/pound. If G_r pounds of refrigerant are circulated per minute and the refrigerating effect per minute is represented by R, then

$$R = G_r(h_1 - h_{f_3}) \quad \text{Btu/minute.} \tag{5.90}$$

Capacity.   The cubic feet per minute (V_1) handled by a compressor operating at N revolutions per minute with a piston displacement (P.D.) per revolution, and drawing in G_r pounds of refrigerant per minute, each pound having a specific volume v_1 cubic feet is

$$V_1 = N \times \text{P.D.} = G_r v_1 \quad \text{feet}^3/\text{minute.} \tag{5.91}$$

NOTE.   This formula assumes no clearance. If the refrigerant is compressed from a suction pressure p_1 pounds absolute per square inch to a discharge pressure p_2 pounds absolute per square inch according to the law $pV^n = $ a constant and a clearance of m percent expressed as a decimal, then

$$V_1 = N \times \text{P.D.} \left[1 + m - m \left(\frac{p_2}{p_1} \right)^{1/n} \right] \quad \text{feet}^3/\text{minute.} \tag{5.92}$$

Tonnage.   One ton of refrigeration is the heat equivalent to the melting of one ton (2000 pounds, 907.2 kg) of ice at 32°F (0°C) in 24 hours. Since one pound of ice melting at 32°F (0°C) will absorb approximately 144 Btu (36.3 kcal), then a ton of refrigeration will absorb 288,000 Btu (72.6 × 10³ kcal) per day or 200 Btu (50.4 kcal) per minute, hence

$$\text{tonnage} = \frac{R}{200} = \frac{G_r(h_1 - h_{f_3})}{200} \quad \text{tons.} \tag{5.93}$$

Power.   The refrigerant enters the compressor with an enthalpy h_1 and leaves with an enthalpy h_2 Btu/pound. If G_r pounds of refrigerant are circulated per minute, the power (P) expressed in foot-pounds per minute for adiabatic compression is

$$P = 778 G_r(h_2 - h_1) \quad \text{foot-pounds/minute.} \tag{5.94}$$

If V_1 cubic feet of refrigerant per minute enter the compressor with a suction pressure p_1 pounds absolute per square inch and is compressed to a discharge pressure

p_2 pounds absolute per square inch according to the law $pV^n = $ a constant, the power (P) is

$$P = 144p_1V_1 \frac{n}{n-1}\left[\left(\frac{p_2}{p_1}\right)^{(n-1)/n} - 1\right] \quad \text{foot-pounds/minute.} \quad (5.95)$$

Coefficient of Performance.   The ratio of the refrigerating effect R to the power $(P/778)$ is called the coefficient of performance (C.P.). Hence, for adiabatic compression.

$$\text{C.P.} = \frac{778R}{P} = \frac{h_1 - h_{f_3}}{h_2 - h_1}, \quad (5.96)$$

and, for polytropic compression

$$\text{C.P.} = \frac{5.40G_r(h_1 - h_{f_3})}{p_1V_1[n/(n-1)][(p_2/p_1)^{(n-1)/n} - 1]}. \quad (5.97)$$

Heat Removed in the Condenser.   If G pounds of refrigerant per minute enter the condenser with an enthalpy h_2 and leave with an enthalpy h_{f_3} Btu/pound, the heat removed per minute in the condenser (Q_c) is

$$Q_c = \frac{W}{778} + R = G(h_2 - h_{f_3}) \quad \text{Btu/minute.} \quad (5.98)$$

Weight of Cooling Water Required.   If Q_c Btu/minute are to be removed in the condenser and the temperatures of the cooling water entering and leaving the condenser are t_c and t_h, respectively, the pounds of cooling water per minute (G_w) required is

$$G_w = \frac{Q_c}{h_{f_h} - h_{f_c}} = \frac{G(h_2 - h_{f_3})}{h_{f_h} - h_{f_c}} \quad \text{pounds/minute.} \quad (5.99)$$

5.7   HEAT TRANSMISSION

Fundamental Equation.   The heat transmitted in engineering apparatus is effected by a combination of the heat transferred by conduction, convection, and radiation. If the temperature is low and the rate of flow of the fluid over the surface is high, the radiation factor is ignored. The fundamental equation for the heat (Q) conducted in time (t), through a material having a thermal conductivity (k) and a surface area (S) which is normal to the flow of heat and of thickness (x) in the direction of the flow of heat with a temperature difference (θ) between its surfaces, is

$$Q = \frac{ckS\theta t}{x} \quad \text{units in time } (t). \tag{5.100}$$

NOTE. Average values of k for various engineering materials expressed in gram-calories per second per square centimeter per centimeter per degree Celsius are given in Chapter 12. The constant c depends on the units of measurement as follows:

Q	S	x	θ	t	c	c^a
gram-calories	cm^2	cm	°C	seconds	1	0.000344
kilogram-calories	meters^2	cm	°C	hours	36,000	12.4
British thermal units	feet^2	inches	°F	hours	2,903	1
joules	cm^2	cm	°C	seconds	4.18	0.00144
joules	feet^2	inches	°F	seconds	851	0.293
kilowatt-hours	meters^2	cm	°C	hours	41.8	0.0144
kilowatt-hours	feet^2	inches	°F	hours	0.851	0.000293

^{a}Values of c if k is expressed in Btu per hour per square foot per inch per degree Fahrenheit.

If the heat is transmitted through a body composed of an inside film, two materials, and an outside film, Equation (5.100), when expressed in English units, becomes

$$Q = \frac{\theta_m}{1/a_1 S_{mf1} + x_1/k_1 S_{m1} + x_2/k_2 S_{m2} + 1/a_2 S_{mf2}} \quad \text{Btu/hour,} \tag{5.101}$$

where θ_m is the mean temperature difference in degrees Fahrenheit between the two fluids while passing over the body, a_1 and a_2 are the inside and outside film coefficients in Btu per hour per square foot per degree Fahrenheit, S_{mf1} and S_{mf2} are the areas of the inside and outside films in square feet, x_1 and x_2 are the thicknesses of the materials in inches, k_1 and k_2 are the conductivities of the materials in Btu per hour per square foot per inch of thickness per degree Fahrenheit, and S_{m1} and S_{m2} are the mean surface areas of the materials.

Mean Temperature Difference.   For building walls, roofs, partitions, and so forth, steam and refrigerating pipes carrying wet or saturated vapor and surrounded by atmospheric air, the heat is assumed to be transmitted from the hot fluid at a uniform temperature (t_1) to a cold fluid at a uniform temperature (t_2). For these cases the mean temperature difference (θ_m) is

$$\theta_m = t_1 - t_2 \quad \text{degrees Fahrenheit.} \tag{5.102}$$

In heat exchangers such as boilers, superheaters, condensers, economizers, liquid and gas heaters or coolers, the temperature of either one or both fluids changes. If the hot fluid enters the apparatus at a temperature t_1 and leaves at t_2 and the contiguous cold fluid temperatures are t_a and t_b, respectively, then

$$\theta_m = \frac{(t_1 - t_a) - (t_2 - t_b)}{\ln\left[(t_1 - t_a)/(t_2 - t_b)\right]} \quad \text{degrees Fahrenheit.} \tag{5.103}$$

NOTE. Equation (5.103) gives the logarithmic mean temperature difference and is applicable only when the overall coefficient of heat transfer (K), the weight (W) of the hot fluid and the weight (w) of the cold fluid and their specific heat (C) and (c), respectively, are approximately constant during the transfer of heat (see also Fig. 5.1).

Mean Surface Area. The most important surfaces encountered in engineering practice are the plane or uniform cross-sectional surface, cylindrical and spherical surfaces. If S_1 and S_2 represent the inside and outside surface areas, respectively, then the mean surface area (S_m) for the plane surface is

$$S_m = \frac{S_2 + S_1}{2} = S \quad \text{feet}^2. \tag{5.104}$$

For the cylindrical surface, the mean surface area (S_m) is

$$S_m = \frac{S_2 - S_1}{\ln \dfrac{S_2}{S_1}} = \frac{2\pi L(r_2 - r_1)}{\ln \dfrac{r_2}{r_1}} \quad \text{feet}^2, \tag{5.105}$$

where L is the length in feet and r_1 and r_2 are the inside and outside radii in feet, respectively.

For the spherical surface, the mean surface area (S_m) is

$$S_m = \sqrt{S_2 S_1} = 4\pi r_1 r_2 \quad \text{feet}^2. \tag{5.106}$$

5.7.1 Heat Exchangers

Heat exchangers are an important type of process equipment widely used in most industries. **Fired heat exchangers** are generally classified as boilers, and typically are governed by the ASME Boiler and Pressure Vessel code (Chapter 12); they are not considered in the following discussion. **Unfired heat exchangers** are often of the **counter flow design** in which the fluid flows are counter to each other, thus assuring that the fluid leaving the exchanger is close in temperature (approach) to the entering fluid. Heat exchangers often use a gas or vapor, frequently steam, as one medium with a liquid as the other. In the process, the vapor is condensed on the exterior of the exchange surface, typically tubing, and the resultant liquid may be further cooled in a drain cooler section. Noncondensing heat exchangers are used for process applications where the temperature exchange occurs above the dew point of the vapor, or where the exchange is between noncondensing media. Heat exchangers range from simple single-pass, straight-tube exchangers to those having multiple passes with U-tube, spiral, or other more complex configurations (Fig. 5.2).

Provision for shell expansion is necessary where heaters operate with significant thermal differences from ambient. For exchangers with the **fluid heads (channel)** at one end only, this is normally accomplished by permitting shell motion away from the channel on rollers or sliding supports. Straight-tube exchangers with channels at each end normally allow one end of the exchanger to float by means of expansion or packed joints in the shell or at the shell-to-channel connection.

General Heat Transfer Cases

Nomenclature:

$$\text{LMTD} = \text{logarithmic mean temperature difference} = \frac{\text{GTD - LTD}}{\text{Loge } \dfrac{\text{GTD}}{\text{LTD}}}$$

TR = temperature rise of cold substance.
TF = temperature fall of hot substance.
LTD + initial temperature difference between hot and cold substances.
FTD = final temperature difference between hot and cold substances.
GTD = greatest temperature difference.
LTD = least temperature difference.

Case I—Hot substance giving up heat with temperature remaining constant to a cold substance absorbing heat with rising temperature.

GTD = ITD
LTD = ITD–TR

This case includes:—steam condensers, ammonia condensers, and boiler feed water heaters.

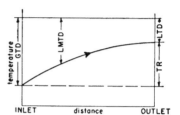

Case II—Hot substance giving up heat with falling temperature to a cold substance absorbing heat with temperature remaining constant.

GTD = ITD
LTD = ITD–TF

This case includes:—steam boilers, ammonia brine coolers, ammonia direct expansion coils in cold storage rooms, evaporator with hot liquid coils or jacket.

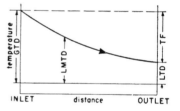

Case III—Two substances both changing temperature, one giving up heat with falling temperature to the other absorbing heat with rising temperature, **parallel flow.**

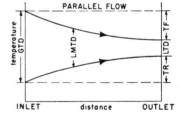

Case IV—Two substances both changing temperature, one giving up heat with falling temperature to the other absorbing heat with rising temperature, **countercurrent flow.**

Cases III & IV include:—steam superheaters and economizers, brine coils in cold storage rooms, compressor intercoolers, cylinder jackets and oil coolers.

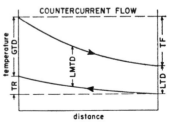

GENERAL HEAT TRANSFER CASES

Fig. 5.1 General heat transfer cases. (From Ref. 2. Copyright by Ingersoll-Rand Company, 1962. Published with permission.)

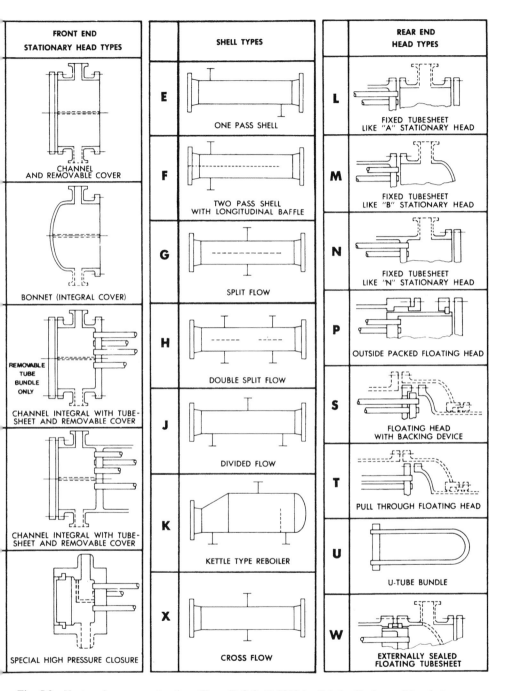

Fig. 5.2 Heat exchanger construction. (From Ref. 3. © 1978 by Tubular Exchange Manufacturers Association.)

Since pressure can be more easily withstood by smaller diameters, the higher-pressure fluid is usually contained on the interior of the internal tubing, with the larger-diameter shell retaining the lower-pressure fluid and thus permitting more economical design. The tubing system is an extension of the channel, which is also constructed to resist the higher pressures.

As a matter of design convenience, manufacturers typically assign pressure ratings to both the channel and shell side of exchangers. These typically are:

Channel		Shell
0–25 psig	(172 kPa)	Normally shell thicknesses are
150 psig	(1034 kPa)	tailored to the operating
300 psig	(2069 kPa)	pressure conditions.
600 psig	(4137 kPa)	
1250 psig	(8619 kPa)	
3000 psig	(20,700 kPa)	

For higher pressures, frequently a special type of locking system is used (in lieu of a conventional blind flange) to provide access to the interior of the channel. These take the form of interrupted bayonet threads, split locking rings, and so forth.

Tubing is connected to the tube sheet normally by **flaring (rolling).** This sealing may also be accomplished by a system of packing, or by tube-to-tube sheet seal welding. Tubing is usually of circular cross section, relatively thin wall (14 to 22 BWG), seamless material. Copper bearing, stainless, or alloy materials are common because of their resistance to corrosion and thus loss of heat transfer capability due to fouling (corrosion product buildup). In some types of exchangers, special tube cross sections with fins or twists are used to promote either internal or external turbulence and thus improve heat transfer coefficients or increase surface area.

Maintenance of a low U factor is important to heat-transfer equipment operation and can be achieved by provision for mechanical or chemical cleaning. During the course of the life of a heat exchanger, some tube leakage (at the tube sheet) can occur. This tube leakage is often repaired by plugging the tube and removing it from service; thus, design of heat exchangers should allow for excess tubing capacity, on the order of 5% or more, depending upon the duty anticipated.

TEMA (Tubular Exchanger Manufacturers Association) standards are widely applied throughout industry. The TEMA standards include thermal and mechanical standards, nomenclature, fabrication, installation, operation and maintenance, and materials specification data, as well as recommended practices.

5.7.2 Building Heat Transfer

Overall Coefficient of Heat Transfer. It is desirable in the solution of engineering problems involving the transfer of heat through typical walls, roofs, partitions, floors, and so on, to use a coefficient of heat transmission that will take into account the effects of conduction, convection, and radiation, together with the type, thickness, and position of the materials, and which may be used with the difference of the temperatures of the fluid temperatures on each side of the composite section. This quantity is termed overall coefficient of heat transfer (U) and is expressed in Btu per hour per square foot of surface area per degree Fahrenheit. The heat transmitted per hour (Q) becomes

$$Q = US\theta_m \quad \text{Btu/hour.} \tag{5.107}$$

NOTE.   Average values of U for the usual building structures are given in Table 5.8.

5.8   AIR AND VAPOR MIXTURES

Specific or Absolute Humidity.   The weight of water vapor per unit volume of space occupied, expressed in grains or pounds per cubic foot, is termed absolute humidity. In order to simplify the solution of problems involving air and vapor mixtures, it is convenient to express the weight of water per cubic foot (d_s) in terms of the weight of dry air per cubic foot (d_a). This ratio has no specific name although the term absolute humidity (ϕ) is often applied to this ratio.

$$\phi = \frac{d_s}{d_a} = \frac{v_a}{v_s} = 0.622 \left(\frac{p_s}{B - p_s} \right) \quad \text{pounds.} \tag{5.108}$$

NOTE.   In Equation (5.108) the perfect gas laws are assumed to hold for both the water vapor and the dry air present in the moisture-laden air. The total pressure of the moisture-laden air (B) expressed in pounds absolute per square inch is assumed to be equal to the sum of the partial pressure exerted by the water vapor (p_s) and the partial pressure exerted by the dry air (p_a), both expressed in pounds absolute per square inch.

Relative Humidity.   The ratio of the actual density of the water vapor in the moisture-laden air (d_s) to the density of saturated vapor (d_{sat}) at the same temperature is termed relative humidity (H). Assuming the perfect gas laws to satisfy this low pressure vapor, then

$$H = \frac{d_s}{d_{\text{sat}}} = \frac{v_{\text{sat}}}{v_s} = \frac{p_s}{p_{\text{sat}}}. \tag{5.109}$$

NOTE.   Although p_s may be determined from Ferrell's or Carrier's equation in terms of the barometric reading, the wet-bulb and dry-bulb temperatures, it is customary to use Equation (5.109) for determining the partial pressure (p_s) and the specific volume (v_s) of the water vapor. In engineering practice the psychometric tables are used for determining the relative humidity (H).

Table 5.8   Overall Coefficients of Heat Transfer U for Building Structures[a] (Expressed in Btu per Hour per Square Foot per Degree Fahrenheit)

Structure	Thickness (inches)		
	8	12	16
Walls			
Brick, without interior plaster	0.50	0.36	0.28
Brick, with interior plaster	0.46	0.34	0.27
Concrete, without interior plaster	0.69	0.54	0.48
Concrete, with interior plaster	0.62	0.49	0.44
Haydite, without interior plaster	0.36	0.26	0.21
Haydite, with interior plaster	0.34	0.24	0.20
Hollow tile, without interior plaster	0.40	0.30	0.25
Hollow tile, with interior plaster	0.38	0.29	0.24
Limestone, without interior plaster	0.71	0.49	0.37
Limestone, with interior plaster	0.64	0.45	0.35
Wood, shingled or clapboarded, with interior plaster			0.25
Stucco, with interior plaster			0.30
Brick veneer, with interior plaster			0.27
Partitions			
4-in. hollow clay tile, plaster both sides			0.40
4-in. common brick, plaster both sides			0.43
4-in. hollow gypsum tile, plaster both sides			0.27
Wood lath and plaster on one side of studding			0.62
Wood lath and plaster on both sides of studding			0.34
Metal lath and plaster on one side of studding			0.69
Partitions			
Metal lath and plaster on both sides of studding			0.39
Plasterboard and plaster on one side of studding			0.61
Plasterboard and plaster on both sides of studding			0.34
2-in. corkboard and plaster on one side of studding			0.12
2-in. corkboard and plaster on both sides of studding			0.063

	4	6	8	10
Floors				
Concrete, no ceiling, and no flooring	0.65	0.59	0.53	0.49
Concrete, plastered ceiling, and no flooring	0.59	0.54	0.50	0.45
Concrete, no ceiling, and terrazzo flooring	0.61	0.56	0.51	0.47
Concrete, plastered ceiling, and terrazzo flooring	0.56	0.52	0.47	0.44
Concrete, on ground, and no flooring	1.07	0.90	0.79	0.70
Concrete, on ground, and terrazzo flooring	0.98	0.84	0.74	0.66
Frame construction, no ceiling, maple or oak flooring on yellow pine sub-flooring on joists				0.34
Frame construction, metal lath and plaster ceiling, maple or oak flooring on yellow pine subflooring on joists				0.35
Frame construction, wood lath and plaster ceiling, maple or oak flooring on yellow pine subflooring on joists				0.24
Frame construction, plasterboard ceiling, maple or oak flooring on yellow pine subflooring on joists				0.24

Table 5.8 (*Continued*)

	2	4	6
Roofs, tar, and gravel			
Concrete, no ceiling, and no insulation	0.82	0.72	0.64
Concrete, no ceiling, and 1-in. rigid insulation	0.24	0.23	0.22
Concrete, metal lath and plaster ceiling, and no insulation	0.42	0.40	0.37
Concrete, metal lath and plaster ceiling, and 1-in. rigid insulation	0.19	0.18	0.18
1-in. wood, no ceiling, and no insulation			0.49
1-in. wood, no ceiling, and 1-in. rigid insulation			0.20
1-in. wood, metal lath and plaster ceiling, and no insulation			0.32
1-in. wood, metal lath and plaster ceiling, and 1-in. rigid insulation			0.16
Metal, no ceiling, and no insulation			0.95
Metal, no ceiling, and 1-in. rigid insulation			0.25
Metal, metal lath and plaster ceiling, and no insulation			0.46

	8	12	16
Roofs, tar, and gravel			
Metal, metal lath and plaster ceiling, and 1-in. rigid insulation			0.19
Wood shingles, rafters exposed			0.46
Wood shingles, metal lath and plaster			0.30
Wood shingles, wood lath and plaster			0.29
Wood shingles, plasterboard and plaster			0.29
Asphalt shingles, rafters exposed			0.56
Asphalt shingles, metal lath and plaster			0.34
Asphalt shingles, wood lath and plaster			0.32
Asphalt shingles, plasterboard and plaster			0.32
Glass			
Single windows and skylights			1.13
Double windows and skylights			0.45
Triple windows and skylights			0.281
Hollow glass tile wall, 6 × 6 × 2-in.-thick blocks			
Wind velocity 15 mph, outside surface; still air, inside surface			0.60
Still air outside and inside surfaces			0.48

Doors	1	1¼	1½	1¾	2	2½	3
Wood	0.69	0.59	0.52	0.51	0.46	0.38	0.33

[a]Correction for exposure:

Direction	North	East	South	West
Multiply U by:	1.3	1.1	1.0	1.2

Dry Bulb Temperature (°F)	Difference Between Wet and Dry Bulb									
	2°	4°	6°	8°	10°	12°	14°	16°	18°	20°
32	79	59	39	20	2	—	—	—	—	—
40	84	68	52	37	22	8	—	—	—	—
45	86	71	59	44	32	19	6	—	—	—
50	87	74	62	50	38	26	16	5	—	—
55	88	76	65	54	43	33	24	14	5	—
60	89	78	68	58	48	39	30	22	13	5
65	90	80	70	61	52	44	35	28	20	12
70	90	81	72	64	56	48	40	32	26	19
75	91	82	74	66	58	51	44	37	30	24
80	92	83	75	68	61	54	47	41	34	29
85	92	84	77	70	63	56	50	44	38	33
90	92	85	78	71	65	58	52	47	41	36
95	93	86	79	72	66	60	54	49	44	39
100	93	86	80	74	68	62	57	52	46	42

NOTE.   The relatively humidity should range between 35 and 45.

Dew Point.   When moisture-laden air is cooled until the temperature reaches that corresponding to the saturation temperature for the partial pressure of the water vapor, condensation or precipitation begins. This temperature is called the dew point.

Determination of Weight of Moisture Precipitated.   In order to precipitate moisture from V cubic feet of moisture-laden air at condition 1 it is necessary to cool the air to the dew point temperature (t_3) for the final condition 2 desired. The pounds of moisture precipitated (M_p) is

$$M_p = \frac{V}{v_{a_1}} (\phi_1 - \phi_2) = \frac{V}{v_{a_1}} \left(\frac{v_{a_1}}{v_{s_1}} - \frac{v_{a_2}}{v_{s_2}} \right) \text{ pounds.} \tag{5.110}$$

NOTE.   The specific volume of the dry air (v_a) may be determined from

$$v_a = \frac{53.34(t + 460)}{144(B - p_s)} \text{ feet}^3/\text{pound,} \tag{5.111}$$

and the specific volume of the water vapor (v_s) may be determined from

$$v_s = \frac{v_{\text{sat}}}{H} \text{ feet}^3/\text{pound.} \tag{5.112}$$

Determination of the Quantity of Heat Removed from the Moisture-Laden Air.   To remove the moisture (M_p), as given in Equation (5.110), it is necessary to supply refrigeration. This refrigeration must cool the dry air and the water vapor in addition to precipitating the moisture, thus the total amount of heat removed (R) in Btu is

$$R = \frac{V}{v_{a_1}} \left[0.241(t_1 - t_3) + \frac{v_{a_2}}{v_{s_2}} (h_{s_1} - h_{s_3}) + \left(\frac{v_{a_1}}{v_{s_1}} - \frac{v_{a_2}}{v_{s_2}} \right) (h_{s_1} - h_{f_3}) \right] \quad \text{Btu.}$$

(5.113)

NOTE. h_s may be assumed to be the same as the enthalpy (h) for the saturated vapor at the same temperature.

Determination of the Heat Added. To precipitate the required moisture from the air at condition 1, it was necessary to cool the air to the dew point temperature (t_3) for condition 2. The saturated air must now be heated to obtain the desired temperature (t_2). This heat must be supplied to the dry air and the water vapor; thus the total amount of heat added (Q) in Btu is

$$Q = \frac{V}{v_{a_1}} \left[0.241(t_3 - t_2) + \frac{v_{a_2}}{v_{s_2}} (h_{s_2} - h_{s_3}) \right] \quad \text{Btu.} \qquad (5.114)$$

REFERENCES

1. J. H. Keenan, F. G. Keyes, et al. *Steam Tables, Thermodynamic Properties of Water,* John Wiley & Sons, Inc., New York, 1969.
2. *Cameron Hydraulic Data,* 18th ed., Ingersoll-Dresser Pumps, Liberty Corner, N.J., 1994.
3. *Standards of Tubular Exchanger Manufacturers Association,* 6th ed., Tubular Exchanger Manufacturers Association, Inc., Tarrytown, N.Y. 1978.

BIBLIOGRAPHY

Dushman, S., *Scientific Foundations of Vacuum Technique,* 2nd ed., John Wiley & Sons, Inc., New York, 1962.

Potter, P. J., *Steam Power Plants.* The Ronald Press Company, New York, 1949.

Skrotzki, B. G. A., and W. A. Vopat, *Power Station Engineering and Economy,* McGraw-Hill Book Company, New York, 1960.

CHAPTER 6
ELECTRICITY AND ELECTRONICS

J. M. SHULMAN

Fellow Engineer (Retired)
Westinghouse Electric Corporation

6.1 ELECTRICAL ENGINEERING TERMS, UNITS, AND SYMBOLS

6.1.1 Standards

In the United States, standards covering design, manufacture, and use of electrical and electronic equipment are published by the Institute of Electrical and Electronic Engineers (IEEE), American National Standards Institute (ANSI), and National Electrical Manufacturers Association (NEMA). Each of these organizations issues a catalog of its standards and updates it at least once a year.[1-3] International standards are published by the International Electro-technical Commission (IEC), located at One Rue de Varembe, Geneva, Switzerland.

IEEE publishes a computer-compiled index of all its standards and all ANSI electrical standards.[4] In its first section the standards are listed by an alphabetical index of key words. The second section is a numerical listing by ANSI and IEEE standard numbers, giving for each the latest date of publication, information for ordering from IEEE, and a complete table of contents. The table of contents enables a reader to quickly determine the possible application of that particular standard.

6.1.2 Terms and Units

One of the IEEE standards is a dictionary of electrical and electronic terms.[5] In addition to terms and definitions previously standardized by IEEE, it now includes many in the ANSI standards and those of the IEC.

The dictionary points out a distinction between use of the terms *electric* and *electrical* as adjectives. **Electric** means containing, producing, arising from, actuated by or carrying electricity, or designed to carry electricity and capable of doing so. **Electrical** means related to, pertaining to or associated with electricity, but not having its properties or characteristics. Examples: electric current, electric motor, electrical engineer, electrical handbook. A note included in the definition of both words states that although some dictionaries indicate they are synonymous, usage in the electrical engineering field has in general been restricted to the meanings in the foregoing definitions. Also, it is recognized that there are borderline cases wherein the usage determines the selection.

The **International System of Units** (SI) is an updated version of the meter–kilogram–second–ampere (MKSA) metric system (which has long been the standard of many countries) and is how being adopted throughout the world. SI units are divided into three classes-base units, supplementary units, and derived units. Base units are those that are regarded as dimensionally independent. They are as follows:

Quantity	Unit	Symbol
Length	meter	m
Mass	kilogram	kg
Time	second	s
Electric current	ampere	A
Thermodynamic temperature	kelvin	K
Amount of substance	mole	mol
Luminous intensity	candela	cd

Supplementary units are those which may be regarded as base units or derived units, as follows:

Quantity	Unit	Symbol
Plane angle	radian	rad
Solid angle	steradian	sr

Derived units are those formed by combining base units, supplementary units, and other derived units according to the algebraic relations linking the corresponding quantities.

In some countries, particularly English-speaking ones, much of the present engineering practice and technical literature uses non-SI units. To encourage and facilitate orderly changeover to metric practice and the use of SI units, a standard on metric practice has been published[6] which contains, in addition to information on base, supplementary, and derived units, complete tables of conversion factors from English system and obsolete metric system units to SI units.

6.1.3   Symbols, Graphics, and Alphabets

A **unit symbol** is a letter or group of letters that may be used in place of the name of the unit. A standard is available[7] that contains a complete alphabetical listing of all SI units and their unit symbols for use in electrical engineering.

Quantity symbols are those used in formulas. A quantity symbol is normally a single letter modified when appropriate by one or more subscripts or superscripts. The standard on quantity symbols[8] lists both quantity and unit symbols used in electrical science and electrical engineering in tables by the following categories: (1) space and time, (2) mechanics, (3) heat, (4) radiation and light, (5) fields and circuits,

(6) electronics and telecommunication, (7) machines and power engineering. Table 8 is a list of symbols for physical constants, Table 9 a list of selected mathematical symbols, and Table 10 an alphabetical list of all quantity symbols.

Device and wiring graphic symbols for use in schematic wiring diagrams of electrical and electronic circuits and in single-line diagrams of electric power circuits have been standardized.[9] Those most frequently used in practice are shown in Fig. 6.1. Figure 6.2 illustrates three types of diagrams that use graphic symbols.

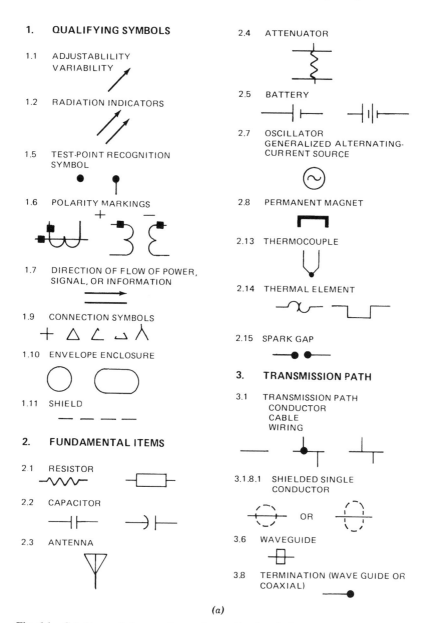

(a)

Fig. 6.1 Graphic symbols most frequently used in electrical and electronics diagrams.

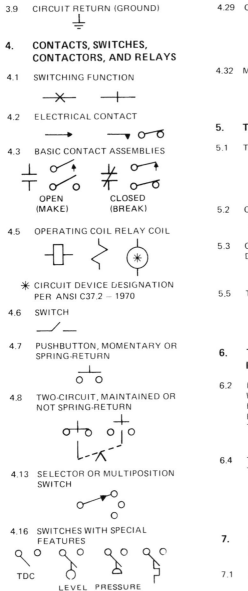

3.9   CIRCUIT RETURN (GROUND)

4.   CONTACTS, SWITCHES, CONTACTORS, AND RELAYS

4.1   SWITCHING FUNCTION

4.2   ELECTRICAL CONTACT

4.3   BASIC CONTACT ASSEMBLIES

OPEN (MAKE)   CLOSED (BREAK)

4.5   OPERATING COIL RELAY COIL

✳ CIRCUIT DEVICE DESIGNATION PER ANSI C37.2 – 1970

4.6   SWITCH

4.7   PUSHBUTTON, MOMENTARY OR SPRING-RETURN

4.8   TWO-CIRCUIT, MAINTAINED OR NOT SPRING-RETURN

4.13   SELECTOR OR MULTIPOSITION SWITCH

4.16   SWITCHES WITH SPECIAL FEATURES

TDC

TIME DELAY   LEVEL ACTUATED   PRESSURE OR VACUUM ACTUATED   TEMPERATURE ACTUATED

4.29   CONTACTOR

4.32   MERCURY SWITCH

5.   TERMINALS AND CONNECTORS

5.1   TERMINALS

5.2   CABLE TERMINATION

5.3   CONNECTOR DISCONNECTING DEVICE

5.5   TEST BLOCKS

6.   TRANSFORMERS, INDUCTORS, AND WINDINGS

6.2   INDUCTOR WINDING REACTOR RADIO FREQUENCY COIL TELEPHONE RETARDATION COIL

6.4   TRANSFORMER TELEPHONE INDUCTION COIL

7.   ELECTRON TUBES AND RELATED DEVICES

7.1   ELECTRON TUBE – COMPONENTS

(b)

Fig. 6.1   (*Continued*)

Prefixes and symbols for decimal multiples and submultiples of SI units are standardized from 10^{18} to 10^{-18}. Those most frequently used in electrical and electronic engineering practice are found in Chapter 12.

SI quantities most frequently used in practice are listed alphabetically in Table 6.1 (page 299), with their quantity symbols, unit names, and unit symbols.

7.3   TYPICAL APPLICATIONS

7.7   NUCLEAR-RADIATION DETECTOR
IONIZATION CHAMBER
PROPORTIONAL COUNTER TUBE
GEIGER-MÜLLER COUNTER TUBE

8.   SEMICONDUCTOR DEVICES

8.2   ELEMENT SYMBOLS

8.5   TYPICAL APPLICATIONS: TWO-
TERMINAL DEVICES

8.6   TYPICAL APPLICATIONS: THREE-
(OR MORE) TERMINAL DEVICES

8.7   PHOTOSENSITIVE CELL

8.8   SEMICONDUCTOR THERMO-
COUPLE

8.9   HALL ELEMENT
HALL GENERATOR

9.   CIRCUIT PROTECTORS

9.1   FUSE

9.3   LIGHTNING ARRESTER
ARRESTER
GAP

9.4   CIRCUIT BREAKER

9.5   PROTECTIVE RELAY

10.   ACOUSTIC DEVICES

10.1   AUDIBLE-SIGNALING DEVICE

**11.   LAMPS AND VISUAL-
SIGNALING DEVICES**

11.2   VISUAL-SIGNALING DEVICE

11.2.1   ANNUNCIATOR

11.2.6   INDICATING LAMP

12.   READOUT DEVICES

12.1   METER INSTRUMENT

A	DB	I	OP	RF	V A
AH	DBM	INT	OSCG	SY	VAR
C	DM	μA	PH	TLM	VARH
CMA	DTR	UA	PI	t^o	VI
CMC	F	MA	PH	THC	VU
CMV	G	NM	RD	TT	W
CRO	GD	OHM	REC	V	WH
		ETC.			

(c)

Fig. 6.1   *(Continued)*

Letter symbols are mainly restricted to the English and Greek alphabets. In print the quantity symbols appear in italic (sloping) type and the unit symbols in roman (upright) type. Quantity symbols are used in both capital and lowercase letter versions. Because of typography variations and use of both capital and lowercase letters, good engineering practice dictates that whenever a possibility of ambiguity or misunderstanding exists in the use of either a quantity or unit symbol, the symbol should be fully identified at the location where it is used.

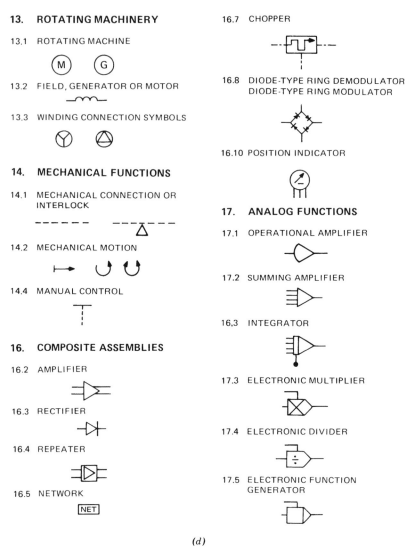

13. ROTATING MACHINERY

13.1 ROTATING MACHINE

13.2 FIELD, GENERATOR OR MOTOR

13.3 WINDING CONNECTION SYMBOLS

14. MECHANICAL FUNCTIONS

14.1 MECHANICAL CONNECTION OR INTERLOCK

14.2 MECHANICAL MOTION

14.4 MANUAL CONTROL

16. COMPOSITE ASSEMBLIES

16.2 AMPLIFIER

16.3 RECTIFIER

16.4 REPEATER

16.5 NETWORK

16.7 CHOPPER

16.8 DIODE-TYPE RING DEMODULATOR
       DIODE-TYPE RING MODULATOR

16.10 POSITION INDICATOR

17. ANALOG FUNCTIONS

17.1 OPERATIONAL AMPLIFIER

17.2 SUMMING AMPLIFIER

16.3 INTEGRATOR

17.3 ELECTRONIC MULTIPLIER

17.4 ELECTRONIC DIVIDER

17.5 ELECTRONIC FUNCTION GENERATOR

(d)

Fig. 6.1 (Continued)

6.2 CIRCUIT ELEMENTS

6.2.1 Circuit Definition

All useful applications of electricity and electronics involve conductors or systems of conductors through which current is intended to flow called circuits. In order for current to flow, one or more voltage sources must be present in the circuit. Voltage sources have many different forms, and circuit currents have magnitudes, directions, and forms depending on voltages. Voltages and currents may be of constant magnitude and direction (continuous dc), varying magnitude and constant direction (varying or pulsating dc), cyclic reversal of direction (ac) or transient (of short duration com-

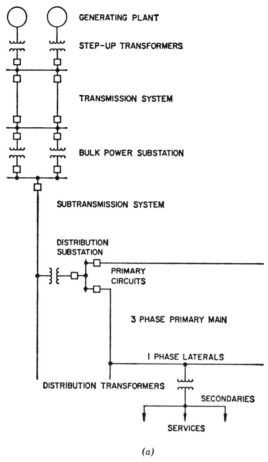

GENERATING PLANT

STEP-UP TRANSFORMERS

TRANSMISSION SYSTEM

BULK POWER SUBSTATION

SUBTRANSMISSION SYSTEM

DISTRIBUTION
SUBSTATION

PRIMARY
CIRCUITS

3 PHASE PRIMARY MAIN

I PHASE LATERALS

DISTRIBUTION TRANSFORMERS

SECONDARIES

SERVICES

(a)

Fig. 6.2 Typical diagrams with graphic symbols: (*a*) single-line diagram of an electric power system; (*b*) electronic circuit with transistors; (*c*) electronic circuit with integrated circuits. [(*a*) Used with permission of the Westinghouse Electric Corporation; (*b* and *c*) used with permission of the American Radio Relay League, Inc.]

pared to the steady-state condition). All circuits contain three linear and bilateral properties that determine the nature of current resulting from voltage, resistance, inductance, and capacitance.

6.2.2 Resistance

The physical properties of a conductor that tend to opposed the flow of current through it, and the conductor dimensions determine its resistance:

$$r = \frac{\rho L}{A},$$ (6.1)

where L is length in the direction of current flow, A is cross-sectional area perpen-

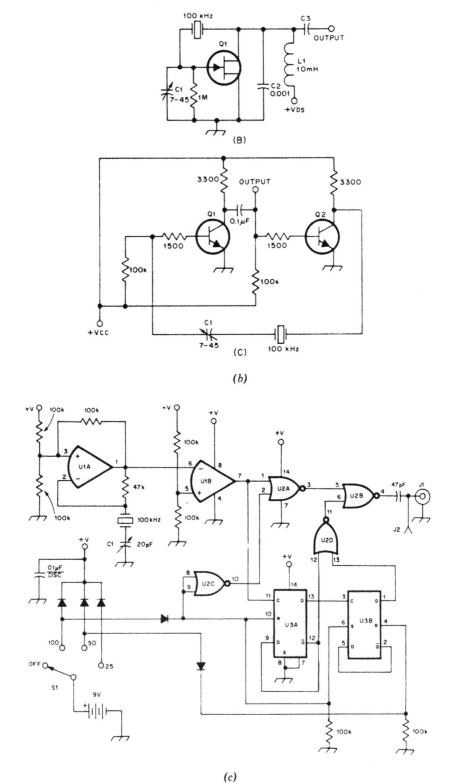

(b)

(c)

Fig. 6.2 (*Continued*)

Table 6.1 Quantities and SI Units Most Frequently Used in Electrical Engineering

Quantity	Quantity Symbol	Unit	Unit Symbol
Admittance	Y	siemens	S
Angular velocity	ω	radian per second	rad/s
Capacitance	C	farad	F
Conductance	G	siemens	S
Conductivity	σ	siemens per meter	S/m
Current	I	ampere	A
Current density	J	ampere per square meter	A/m²
Electric charge	Q	coulomb	C
Electric field strength	E	volt per meter	V/m
Electric flux	Ψ	coulomb	C
Electric flux density	D	coulomb per square meter	C/m²
Electromotive force	E	volt	V
Energy	W	joule	J
Force	F	newton	N
Frequency	f	hertz	Hz
Impedance	Z	ohm	Ω
Inductance	L	henry	H
Inductance, mutual	M	henry	H
Length	L	meter	m
Magnetic field strength (also called magnetizing force)	H	ampere per meter	A/m
Magnetic flux	Φ	weber	Wb
Magnetic flux density	B	tesla	T
Magnetomotive force	$\mathcal{F}$	ampere	A
Mass	m	kilogram	kg
Permeability	μ	henry per meter	H/m
Permeance	$\mathcal{P}$	henry	H
Permittivity	ϵ	farad per meter	F/m
Phase angle	Θ	radian	rad
Potential, potential difference	V	volt	V
Power, reactive	Q	var	var
Power, real	P	watt	W
Pressure	p	newton per square meter	N/m²
Reactance	X	ohm	Ω
Reluctance	$\mathcal{R}$	reciprocal henry	H⁻¹
Resistance	R	ohm	Ω
Resistivity	ρ	ohm meter	Ω·m
Susceptance	B	siemens	S
Temperature	T	degrees Celsius	°C
Time	t	second	s
Voltage	V	volt	V
Wavelength	λ	meter	m
Work	W	joule	J

dicular to flow, and ρ is a constant for the material, called resistivity or specific resistance. Rho is numerically equal to R when L and A are equal to one. For L in meters (m) and A in square meters (m²), the unit for ρ is ohm meters.

Resistivity of metallic conductors varies linearly between approximately 100 and $-200°C$ (Fig. 6.3). It approaches zero nonlinearly between $-200°C$ and absolute zero temperature $-273.13°C$. Since most practical work is in the linear range, extending

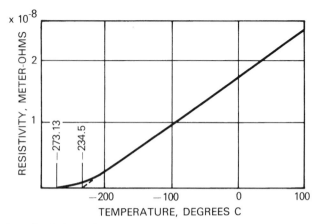

Fig. 6.3 Variation of resistivity of copper with temperature. (From H. H. Skilling, *Electrical Engineering Circuits,* 2nd ed. Copyright © 1965 by John Wiley & Sons, Inc., Reproduced with permission.)

the linear portion of the curve to the abscissa yields a value of temperature T, on which the rate of change of resistivity α_1 at any temperature t_1 higher than T can be calculated as

$$\alpha_1 = \frac{1}{T + t_1}. \tag{6.2}$$

T is 234.5 for standard annealed copper. If t_1 is $+20°C$, $\alpha_1 = 1/254.5 = 0.00393$. α_1 is called the temperature coefficient of resistivity (or resistance) at temperature t_1, which must be stated. Table 6.2 lists values of resistivity, temperature coefficient of resistivity, and reference temperature T for copper and other commonly used conductors.

If R_1 and α_1 are known at any temperature t_1 in the linear range, R_2 at any temperature t_2 also in the linear range can be found from the formula

$$R_2 = R_1[1 + \alpha_1(t_2 - t_1)]. \tag{6.3}$$

If reference temperature T is known instead of α_1, R_2 can be found from the formula

$$R_2 = R_1\left(\frac{T + t_2}{T + t_1}\right). \tag{6.4}$$

Most resistance–temperature calculations in practice are for copper, for which $T = 234.5$, a figure easily memorized, so Equation (6.5) is more frequently used.

$$L = \frac{N\Phi}{I} \quad \text{henrys.} \tag{6.5}$$

American Wire Gauge (B&S) dimensional and resistance data for standard annealed copper and for hard-drawn aluminum wire are given in Tables 6.3 and 6.4.

Table 6.2 Resistivity of Metals and Temperature Coefficients of Resistivity

Metal	Resistivity at 20 °C (Ω m)	Conductivity (% of Standard Annealed Copper)	Temperature Coefficient of Resistivity per Degree Celsius at 20° C	Reference Temperature (T) in Equations (6.2) and (6.4)
Copper, standard annealed	1.7241×10^{-8}	100	0.00393	234.5
Coper, soft-drawn	1.76×10^{-8}	98	0.00385	240
Copper, medium hard-drawn	1.76–1.78×10^{-8}	97–98	0.00383 av	241
Copper, hard-drawn	1.77×10^{-8}	97.3	0.00382	242
Aluminum, 99.97% pure	2.66×10^{-8}	64.6	0.00427	214
Aluminum, average commercial hard-drawn	2.83×10^{-8}	61.0	0.00403	228
Brass, typical	7–8×10^{-8}	21–25	0.002	480
Gold	2.44×10^{-8}	70.8	0.0034	274
Iron, pure	10×10^{-8}	17	0.005	180
Platinum	10×10^{-8}	17	0.003	313
Silver	1.63×10^{-8}	106	0.0038	243
Steel, carbon, medium to hard	20–50×10^{-8}	3–9	0.002–0.005	480–180
Tin	11.5×10^{-7}	15	0.0042	218
Tungsten	5.51×10^{-8}	31	0.0045	202
Zinc	5.8×10^{-8}	30	0.0037	250

Source: H. H. Skilling, *Electrical Engineering Circuits,* 2nd ed., Copyright © 1965 by John Wiley & Sons, Inc. Reproduced with permission.

6.2.3 Inductance and Ferromagnetic Circuits

A varying current flowing in a conductor creates a magnetic field around the conductor, which in turn creates an electromotive force (voltage) in the conductor and also in any other conductor close enough to be in the magnetic field. **Self-inductance** is the property of the conductor that produces an induced voltage in itself with changing current. The term "inductance" alone means self-inductance. When a varying current in one conductor induces a voltage in a neighboring conductor the effect is called **mutual inductance.**

Amount of inductance is a function of conductor configuration, dimensions, and magnetic characteristics of the medium surrounding it. Inductance is increased when a conductor is wound into the form of a coil. It is further increased by a large magnitude when the medium for the magnetic field inside the coil is magnetic material instead of free space or air.

A coil of N turns has an inductance of 1 henry if a current of 1 ampere in it produces a flux of 1 weber:

Table 6.3 Wire Table, Standard Annealed Copper: American Wire Gauge (B&S), English Units

Gauge No. (AWG)	Diameter in Mils at 20°C	Cross Section at 20°C		Ohms per 1000 ft^a at 20°C (= 68°F)	Pounds per 1000 ft	Feet per Pound	Feet per Ohm^b at 20°C (= 68°F)	Ohms per Pound at 20°C (= 68°F)	Pounds per Ohm at 20°C (= 68°F)
		Circular Mils	Square Inches						
0000	460.0	211,600	0.1662	0.049 01	640.5	1.561	20,400	0.000 076 52	13,070
000	409.6	167,800	0.1318	0.061 80	507.9	1.968	16,180	0.000 1217	8,219
00	364.8	133,100	0.1045	0.077 93	402.8	2.482	12,830	0.000 1935	5,169
0	324.9	105,500	0.082 89	0.098 27	319.5	3.130	10,180	0.000 3076	3,251
1	289.3	83,690	0.065 73	0.1239	253.3	3.947	8,070	0.000 4891	2,044
2	257.6	66,370	0.052 13	0.1563	200.9	4.977	6,400	0.000 7778	1,286
3	229.4	52,640	0.041 34	0.1970	159.3	6.276	5,075	0.001 237	808.6
4	204.3	41,740	0.032 78	0.2485	126.4	7.914	4,025	0.001 966	508.5
5	181.9	33,100	0.026 00	0.3133	100.2	9.980	3,192	0.003 127	319.8
6	162.0	26,250	0.020 62	0.3951	79.46	12.58	2,531	0.004 972	201.1
7	144.3	20,820	0.016 35	0.4982	63.02	15.87	2,007	0.007 905	126.5
8	128.5	16,510	0.012 97	0.6282	49.98	20.01	1,592	0.012 57	79.55
9	114.4	13,090	0.010 28	0.7921	36.63	25.23	1,262	0.019 99	50.03
10	101.9	10,380	0.008 155	0.9989	31.43	31.82	1,001	0.031 78	31,47
11	90.74	8,234	0.006 467	1.260	24.92	40.12	794.0	0.050 53	19.79
12	80.81	6,530	0.005 129	1.588	19.77	50.59	629.6	0.080 35	12.45
13	71.96	5,178	0.004 067	2.003	15.68	63.80	499.3	0.1278	7.827
14	64.08	4,107	0.003 225	2.525	12.43	80.44	396.0	0.2032	4.922
15	57.07	3,257	0.002 558	3.184	9.858	101.4	314.0	0.3230	3.096
16	50.82	2,583	0.002 028	4.016	7.818	127.9	249.0	0.5136	1.947
17	45.26	2,048	0.001 609	5.064	6.200	161.3	197.5	0.8167	1.224
18	40.30	1,624	0.001 276	6.385	4.917	203.4	156.6	1.299	0.7700

19	35.89	1,288	0.001 012	8.051	3.899	256.5	124.2	2.065	0.4843
20	31.96	1,022	0.000 802 3	10.15	3.092	323.4	98.50	3.283	0.3046
21	28.46	810.1	0.000 636 3	12.80	2.452	407.8	78.11	5.221	0.1915
22	25.35	642.4	0.000 504 6	16.14	1.945	514.2	61.95	8.301	0.1205
23	22.57	509.5	0.000 400 2	20.36	1.542	648.4	49.13	13.20	0.075 76
24	20.10	404.0	0.000 317 3	25.67	1.223	817.7	38.96	20.99	0.047 65
25	17.90	320.4	0.000 251 7	32.37	0.9699	1,031	30.90	33.37	0.029 97
26	15.94	254.1	0.000 199 6	40.81	0.7692	1,300	24.50	53.06	0.018 85
27	14.20	201.5	0.000 158 3	51.47	0.6100	1,639	19.43	84.37	0.011 85
28	12.64	159.8	0.000 125 5	64.90	0.4837	2,067	15.41	134.2	0.007 454
29	11.26	126.7	0.000 099 53	81.83	0.3836	2,607	12.22	213.3	0.004 688
30	10.03	100.5	0.000 078 94	103.2	0.3042	3,287	9.691	339.2	0.002 948
31	8.928	79.70	0.000 062 60	130.1	0.2413	4,145	7.685	539.3	0.001 854
32	7.950	63.21	0.000 049 64	164.1	0.1913	5,227	6.095	857.6	0.001 166
33	7.080	50.13	0.000 039 37	206.9	0.1517	6,591	4.833	1,364	0.000 7333
34	6.305	39.75	0.000 031 22	260.9	0.1203	8,310	3.833	2,168	0.000 4612
35	5.615	31.52	0.000 024 76	329.0	0.095 42	10,480	3.040	3,448	0.000 2901
36	5.000	25.00	0.000 019 64	414.8	0.075 68	13,210	2.411	5,482	0.000 1824
37	4.453	19.83	0.000 015 57	523.1	0.060 01	16,660	1.912	8,717	0.000 1147
38	3.965	15.72	0.000 012 35	659.6	0.047 59	21,010	1.516	13,860	0.000 072 15
39	3.531	12.47	0.000 009 793	831.8	0.037 74	26,500	1.202	22,040	0.000 045 38
40	3.145	9.888	0.000 007 766	1,049	0.029 93	33,410	0.9534	35,040	0.000 028 54

Source: O. W. Eshbach and M. Sounders, *Handbook of Engineering Fundamentals*, 3rd ed., Copyright © 1975 by John Wiley & Sons, Inc. Reproduced with permission.

[a]Resistance at the stated temperatures of a wire whose length is 1000 ft at 20° C.
[b]Length at 20°C of a wire whose resistance is 1 Ω at the stated temperatures.

Table 6.4   Wire Table, Aluminum: Hard-Drawn Aluminum Wire at 20°C (68°F), American Wire Gauge (B&S), English Units

| Gauge No. | Diameter (mils) | Cross Section | | Ohms per 1000 ft | Pounds per 1000 ft | Pounds per Ohm | Feet per Ohm |
		Circular Mils	Square Inches				
0000	460	212,000	0.166	0.0804	195	2,420	12,400
000	410	168,000	0.132	0.101	154	1,520	9,860
00	365	133,000	0.105	0.128	122	957	7,820
0	325	106,000	0.0829	0.161	97.0	602	6,200
1	289	83,700	0.0657	0.203	76.9	379	4,920
2	258	66,400	0.0521	0.256	61.0	238	3,900
3	229	52,600	0.0413	0.323	48.4	150	3,090
4	204	41,700	0.0328	0.408	38.4	94.2	2,450
5	182	33,100	0.0260	0.514	30.4	59.2	1,950
6	162	26,300	0.0206	0.648	24.1	37.2	1,540
7	144	20,800	0.0164	0.817	19.1	23.4	1,220
8	128	16,500	0.0130	1.03	15.2	14.7	970
9	114	13,100	0.0103	1.30	12.0	9.26	770
10	102	10,400	0.008 15	1.64	9.55	5.83	610
11	91	8,230	0.006 47	2.07	7.57	3.66	484
12	81	6,530	0.005 13	2.61	6.00	2.30	384
13	72	5,180	0.004 07	3.29	4.76	1.45	304
14	64	4,110	0.003 23	4.14	3.78	0.911	241
15	57	3,260	0.002 56	5.22	2.99	0.573	191
16	51	2,580	0.002 03	6.59	2.37	0.360	152
17	45	2,050	0.001 61	8.31	1.88	0.227	120
18	40	1,620	0.001 28	10.5	1.49	0.143	95.5

19	36	1,290	0.001 01	13.2	1.18	0.0897	75.7
20	32	1,020	0.000 802	16.7	0.939	0.0564	60.0
21	28.5	810	0.000 636	21.0	0.745	0.0355	47.6
22	25.3	642	0.000 505	26.5	0.591	0.0223	37.8
23	22.6	509	0.000 400	33.4	0.468	0.0140	29.9
24	20.1	404	0.000 317	42.1	0.371	0.008 82	23.7
25	17.9	320	0.000 252	53.1	0.295	0.005 55	18.8
26	15.9	254	0.000 200	67.0	0.234	0.003 49	14.9
27	14.2	202	0.000 158	84.4	0.185	0.002 19	11.8
28	12.6	160	0.000 126	106	0.147	0.001 38	9.39
29	11.3	127	0.000 099 5	134	0.117	0.000 868	7.45
30	10.0	101	0.000 078 9	169	0.0924	0.000 546	5.91
31	8.9	79.7	0.000 062 6	213	0.0733	0.000 343	4.68
32	8.0	63.2	0.000 049 6	269	0.0581	0.000 216	3.72
33	7.1	50.1	0.000 039 4	339	0.0461	0.000 136	2.95
34	6.3	39.8	0.000 031 2	428	0.0365	0.000 085 4	2.34
35	5.6	31.5	0.000 024 8	540	0.0290	0.000 053 7	1.85
36	5.0	25.0	0.000 019 6	681	0.0230	0.000 033 8	1.47
37	4.5	19.8	0.000 015 6	858	0.0182	0.000 021 2	1.17
38	4.0	15.7	0.000 012 3	1,080	0.0145	0.000 013 4	0.924
39	3.5	12.5	0.000 009 79	1,360	0.0115	0.000 008 40	0.733
40	3.1	9.9	0.000 007 77	1,720	0.0091	0.000 005 28	0.581

Source: O. W. Eshbach and M. Souders, *Handbook of Engineering Fundamentals*, 3rd ed. Copyright © 1975 by John Wiley & Sons, Inc. Reproduced with permission.

The **magnetic permeability** of free space is $4\pi \times 10^{-7}$. One weber is equal to 10^8 flux lines. Air-core inductors therefore have inductances very low compared with iron-core inductors. The latter, however, are not linear, so air-core inductors must be used when linearity is a requirement.

The following formulas give inductances for coils and other conductor configurations in air most frequently encountered in engineering practice.

The **self-inductance** (L) per centimeter axial length of the turns near the center of an air solenoid, A square centimeters in sectional area, wound uniformly with n turns per centimeter length (dimensions of sectional area negligible compared with the axial length):

$$L = 12.6n^2A \times 10^{-9} \text{ henrys.} \tag{6.6}$$

NOTE. If the solenoid is filled completely with a medium of constant permeability μ, the self-inductance per centimeter length is $12.6n^2\mu A \times 10^{-9}$ henrys, and if filled partially throughout its length with a medium of constant permeability μ and B square centimeters in constant sectional area, the self-inductance per centimeter length is $12.6n^2(\mu B + A - B) \times 10^{-9}$ henry.

The **self-inductance** (L) of a single-layer short solenoid of N turns, l centimeters in axial length, and r centimeters in radius (l small compared with r):

$$L = 12.6rN^2 \left[\ln \frac{8r}{l} - \frac{1}{2} + \frac{l^2}{32r^2}\left(\ln \frac{8r}{l} + \frac{1}{4} \right) \right] \times 10^{-9} \text{ henrys.} \tag{6.7}$$

The **self-inductance** (L) of a multiple-layer short solenoid of N turns, l centimeters in axial length, R centimeters in external radius, and r centimeters in internal radius (l small compared with R or r):

$$L = 12.6aN^2 \left[\ln \frac{8a}{b}\left(1 + \frac{3b^2}{16a^2} \right) - \left(2 + \frac{b^2}{16a^2} \right) \right] \times 10^{-9} \text{ henrys.} \tag{6.8}$$

NOTE. $a = (r + r)/2$ and $b = 0.2235(l + R - r)$. ln equals $\log_e$.

The **self-inductance** (L) of a toroidal coil wound uniformly with a single layer of N turns on a surface generated by the revolution of a circle r centimeters in radius about an axis R centimeters from the center of the circle:

$$L = 12.6N^2(R - \sqrt{R^2 - r^2}) \times 10^{-9} \text{ henrys.} \tag{6.9}$$

The **self-inductance** $(L$ of a toroidal coil of rectangular section, r and R centimeters in internal and external radius, respectively, sides of section $(R - r)$ centimeters and l centimeters, respectively, and wound uniformly with a single layer of N turns:

$$L = 2N^2 l \ln \frac{R}{r} \times 10^{-9} \text{ henrys.} \qquad (6.10)$$

The **self-inductance** (L) per centimeter length of one of two parallel straight, cylindrical wires, each r centimeters in radius, their axes d centimeters apart, and conducting the same current in opposite directions (distance d small compared with the length of the wires):

$$L = \left(2 \ln \frac{d}{r} + 0.5 \right) \times 10^{-9} \text{ henrys.} \qquad (6.11)$$

NOTE. The self-inductance of each wire per mile is $0.08047 + 0.7411 \log (d/r)$ millihenrys, and for two wires is twice as great. Equation (6.11) also gives the self-inductance per centimeter length of one of three wires, located at the vertices of an equilateral triangle (d is the distance between the axes of any two wires, provided the algebraic sum of the instantaneous currents conducted respectively by the three wires in the same direction equals zero).

The **self-inductance** (L) per mile length of one of three unsymmetrically spaced but completely transposed wires, each r inches in radius, with axial spacings of d_{12}, d_{23}, and d_{13} inches, and with the algebraic sum of the instantaneous currents conducted respectively by the three wires in the same direction equal to zero:

$$L = 0.08047 + 0.7411 \log \frac{\sqrt[3]{d_{12}d_{23}d_{13}}}{r} \text{ millihenrys.} \qquad (6.12)$$

NOTE. A completely transposed three-wire circuit is one in which each wire occupies each position for one-third of the distance.

The **self-inductance** (L) per centimeter length of two straight cylindrical concentric wires of equal section conducting the same current in opposite directions, the inner radius of the outer conductor being b centimeters and the radius of the solid inner conductor being c centimeters:

$$L = \left(2 \ln \frac{b}{c} + \frac{1}{2} + \frac{c^2}{3b^2} - \frac{c^4}{12b^4} + \frac{c^6}{30b^6} - \cdots \right) \times 10^{-9} \text{ henrys.} \qquad (6.13)$$

The **self-inductance** (L) of a single circular turn of wire of circular section, the mean radius of the turn being R centimeters and the radius of the section r centimeters:

$$L = 12.6R \left[\left(1 + \frac{r^2}{8R^2} \right) \ln \frac{8R}{r} \frac{r^2}{25R^2} - 1.75 \right] \times 10^{-9} \text{ henrys.} \qquad (6.14)$$

The **mutual inductance** (M) of two coils in which a current of I_1 amperes in one establishes a flux of Φ_2 webers through the N_2 turns of the other:

$$M = \frac{N_2\Phi_2}{I_1} \quad \text{henrys.} \tag{6.15}$$

The **mutual inductance** (M) of two parallel circular coaxial turns, each r centimeters in radius and their planes d centimeters apart (d small compared with r):

$$M = 12.6r\left[\ln\frac{8r}{d}\left(1 + \frac{3d^2}{16r^2}\right) - \left(2 + \frac{d^2}{16r^2}\right)\right] \times 10^{-9} \text{ henrys.} \tag{6.16}$$

The **mutual inductance** (M) of two concentric solenoids, the exterior of N_1 turns and length l centimeters and the interior of N_2 turns and sectional area A_2 square centimeters (the axial length of the interior solenoid small compared with the axial length of the exterior solenoid):

$$M = \frac{12.6N_1N_2A_2}{l} \times 10^{-9} \text{ henrys.} \tag{6.17}$$

The **self-inductance** (L) of two series connections of self-inductance L_1 and L_2 henrys, respectively, and **mutual inductance** M_{12} henrys:

$$L = L_1 + L_2 \pm 2M_{12} \quad \text{henrys.} \tag{6.18}$$

NOTE.   The sign is $+$ when the mutual fluxes are in conjunction and is $-$ when the mutual fluxes are in opposition. The mutual inductance (M) of two series connections of self-inductance L_1 and L_2 henrys, respectively, with K percent coupling is given by $M = K(L_1L_2)^{1/2}$ henrys. The coefficient of coupling K is expressed as a decimal fraction representing the ratio of the flux caused by one coil to that which links the other.

The **self-inductance** (L) of several coils of self-inductances L_1, L_2, L_3, wound on the same core with 100% coupling in conjunction ($+$ sign) or opposition ($-$ sign):

$$L = (\sqrt{L_1} \pm \sqrt{L_2} \pm \sqrt{L_3} \pm \cdots)^2 \quad \text{henrys.} \tag{6.19}$$

Relative to free-space permeability, ferromagnetic substances, including iron, steel, nickel, cobalt, and magnetic alloys, have very high permeability. For example, the relative permeability of transformer-quality steel may exceed 2000. This means a given coil with a closed-loop magnetic core of steel instead of an air core will have an inductance approximately 2000 times that given by the air-core formula. However, relative permeabilities of ferromagnetic substances vary greatly with flux density and may also be affected by heat treatment, previous magnetic history, and impurities introduced to reduce losses.

Magetization curves of commercially used forms of iron and steel for magnetic circuits are shown in Fig. 6.4. Relative permeabilities of cast steel, cast iron, transformer hot-rolled steel, and some specially developed high-permeability alloys as a function of flux density are shown in Fig. 6.5.

Toroidal cores with ferromagnetic properties are manufactured of powdered iron and ferrite materials with a range of relative permeabilities from 1 to 8000. Although designed and used primarily for high-frequency and very-high-frequency applications,

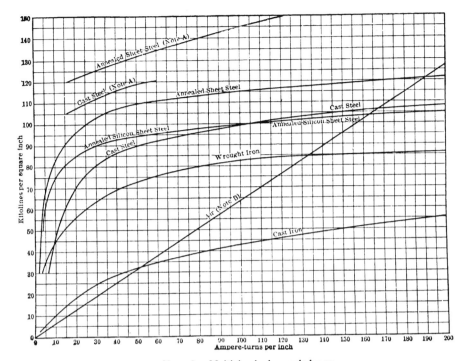

NOTE A.   Multiply abscissa scale by 10.
NOTE B.   Multiply abscissa scale by 200.

Fig. 6.4   Magnetization curves of iron and steel.

toroidal cores are also being used in power transformers at 50 and 60 Hz. Tables 6.5 and 6.6 give information for making inductances using commercially available cores.

In a closed loop ferromagnetic circuit magnetic flux (Φ) is established by a **magnetomotive force** ($\mathfrak{F}$) and limited in magnitude by **reductance** ($\mathfrak{R}$), a property of the ferromagnetic material. The relationship is similar to that of Ohm's law for electric circuits:

$$\Phi = \frac{\mathfrak{F}}{\mathfrak{R}}. \tag{6.20}$$

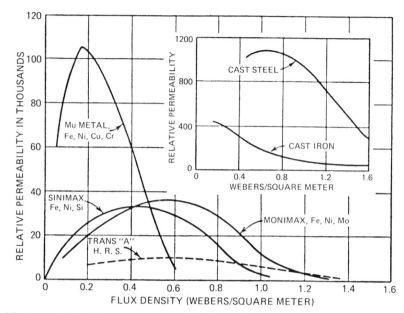

Fig. 6.5 Permeability of ferromagnetic materials (From O. W. Eshbach and M. Souders, *Handbook of Engineering Fundamentals,* 3rd ed. Copyright © 1975 by John Wiley & Sons, Inc. Reproduced with permission.)

The **magnetomotive force** ($\mathcal{F}$) due to N turns of conductor each conducting I amperes:

$$\mathcal{F} = NI \quad \text{ampere turns.} \tag{6.21}$$

The **reluctance of a ferromagnetic path** L meters long and A square meters in cross section with relative permeability:

$$\mathcal{R} = \frac{L}{\mu A} \quad \text{reciprocal henrys.} \tag{6.22}$$

From Equation (6.20), reluctance is also equal to the number of ampere turns per weber:

$$\mathcal{R} = \frac{\mathcal{F}}{\Phi}. \tag{6.23}$$

The **Permeance** ($\mathcal{P}$) of the magnetic circuit is the reciprocal of reluctance:

$$\mathcal{P} = \frac{\mu A}{L} = \frac{\Phi}{\mathcal{F}} \quad \text{henrys.} \tag{6.24}$$

Because μ varies with flux density in ferromagnetic circuits, data on ferromagnetic materials are always given in terms of flux density (B) as a function of magnetic

Table 6.5 Powdered Iron Toroidal Core Data

Powered-Iron Toroidal Cores—A_L Values (μH/100 Turns)[a]

Core Size	41-Mix Green $\mu = 75$	3-Mix Gray $\mu = 35$ 0.05–0.5MHz	15-Mix Red and White $\mu = 25$ 0.1–2 MHz	1-Mix Blue $\mu = 20$ 0.5–5 MHz	2-Mix Red $\mu = 10$ 1–30 MHz	6-Mix Yellow $\mu = 8$ 10–90 MHz	10-Mix Black $\mu = 6$ 60–150 MHz	120Mix Green and White $\mu = 3$ 100–200 MHz	0-Mix Tan $\mu = 1$ 150–300 MHz
T-200	755	360	NA[b]	250[c]	120	100[c]	NA	NA	NA
T-184	1640	720	NA	500[c]	240	195	NA	NA	NA
T-157	970	420	360[c]	320[c]	140	115	NA	NA	NA
T-130	785	330	250[c]	200	110	96	NA	NA	NA
T-106	900	405	345[c]	325[c]	135	116	NA	NA	19.0[c]
T-94	590	248	200[c]	160	84	70	58	32	10.6
T-80	450	180	170	115	55	45	32[c]	22	8.5
T-68	420	195	180	115	57	47	32	21	7.5
T-50	320	175	135	100	49[c]	40	31	18	6.4
T-44	229	180	160	105	52[c]	42	33	NA	6.5
T-37	308	120[c]	90	80	40[c]	30	25	15	4.9
T-30	375	140[c]	93	85	43	36	25	16	6.0
T-25	225	100	85	70	34	27	19	13	4.5
T-20	175	90	65	52	27	22	16	10	3.5
T-16	130	61	NA	44	22	19	13	8	3.0
T-12	112	60	50[c]	48	20[c]	17[c]	7.5	3.0	

Table 6.5 (*Continued*)

Number of Turns versus Wire Size and Core Size
(Approximate Maximum of Turns—Single-Layer, Wound Enameled Wire)

Wire Size	T-200	T-130	T-106	T-94	T-80	T-68	T-50	T-37	T-25	T-12
10	33	20	12	12	10	6	4	1		
12	43	25	16	16	14	9	6	3		
14	54	32	21	21	18	13	8	5	1	
16	69	41	28	28	24	17	13	7	2	
18	88	53	37	37	32	23	18	10	4	1
20	111	67	47	47	41	29	23	14	6	1
22	140	86	60	60	53	38	30	19	9	2
24	177	109	77	77	67	49	39	25	13	4
26	223	137	97	97	85	63	50	33	17	7
28	281	173	123	123	108	80	64	42	23	9
30	355	217	154	154	136	101	81	54	29	13
32	439	272	194	194	171	127	103	68	38	17
34	557	346	247	247	218	162	132	88	49	23
36	683	424	304	304	268	199	162	108	62	30
38	875	544	389	389	344	256	209	140	80	39
40	1103	687	492	492	434	324	264	178	102	51

Physical Dimensions[a]

Core Size	Outer Diameter (in.)[e]	Inner Diameter (in.)	Height (in.)	Cross-Section Area (cm²)	Mean Length (cm)	Core Size	Outer Diameter (in.)	Inner Diameter (in.)	Height (in.)	Cross-Section Area (cm²)	Mean Length (cm)
T-200	2.000	1.250	0.550	1.330	12.97	T-50	0.500	0.303	0.190	0.121	3.20
T-184	1.840	0.950	0.710	2.040	11.12	T-44	0.440	0.229	0.159	0.107	2.67
T-157	1.570	0.950	0.570	1.140	10.05	T-37	0.375	0.205	0.128	0.070	2.32
T-130	1.300	0.780	0.437	0.733	8.29	T-30	0.307	0.151	0.128	0.065	1.83
T-106	1.060	0.560	0.437	0.706	6.47	T-25	0.255	0.120	0.096	0.042	1.50
T-94	0.942	0.560	0.312	0.385	6.00	T-20	0.200	0.088	0.067	0.034	1.15
T-80	0.795	0.495	0.250	0.242	5.15	T-16	0.160	0.078	0.060	0.016	0.75
T-68	0.690	0.370	0.190	0.196	4.24	T-12	0.125	0.062	0.050	0.010	0.74

Source: The Radio Amateur's Handbook, 60th ed., American Radio Relay League, Newington, Conn., 1983.

[a]Turns = 100 $(L_{\mu H}/A_L \text{ value})^{1/2}$. All frequency figures optimum.

[b]NA, not available in that size.

[c]Updated values (1979) from Micrometals, Inc.

[d]Inches × 25.4 = mm.

[e]Courtesy of Amidon Assoc., North Hollywood, Calif., and Micrometals, Inc.

Table 6.6   Ferrite Toroidal Coil Data

Ferrite Toroids—A_L Values (mH/1000 Turns)[a] Enameled Wire

Core Size	63-Mix $\mu = 40$	61-Mix $\mu = 125$	43-Mix $\mu = 950$	72-Mix $\mu = 2000$	75-Mix $\mu = 5000$
FT-23	7.9	24.8	189.0	396.0	990.0
FT-37	17.7	55.3	420.0	884.0	2210.0
FT-50	22.0	68.0	523.0	1100.0	2750.0
FT-82	23.4	73.3	557.0	1172.0	2930.0
FT-114	25.4	79.3	603.0	1268.0	3170.0

Number turns $= 1000 \sqrt{\text{desired } L(\text{mH})} \div A_L$ value (above)

Ferrite Magnetic Properties

Property	Unit	63-Mix	61-Mix	43-Mix	72-Mix	75-Mix
Initial perm. (μ_1)		40	125	950	2000	5000
Maximum perm.		125	450	3000	3500	8000
Saturation flux density at 13 oer	Gauss	1850	2350	2750	3500	3900
Residual flux density	Gauss	750	1200	1200	1500	1250
Curie temp.	°C	500	300	130	150	160
Vol. resistivity	ohm/cm	1×10^8	1×10^8	1×10^5	1×10^2	5×10^2
Opt. freq. range	MHz	15–25	0.2–10	0.01–1	0.001–1	0.001–1
Specific gravity		4.7	4.7	4.5	4.8	4.8
Loss factor	$\dfrac{1}{uO}$	9.0×10^{-5} @ 25 MHz	2.2×10^{-5} @ 2.5 MHz	2.5×10^{-5} @ 0.2 MHz	9.0×10^{-6} @ 0.1 MHz	5.0×10^{-6} @ 0.1 MHz
Coercive force	Oer.	2.40	1.60	0.30	0.18	0.18
Temp. coeff. of initial perm.	%/°C 20–70°C	0.10	0.10	0.20	0.60	

Ferrite Toroids—Physical Properties[b]

Core Size	OD	ID	Hgt	A_e	l_e	V_e	A_s	A_w
FT-23	0.230	0.120	0.060	0.00330	0.529	0.00174	0.1264	0.01121
FT-37	0.375	0.187	0.125	0.01175	0.846	0.00994	0.3860	0.02750
FT-50	0.500	0.281	0.188	0.02060	1.190	0.02450	0.7300	0.06200
FT-82	0.825	0.520	0.250	0.03810	2.070	0.07890	1.7000	0.21200
FT-114	1.142	0.748	0.295	0.05810	2.920	0.16950	2.9200	0.43900

OD = Outer diameter (inches)[c] A_e = Effective magnetic cross-sectional area (inches²)
ID = Inner diameter (inches) l_e = Effective magnetic path length (inches)
Hgt = Height (inches) V_e = Effective magnetic volume (inches³)
A_w = Total window area (inches²) A_s = Surface area exposed for cooling (inches²)

Source: The Radio Amateur's Handbook, 60th ed., American Radio Relay League, Newington, Conn., 1983.
[a] Number of turns $= 1000 \, (L_{(m/H)} A_L \text{ value})^{1/2}$.
[b] Courtesy of Amidon Assoc., North Hollywood, Calif.
[c] Inches $\times 25.4 =$ mm.

field strength (also called magnetizing force). (*H*). From Equations (6.22) and (6.23), $B = \mu H$ tesla.

Much current technical literature and data on ferromagnetic materials uses CGS metric units or English units. Conversion factors to SI units for magnetic flux (Φ), magnetomotive force ($\mathcal{F}$), magnetic flux density (B), and magnetic field strength (H are as follows:

To convert from:	To:	Multiply by:
(Φ) line (English)	weber	10^{-8}
(Φ) maxwell (CGS)	weber	10^{-8}
($\mathfrak{F}$) ampere-turn (English)	ampere-turn	1.0
($\mathfrak{F}$) gilbert (CGS)	ampere-turn	0.796
(B) line per square inch (English)	tesla	1.55×10^{-5}
(B) gauss (CGS)	tesla	10^{-4}
(H) ampere-turn per inch (English)	ampere-turn per meter	39.4
(H) oersted (CGS)	ampere-turn per meter	79.6

The **energy of a magnetic field** (W) established by a circuit with inductance L henrys conducting a current I amperes:

$$W = \frac{1}{2}LI^2 \quad \text{joules.} \tag{6.25}$$

When flux is caused by changing current i, an increase of i converts electric to magnetic energy and a decrease of i converts magnetic to electric energy:

$$W = N \int i\,d\Phi \times 10^{-8} \text{ joule.} \tag{6.26}$$

6.2.4 Capacitance

Two conductors across which a voltage is applied, separated by an insulating medium, form a capacitor. Quantitatively the capacitance is the amount of charge per volt which accumulates at the boundaries between conductors and insulating medium. With charge (Q) in coulombs and V in volts,

$$C = \frac{Q}{V} \quad \text{farads.} \tag{6.27}$$

The **dielectric constant** (k) of the insulating medium in a capacitor is the ratio of electric flux density D to electric field strength E, relative to the ratio for an insulating medium of free space equal to one. Table 6.7 lists relative dielectric constants of insulating mediums (dielectrics) used in capacitors. Table 6.8 lists the resistivity and temperature coefficient of resistance of conductors in addition to those in Table 6.2. Table 6.9 lists the resistivity and dielectric constant of insulating materials in addition to those in Table 6.7.

Equations for calculating common capacitor configurations and lines most used in practice follow.

The **capacitance** (C) **of a parallel-plate capacitor** in which the positive and negative charges are each distributed uniformly over a surface are of A square cen-

Table 6.7   Relative Dielectric Constant (Permittivity)[a]

Air	1.0006	Glass, crow (window)*	6
Hydrogen	1.0003	Mica, good quality*	7
Water, pure	78	Porcelain*	6.5
Alcohol, ethyl	25.7	Rubber, vulcanized*	3
Oil, petroleum*	2.1	Teflon (all frequencies)	2.1
Paraffin wax*	2.25	Polystyrene (all frequencies)	2.55
Paper, dry*	3.5	Polyethylene (all frequencies)	2.26
Paper, oiled*	3.5	Titanium oxides and titanates*	10–10,000

Source: H. H. Skilling, *Electrical Engineering Circuits,* 2nd ed., Copyright © 1965 by John Wiley & Sons, Inc. Reproduced with permission.

[a]$K = \epsilon/\epsilon_0$; values given for 20°C and atmospheric pressure; frequency less than 1 MHz unless otherwise indicated. Substances marked * are quite variable, and K may differ between samples by 10 to 20% or even more.

timeters, the uniform distance between the oppositely charged surfaces is d centimeters, and the medium between the oppositely charged surfaces is of dielectric constant k (d is assumed to be small compared with all other dimensions):

$$C = \frac{kA}{36\pi d \times 10^5} \quad \text{microfarads.} \tag{6.28}$$

The **capacitance** (C) **of two concentric spheres,** the inner, r_1 centimeters in external radius, the outer, r_2 centimeters in internal radius, and separated by a medium of dielectric constant k:

$$C = \frac{r_1 r_2 k}{9(r_2 - r_1) \times 10^5} \quad \text{microfarads.} \tag{6.29}$$

The **capacitance** (C) **of two coaxial cylinders** per centimeter axial length, the inner, r_1 centimeters in external radius, the outer, r_2 centimeters in internal radius, and separated by a medium of dielectric constant k (ln equals $\log_e$):

$$C = \frac{k}{18} \ln(r_2/r_1) \times 10^5 \quad \text{microfarads.} \tag{6.30}$$

NOTE.   The capacitance per mile is $0.03882k/\log(r_2/r_1)$ microfarads.

The **capacitance** (C) **of two parallel cylinders** per centimeter length, each cylinder r centimeters in radius, their centers separated by a distance of d centimeters, and immersed in a medium of dielectric constant k (r small compared with d and all dimensions small compared with distance to surrounding objects):

$$C = \frac{k}{36 \ln(d/r) \times 10^5} \quad \text{microfarads.} \tag{6.31}$$

NOTE.   The capacitance per mile is $(1.941k \times 10^{-2})/\log(d/r)$ microfarads. The capacitance per conductor (to neutral) of a balanced three-phase transmission line with conductors located at the vertices of an equilateral triangle equals $(3.882k \times 10^{-2}/\log(d/r)$ microfarads per mile.

Table 6.8 Resistivity (ρ) and Temperature Coefficient of Resistance (α) of Certain Conductors[a]

Material	ρ ($\mu\Omega/cm^3$)	α	Material	ρ ($\mu\Omega/cm^3$)	α
Aluminum	2.688	0.00403	Mercury	95.8	0.00089
Antimony	39.1 (0°C)	0.0036	Molybdenum	5.08 (0°C)	0.0047 (0–100°C)
Barium	9.8	0.0033	Monel metal	42	—
Beryllium	10.1	—	Nickel	7.8	0.00537 (20–100°C)
Bismuth	120	0.004	Osmium	9.5	0.0033
Calcium	4.59	0.00364 (0–600°C)	Palladium	9.83 (0°C)	0.003
Carbon	3500 (0°C)	−0.009	Platinum	11	—
Cerium	78	—	Potassium	6.1 (0°C)	0.0055 (0°C)
Cesium	19 (0°C)	—	Rhodium	5.11 (0°C)	0.0043 (0°C)
Chromium	2.6 (0°C)	—	Silver	1.629 (18°C)	0.0038
Cobalt	9.7	0.00658 (0–100°C)	Sodium	4.3 (0°C)	0.0054
Copper	1.724	0.00393	Strontium	24.8	—
Gold	2.44	0.0034	Tantalum	15.5	0.0031
Graphite	800 (0°C)	—	Tellurium	2×10^5	—
Iron	9.8	0.0065 (0–100°C)	Thallium	17.6 (0°C)	0.0040 (0°C)
Iron, cast	79–104	—	Thorium	18	0.0021 (20–1800°C)
Lead	22.0	0.0039	Tin	11.5	0.0042
Lithium	8.55 (0°C)	0.0047 (0°C)	Titanium	3.0	—
Magnesium	4.46	0.0040	Tungsten	5.5	0.0047 (0–100°C)
Manganese	5	—	Zinc	5.75 (0°C)	0.0037

[a]Temperature is 20°C unless otherwise specified.

Table 6.9 Resistivity (ρ) and Dielectric Constant (k) of Certain Insulators at Room Temperature

Material	ρ (MΩ/cm^3)	k	Material	ρ (MΩ/cm^3)	k
Alcohol, ethyl	0.3	5.0–5.46	Oil, olive	5×10^6	3.11
Alcohol, methyl	0.14	31.2–35.0	Oil, paraffin	10^{10}	—
Amber	5×10^{10}	—	Oil, petroleum	2×10^{10}	2.13
Amylacetate	—	4.81	Paper	10^4–10^9	1.7–3.8
Asbestos paper	1.6×10^5	2.7	Paraffin	5×10^{10}–5×10^{12}	1.9–2.3
Asphalt	—	2.7	Porcelain	3×10^8	4.4
Bakelite	10^5–10^{10}	4.5–5.5	Quartz	10^8–5×10^{12}	4.7–5.1
Beeswax	6×10^8	—	Rosin	7×10^9–5×10^{10}	2.5
Cellophane	—	8	Rubber, hard	3×10^{10}–10^{12}	2.0–3.5
Celluloid	2×10^4	13.3	Sealing wax	10^9–8×10^9	—
Cellulose acetate	—	5	Selenium	0.06	6.1–7.4
Glass	5×10^5–10^{10}	5.5–9.1	Shellac	10^{10}	3.0–3.7
Glycerine	—	56.2	Silica, fused	10^8–10^{13}	3.5–3.6
Gutta percha	3×10^4	2.9	Slate	10^2–10^4	6.6–7.4
Ice	720	86	Sulfur	8×10^9–10^{11}	2.9–3.2
Ivory	200	—	Turpentine	—	2.23
Marble	10^3–10^5	8.3	Water, distilled	0.5	81
Mica	4×10^7–2×10^{11}	5–7	Wood, paraffined	3×10^4–4×10^7	4.1

The **capacitance** (C) **to neutral** per mile of one conductor of a balanced three-phase transmission line with unsymmetrical spacing but completely transposed, d_{12}, d_{23}, and d_{13} being the axial spacings in inches and r being the conductor radius in inches:

$$C = \frac{3.882 \times 10^{-2}}{\log \sqrt[3]{d_{12}d_{23}d_{13}/r}} \quad \text{microfarads.} \tag{6.32}$$

NOTE.   See note following Equation (6.12).

The **total capacitance** (C_0) **of several parallel capacitors** of capacitance C_1, C_2, and C_3 farads, respectively:

$$C_0 = C_1 + C_2 + C_3 \text{ farads.} \tag{6.33}$$

The **total capacitance** (C_0) **of several series capacitors** of capacitance C_1, C_2, and C_3 farads, respectively:

$$C_0 = \frac{1}{1/C_1 + 1/C_2 + 1/C_3} \quad \text{farads.} \tag{6.34}$$

Capacitors used in electric power applications are rated in vars (volt amperes reactive) or kilovars instead of farads or microfarads because their useful output is reactive volt amperes at the alternating current frequency of the power system. If C is capacitance in microfarads, V is system rms volts, and f is system frequency in hertz,

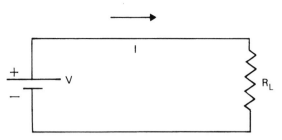

Fig. 6.6   Simplest series electric circuit.

$$\text{kvar} = 2\pi f C V^2 \times 10^{-9}. \tag{6.35}$$

The **energy** *(W)* **stored in a capacitor** of C farads at potential V volts:

$$W = \tfrac{1}{2}CV^2 \quad \text{joules.} \tag{6.36}$$

6.3   DIRECT-CURRENT CIRCUITS

6.3.1   Ohm's Law

In a circuit where a source of constant voltage V and a load resistance R_L are connected by two conductors (Fig. 6.6), the current is proportional to voltage and inversely proportional to resistance. When any two of the three quantities are known the third can be found from the equations

$$I = V/R \text{ amperes,} \qquad R = V/I \text{ ohms,} \qquad V = IR \text{ volts.} \tag{6.37}$$

A practical circuit always contains resistance in the source R_s and in the conductors R_C in addition to the load resistance R_L (Fig. 6.7). If R_s and R_C are significant in magnitude compared with R_L,

$$I = \frac{V}{R_L + R_s + R_C} \quad \text{amperes.} \tag{6.38}$$

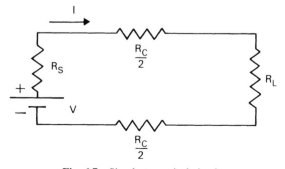

Fig. 6.7   Simplest practical circuit.

6.3.2 Series and Parallel Connections

Figures 6.6 and 6.7 are the simplest possible form of a circuit with a single series path. If a series path has more than one voltage source, the total voltage in the path is their algebraic sum; if more than one resistance, the total resistance is their numerical sum.

The **Equivalent resistance** (R_p) of a parallel circuit the respective branches of which have resistances R_1, R_2, R_3, and so on, and contain no voltage, each resistance in ohms:

$$\frac{1}{R_p} = \frac{1}{R_1} + \frac{1}{R_2} + \frac{1}{R_3}, \text{ etc.} \quad \text{siemens.} \tag{6.39}$$

When there are only two branches, Equations (6.39) reduces to

$$R_p = \frac{R_1 R_2}{R_1 + R_2} \quad \text{ohms.} \tag{6.40}$$

When there are n parallel paths of equal resistance R_1 ohms,

$$R_p = \frac{R_1}{n} \text{ ohms.} \tag{6.41}$$

6.3.3 Kirchhoff's Laws

A circuit containing three or more parallel paths is called a **network.** It is possible to determine the currents in each branch of simple networks by changing parallel paths to equivalent series paths, but a more systematic approach applicable to all networks is to set up equations based on **Kirchhoff's** two **laws:**

1. The sum of voltage sources (rises) and voltage drops around any closed series path of a network equals zero.
2. At any current junction in a network the total of individual currents leaving the junction is equal to the total of currents flowing toward the junction.

Figure 6.8 shows a part circuit with voltage source V, resistance R, and current I_{AB} flowing in the direction A to B. In using Kirchhoff's first law it is necessary to consistently distinguish a voltage rise from a voltage drop by algebraic sign

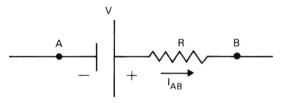

Fig. 6.8   Part circuit containing voltage source and resistance.

$$V_{AB} = +V - I_{AB}R_{AB} \quad \text{volts.} \tag{6.42}$$

The sign of the voltage source or current is positive when acting in the direction shown in Fig. 6.8, and is negative when acting in the opposite direction. When V_{AB} is positive it is called a potential rise from A to B, and when V_{AB} is negative it is called a potential drop from A to B. If V is zero, $V_{AB} = -I_{AB}R_{AB}$ volts, and if either I_{AB} or R_{AB} is zero, $V_{AB} = +V$ volts.

The **current** (I_{AB}) in Fig. 6.8 flowing from A toward B:

$$I_{AB} = \frac{+V - V_{AB}}{R_{AB}} \quad \text{amperes.} \tag{6.43}$$

The direction of current is determined by its sign, a positive sign indicating flow from A to B and a negative sign a flow from B to A. When V_{AB} = zero, I_{AB} = $+V/R_{AB}$, and when V = zero, $I_{AB}R_{AB} = -V_{AB}$.

The **total current** (I_0) flowing toward a junction from which currents I_1, I_2, I_3, and so on, flow away, all currents being measured in amperes:

$$I_0 = I_1 + I_2 + I_3 + \cdots \quad \text{amperes.} \tag{6.44}$$

In the network of Fig. 6.9 the magnitudes of currents, voltage sources and resistances are indicated by the symbols I, V, and R, respectively, and directions of currents and voltage sources are indicated by arrows and $+/-$ signs, respectively, any unknown direction being assumed arbitrarily. The Kirchhoff's law equations are

$$1.\ +V_1 - I_1R_1 + I_2R_2 + V_2 = 0, \tag{6.45}$$
$$2.\ +V_1 - I_2R_1 - I_3R_3 - V_3 = 0, \tag{6.46}$$
$$3.\ I_1 + I_2 - I_3 = 0. \tag{6.47}$$

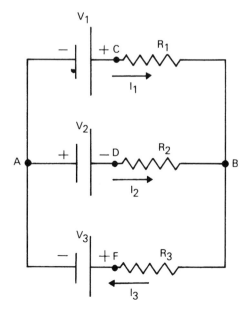

Fig. 6.9 Network with three branches.

The magnitude and direction of each current may be determined by solving the simultaneous equations, a positive value of current indicating the same direction and a negative value indicating the opposite direction to that assumed in the figure.

6.3.4   Power and Energy in a Direct-Current Circuit

The **power** (P) delivered to or from a part circuit conducting a current of I amperes and across which the voltage is V volts:

$$P = VI \text{ watts.} \tag{6.48}$$

A voltage rise in the direction of the current indicates power delivered from the part circuit, and a voltage drop in the direction of the current indicates power delivered to the part circuit.

The **power** (P) delivered to a part-circuit of R ohms resistance containing no voltage source and conducting a current of I amperes:

$$P = I^2R \quad \text{watts.} \tag{6.49}$$

The **energy** (W) delivered to or used by a circuit operating at power P watts for time t hours.

$$W = VIt = I^2Rt \quad \text{watt-hours.} \tag{6.50}$$

6.4   TRANSIENT VOLTAGES AND CURRENTS

6.4.1   Definition of Transient State

Because of the presence of inductance and capacitance in electric circuits, whether intended or not intended, whenever an abrupt change occurs in voltage, current or impedance, transient conditions exist in one or more of these during the time interval between the initial change and the establishment of a steady-state condition. Figures 6.10 and 6.11 illustrate the transient waveforms that occur, for example, when a switch is closed, applying voltage to various combinations of resistance, capacitance, and inductance. Transients also occur when current flow in a circuit is interrupted by the opening of a switch. In the terminology of power systems, large transient increases in voltage or current during switching operations or accidental short circuits are called **surges.**

6.4.2   *RL* Circuits

The **current** (i) flowing in a series circuit of R ohms resistance and L henrys self-inductance t seconds after a constant emf of E volts is impressed upon the circuit.*

---

*The value of ϵ throughout is 2.718.

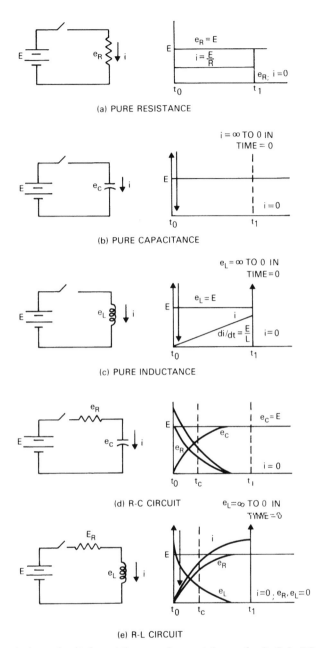

Fig. 6.10 Transients on circuit close at time t_0 and open at time t_1, for R, C, L, RC, and RL circuits. (Used with permission of Westinghouse Electric Corporation.)

$$i = \frac{E}{R} (1 - \epsilon^{-Rt/L}) + I\epsilon^{-Rt/L} \quad \text{amperes.} \tag{6.51}$$

NOTE. I is the current in amperes flowing in the circuit at the instant before the emf is impressed. It is a positive quantity if flowing in conjunction with and is a negative quantity if flowing in opposition to the emf.

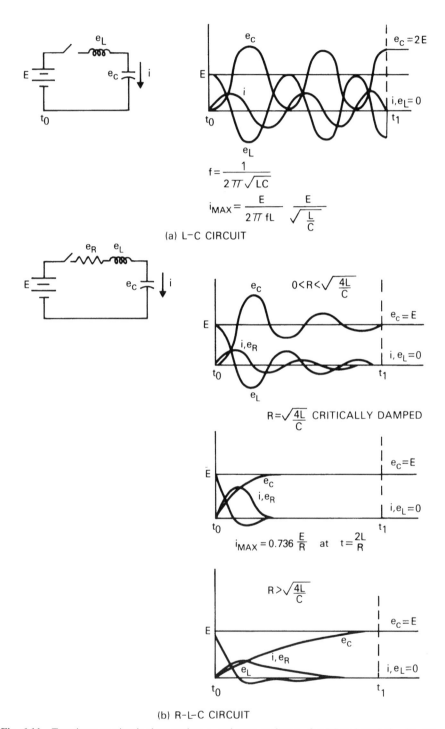

$$f = \frac{1}{2\pi\sqrt{LC}}$$

$$i_{MAX} = \frac{E}{2\pi fL} \quad \frac{E}{\sqrt{\frac{L}{C}}}$$

(a) L–C CIRCUIT

$$0 < R < \sqrt{\frac{4L}{C}}$$

$$R = \sqrt{\frac{4L}{C}} \quad \text{CRITICALLY DAMPED}$$

$$i_{MAX} = 0.736\frac{E}{R} \quad \text{at} \quad t = \frac{2L}{R}$$

$$R > \sqrt{\frac{4L}{C}}$$

(b) R–L–C CIRCUIT

Fig. 6.11 Transients on circuit close at time t_0 and open at time t_1, for LC and RLC circuits. (Used with permission of Westinghouse Electric Corporation.)

The **current** (i) flowing in a series circuit of R ohms resistance and L henrys self-inductance t seconds after its source of emf is short-circuited, the current flowing in the circuit at the instant before the sort-circuit being I amperes:

$$i = I\epsilon^{-Rt/L} \quad \text{amperes.} \tag{6.52}$$

6.4.3 RC Circuits

The **current** (i) flowing in a circuit of R ohms resistance and C farads series capacitance t seconds after a constant emf of E volts is impressed upon the circuit:

$$i = \left(\frac{E - V}{R}\right)\epsilon^{-t/RC} \quad \text{amperes.} \tag{6.53}$$

NOTE. V is the potential across the capacitor at the instant before the emf is impressed. It is a positive quantity if acting in opposition to, and a negative quantity if acting in conjunction with, the impressed emf.

The **potential** (v) across the capacitor at any time t under the conditions stated in Equation (6.53):

$$v = E(1 - \epsilon^{-t/RC} + V\epsilon^{-t/RC} \quad \text{volts.} \tag{6.54}$$

The **current** (i) flowing in a circuit of R ohms resistance and C farads series capacitance t seconds after its source of emf is short-circuited, the potential across the condenser at the instant before the short-circuit being V volts:

$$i = \frac{V}{R}\epsilon^{-t/RC} \quad \text{amperes.} \tag{6.55}$$

The **potential** (v) across the capacitor at any time t under the conditions stated in Equation (6.55):

$$v = V\epsilon^{-t/RC} \quad \text{volts.} \tag{6.56}$$

6.4.4 RLC Circuits

The **current** (i) flowing in a circuit of R ohms resistance, L henrys self-inductance, and C farads series capacitance t seconds after a constant emf of E volts is impressed upon the circuit, the potential across the capacitor and the current flowing in the circuit at the instant before the emf is impressed being V volts and I amperes, respectively.

CASE I. $R^2C > 4L$

$$i = \left[\frac{E - V - aLI}{(b - a)L}\right]\epsilon^{-at} - \left[\frac{E - V - bLI}{(b - a)L}\right]\epsilon^{-bt} \quad \text{amperes.} \tag{6.57}$$

CASE II. $R^2C = 4L$

$$i = \left\{ I + \left[\frac{2(E - V) - RI}{2L} \right] t \right\} \epsilon^{-Rt/(2L)} \quad \text{amperes.} \tag{6.58}$$

CASE III. $R^2C < 4L$

$$i = \left[\frac{2(E - V) - RI}{2\omega_1 L} \sin \omega_1 t + I \cos \omega_1 t \right] \epsilon^{-Rt/2L} \quad \text{amperes.} \tag{6.59}$$

NOTE. $a = \dfrac{RC - (R^2C^2 - 4LC)^{1/2}}{2LC}$, $b = \dfrac{RC + (R^2C^2 - 4LC)^{1/2}}{2LC}$, and $\omega 1 = \dfrac{(4LC - R^2C^2)^{1/2}}{2LC}$. The current ($I$) is positive when flowing in the same direction as the impressed emf and the sign of the potential (V) is obtained as in Equation (6.52).

The **current** (i) flowing in a circuit of R ohms resistance, L henrys self-inductance, and C farads series capacitance t seconds after its source of emf is short-circuited, the potential across the capacitor and the current flowing in the circuit at the instant before the short-circuit being V volts and I amperes, respectively.

NOTE.   Write Equation (6.57), (6.58), or (6.59), making E zero in each case.

The **current** (i) flowing in a circuit of R ohms resistance, L henrys self-inductance, and C farads series capacitance t seconds after a sinusoidal emf, $e = E_m \sin (\omega t + \alpha)$ volts, is impressed upon the circuit, the potential across the capacitor and the current flowing in the circuit at the instant before the emf is impressed being V volts and I amperes, respectively.

CASE I. $R^2C > 4L$

$$i = G\epsilon^{-at} - H\epsilon^{-bt} + \frac{E_m}{Z} \sin(\omega t + \alpha - \theta) \quad \text{amperes.} \tag{6.60}$$

CASE II. $R^2C = 4L$

$$i = (J + Kt)\epsilon^{-Rt/2L} + \frac{E_m}{Z} \sin(\omega t + \alpha - \theta) \quad \text{amperes.} \tag{6.61}$$

CASE III. $R^2C < 4L$

$$i = (M \sin \omega_1 t + N \cos \omega_1 t)\epsilon^{-Rt/2L} + \frac{E_m}{Z} \sin(\omega t + \alpha - \theta) \quad \text{amperes.} \tag{6.62}$$

NOTE.

$$G = \frac{E_m \sin \alpha - V - aLI - (E_m L/Z) [b \sin (\alpha - \theta) + \omega \cos (\alpha - \theta)]}{(b - a)L}$$

$$H = \frac{E_m \sin \alpha - V - bLI - (E_m L/Z) [a \sin (\alpha - \theta) + \omega \cos (\alpha - \theta)]}{(b - a)L}$$

$$J = I - \frac{E_m}{Z} \sin (\alpha - \theta) \qquad \omega_1, a, \text{ and } b \text{ as in Equation (6.59).}$$

$$K = \frac{1}{L} \left\{ E_m \sin \alpha - V - \frac{RI}{2} - \frac{E_m}{Z} \left[\frac{R}{2} \sin (\alpha - \theta) + L\omega \cos(\alpha - \theta) \right] \right\}.$$

$$M = \frac{K}{\omega_1}. \qquad N = J. \qquad \omega = 2\pi f. \qquad \theta = \cos^{-1} \frac{r}{Z}.$$

$$Z = \sqrt{R^2 + \left(\omega L - \frac{1}{\omega C} \right)^2}.$$

$\alpha = \sin^{-1}(e/E_m)$, where e equals the algebraic value of the sinusoidal emf at the instant that it is impressed on the circuit. The current (I) is positive if flowing in the same direction as the impressed emf and the potential (V) is positive if acting in opposition to the impressed emf, both at time (t) equals zero. If the circuit contains no capacitor, the series capacitance is infinite and $a = 0$, $b = R/L$, and $Z = (R^2 + \omega^2 L^2)^{1/2}$.

6.5 ALTERNATING-CURRENT CIRCUITS

6.5.1 Types of Alternating Currents

The term alternating current refers to a current that reverses at regularly recurring intervals of time and that has alternately positive and negative values. An alternating current may be either sinusoidal, in which case it has only one frequency component, called the fundamental; or nonsinusoidal, in which case it has a fundamental frequency plus one or more harmonics. A harmonic is a sinusoidal component having a frequency an integral multiple of the fundamental.

6.5.2 Sinusoidal Voltages and Currents

The electromotive force (e) of an emf of maximum value E_m volts and angular velocity ω radians per second at any harmonic time t seconds:

$$e = E_m \sin \omega t \text{ volts.} \tag{6.63}$$

NOTE. A cycle is a single sequence of values from zero to positive maximum to zero to negative maximum to zero. The frequency (f) is the sequence rate in cycles per second (H_z). The angular velocity (ω) in radians per second equals 2π times the frequency (f). The time (t) is the time in seconds measured from the instant when the value is zero and is increasing to a positive maximum. When an emf is indicated by the expression $e = E_m \sin(\omega t + \alpha)$, time is measured from the instant when $e = E_m \sin \alpha$.

The **current** (i) flowing at emf time t seconds in a circuit of R ohms resistance, L henrys self-inductance, and C farads series capacitance, upon which an emf, $e = E_m \sin \omega t$, is impressed:

$$i = \frac{E_m}{\sqrt{R^2 + (L\omega - 1/C\omega)}} \sin\left[\omega t - \tan^{-1}\left(\frac{l\omega - 1/C\omega}{R}\right)\right] \quad \text{amperes.} \quad (6.64)$$

NOTE.   It is assumed that the emf has been impressed upon the circuit long enough to produce a current.

The **maximum current** (I_m) flowing in a circuit under the conditions stated in Equation (6.64):

$$I_m = \frac{E_m}{\sqrt{R^2 + (L\omega - 1/C\omega)^2}} \quad \text{amperes.} \quad (6.65)$$

The **effective** or **root-mean-square emf** (E) of an emf, $e = E_m \sin \omega t$ volts:

$$E = \frac{E_m}{\sqrt{2}} \quad \text{volts.} \quad (6.66)$$

NOTE.   The effective current (I) of a current equals $I_m/2^{1/2}$ amperes.

The **average emf** (E_a) of an emf, $e = E_m \sin \omega t$ volts:

$$E_a = \frac{2E_m}{\pi} \quad \text{volts.} \quad (6.67)$$

NOTE.   The average current (I_a) of a current equals $2I_m/\pi$ amperes.

The **form factor** ($f.f.$) and **amplitude factor** ($a.f.$), respectively, of an emf, $e = E_m \sin \omega t$ volts:

$$f.f. = \frac{E}{E_a} = 1.11. \quad (6.68)$$

$$a.f. = \frac{E_m}{E} = 1.414. \quad (6.69)$$

NOTE.   The form factor ($f.f.$) and amplitude factor ($a.f.$), respectively, of a current are $I/I_a = 1.11$ and $I_m/I = 1.414$.

6.5.3   Reactance and Impedance

The **Reactance** (X) of a circuit of L henrys self-inductance and C farads series capacitance when conducting a current of ω radians per second angular velocity:

$$X = L\omega - \frac{1}{C\omega} \quad \text{ohms.} \tag{6.70}$$

NOTE. $L\omega$ is called the inductive reactance and $1/(C\omega)$ the capacitive reactance of the circuit, each measured in ohms.

The **impedance** (Z) of a circuit of R ohms resistance and X ohms series reactance:

$$Z = \sqrt{R^2 + X^2} \quad \text{ohms.} \tag{6.71}$$

The **phase angle** (θ) of a circuit of R ohms resistance and X ohms series reactance:

$$\theta = \tan^{-1}\frac{X}{R}. \tag{6.72}$$

NOTE.   The phase angle (θ) of a circuit in radians divided by the angular velocity (ω) of the conducted current in radians per second equals the time t in seconds by which the current lags or leads the emf. A positive value of X/R indicates a lagging current, and a negative value a leading current.

The **power factor** (PF) of a part circuit of R ohms resistance, Z ohms impedance, and phase angle θ containing no generated emf:

$$p.f. = \frac{R}{Z} = \cos\,\theta. \tag{6.73}$$

The **total reactance** (X_s) of a series circuit the respective parts of which have reactances of $X_1, X_2, X_3, \ldots$, ohms:

$$X_s = X_1 + X_2 + X_3 + \cdots \quad \text{ohms.} \tag{6.74}$$

NOTE.   The addition is algebraic, inductive reactance being positive and capacitive reactance negative.

The **total impedance** (Z_s) of a series circuit of R_s ohms total resistance and X_s ohms total reactance. See Equation (6.71).

NOTE.   The total impedance of a series circuit does not equal the sum of the impedances of its respective parts unless the ratio of reactance to resistance in each part is the same and the net reactances are of the same sign.

6.5.4   Power and Reactive Power

The **Power** (P) delivered to or from a part circuit conducting an effective current of I amperes across which the effective potential rise in the direction of the current is V volts, the phase angle between the current and the potential rise being θ:

$$P = VI \cos\,\theta \quad \text{watts.} \tag{6.75}$$

NOTE. Positive power indicates net power delivered from, and negative power indicates net power delivered to the part circuit.

The **reactive power** (Q) under the conditions stated in Equation (6.75):

$$Q = VI \sin \theta \quad \text{vars.} \tag{6.76}$$

NOTE. Leading reactive power is considered by convention to be positive. Lagging reactive power is considered negative.

Volt-amperes (*VA*), or apparent power, under the conditions stated in Equation (6.75) is the product of V volts and I amperes (*VI*).

The **power** (*P*) delivered to a part circuit of R ohms resistance conducting an effective current of I amperes and containing no generated emf:

$$P = I^2R \quad \text{watts.} \tag{6.77}$$

NOTE. The net power delivered to a reactance is zero.

The **reactive power** (Q) delivered to a part circuit of X ohms reactance, conducting an effective current of I amperes, and containing no generated emf:

$$Q = I^2X \quad \text{vars.} \tag{6.78}$$

NOTE. Q is positive for a capacitive reactance and negative for an inductive reactance. The reactive power delivered to a resistance is zero.

The **volt-amperes** (VA) delivered to a part circuit of Z ohms impedance conducting an effective current of I amperes and containing no generated emf:

$$\text{VA} = I^2Z \quad \text{volt-amperes.} \tag{6.79}$$

The **volt-amperes** (VA) corresponding to a power of P watts and a reactive power of Q vars:

$$\text{VA} = \sqrt{P^2 + Q^2}. \tag{6.80}$$

The **effective vector expression** (**E**) and (**I**) for an emf, $e = E_m \sin(\omega t + \alpha)$ volts, and a current, $i = I_m \sin(\omega t - \beta)$ amperes:

$$\mathbf{E} = \left(\frac{E_m}{\sqrt{2}} \cos \alpha + \mathbf{j} \frac{E_m}{\sqrt{2}} \sin \alpha \right) = \frac{E_m}{\sqrt{2}} \underline{/\alpha} \quad \text{volts.} \tag{6.81}$$

$$\mathbf{I} = \left(\frac{I_m}{\sqrt{2}} \cos \beta - \mathbf{j} \frac{I_m}{\sqrt{2}} \sin \beta \right) = \frac{I_m}{\sqrt{2}} \underline{/-\beta} \quad \text{amperes.} \tag{6.82}$$

NOTE. In symbolic notation, the horizontal component of a vector is without prefix and its sign is + to the right and − to the left of the Y axis; the vertical component is designated by the prefix **j** and its sign is + above and − below the X axis. In some mathematical operations the symbol **j** has the value $(-1)^{1/2}$. The symbols $\underline{/\alpha}$ and $\underline{/-\beta}$

indicate vectors making the angles $+\alpha$ and $-\beta$, respectively, with the X axis and having magnitudes given by the quantity preceding the symbols. This is known as the polar form of the vector expression.

6.5.5 Vector and Symbolic Expressions

The **vector electromotive force** ($\mathbf{E}_{AD}$) in a circuit the constituent parts of which contain the vector emf's $\mathbf{E}_{AB}$, $\mathbf{E}_{BC}$, and $\mathbf{E}_{CD}$ volts:

$$\mathbf{E}_{AD} = \mathbf{E}_{AB} + \mathbf{E}_{BC} + \mathbf{E}_{CD} \quad \text{volts.} \tag{6.83}$$

NOTE. Each vector emf must be referred to the same axis of reference. The subscripts in each case indicate the direction of emf rise.

The **vector current** ($\mathbf{I}_{BA}$) flowing from B toward a junction A from which the vector current $\mathbf{I}_{AC}$, $\mathbf{I}_{AD}$, and $\mathbf{I}_{AF}$ amperes flow away:

$$\mathbf{I}_{BA} = \mathbf{I}_{AC} + \mathbf{I}_{AD} + \mathbf{I}_{AF} \quad \text{amperes.} \tag{6.84}$$

The **electromotive force equivalent** (E) of a vector emf, $\mathbf{E} = a + jb$ volts:

$$E = \sqrt{a^2 + b^2} \quad \text{volts.} \tag{6.85}$$

NOTE. In polar form, $\mathbf{E} = (a^2 + b^2)^{1/2} \underline{/\tan^{-1}(b/a)}..$ The current equivalent (I) of a vector current $\mathbf{I} = (c + jd)$ amperes is $(c^2 + d^2)^{1/2}$ amperes, and in polar form $I = (c^2 + d^2)^{1/2} \underline{/\tan^{-1}(d/c)}..$

The **symbolic expression** ($\mathbf{Z}$) for the impedance of a circuit of R ohms resistance and X ohms reactance:

$$\mathbf{Z} = (R + jX) \quad \text{ohms.} \tag{6.86}$$

NOTE. The resistance component has no prefix and is always $+$; the reactance component has the prefix j, a $+$ sign indicating net inductive reactance and a $-$ sign net capacitive reactance. In polar form, $\mathbf{Z} = (R^2 + X^2)^{1/2} \underline{/\tan^{-1}(X/R)}..$

The **symbolic impedance** ($\mathbf{Z}_{AD}$) between the ends A and D of a part circuit containing several series parts of symbolic impedance $\mathbf{Z}_{AB}$, $\mathbf{Z}_{BC}$, and $\mathbf{Z}_{CD}$ ohms, respectively:

$$\mathbf{Z}_{AD} = \mathbf{Z}_{AB} + \mathbf{Z}_{BC} + \mathbf{Z}_{CD} \quad \text{ohms} \tag{6.87}$$
$$= (R_{AB} + R_{BC} + R_{CD}) + j(X_{AB} + X_{BC} + X_{CD}) \quad \text{ohms.}$$

The **vector current** ($\mathbf{I}$) flowing in the direction of an emf rise, $\mathbf{E} = (a + jb)$ volts acting in a circuit of symbolic impedance $\mathbf{Z} = (r + jx)$ ohms:

$$\mathbf{I} = \frac{a + jb}{r + jx} \quad \text{amperes.} \tag{6.88}$$

NOTE. To rationalize, multiply both numerator and denominator by the denominator with the sign of its j term reversed. We then have

$$I = \frac{(a + jb)(r - jx)}{(r + jx)(r - jx)} = \frac{ar - j^2bx + jbr - jax}{r^2 - j^2x^2}.$$

In this operation, $j = (-1)^{1/2}$ or $j^2 = -1$. Hence

$$I = \frac{(ar + bx) + j(br - ax)}{r^2 + x^2} = \frac{ar + bx}{r^2 + x^2} + j\left(\frac{br - ax}{r^2 + x^2}\right).$$

Alternatively, in polar form,

$$I = \sqrt{a^2 + b^2}\ \frac{\underline{/\tan^{-1}(b/a)}}{\sqrt{r^2 + x^2}\ \underline{/\tan^{-1}(x/r)}} = \frac{\sqrt{a^2 + b^2}}{\sqrt{r^2 + x^2}}\ \Big/\tan^{-1}\frac{b}{a} - \tan^{-1}\frac{x}{r}.$$

The **vector potential rise** ($\mathbf{V}_{AB}$) between the ends A and B of a part circuit of symbolic impedance $\mathbf{Z}_{AB}$ ohms conducting a current of vector value $\mathbf{I}_{AB}$ amperes and containing an emf rise of vector value $\mathbf{E}_{AB}$ volts:

$$\mathbf{V}_{AB} = +\mathbf{E}_{AB} - \mathbf{I}_{AB}\mathbf{Z}_{AB}\quad\text{volts.}\tag{6.89}$$

NOTE. If

$$\mathbf{E}_{AB} = a + jb,\ \mathbf{I}_{AB} = c + jd,\text{ and }\mathbf{Z}_{AB} = r + jx,$$

$$\mathbf{V}_{AB} = a + jb - (c + jd)(r + jx)$$

$$= a + jb - cr - j^2dx - jcx - jdr$$

and since $j^2 = -1$,

$$\mathbf{V}_{AB} = (a - cr + dx) + j(b - cx - dr).$$

The **vector potential rise** ($\mathbf{V}_{AD}$) between the ends A and D of a part-circuit containing several series parts across which the respective vector potential rises are $\mathbf{V}_{AB}$, $\mathbf{V}_{BC}$, and $\mathbf{V}_{CD}$ volts:

$$\mathbf{V}_{AD} = \mathbf{V}_{AB} + \mathbf{V}_{BC} + \mathbf{V}_{CD}\quad\text{volts.}\tag{6.90}$$

The **power** (P) delivered to or from a part-circuit conducting a vector current $\mathbf{I} = (c + jd)$ amperes and across which the vector potential rise in the direction of the current is $\mathbf{V} = (a + jb)$ volts:

$$P = (ac + bd)\quad\text{watts.}\tag{6.91}$$

NOTE. The signs of a, b, c, and d are preserved. Positive power indicates power delivered from, and negative power indicates power delivered to, the part circuit. The power does not equal $(a + jb)(c + jd)$.

The **reactive power** (Q) under the conditions stated in Equation (6.91):

$$Q = (ad - bc) \quad \text{vars.} \tag{6.92}$$

NOTE. The signs of a, b, c, and d should be preserved. Leading reactive power is positive; lagging reactive power is negative.

The **conductance** (G) and **susceptance** (B) of a branch of R ohms resistance, X ohms series reactance, and Z ohms impedance:

$$G = \frac{R}{R^2 + X^2} = \frac{R}{Z^2} \quad \text{siemens.} \tag{6.93}$$

$$B = \frac{X}{R^2 + X^2} = \frac{X}{Z^2} \quad \text{siemens.} \tag{6.94}$$

The **admittance** (Y) of a branch of Z ohms impedance, G siemens conductance, and B siemens susceptance:

$$Y = \frac{1}{Z = \sqrt{G^2 + B^2}} \quad \text{siemens.} \tag{6.95}$$

The **total conductance** (G_0) of several parallel branches of G_1, G_2, and G_3 siemens conductance, respectively:

$$G_0 = G_1 + G_2 + G_3 \quad \text{siemens.} \tag{6.96}$$

The **total susceptance** (B_0) of several parallel branches of B_1, B_2, and B_3 siemens susceptance, respectively:

$$B_0 = B_1 + B_2 + B_3 \quad \text{siemens.} \tag{6.97}$$

NOTE. The addition is algebraic, inductive susceptance being positive and capacitive susceptance negative.

The **total admittance** (Y_0) of several parallel branches of total conductance G_0 siemens and total susceptance B_0 siemens. See Equation (6.95).

NOTE. The total admittance of a parallel circuit does not equal the sum of the admittances of the respective branches unless the ratio of susceptance to conductance in each branch is the same and the net susceptances are of the same sign.

The **phase angle** (θ) of a circuit of G siemens conductance and B siemens susceptance.

$$\theta = \tan^{-1} \frac{B}{G}. \tag{6.98}$$

The **power factor** (PF) of a part circuit of G seimens conductance and Y siemens admittance, containing no generated emf:

$$\text{PF} = \frac{G}{Y}. \tag{6.99}$$

The **resistance** (R) and **reactance** (X) of a circuit of G siemens conductance, B siemens susceptance, and Y siemens admittance:

$$R = \frac{G}{G^2 + B^2} = \frac{G}{Y^2} \quad \text{ohms.}$$

$$X = \frac{B}{G^2 + B^2} = \frac{B}{Y^2} \quad \text{ohms.} \tag{6.100}$$

The **impendance** (Z) of a circuit of Y siemens admittance:

$$Z = \frac{1}{Y} \quad \text{ohms.} \tag{6.101}$$

The **symbolic expression** ($\mathbf{Y}$) for the admittance of a circuit of G siemens conductance and B siemens susceptance:

$$\mathbf{Y} = (G - \mathbf{j}B) \quad \text{siemens.} \tag{6.102}$$

NOTE. In polar form, $\mathbf{Y} = \sqrt{G^2 + B^2} \; \underline{/\tan^{-1}(-B/G)}$.

The **symbolic admittance** ($\mathbf{Y}_0$) of a parallel circuit containing several branches of symbolic admittance $\mathbf{Y}_1$, $\mathbf{Y}_2$, and $\mathbf{Y}_3$ siemens, respectively:

$$\mathbf{Y}_0 = \mathbf{Y}_1 + \mathbf{Y}_2 + \mathbf{Y}_3 \quad \text{siemens.} \tag{6.103}$$

The **vector current** ($\mathbf{I}$) flowing in the direction of an emf rise $\mathbf{E}$ acting in a circuit of $\mathbf{Y}$ siemens symbolic admittance:

$$\mathbf{I} = \mathbf{EY} \quad \text{amperes.} \tag{6.104}$$

6.5.6   Nomenclature for Three-Phase Circuits

Line emf or voltage, E_1 volts; phase emf or voltage, E_p volts; line current I_1 amperes; phase current, I_p amperes; phase angle between phase voltage and phase current, Θ_p. In Equations (6.109) and (6.111), $\mathbf{E}_a$, $\mathbf{E}_b$, and $\mathbf{E}_c$ are any three voltage vectors that may exist in a three-phase system, such as voltages to neutral or to ground, line-to-line voltages, and induced voltages. Similarly, $\mathbf{I}_a$, $\mathbf{I}_b$, and $\mathbf{I}_c$ in Equations (6.110) and (6.112) may be line currents, phase currents, the currents in a Δ-connected winding, and so on. The subscripts 1, 2, and 0 in Equations (6.109) to (6.112) denote, respectively, positive-, negative-, and zero-sequence components. Positive phase rotation, ABC in counterclockwise direction.

Conditions for balanced three-phase circuit: all phase currents, phase emf's, and phase voltages, respectively, equal and differing in phase by 120°. Conditions for unbalanced three-phase circuit: phase currents, phase emf's, or phase voltages, respectively, unequal or not differing in phase by 120°.

6.5.7 Three-Phase Circuits

Balanced Y-connected branches (Fig. 6.12)

$$E_1 = \sqrt{3}\, E_p.$$

$$I_1 = I_p.$$

$$\mathbf{E}_{OA} + \mathbf{E}_{OB} + \mathbf{E}_{OC} = 0. \tag{6.105}$$

$$\mathbf{E}_{AB} + \mathbf{E}_{BC} + \mathbf{E}_{CA} = 0.$$

$$\mathbf{E}_{AB} = \mathbf{E}_{OB} - \mathbf{E}_{OA} = \sqrt{3}\, \mathbf{E}_{OB} \; \underline{/30°} = \sqrt{3}\, \mathbf{E}_{OC} \; \underline{/90°}.$$

$$\mathbf{I}_{OA} + \mathbf{I}_{OB} + \mathbf{I}_{OC} = 0.$$

Balanced Δ-connected branches (Fig. 6.13)

$$E_1 = E_p.$$

$$I_1 = \sqrt{3}\, I_p. \tag{6.106}$$

$$\mathbf{E}_{AB} + \mathbf{E}_{BC} + \mathbf{E}_{CA} = 0.$$

$$\mathbf{I}_{AB} + \mathbf{I}_{BC} + \mathbf{I}_{CA} = 0.$$

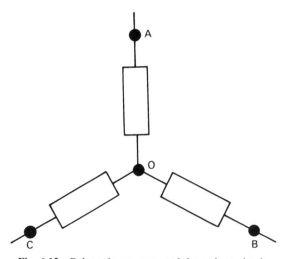

Fig. 6.12   Balanced wye-connected three-phase circuit.

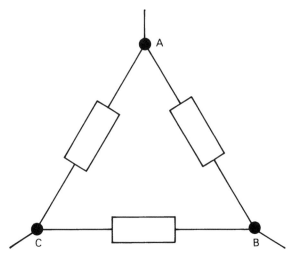

Fig. 6.13   Balanced delta-connected three-phase circuit.

$$\mathbf{I}'_{AA} = \mathbf{I}_{CA} - \mathbf{I}_{AB} = \sqrt{3}\,\mathbf{I}_{CA}\,\underline{/30^\circ} = \sqrt{3}\,\mathbf{I}_{BC}\,\underline{/90^\circ}.$$ (6.106a)

$$\mathbf{I}'_{AA} = \mathbf{I}'_{BB} + \mathbf{I}'_{CC} = 0.$$

Unbalanced Y-connected branches

$$\mathbf{E}_{AB} + \mathbf{E}_{BC} + \mathbf{E}_{CA} = 0.$$

$$\mathbf{E}_{AB} = \mathbf{E}_{OB} - \mathbf{E}_{OA}.$$ (6.107)

$$\mathbf{I}_{OA} + \mathbf{I}_{OB} + \mathbf{I}_{OC} = 0.$$

NOTE.   If there is a current flowing out of the point O in a neutral connection, it must be added vectorially to the left-hand side of the last equation.

Unbalanced Δ-connected branches

$$\mathbf{E}_{AB} + \mathbf{E}_{BC} + \mathbf{E}_{CA} = 0.$$ (6.108)

$$\mathbf{I}'_{AA} = \mathbf{I}_{CA} - \mathbf{I}_{AB}.$$

Symmetrical components of voltage in an unbalanced three-phase system:

$$\mathbf{E}_{a1} = \tfrac{1}{3}(\mathbf{E}_a + \mathbf{E}_b\underline{/120^\circ} + \mathbf{E}_c\underline{/120^\circ}).$$

$$\mathbf{E}_{a2} = \tfrac{1}{3}(\mathbf{E}_a + \mathbf{E}_b\underline{/120^\circ} + \mathbf{E}_c\underline{/120^\circ}).$$ (6.109)

$$\mathbf{E}_0 = \tfrac{1}{3}(\mathbf{E}_a + \mathbf{E}_b + \mathbf{E}_c).$$

Symmetrical components of current in an unbalanced three-phase system:

Replace $\mathbf{E}$ in Equation (6.109) by $\mathbf{I}$.   (6.110)

Three-phase voltages in terms of the symmetrical components of voltage.

$$\mathbf{E}_a = \mathbf{E}_{a1} + \mathbf{E}_{a2} + \mathbf{E}_0.$$

$$\mathbf{E}_b = \mathbf{E}_{a1}\underline{/120°} + \mathbf{E}_{a2}\underline{/120°} + \mathbf{E}_0.$$ (6.111)

$$\mathbf{E}_c = \mathbf{E}_{a1}\underline{/120°} + \mathbf{E}_{a2}\underline{/120°} + \mathbf{E}_0.$$

Three-phase currents in terms of the symmetrical components of current:

Replace $\mathbf{E}$ in Equation (6.111) by $\mathbf{I}$.   (6.112)

Y-connected impedances which are equivalent to a given set of Δ-connected impedances so far as conditions at the terminals are concerned (see Fig. 6.14):

$$\mathbf{Z}_A = \frac{\mathbf{Z}_{CA}\mathbf{Z}_{AB}}{\mathbf{Z}_{AB} + \mathbf{Z}_{BC} + \mathbf{Z}_{CA}}.$$

$$\mathbf{Z}_B = \frac{\mathbf{Z}_{AB}\mathbf{Z}_{BC}}{\mathbf{Z}_{AB} + \mathbf{Z}_{BC} + \mathbf{Z}_{CA}}.$$ (6.113)

$$\mathbf{Z}_C = \frac{\mathbf{Z}_{BC}\mathbf{Z}_{CA}}{\mathbf{Z}_{AB} + \mathbf{Z}_{BC} + \mathbf{Z}_{CA}}.$$

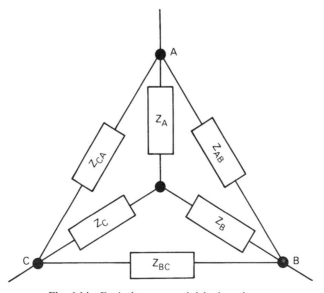

Fig. 6.14   Equivalent wye and delta impedances.

NOTE. If the impedances are balanced, the Y-connected impedances are ⅓ of the Δ-connected impedances.

Δ-connected impedances that are equivalent to a given set of Y-connected impedances so far as conditions at the terminals are concerned (see Fig. 6.14):

$$Z_{AB} = \frac{Z_A Z_B + Z_B Z_C + Z_C Z_A}{Z_C}$$

$$Z_{BC} = \frac{Z_A Z_B + Z_B Z_C + Z_C Z_A}{Z_A} \qquad (6.114)$$

$$Z_{CA} = \frac{Z_A Z_B + Z_B Z_C + Z_C Z_A}{Z_B}$$

NOTE. If the impedances are balanced, the Δ-connected impedances are three time the Y-connected impedances.

The **power** (*P*) delivered to or from a balanced three-phase line:

$$P = \sqrt{3} E_1 I_1 \cos \theta_p \quad \text{watts.} \qquad (6.115)$$

The **power factor** (PF) of a balanced three-phase load:

$$\text{PF} = \cos \theta_p. \qquad (6.116)$$

The **power** (*P*) delivered to or from an unbalanced three-phase line:

$$P = E_{p1} I_{p1} \cos \theta_{p1} + E_{p2} I_{p2} \cos \theta_{p2} + E_{p3} I_{p3} \cos \theta_{p3} \quad \text{watts.} \qquad (6.117)$$

NOTE. The subscripts 1, 2, and 3 here denote the three phases.

The **power factor** (PF) of an unbalanced three-phase load:

$$\text{PF} = \frac{P}{E_{p1} I_{p1} + E_{p2} I_{p2} + E_{p3} I_{p3}}. \qquad (6.118)$$

NOTE. The subscripts 1, 2, and 3 here denote the three phases.

The **power** (*P*) measured by two wattmeters connected in a three-phase line as shown in Fig. 6.15:

$$P = P_A \pm P_B \quad \text{watts.} \qquad (6.119)$$

NOTE. To determine use of + or − sign, break connection of potential coil of wattmeter A at line C and connect to line B. A wattmeter deflection on scale indicates the use of the + sign and a deflection off scale indicates the use of the − sign.

The **power** (P_A) and (P_B), respectively, measured by two wattmeters connected in a three-phase balanced line as shown in Fig. 6.15:

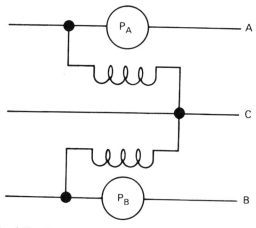

Fig. 6.15 Three-phase power measured by two wattmeters.

$$P_A = E_1 I_1 \cos(30° - \theta_p) \quad \text{watts.}$$
$$\tag{6.120}$$
$$P_B = E_1 I_1 \cos(30° + \theta_p) \quad \text{watts.}$$

The **phase angle** (θ) of a balanced three-phase load when two wattmeters connected as shown in Fig. 6.15 measure P_A and P_B watts, respectively:

$$\theta = \tan^{-1} \sqrt{3} \, \frac{P_A - P_B}{P_A + P_B}. \tag{6.121}$$

NOTE.   The additions and substractions are algebraic.

6.5.8   Nonsinusoidal Alternating Voltages and Currents

Any nonsinusoidal voltage or current waveform can be resolved into fundamental frequency sine wave and its harmonics by a Fourier's series. If y is the amplitude of a nonsinusoidal wave at time t; Y_1, $Y_2 \cdots$, Y_n the maximum amplitude of the 1st, 2nd, $\cdots$, nth harmonics; θ_1, $\theta_2, \cdots$, θ_n the angles that determine relative phase of the harmonics; and $\omega(2\pi f)$ the angular velocity of the fundamental wave corresponding to f hertz,

$$y = Y_1 \sin(\omega t + \theta_1) + Y_2 \sin(2\omega T + \theta_2) + \cdots + Y_n \sin(n\omega t + \theta_n). \tag{6.122}$$

The **effective** or **root-mean-square value** (E) of a periodically time-varying emf or voltage $e = f(t)$ volts having a period of T seconds:

$$E = \sqrt{\frac{1}{T} \int_0^T e^2 \, dt} \text{ volts.} \tag{6.123}$$

The **effective** or **root-mean-square value** (i) of a periodically time-varying current $i = f(t)$ amperes having a period of T seconds:

$$I = \sqrt{\frac{1}{T} \int_0^T i^2 \, dt} \text{ amperes.} \tag{6.124}$$

The **electromotive force** (e) of a nonsinusoidal emf or voltage at any time t seconds:

$$e = E_{m1} \sin(\omega t + \theta_1) + E_{m3} \sin(3\omega t + \theta_3)$$
$$+ E_{m5} \sin(5\omega t + \theta_5) + \cdots \text{ volts.} \tag{6.125}$$

NOTE. $E_{m1}, E_{m3}, E_{m5}, \cdots$ represent the maximum values of the first, third, fifth, and so on, harmonics, and $\theta_1, \theta_2, \theta_3, \cdots$ their respective phase angles with a common axis of reference. The angular velocity ω is that of the fundamental or first harmonic. Alternators do not normally generate even harmonics of voltage. Some nonsinusoidal voltages, such as the outputs of rectifiers, however, may contain odd or even harmonics and average or dc components. In such cases the terms $E_0 + E_{m2} \sin(2\omega t + \theta_2) + E_{m4} \sin(4\omega t + \theta_4) + \cdots$ should be added to the right-hand member of Equation (6.125).

The **current** (i) at any time t seconds flowing in a circuit of R ohms resistance, L henrys self-inductance, and C farads series capacitance upon which a nonsinusoidal emf of the form stated in Equation (6.125) is impressed:

$$i = I_{m1} \sin(\omega t + \theta_1') + I_{m3} \sin(3\omega t + \theta_3')$$
$$+ I_{m5} \sin(5\omega t + \theta_5') + \cdots \text{ amperes.} \tag{6.126}$$

NOTE

$$I_{m1} = \frac{E_{m1}}{\sqrt{R^2 + (L\omega - 1/C\omega)^2}}, \quad \theta_1' = \theta_1 - \tan^{-1}\left(\frac{L\omega - 1/C\omega}{R}\right),$$

$$I_{m3} = \frac{E_{m3}}{\sqrt{R^2 + (3L\omega - 1/3C\omega)^2}}, \quad \theta_3' = \theta_3 - \tan^{-1}\left(\frac{3L\omega - 1/3C\omega}{R}\right),$$

$$I_{m5} = \frac{E_{m5}}{\sqrt{R^2 + (5L\omega - 1/5C\omega)^2}}, \quad \theta_5' = \theta_5 - \tan^{-1}\left(\frac{5L\omega - 1/5C\omega}{R}\right).$$

The **effective emf** (E) of a nonsinusoidal emf of the form stated in Equation (6.125):

$$E = \sqrt{\frac{(E_{m1})^2 + (E_{m3})^2 + (E_{m5})^2}{2}} \text{ volts.} \tag{6.127}$$

NOTE. The effective value of a nonsinusoidal potential or current is obtained in the same manner.

The **power** (P) delivered to a part circuit conducting a nonsinusoidal current of the form stated in Equation (6.126) and upon which is impressed a nonharmonic emf of the form stated in Equation (6.125):

$$P = \frac{E_{m1}I_{m1}}{2} \cos(\theta_1 - \theta_1') + \frac{E_{m3}I_{m3}}{2} \cos(\theta_3 - \theta_3')$$
$$+ \frac{E_{m5}I_{m5}}{2} \cos(\theta_5 - \theta'5) + \cdots \text{ watts.} \tag{6.128}$$

The **power factor** (PF) of a part circuit conducting a nonsinusoidal current of effective value I amperes and absorbing energy at a rate of P watts, the effective value of the nonsinusioidal potential between its ends being V volts:

$$PF = \frac{P}{VI}. \qquad (6.129)$$

The **sinusoidal emf and current** equivalent to the nonsinusoidal forms stated in Equations (6.125) and (6.126):

$$e = \sqrt{2}\, E \sin \omega \quad \text{volts}; \qquad (6.130)$$
$$i = \sqrt{2}\, I \sin(t \pm \cos^{-1} PF) \quad \text{amperes}.$$

NOTE. PF is the power factor of the circuit upon which the nonharmonic emf is impressed.

The **resistance** (R) of a part-circuit containing no source of generated emf, conducting an effective current of I amperes, and absorbing energy at a rate of P watts:

$$R = \frac{P}{I^2} \quad \text{ohms}. \qquad (6.131)$$

The **impendance** (Z) of a part-circuit containing no source of generated emf, conducting an effective current of I amperes, and across which the potential is V volts:

$$Z = \frac{V}{I} \quad \text{ohms}. \qquad (6.132)$$

The **reactance** (X) of a part circuit of R ohms resistance and Z ohms impendance:

$$X = \sqrt{Z^2 - R^2} \quad \text{ohms}. \qquad (6.133)$$

NOTE. The reactance of a part circuit to a nonsinusoidal current does not equal $[L\omega - (1/C\omega)]$ ohms.

6.6 DIRECT-CURRENT GENERATORS AND MOTORS

6.6.1 Rotating Masses

A generator is accelerated to its rated speed by applying mechanical torque (T) to its shaft; a motor is accelerated by creating mechanical torque at its shaft. For either machine, where I is total inertia of the rotating mass and α is angular acceleration,

$$\alpha = \frac{T}{I}. \qquad (6.134)$$

Units for α, T, and I in the English system, which is still in common use, can be converted to SI system units.

To Convert from:		To:	Multiply by:
α	rad/s	rad/s^2	1.0
T	lbf-ft	newton meter (N·m)	1.356
I	lb-ft^2	kilogram meter^2 (kg·m^2)	0.0421

The **time** (t) to accelerate a rotating mass from standstill to rated speed n revolutions per minute when it has inertia I and average torque T_A during the accelerating interval:

$$t = \frac{0.105In}{gT_A} \quad \text{seconds.} \tag{6.135}$$

I, g, and T_A can be either metric or English units:

	I	g	T_A
Metric	kg·m^2	9.807 m/s^2	N·m
English	lb-ft^2	32.17 ft/s^2	lbf-ft

6.6.2   Mechanical Shaft Power

The **mechanical shaft power,** applied to a shaft in the case of a generator, or taken from the shaft of a motor at rated speed n revolutions per minute:

$$\text{For } T \text{ in N·m, } P = \frac{nT}{5190} \quad \text{kilowatts.} \tag{6.136}$$

$$\text{For } T \text{ in lbf-ft, } P = \frac{nT}{5250} \quad \text{horsepower.} \tag{6.137}$$

All the equations in Sections 6.6, and in Section 6.7 on alternating current rotating machines, relate to the conversion of mechanical shaft power to electric power or vice versa, and are applicable to either a generator or a motor.

6.6.3   DC Machine Nomenclature and Units of Measurement

Emf generated in armature, E volts; terminal voltage, V volts; armature current, I amperes; line current, I_l amperes; shunt field current, I_f amperes; series field current, I_s amperes; armature resistance between brushes, R ohms; shunt field resistance including rheostat, R_f ohms; series field resistance including shunt, R_s ohms; number of poles, p; shunt field turns per pole, N_f; series field turns per pole, N_s; number of armature paths between terminals, m; number of armature conductors, Z; magnetic flux per pole, Φ webers; armature speed, n revolutions per minute; armature torque, T newton meters.

6.6.4 DC Machine Characteristics

The **electromotive force** (E) generated in the armature of a dc machine:

$$E = \frac{p\Phi Zn}{60m} \quad \text{volts.} \tag{6.138}$$

The **shunt field current** (I_{fd}) equivalent to the demagnetizing magnetomotive force of the armature per pole when the armature current is I amperes and the brushes are shifted through an angle of θ space degrees from the neutral plane to improve commutation:

$$I_{fd} = \frac{ZI\theta}{360N_f m} \quad \text{amperes.} \tag{6.139}$$

The **shunt field current** (I_{fs}) equivalent to the magnetomotive force of the series turns per pole:

$$I_{fs} = \frac{N_s}{N_f} I_s \quad \text{amperes.} \tag{6.140}$$

The **net field current** (I_{fn}) at any load:

$$I_{fn} = I_f - I_{fd} \pm I_{fs} \quad \text{amperes.} \tag{6.141}$$

NOTE. The sign before I_{fs} is $+$ for a cumulative and $-$ for a differential compound machine.

The **terminal voltage** (V) of a shunt machine when the armature current is I amperes and the generated emf is E volts:

$$V = E \pm IR \quad \text{volts.} \tag{6.142}$$

NOTE. The sign before IR is $+$ for a motor and $-$ for a generator. In a series or long-shunt compound machine, $V = E \pm I(R + R_s)$ volts, and in a short-shunt compound machine, $V = E \pm IR \pm I_s R_s$ volts.

The **armature speed** (n) when the generated emf is E volts:

$$n = \frac{60Em}{p\Phi Z} \quad \text{revolutions/minute.} \tag{6.143}$$

The **armature torque** (T) of a dynamo when the armature current is I amperes:

$$T = \frac{0.159Z\Phi Ip}{m} \quad \text{newton meters.} \tag{6.144}$$

The **rotational losses** (P_r) of a machine that, operated as a shunt motor at no load with a voltage between brushes of V volts, takes an armature current of I amperes:

$$P_r = VI - I^2R \quad \text{watts}. \tag{6.145}$$

NOTE.   To determine the rotational losses corresponding to a definite load, the machine operated as a shunt motor at no load must be run at the same speed and with the same generated emf as when running at the definite load.

The **copper losses** (P_k) at any load:

$$\text{Shunt field: } P_f = I_f^2 R_f = VI_f \quad \text{watts}. \tag{6.146}$$

$$\text{Series field: } P_s = I_s^2 R_s \quad \text{watts}. \tag{6.147}$$

$$\text{Armature: } \quad P_a = I^2R \quad \text{watts}. \tag{6.148}$$

The **power input** (P_i) **to a generator** at any load:

$$\begin{aligned} P_i &= EI + P_r \quad \text{watts} \\ &= P_o + P_k + P_r \quad \text{watts} \\ &= 0.1420nT \quad \text{watts} \\ &= 1.903nT \times 10^{-4} \quad \text{horsepower}. \end{aligned} \tag{6.149}$$

The **power output** (P_o) **of a generator** at any load:

$$P_o = VI_1 \quad \text{watts}. \tag{6.150}$$

The **power input** (P_i) **of a motor** at any load:

$$P_i = VI_1 \quad \text{watts}. \tag{6.151}$$

The **power output** (P_o) **of a motor** at any load:

$$\begin{aligned} P_o &= EI - P_r \quad \text{watts} \\ &= P_i - P_k - P_r \quad \text{watts} \\ &= 0.1420nT \quad \text{watts} \\ &= 1.903nT \times 10^{-4} \quad \text{horsepower}. \end{aligned} \tag{6.152}$$

The **efficiency** (η) of a machine at any load:

$$\eta = \frac{P_o}{P_i} = \frac{P_o}{P_o + P_k + P_r} = \frac{P_i - P_k - P_r}{P_i}. \tag{6.153}$$

6.7   ALTERNATING-CURRENT MACHINES

6.7.1   Synchronous Machines

All equations in this subsection apply to synchronous generators, motors, or condensers unless indicated otherwise. Sinusoidal voltages and currents are assumed, and their magnitudes expressed by effective values. Three-phase machines with wye-connected windings are assumed with all machine impedances, voltages, and currents being phase values for a wye connection. The **terminal voltage** of a wye-connected machine is $(3)^{1/2}$ times phase voltage.

6.7.2   AC Synchronous Machine Characteristics

The **frequency** (f) of the voltage generated in a synchronous machine having p poles, the speed of the rotating magnetic field being n revolutions per minute:

$$f = \frac{pn}{120} \quad \text{hertz.} \tag{6.154}$$

The **synchronous internal voltage or excitation voltage** (E_i) generated in the armature of a synchronous machine:

$$E_i = 4.44 fN\Phi \, \cos\frac{\beta}{2} \left(\frac{\sin\,(m\alpha/2)}{m \, \sin\,(\alpha/2)} \right) \quad \text{volts.} \tag{6.155}$$

NOTE. N is the number of armature series turns per phase or one-half the number of series conductors on the armature divided by the number of phases, Φ is the main field flux per pole in webers, β is the pitch deficiency or the difference in electrical degrees between the pole pitch (180°) and the coil pitch, m is the number of slots per pole per phase, and α is the angle between adjacent slot centers in electrical degrees. Electrical degrees equal space degrees multiplied by $p/2$. This equation assumes that the arrangement of the winding before each pole is the same and hence that there are an integral number of slots per pole per phase. If ϕ_r, the resultant air-gap flux per pole corresponding to the mmf R in Equation (6.158), is used instead of ϕ, the voltage given by Equation (6.155) is the air-gap voltage E_a.

The **field magnetomotive force** (F) in a cylindrical-rotor machine:

$$F = \frac{4}{\pi} N_f I_f \frac{\sin\,n_f\alpha_f/2}{n_f \, \sin\,\alpha_f/2} \quad \text{ampere-turns/pole.} \tag{6.156}$$

NOTE. F is the maximum value of the fundamental field mmf. N_t is the number of field turns per pole carrying the current I_f amperes per turn. n_t is the number of rotor slots per pole, and α_f is the electrical angle between adjacent slots in a belt. Equation (6.156) assumes that the field winding is a regular, distributed, full-pitch winding.

The **armature magnetomotive force** (A) in a cylindrical-rotor machine:

$$A = 0.90 K N_a I \quad \text{ampere-turns/pole.} \tag{6.157}$$

NOTE. K equals $\cos (\beta/2)\{\sin(m\alpha/2)/[m \sin(\alpha/2)]\}$ as explained in Equation (6.155). N_a is the number of armature series turns per pole. I is the armature current. A may be obtained in equivalent field amperes by dividing by N_f.

The **resultant magnetomotive force** (R) in a machine with a field magnetomotive force F ampere-turns per pole and an armature magnetomotive force A ampere turns per pole.

$$R = F + A \quad \text{ampere-turns/pole.} \tag{6.158}$$

NOTE.   The addition must be made vectorially. See Equations (6.155), (6.161), and (6.162).

Characteristic curves of a synchronous machine are shown in Fig. 6.16. Data are plotted as follows: OCC (open-circuit characteristic), terminal voltage at no load versus field current; the air-gap line is drawn tangent to the lower part of OCC; SCC (short-circuit characteristic), armature current with the armature terminals short-circuited versus field current; ZPF (zero-power-factor characteristic), terminal voltage versus field current with the armature supplying a constant current I_0 to a ZPF lagging load at its terminals, I_0 usually being taken equal to rated armature current.

The **Potier triangle** cde (Fig. 6.16) is obtained as follows: Choose a point d well up on the curved part of ZPF; draw db parallel to and equal to ao; draw be parallel to the air-gap line; draw ex perpendicular to db; cde is then the Potier triangle.

The **Potier reactance** (X_p) from Potier triangle (see Fig. 6.16):

$$X_p = \frac{ec \text{ in volts}}{I_0 \text{ in amperes}} \quad \text{ohms.} \tag{6.159}$$

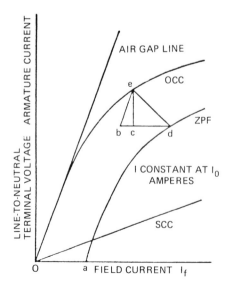

Fig. 6.16 No-load and short-circuit characteristics of a synchronous machine.

NOTE. I_0 is the constant armature current for which ZPF is drawn. X_p is very nearly equal to, and is frequently used for, the armature leakage reactance X_a.

The **armature magnetomotive force** (A) corresponding to the armature current $I = I_0$ from Potier triangle (see Fig. 6.16):

$$A = cd \text{ equivalent field amperes.} \tag{6.160}$$

NOTE. To obtain A in ampere-turns per pole as in Equation (6.157), substitute cd for I_f in Equation (6.156). A for any other value of I may be obtained by direct proportion.

The **general-method vector diagram** for a cylindrical-rotor machine with armature current I, terminal voltage V, and phase angle θ between I and V. Figure 6.17 is drawn for generator operation with a lagging power-factor load. All voltage vectors are voltage rises. For motor operation, I and the voltage drop vector $-V$ are θ degrees out of phase.

Air-gap voltage E_a is given by

$$\mathbf{E}_a = \mathbf{V} + \mathbf{I}(r + \mathbf{j}X_a), \tag{6.161}$$

in which the voltages $\mathbf{E}_a$ and $\mathbf{V}$ are rises and r is the armature resistance.

A is obtained from Equation (6.160) or (6.157).

R is obtained in equivalent field amperes by entering OCC (Fig. 6.16) with E_a volts and reading the corresponding field current.

The **field current** (I_f) required in a cylindrical-rotor machine with armature current I, terminal voltage V, and phase angle θ between I and V.

$$I_f = \text{magnitude of } (\mathbf{R} - \mathbf{A}) \quad \text{amperes.} \tag{6.162}$$

NOTE. R and A must be in equivalent field amperes in Equation (6.162); if they are in ampere-turns per pole, use Equation (6.156) to convert ampere-turns per pole into equivalent field amperes. See Fig. 6.17. R and A obtained as in Equation (6.161).

The **excitation voltage** (E_i) under conditions in Equation (6.162): Enter OCC (Fig. 6.16) with I_f from Equation (6.162) and read the corresponding voltage.

For **salient-pole machines** the methods of Equation (6.156) to (6.162) will give approximately correct results. In Equation (6.156), F should be taken as $N_f I_f$, and in

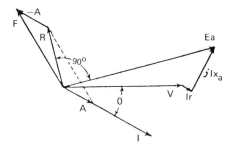

Fig. 6.17 Voltage and current phasor diagram of a cylindrical-rotor synchronous machine.

Equation (6.157) the factor 0.75 should be used instead of 0.9. For more accurate results the Blondel two-reaction method should be used.

The **total power output** (P_o) of a cylindrical-rotor synchronous machine with excitation voltage E_i volts, terminal voltage V volts, angle δ electrical degrees between the vectors $\mathbf{E}_i$ and $\mathbf{V}$, and synchronous reactance X_s ohms:

$$P_o = \frac{3VE_i}{X_s} \sin \delta \quad \text{watts.} \tag{6.163}$$

NOTE. Losses are neglected. X_s should be properly adjusted for saturation. The maximum power output is $(3VE_i)/X_s$ watts; it may not be attainable without loss of synchronism. ($3VE_i/X_s$ watts gives the breakdown or pull-out power for a cylindrical-rotor motor connected directly to a power system of capacity large compared to that of the motor.

The **total power output** (P_o) of a salient-pole synchronous machine with excitation voltage E_d volts, terminal voltage V colts, angle δ electrical degrees between the vectors $\mathbf{E}_d$ and $\mathbf{V}$, direct-axis synchronous reactance X_d ohms, and quadrature-axis synchronous reactance X_q ohms:

$$P_o = 3 \left[\frac{VE_d}{X_d} \sin \delta + \frac{V^2(X_d - X_q)}{2X_d X_q} \sin 2\delta \right] \quad \text{watts.} \tag{6.164}$$

NOTE. Losses are neglected. The reactances should be properly adjusted for saturation. The maximum value of P_o, as δ varies, may not be attainable without loss of synchronism. This maximum value give the breakdown or pull-out power for a salient-pole motor connected directly to a power system of capacity large compared to that of the motor.

The **efficiency** (η) of a synchronous machine when the output is P_o watts:

$$\eta = \frac{P_o}{P_o + P_a + P_c + P_s + P_{fw} + P_f}. \tag{6.165}$$

NOTE. All powers are total three-phase powers. P_a, the armature copper loss, equals three times the armature current squared times the ohmic resistance per phase. P_c is the open-circuit core loss (hysteresis and eddy-current losses). To determine P_c, enter a curve of open-circuit core loss versus open-circuit voltage with a voltage equal to the vector sum of the terminal voltage plus the armature resistance drop, and read the corresponding value of P_c. P_s is the stray-load loss (skin effect and eddy-current losses in the armature conductors plus local core losses due to armature leakage flux) and may be found from a curve of stray-load loss versus armature current. P_{fw} is the friction and windage loss, and P_f is the field copper loss.

6.7.3 Induction Machines

Three-phase wye-wound machines are assumed throughout and unless indicated otherwise each formula applies to a generator or a motor. All rotor resistances and reactances are referred to the stator. All impedances, voltages, and currents are phase values for a wye connection. All values of power P are total three-phase powers.

6.7.4 AC Induction Machine Characteristics

The **equivalent effective resistance** (R_i) of an induction machine:

$$R_1 = \frac{P_i}{3I_1^2} \quad \text{ohms.} \tag{6.166}$$

NOTE. P_i is the power input on blocked-rotor and I_i is the stator blocked-rotor current.

The **equivalent impedance** (Z_1) of an induction machine:

$$Z_1 = \frac{V_1}{I_1} \quad \text{ohms.} \tag{6.167}$$

NOTE. V_1 is the stator terminal voltage during blocked-rotor and I_1 is the stator blocked-rotor current.

The **equivalent reactance** (X_1) of an induction machine:

$$X_1 = \sqrt{Z_1^2 - R_1^2} \quad \text{ohms.} \tag{6.168}$$

NOTE. Z_1 and R_1 are determined as in Equations (6.167) and (6.168).

The **rotor resistance** (r_2) of an induction machine referred to the stator:

$$r_2 = T_1^2 r_2' \quad \text{ohms.} \tag{6.169}$$

NOTE. r_2' is the actual rotor resistance and T_1 is the ratio of transformation from stator to rotor or the ratio of the emf's induced in the stator and rotor, respectively, during blocked-rotor.

The **rotor leakage reactance** (x_2) of an induction machine referred to the stator:

$$x_2 = T_1^2 x_2' \quad \text{ohms.} \tag{6.170}$$

NOTE.   Read note to Equation (6.169), substituting x_2' for r_2' and reactance for resistance.

The **equivalent effective resistance** (R_1) of an induction machine or r_{1e} ohms effective stator resistance and r_{2e} ohms effective rotor resistance referred to the stator:

$$R_1 = r_{1e} + r_{2e} \quad \text{ohms.} \tag{6.171}$$

The **equivalent reactance** (X_1) of an induction machine of x_1 ohms stator leakage reactance and x_2 ohms rotor leakage reactance referred to the stator:

$$X_1 = x_1 + x_2 \quad \text{ohms.} \tag{6.172}$$

The **synchronous speed** (n_1) of an induction machine having p poles, the frequency of the impressed voltage being f hertz:

$$n_1 = \frac{120f}{p} \quad \text{revolutions/minute.} \tag{6.173}$$

The **slip** (s) of an induction machine of synchronous speed (n_1) revolutions per minute when the rotor speed is n_2 revolutions per minute:

$$s = 1 - \frac{n_2}{n_1}. \tag{6.174}$$

The **equivalent circuit** of an induction motor (Fig. 6.18), r_{1e} and r_{2e} as in Equation (6.171). x_1 and x_2 as in Equation (6.172). Z_n, the exciting impedance, and its components r_n and x_n may be determined from a no-load run. The determination is similar to that for R_1, Z_1, and X_1 in Equations (6.166), (6.167), and (6.168) except that the no-load voltage and current and the no-load power input minus friction and windage losses must be used. $r_{2e} [(1 - s)/s]$ is the resistance equivalent for the rotational power, which is the shaft power output plus friction and windage and a small core loss.

The **induced stator emf** (E_n) of an induction machine at no load:

$$E_n = V_1 - I_n \sqrt{r_{1e}^2 + x_1^2} \quad \text{volts.} \tag{6.175}$$

NOTE. V_1 is the stator terminal voltage, I_n the no-load line current, r_{1e} and x_1 as in Equation (6.171) and (6.172).

The **rotor current** (I_2) referred to stator of an induction machine at slip (s):

$$I_2 = \frac{E_n}{\sqrt{(r_{1e} + r_{2o}/s)^2 + (X_1)^2}} \quad \text{amperes.} \tag{6.176}$$

NOTE. For a wound-rotor induction motor with an effective external resistance R_e ohms per phase referred to the stator, substitute ($r_{2e} + R_e$) for r_{2o}. To find the starting rotor current of an induction motor, make s equal one.

The **stator current** (I_1) of an induction machine at slip (s):

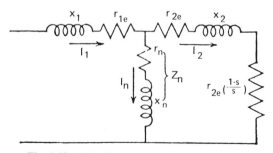

Fig. 6.18   Induction machine equivalent circuit.

$$I_1 = \sqrt{I_2^2 + I_n^2 + 2I_2I_n \sin \alpha} \quad \text{amperes.} \quad (6.177)$$

NOTE. I_2 is determined by Equation (6.176), I_n is the no-load current, and $\alpha = \sin^{-1} ($ p.f. at no load$) + \tan^{-1}(sx_2/r_{2o})$. For a wound-rotor motor, substitute $(r_{2e} + R_e)$ for r_{2o} in the expression for α. The starting stator current is given by making s equal one, the starting rotor current being determined as indicated in Equation (6.176).

The **power output** (P_o) of an induction machine at slip (s):

$$P_o = 3I_2^2r_{2o}\left(\frac{1-s}{s}\right) - P_{fw} \quad \text{watts.} \quad (6.178)$$

NOTE. For a wound-rotor motor, r_{2o} should be increased as indicated in Equation (6.176). When the slip is negative, P_o is negative and gives the power input to an induction generator. P_{fw} is the friction and windage loss.

The **power input** (P_i) to an induction machine at slip (s):

$$P_i = 3\frac{I_2^2r_{2o}}{s} + 3I_1^2r_{1e} + P_n - 3I_n^2r_{1e} \quad \text{watts.} \quad (6.179)$$

NOTE. P_n is the power in watts taken at no load. For a wound-rotor motor, r_{2o} should be increased as indicated in Equation (6.176). When the slip is negative, P_i is negative and gives the power output of an induction generator.

The **output torque** (T) of an induction motor at slip (s):

$$T = 0.239\frac{PI_2^2r_{2o}}{fs} - T_{fw} \quad \text{newton meters.} \quad (6.180)$$

NOTE. Read note on r_{2o} for a wound-rotor motor and s under starting conditions following Equation (6.176). T_{fw} is the friction and windage torque. If P_{fw} is known in watts, T_{fw} equals $\{0.0796pP_{fw}/[f(1-s)]\}$ newton meters.

The **slip** (s) of an induction motor at any stated load:

$$s = \frac{r_{2o}[(3E_n^2/P_o) - 2r_{1e} - 2r_{2o}]}{[(3E_n^2/P_o) - 2r_{1e} - 2r_o]^2 - Z_1^2} \quad \text{approximately.} \quad (6.181)$$

NOTE. Read note on r_{2o} for a wound-rotor motor following Equation (6.176).

The **slip** (s) of an induction generator at any stated load:

$$s = \frac{r_{2o}(3E_n^2/P_o)}{(3E_n^2/P_o) - Z_1^2 + r_{1e}(3E_n^2/P_o)} \quad \text{approximately.} \quad (6.182)$$

The **efficiency** (η) of an induction machine:

$$\eta = \frac{P_o}{P_i}. \quad (6.183)$$

6.8   ELECTRIC MOTOR APPLICATIONS

Widely used as prime movers, for uninterrupted power supply, and for power factor correction, **electric motors** are available in a wide variety of sizes and types for virtually any application.[10,11] Lower-horsepower applications, typically up to 5 hp, favor the use of single-phase motors, while higher-horsepower applications favor multi (typically three)-phase motors. Motors over 250 hp are usually produced to order with designs tailored to particular service applications. Motors below 250 hp are normally produced in accordance with NEMA standards and are widely stocked as standard motors. Motors are rated with a service factor of 1.0 or 1.15, which permits the motor to be run continuously at 100% or 115% of its nameplate rating.

6.8.1   Motor Classification

Motors are broadly classified as either **alternating-current** or **direct-current** types, with further subclassifications depending on the armature type, such as **squirrel cage, wound rotor,** and so on, enclosure type, voltage level, insulation level, or often for specific service, such as **synchronous motors** and **condensers.** Manufacturers may further classify motors into categories depending on their internal design features, which are often proprietary. All motors are rated based on heating and heat dissipation; as a result, motor applications in warm environments, above 40°C, or at altitudes above 3000 ft, will normally require derating of the motor.

6.8.2   Alternating-Current Motors

Alternating-current motors have a **synchronous speed** depending on the numbers of poles in the motor stator and the frequency of the system on which they are installed. This speed is given by formula (6.173).

Under full load, induction motors will typically **slip** from 1 to 3% [small (up to 2 hp) motors may have slips up to 5%], with the slip as given by formula (6.174).

All motors will exhibit a speed–torque characteristic in general conformance with the curve shown in Fig. 6.19. Of interest are several characteristics: rated torque at full load, **locked rotor** or starting torque, pull-up torque, and breakdown torque.

Induction Motors

Squirrel-Cage Motors.   By far the most common type of motor used is the **squirrel-cage** induction motor. A magnetic flux is developed by a series of coils in the stationary outer frame of the motor, the stator. On large motors the rotor consists of a series of bars and end rings, usually copper but sometimes aluminum, which are connected to each other at the end rings. This creates a short circuit in the rotor which interacts with the magnetic flux of the stator and produces a torque. The torque and hence the power output increases with the degree of slip to a maximum value at a breakdown point called the **pull-out torque.** The required speed–torque characteristics of the motor can be tailored to the particular application by designing a rotor cage of suitable resistance. Because of the simplicity of this design, reliability is high, maintenance requirements relatively low, and cost economical for most applications.

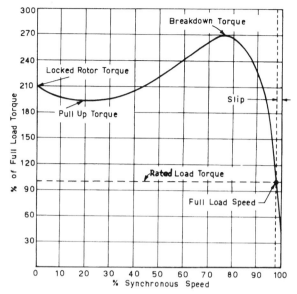

Fig. 6.19 Generalized speed–torque curve. (Courtesy of Lincoln Electric Company, Motor Division.)

Motor **efficiencies** are normally developed from test data in accordance with the IEEE Standard 112 and range from slightly above 70% for motors up to about 2 hp to 90 to 94% for motors of 200 to 250 hp. **Locked rotor** starting current is about six time full-load current and results in a stator winding heating effect of 36 times normal. A motor can sustain locked rotor currents for only brief periods, typically 10 to 15 seconds. Some rules of thumb for motors may be helpful: at 1800 rpm a squirrel-cage motor typically develops 3 lb-ft per hp; at 1200 rpm, 4.5 lb-ft; while at 230 V a three-phase motor draws 2.5 A per hp; at 460 V, 1.25 A; at 575 V, 1.0 A.

Squirrel-cage motors up to 250 hp are normally produced in accordance with **NEMA (National Electrical Manufacturers Association)** standards and are widely stocked as standard motors designated by frame sizes for each type of motor enclosure. They are typically available as NEMA Design Class B (temperature rise) for 40°C maximum ambient temperature, with a service factor of 1.15, which permits the motor to be run continuously, in this example, at a 15% greater load than its basic rating. A frame number (size) in the smaller sizes, say to 20 hp, may be used for several horsepower ratings. Frame numbers 56 through 445 have torque–speed curves as shown in Fig. 6.20. NEMA letter designations B, C, and D are the most common torque designs. The different letter designations refer to standardized rotor resistance design and starting torque. Their characteristics are shown in Table 6.10 and Fig. 6.21. Many motors available in the United States are now produced in metric frame sizes, conforming to IEC (International Electrotechnical Commission) standards. Motor nameplates typically carry a letter designation that specifies the locked rotor kVA drawn by the motor (Table 6.11).

NEMA motors designed for 60-Hz operation at 230/460 V are often used on 50-Hz systems. For this case, horsepower developed will be approximately 0.9 × nameplate horsepower, speed will be 0.83 × nameplate rpm, service factor will be 1.00, and the motor should be protected against an ampere draw of 1.1 × nameplate amperes:

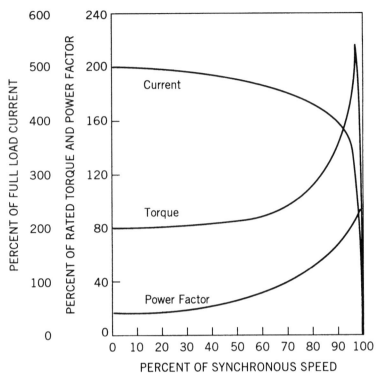

Fig. 6.20 Typical motor characteristics for squirrel-cage motors. (Courtesy of Westinghouse Motor Company.)

Table 6.10 NEMA Torque Designs for Polyphase Motors

NEMA Design[a]	Starting Current	Locked Rotor	Breakdown Torque	Percent Slip	Applications
B	Medium	Medium torque	High	Max. 5	Normal starting torque for fans, blowers, rotary pumps, unloaded compressors, some conveyors, metal cutting machine tools, miscellaneous machinery; constant load speed
C	Medium	High torque	Medium	Max. 5	High inertia starts such as large centrifugal blowers, flywheels, and crusher drums; loaded starts such as piston pumps, compressors, and conveyors; constant load speed
D	Medium	Extra-high torque	Low		Very high inertia and loaded starts; also, considerable variation in load speed
				5 or more	Punch presses, shears, and forming machine tools; cranes, hoists, elevators, and oil well pumping jacks

Source: Courtesy of the Lincoln Electric Company, Motor Division.
[a]NEMA Design A is a variation of Design B, having higher locked rotor current.

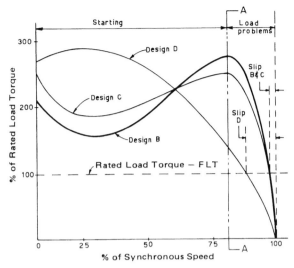

Fig. 6.21 Speed–torque curves for NEMA motors. (Courtesy of the Lincoln Electric Company, Motor Division.)

Commonly Used Ratings. The more widely used and commonly available horsepower **ratings** of squirrel-cage motors are shown in Table 6.12. Motors larger than 250 hp are custom designed, and it is possible to obtain motors with speed–torque characteristics closely tailored to special applications.

Multispeed motors, usually two-speed, are available to closely match the torque requirements of the driven equipment. With single winding motors, only speeds of 50% and 100% were formerly available. In one current version, the **pole amplitude modulation technique,** a squirrel-cage motor has a single winding but through an external speed-changing switch permits selection of one of two predetermined speeds by means of different connection points on the winding. This permits up to a 40%

Table 6.11   Locked Rotor kVA/hp

Code Letter	kVA/hp[a]	Code Letter	kVA/hp[a]
A	0–3.15	L	9.0–10.0
B	3.15–3.55	M	10.0–11.2
C	3.55–4.0	N	11.2–12.5
D	4.0 –4.5	P	12.5–14.0
E	4.5 –5.0	R	14.0–16.0
F	5.0 –5.6	S	16.0–18.0
G	5.6 –6.3	T	18.0–20.0
H	6.3–7.1	U	20.0–22.4
J	7.1–8.0	V	22.4 and up
K	8.0 –9.0		

Source: Courtesy of the Lincoln Electric Company, Motor Division.

[a]Locked rotor kVA/hp range includes the lower figure up to, but not including, the higher figure.

**Table 6.12 Squirrel-Cage
Motors Commonly Available,
¼ to 250 hp**

⅓	5	25	100
½	7½	30	125
1	10	40	150
1½	15	50	200
2	20	60	250
3		75	

reduction in rotor heating for starting high-inertial loads and may also have an advantage where energy charges are high and the driven equipment operates at a lower speed for a significant percentage of time. Performance data for this type arrangement are shown in Fig. 6.22.

In a variation of this, the speed of the motor can be changed over a relatively wide range by changing the frequency of the ac power fed to the machine. In this case, line ac power is converted to dc power by **thryistors,** the power then being inverted to ac by a second set of thryistors and at a frequency called for by a solid-state digital control system (see Fig. 6.23). The advantages of such an arrangement include a softer start for the motor, lower operating costs, and longer life. These are somewhat offset by the more complicated control system and its higher initial cost.

Induction motors driven above synchronous speed by a separate prime mover become **induction generators** and are often used in the petrochemical industry with steam turbine drivers for waste heat recovery. Great care needs to be taken with their control systems, however, as loss of load can immediately cause the prime mover to overspeed, often reaching 200% of design speed in 5 seconds.

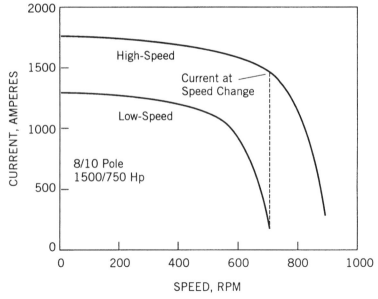

Fig. 6.22 Pole amplitude modulation motor. (Courtesy of Westinghouse Motor Company.)

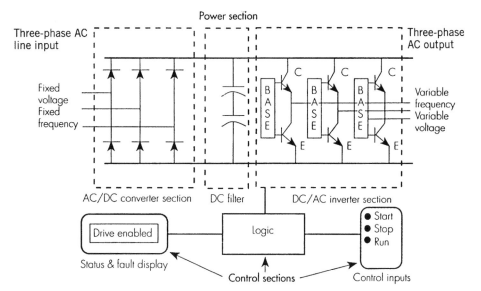

Fig. 6.23 Thyristor frequency control system. (Courtesy of Westinghouse Motor Company.)

Wound-Rotor Motors.   As their name implies, **wound-rotor motors** consist of a rotor with coil windings running within a stator which also has coil windings. The rotor windings are led to a set of slip rings at the end of the shaft, where carbon brushes complete an electrical circuit to an external system. This permits the resistance of the rotor coil circuit to be varied, thus changing the motor speed–torque characteristics. Figure 6.24 illustrates the general speed–torque characteristics of wound motors. These motors are useful in variable-speed applications, where the motor torque can be matched to the load to produce the speed desired.

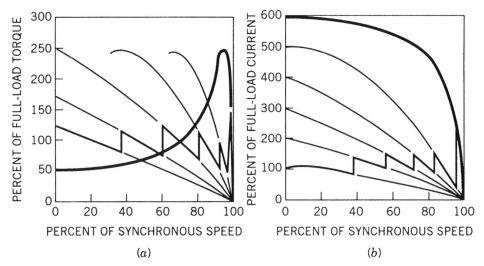

Fig. 6.24 Typical wound-rotor motor characteristics. (Courtesy of Westinghouse Motor Company.)

Rotor resistance is normally controlled by an automated rheostate based on inputs from the driven device, although in some designs the rotor resistance is rarely changed. Because of the more complex construction of wound-rotor machines, they are more expensive than squirrel-cage machines and require somewhat more maintenance. For many applications this is more than offset by their speed matching performance.

Motor Starting.   **Motor starting** requires that sufficient torque be available to accelerate the load to running speed without damaging the motor due to excessive or prolonged heat. In smaller sizes (to 200 hp) where load inertia is not a problem, this is usually accomplished by **across-the-line starters** where full line voltage is applied to the motor terminals. This has the advantage of simplicity but creates a large starting current **inrush,** typically 600 to 800% of full-load current. The large current flow continues until the motor approaches 90% of synchronous speed. This can be a problem with motors that have a high starting inertia or are driving high-inertial loads and take some time to come up to speed because of the heating caused by the high current flow. High inertia favors an oversized motor design and in some cases, special rotor bar material. Inrush becomes a concern on smaller power supply systems, where there may be a significant reduction in voltage. If there is any doubt, calculations should be run to determine probable reduced terminal voltage due to inrush.

To avoid this problem, **reduced-voltage starting** is used almost universally for larger motors to reduce starting inrush. Since the starting torque is reduced as a function of the square of the voltage (e.g., 80% voltage yields 64% starting torque) in cases where the starting torque requirements are high, steps must be taken to reduce the starting torque requirements of the driven equipment. For example, closing the discharge valve of a mixed flow pump or the inlet vanes of a large fan will often reduce the required torque and permit load acceleration within safe limits. Table 6.13 presents a comparison of commonly used starting methods, while Table 6.14 indicates the typical load–torque requirements for several common loads. Where there is any concern, the motor manufacturer and the supplier of the driven equipment should be consulted.

Synchronous Motors

Synchronous motors are usually found in larger sizes, typically 2000 hp or higher. They may be economical where the horsepower requirement is larger than the motor speed (e.g., 2100 hp at 720 rpm). Common upper limits for speed are 1800 or 1200 rpm. Somewhat similar in construction to a wound-rotor motor, the rotor contains field coils which are connected to a source of dc excitation. The excitation causes a strong magnetic field, which results in the rotor locking into the rotating field of the stator and rotating in synchronism with it. Thus the two principal criteria for a synchronous motor are the number of poles on the rotor, which establishes motor speed, and the method and strength of **excitation,** which establishes the pullout torque of the motor. Synchronous motors can be designed for a wide range of starting, pull-in, and pull-out torques, to suit virtually any application.

Table 6.13 Comparison of Starting Methods[a]

Starting Method	Operation	Starting Current (% of locked rotor current)	Starting Torque (% of locked rotor torque)	Open or Closed Transition	Basic Characteristics		
					Advantages		Limitations
Across-the-line	Initially connects motor directly to power lines	100	100	None	1. Lowest cost 2. Highest starting torque 3. Used with any standard motor 4. Least maintenance		1. High starting current 2. High starting torque 3. May shock driven machine
Primary resistance-reduced voltage	Inserts resistance units in series with motor during first step(s)	50–80	25–64	Closed	1. Smoothest starting 2. Least shock to driven machine 3. Most flexible in application 4. Used with any standard motor		1. High power loss because of heating resistors 2. Heat must be dissipated 3. Low torque per ampere input 4. Highest cost
Autotransformer reduced voltage	Uses autotransformer to reduce voltage applied to motor	Tap: 50% 25 65% 42 80% 64	25 42 64	Closed	1. Best for hard to start loads 2. Adjustable starting torque 3. Used with any standard motor 4. Less strain on motor		1. May shock driven machine 2. High cost
Wye–delta or star-delta	Starts motor with windings wye connected, then reconnects them in delta connection for running	33	33	Open or closed	1. Medium cost 2. Low starting current 3. Low starting torque 4. Less strain on motor		1. Low starting torque 2. Requires delta wound motor
Part Winding	Starts motor with only part of windings connected, then adds remainder for running	70–80	50–60 Minimum pull-up torque 35% of full load torque	Closed	1. Low cost 2. Popular method for medium starting torque applications 3. Low maintenance		1. Not good for frequent starts 2. May require special wound motor 3. Low pull-up torque 4. May not come up to speed on first step when started with load applied

Source: Courtesy of Lincoln Electric Company, Motor Division.

[a]The reduced starting torque (LRT) for the various reduced-voltage starting methods can prevent proper starting of high-inertia loads and must be considered when sizing motors and choosing starters.

Table 6.14   Typical Load–Torque Requirements

| | | Torque as Percentage of Motor Full-Load Torque | |
		Locked Rotor	Pullout
Item	Application		
1	Banbury mixers	125	250
2	Blowers, positive displacement, rotary—bypassed for starting	60	205
3	Chippers—starting empty	100	225
4	Compressors, centrifugal— starting with inlet or discharge value closed	60	205
5	Crushers—starting loaded, ball or rod mills	150–200	250
6	Fans, centrifugal—starting with inlet or discharge valve closed	60	205
7	Grinders, pulp–starting unloaded	60	205
8	Pulverizers	150–200	250
9	Pumps, axial flow, fixed blade— starting with casing dry	60	205
10	Pumps, centrifugal—starting with casing filled, discharge closed	60	205
11	Pumps, mixed flow—starting with casing filled, discharge closed	60	205
12	Refiners, pulp—including Jordans	60	205
13	Rubber mills	100–150	225

Source: Courtesy of Westinghouse Motor Company.

A common construction of synchronous motors uses slip rings and carbon brushes to provide dc power to the rotor with static excitation from an external transformer and rectifier unit. For high-horsepower applications, multiple-brush rigging for excitation is sometimes used. Another widely used method employs a dc generator connected to (often integral with) the motor shaft, with brushes used to transfer dc power to the rotor coils. Perhaps the most popular method uses an ac generator mounted on the same shaft with solid-state rectifier assemblies also mounted on the same shaft, thus avoiding the need for brushes. This arrangement tends to be more reliable, reduces maintenance requirements, and permits utilization in some hazardous atmospheres.

A major advantage of synchronous motors is their ability to correct **power factor.** By increasing excitation beyond that needed merely to keep the machine in synchronism, it is possible to develop a leading power factor usually up to 0.8. Excitation is normally steplessly controlled with a voltage regulator and power factor controller to maintain a constant power factor. Alternatively, if a constant power output is required, the power factor will vary inversely with the load. This can be a major advantage on systems with lagging power factors and can lead to substantial savings on capacity factor charges.

In more recent years the development of variable-speed synchronous motors using adjustable-frequency control technology has partially displaced dc motors as drivers for high-torque, low-speed applications, as low as 400 rpm.

Synchronous condensers are synchronous motors that are not connected to an external load and, typically through solid-state microprocessors, provide excitation to run at zero power factor, thus supplying reactive power (kvar) to the line. Because there are no switching transients as with fixed capacitors, which are frequently used for power factor correction, synchronous condensers are often preferred. These are often three- or six-phase machines.

Motor Starting. Synchronous motors are started as induction motors and when within about 5% of synchronous speed, field power is applied and the rotor locks into synchronization. The motor will stay in synchronization until the applied load torque exceeds the pull-out torque. At this point the motor will pull out of synchronization and drop off in speed; this will be accompanied by extreme heating and the motor will be destroyed unless taken off line promptly.

6.8.3 Direct-Current Motors

Direct current motors have excellent speed-control characteristics and are used on precise applications such as newspaper printing presses, extruders, runout tables, and for similar uses. Dc motors are available as **shunt, series,** and **compound wound** types. They are normally employed either where dc power is available, such as battery-powered applications or occasionally, ships or aircraft. They can provide precise speed control over a wide range of speeds, up to 4:1 in the same motor. In general, as surrent is increased, torque increases almost linearly; thus for, say, 150% of full-load current, 150% or more torque is produced.

Shunt motors normally have the rotor windings connected in parallel with the stator, while series motors, as their name implies, have the rotor windings connected in series with the stator. The compound wound motor has its rotor windings connected, both in series and parallel with stator, incorporating windings of both the shunt and series types.

6.8.4 Motor Enclosures

Motor enclosures are available in accordance with standard NEMA designations. For larger motors, the types of enclosures available will vary slightly from one manufacturer to another. Motor enclosures most commonly used are Type I weather protected (WPI), Type II weather protected (WPII) with three-directional change in cooling air direction, totally enclosed fan cooled (TEFC), totally enclosed pipe ventilated (TEPV) using ventilation ducting, totally enclosed air-to-air cooled (TEAAC), and totally enclosed water-to-air cooled (TEWAC) using a water-to-air heat exchanger. Enclosures are available in the types mentioned above for vertical applications as well. RTDs, space heaters, and winding heaters are available as well. Winding heaters utilize a small current in one phase of the motor windings and are normally supplied by the motor control manufacturer.

6.8.5 Motor Insulation

Motor insulation type is normally designated by thermal endurance, most commonly using the NEMA letter classifications defined in Table 6.15.

For larger motors, most manufacturers use some sort of epoxy-rich taping system, often proprietary, for both the rotor and stator coils and their leads to provide a highly impervious, long-lived insulation system. These may be vacuum or pressure applied to develop a continuous void-free barrier and are often baked to provide a durable, long-lived surface. They normally provide excellent protection against dust, moisture, and similar materials. They will vary in their resistance to airborne abrasives, chemicals, and in particular to acids, both strong and weak. It is important to establish with the motor manufacturer the insulation system that will be provided, its resistance to the various environmental factors present, and warranties that may apply.

6.8.6 Bearings and Lubrication

Small motors will normally be equipped with antifriction **bearings** having both radial and, in the case of vertical motors, significant thrust capability. These bearings often carry an AFBMA B-10 life of 40,000 to 50,000 hours when the motor is operated in a cool, clean environment. Larger motors, over 200 hp, will often be supplied with split-sleeve bearings of the babbit/oil-lubricated variety. They may have provisions for oil jacking of the motor shaft to facilitate starting, and while most have gravity-ring-type **oiling systems,** in larger sizes or for critical service may have special oil pumps and positive-pressure oiling systems. Most motors of 1000 hp and larger will have one bearing insulated to avoid pitting of bearings from voltages induced in the shafting. In addition to their split-sleeve bearings, many larger motors will have either spherical roller bearings or larger plate-type thrust bearings to handle nonstandard

Table 6.15   NEMA Motor Insulation Classifications

Class A—A Class A insulation system is one which by experience or accepted test can be shown to have suitable thermal endurance when operated at the limiting Class A temperature of 105°C. Typical materials used include cotton, paper, cellulous acetate films, enamel-coated wire, and similar organic materials impregnated with suitable substances.

Class B—A Class B insulation system is one which by experience or accepted tests can be shown to have suitable thermal endurance when operated at the limiting Class B temperature of 130°C. Typical materials include mica, glass fiber, asbestos and other materials, not necessarily inorganic, with compatible bonding substances having suitable thermal stability.

Class F—A Class F insulation system is one which by experience or accepted test can be shown to have suitable thermal endurance when operating at the limiting Class F temperature of 155°C. Typical materials include mica, glass fiber, asbestos and other materials, not necessarily inorganic, with compatible bonding substances having suitable thermal stability.

Class H—A Class H insulation system is one which by experience or accepted test can be shown to have suitable thermal endurance when operated at the limiting Class H temperature of 180°C. Typical materials used include mica, glass fiber, asbestos, silicone elastomer, and other materials, not necessarily inorganic, with compatible bonding substances, such as silicone resins, having suitable thermal stability.

Source: Courtesy of Lincoln Electric Company, Motor Division.

thrust loads. Vertical applications will include thrust bearings whose capacity should be coordinated with the driven equipment requirements to assure adequacy.

6.8.7 Motor Selection

For all but the most conventional uses it is important to consult with the motor and driven equipment manufacturers to assure the correct motor selection for the particular application. With the widespread use of computerized design programs, manufacturers are readily able to modify their standard designs to suit the specific requirements of virtually any application. The items of concern include:

- Load horsepower required under the various operating conditions
- Starting and running torque requirements of the load
- Load inertia, Wk^2
- Multispeed requirements
- Axial or thrust bearing loadings
- For large motors, the source of power (i.e., starting inrush and likely voltage drop)
- For medium-size and large motors, frequency of starting
- Installation factors, foundation, crane capacity required, clearances, and so on
- Ambient conditions, such as temperature, altitude, dust, abrasives, cooling medium available, and so on
- Special requirements for specific applications, such as power factor correction, noise-level limitations, vibration, and so on

6.9 TRANSFORMERS

6.9.1 Transformer Characteristics

The **electromotive force** (E) induced in N turns linked by a flux, $\phi = \Phi_m \sin 2\pi ft$ webers:

$$E = 4.44 Nf\Phi_m \quad \text{volts.} \tag{6.184}$$

The **core loss** (P_c) of a transformer at any load:

$$P_c = P_h + P_e \quad \text{watts.} \tag{6.185}$$

NOTE. P_h is the hysteresis loss and P_e the eddy-current loss in the magnetic circuit of the transformer in Equations (6.186) and (6.187).

The **hysteresis loss** (P_h) in a medium in which a variable magnetic flux of maximum density B_m teslas changes from positive to negative to positive maximum f times per second:

$$P_h = \eta f B_m^{1.6} \times 10^{-3} \quad \text{watts/centimeter}^3. \tag{6.186}$$

NOTE.   This is an empirical equation and in some cases the exponent of B_m may differ appreciably from 1.6. The hysteresis loop must be symmetrical with no re-entrant loops. The hysteresis coefficient (η) varies in different materials as follows: cast iron, 0.012; cast steel, 0.005; hipernik (50 N), 0.00015; low-carbon sheet steel, 0.003; permalloy (78 N), 0.0001; pure Norway iron, 0.002; silicon sheet steel, 0.00046 to 0.001. Equation (6.186) does not apply to the hysteresis loss in iron rotated in a magnetic field. In the latter case at low flux densities the loss may be twice as much as that due to an alternating flux, but declines in value as the flux density increases. For soft iron the loss by either process will be about the same at 1.50 teslas, and at 2.00 teslas the loss due to rotation is practically zero.

The **eddy-current loss** (P_e) in thin laminations placed in a sinusoidally varying magnetic flux:

$$P_e = \frac{1.64(tfB_m)^2}{p \times 10^8} \quad \text{watts/centimeter}^3. \tag{6.187}$$

NOTE. t is the thickness of the laminations in centimeters, f is the frequency of flux variation in hertz, B_m is the maximum flux density in teslas, and ρ is the resistivity of the laminations in ohms per cubic centimeter.

The **ratio of transformation** (T_1) from primary to secondary of a transformer wound with two coils of N_1 (primary) and N_2 (secondary) turns, respectively:

$$T_1 = \frac{N_1}{N_2}. \tag{6.188}$$

NOTE. T_1 equals the ratio E_1/E_2 of the emf's induced, respectively, in the primary and secondary coils and equals approximately the ratio V_1/V_2 of terminal voltages of the primary and secondary coils or the ratio I_2/I_1 of the secondary and primary currents. The ratio of transformation (T_2) from secondary to primary equals $1/T_1$.

The **magnetizing current** (I_m) in a coil of N turns wound on a magnetic circuit of uniform maximum permeability (μ), l centimeters in mean length, A square centimeters in mean section, and conducting a flux, $\phi = \Phi_m \sin 2\pi ft$ webers:

$$I_m = \frac{10\Phi_m l \times 10^{-8}}{4\pi N\mu A\sqrt{2}} \quad \text{amperes (approx.).} \tag{6.189}$$

The **core-loss current** (I_c) in a coil containing an induced emf of E volts and wound on a magnetic circuit in which the core loss is P_c watts.

$$I_c = \frac{P_c}{E} \quad \text{amperes.} \tag{6.190}$$

The **no-load current** (I_n) taken by a transformer which requires a magnetizing current of I_m amperes and a core-loss current of I_c amperes.

$$I_n = \sqrt{I_m^2 + I_c^2} \quad \text{amperes.} \tag{6.191}$$

The **equivalent resistance** (R_1) and **equivalent reactance** (X_1) between the primary terminals of a transformer which has a primary resistance of r_1 ohms, a primary leakage reactance of x_1 ohms, a secondary resistance of r_2 ohms, a secondary leakage reactance of x_2 ohms, and primary to secondary ratio of transformation of T_1.

$$R_1 = r_1 + T_1^2 r_2 \quad \text{ohms.} \tag{6.192}$$

$$X_1 = x_1 + T_1^2 x_2 \quad \text{ohms.} \tag{6.193}$$

NOTE. The equivalent resistance and reactance, respectively, between the secondary terminals is given by $R_2 = r_2 + T_2^2 r_1$ ohms and $X_2 = x_2 + T_2^2 x_1$ ohms. The equivalent impedance in each case equals $(R^2 + X^2)^{1/2}$ ohms and $Z_1 = T_1^2 Z_2$ ohms.

The **equivalent resistance** (R_1) between the primary terminals of a transformer which, with short-circuited secondary, absorbs P_i watts with a primary current of I_1 amperes:

$$R_1 = \frac{P_i}{I_1^2} \quad \text{ohms.} \tag{6.194}$$

The **equivalent impedance** (Z_1) between the primary terminals of a transformer which, with secondary short-circuited and with V_1 volts between the primary terminals, takes a primary current of I_1 amperes:

$$Z_1 = \frac{V_1}{I_1} \quad \text{ohms.} \tag{6.195}$$

The **equivalent reactance** (X_1) between the primary terminals of a transformer of equivalent resistance (R_1) ohms and equivalent impedance (Z_1) ohms between the primary terminals:

$$X_1 = \sqrt{Z_1^2 - R_1^2} \quad \text{ohms.} \tag{6.196}$$

The **primary voltage** (V_1) of a transformer of ratio of transformation (T_1), equivalent resistance and reactance, respectively, between secondary terminals (R_2) and (X_2) ohms, secondary terminal voltage (V_2) volts, secondary current (I_2) amperes, and power factor of the load on the secondary ($\cos\theta_2$):

$$V_1 = T_1 \sqrt{(V_2 \cos\theta_2 + I_2 R_2)^2 + (V_2 \sin\theta_2 \pm I_2 X_2)^2} \quad \text{volts.} \tag{6.197}$$

NOTE. The sign before $I_2 X_2$ is $+$ for zero or lagging current phase and $-$ for leading current phase.

The **voltage regulation** (V.R.) of a transformer at any load, V_1, V_2, and T_1 as in Equation (6.197):

$$\text{V.R.} = \frac{V_1 - T_1 V_2}{T_1 V_2}. \tag{6.198}$$

The **efficiency** (η) of a transformer at any load:

$$\eta = \frac{I_2 V_2 \cos \theta_2}{I_2 V_2 \cos \theta_2 + I_2^2 R_2 + P_c}. \tag{6.199}$$

6.10 AC POWER TRANSIMISSION

6.10.1 Conventions

Three-phase power networks are assumed throughout. All apparatus is assumed to be wye-connected, and all impedances and admittances are for one phase or line on this basis. All currents are line currents (phase currents for a wye-connection). Unless otherwise stated, all voltages except in Equations (6.200) and (6.201) are line-to-neutral voltages (phase voltages for a wye-connection). In Equations (6.200) and (6.201) line-to-line voltages are used.

6.10.2 Per Unit Quantities

Ohms's law and **Kirchoff's laws** apply in circuits whether V, I, and Z are expressed in their absolute units or in per unit (percent/100) of a base value. The base may be chosen arbitrarily, but it must be consistent throughout any set of computations. For electrical equipment it is customary to use the equipment ratings as the base, that is, $V_{rated} = 1.0$ per unit voltage, $I_{rated} = 1.0$ per unit current, and rated kVA (kilovolt-amperes) = 1.0 per unit kVA. In a power network containing transformers, the same voltage, current, and kVA bases consistent with the transformer ratios must be used throughout.

Impedance and **admittance** are expressed in per unit of base quantities V_{rated}/I_{rated} and I_{rated}/V_{rated}, respectively. For changing from one base quantity to another, per unit impedances are directly proportional to the kVA base and inversely proportional to the square of the voltage base. Per unit admittances are inversely proportional to the kVA base and directly proportional to the square of the voltage base.

The **per unit impedance** (Z_{pu}) in terms of the impedance Z in ohms in one phase of a three-phase system, expressed on a three-phase base of kVA kilovolt-amperes and a line-to-line base voltage of V_b volts:

$$Z_{pu} = \frac{Z \times kva}{V_b^2} \times 10^3 \text{ per unit.} \tag{6.200}$$

The **per unit admittance** (Y_{pu}) in terms of the admittance Y in siemens in one phase of a three-phase system, expressed on a three-phase base of kva kilovolt-amperes and a line-to-line voltage base of V_b volts:

$$Y_{pu} = \frac{YV_b^2}{kva} \times 10^{-3}. \tag{6.201}$$

In a single-phase power system, with base kVA equal to rated kVA and base voltage equal to rated kV, the following relationships apply:

Base current in amperes = base kVA/base voltage.

Base impedance in ohms $= \dfrac{\text{base voltage} \times 1000}{\text{base current}}$

$= \dfrac{\text{base voltage}^2 \times 1000}{\text{base kVA}}.$

Per unit voltage = actual voltage/ base voltage.

Per unit current = actual current/base current.

Per unit impedance = actual impedance/base impedance

Per unit kVA $= \dfrac{\text{actual impedance} \times \text{base kVA}}{\text{base voltage}^2 \times 1000}.$

Per unit kVA = actual kVA/base kVA.

In a three-phase system, with base kVA equal to total three-phase kVA, and base voltage the phase to neutral voltage in kV:

Base current in amperes $= \dfrac{\text{base kVA}}{3 \times \text{base voltage}}.$

Base impedance in ohms $= \dfrac{\text{base voltage} \times 1000}{\text{base current}}$

$= \dfrac{3 \times \text{base voltage}^2 \times 1000}{\text{base kVA}}.$

Per unit voltage = actual voltage/ base voltage.

Per unit current = actual current/base current.

Per unit impedance = actual impedance/base impendace

$= \dfrac{\text{actual impedance} \times \text{base kVA}}{3 \times \text{base voltage}^2 \times 1000}.$

Per unit kVA = actual kVA/base kVA.

To convert from one kVA base, KVA_1, to another, kVA_2:

Per unit impedance on base $\text{kVA}_2 = \dfrac{\text{kVA}_2}{\text{kVA}_1} \times$ per unit impedance on base kVA_1.

Per unit current on base $\text{kVA}_2 = \dfrac{\text{kVA}_1}{\text{kVA}_2} \times$ per unit current on base kVA_1.

Per unit kVA on base $\text{kVA}_2 = \dfrac{\text{kVA}_1}{\text{kVA}_2} \times$ per unit kVA on base kVA_1.

To convert from one kV base, KV_1, to another, kV_2:

Per unit impedance on base $\text{kV}_2 = \left(\dfrac{\text{kV}_1}{\text{kV}_2}\right)^2 \times$ per unit impedance on base kV_1.

Per unit current on base $\text{kV}_2 = \dfrac{\text{kV}_2}{\text{kV}_1} \times$ per unit current on base kV_1.

Per unit voltage on base $\text{kV}_2 = \dfrac{\text{kV}_1}{\text{kV}_2} \times$ per unit voltage on base kV_1.

6.10.3   Transmission Line Characteristics

The **voltage** (E_s) at the sending end of a transmission line of R ohms resistance per conductor and X ohms reactance per conductor conducting a current of I amperes, the voltage at the receiving end being E_r volts and the phase angle between the receiving-end voltage and the line current being θ_2:

$$E_s = \sqrt{(E_r \cos \theta_r + IR)^2 + (E_r \sin \theta_r \pm IX)^2} \text{ volts.} \tag{6.202}$$

NOTE.   See vector diagram, Fig. 6.25. When I lags or is in phase with E_r, the sign before IX is $+$, and when I leads E_r, the sign before IX is $-$. Equation (6.202) neglects capacitance. For transmission lines longer than 40 miles, the capacitance should be included. For cables longer than about 2 miles the capacitance should be included. See Equations (6.207), (6.216), and (6.218).

For application to single-phase lines, use $2R$ and $2X$ in place of R and X.

The **phase angle** (θ_s) between the sending-end voltage and the line current under the conditions stated in Equation 6.202:

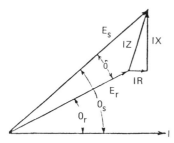

Fig. 6.25   Vector diagram of transmission line voltages and line current.

$$\theta_s = \tan^{-1} \frac{E_r \sin \theta_r \pm IX}{E_s \cos \theta_s + IR}.$$ (6.203)

NOTE. The power factor at the sending end is $\cos \theta_s$.

The **power input** (P_s) to the sending end of a transmission line with δ degrees angular displacement between sending- and receiving-end voltages, the line having an impedance Z ohms and an impedance angle $\theta = \tan^{-1}(X/R)$:

$$P_s = \frac{E_s^2}{Z} \cos \theta - \frac{E_s E_r}{Z} \cos (\delta + \theta) \quad \text{watts.}$$ (6.204)

NOTE. δ is positive if E_s leads E_r. Capacitance is neglected. P_s will be total three-phase power if line-to-line voltages are used, and power per phase if line-to-neutral voltages are used.
   The maximum value is $[(E_s E_r)/Z] - (E_s^2/Z) \cos \theta$; this maximum value may not be attainable without instability in the system.

The **power output** (P_r) at the receiving end of a transmission line under the conditions in Equation (6.204):

$$P_r = \frac{E_s E_r}{Z} \cos (\delta - \theta) - \frac{E_r^2}{Z} \cos \theta \quad \text{watts.}$$ (6.205)

NOTE. Read first paragraph of note to Equation (6.204), substituting P_r for P_s. The maximum value of P_r is $[(E_s E_r)/Z] - (E_r^2/Z) \cos \theta$; this maximum value may not be attainable without instability in the system.

The **efficiency** (η) of a transmission line under the conditions stated in Equation (6.202):

$$\eta = \frac{E_r I \cos \theta_r}{E_r I \cos \theta_r + I^2 R} = \frac{E \cos \theta_r}{E_s \cos \theta_s} = \frac{P_r}{P_s}.$$ (6.206)

Nominal π (Fig. 6.26) and **nominal T** (Fig. 6.27) equivalent circuits for a transmission line of length l miles having an impedance z ohms per mile composed of resistance r ohms per mile and reactance x ohms per mile, and an admittance y siemens per mile composed of a leakage conductance g siemens per mile and a capacitive susceptance of b siemens per mile.

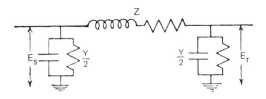

Fig. 6.26 Nominal pi equivalent circuit for a transmission line.

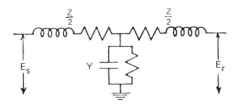

Fig. 6.27 Nominal T equivalent circuit for a transmission line.

$$Z = zl \quad \text{ohms}$$
$$= (r + jx)l \quad \text{vector ohms.} \tag{6.207}$$
$$Y = yl \quad \text{siemens}$$
$$= (g + jb) \quad \text{vector siemens.} \tag{6.208}$$

NOTE.   The admittance Y is almost always purely capacitive, leakage usually being negligible. Nominal π and T circuits are usually used for transmission lines of lengths between about 40 and 100 miles. Below 40 miles the capacitance may be neglected; above 100 miles the equivalent π or T should be used [see Equations (6.216) and (6.218))]. For cables, the nominal π and T circuits are usually used for lengths between about 2 and 5 miles. Below 2 miles the capacitance may be neglected; above 5 miles the equivalent π or T should be used.

Long Transmission Line.   Nomenclature: length of line, l miles. Resistance, inductive reactance, capacitance, and leakage conductance, respectively, r ohms per mile, x ohms per mile, c farads per mile, and g siemens per mile; all per conductor. Impedance, $\mathbf{z}$ vector ohms per mile. Admittance, $\mathbf{y}$ vector siemens per mile. In the following equations, $\mathbf{z}$ and $\mathbf{y}$ are vectors equal, respectively, to $r + jx$ and $g + j\omega c$. Sending-end voltage and current, respectively, $\mathbf{E}_s$ vector volts and $\mathbf{I}_s$ vector amperes. Receiving-end voltage and current, respectively, $\mathbf{E}_r$ vector volts and $\mathbf{I}_r$ vector amperes. In the following equations, $\mathbf{E}_s$, $\mathbf{I}_s$, $\mathbf{E}_r$, and $\mathbf{I}_r$ are vector quantities with a common, arbitrary reference axis.

The **propagation constant** (α) of a transmission line:

$$\alpha = \sqrt{\mathbf{zy}} = \sqrt{(r + jx)(g + j\omega c)} \quad \text{hyps/mile.} \tag{6.209}$$

NOTE.   The real part of α is called the attenuation constant. The imaginary part is called the wavelength constant, phase constant, or velocity constant.

The **surge impedance** or **characteristic impedance** (Z_0) of a transmission line:

$$Z_0 = \sqrt{\frac{\mathbf{z}}{\mathbf{y}}} = \sqrt{\frac{r + jx}{g + j\omega c}} \quad \text{vector ohms.} \tag{6.210}$$

The **hyperbolic angle** (θ) of a transmission line:

$$\theta = \alpha l \quad \text{numerics.} \tag{6.211}$$

The **current** (i) and **voltage** (e) at a point on the line distant x miles from the receiving end in terms of receiving-end voltage and current:

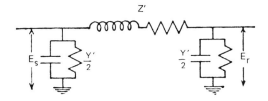

Z'

$\dfrac{Y'}{2}$ $\dfrac{Y'}{2}$

E_s E_r

Fig. 6.28 Equivalent pi circuit for a long transmission line.

$$i = \mathbf{I}_r \cosh \alpha x + \frac{\mathbf{E}_r}{Z_0} \sinh \alpha x \quad \text{vector amperes.} \qquad (6.212)$$

$$e = \mathbf{E}_r \cosh \alpha x + \mathbf{I}_r Z_0 \sinh \alpha x \quad \text{vector volts.} \qquad (6.213)$$

NOTE. To obtain $\mathbf{I}_s$ and $\mathbf{E}_s$, substitute θ for αx.

The **current** (i) and **voltage** (e) at a point on the line distant x miles from the receiving end in terms of sending-end voltage and current:

$$i = \mathbf{I}_s \cosh(l - x)\alpha - \frac{\mathbf{E}_s}{Z_0} \sinh(l - x)\alpha \quad \text{vector amperes.} \qquad (6.214)$$

$$e = \mathbf{E}_s \cosh(l - x)\alpha - \mathbf{I}_s Z_0 \sinh(l - x)\alpha \quad \text{vector volts.} \qquad (6.215)$$

NOTE. To obtain $\mathbf{I}_r$ and $\mathbf{E}_r$, substitute θ for $(l - x)\alpha$.

The **equivalent + circuit** for a long transmission line. (see Fig. 6.28):

$$Z' = zl \frac{\sinh \theta}{\theta} \quad \text{vector ohms.} \qquad (6.216)$$

$$Y' = yl \frac{\tanh (\theta/2)}{\theta/2} \quad \text{vector siemens.} \qquad (6.217)$$

NOTE. Equivalent π and T circuits will represent exactly the performance of smooth transmission lines at their terminals under steady-state conditions. For transmission lines of lengths below about 100 miles, the nominal π and T circuits [see Equations (6.207) and (6.208)] may be used unless very precise results are desired.

The **equivalent T circuit** for a long transmission line (see Fig. 6.29):

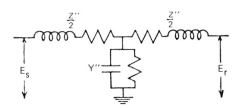

$\dfrac{Z''}{2}$ $\dfrac{Z''}{2}$

E_s Y'' E_r

Fig. 6.29 Equivalent T circuit for a long transmission line.

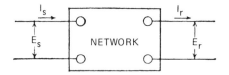

Fig. 6.30 Voltages and currents at sending and receiving ends of a network.

$$Z'' = zl \frac{\tanh (\theta/2)}{\theta/2} \quad \text{vector ohms.} \tag{6.218}$$

$$Y'' = yl \frac{\sinh \theta}{\theta} \tag{6.219}$$

NOTE.   See note to Equations (6.216) and (6.217).

General circuit constants (*A*, *B*, *C*, and *D*) are often used to express the steady-state performance of a network consisting of any combination of constant impedances to which power is supplied at one point and received at another. See Fig. 6.30 and note to Equations (6.220) and (6.221).

The **sending-end voltage** (E_s) and **current** (I_s) in terms of the receiving-end voltage $\mathbf{E}_r$ volts and current $\mathbf{I}_r$ amperes for the network of Fig. 6.30:

$$\mathbf{E}_s = A\mathbf{E}_r + B\mathbf{I}_r \quad \text{vector volts.} \tag{6.220}$$

$$\mathbf{I}_s = C\mathbf{E}_r + D\mathbf{I}_r \quad \text{vector amperes.} \tag{6.221}$$

NOTE.   All quantities are vector quantities. These are the defining equations for general circuit constants. *A* and *B* are found for a given network by obtaining from circuit equations an expression for $\mathbf{E}_s$ in terms of $\mathbf{E}_r$ and $\mathbf{I}_r$; *A* is then the coefficient of $\mathbf{E}_r$ in this expression, and *B* is the coefficient of $\mathbf{I}_r$. *C* and *D* are found similarly from the expression for $\mathbf{I}_s$ in terms of $\mathbf{E}_r$ and $\mathbf{I}_r$. Thus, for a network consisting of a series impedance Z, $\mathbf{E}_s = \mathbf{E}_r + Z\mathbf{I}_r$, and $\mathbf{I}_s = \mathbf{I}_r$; hence for this network, $A = 1$, $B = Z$, $C = 0$, $D = 1$.

The **receiving-end voltage** (E_r) and **current** (I_r) in terms of the sending-end voltage $\mathbf{E}_s$ volts and current $\mathbf{I}_s$ amperes for the network of Fig. 6.30:

$$\mathbf{E}_r = D\mathbf{E}_s - B\mathbf{I}_s \quad \text{vector volts.} \tag{6.222}$$

$$\mathbf{I}_r = -C\mathbf{E}_s + A\mathbf{I}_s \quad \text{vector amperes.} \tag{6.223}$$

NOTE.   All quantities are vector quantities.

6.11   ELECTRICAL PROTECTION

6.11.1   Protection Principles

Good engineering practice dictates that in the design, operation, and maintenance of electrical equipment, natural and man-made events that can cause danger to people and damage to property must be planned for and protected against. Protection is the process of sensing an abnormal condition, applying logic to determine how and when to take corrective action, and taking the corrective action at the optimum time.

6.11.2 Abnormal Conditions

Abnormal electrical conditions are those during which measurable quantities being monitored, such as voltage, current, temperature, real power, reactive power, and phase angle, go outside of assigned safe limits. Common causes of abnormal conditions are overloads, switching operations, insulation breakdowns, and accidental short circuits or open circuits. Lightning is a frequent cause on high voltage power systems. It is not always possible to accurately categorize an abnormal condition in terms of whether it is a cause or an effect. An initiating event itself is sometimes an effect of an obscure cause and it may result in a sequence of other events which might be termed either causes or effects.

6.11.3 Speed of Response for Corrective Action

Some abnormal conditions such as high-current short circuits and high transient voltages require corrective action as soon as possible after the initiating event, in an elapsed time of the order of milliseconds or even microseconds. Electromechanical protection devices such as circuit breakers and relays are considered **instantaneous** if they operate in 1 to 3 cycles (1 cycle = 16.7 milliseconds for 60-hertz power). Solid-state protection devices can operate at microseconds with precise timing. Time delays can be introduced in both types either electronically or by auxiliary time delay mechanical devices.

6.11.4 Protection Coordination

Electrical protection is applied to both individual components and to systems. Coordination is a method of protection for a system in which the failure of any individual component will isolate that component only from the system and allow the rest of the system to continue to operate normally. In practice it requires that the **disconnect of power** at devices farthest from the power source will always take place before disconnect of power closer to the source, and where a protective device is fed from more than one source it will be isolated from all its sources before any disconnects occur closer to the source.

6.11.5 Protection Devices

Devices used for protection of electric power equipment were assigned **device numbers** based on their definition and function by IEEE, and these numbers are incorporated into ANSI Standard C37.2-1979 for power switchgear assemblies. Table 6.16 is an alphabetical index of these standard devices by name or function. In addition to the standard device number, it indicates whether the device is used primarily for **protection, indication, control,** or as **auxiliary equipment.** A numerical list of these

Table 6.16   IEEE/ANSI Standard C37.2-1979 Protection Devices with Assigned Device Numbers

Device Name or Function	No.	Application[a]	Device Name or Function	No.	Application[a]
Accelerating	18	C	Jogging	66	C
Alarm	74	I	Level	63, 71	I P
Annunciator	30	I,	Line	89	C
Atmospheric condition	45	I, P	Lockout	30, 86	P
Auxiliary motor, M-G	88	A	Manual transfer	43	C
Balance, current	61	P	Master contactor	4	C, P
Balance, voltage	60	P	Master element	1	C
Bearing	38	P	Mechanical condition	39	P
Blocking	68	P	Motor, M-G, auxiliary	88	A
Brush operating	35	C	Notching	66	C
Carrier	85	P	Operating mechanism	84	C
Checking	3	C, I	Out-of-step	78	P
Circuit breaker, ac	52	C, P	Overcurrent, ac	50, 51	P
Circuit breaker, anode	7	P	Overcurrent, dc	76	P
Circuit breaker, dc	54, 72	C, P	Overcurrent, directional	67	P
Circuit breaker, field	41	C, P	Overspeed	12	P
Circuit breaker, running	42	C	Overvoltage	59	P
			Permissive control	69	C, P
Contactor	4	C, P	Phase angle	78	P
Control power	8	C, P	Phase reversal, balance	46	P
Current balance	61	P	Phase sequence voltage	47	P
Current, directional	67	P	Pilot wire	85	P
Current, instantaneous	50	P	Polarity	36	P
Current, phase rev., bal.	46	P	Position	33	C
Current, time	51	P	Position changing	75	A
Current, under	37	P	Power, control	8	C, P
Decelerating	18	C	Power, directional	32, 92	P
Differential	87	P	Power factor	55	P
Directional current	67	P	Power, under	37	P
Directional power	32, 92	P	Pressure	63	P
Directional voltage	91, 92	P	Pulse transmitter	77	C, I
Discharge	17	P	Reclose, ac	79	P
Distance	21	P	Reclose, dc	82	P
Equalizer	22	C	Rectifier	35	C
Excitation	31, 53	C	Rectifier misfire	53	P
Field application	56	C	Regulate	90	C
Field breaker	41	C, P	Resistance, rheostat	70	C
Field change	93	C	Resistor contactor	73	C
Field excitation	40	P	Reverse	9	C
Flow	63, 80	C, P	Reverse phase, current	46	P
Frequency	81	P	Sealing, mechanical	39	P
Governor	65	C	Select	43, 83	C
Ground	64	P	Sequence change	10	C
Grounding	57	P	Sequence, incomplete	48	C, P
Incomplete sequence	48	C, P	Sequence motor operated	34	C
Interlocking	3	C, I			
Isolating	29	A	Sequence, unit starting	44	C

Table 6.16 (*Continued*)

Device Name or Function	No.	Application[a]	Device Name or Function	No.	Application[a]
Short circuit	57	P	Transition	19	C
Shunting	17	P	Trip, trip-free	94	P
Slip ring shorting	35	A	Unbalance, current	46	P
Speed	12, 14	P	Unbalance, voltage	47	P
Speed matcher	15	C	Undercurrent	37	P
Starting circuit breaker	6	C	Underpower	37	P
Start-run contactor	19	C	Underspeed	14	P
Stop	5	P	Undervoltage	27	P
Synchronizing, sync. check	13, 25	C, P	Unit sequence	10	C
			Valve, electrically operated	20	A
Temperature	23	C	Vibration	39	P
Temperature	26, 49	P	Voltage balance	60	P
Time delay	2, 62	C	Voltage, directional	91, 92	P
Transfer	43, 83	C	Voltage, under	27	P

[a]A, auxillary equipment; C, control; I, indication; P, protection.

device numbers is included in Chapter 10. These numbers are used in schematic diagrams, connection diagrams, specifications, and instruction literature of electric power equipment, with appropriate suffix letters when necessary to clarify functions they perform.

6.11.6 Voltage and Current Transformers

Protection devices that monitor voltage, current, or other quantities derived from voltage and current, are designed for 120 rated volts and 5 rated amperes. In power circuits with higher voltages and currents, transformers are connected between the power circuit and the device input circuits as shown in Fig. 6.31.

The **current transformer** is connected in series with the power line and transforms its nominal value of current to 5 A. The **voltage transformer** is connected across the power line and transforms its nominal voltage to 120 V. To keep the secondary circuits safe, the secondary circuits of both voltage and current transformers must be grounded as shown. The **polarity markers** (black circles) and arrows indicate relative instantaneous directions of current in the transformer windings. They are important for correct operation of devices having more than one voltage or current input. They are always used in a manner that keeps direction of current flow in the secondary-connected device the same as if the primary circuit were itself connected to the device at that location. Data on standard current transformers are given in Tables 6.17, 6.18, and 6.19.

The term **instrument transformer** is applied to these transformers in practice, whether they are used with instruments or with protective devices (relaying). The term **burden** is used with instrument transformers to differentiate this type of load from a primary circuit power load.

Current transformers used for relaying have an accuracy class designation consisting of the letter C or T followed by a number 10, 20, 50, 100, 200, 400, or 800.

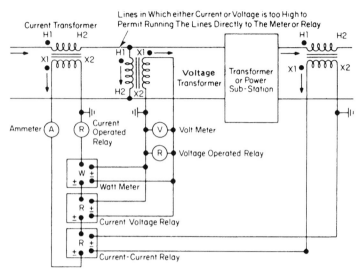

Fig. 6.31 Voltage and current (instrument) transformer connections. (Used with permission of Westinghouse Electric Corporation.)

Table 6.17   Standard Current Transformer Ratings, Single and Double Ratio

Single Ratio (A)	Double Ratio with Series, Parallel Primary Windings (A)	Double Ratio with Taps in Secondary Winding (A)
10/5	25 × 50/5	25/50/5
15/5	50 × 100/5	50/100/5
25/5	100 × 200/5	100/200/5
40.5	200 × 400/5	200/400/5
50/5	400 × 800/5	300/600/5
75/5	600 × 1,200/5	400/800/5
100/5	1,000 × 2,000/5	600/1,200/5
200/5	2,000 × 4,000/5	1,000/2,000/5
300/5		1,500/3,000/5
400/5		2,000/4,000/5
600/5		
800/5		
1,200/5		
1,500/5		
2,000/5		
3,000/5		
4,000/5		
5,000/5		
6,000/5		
8,000/5		
12,000/5		

Source: Courtesy of IEEE, New York. © 1975 IEEE.

Table 6.18 Standard Current Transformer Ratings, Multiratio Bushing Type

Current Ratings (A)		Secondary Taps	Current Ratings (A)		Secondary Taps
600/5	50/5	X2-X3	2000/5	300/5	X3-X4
	100/5	X1-X2		400/5	X1-X2
	150/5	X1-X3		500/5	X4-X5
	200/5	X4-X5		800/5	X2-X3
	250/5	X3-X4		1100/5	X2-X4
	300/5	X2-X4		1200/5	X1-X3
	400/5	X1-X4		1500/5	X1-X4
	450/5	X3-X5		1600/5	X2-X5
	500/5	X2-X5		2000/5	X1-X5
	600/5	X1-X5	3000/5	1500/5	X2-X3
1200/5	100/5	X2-X3		2000/5	X2-X4
	200/5	X1-X2		3000/5	X1-X4
	300/5	X1-X3	4000/5	2000/5	X1-X2
	400/5	X4-X5		3000/5	X1-X3
	500/5	X3-X4		4000/5	X1-X4
	600/5	X2-X4	5000/5	3000/5	X1-X2
	800/5	X1-X4		4000/5	X1-X3
	900/5	X3-X5		5000/5	X1-X4
	1000/5	X2-X5			
	1200/5	X1-X5			

Source: Courtesy of IEEE, New York. © 1975 IEEE.

C means the percent **ratio error** can be calculated; T means it has been determined by test. The number is the secondary terminal voltage the transformer will deliver to a standard burden as listed in Table 6.19 at 20 times normal secondary current without exceeding a 10% ratio error. Also, the ratio error should not exceed 10% at any current from 1 to 20 times rated current at any lesser burden. For example, a transformer rated C100 means the ratio error can be calculated and will not exceed 10% at any current from 5 to 100 amperes with a connected burden of 1 ohm.

Table 6.19 Standard Burden for Current Transformers

Standard Burden Designation	Characteristics		Characteristics for 60 Hz and 5 A Secondary Current		
	Resistance (Ω)	Inductance (mH)	Impedance (Ω)	Apparent Power[a] (VA)	Power Factor
B-0.1	0.09	0.116	0.1	2.5	0.9[b]
B-0.2	0.18	0.232	0.2	5.0	0.9[b]
B-0.5	0.45	0.580	0.5	12.5	0.9[b]
B-1	0.5	2.3	1.0	25	0.5[c]
B-2	1.0	4.6	2.0	50	0.5[c]
B-4	2.0	9.2	4.0	100	0.5[c]
B-8	4.0	18.6	8.0	200	0.5[c]

Source: Courtesy of IEEE, New York. © 1975 IEEE.

[a]At 5 A; note that VA = I^2Z, or 2 Ω at 5 A = 2×5^2 = 50 VA.
[b]Usually considered metering burdens, but data sheets may give metering accuracies at B-1.0 and B-2.0.
[c]Usually considered relaying burdens.

Figure 6.32 shows typical excitation current/voltage characteristics for current transformers with ratios 50/5 to 600/5. In some applications where the ratio is low and burden is high, such as in detection of small ground-fault currents, the ratio error is high, and it is necessary to add exciting current to burden current to determine the primary current at which a relay will operate. For example, a sensitive overcurrent relay with a minimum operating current of 0.25 A has a burden impedance of about 7.0 Ω. When used with the 50/5 ratio transformer of Fig. 6.32 its secondary resistance plus lead resistance adds about 0.1 Ω. Secondary voltage required to provide 0.25 A burden current is 7.1 × 0.25 = 1.8 V. From Fig. 6.32, exciting current at 1.8 V is 0.38 A and total secondary current is 0.63 A. Total primary current is 10 × 0.63 = 6.3 A to operate the relay, not 2.5 A which would be the value based on the turns ratio if there were no ratio error.

Voltage transformers, also called **potential transformers** (although the term *voltage transformer* is now preferred), have primary windings rated at nominal system voltages per Table 6.20 and secondary windings normally rated 120 V to match the voltage rating of most protective devices. **Accuracy classifications** of voltage transformers are stated by a number from 0.3 to 1.2, representing the percent ratio correction to obtain a true ratio. These accuracies are intended mainly for metering or indication purposes and are high enough that any standard voltage transformer is adequate for protective relaying purposes as long as it is applied within its thermal and voltage limits.

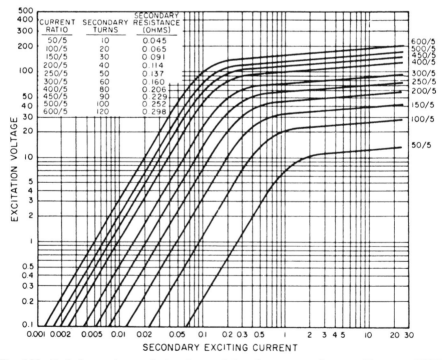

Fig. 6.32 Typical secondary excitation characteristics of current transformers. (Courtesy of IEEE, New York. © 1975 IEEE.)

Table 6.20 Voltage Classes and Nominal System Voltages

Voltage Class	Nominal System Voltage (volts)[a] Two-Wire	Three-Wire	Four-Wire	Maximum System Voltage	Reference
Low voltage					
Single-phase systems	(120)	120/240		127 or 127/254	IEEE Std 100-1972 (ANSI C42.100-1972) defines low-voltage system as "less than 750 V." It is now being proposed as "maximum rms ac voltage of 1000 V."
Three-phase systems			208Y/120	220	
		(240)	240/120	254	
		(480)	480Y/277	508	
		(600)		635	
Medium voltage		(2,400)		2,540	These voltages are listed in ANSI C84.1-1970 but are not identified by voltage class.
		4,160		4,400	
		(4,800)		5,080	
		(6,900)		7,260	
			12,470Y/7,200	13,970	
			13,200Y/7,620	13,200	
		13,800	(13,800Y/7,970)	13,970	
				14,520	
		(23,000)		24,340	Note that additional voltages in this class are listed in ANSI C84.1-1970 but are not included because they are oriented primarily to electric-utility system practice.
			24,940Y/14,400	26,400	
		(34,500)	34,500Y/19,920	36,510	
		(46,000)		48,300	
		(69,000)		72,500	
High voltage		115,000		121,000	ANSI C84.1-1970 identifies these as "higher voltage three-phase systems" (as well as 46 kV and 69 Kv). ANSI C92.2-1974
		138,000		145,000	
		(161,000)		169,000	
		230,000		242,000	
Extrahigh voltage		345,000		362,000	
		500,000		550,000	
		735,000–765,000		800,000	

Source: Courtesy of IEEE, New York. © 1975 IEEE.

[a] System voltages shown without parentheses are preferred.

6.11.7   Fuses

The simplest and oldest of protection devices, fuses, are now standardized in low, medium, and high voltage ratings (Table 6.21). They are used either for primary protection or for backup protection coordinated with circuit breakers. Low-voltage fuses are made in plug and cartridge housings up to 120 V and in cartridges up to 600 V. High-voltage power fuses have insulating tube cartridge housings. At voltages up to 34.5 kV they are made with replaceable fusible elements, above 34.5 kV the entire assembly is replaced after operation.

Current-limiting fuses have fast-melting characteristics which cause interruption of fault current before it reaches its peak value on the first cycle of a short circuit. If a power circuit has total available fault current higher than the interrupting capability of its circuit breaker, current-limiting fuses used on the source side of the breaker will limit fault current and interrupt it before the breaker operates, thus preventing possible damage to the breaker.[1]

6.11.8   Synchronous Generator Protection

A typical generator protection scheme is shown in Fig. 6.33. The devices shown dashed are optional for small or low-voltage machines. Additional devices used for protection of large power station generators are discussed in Chapter 6 of Ref. 12.

6.11.9   Power Bus Protection

Power buses connected through circuit breakers to sources or as feeders to loads require protection against short circuits by the fastest possible means of sensing and breaker opening. Moreover, the protective relays must differentiate between a short circuit on the bus itself and one on the source or feeder, and operate the breaker only on the former. To accomplish this, special types of **differential relays** (device 87) with multiple **restraint windings** connected (Fig. 6.34) to current transformers on each source or feeder group are used. Under normal operation, the restraint winding currents are balanced and no current flows through the relay operating winding. A fault on the bus disturbs the balance and causes the relay to operate; a fault on the source or load side of any current transformer does not do so.

An alternative differential protection system using a sensitive **voltage relay** and **linear couplers** is shown in Fig. 6.35. Linear couplers are air-core bushing-type transformers wound on nonmagnetic toroidal cores, designed to produce 5 secondary volts per 1000 amperes of primary current. Under normal operation or external fault conditions, the secondary voltages in all linear couplers add up to zero; for a fault on the bus, their sum is a net voltage which operates the relay. Advantages of this system are that the linear couplers are not responsive to dc transients or saturation effects of iron, which sometimes cause false operation of differential relays using iron-core current transformers.

6.11.10   Transformer Protection

Primary **fuses** and a secondary **circuit breaker** operated by phase and ground (neutral) overcurrent relays (Fig. 6.36) form a satisfactory protection system for most

Table 6.21 Standard Fuse Ratings

Low Voltage

0-600 A			601-6000 A
250 V or Less (A)	300 V or Less (A)	600 V or Less (A)	600V or Less (A)
0–30	0–15	0–30	601–800
31–60	16–20	31–60	801–1200
61–100	21–30	61–100	1201–1600
101–200	31–60	101–200	1601–2000
201–400		201–400	2001–2500
401–600		401–600	2501–3000
			3001–4000
			4001–5000
			5001–6000

High Voltage

Nominal Rating (kV)	Maximum Continuous Current (A)	Maximum Three-Phase Symmetrical Interrupting Rating (MVA)
	Boric-Acid Power Fuses, Refillable	
2.4	200, 400, 720	155
4.16	200, 400, 720	270
7.2	200, 400, 720	325
14.4	200, 400, 720	620
23	200, 300	750
34.5	200, 300	1000
	Boric-Acid Power Fuses, Nonrefillable	
34.5	100, 200, 300	2000
46	100, 200, 300	2500
69	100, 200, 300	2000
115	100, 250	2000
138	100, 250	2000
	Outdoor Expulsion-Type Power Fuses	
7.2	100, 200, 300, 400	162
14.4	100, 200, 300, 400	406
23	100, 200, 300, 400	785
34.5	100, 200, 300, 400	1174
46	100, 200, 300, 400	1988
69	100, 200, 300, 400	2350
115	100, 200,	3110
138	100, 200,	2980
161	100, 200,	3480
	Current-Limiting Power Fuses	
2.4	100, 200, 450	155–210
2.4/4.16Y	450	360
4.8	100, 200, 300, 400	310
7.2	100, 200	620
14.2	50, 100, 175. 200	780–2950
23	50, 100	750–1740
34.5	40, 80	750–2600

Source: Courtesy of IEEE, New York. © 1975 IEEE.

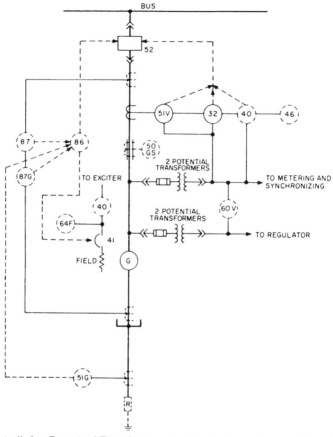

3 Voltage-Controlled or Restrained Time Over-current Relays (Device 51V)

1 Time Overcurrent Relay (Device 51G) (use if generator neutral is grounded)

1 Instantaneous Overcurrent Relay (Device 50GS) (use if generator neutral is not grounded)

1 Power Directional Relay (Device 32) (may be omitted if protective function is included with steam turbine)

1 Stator Impedance or Loss of Field Current Relay (Device 40)

1 Negative Phase Sequence Current Relay (Device 46)

1 Field Circuit Ground Detector (Device 64F)

1 Potential Transformer Failure Relay (Device 60V)

1 Lockout Relay (Device 86) (hand reset)

3 Fixed or Variable Percent Differential Relays (Device 87)

1 Current-Polarized Directional Relay (Device 87G)

Fig. 6.33 Synchronous generator protection. (Courtesy of IEEE, New York. © 1975 IEEE.)

small power transformers. On large transformers, the additional protection cost of a primary breaker, differential relay system (87T), and pressure switch (63) (Fig. 6.37) is recommended because of the very high outage and repair costs of an internal fault. Transformers are often paralleled in practice. A tie circuit breaker is used to isolate parallel secondary circuits for a fault on either side. Protection diagrams for these arrangements are shown in Chapter 8 of Ref. 12.

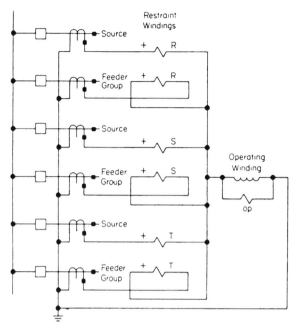

Fig. 6.34 Power bus protection with iron-core current transformers and differential current relay. (Reproduced by permission of the Relay-Instrument Division, Westinghouse Electric Corporation.)

6.11.11 Motor Protection

Ac motors vary in ratings from fractional horsepower to tens of thousands of horse-power and in speeds from 3600 revolutions per minute down to 180 or less. Physical size of a motor is directly proportional to horsepower rating and inversely proportional to speed. Standards for motors separate them into three size categories; fractional horsepower, medium (integral horsepower to approximately 1500 hp at 3600 rpm), and large (over 1500 hp).

Fractional-horsepower motors can be adequately protected by a **temperature-sensitive thermostatic switch** embedded in the end windings and connected in series with the power source, if they operate from single-phase power.

Three-phase fractional-horsepower and medium ac motors in most applications are protected by the combination of components called **motor control** shown in Fig. 6.38. Where control of many motors is desired from a single location, units of motor control consisting of these components are grouped in separate enclosure cubicals and the groups are called *motor control centers*. Function of the circuit breaker is to protect against short-circuit currents in the motor or in the power leads connecting to the motor. The three-pole **magnetic contactor M** applies and disconnects power to the motor. Its auxiliary contacts operate one or more indicator lights to show operating conditions. The **overload relays** open the contactor circuit at preset conditions of overcurrent and time.

Where reliability of motors is important enough to warrant more complete and more sophisticated protection than provided by motor control, **relay protection** for

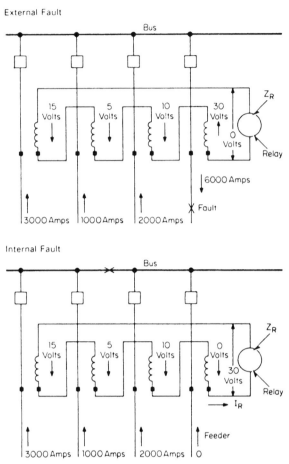

Fig. 6.35 Bus protection with air-core linear coupler sensors and sensitive voltage relay. (Reproduced by permission of the Relay-Instrument Division, Westinghouse Electric Corporation.)

operating circuit breakers is used as shown in Fig. 6.39, up to about 1500 hp. This scheme provides **undervoltage** and **phase-sequence voltage protection** and sensitive **ground-fault protection** in addition to **phase overcurrent** and **fault protection.**

For large motors above 1500 hp the complete complement of relay protection of Fig. 6.40 is typical. **Differential current protection** (device 87) for very fast **internal fault breaker opening** and **resistance temperature detectors** (RTDs) in the stator windings for accurate alarm and protection on overloads are used in addition to the protective devices on smaller motors.

Protection of a large motor during a long acceleration period may be required in addition to protection while running. The time required for a motor to accelerate from zero to full speed is proportional to total inertia of the motor rotor, coupling, and load, and inversely proportional to average torque during the starting period, as indicated by Equation (6.135). Motor standards define a maximum value of total inertia for a given speed and horsepower rating that can safely be accelerated by a motor of standard electrical design without injurious temperature rise. In present

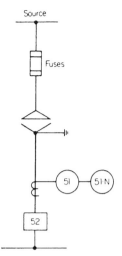

Fig. 6.36 Transformer protection with primary fuses and secondary breaker. (Reproduced by permission of the Relay-Instrument Division, Westinghouse Electric Corporation.) *Note:* Refer to Table 10.1, for identification of numbered devices in this and subsequent figures.

practice, large motors are frequently required to accelerate loads with higher total inertia than these maximum values. To increase the physical size of a motor for the sole purpose of starting a high-inertia load may be uneconomical from the standpoints of both higher first cost and lower efficiency at its rated load. Therefore, in applications of motors driving high-inertia loads, special attention should be given to **starting protection.**

If the starting period is equal to or greater than the maximum permissible locked-rotor time for safe temperature rise, starting protection requires, in addition to an

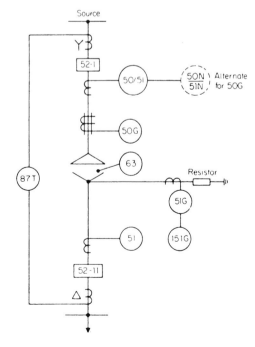

Fig. 6.37 Transformer protection with primary and secondary breakers and differential relay. (Reproduced by permission of the Relay-Instrument Division, Westinghouse Electric Corporation.)

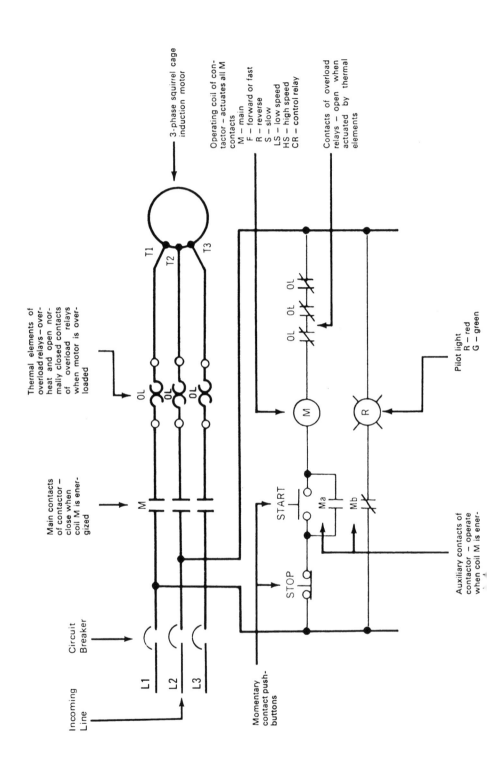

3-phase squirrel cage induction motor

Operating coil of contactor – actuates all M contacts

M – main
F – forward or fast
R – reverse
S – slow
LS – low speed
HS – high speed
CR – control relay

Contacts of overload relays – open when actuated by thermal elements

Thermal elements of overload relays – overheat and open normally closed contacts of overload relays when motor is overloaded

Pilot light
R – red
G – green

Main contacts of contactor – close when coil M is energized

Auxiliary contacts of contactor – operate when coil M is ener-

Circuit Breaker

Incoming Line

Momentary contact pushbuttons

L1
L2
L3

OL

M

T1
T2
T3

OL OL OL

M

R

START

STOP

Ma

Mb

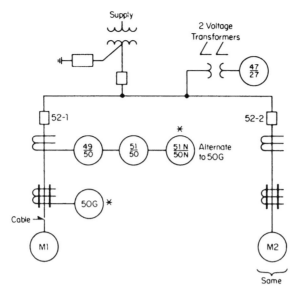

Fig. 6.39 Relay protection for motors with ratings up to 1500 hp. Reproduced by permission of the Relay-Instrument Division, Westinghouse Electric Corporation.)

overcurrent relay set for maximum permissible locked-rotor time, a means of disabling (blocking) this relay at a time during the starting period when the rotor is known to be in motion. Two methods of detecting rotor motion during start and blocking the overcurrent relay as a result of this motion are in current use: (1) a **mechanical speed switch** coupled to the motor shaft with contacts set to open at a predetermined speed, and (2) an **impedance relay** responsive to the combination of motor voltage, current, and phase angle. The impedance relay has closed contacts at zero and low speeds that open at a set point represented by higher impedance and lower phase angle as the speed increases.

Voltage at the motor terminals during the starting period is another important consideration in motor starting protection because it affects the starting time. Large induction motors have **inrush currents** of 400 to 700%, which decrease slowly until the rotor is almost up to full speed. At any time during the starting period, torque is proportional to the square of the voltage. If the voltage remains low during a large part of the period, the starting time increases as the square of the voltage drop. For example, at 70% voltage, the average torque is only about half of normal and the starting time is approximately doubled.

6.11.12 Transient Protection

Since abnormal short-time transient voltages and currents can occur in all high- and low-voltage circuits even under normal operating conditions, whenever the possibility exists for their causing malfunction or damage, protection measures must be taken to prevent or at least mitigate these effects.

In high-voltage power circuits the two most common causes of **transient surges** are **lightning** and **switching** operations. **Lightning arresters** are used between high-

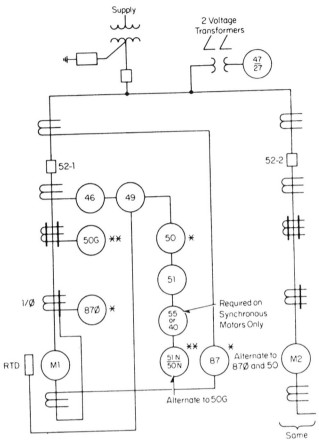

Fig. 6.40   Relay protection for large motors above 1500 hp. (Reproduced by permission of the Relay-Instrument Division, Westinghouse Electric Corporation.)

voltage terminals and ground on outdoor electrical equipment. These are nonlinear resistors made of zinc oxide blocks that act as insulators at the rated line voltage but pass very high currents during voltage surges. To reduce the magnitude of voltage surges during switching operations in high-voltage circuits, resistors are inserted across the breaker during the switching operation to reduce the rate of change of current during the operation. Detailed descriptions of up-to-date devices and practices for surge protection of high voltage equipment can be found in Ref. 13.

From the standpoint of protection of equipment from malfunction, low-voltage devices must be protected against transients caused either within their own circuits or by **electromagnetic** or **electrostatic coupling** to high-voltage circuits. One or more of the following protective measures are applied in relaying and control circuits to guard against malfunction and damage due to transients: (1) **physical separation of coupled circuits** to reduce mutual inductance and mutual capacitance, (2) **series inductance** to slow the rate of change of current surges, (3) **parallel capacitance** or **resistance-capacitance** combinations to slow the rate of rise of voltage surges, (4) **parallel zener (clamping) diodes** to limit magnitude of voltage surges, (5) insertion of **resistance** across breaker or relay contacts during operation, (6) **shielding of**

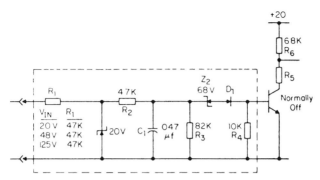

Fig. 6.41 Typical filter for transient protection of solid-state relays. (Reproduced by permission of the Relay-Instrument Division, Westinghouse Electric Corporation.)

control leads to reduce capacitive coupling to other circuits, (7) **twisting of control leads** to reduce common mode coupling to other circuits, (8) use of **surge protection package circuits,** also called *filters* or *buffers,* in the input lead circuits of all solid-state relays. A typical such filter is shown in Fig. 6.41. For a detailed discussion of surge protection in power system relaying refer to Chapter 4 in Ref. 12.

6.12 EMERGENCY POWER SYSTEMS

6.12.1 Reliability of Electric Utilities

Every means available to electric utilities is used in the design and operation of their distribution, transmission, and generation systems to give the highest possible reliability of service. **Parallel** or **looped feeders** provide alternate power paths from sources to consumers wherever possible. Low-voltage distribution systems are interconnected into **networks** to provide multiple power paths to commercial and industrial buildings in downtown and other areas where these buildings are concentrated. Protective devices restore power almost immediately in many cases where a fault occurs in a distribution system and can be isolated from the rest of the system. However, the needs of some consumers are so critical that the loss of power for even a very short time interval cannot be tolerated. For example, a hospital cannot have loss of power for more than a few seconds without possibly endangering someone's life. Industries such as those manufacturing semiconductors and photo processing require constant temperatures maintained by electric heat and electric control, where even a short power failure can be extremely costly. Communication services, particularly telephone systems, are another example. And one of the most critical applications where continuous power is essential is computers, where power interruptions as short as a few cycles (1 cycle = 0.0167 second) can cause serious problems.

6.12.2 Standby Power Sources

If power loss of a few seconds or a few minutes at most can be tolerated, an **alternative source,** normally not in use until the prime (utility) source fails and operating completely independent of the prime source, is used. The most common type of

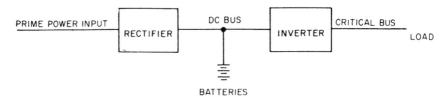

Fig. 6.42 Nonredundant uninterruptible power supply. (Courtesy of IEEE, New York. © 1975 IEEE.)

standby power source is an engine-driven generator set. On loss of voltage at the load bus a normally closed transfer switch disconnects the load from the utility source and initiates startup of the generator set. When rated voltage exists at the generator terminals a normally open switch connects it to the load. Most diesel and gasoline engine-driven generators can restore power in less than a minute.

6.12.3 Uninterruptible Power Supplies

To meet the operating needs of computers and critical manufacturing processes that cannot tolerate interruptions of even a few cycles, power supplies that operate in parallel with electric utility sources and supply their load without interruption when the utility source fails have become widely used during the past two decades. Most **uninterruptible power supplies** (UPSs) have batteries operating an inverter connected to the critical load at all times, the batteries being charged by a rectifier from the prime power source as shown in Fig. 6.42. Where utmost reliability is required, two or more redundant systems of rectifiers and inverters are connected in parallel with static selector switches on their load side as shown in Fig. 6.43. A UPS system with static selector switches connecting to either the utility prime source or to a rectifier, battery, and inverter source is shown in Fig. 6.44. Figure 6.45 goes one step further in capability for long-time emergency operation by providing alternate power

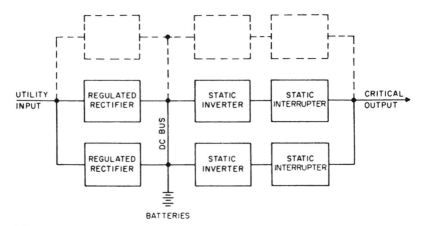

Fig. 6.43 Redundant uninterruptible power supply. (Courtesy of IEEE, New York. © 1975 IEEE.)

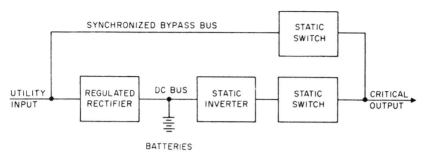

Fig. 6.44 Uninterruptible power supply with static switch bypass to prime source. (Courtesy of IEEE, New York. © 1975 IEEE.)

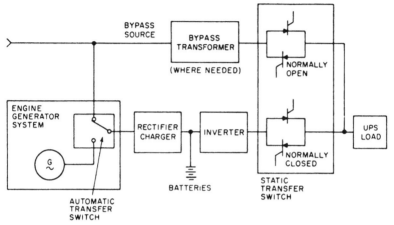

Fig. 6.45 Uninterruptible power supply with static switch bypass to prime source and automatic transfer switch to standby source.

to charge the batteries from an engine-driven generator. Reference 14 describes the preceding UPS systems and other emergency power systems in more detail.

6.13 ELECTRONICS

6.13.1 Definition and Scope

Electronics is the specialized field within electrical engineering that deals with electron devices and their utilization. An **electron device** is one in which current flows principally by the movement of electrons through a **vacuum, gas,** or **semiconductor.** In semiconductors, current flow may occur either by electron movement in the negative-to-positive direction, or by the apparent movement of electron sites, called holes, in the opposite direction.

6.13.2 Electronic Applications

The term *electronic* describes circuits, systems, and items of equipment utilizing electron devices, for example, electronic alarm and electronic voltage regulator.

6.13.3   Electronics Engineering

Branches of electrical engineering that deal primarily with electronics are radio, television, data communication, data and word processing, solid-state control and indication, and audio systems. Nearly every other engineering discipline becomes involved in electronics technology today because of the proliferation of electronic devices and techniques in other fields.

6.14   ELECTRONIC DEVICES

6.14.1   Vacuum and Gas Devices

Vacuum tubes contain a thermionic emitter of electrons, the **filament** or **cathode,** and an electron-collecting element called the **anode.** A tube with only these **two elements** forms a **diode** capable of rectification and other diode functions. **Triodes, tetrodes, pentodes,** and other multigrid configurations contain one or more grids between cathode and anode to control electron flow, thus making them capable of amplification, switching, and generation of ac voltages and currents by oscillation. Tubes containing gases, with or without thermionic emitters, in which the gas ionizes and allows current flow from anode to cathode are used as **rectifiers, voltage regulators,** and **light generators.**

Prior to the invention of the transistor in 1947, vacuum and gas tubes were used in nearly all electronic applications. In present technology they have been largely superseded by **semiconductor devices,** which in most cases perform the same functions better and at less cost. Two exceptions are high-power vacuum tubes and gas-tube lighting devices. Up-to-date reference literature on these is available and also on small vacuum tubes because of the large number of them still in use.[15,16]

6.14.2   Semiconductor Diode

When **impurities** are added to the semiconductor materials **silicon** and **germanium** in a process called **doping,** they acquire either an **excess** of **electrons** and become **N-type material,** or they acquire a **deficiency** of **electrons** and become **P-type material,** depending on the type of impurity added. A **diode** is formed by a surface or a point contact junction between P-type and N-type materials. When voltage is applied across it, positive to P-type and negative to N-type, current flows freely across the junction, labeled I_F in Fig. 6.46. When voltage is applied in the reverse direction, only a small leakage current flows until a breakdown voltage $-B_{VR}$ causes an avalanche current $-I_R$, as shown in Fig. 6.47.

Diodes are marked or labeled so as to indicate the direction of currrent flow through them (Fig. 6.48). This is opposite to the direction of electron flow, and is in the same direction as the flow of electron sites or holes.

The most common applications of diodes are their use as ac power rectifiers (see Figs. 6.49 through Fig. 6.53. Many other diode applications have resulted from unique properties of some types, as listed in Table 6.22.

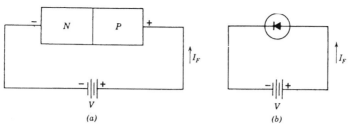

Fig. 6.46 Forward current in a diode. (*a*) Physical layout, (*b*) circuit symbol. (From E. N. Lurch, *Fundamentals of Electronics*. Copyright © 1981 by John Wiley & Sons, Inc. Reproduced with permission.)

6.14.3 Junction Transistor

A thin section of P-type material sandwiched between two thicker sections of N-type material forms an **NPN junction transistor** (Fig. 6.54*a*). The collector, base, and emitter are labeled *C*, *B*, and *E*, respectively. Figure 6.54*c* shows a PNP transistor with an N section between two P sections, and the polarity of V_{BB} and V_{CC} reversed. Because junction transistors operate both by electron flow from negative to positive and by hole flow from positive to negative, they are also called **bipolar junction transistors** (BJTs).

Junction transistors have collector current I_C larger than base current I_B, and emitter current:

$$I_E = I_B + I_C. \qquad (6.224)$$

Dc beta (β_{dc}) is defined at any operating point by

$$\beta_{dc} = \frac{I_C}{I_B}. \qquad (6.225)$$

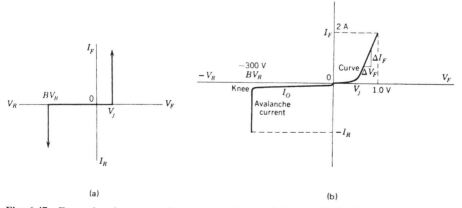

Fig. 6.47 Forward and reverse voltage–current characteristics: (*a*) ideal diode; (*b*) actual diode. (From E. N. Lurch, *Fundamentals of Electronics*. Copyright © 1981 by John Wiley & Sons, Inc. Reproduced with permission.)

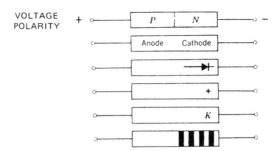

Fig. 6.48 Diode markings and voltage polarity for forward current flow. (From E. N. Lurch, *Fundamentals of Electronics.* (Copyright © 1981 by John Wily & Sons, Inc. Reproduced by permission.)

This ratio is equal to current gain when the transistor is used as a dc amplifier. **Dc alpha** (α_{dc}) is defined by

$$\alpha_{dc} = \frac{I_C}{I_E}. \tag{6.226}$$

When transistors are used as ac amplifiers, three circuits are used, with different gain, input resistance, and phase shift characteristics: common-emitter, common-collector, and common-base.

Ac current gain (β_{ac}) is defined as the change in collector current caused by a given change in base current at the operating point:

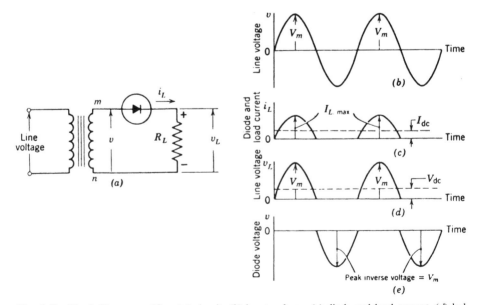

Fig. 6.49 The half-wave rectifier: (*a*) circuit, (*b*) input voltage, (*c*) diode and load current, (*d*) lod voltage, (*e*) diode voltage. (From E. N. Lurch, *Fundamentals of Electronics.* (Copyright © 1981 by John Wiley & Sons, Inc. Reproduced with permission.)

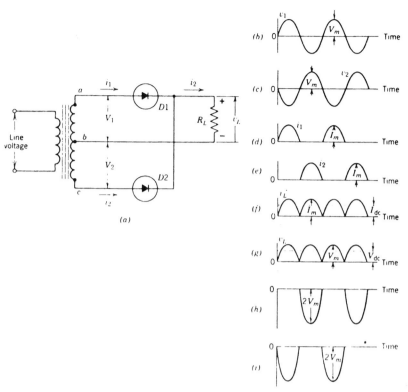

Fig. 6.50 Full-wave rectifier: (*a*) circuit; (*b*) voltage applied to D1; (*c*) voltage applied to D2; (*d*) current in D1; (*e*) current in D2; (*f*) load current; (*g*) load voltage; (*h*) voltage across D1; (*i*) voltage across D2. (From E. N. Lurch, *Fundamentals of Electronics.* (Copyright © 1981 by John Wiley & Sons, Inc. Reproduced with permission.)

$$\beta_{ac} = \frac{\Delta I_C}{\Delta I_B}. \tag{6.227}$$

Ac alpha (α_{ac}) is defined as the ratio of change in collector current to change in emitter current:

$$\alpha_{ac} = \frac{\Delta I_C}{\Delta I_E}. \tag{6.228}$$

Circuits and waveforms of a **common-emitter amplifier, common-collector amplifier,** and **common-base amplifier** are shown in Figs. 6.55, 6.56, and 6.57, respectively. Table 6.23 compares typical values of gain, input resistance, and phase shift of these amplifier circuits.

6.14.4 Junction Field-Effect Transistor

The **junction field-effect transistor** (JFET) has a high input resistance in the order of megohms and operates as an amplifier with negligible base current, thus overcom-

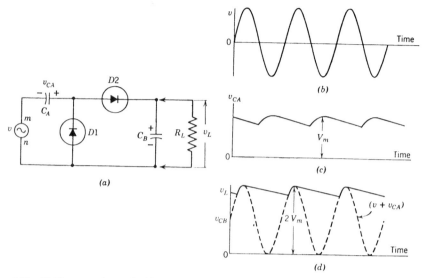

Fig. 6.51 Half-wave voltage doubler: (*a*) circuit; (*b*) line voltage; (*c*) waveform across C_A; (*d*) waveform across C_B. (From E. N. Lurch, *Fundamentals of Electronics*. Copyright © 1981 by John Wiley & Sons, Inc. Reproduced with permission.)

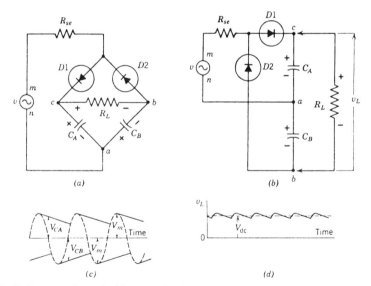

Fig. 6.52 Full-wave voltage doubler: (*a*) circuit; (*b*) alternate from of circuit layout; (*c*) voltage waveform across C_A and C_B; (*d*) output voltage waveform. (From E. N. Lurch, *Fundamentals of Electronics*. (Copyright © 1981 by John Wiley & Sons, Inc. Reproduced with permission.)

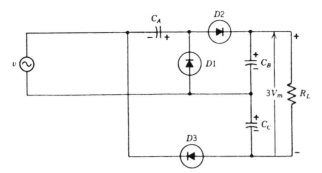

Fig. 6.53 Voltage tripler. (From E. N. Lurch, *Fundamentals of Electronics.* (Copyright © 1981 by John Wiley & Sons, Inc. Reproduced with permission.)

ing one of the most important limitations of the junction transistor, and operating in a manner closely approximating that of a vacuum tube. Construction of an N-channel JFET is shown in Fig. 6.58; a P-channel unit has P-type material channels, an N-type gate, and the polarities of V_{GG} and V_{DD} reversed.

Table 6.22   Semiconductor Diode Applications

Application	Diode Name or Type	Diode Properties
Amplifier, radio frequency	Gunn (tunnel)	Negative resistance at UHF and microwave frequencies
Capacitance, variable	Varactor	C variable with voltage
Frequency multiplier	PIN	Charge storage, variable R
Light source	Light-emitting (LED)	Light emission
Light to electric power	Photo-voltaic	Dc voltage and current proportional to light
Light sensor	Photo	Resistance variable with light
Noise source	Noise	Random noise generation
Oscillator	Gunn	Negative resistance at UHF and microwave frequencies
Oscillator frequency control	Varactor	C variable with voltage
Rectifier, ac power or signal	Silicon	High or lower power, usuable at high temperatures
Rectifier, signal	Germanium	Low forward voltage drop
Switching, high speed	PIN	Low forward voltage drop, variable resistance
Switching, UHF and microwave	Hot carrier (HCD)	Low capacitance, high efficiency
Transient suppression	Zener	Unidirectional voltage limiting
Voltage reference	Zener	Constant voltage with variable current
Voltage regulator	Zener	Constant voltage with variable current

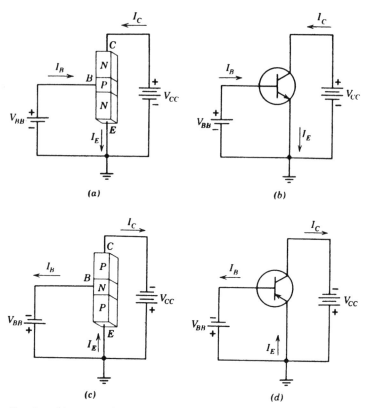

Fig. 6.54 Transistor bias connections: (*a*) and (*b*) NPN transistor; (*c* and (*d*) PNP transistor. (From E. N. Lurch, *Fundamentals of Electronics*. (Copyright © 1981 John Wiley & Sons, Inc., Reproduced with permission.)

6.14.5 Metal–Oxide Semiconductor Field-Effect Transistor

A different type of field-effect transistor, in which the gate is insulated from the channel by a very thin layer of glass (SiO$_2$), is shown in Fig. 6.59. The **metal–oxide semiconductor field-effect transistor** (MOSFET) is also called an **insulated gate field-effect transistor** (IGFET). The advantage of the insulated gate is an order of magnitude increase in input resistance, from around 10 megohms to 100 MΩ. A disadvantage of this type is that because the gate is so sensitive to static or stray voltage, the insulating glass layer can easily be punctured and the transistor destroyed. The most recently designed MOSFETs have internal protective diodes between gate and source (Fig. 6.60). Those without this feature must have all terminals short-circuited by grounding rings during handling, and the rings removed after the unit is wired into its circuit.

6.14.6 Unijunction Transistor

This is a three-terminal semiconductor device consisting of a P-material bar imbeded into an N-material block, with ohmic metallic contacts called base 1 and base 2

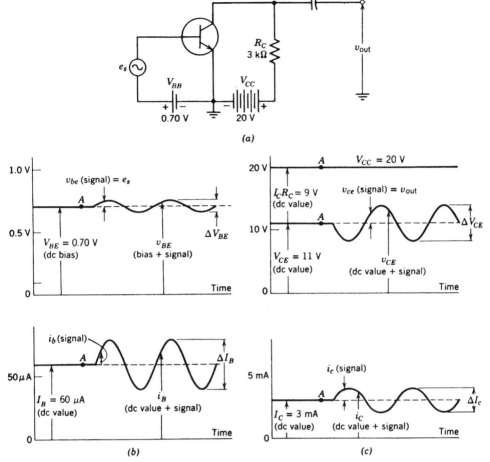

Fig. 6.55 Common-emitter amplifier: (*a*) circuit; (*b*) base waveforms; (*c*) collector waveforms. (From E. N. Lurch, *Fundamentals of Electronics*. (Copyright © 1981 John Wiley & Sons, Inc. Reproduced with permission.)

welded to the N block without creating P–N junctions (Fig. 6.60*a*). When voltage is applied across the two bases, the emitter current voltage characteristic exhibits negative resistance over a portion of its range, that is, an increase in emitter current causes a decrease in emitter voltage. The negative resistance characteristic can be used in circuits of oscillators, timing circuits, triggering circuits, and voltage- and current-sensing applications. Its most common use is in relaxation oscillators.

6.14.7 Thyristor

The term **thyristor** was first used to describe a basic four-layer, three-junction diode that had two stable states of operation. It was also called **Shockley diode** and **reverse blocking diode thyristor.** Figure 6.61 shows its construction and characteristics. It is used in high-power applications comparable to those for the unijunction transistor.

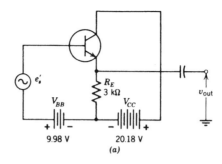

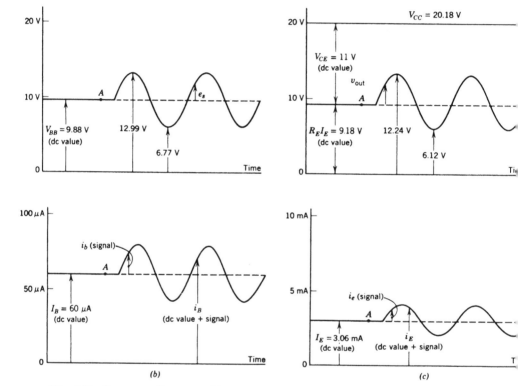

Fig. 6.56 Common-collector amplifier: (*a*) circuit; (*b*) input waveforms; (*c*) output waveforms. (From E. N. Lurch, *Fundamentals of Electronics*. (Copyright © 1981 John Wiley & Sons, Inc. Reproduced with permission.)

A light-activated switch (LAS) is a thyristor triggered by incident light instead of by voltage change.

6.14.8 Silicon-Controlled Rectifier

A thyristor with a gate that shown in Fig. 6.62 is called a **silicon-controlled rectifier** (SCR), although in practice the term **thyristor** is becoming commonly used for either.

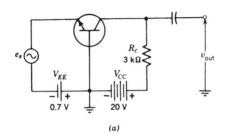

(a)

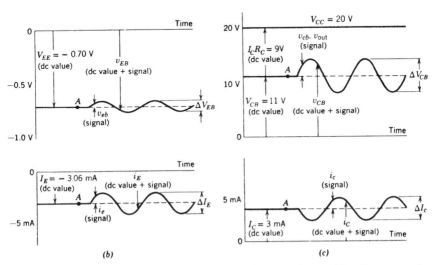

(b)   *(c)*

Fig. 6.57 Common-base amplifier: *(a)* circuit; *(b)* emitter waveforms; *(c)* collector waveforms. (From E. N. Lurch, *Fundamentals of Electronics.* (Copyright © 1981 John Wiley & Sons, Inc. Reproduced with permission.)

SCRs have wide application in the control of ac power, power switching, dc to ac inverters, dc to dc converters, and var control in ac power transmission.

6.14.9 Bidirectional Thyristor

A thyristor can be made with the structure shown in Fig. 6.63 with or without a gate, so as to make it operable in both directions of applied ac voltage. This device is also called a **bidirectional dipole thyristor,** or diac.

6.14.10 Triac

A **triac** is a form of SCR that can be triggered into conduction with polarity of either gate voltage or anode voltage (Fig. 6.64). These characteristics make it very useful as a transient surge protector in ac power circuits, and in the control of ac power to a load without rectification.

Table 6.23   Comparison of Basic Amplifier Circuits

	Common Emitter	Common Collector (Emitter Follower)	Common Base
Current gain, A_i	50	51	$0.98 \approx 1$
Voltage gain, A_v	60	1	60
	Out of phase	In phase	In phase
Power gain, A_p	3000	51	$58.8 \approx 60$
Input resistance, r_{in}	250 Ω	155,500 Ω	49 Ω
Phase shift	180°	0°	0°

Source: E. N. Lurch, *Fundamentals of Electronics.* Copyright © 1981 by John Wiley & Sons, Inc. Reproduced with permission.

6.14.11   Integrated Circuits

Combinations of diodes, transistors, and the passive electric circuit components (resistance, inductance, and capacitance) can be formed and connected together by deposition processes to form **integrated circuits** (ICs) on a silicon chip less than a square centimeter in size. Examples of ICs for linear and digital electronic applications are shown in Figs. 6.65 and 6.66. Standard ICs for digital applications are shown in Table 6.24. Large-scale integration (LSI) is the process by which IC blocks or modules with individual functions are combined on a larger silicon chip to form complete integrated assemblies of circuits, such as computer memories.

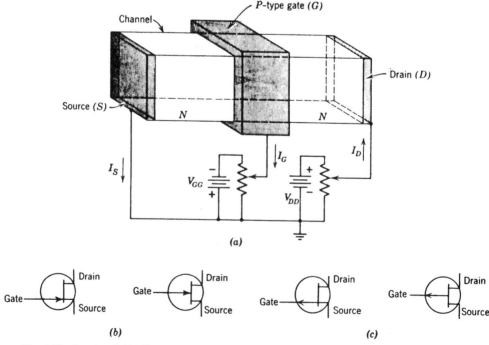

Fig. 6.58   Junction field-effect transistor: (*a*) construction; (*b*) symbol for N-channel JFET; (*c*) symbol for P-channel JFET. (From E. N. Lurch, *Fundamentals of Electronics.* (Copyright © 1981 John Wiley & Sons, Inc. Reproduced with permission.)

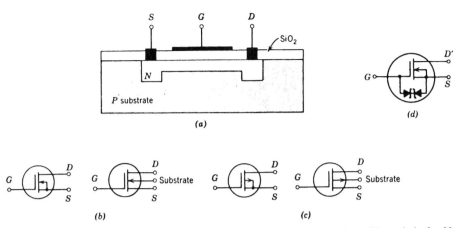

Fig. 6.59 Metal–oxide silicon field-effect transistor: (*a*) cross-section view; (*b*) symbols for N-channel MOSFET; (*c*) symbols for P-channel MOSFET; (*d*) MOSFET with internal Zener diode protection (From E. N. Lurch, *Fundamentals of Electronics*. (Copyright © 1981 John Wiley & Sons, Inc. Reproduced with permission.)

6.14.12 Packaging of Electronic Chips[17,18]

Packaging of **electronic chips** provides for their easy handling, offers a convenient way to make electrical connections, in some cases assists in heat removal, and protects the chip from corrosion due to the ambient environment. For some applications chips are mounted directly on **printed circuit broad,** or **substrate,** using either the **flip chip** or **wafer bumping** process. For most applications however, the chip is mounted in a plastic protective package of which there are several types, such as quad flat packs with 80 pins on each of four sides (often used for **microprocessors**), pin grids of 196 pins in a 14 × 14 array (somewhat expensive and not widely used), and less expensive widely used dual-in-line packages, which typically come with 6, 8, 14, 16, 20, 28, 40, or more pins.

Packaging for electronic chips take the form of either **through-hole** (with long leads that go through holes in the printed circuit board) or **surface-mount** designs (with shorter leads). The former offers a somewhat simpler mounting, while the later are smaller and lighter and can provide a higher density of elements since available

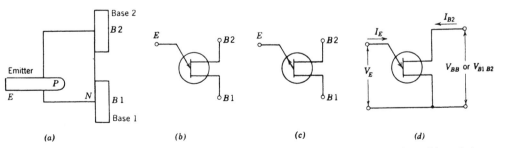

Fig. 6.60 Unijunction transistor: (*a*) construction; (*b*) symbol for UJT with N-type base; (*c*) symbol for UJT with P-type base; (*d*) nomenclature. (From E. N. Lurch, *Fundamentals of Electronics*. (Copyright © 1981 John Wiley & Sons, Inc. Reproduced with permission.)

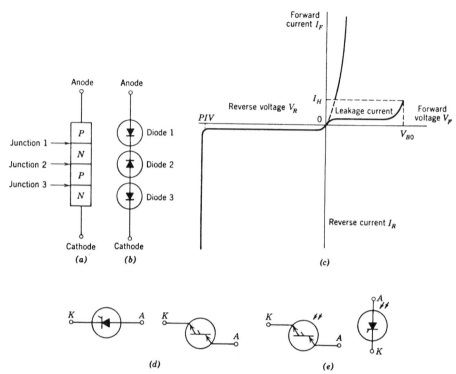

Fig. 6.61 Fundamental thyristor (NPNP transistor): (a) construction; (b) diode model; (c) breakdown characteristics; (d) symbols; (e) symbols for light-activated switch (LAS). (From E. N. Lurch, *Fundamentals of Electronics*. (Copyright © 1981 John Wiley & Sons, Inc. Reproduced with permission.)

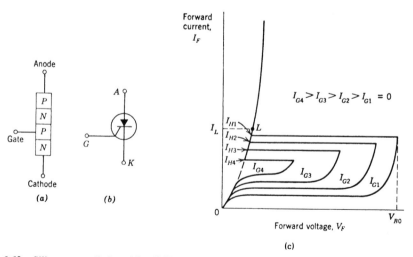

Fig. 6.62 Silicon-controlled rectifier (SCR): (a) construction; (b) symbol; (c) characteristics. (From E. N. Lurch, *Fundamentals of Electronics*. (Copyright © 1981 John Wiley & Sons, Inc. Reproduced with permission.)

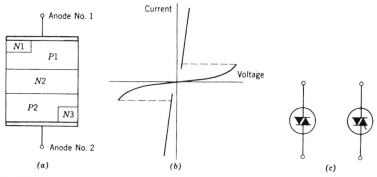

Fig. 6.63 Bidirectional thyristors: (*a*) layer structure; (*b*) characteristics; (*c*) symbols. (From E. N. Lurch, *Fundamentals of Electronics*. (Copyright © 1981 John Wiley & Sons, Inc. Reproduced with permission.)

board space is used more efficiently. Several variations of the surface-mount configuration are available which multiply the chip density, including systems that permit vertical stacking with extended leads.

For packaging, the chip is first bonded to the package base using **epoxy** (conductive or nonconductive) or **polyimide** or a similar cement. For power devices the die attachment uses soft solder or a **eutectic** (for best reliability) **solder** die attachment; for others, the die is mounted on a paddle part of the leadframe. Fine wires, typically aluminum, are bonded to the contact pads of the chip and to the pin contacts on the package. The chip is then sealed inside the package using a plastic, metal, or ceramic seal system. Next, the packages are labeled with the chip type and usually, a production data code. Following this, final testing is performed.

A wide variety of packing materials for integrated circuits and electronic chips have been developed. These include plastic, and **hermetic** (ceramic and metallic)

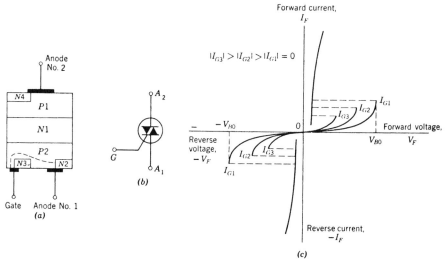

Fig. 6.64 Triac: (*a*) cross section; (*b*) symbol; (*c*) characteristics. (From E. N. Lurch, *Fundamentals of Electronics*. (Copyright © 1981 John Wiley & Sons, Inc. Reproduced with permission.)

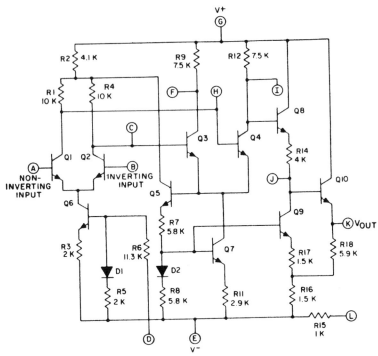

Fig. 6.65 Integrated-circuit operational amplifier for linear application. (Courtesy of RCA Solid State.)

materials and are available for both through-hole and surface-mounted applications. Hermetic packages prevent gases and liquids from entering the package cavity and are able to withstand higher temperatures than can equivalent plastic packages. The primary materials used in a plastic package are a leadframe, die-attached material, bond wire, mold compound, and a lead finish. The leadframe provides increased thermal conductivity and tends to improve life and performance.

Hermetic packages are of three types: (1) multilayer ceramic, (2) pressed ceramic, and (3) metal can. For **multilayer ceramic packages** ceramic tape layers are metal-

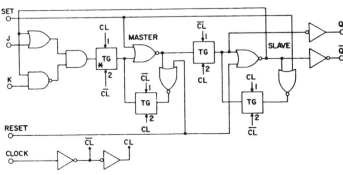

Fig. 6.66 Type J-F flip-flop integrated-circuit block for digital applications. (Courtesy of RCA Solid State.)

Table 6.24   Standardized Complementary Metal–Oxide Semiconductor Integrated Circuits for Digital Applications

---

Gates

CD4000A	Dual 3-input NOR gate plus inverter
CD4001A	Quad 2-input NOR gate
CD4002A	Dual 4-input NOR gate
CD4011A	Quad 2-input NAND gate
CD4012A	Dual 4-input NAND gate
CD4019A	Quad AND-OR select gate
CD4023A	Triple 3-input NAND gate
CD4025A	Triple 3-input NOR gate
CD4030A	Quad exclusive-OR gate
CD4037A	Triple AND-OR biphase pairs
CD4048A	Expandable 8-input gate

Flip-Flops

CD4013A	Dual D with set/reset capability
CD4027A	Dual J-K with set/reset capability
CD4047A	Monostable/astable multivibrator

Latches

CD4042A	Quad clocked D latch
CD4043A	NOR R/S latch (three output states)
CD4044A	NAND R/S latch (three output states)

Arithmetic Devices

CD4008A	Four-bit adder, parallel carry-out
CD4032A	Triple serial adder, internal carry (negative logic)
CD4038A	Triple serial adder, internal carry (positive logic)
CD4057A	LSI 4-bit arithmetic logic unit

Buffers

CD4009A	Hex inverting type
CD4010A	Hex noninverting type
CD4041A	Quad inverting and noninverting type
CD4049A	Hex buffer/converter (inverting)
CD4050A	Hex buffer/converter (noninverting)

---

lized, laminated, and fired to create the package body. The metallized areas are then brazed to the package body. The metallized areas of the package are than electroplated (usually, nickel followed by gold). After assembly, the hermetic seal is achieved by soldering a metal lid onto the metallized and plated seal ring. These packages are often referred to as **solder seal packages.** The multilayer construction allows the package designer to incorporate electrical enhancements within the package body; for example, power and ground planes to reduce inductance, shield planes to reduce crosstalk, and controlled characteristic impedance of signal lines have been incorporated into several variations of multilayer ceramic packages.

Pressed ceramic packages are usually a three-part construction consisting of a base, lid, and leadframe. The base and lid are manufactured by pressing ceramic

Table 6.24   Standardized Complementary Metal–Oxide Semiconductor Integrated Circuits for Digital Applications

	Complementary Pairs
CD4007A	Dual complementary pair plus inverter
	Multiplexers and Decoders
CD4016A	Quad bilateral switch
CD4028A	BCD-to-decimal decoder
CD4066A	Quad bilateral switch
	Counters
CD4017A	Decade counter/divider plus 10 decoded decimal outputs
CD4018A	Presettable divide-by-N counter
CD4020A	14-Stage ripple counter
CD4022A	Divide-by-8 counter/divider, 8 decoded outputs
CD4024A	7-Stage ripple counter
CD4026A	Decade counter/divider, 7-segment display output
CD4029A	Presettable up/down counter, binary or BCD-decade
CD4033A	Decade counter/divider, 7-segment display output
CD4040A	12-Stage binary/ripple counter
CD4045A	21-Stage ripple counter
	Shift Registers
CD4006A	18-Stage static shift register
CD4014A	8-Stage synch shift register, parallel-in/serial-out
CD4015A	Dual 4-stage shift register, serial-in/parallel-out
CD4021A	8-Stage asynchronous shift register, parallel-in/serial-out
CD4031A	64-Stage static shift register
CD4034A	Parallel-in/parallel-out shift register (3 output states)
CD4035A	4-Bit parallel-in/parallel-out shift register, J-$\overline{\text{K}}$ in, true-comp. out
	Phase-Locked Loop
CD4046A	Micropower phase-locked loop
	Memories
CD4036A	4-Word by 8-bit RAM (binary addressing)
CD4039A	4-Word by 8-bit RAM (word addressing)
CD4061A	256-Word by 1-bit state RAM
	Drivers
CD4054A	4-Line liquid-crystal-display driver
CD4055A	BCD to 7-segment decoder/driver
CD4056A	BCD to 7-segment decoder/driver

Source: Courtesy of RCA Solid State.

powder into the desired shape and then firing. Glass paste is then screened onto the fired base and lid and the glass paste is fired. During package assembly a separate leadframe is embedded in the base glass. The hermetic seal is formed by melting the lid glass over the base–leadframe combination. This seal method is referred to as a **frit seal,** and the package is often called a **glass frit seal package.** Pressed ceramic packages are typically lower in cost than multilayer packages; however, their simplier construction does not allow for many electrical enhancements.

Metal can packages consist of a metal base with leads exited through a glass seal. This seal can be a compression or matched seal. After device assembly in the package, a metal lid (or can) is resistance welded to the metal base, forming the hermetic seal. Metal can packages usually have a low lead count (less than 24) and are low in cost. Some types have very low thermal resistance. Metal can packages are used in both linear and hybrid applications.

Hermetically sealed packages of metal or ceramic, with their ability to withstand higher heat (above about 160°C), are used extensively for military and some special commercial applications and ensure high reliability and high performance. Plastic (epoxy) packaging can also be quite stable and moisture resistant, and with typical transition temperatures of about 155°C performs almost as well at temperatures of 150°C or lower. Broadly, hermetic packaging tends to yield performance about twice as good as that of plastic packaging, which seems a significant increase in capability. However, plastic packaged chips are currently somewhere in the range of 10 to 100 times more reliable than the best metal or ceramic chips of just a few years ago. Thus for applications where temperature is not a problem, plastic packaging is entirely suitable. Plastic packaging is used for virtually all commercial applications.

6.15   PRACTICAL ELECTRONIC CIRCUITS

Examples of circuits frequently used in practice can be found in Ref. 19.

REFERENCES

1. *ANSI Catalog of American National Standards,* American National Standards Institute, New York 1982.
2. *IEEE Standards Catalog and Standards Listing,* Institute of Electrical and Electronic Engineers, New York, 1982.
3. *NEMA Standards Publications,* National Electrical Manufacturers Association, Washington, D.C., 1982.
4. *Quick Reference to IEEE Standards (QRIS),* Institute of Electrical and Electronic Engineers, New York, 1980.
5. *IEEE Standard Dictionary of Electrical and Electronics Terms,* ANSI/IEEE Std. 100-1977, Institute of Electrical and Electronic Engineers, New York, 1977.
6. *Metric Practice,* ANSI Std. Z 210.1-1976, IEEE Std. 268-1976, ASTM Std. E 380-76, 1976.
7. *IEEE Standard Letter Symbols for Units of Measurement,* ANSI/IEEE Std. 260-1978, Institute of Electrical and Electronic Engineers, New York, 1978.
8. *Letter Symbols for Quantities Used in Electrical Science and Electrical Engineering,* ANSI/IEEE Std. 280-1968, Institute of Electrical and Electronic Engineers, New York, 1968.
9. *Electrical and Electronics Graphic Symbols and Reference Designations,* Std. 76-ANSI/IEEE Y32E, Institute of Electrical and Electronic Engineers, New York, 1976.

10. *Application Technology Bulletins, D, D, 2T,* Lincoln Electric Company, Cleveland, Ohio, 1991.
11. *Westinghouse Large Induction Motors,* Publication B236, Westinghouse Motor Company, Round Rock, Texas.
12. *Applied Protective Relaying,* Westinghouse Electric Corp., Coral Springs, Fla., 1979.
13. *Surge Protection of Power Systems,* Westinghouse Electric Corp. Pittsburgh, Pa. 1971.
14. *IEEE Recommended Practice for Emergency and Standby Power Systems for Industrial and Commercial Application,* IEEE Std. 446-1980, Institute of Electrical and Electronic Engineers, New York, 1980.
15. *Receiving Tube Manual, RC-30,* RCA Corporation, Camden, N.J. 1975.
16. *Transmitting Tubes,* Technical Manual TT, RCA Corporation, Camden, N.J.
17. "Chip Packaging," R. Pease, National Semiconductor Corporation, Santa Clara, Calif., private communication, 1996.
18. "Chip Packaging," P. Rissman, Hewlett-Packard, Palo Alto, Calif. private communication, 1996.
19. *Solid-State Devices Manual, SC-16,* RCA Solid-State Division, Somerville, N.J., 1975.

BIBLIOGRAPHY

Applied Protective Relaying, Westinghouse Electric Corp., Coral Springs, Florida, 1979.

Baumeister, T., E. A., Avallone, and T. Baumeister III, *Mark's Standard Handbook for Mechanical Engineers,* 8th ed. McGraw-Hill Book Company, New York, 1978.

Bonebreak, R. L., *Practical Techniques of Electronic Circuit Design,* John Wiley & Sons, Inc., New York, 1982.

Cahill, S. J., *Digital and Microprocessor Engineering,* John Wiley & Sons, Inc., New York, 1982.

Cowles, L. G., *Transistor Circuits and Applications,* 2nd ed., Prentice Hall, Upper Saddle River, N. J., 1974.

DeSa, A., *Principles of Electronic Instrumentation,* John Wiley & Sons, Inc., New York, 1981.

Distribution Systems Reference Book, Westinghouse Electric Corp., Pittsburgh, Pa., 1965.

Electrical Transmission and Distribution Reference Book, Westinghouse Electric Corp., Pittsburgh, Pa., 1964.

Eshbach, O. W., and M., Souders, *Handbook of Engineering Fundamentals,* 3rd ed., John Wiley & Sons, Inc., New York, 1975.

Faber, R. B. *Applied Electricity and Electronics for Technology,* 2nd ed., John Wiley & Sons, Inc., New York, 1982.

Fink, D. G., and H. W., Beaty, *Standard Handbook for Electrical Engineers,* 11th ed., McGraw-Hill Book Company, New York, 1978.

Ginsberg, G. L., *A User's Guide to Selecting Electronic Components,* John Wiley & Sons, Inc., New York, 1981.

Howes, M. J., and D. V., Morgan, *Large Scale Integration Devices, Circuits and Systems,* John Wiley & Sons, Inc., New York, 1981.

Krauss, H. L., C. W. Bostian, and F. H., Raab, *Solid State Radio Engineering,* John Wiley & Sons, Inc., New York, 1980.

Krutz, R. L., *Microprocessors and Logic Design,* John Wiley & Sons, Inc., New York, 1980.

Lurch, E. N., *Fundamentals of Electronics, 3rd edition,* John Wiley & Sons, Inc., New York, 1981.

Sen, P. C., *Thyristor DC Drives,* John Wiley & Sons, Inc., New York, 1981.

Sessions, K. W., *IC Schematic Sourcemaster,* John Wiley & Sons, Inc., New York, 1978.

Seymour, J., *Electronic Devices and Components,* John Wiley & Sons, Inc., New York, 1981.

Skilling, H. H., *Electrical Engineering Circuits,* 2nd edition, John Wiley & Sons, Inc. New York, 1965.

Sonde, B. S., *Introduction to System Design Using Integrated Circuits,* John Wiley & Sons, Inc., New York, 1980.

Surge Protection of Power Systems, Westinghouse Electric Corp., Pittsburgh, Pa., 1971.

Sze, S. M., *Physics of Semiconductor Devices,* John Wiley & Sons, Inc., New York, 1981.

Wildi, T., *Electrical Power Technology,* John Wiley & Sons, Inc., New York, 1981.

CHAPTER 7
CONTROLS

7.1 CONTROL THEORY

Control theory deals with methods for changing an output (or process) based on preestablished performance values (**set points**). A control loop can be generalized as shown in Fig. 7.1. The typical response of a controlled system will vary depending on the type of control exercised.

Automatic control can be broadly categorized into the following types:

- On–off
- Proportional (P)
- Integral (I) (also called reset control)
- Proportional plus reset (PI)
- Proportional plus integral plus derivative (PID) or error rate

7.1.1 On–Off Control

An **on–off controller** acts by applying full corrective action whenever an error is detected. It has only two states, one of which corrects positive deviations of the system, the other negative. The control signal applied is thus either 100% or zero. A system controlled in this way is always in a transient state, and oscillation between limits will always occur. The controller compensates for disturbances, applying its full signal when its set point is reached. The system performance will be dependent on the rate of system deviation (in the *off* condition) and the rate of system recovery after control signal initiation (the *on* condition). Figure 7.2 shows a typical time response of a system under on–off control. A common example is the furnace control in a home heating system.

7.1.2 Proportional Control

Proportional or **throttling control** (Fig. 7.3) is the simplest linear control, in which output of the controller is a linear algebraic function of its input. A change in the measured variable produces a proportionate change in the control signal. Thus

$$\text{control output} = K \times E \tag{7.1}$$

where K is a proportional constant and E is the error. Since the difference between

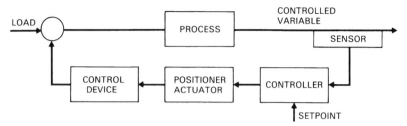

Fig. 7.1 Block diagram of a generalized control loop.

the measured variable and its set point controls the magnitude of the correction signal, once correction is under way, this signal is reduced, leading to progressively smaller correction signals; the difference between the measured variable and the set point is called **proportional offset:**

$$\text{proportional offset} = SP - MV \tag{7.2}$$

where SP is the set point and MV is the measured variable. The ratio of control signal change to measured variable change is called the **gain of the controller.** The **proportional band** of the controller is defined as 100 times the percent input change divided by the percent output change. Thus a gain of 0.1 is equivalent to 1000% proportional band, a gain of 1.0 is 100% proportional band, and so on. Low gain is equivalent to a wide proportional band, high gain to a narrow proportional band (Fig. 7.4).

7.1.3 Integral Control

The effect of proportional offset can be eliminated by using a control system in which the measured error signal is continuously integrated with respect to time. Recent-model, commercially available controllers with integral action, although not perfect

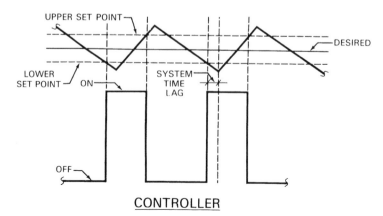

Fig. 7.2 On–off control.

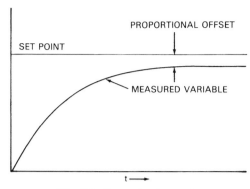

Fig. 7.3 Proportional control.

integrators, can closely approach a true integral for the periods of time under concern. The gain at very low frequencies is limited and commonly called the **reset gain** of the controller. Too low a value for this figure is undesirable because it can lead to **static offset.** Proportional control can be fast acting. However, it does give static offset. Integral control, on the other hand, minimizes static offset but is sluggish because of its attenuation of controller output at higher frequencies. A mode of control that can be used to give the best features of both is proportional plus integral.

7.1.4 Proportional Plus Integral Control

PI control is probably the most commonly used mode of control. Its principal disadvantage is that extended open-loop operation can lead to comparatively large overshoot in the integral part of the action through the effect known as *reset windup.* All modern controllers are designed to minimize this effect. PI control can be expressed

$$\text{controller output} = K \times E + K_I \times E \int dt \qquad (7.3)$$

where K_I is the constant of integration (adjustable). A typical time response of proportional plus integral (PI) controller is shown in Fig. 7.5, which also indicates how

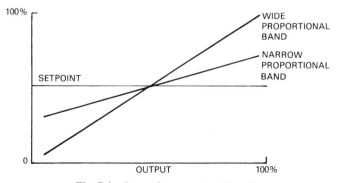

Fig. 7.4 Output for proportional bands.

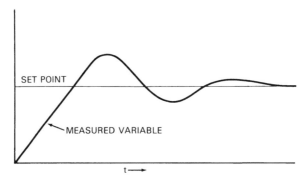

Fig. 7.5   Proportional plus integral control, "pure action."

the pure-action controllers would behave under the same stimulus; Fig. 7.6 shows their actual response.

7.1.5   Proportional Plus Integral Plus Derivative Control

In a third linear control mode used occasionally the output of the controller is proportional to the rate of change of the error signal. Thus, if an error begins to develop, a controller with *rate action* can anticipate and begin to take action before the error becomes large. If the action were perfect, the gain would increase steadily with the frequency, and the phase lead would always be 90°. This sort of control action would be highly undesirable, since high-frequency noise would be excessively amplified. Derivative controllers must, therefore, be modified to limit their action at high frequencies.

A pure derivative controller would not be a controller at all, since there is no unique relationship between error and control action. Derivative action is therefore used only to augment proportional or proportional-plus-integral controllers to improve their high-frequency response:

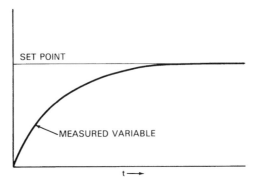

Fig. 7.6   Proportional plus integral control, "actual."

$$\text{controller output} = KE + K_I E \int dt + K_d \frac{de}{dt} \qquad (7.4)$$

where K_d is a rate constant (adjustable).

7.2 PROCESS CONTROL

Control systems can be modeled mathematically and the required characteristics of the control elements thus determined. The process, however, can be laborious and subject to inaccuracies in predicting performance and characteristics of the process. Thus, for **process control,** the **sensing elements, transmitters,** and **final control elements** are selected based on the process requirements. Off-the-shelf controllers are used, and the system is tuned during startup of the process to provide the mode or combinations of modes of control needed. With the recent advent of microprocessors and digital control, this has been greatly facilitated. Most commercially available controllers have all the control modes described previously either switch selectable, as modules, or by program selection.

Some **typical applications** of control modes to industrial processes are:

P	Blending to meet a desired composition
I	Speed control of centrifugal machines
	Liquid flow when a fast actuation is needed
PI	Majority of industrial processes
	Gas pressure
	Liquid flow with slow actuation
	Concentration controllers
PID	Temperature control where sensing lag is significant compared to process lag

7.2.1 Control Systems

Plant control systems can be visualized as a pyramidal hierarchy of control (Fig. 7.7) truncated, depending on the degree of centralized control desired.

Hierarchy of Control

Item	Level
Local/hand	Component
Subautomation (local loop)	Subsystem
Automation	System
Integrated plant control	Integrated plant

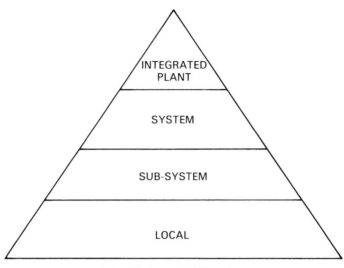

Fig. 7.7 Pyramidal hierarchy.

All the levels of controls are compatible with and typically form part of some degree of an integrated plant control system. The degree of central control can be varied widely depending on process or economic requirements.

Figure 7.8 and the table that follows it indicate broadly the degree of process control capability readily achievable both as **local** and **centralized control.** Control technology is subject to high rates of change (and obsolescence), as witnessed by the recent developments in computer chip capability, cathode ray tube (CRT), and communication technology. Further, as processes have become more complex, instrumentation and control requirements have increased.

Microprocessor-based control systems offer an attractive solution to many control problems because it is possible to **analog** (represent or duplicate) any of the P, PI, PD, or PID control modes by software programming and thus modify the control mode with minimum physical changes. They have the added advantage of low cost, thus permitting the incorporation of redundant microprocessors to improve reliability. With microprocessors it is possible to analog all elements in a single controller whose mode can be modified by software changes only; in some instances, several controllers can be housed in one case.

7.2.2 Degree of Automatic Control

The **degree of automatic control** applied depends on the process requirements, which may range from batch mode to continuous applications. Considerations include the length of time for batch processes, the economic consequences of outage time, and accuracy and degree of control required. Other significant factors include the response time required, the control limits, and the effect of variation within that for the value (or quality) of the product being produced. Other factors include system lag, the time required for control to take effect (a function of the combined response times of the control system and the process), and the precision with which control

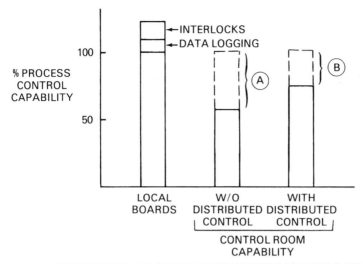

LIMITATIONS: (A) SPACE; INFORMATION HANDLING CAPABILITY
(B) ECONOMICS

Fig. 7.8   Local and centralized control.

Without Distributed Control	With Distributed Control
Advantages	*Advantages*
Lower capital cost	Lower installation cost
	Few operators, more control and data in central control room
Less sophisticated personnel requirements	Adaptive control strategies easier to implement
	Process control changes implemented largely by software changes (interactive systems)
Disadvantages	*Disadvantages*
Controls/data breadth limited by human capability	Higher capital cost
More operating personnel required	Information handling in control room greater and possibly more complex
Control changes cumbersome	More sophisticated maintenance required
	Some applications may require backup computer
	Increased operator training and technical support requirements

can be exercised. Because of lag time, sensitive systems frequently require antici-patory control.

Today, electronic systems are specified almost exclusively for large control sys-tems. Control systems are also classified as **analog** or **direct digital control** (DDC) types. An important and widely used subset of direct digital control is distributed digital control.

7.2.3   Analog Control

Analog control utilizes a continuous signal with respect to both value and time, whereas **digital control** normally operates in binary digits or bits, each bit equaling either 0 or 1. Typically, analog transmitters produce a continuous signal directly proportional to the variable; thus for digital control applications, analog-to-digital (A/D) converters are normally provided. Similarly, digital-to-analog converters (D/A) are used to adapt digital signals to analog process control units. With the extensive development and speed of digital data transmission and processing most analog control is actually exercised through digitized data.

7.2.4 Analog Application Considerations

These systems typically utilize 3 to 15 psig (48 to 103 kPa) air or low current (4 to 20 or 10 to 50 mA dc) as the variable signal. Analog control is simple and relatively easy to understand because of the limited number of process variables involved. Similarly, maintenance of the hardware circuitry is relatively simple. Analog controllers performing only a single function can achieve control very close to the set point. Analog instrumentation may be used to implement advanced control (i.e., control in terms of the final objective, such as energy optimization) rather than merely controlling flows, pressures, temperatures, and so on. The adjustability of the control modes are limited, however, and typically are as follows:

- *Proportional band:* 1 to 1000% (gain: 100 to 0.1)
- *Integral:* 0.02 to 100 minutes (50 to 0.01 minutes/repeat)
- *Derivative:* 0 to 10 minutes

The limitations are determined by controller pneumatic bellows sizing or electronic controller capacitor capability. Signals can also be provided at 1 to 5 V dc, ± 10 V dc, and on occasion ± 10, 20, and 50 mV dc.

Calculation accuracy for each element is typically 0.25 to 0.5%; thus, a multi-element calculation may have an overall accuracy of 2 to 4%, which is usually sufficient but may not be adequate for specialized applications. Analog systems tend to require more components, thus lowering overall reliability; and modifications to equipment are awkward once the equipment is installed. Providing compensation for transmitter failure is difficult and adds considerable complications to systems.

7.2.5 Direct Digital Control[1]

In **direct digital control** (DDC) a central computer performs the entire control function; the computer contains the control function or **algorithm** for all controls in the system. Field data in the form of analog inputs and outputs are transmitted to and from this central computer via "data highways." Because of the large number of data manipulations required, fast computers are necessary. Figure 7.9 shows direct digital control of a process flow. The computer receives an analog signal from the flow transmitter, converts this signal into a numerical value, compares this value with a set point, calculates the proper output signal, converts the result into an analog value, and sends a signal corresponding to this value to the control device. Indication of the value of the flow is performed by a computer-operated display device such as a cathode ray tube.

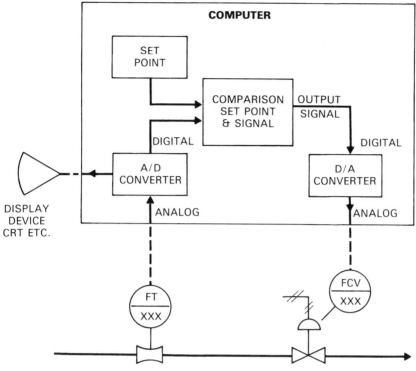

Fig. 7.9 Typical DDC control loop.

7.2.6 DDC Application Considerations[1]

The control modes and their settings are part of the microprocessor controller software; thus the combinations that may be used, proportional band, and reset and rate times are relatively unlimited. A digital computer system can take into account a significantly larger number of variables than an equivalent analog system. Calculation accuracy is better than in analog systems because the calculation is limited only by the accuracies of the transmitter and the D/A converters. Transmitter failure can be isolated from system performance such that it does not shut down the process but merely removes the element from control. Computer interface stations, which hold the last computer-calculated signal in the event of computer failure, are usually provided as analog backup devices between the computer and the controller. The pulse converter (D/A) may be considered as a computer manual (C/M) device, since it includes a manual operation mode allowing the operator to bypass the computer. To allow smooth transfer between local and supervisory control, the set-point output of the interface station can be fed back to the computer as an analog input.

A digital system can normally make on-line changes in the control strategy without shutting down the process, since control strategy is a function of computer software. Optimization algorithms involve many variables and calculations that are feasible only on a digital computer. DDC systems normally utilize CRTs as a display device. Interface with the computer by these displays improves and simplifies operator management of the control system. Since loss of computer also means the loss of any computer-operated display device, each important interface station should also contain a backup process indicator.

Large DDC systems tend to be slower than analog systems in execution. Thus the last loop variations from set point can be greater than for analog systems. DDC systems are more complex, difficult to understand, and difficult to maintain than analog systems. They require maintenance by technicians with an understanding primarily of electronics. Modularized software largely eliminates the need for reprogramming and permits relatively easy modifications after initial installation.

7.2.7 Distributed Digital Control[1]

With the recent development of economical, dependable microprocessors and miniaturized computers, it is increasingly practical to utilize distributed digital control for systems having from only 1 loop to those with 1000 or more loop applications. **Distributed digital control** distributes the control intelligence to devices at or near the point of control, and only exception and special requests are transmitted to and from the central control computer. As a result, slower **data highways** are satisfactory because of the lower quantity of data being handled. Distributed control highways can interfere with pneumatic, hydraulic, or electronic local control systems of the analog type, eliminating the need for all controls to be the direct digital type. For distributed digital control systems, large central processors are usually needed only for data reduction and evaluation but not for control.

Distributed digital control permits substantial savings in cable quantities, raceways, and so on, because it allows the process controllers to be placed close to the process; permits ready modification of control mode, loop performance, and display systems over the life of the plant; and broadly simplifies the problem of central control. Care must be given in the initial planning to space allocation, ambient environmental conditions, and data highway type and routings. Changes, loop additions, and other modifications can readily be accomplished, providing flexibility in application.

Both distributed and direct digital control can benefit from the use of data highways. All communication in these control systems is by a **single coaxial cable** (rather than multipair cables). The system can readily be expanded but has length limitations on the order of 3000 ft (900 m) between the computer and the farthest component. Usual points of connection with the data highway are **multiplexers** or **data terminals.** Some manufacturers offer prefabricated cables to simplify installation. In addition to the savings in wiring, its space requirements, and so on, data highways are relatively immune to electrical noise.

7.2.8 Functional Distribution Concepts

With small modularized microprocessor-based systems it is possible to add input/output modules, CRTs, and so on, and thus expand the capability readily. These modularized units are then interconnected through the data highways to provide an integrated control system. The near-term trends appear to be in the direction of heuristic (learning) control, more sophisticated networking, and improved diagnostic capability. Problems remain in the areas of equipment component standardization, commonality to permit different vendors' products to be used in a common system, and the availability of extensive, complete system documentation, communication checks, and to a lesser extent, system security (reliability).

7.3   CONTROL SYSTEMS EQUIPMENT

Control systems equipment includes sensors, transmitters, indicators, recorders, controllers, process computers, and final control elements. Frequently, functions are grouped in a single unit, and with the advent of microprocessors, this trend is accelerating.

7.3.1   Sensors

Sensors are available in two basic types: those in **direct** contact with the fluid for normal applications and the **indirect** (noncontact) type for applications that involve abrasive or corrosive materials, materials that tend to build up, or where leakage or contamination are undesirable. A frequently used system injects a **purge liquid** (or gas) and reads differentials between it and the measured variable, thus avoiding contact between the fluid and the measuring system. Typical **commercial sensor accuracy** is ±0.5% of range, with **precision sensors** available with accuracies of ±0.2% (or better) of the range.

Pressure sensors for analog systems tend to be of the bourdon tube type or in some cases diaphragm (bellows) type suitable for pressures up to the 50,000 psi (35×10^4 kPa) range. Electronic control systems use capacitance, inductance, and resonant wire primary sensing elements, due to their improved accuracy and direct conversion from measured variable to the process signal. Differential pressure sensors normally use coupled diaphragms.

Temperature sensors commonly used are bimetallic or filled fluid systems utilizing liquid, vapor, or gas. These can handle temperatures from −400 to +1400°F (−240 to 760°C). For high temperatures to +2700°F (1500°C), **thermocouples** are commonly used, whereas **resistance temperature detectors** (RTDs) are frequently used for midrange temperatures (−300 to +1200°F; −185 to 650°C). For electronic control systems, thermocouples and RTDs are used for virtually all temperature ranges.

Level can be determined by displacers or floats (either fixed or operating through a linkage or by stationary probes that electrically ground out when liquid levels reach them), differential pressure cells, or capacitance devices. Pressure cells are also frequently used with bubblers (measuring backpressure) for fluids that plug or coat normal sensors.

Density and **thickness** can be determined by absorption of radiation using a low-level source located on one side of a pipe (or barrier) and a sensor on the other. Where piping is carrying the fluid to be measured, rather careful calibration is needed to compensate properly for pipe wall thickness absorption, which is usually greater than changes in fluid density. Thickness can also be measured indirectly, although easily, by reference to a specific datum such as roll setting or by direct measuring devices such as calipers.

Flowmeters are available as turbine (bladed), target (paddle), vortex shedding, magnetic flow tube, or orifice types.

Other more specialized sensors are available for conductivity, relative humidity, sound intensity, vibration, various types of gas analysis (including gas chromatograph and infrared), pH, and similar variables.

7.3.2 Transmitters

Transmitters are commercially available either in combination with sensors or as separate units. Depending on the control mode employed, they can be obtained with 3- to 15-psi (48- to 103-kPa) pneumatic, 4- to 20- or 10- to 50-mA dc, outputs. Other versions also available yield 20- to 100-kPa signals. Transmitter signals are essentially proportional to the change in the measured variable. **Converters** are also available to convert pneumatic to electronic signals [pressure to current (P/I) and vice versa (I/P)], thus permitting adaptability and interchange of sensors.

7.3.3 Indicators

Indicators are sometimes included as an integral part of the sensor (local indicators), but remote indicator dials, vertical scales, counters, and so on, are available for mounting at other locations. They can accept mechanical, pneumatic, or electronic inputs and are available in a wide variety of types and sizes.

7.3.4 Recorders

Recorders are usually of the **circular** or **strip-chart type,** almost all today of the electronic type (both chart and computerized). Able to accept a variety of inputs, they are usually panel mounted and can include controllers.

7.3.5 Controllers

Controllers are available in designs similar to recorders but are frequently intended for panel mounting; thus they are often of the vertical scale type. To facilitate reading and reducing reading error, matching scale pointers are frequently utilized.

They are compatible with a wide variety of inputs and can be obtained with any combination of control modes desired, although, more recently, PID controllers with switchable mode selection are most widely used.

Ratio controllers, autoselector systems, cascade controllers, and computing devices are also available with functions that can be tailored to specific control requirements.

7.3.6 Scanners

Scanners are used to monitor multiple sensors and alarm when out-of-range conditions are detected. Frequently, they are high-speed type, some models scanning thousands of points per second. Although their functions are somewhat similar, scanners are considered distinct from **multipoint recorders,** which may have alarms. Some examples of scanners include fire eye, smoke detectors, and RTD or thermocouple scanning systems for bearing and electrical winding protection.

7.3.7   Trend Indicators

Trend indicators may be continuous or noncontinuous operating instruments. The noncontinuous type is normally wired to start at a preset level of the variable or upon command. Their purpose is to establish or display a trend of the variable and thus provide operators with process direction changes in advance of reaching process control limits. Frequently, they include some internal reduction (mathematical) capability. They are typically of the strip-chart variety in board-mounted cases up to 4 in. wide.

7.3.8   Annunciators

Annunciators are available in a variety of types producing audible, visual, or other types of signals. For control rooms, the **drop window** type is frequently used. When actuated, it provides a flashing light behind a translucent engraved window together with an audible signal. This continues until the operator silences the audible signal; the light at that time remains on, but no longer flashes. Subsequently, the light is extinguished when the alarm signal is terminated.

Normally, the annunciator windows are arranged in groups above the uppermost operating controls; large control rooms may have several hundred windows. The windows are engraved with the name of the problem (e.g., high bearing temperature, low water level) and are available in a variety of colors, although translucent white or cream is most common. When annunciators are specified, it is considered good practice to provide from 10 to 25% spares (unassigned) for future needs.

7.3.9   Data Loggers

Data loggers are normally of the printing type. Frequently, they operate noncontinuously, provide a record of data or events, and are triggered to function as event recorders. Normally electronic, they are capable of operating and/or recording data at high speed. They can accept input from a wide variety of sensors.

7.3.10   Accessories

A family of accessories is available for instrument systems, including alarms, ratio controllers, and computing devices. Alarms are available with most instruments and usually consist of electrical contacts whose settings are adjustable to operate at selected process signal levels.

7.3.11   Enclosures

Enclosures are available for most types of instruments for use in outdoor, corrosive, explosive, or other special environments. In some cases specially treated purge air systems are provided to pressurize instrument cases to avoid contamination. Enclosures provide mechanical and electrical protection for the operator and equipment. Brief descriptions of the more common types of enclosures follow. NEMA standards are used to designate the type of enclosure used. (See NEMA Standards Publication No. 250-1979 for more comprehensive descriptions, definitions, and/or test criteria.)

NEMA Type 1: For Indoor Use

Suitable for most applications where unusual service conditions do not exist and where a measure of protection from accidental contact with enclosed equipment is required. Designed to meet tests for rod entry and rust resistance.

NEMA Type 3R: For Outdoor Use

Primarily intended for applications where falling rain, sleet, or external ice formation are present. Gasketed cover. Designed to meet tests for rain, rod entry, external icing, and rust resistance.

NEMA Type 4: For Indoor or Outdoor Use

Provides a measure of protection from splashing water, hose-directed water, and wind-blown dust or rain. Constructed of sheet steel or stainless steel with gasketed cover. Designed to meet tests for hosedown, dust, external icing, and rust resistance.

NEMA Type 4X: Nonmetallic for Indoor or Outdoor Use

Corrosion-resistant glass polyester or similar construction. Excellent for applications where wind-blown dust, rain, hose-directed water, or splashing water are present. Designed to meet tests for hosedown, dust, external icing, and corrosion resistance.

NEMA Type 7: For Hazardous Gas Locations

For use in Class 1, Group C or D indoor locations as defined in the National Electrical Code, NEMA Type 7 enclosures must withstand the pressure generated by explosion of internally trapped gases and be able to contain the explosion so that gases in the surrounding atmosphere are not ignited. Under normal operation, the surface temperature of the enclosure must be below the point where it could ignite explosive gases present in the surrounding atmosphere. Designed to meet explosion, temperature, and hydrostatic design tests.

NEMA Type 9: For Hazardous Dust Locations

For use in Class II, Group E, F, or G indoor locations as defined in the National Electrical Code. Heat-generating devices within the enclosure are designed to maintain the surface temperature of the enclosure below a point where it could ignite the dust–air mixture in the surrounding atmosphere or cause discoloration of surface dust. These enclosures are designed to meet tests for dust penetration, temperature, and aging of gaskets (if used).

NEMA Type 12: For Indoor Use

Provides a degree of protection from dripping liquids (noncorrosive), falling dirt, and dust. Designed to meet tests for drip, dust, and rust resistance.

NEMA Type 13: For Indoor Use

Provides dust protection and protection against water, oil, or noncorrosive coolant spray. Designed to meet tests for oil exclusion and rust resistance.

NOTE. Enclosures are not designed to protect equipment against condensation, corrosion, icing, or contamination that can occur inside the enclosure or enter through a conduit or unsealed opening. Provisions to safeguard against such conditions must be made by the user.

7.4 APPLICATION CONSIDERATIONS

7.4.1 Automatic Data Collection

Automatic data collection (ADC) technologies include:

Conventional data acquisition	Bar coding
Optical character recognition	Radio-frequency data communication
Smart cards	Voice recognition
Biometrics	Magnetic stripe
Optical cards	Radio-frequency identification (RFID)
Touch memory	Machine vision
Two-dimensional scanners	

Among those most widely used are conventional data acquisition, bar coding, machine vision, optical character recognition, voice recognition, and magnetic strip technology.

Conventional data acquisition systems are available in a wide variety of forms. They can acquire, reduce, and display information in a variety of ways, depending on the type of control and data needed. Using a dedicated computer, usually internal to the system, they have a wide range of capability. A typical unit contains a range of signal conditioners to measure voltage, strain, pressure, temperature, and other parameters. With **digital signal processing** (DSP) capability, extensive signal filtering is available. High-speed data networks permit (by multiplexing) several hundred data channels to input simultaneously, with sample rates of 5 kHz per individual channel. Display capability includes CRTs, strip charts, alphanumeric, LCD, and so on. High-capacity internal data storage is available with capability to download to PCs for further data processing. Suitable software packages permit signal conditioning, monitoring, data acquisition, and data review.

Bar coding[2-5] utilizes one of three types of graphic patterns to store information, which may be numeric, alpha and numeric, or full **ASCII (American Standard Code**

for Information Interchange). Linear symbologies utilize a single row of bars and spaces, while **stacked** (multiscan) symbologies have several rows of bars and spaces. **Matrix** symbologies are regular polygonal arrays of data cells, plus "finder" and orientation structures that typically are read from two-dimensional images. Examples of each of these three symbologies are shown in Fig. 7.10.

Linear and stacked bar codes may have equal or varying width and spacing of either the bars or spaces, or both. Bar code design and density depend on the narrowest bar and space used, whose dimensions in turn are limited by the resolution of both the printing process used to create them and the scanning process used to read them. Low-density codes use narrow element widths of 20 mils and higher, while for medium-density codes these are about 10 mils, and for high-density codes, about 7 mils. Codes can have accuracy check features built into them. The matrix symbologies generally represent data with dark modules representing binary "1"s and light modules representing "0"s. Both stacked and matrix symbols can contain about 100 times more information than can linear bar codes.

Scanning devices can be either manual or automatic, fixed or portable. They can be in the form of pens, guns, or slot scanners. The automatic devices utilize oscillating or rotating mirror arrays to move the light source across the code. The decoder reads the reflected light as a "1" and the absorbed light as a "0," thus digitizing the signal. Codes can be read at up to 35 scans per second. Autodiscrimination of up to 15 symbologies is usually incorporated into the device. They have the capability of reading in intense ambient-light conditions, up to and including direct sunlight. Scanning is performed by a moving-band **helium–neon laser** (633 nm wavelength), laser diode (670 nm wavelength), **infrared** (>700 nm wavelength), visible red light (630 nm wavelength), or a linear **charged-coupled device** (CCD) to read the reflected or absorbed light from the code. Gun-type linear readers have a depth of field up to 40 in. (some special types up to 20 ft or more), with angular capability, both vertically and laterally, of up to 60°. Manually scanned wand readers generally work only in contact with the symbol.

Linear

Stacked

Matrix
(a)

Fig. 7.10 Bar code examples: (a) bar code types; (b) linear bar code examples; (c) matrix bar code examples. (Courtesy of MACtec-Morgan Adhesives Co., Roll Products Division.)

00 0 0012345 6666666666 7

CODE 128: Complex Product Identification

Benefits of Code 128 include:

- Uses least amount of label space for messages of 6 or more characters.
- Easiest code to scan.
- Highest message integrity (accuracy).

CODE 39 is the most frequently used bar code in **Industrial Applications.** Standard for **Vendor Parts Labeling** in the **Automotive Industry.**

A12345678B

CODABAR: codabar is used in **Specialized Applications** such as blood bag labeling, library systems and airbill tracking.

(b) (Continued)

Fig. 7.10

429

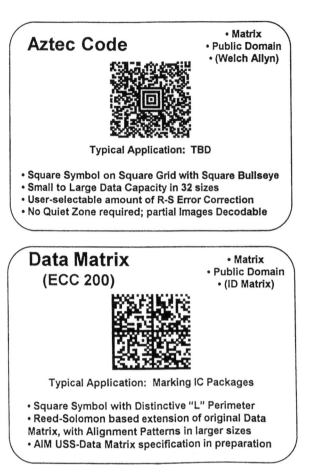

(c)

Fig. 7.10   (Continued)

Linear charge-coupled devices have been developed to provide high speed (throughput) and to avoid the resolution problems of moving-beam methods but have limited field widths. As a result, they require mounting close to the application being read, typically from 5 in. to 2 to 3 ft, depending on the lense system used to focus. Frequently, they use a CCD array of 2048 photoreceptors, generating an analog signal that is sent to a digitizer for processing.

Readers based on two-dimensional CCD image sensors are needed to read matrix bar code symbols. These CCDs are similar to the 600 × 400 pixel devices used in camcorders and security cameras, with lensing that typically provides a 4 to 10-in. depth of field. Image-based decoding also requires substantially more computing power than does linear scan–based decoding.

Machine vision[6] systems utilize several different methodologies: pattern matching (frequently using correlation and feature finding), dimensional inspection, guidance and machine control, and flaw detection (lighting and optics are involved here). Machine vision normally utilizes CCDs to acquire the significant dimensional or outline measurements of the object. Typical configurations use 512 × 480 to 1024 × 1024 pixel arrays for each of several cameras providing input to a host computer, which processes and then compares the data to its library and provides a readout of the data. The library data can include specific physical data such as template, line, curve, and arc data, camera gain, grayscale (128 or 256 gray levels) for a few applications, color capability, and a variety of pixel statistical data to facilitate identification. Image manipulation, including **morphing,** erosion and dilation, addition and subtraction, and similar features, are normally available. Color discrimination capability can also be included. Data input rates between 10 and 58.9 megabytes per second from each CCD are typical, and identification of simple parts are achieved in less than 50 msec, while complex parts in one plane require 2 sec. For the most complex parts, where color and color scale differences are involved, identification time ranges up to 10 sec. This speed can be increased by a factor of 10 to 100, depending on the processing hardware utilized. Host computers include the **heuristic** ability to facilitate later searches.

Machine vision systems are frequently combined with control systems for manipulation of items for processing, such as robotic control and machine control. Control systems increasingly utilize force, torque sensing, and measuring capability (including feedback loops), all controlled by the host computer.

Optical character recognition[5] (OCR) represents a somewhat specialized subset of the scanners described above. OCR is generally divided into data and text applications. **Data capture readers** utilize the technology above with specific machine-readable font, which can expected to be 100% accurate (one error in 3×10^6 characters) through the use of a **check digit verification** (CDV), where an additional character is added to the data string. Without CDV accuracies of one error in 2×10^4 characters is readily achieved. OCR scanners are normally used to read alpha, alphanumeric, or bar code fonts. They are frequently used in high-speed throughput systems such as bank check clearing. Usually, auto discrimination is programmed into the scanner or its reader, which may be integral with the scanner. OCR has high speed with perfect accuracy but limited flexibility. Figure 7.11 illustrates the typical group of fonts which these devices are able to read.

For text applications, the scanner must distinguish between all fonts, including different sizes and bold, italic, and similar differences in typefaces. **Text recognition software** is available with sufficient flexibility to read most text but with accuracies on the order of 99% and speed considerably slower than that of data capture readers. **Neural network** algorithms can read widely different text fonts and forms and offers great promise for further development and wide use. **Handprint recognition,** called **intelligent character recognition** (ICR), with its ability to read cursive handwriting, is undergoing increasing improvement and is likely to undergo significant improvements in the near term.

Voice data collection (also called **voice data entry** or **voice recognition**) utilizes a voice-processing module that compares the voice analog signal from the speaker to preprogrammed patterns (words). This permits interactive communication. The systems available are classified into several different types. **Speaker-independent systems** require the speaker to sound words in a uniform way among several users. These systems typically have a limited vocabulary of about 20 words but allow different persons to provide input. **Speaker-dependent systems** are those where the operator trains the system to recognize his or her voice. They are less sensitive to external noise and other voices and can handle non-English input. They are normally

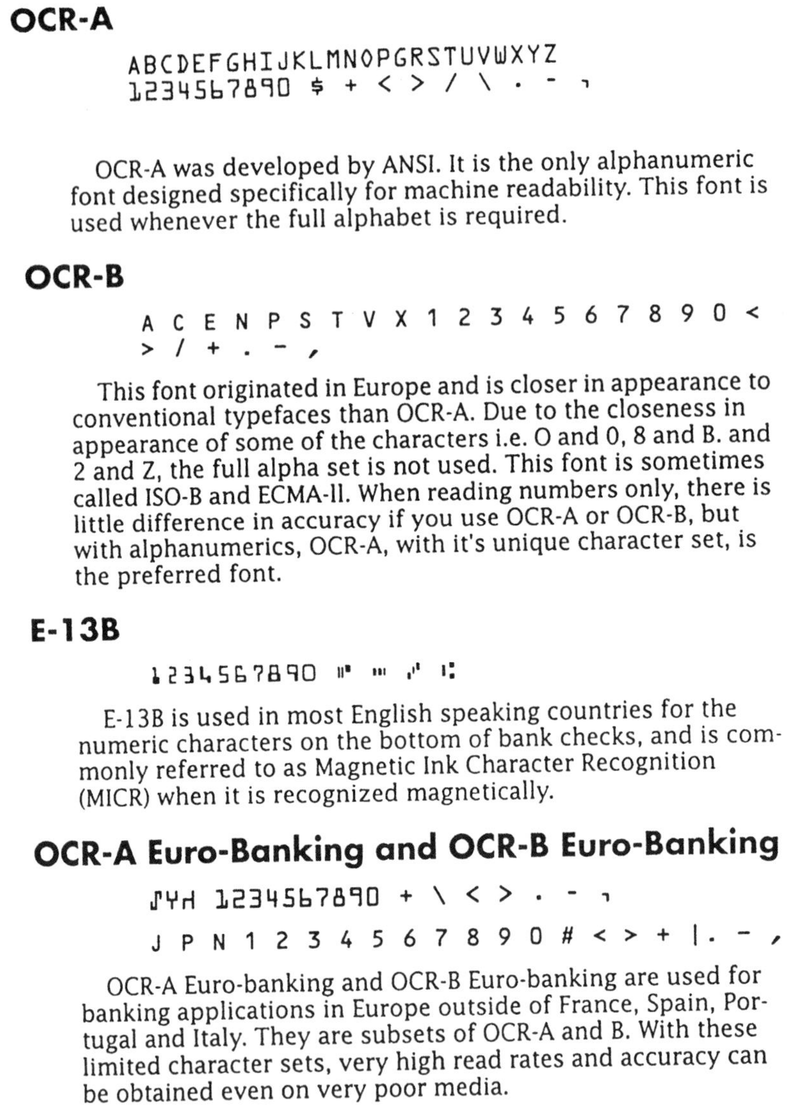

OCR-A

ABCDEFGHIJKLMNOPGRSTUVWXYZ
1234567890 $ + < > / \ . - ,

OCR-A was developed by ANSI. It is the only alphanumeric font designed specifically for machine readability. This font is used whenever the full alphabet is required.

OCR-B

A C E N P S T V X 1 2 3 4 5 6 7 8 9 0 <
> / + . - ,

This font originated in Europe and is closer in appearance to conventional typefaces than OCR-A. Due to the closeness in appearance of some of the characters i.e. O and 0, 8 and B. and 2 and Z, the full alpha set is not used. This font is sometimes called ISO-B and ECMA-ll. When reading numbers only, there is little difference in accuracy if you use OCR-A or OCR-B, but with alphanumerics, OCR-A, with it's unique character set, is the preferred font.

E-13B

1234567890 ⑂ ⑃ ⑄ ⑅

E-13B is used in most English speaking countries for the numeric characters on the bottom of bank checks, and is commonly referred to as Magnetic Ink Character Recognition (MICR) when it is recognized magnetically.

OCR-A Euro-Banking and OCR-B Euro-Banking

JYH 1234567890 + \ < > . - ,

J P N 1 2 3 4 5 6 7 8 9 0 # < > + |. - ,

OCR-A Euro-banking and OCR-B Euro-banking are used for banking applications in Europe outside of France, Spain, Portugal and Italy. They are subsets of OCR-A and B. With these limited character sets, very high read rates and accuracy can be obtained even on very poor media.

Fig. 7.11   Optical character reader fonts. (Courtesy of Caere Corporation Automated Data Entry.)

limited to use by one or only a few persons. **Discrete recognition systems** require a pause between words and that numbers be broken down into digits (e.g., 3–8–5–3). **Continuous recognition systems** do not require absolute precision in speech, and some systems use internal grammar programs to allow for different meanings of the same word (phonetic) sound or similar-sounding words.

Magnetic stripe[7] is the most widely used ADC method. This utilizes **digital** encoding on magnetic media of iron oxide, chrome oxide, cobalt, or other metallic film, typically in tracks $\frac{1}{10}$ in. in width. An important feature is that data can be reencoded. Because of the need to align the tracks with the readout device, media are normally configured in a linear pattern. Reading the stripe has two criteria: The reading head must contact the stripe, and relative motion (**swipe**) between the head and the stripe must be maintained, either manually or by a power-driven device. Readers can generally read between 4 and 55 in./sec. A reader encoder can perform error checking, and its output in **ASCII characters** permits ready adaptation and input to user programs. Standards exist for many track locations.

Cards are of two principal types, financial, as with credit cards, and access control cards, used where entry is restricted. Credit card media are generally low-coercivity encoded, typically with field strengths from 300 to 600 **oersteds** (Oe), while physical access cards encoding are generally between 1750 and 4000 Oe. Cards from 2700 to 4000 Oe are preferred because common magnets often have field strengths near 2000 Oe. Media using higher strengths are a bit more difficult to encode or erase but are more resistant to inadvertent damage.

Usable worldwide and best known are **ISO tracks** 1, 2, and 3 on credit cards holding 79 **alphanumeric** characters on track 1, 40 numeric characters on track 2, and 107 numeric characters on track 3. Tracks 1 and 3 are encoded at 210 bits per inch, and track 2 is encoded at 75 bits per inch. Track 1, developed by the International Air Transport Association (IATA), contains alphanumeric information for automation of airline ticketing or other transactions where a reservation database is accessed. Track 2, was developed by the American Bankers Association (ABA), contains numeric information for automation of financial transactions. This track is also used by most systems that require an identification number and a minimum of other control information. Track 3, developed by the thrift industry, contains information some of which is intended to be updated (rerecorded) with each transaction (e.g., cash dispensers which operate off-line).

Access control cards use different variations of the ISO standard, depending on the company and the need. Many applications exist for nonstandard use of the magnetic stripe card where data content, format, and densities can be changed to fit applications such as inventory, insurance, and medical uses.

7.4.2 Reliability and Redundancy

Reliability is critical to many process control applications. It is usually not practical to provide duplicate *control* devices, and further, these devices have proven to be relatively troublefree and in themselves highly dependable. **Sensors** and **transducers** tend to be the least reliable components in the loop, due mainly to environment. To achieve reliability, critical applications frequently have those components duplicated to provide reliability through **redundancy.** Special requirements for reliability are frequently achieved by hard-wired control circuits (rather than using electronic controls running through computers), the use of backup computers where the computer is used in an interactive control mode, and in some cases by redundant control systems where it is absolutely critical that one of the systems functions.

7.4.3 Fail-Safe Requirements

Fail-safe control systems assure that failure of a device or a control element does not adversely affect the operation of a system nor cause a hazardous condition. These can be broadly described as **stored energy systems,** those depending on **natural forces** or those in which the process is **self-compensating.**

Stored energy systems in many cases use springs, pneumatic fluid under pressure (such as bottled high-pressure gases), or in some cases explosive (chemical energy) actuators such as explosive valves for critical services. Other stored energy systems may use battery packs or chemical reactions to cause a control action. Examples of these are emergency lighting systems and the common soda–acid fire extinguisher.

Natural force systems frequently use gravity, for example, those systems in which a weight may be available but prevented from acting by a fusible link, as in fire doors.

Self-compensating systems are those in which the system tends to be self-regulating. An example of this is the float on a level control valve that uses the buoyant force of the controlled liquid acting on the float to provide the force necessary to operate the valve itself.

7.4.4 Requirements for Local Control

It is a matter of good practice to provide for **local control** wherever practical. In some cases the governing codes, frequently the National Electrical Code, will require that certain control devices be located within sight of the process or within sight of the operator so that in the event of a malfunction the energy source can be rapidly disconnected. The degree to which local control is incorporated depends not only on code requirements but also on the funding available for allocation to centralized process control. It may also depend on convenience when facilities are spread over a wide area, thus rendering local control inconvenient. Where processes are controlled from a remote location, it is good safety practice and often a legal requirement to provide sufficient warning signs, guards, and similar devices to avoid hazard to personnel or equipment in the area.

7.4.5 Interrelationships

Typically, basic plant service systems such as compressed air and service water are automated and controlled separately from other processes. To minimize operational difficulty, **control interconnections** between service systems and process systems should be minimized. Peripheral and secondary process systems are automated next, and, finally, the main process systems are themselves automated. This separation minimizes the potential for major difficulties because the building-block approach proceeds from those systems that are less complex and have the least impact on the process to those that are more sophisticated and have the greatest impact.

7.4.6 Interlocks

Although **interlocks** are used to protect downstream equipment and components, their control typically cascades upstream. Examples of this are conveyors, hydraulic controls, and level controls. These interlocks may be mechanical, electrical, hydraulic, or electronic.

7.5   CONTROL VALVES[8–10]

Control valve bodies are often at least one size smaller than the line in which they are installed, and it is not uncommon for them to be two sizes smaller. It is not good practice to install control valves with bodies smaller than two sizes less than the connecting lines unless a stress analysis is performed to ensure that the valve body is not overstressed due to thermal transients or dynamic forces.

To fit an accepted definition of a control valve, the valve body must be linked with an **actuator** mechanism which receives, and acts upon, a signal from the controlling system. One of the more common types of actuator uses a **spring/diaphragm** combination. Control valves are frequently provided with **shut-off (block) valves** on each side to permit them to be isolated for maintenance, and if continuity of service is important, a bypass globe pattern valve is also provided. Many control valves employ a **characterized plug** or **cage,** that is, a plug or cage that is specially contoured to provide some specific relationship between valve stem (and hence plug) position and valve flow area. To avoid **wire drawing** (seat damage) due to throttling, seats, and in some cases plugs, are hardened or hard surfaced. A typical control valve is shown in Fig. 7.12.

7.5.1   Valve Characterization

The **flow characteristics** of a control valve is defined as the relationship that exists between the **flow rate** through the valve and the **plug travel** from zero to 100%. The flow rate through the valve is also a function of the pressure drop across the valve. Since the pressure drop in an actual installation is usually not constant and ordinarily is different for each application, the recognized way to compare one valve with another is under similar conditions of constant pressure drop. When the pressure drop is held constant, the relationship of flow rate to plug travel is known as the **inherent flow characteristic.** There is also an **installed flow characteristic,** which is obtained under the varying pressure drop condition of an actual installation.

Although there are numerous inherent flow characteristics, only four of the more common ones are discussed here: quick-opening, linear, equal-percentage, and modified parabolic. They are shown graphically in Fig. 7.13.

The **quick-opening flow characteristic** provides maximum change in flow rate at low plug travels, with a fairly linear relationship. Additional increases in plug travel give sharply reduced changes in flow rate, and when the plug nears the wide-open position, the change in flow rate approaches zero. This characteristic is used primarily for on–off service where the flow must be established quickly when the valve begins to open. A typical example is in relief valve applications. In many applications it is possible to use a quick-opening characteristic where a linear characteristic would normally be specified, because the quick-opening characteristic is essentially linear up to about 70% of the maximum flow rate.

The **ideal linear flow characteristic** curve indicates a flow rate directly proportional to the plug travel. This proportional relationship produces a characteristic with a constant slope, so with constant pressure drop, the valve gain is the same at all flow rates. A linear characteristic is commonly specified for liquid level control and for certain flow control applications requiring constant gain.

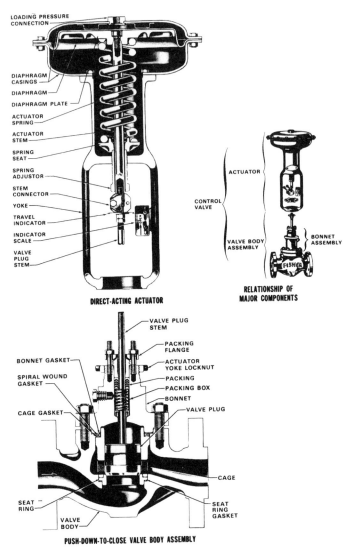

Fig. 7.12 Typical control valve assembly. (Courtesy of Fisher Controls International, Inc.)

With the **ideal equal-percentage flow characteristic,** equal increments of plug travel produce flow rate changes, which are equal percentages of the existing flow. The change in flow rate is always proportional to the flow rate that exists just before the change in plug position is made. When the plug is near its seat and the flow rate is small, the change in flow rate will be small; with a large flow rate, the change in flow rate will be large. These are generally used on pressure-control applications, and on other applications where a large percentage of the pressure drop is normally absorbed by the system itself with only a relatively small percentage available at the control valve. Equal-percentage valve characteristics should also be considered where highly varying pressure drop conditions can be expected.

The **modified parabolic flow characteristic** curve falls between the linear and equal-percentage characteristics. It provides close throttling action at low valve-plug

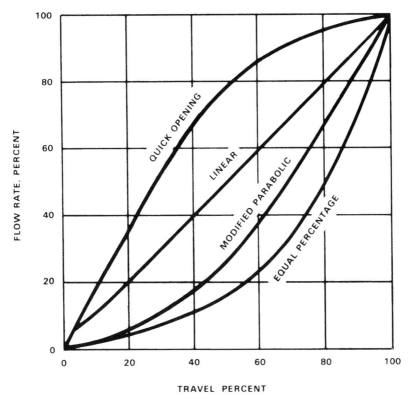

Fig. 7.13 Flow characteristic curves. (From Ref. 9. Courtesy of Fisher Controls International, Inc.)

travel and approximately linear characteristics for upper portions of plug travel. Linear or equal-percentage characteristics can be substituted with little change in performance.

Two coefficients are used for sizing and comparing control valve characteristics: C_v, valve sizing coefficient for liquids; and C_g, valve sizing coefficient for gases.

7.5.2 Liquid Flow[9]

$$Q = C_v \sqrt{\frac{\Delta P}{G}}, \tag{7.5}$$

where Q is the flow (gpm), ΔP is the pressure differential across valve (psi), G is the specific gravity of fluid (water = 1.0), and C_v is the valve flow coefficient (includes effects of flow area, discharge coefficient, etc.). C_v equals the number of U.S. gallons of water that will flow through the valve in 1 minute when the water temperature is 60°F (15.6°C) and the pressure differential across the valve is 1 psi (6.89 kPa). Thus C_v provides an index for comparing the liquid flow capacities of different types of valves under a standard set of conditions.

Control valves are available in **high** and **low (pressure) recovery** designs. **High recovery** is achieved by streamlining the internal design of the valve (e.g., high recovery—a V-notch ball valve; low recovery—a globe pattern valve).

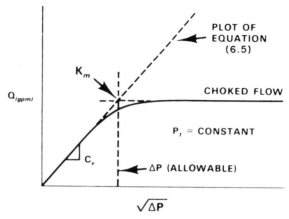

Fig. 7.14 Fow curve showing C_v and K_m. (From Ref. 9. Courtesy of Fisher Controls International, Inc.)

Choked flow is a condition where the flow of a fluid through a control valve remains constant when the pressure differential across the valve increases. Figures 7.14 and 7.15 show this condition. Although the exact mechanisms of the choking process are currently unknown, the presence of a certain amount of vapor in the liquid is associated with this phenomenon.

The point at which choking will occur is given by the following equation:

$$\Delta P_{\text{allow}} = K_m(P_i - r_c P_v) \qquad (7.6)$$

where ΔP_{allow} is the maximum allowable differential pressure for sizing purposes (psi), K_m is the valve recovery coefficient from the manufacturer's literature, P_i is the body inlet pressure (psia), r_c is the critical pressure ratio (determined from Figs. 7.16 and 7.17), and P_v is the vapor pressure of the liquid at body inlet temperature (psia).

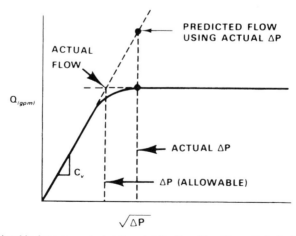

Fig. 7.15 Relationship between actual ΔP and ΔP allowable. (From Ref. 9. Courtesy of Fisher Controls International, Inc.)

(Vapor pressures and critical pressures for many common liquids are provided in Chapter 12.) After calculating ΔP_{allow}, substitute it into the basic liquid sizing equation $Q = C_v(\Delta P/G)^{1/2}$ to determine either Q or C_v. If the actual ΔP is less than ΔP_{allow}, the actual ΔP should be used in the sizing equation.

Use the curve in Fig. 7.16 for water. Enter on the abscissa at the water vapor pressure at the valve inlet. Proceed vertically to intersect the curve. Move horizontally to the left to read the critical pressure ratio r_c on the ordinate.

Use the curve in Fig. 7.17 for liquids other than water. Determine the vapor pressure/critical pressure ratio by dividing the liquid vapor pressure at the valve inlet by the critical pressure of the liquid. Enter on the abscissa at the ratio just calculated and proceed vertically to intersect the curve. Move horizontally to the left and read the critical pressure ratio r_c on the ordinate.

7.5.3 Gaseous Flow[14]

Sizing of control valves for gases is more complex due to problems of critical flow and gas compressibility. One prominent manufacturer has developed the following empirical formula to reflect valve performance over its entire pressure range and for either high- or low-recovery valves. Because of the use of several coefficients this formula must be used in conjunction with performance data from the manufacturer. This requirement is similar to but varies among the manufacturers since valve performance is sensitive to valve body design (C_g) as well as the gas characteristics for the specific application.

$$Q = 1.06 \sqrt{d_i P_i} \, C_g \sin \left(\frac{3417}{C_1} \sqrt{\frac{\Delta P}{P_1}} \right)_{deg} \qquad (7.7)$$

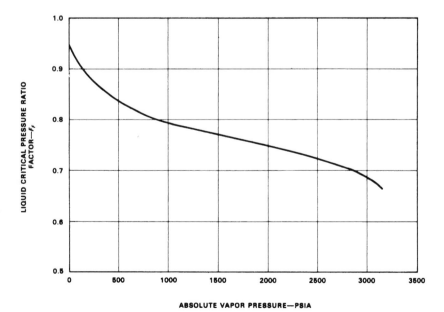

Fig. 7.16   Critical pressure ratios for water. (From Ref. 8. Courtesy of Fisher Controls International, Inc.)

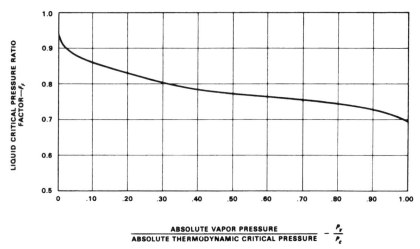

Fig. 7.17   Critical pressure ratios for liquids other than water. (From Ref. 8. Courtesy of Fisher Controls International, Inc.)

where $Q_{lb/hr}$ is the gas, steam, or vapor flow (lb/hr), d_i is the inlet gas density (lb/ft³), and C_1 ratio of gas to liquid flow coefficients, range 16 to 37. (High-recovery valves have low C_1 values, low-recovery valves have high C_1 values.)

7.6   LASERS*

Laser (*light amplification by stimulated emission of radiation*) technology is widely used for geometric positioning, communication, marking and cutting, materials processing, heating, welding, medical uses of a wide variety, reading of CD information, surveying, and similar activities.[11-13] Major laser classifications are gas, solid-state, ion, free-electron, and organic dye lasers. The material being lased can be in the solid, liquid, or gaseous state. The energy of a light source is focused on a material whose electron energy level is amplified to high intensity by stimulated emission. As the energy level decays to a steady state, some portion of the higher-level energy (by natural decay) is given off as **photons** of coherent, laser radiation. Commonly, the light source and the lased material are contained within a highly reflective shroud or cavity which reflects and concentrates the light energy on the lased material. With a mirror placed at each end of the cavity or shroud (**resonator** or **cavity mirrors**) and the reflectance of the cavity, the photons released by the lased material are reflected internally, causing other photons to be emitted and multiplying the effect. One of the end mirrors is coated to leak about 10% of the amplified light, which becomes the laser beam. While highly inefficient in overall energy conversion, with typical inputs of hundreds of watts to produce perhaps 1 W of output, the ability to produce a highly collimated and often high-energy beam is their principal and unique advantage.

---

*Abstracted from *Understanding Laser Technology,* 2nd ed., by C. Breck Hitz (1988, PennWell Publishing Co.). Used with permission.

The color of laser radiation or light is normally expressed in terms of the wavelength of the laser light in **nanometers** (nm), 1 nanometer being one billionth (e.g., 1×10^{-9}) of a meter, or 10 angstroms. Laser energy frequently is in or near the **optical** portion of the electromagnetic spectrum and ranges from 100 to 1000 nm. The term **micron,** μ (1 micron equals 1000 nm), is often also used to describe laser wavelengths. Laser radiation can be tuned to specific wavelengths by the use of optical filters or by the choice of laser materials, including the addition of specific elements **(doping)** to the light source (lased material). See Fig. 7.18.

Lasers are available as **pulsed** or **continuous** light sources. Normally, the light source provides continuous excitation to the laser material and the flow of energy from the laser is continuous. For certain applications where more power is required, a **Q-switch** is employed to release the energy in pulses; for example, a pulsed 100-W laser can produce 120 kW of peak power for a duration (pulse width) of 170 nsec. This amplification of three orders of magnitude (1000) is fairly average for pulsed laser systems. Alternatively, some lasers use **gas discharge lamps,** either dc arc lamps for continuous **pumping,** or **flashlamps** for pulsed pumping. Flashlamps discharge high current pulses (hundreds of amperes) with durations ranging from $\frac{1}{10}$ to 20.0 msec. Arrangements of mirrors and lenses are used (internally) to generate, maintain, and amplify the laser discharge beam.

Focusing, reflecting, or redirecting the laser beam will cause some of it to be absorbed by the target object, resulting in heating of the surface. If laser radiation is short (i.e., in the ultraviolet or blue region), **photochemical effects** occur in addition to the heating effect. Focusing concentrates the light beam, increasing its power density, and permits laser use in industry for such tasks as communication (fiber optics), cutting of metals and marking (often by actual vaporization of minute amounts of the target material), and welding.

7.6.1 Laser Types

Of all the lasers, one of the most common is the **gaseous HeNe laser.** In this same family, the **CO_2 laser** is also widely used. The HeNe operates at 632.8-nm wave-

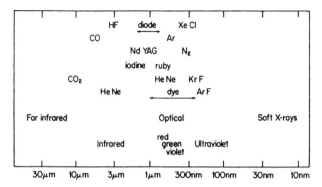

Fig. 7.18 Near-optical regions of the electromagnetic spectrum, with types of lasers located approximately at their operating wavelength or wavelengths. (From J. H. Eberly and P. W. Milonni, *Lasers.* Copyright © 1988 by John Wiley & Sons, Inc. Reprinted with permission.)

length, with average power outputs from a fraction of 1 to approximately 50 mW, while the CO_2 operates at 10.6-μm wavelength with average power outputs from milliwatts to over 20 kW. These higher-power outputs are particularly important for industrial applications such as metal cutting, heat treating, and welding.

Solid-state lasers (usually water cooled) typically are those generated by **Nd:YAG** (neodymium-doped yttrium aluminum garnet) and **Nd:glass,** both operating at 1.06-μm wavelength. For certain medical, military, scientific, and communications applications, erbium, holmium, chromium-doped sapphire (ruby), titanium-doped sapphire, and alexandrite solid-state lasing media are used. In addition, many other media are available, depending on the output wavelength desired.

Semiconductor diode lasers are the most widely used of lasers. They have the advantages of small size, are able to withstand rough handling, and are cheap. Two types found most widely are the GaAs/AlGaAs, with an output of about 800 nm, and GaAsP/InGaAsP, with an output between 1300 and 1550 nm. With the lack of a focusing structure, the beam width tends to be broader than other lasers, and since the geometry of the diode output is temperature dependent, they have less precision than do other laser types. Despite those characteristics and with an output as high as 1 W, they have found wide use as internal pumps for Nd:YAG and other solid-state lasers and for short-distance fiber optic systems.

Ion lasers are considered among the highest-powered continuous wave lasers currently available. They have power outputs as high as 30 W. The two common types are argon, with a blue line at 488 nm and a green line at 514.5 nm, and krypton, with a red line at 647.1 nm. The krypton laser can also produce blue, yellow, and green light. These lasers normally employ water cooling because of their high current density. They produce outputs between milliwatts and tens of watts.

Free-electron lasers have great development potential, with both tunability and the potential to produce large amounts of power. Operating on a slightly different principle from that of other lasers, they produce their energy by a form of particle acceleration produced by an alternating magnetic field. While they have the theoretical capability to generate very large quantities of power, they are very large machines, on the order of $6 \times 6 \times 15$ ft in size.

Organic dye lasers are a special class of lasers that use a large molecule organic dye in a liquid solvent. These lasers typically are optically pumped either by another laser or by a flashlamp. The gain bandwidth of this class of laser is quite wide, on the order of 100 nm, and can be tuned by prisms, gratings, and such. Their output, on the order of several tens of watts, and their tunable characteristic make them particularly suitable for photochemistry and spectroscopy.

7.6.2 Laser Safety

Laser beam energy tends to cause primarily thermal heating effects, and damage to unprotected skin is a significant concern. Similarly, since the human eye amplifies and concentrates light about 100 times, energy streams of laser light, which can have high power density levels, if aligned with or quite near the axis of the eye can be extremely damaging to vision and can result in blindness. As a result, the **ANSI Standard Z136 series** have been established to provide safety standards for lasers with wavelengths between 180 nm and 1 mm, based on the energy level and duration of exposure to the radiation.

The basis of the hazard classification scheme is the ability of the primary laser beam or reflected primary laser beam to cause biological damage to the eye or skin during intended use. For example, a Class 1 laser is considered to be incapable of producing damaging radiation levels during operation and maintenance and is, therefore, exempt from any control measures or other forms of surveillance. Class 2 lasers (low-power) are divided into two subclasses, 2 and 2a. A Class 2 laser emits in the visible portion of the spectrum (0.4 to 0.7 μm) and eye protection is normally afforded by the aversion response, including the blink reflex. Class 3 lasers (medium-power) are divided into two subclasses, 3a and 3b. A Class 3 laser may be hazardous under direct and specular reflection viewing conditions, but the diffuse reflection is usually not a hazard. A Class 3 laser is normally not a fire hazard. A Class 4 laser (high-power) is a hazard to the eye or skin from the direct beam and sometimes from a diffuse reflection and also can be a fire hazard. Class 4 lasers may also produce laser-generated air contaminants and hazardous plasma radiation.

It must be recognized that the classification scheme relates specifically to the laser product and its potential hazard, based on operating characteristics. However, the conditions under which the laser is used, the level of safety training of those using the laser, and other environmental and personnel factors are important considerations in determining the full extent of safety control measures. Since such situations require informed judgments by responsible persons, major responsibility for such judgments is assigned to a person with the requisite authority and responsibility, the laser safety officer (LSO).

Laser hazard classification based on power density, W/cm^3, has established four levels, Classes 1 through 4. These limits, with a description of the safety levels, are shown in Table 7.1. Since all commercial laser systems have enclosures, with some form of safety shutter or cutoff when opened, concerns for laser light safety applies principally to open laser systems as found in laboratory or developmental work. In

Table 7.1 Maximum Permissible Exposure Limits

Class 1—A Class 1 laser is considered safe based on current medical knowledge. This class includes all lasers or laser systems that cannot emit levels of optical radiation above the exposure limits for the eye under any exposure conditions inherent in the design of the laser product. There may be a more hazardous laser embedded in the enclosure of a Class 1 product, but no harmful radiation can escape the enclosure.

Class 2—A Class 2 laser or laser system can emit a visible laser beam which because of its brightness will be too dazzling to stare into for extended periods. Momentary viewing is not considered hazardous since the upper radiant power limit on this type of device is 1 mW, which corresponds to the total beam power entering the eye for a momentary exposure of 0.25 second that is safe (i.e., the MPE for a 0.25-second exposure).

Class 3—A Class 3 laser or laser system can emit any wavelength, but it cannot produce a diffuse (scattered) reflection hazard unless focused or viewed for extended periods at close range. It is also not considered a fire hazard or serious skin hazard. Any continuous wave (CW) laser that is not Class 1 or Class 2 is a Class 3 device if its output power is 0.5 W or less. Since the output beam of such a laser is definitely hazardous for intrabeam viewing, control measures center on eliminating this possibility.

Class 4—A Class 4 laser or laser system is one that exceeds the output limits (accessible emission limits, AELs) of a Class 3 device. As would be expected, these lasers may be either a fire or skin hazard or a diffuse reflection hazard. Very stringent control measures are required for a Class 4 laser or laser system.

Source: Copyright Laser Institute of America, Orlando, Fla. Adapted with permission from American National Standard Z136.1 (1986).

fact, the major concern in utilizing commercial laser systems is hazard from the electrical supply system.

7.7   HUMAN–MACHINE INTERFACE[14–17]

Limitations of Humans

- *Accuracy:* susceptible to constant and variable errors.
- *Speed:* time required for decision and movement.
- *Force:* depends on body member in use and fatigue.
- *Computing:* slow, inaccurate (limited to single integration and differentiations).
- *Decision making:* optimum strategy not always used; preservation exists.
- *Information input rate:* susceptible to overloading; stress and boredom affect performance.

Advantages of Humans

- *Detection:* can detect a wide range of signals, strong to weak.
- *Perception:* can see through complex situations, have constancy, and can detect signals through noise.
- *Flexibility:* can shift rapidly in attention, can revert to alternate modes of operation.
- *Judgment:* have inductive reasoning, incidental intelligence, "hunches."
- *Reliability:* satisfactory performance under adverse conditions is possible; can perform when parts are out of order; performance good when highly motivated.

Limitations of Machines

- Maintenance is required.
- Monitoring is required.
- Decision making is limited.

Advantages of Machines

- Speed is possible.
- Accuracy is possible.
- Short-term memory.
- Simultaneous activities are possible.
- Complex problems can be handled.
- Good for repetitive tasks.
- Do not tire.

A human being is more desirable than an automatic regulator only when the eye and brain take an essential part in the regulation process (e.g., in flying an aircraft in difficult situations or driving a car in traffic). However, regulation of temperature, sound volume, or physical quantities can be performed better by an automatic device.

Humans have two faculties that cannot be matched by technical transmission systems:

- The human eye is extremely efficient, especially in visual acuity.
- The connection between eye and brain makes possible a logical discrimination of things that are seen.

7.8   PANEL DESIGN

Panel design must consider the characteristics of the human body and the functions to be exercised.

7.8.1   General Specifications

Height.   For optimum results, locate panel devices below eye level but not lower than a sight line 30° below horizontal. Easy eye movements are from horizontal to 30° below.

Distance.   Optimum reading radius is 18 to 22 in. (0.4 to 0.6 m). Standard displays are designed to be read at a 28-in. (0.7-m) radius from the eye but distance can be greatly increased if the display is designed accordingly and reach is not a factor. Minimum acceptable reading distance is 13 in. (0.3 m). Normal reading distance for standard cathode ray tubes is 14 to 18 in. (0.35 to 0.46 m).

Compactness.   Avoid extended displays to minimize scanning.

Priority.   Locate important and frequently used instruments nearest the normal line of sight, which varies for the seated or standing posture.

Standardization.   Primary instruments should be located in standard arrangement from one situation to another to prevent error in reading.

Associated Manual Controls.   Associated knobs and switches should be located below the instrument or to the right to prevent visual interference due to hand movements.

7.8.2   Panels

Panels are frequently laid out to indicate the process being controlled. In the arrangement called **mimic panels,** main process flow paths and equipment are indicated in schematic form, and instruments, status lights, and so on, are located approximately matching the process locations of the sensors, control elements, and so on. **Semi-**

mimic panels are a more generalized layout concept and are usually substantially cheaper to produce. For processes subject to change, several manufacturers offer a modular grid mimic panel where grid elements can be changed to suit process changes. More recently, the expanded use of CRTs to display process diagrams, including in some cases flow rates and temperatures, has reduced the interest in mimic panels. The programming costs and complications must be considered when making the decision to use CRTs for this purpose.

CRT Consoles.   Several manufacturers have begun offering CRT consoles as an alternate (small systems) or supplement (large systems) to conventional control panels. The CRTs partially or completely replace the panels, switches, annunciators, and so forth, and utilize one or more color CRTs with an associated typewriter-style keyboard. The CRTs display either the manufacturer's standard or customized formats and graphics. The displays are in color and can incorporate system graphics, performance and status data in bar graph form, and flashing (alarm) indications. They can also permit selection of level of detail displayed as well as specific parameters of interest. Through the associated keyboard, set points can be changed, motors started, valves positioned, and so on. In some cases, **interactive graphics** utilizing either light pens, a mouse, or touch pads are provided. These devices permit the operator to actuate components displayed on the CRT. They can provide overview data, trend data, and permanent records via line printers. Although providing major space savings and improving operator span of control, CRT consoles are more costly than conventional instrumentation and may require more programming and operator training.

7.8.3   Instruments

Kinds	Reading Error (%)
Counters	0.4
Open reading dials	0.5
Round dials	10.9
Semicircular dials	16.6
Horizontal dials	27.5
Vertical dials	35.5

Preference.   Direct-reading counters and circular dials are best; all linear scales lead to errors.

7.8.4   Round Dials

- Standard dial face diameters of 2¾ to 3 in. (70 to 76 mm) are recommended.
- A fixed dial with a movable pointer is superior to a moving face and a fixed pointer, which is not readable during rapid motion.
- All dials should increase from zero in a clockwise direction.

- Locate the zero value at the 9 to 12 o'clock positions. Locate the zero value at the 12 o'clock position if the values go from plus to minus.
- Dials with the fewest markings can be read the most rapidly.
- Scales numbered by intervals of 1, 10, 100, and so on, and subdivided by 10 graduation intervals are superior to other acceptable scales. Avoid irregular scales, except logarithmic and special scales.
- Except for locations higher than the head, titles should be placed above instruments and manual controls to prevent visual interference.

7.8.5 Pictorial Indicators

Pictorial indicators are used to assist in interpretation of special relationships [e.g., to symbolize aircraft attitude (i.e., orientation in space), pitch, and roll].

7.8.6 Signal Lights

- Alerting with lights.
- Signal light colors:
    Red—critical, malfunction
    Amber—cautionary
    White—general status
    Green—safe
    Yellow—caution
- Warning lights for critical functions must be within 30° of the normal line of sight. A central warning light may be used to direct attention to another panel area.
- Keep number and colors to a minimum.
- Locate on or near the associated control to facilitate appropriate action.
- Normal operating signals should be arranged in patterns or superimposed on diagrams compatible with the equipment they symbolize.

7.8.7 Cathode Ray Tubes

Scope Size (Diagonal). The standard scope size is 5 in. (125 mm) minimum; 19 in. (480 mm) typical; 25 in. (623 mm) maximum.

Scope Position. The viewing distance is 12 to 18 in. (300 to 450 mm). For 25-in. (635-mm) screens, distances to 6 ft (1.8 m) are satisfactory. The screen should be perpendicular to the normal line of sight or not greater than 30° off.

7.8.8 Audible Signals

Buzzer. A buzzer is used in quiet locations and commands attention without causing undue alarm (e.g., individual operator alert).

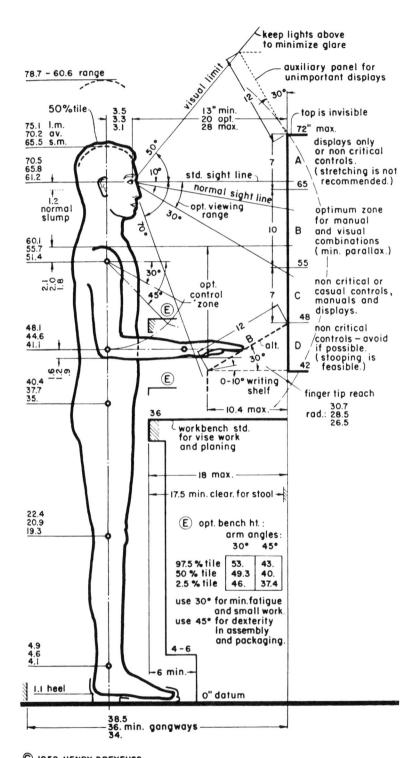

Fig. 7.19 Anthropometric data, adult male standing at control board. (From Ref. 16. © 1959, 1960, 1967, by Henry Dreyfuss & Associates. Reprinted by permission of The Whitney Library of Design.)

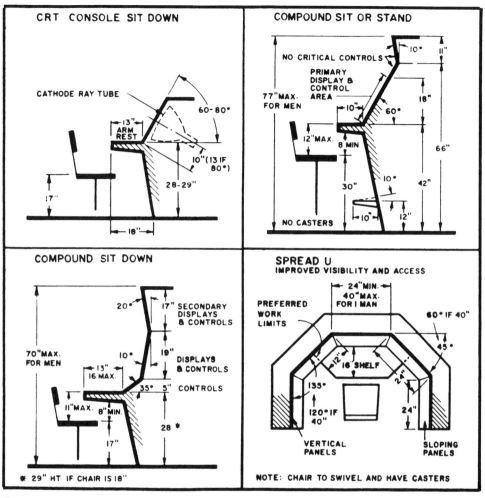

Fig. 7.20 Preferred console dimensions.[15]

Bells. Bells produce a penetrating (low-frequency) noise. Abrupt onset demands attention and fast response (e.g., fire alarm).

Chimes. Chimes are used for nonurgent actions. Chimes do not cause undue alarm.

Tones. Tones can be used over an electric intercom system.

Recommended Design Practice

- Warning signals should be at least 10 dB above noise level.
- Signals should not exceed 110 dB unless ear protectors are used. Preferred frequency, 200 to 5000 Hz (sensitive range of hearing).

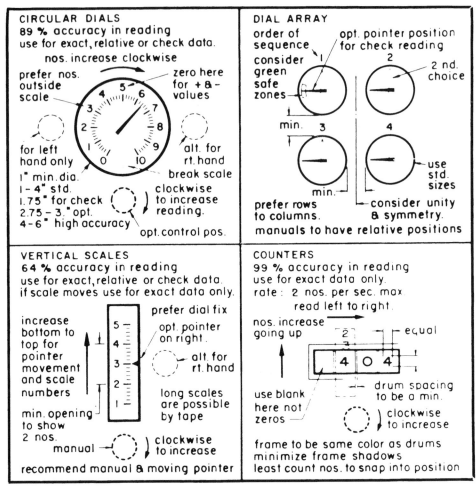

CIRCULAR DIALS
89 % accuracy in reading
use for exact, relative or check data.
    nos. increase clockwise
prefer nos.   zero here
outside   for + & -
scale   values
                                 alt. for
for left   rt. hand
hand only   break scale
1" min. dia.
1 - 4" std.   clockwise
1.75" for check   to increase
2.75 - 3." opt.   reading.
4 - 6" high accuracy
                                 opt. control pos.

DIAL ARRAY
order of   opt. pointer position
sequence   for check reading
consider   2nd.
green   choice
safe
zones
    min.
prefer rows   use
to columns.   std.
                                   sizes
                        consider unity
                        & symmetry.
manuals to have relative positions

VERTICAL SCALES
64 % accuracy in reading
use for exact, relative or check data.
if scale moves use for exact data only.
                        prefer dial fix
increase   opt. pointer
bottom to   on right.
top for
pointer   alt. for
movement   rt. hand
and scale
numbers   long scales
                        are possible
min. opening   by tape
to show
2 nos.
     manual   clockwise
                        to increase
recommend manual & moving pointer

COUNTERS
99 % accuracy in reading
use for exact data only.
rate : 2 nos. per sec. max
    read left to right.
nos. increase
going up   equal

use blank   drum spacing
here not   to be a min.
zeros   clockwise
                          to increase
frame to be same color as drums
minimize frame shadows
least count nos. to snap into position

© 1960 HENRY DREYFUSS

Fig. 7.21 Basic display data. (From Ref. 16.)

7.8.9 Anthropometric Data

Standard anthropometric data together with information on displays and panel design
are shown in Fig. 7.19.

7.8.10 Consoles

Selected dimensional data on consoles are shown in Fig. 7.20.

7.8.11 Basic Display Data

Orientation and other data on basic displays are shown in Fig. 7.21.

REFERENCES

1. C. Warren, *A Comparison of Analog and Digital Control Techniques,* Fisher Controls International, Marshalltown, Iowa, 1979.

2. A. Longacre, "Bar Coding," Welch Allyn, Inc., Skaneateles Falls, N.Y., Private communication, 1996.

3. J. Feinstein, *Basic Bar Code Technology,* MACtec-Morgan Adhesives Co., Stow, Ohio, 1993.

4. S. Itkin, *Basic Bar Coding, Scan-Tech 93,* Symbol Technologies, Inc., Bohemia, N.Y., 1993.

5. *Guide to Bar Code and OCR Technology,* Caere Corp., Los Gatos, Calif., 1988–94.

6. J. Dulchinos, *Machine Vision,* Adept Technologies, San Jose, Calif., private communication, 1996.

7. M. Guillen, *Magnetic Stripe,* American Magnetics Corp., Carson, Calif., private communication, 1996.

8. *Control Valve Handbook,* 2nd ed., Controls International, Inc., Marshalltown, Iowa, 1977.

9. *Technical Monographs 29, 30, and 31,* Fisher Controls International, Marshalltown, Iowa.

10. J. W. Hutchinson, *ISA Handbook of Control Valves,* 2nd ed., Instrument Society of America, Pittsburgh, Pa., 1976.

11. Peter W. Milonni and Joseph H. Eberly, *Lasers,* John Wiley & Sons, Inc., New York, 1988.

12. *Industrial Strength Laser Marking, Turning Photons into Dollars,* Excel Control Laser, Orlando, Fla., 1992.

13. C. B. Hitz, *Understanding Laser Technology,* 2nd ed., PenWell Publishing Company, Tulsa, Okla., 1989.

14. W. E. Woodson and D. W. Conover, *Human Engineering Guide for Equipment Designers,* University of California Press, Berkeley, Calif., 1964.

15. E. J. McCormick, *Human Factors Engineering,* 3rd ed., McGraw-Hill Book Company, New York, 1970.

16. H. Dreyfus, *The Measure of Man,* Library of Design, Watson-Guptill Publications, New York, 1967.

17. J. Hammond, *Understanding Human Engineering: An Introduction to Ergonomics,* David and Charles, Inc., North Pomfret, Vt., 1979.

CHAPTER 8
ECONOMICS/STATISTICS

8.1 INTEREST

Interest is the time value of money, often referred to as the **cost of money.**[1-3]

8.1.1 Simple Interest

Simple interest assumes that the debt (principal) is repaid at the end of the period with interest earned over the period. Thus

$$I = Pni, \tag{8.1}$$

where I is the interest and P is the principal, and

$$T = P(I + ni), \tag{8.2}$$

where n is the time (usually number of years), i is the interest rate per unit time (usually per year), and T is the total amount due.

8.1.2 Compound Interest

Compound interest assumes that the periodic interest payments are added to the principal and subsequent interest is earned on the resulting larger principal:

$$P \text{ at end of period} = P(1 + i)^n. \tag{8.3}$$
$$I \text{ earned during period} = P(1 + i)^{n-1}i. \tag{8.4}$$
$$T \text{ due at end of period} = P(1 + i)^n. \tag{8.5}$$

As a matter of convenience, interest rates compounded daily are often based on a 360-day year to permit simplified conversion to annual rates, rather than using a 365-day year. Also, extensive tables are available for interest and principal values (factors).

A useful rule of thumb for quickly estimating the effect of compound interest is the **rule of 72.** If 72 is divided by the interest rate, the answer is a very close approximation of the number of periods required to double the principal. Thus at 6% annual interest, compounded, the principal doubles in 12 years, at 8%, 9 years, and so on.

453

8.1.3 Present Worth

Present worth is the value of a principal to be paid (or redeemed) at a future date, the principal bearing compound interest during the intervening time:

$$P_w = \frac{1}{(1 + i)^n},$$

(8.6)

where P_w is the present worth of the principal. (Tables of P_w are given in Section 8.9.)

8.1.4 Discounted Value

The **discounted value** is the present value of a principal to be redeemed or paid at a future date. The **discount rate** (d) is the discount on the principal of 1 over one unit of time. Thus

$$i = \frac{d}{1 - d},$$

(8.7)

where d is the discount rate and i is the effective interest rate.

8.1.5 Annuities

An **annuity** is a series of equal periodic transactions that earn interest. The calculation of annuities can involve either deposits or withdrawals.

Amount of Annuity

$$A_n = \frac{(1 + i)^n - 1}{i},$$

(8.8)

where A_n is the amount of the annuity. (Tables of A_n are given in Section 8.9.)

Present Worth of an Annuity

$$P_w = \frac{1 - (1 + i)^{-n}}{i}.$$

(8.9)

(Tables of P_w are given in Section 8.9.)

8.2 DEPRECIATION

Depreciation represents a charge reflecting the deterioration in a capital asset.[1-3] In business this permits accumulation of funds to replace assets as they wear out. Another form of deterioration is **obsolescence,** which represents a limited economic life of an asset due to technological change. Both of these can be provided for by setting aside funds usually called *reserves for depreciation* to permit replacement of the asset. Allowable periods of write-off for depreciation permitted by the Internal Revenue Code do not necessarily reflect the actual economic life of the asset, but may be structured to encourage investment or plant modernization as a matter of governmental policy. Depreciation allowances may be calculated in several ways. (Note that none of these formulas make allowance for the effect of inflation; i.e., the new asset costs more than the asset being replaced. To some extent sinking funds that earn interest can compensate for the effect of inflation, however.)

8.2.1 Straight Line

The simplest, most widely used formula, the **straight-line** method, assumes uniform depreciation over the economic life of the asset:

$$d = \frac{C - C_S}{L},$$

(8.10)

where d is the depreciation per year, C is the original cost, C_S is the salvage value, and L is the life (years).

8.2.2 Sinking Fund

The **sinking fund** method assumes that a sinking fund receiving periodic payments and earning interest is established. Thus

$$d = (C - C_S) \frac{1}{S_L}.$$

(8.11)

8.2.3 Accelerated Depreciation

Accelerated depreciation assumes that the asset declines more rapidly in value during its early years. Several types are commonly used: for example, double declining and sum of the digits. **Double-declining depreciation** utilizes deductions of twice the straight-line amount applied in each year to the remaining balance, thus accelerating the amount of depreciation deducted during the early life of the asset. For an asset with a 10-year life, the depreciation deducted is 20% the first year, 16% (20% of the remaining 80%) the second year, and so on.

The **sum-of-the-digits** method utilizes the life of the asset as the numerator and the sum of the digits of the life of the asset as the denominator. Thus for a 10-year life, the sum of the digits 1 through 10 is 55, and the depreciation taken the first year is 10/55, the second year, 9/55, and so on:

$$d = \frac{n'}{n[(n + 1)/2]},$$

(8.12)

where d is the depreciation taken in year n', n' is the particular year, and n is the life of asset remaining (number of years).

8.3 STATISTICS

8.3.1 Measures of Central Tendency

The two most useful measures of central tendency are the arithmetic mean and the median. The **arithmetic mean**

$$\bar{x} = \sum \frac{x}{n}, \tag{8.13}$$

where $\bar{x}$ is the arithmetic mean, x are the individual observed values, and n is the number of values observed.

The **median** is the individual observed value that has an equal number of observed values above and below it; that is, it is the middle observed value in a series. In small or skewed statistical populations, it may vary widely from the arithmetic mean, whereas in large nonskewed populations it will closely approach the arithmetic mean. **Population** is the term applied to the entire group from which characteristics will be determined.

8.3.2 Measures of Dispersion

Observations will tend to cluster around the arithmetic mean with the degree of their dispersion defined as the **standard deviation:**

$$s = \sqrt{\frac{\Sigma(x - \bar{x})^2}{n - 1}}, \tag{8.14}$$

where s is the standard deviation. For normal (nonskewed) distributions (Fig. 8.1), a predictable proportion of the values will fall within a range of standard deviations:

Range	Percent of Values Included
$-1s$ to $+1s$	68.26
$-1.96s$ to $+1.96s$	95.0
$-2s$ to $+2s$	95.46
$-3s$ to $+3s$	99.73

The **coefficient of variation** is a useful measure to compare dispersion of data from different sources or bases. This can be particularly useful when establishing correlation:

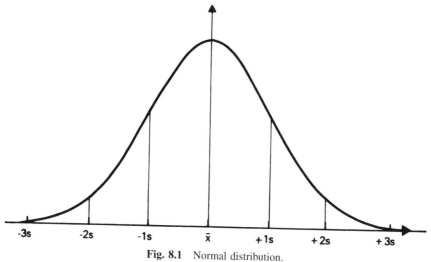

Fig. 8.1 Normal distribution.

$$V = \frac{s}{\bar{x}}, \tag{8.15}$$

where V is the coefficient of variation.

8.3.3 Sampling

Sampling provides a method by which the characteristics of a population can be predicted by examination of only a small portion of the population. To achieve this it is necessary that the sample be representative (usually achieved by random sampling). Where all items in the population are examined, the sample is 100%. This is sometimes also referred to as *exhaustive sampling.*

Sampling frequently uses a single sample whose size varies depending on the size of the population (or lot) and the **confidence level** desired. The confidence level desired will have the larger effect on sample size required. Predetermined acceptance levels are compared to the sample characteristics to establish acceptability or rejectability of the lot.

Where a **single sample** indicates noncompliance with acceptable criteria, **double** (or **multiple**) **sampling** is frequently employed. This requires larger progressive samples with a reduced proportion of nonacceptable items permitted.

After determining the **acceptable quality level** (AQL) (the level of defects acceptable) together with the **confidence level** (CL) of achieving or bettering that level, standard published tables of acceptance sampling plans can be referred to for lot sizes, sample sizes, acceptance and reject quantities for each, and so on. See Section 8.4.

8.3.4 Histogram

A **histogram** is a powerful diagnostic tool to establish causes of events. Where events are grouped by cause (or type) and the number of occurrences of each are plotted as

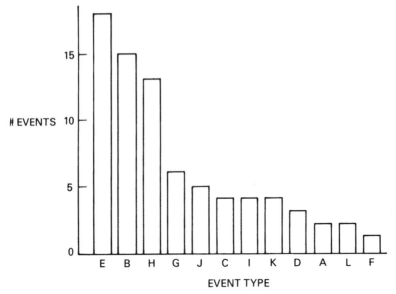

Fig. 8.2   Histogram.

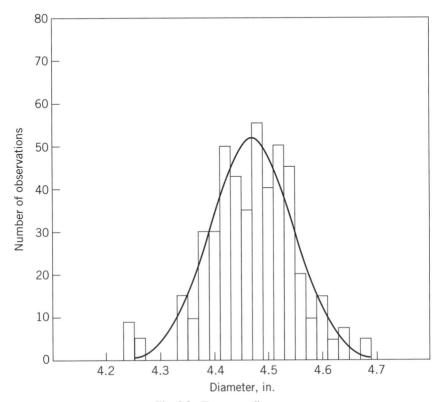

Fig. 8.3   Frequency diagram.

a simple bar graph, the diagram will clearly establish those items that occur most frequently and that deserve first attention. As is apparent in Fig. 8.2, event types E, B, an H are the major repetitive ones and should be given first attention. Of the 80 events, 46 are of only three types, while the remaining 34 comprise 10 different event types. The **Pareto principle,** which describes this, states that only a few causes account for the majority of the events. When testing data, a rule of thumb often used is "20% of the causes generate 80% of the events" the **80–20 rule.**

8.3.5   Frequency Diagram

Frequency diagrams are used to determine the characteristics of a population of data.[4] By plotting the number of observations against a variable characteristic such as size, height, or a machined dimension, it is readily possible to develop a curve that can be analyzed mathematically. Where the process values or variations are distributed relatively uniformly around a mean value, the curve takes on a normal distribution. For this case determination of standard deviation, coefficient of variation and similar predictions of the entire population of data can readily be performed. Figure 8.3 depicts a frequency diagram with a normal distribution superimposed on it. For normal distributions, as the number of observations is increased, the data will match the normal curve more and more closely.

8.4   STATISTICAL PROCESS CONTROL

Statistical process control (SPC) is a system to control the output of a repetitive process using sampling and statistical formulae to determine deviations from specified performance, identify trends, and areas for corrective action.[4–6] SPC utilizes both mathematical and graphical methodology to achieve its results. Although originally developed for manufacturing operations, it has found application in other industries as well.

Much of the basis for SPC lies in the **normal distribution curve** (Fig. 8.1), an important **measure of central tendency.** Where populations (of data) exhibit equal deviations on both sides of the mean, the distribution is said to be normal, permitting characteristics of the entire **population** to be inferred from small samples of that population. Most manufacturing processes produce normal distributions, and this methodology is widely used.

The degree of variance of a distribution is measured by its **standard deviation,** Equation (8.14). The greater the degree of process control, the less the standard deviation; thus improvements in process control always yield output that conforms more closely and shows less variation between output elements. **Kurtosis** measures the tightness of the normal curve and is greater for those curves with smaller standard deviations.

8.4.1   Process Capability

The terms **capability of process** (CP) and **capability index** (CpK) describe the degree of process control potentially available. CP provides a measure of dispersion, while CpK provides a measure of both dispersion and centeredness. Their formulas are given below. Numerous computer programs are available to reduce data and present them in the form noted above.

$$CP = \frac{\text{tolerance}}{6\sigma}$$

$$CpK = \text{the lesser of } \frac{USL - X \text{ bar}}{3\sigma} \quad \text{or} \quad \frac{(X \text{ bar} - LSL)}{3\sigma} \tag{8.16}$$

where USL is the upper specification limit, LSL is the lower specification limit, and σ the standard deviation.

Figure 8.4 indicates how the values of CpK change for various types of standard distributions. Since CpK is the relationship of the actual curve to a normal curve of ± 3 standard deviations, a value of 1.00 indicates a normal curve with 99.73% of the values conforming. Values of CpK > 1.0 indicate tighter curves with smaller standard deviations, while lower values indicate higher standard deviations. Values of CpK < 1.0 indicate a distribution not centered on the specification midpoint and with values beyond either the upper or lower specification limit.

Many purchasers are now requiring CpK values of 1.33 or 2.00. A value of 1.33 indicates that the difference between the mean and the specification limits is ± 4

Table II   Single sampling plans for normal inspection (Master table)

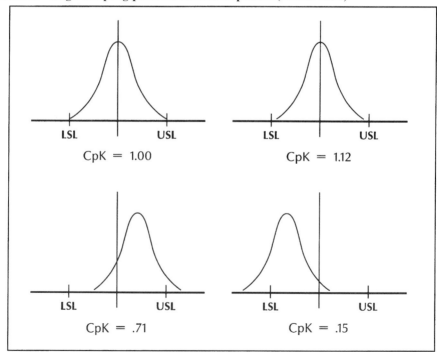

Fig. 8.4   Capability index examples. (Courtesy of DataMyte Business of Rockwell Automation. Copyright Allen-Bradley Company, Inc., 1995. Used with permission.)

standard deviations and that 99.994% of the material is conforming. Values higher than 1.33 can be justified only by a rigorous cost/benefit analysis as conformance becomes increasingly difficult and may not be economically justified.

It is important to note that manufacturers of many complex products such as integrated circuits are increasingly striving for **6 sigma quality,** which is equivalent to a CpK value of 2.00. This approach uses a ±6-sigma band (12 sigma) between the upper and lower specification limits and permits a 1.5-sigma drift of the mean in response to normal manufacturing variations while limiting the defect rate to 3.4 parts per million.

For convenience, many SPC systems use precalculated values for **accept/reject** quantities, based on **lot size, sample size,** and the **acceptable quality level** (maximum number of defects per hundred) permitted by the user. MIL-STD 105D, "Sampling Procedures and Tables for Inspection by Attributes," formerly used, has been withdrawn from use and superseded by **ASQC Z1.4** (same title), which corresponds to it directly. This is perhaps the most widely used sampling standard. The tables are from lot sizes as small as 2 to those over 500,000 and for a variety of situations, including normal, tightened, and reduced inspections. Figure 8.5 is an example of one of these tables.

8.4.2 Control Charts

Control charts, also called **X bar/R charts** (or **Shewart control charts**) (Fig.8.6), are widely used for process control. Normally, the two are used together. X bar represents the average value of several measurements plotted against the **upper** and **lower control limits** for the measurement, while R represents the widest variance within X bar between successive readings. For small production quantities, X is often used in lieu of X bar. The distance between the upper and lower control limits are often set at ±3 sigma. This provides assurance that these limits include 99.73% of all the values observed. In some cases the upper and lower control limits are reduced, providing tighter control on the process. Because these represent control limits on the process, considerable care should be taken to correlate them to the tolerances permitted for the particular attribute being measured. In no case should the control limits exceed the allowable variation (i.e., the tolerance).

R represents the **range** or difference between the largest and smallest readings, within the series of measurements averaged into X bar, between successive readings. R indicates both whether the measured value is within control limits and whether there is a trend toward one of the limits, as would occur, for example, with progressive tool wear. The pattern of range measurements may also indicate whether the measured values meander in a random way, as with a machine tool spindle bearing that is not sufficiently rigid.

8.4.3 Skewed Distributions

Skewed distributions are those where the tails of the curve are not even. A positive skew occurs when the tail of the curve extends toward the higher values, while the tail of a negative skew extends toward the lower values (see Fig. 8.7). A convenient

Table II-A.   Single Sampling Plans for Normal Inspection (Master table)

Sample size code letter	Sample size	\ 0.010	0.015	0.025	0.040	0.065	0.10	0.15	0.25	0.40	0.65	1.0	1.5	2.5	4.0	6.5	10	15	25	40	65	100	150	250	400	650	1000
		Ac Re	Ac Re	Ac Re	Ac Re	Ac Re	Ac Re	Ac Re	Ac Re	Ac Re	Ac Re	Ac Re	Ac Re	Ac Re	Ac Re	Ac Re	Ac Re	Ac Re	Ac Re	Ac Re	Ac Re	Ac Re	Ac Re	Ac Re	Ac Re	Ac Re	Ac Re
A	2	↓	↓	↓	↓	↓	↓	↓	↓	↓	↓	↓	↓	↓	↓	↓	↓	0 1	1 2	2 3	3 4	5 6	7 8	10 11	14 15	21 22	30 31
B	3	↓	↓	↓	↓	↓	↓	↓	↓	↓	↓	↓	↓	↓	↓	↓	0 1	1 2	2 3	3 4	5 6	7 8	10 11	14 15	21 22	30 31	44 45
C	5	↓	↓	↓	↓	↓	↓	↓	↓	↓	↓	↓	↓	↓	↓	0 1	1 2	2 3	3 4	5 6	7 8	10 11	14 15	21 22	30 31	44 45	↑
D	8	↓	↓	↓	↓	↓	↓	↓	↓	↓	↓	↓	↓	↓	0 1	1 2	2 3	3 4	5 6	7 8	10 11	14 15	21 22	30 31	44 45	↑	↑
E	13	↓	↓	↓	↓	↓	↓	↓	↓	↓	↓	↓	↓	0 1	1 2	2 3	3 4	5 6	7 8	10 11	14 15	21 22	30 31	44 45	↑	↑	↑
F	20	↓	↓	↓	↓	↓	↓	↓	↓	↓	↓	↓	0 1	1 2	2 3	3 4	5 6	7 8	10 11	14 15	21 22	30 31	44 45	↑	↑	↑	↑
G	32	↓	↓	↓	↓	↓	↓	↓	↓	↓	↓	0 1	1 2	2 3	3 4	5 6	7 8	10 11	14 15	21 22	30 31	44 45	↑	↑	↑	↑	↑
H	50	↓	↓	↓	↓	↓	↓	↓	↓	↓	0 1	1 2	2 3	3 4	5 6	7 8	10 11	14 15	21 22	30 31	44 45	↑	↑	↑	↑	↑	↑
J	80	↓	↓	↓	↓	↓	↓	↓	↓	0 1	1 2	2 3	3 4	5 6	7 8	10 11	14 15	21 22	30 31	44 45	↑	↑	↑	↑	↑	↑	↑
K	125	↓	↓	↓	↓	↓	↓	↓	0 1	1 2	2 3	3 4	5 6	7 8	10 11	14 15	21 22	30 31	44 45	↑	↑	↑	↑	↑	↑	↑	↑
L	200	↓	↓	↓	↓	↓	↓	0 1	1 2	2 3	3 4	5 6	7 8	10 11	14 15	21 22	30 31	44 45	↑	↑	↑	↑	↑	↑	↑	↑	↑
M	315	↓	↓	↓	↓	↓	0 1	1 2	2 3	3 4	5 6	7 8	10 11	14 15	21 22	30 31	44 45	↑	↑	↑	↑	↑	↑	↑	↑	↑	↑
N	500	↓	↓	↓	↓	0 1	1 2	2 3	3 4	5 6	7 8	10 11	14 15	21 22	30 31	44 45	↑	↑	↑	↑	↑	↑	↑	↑	↑	↑	↑
P	800	↓	↓	↓	0 1	1 2	2 3	3 4	5 6	7 8	10 11	14 15	21 22	30 31	44 45	↑	↑	↑	↑	↑	↑	↑	↑	↑	↑	↑	↑
Q	1250	↓	↓	0 1	1 2	2 3	3 4	5 6	7 8	10 11	14 15	21 22	30 31	44 45	↑	↑	↑	↑	↑	↑	↑	↑	↑	↑	↑	↑	↑
R	2000	↓	0 1	1 2	2 3	3 4	5 6	7 8	10 11	14 15	21 22	30 31	44 45	↑	↑	↑	↑	↑	↑	↑	↑	↑	↑	↑	↑	↑	↑

Acceptable Quality Levels (normal inspection)

↓ = Use first sampling plan below arrow. If sample size equals, or exceeds, lot or batch size, do 100 percent inspection.

↑ = Use first sampling plan above arrow.

Ac = Acceptance number.

Re = Rejection number.

Fig. 8.5   Typical table ASQC Z1.4. (Reprinted with permission from American Society for Quality Control Standard Z1.4, 1993.)

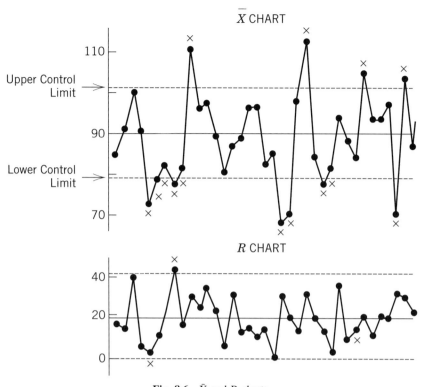

Fig. 8.6 $\bar{X}$ and R charts.

way to analyze such distributions is to calculate the CpK value for each half of the curve (above and below the maximum value). From this the percent of nonconformances for each can be determined and treated accordingly.

8.5 PROBABILITY

Probability establishes the mathematical likelihood of an event occurring. The probability of a single event is expressed by a number between 1.0 (certainty) and 0.0 (impossibility). The probability of a series of events occurring is the product of their individual probabilities:

$$P_s = P_a \cdot P_b \cdot P_c \cdots P_n, \qquad (8.17)$$

where P_s is the probability of the series of events and P_a, P_b, P_c, and P_n are the probabilities of the individual events.

The probability of a series of events can be greatly increased by identifying and improving those items having low probability. Alternatively, the probability or performance of a system of components can be improved by establishing parallel circuits:

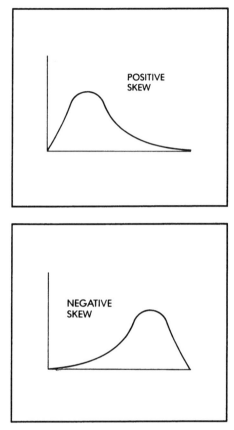

Fig. 8.7 Skewed distributions. (Courtesy of DataMyte Business of Rockwell Automation. Copyright Allen-Bradley Company, Inc., 1995. Used with permission.)

$$P_s = 1 - (1 - P_a)(1 - P_b) \cdots (1 - P_n), \tag{8.18}$$

where P_s is the probability of the system. Thus for a **series system** (Fig. 8.8) with a low-probability (reliability) component, for example:

$$P_a = 0.98$$
$$P_b = 0.88$$
$$P_c = 0.65$$
$$P_d = 0.98,$$

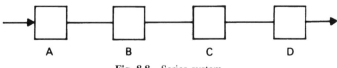

Fig. 8.8 Series system.

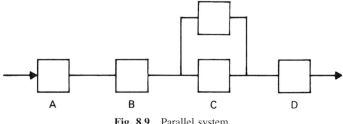

Fig. 8.9 Parallel system.

$$P_s = (0.98)(0.88)(0.65)(0.98) = 0.549. \tag{8.19}$$

Duplicating the low-reliability component in **parallel** (Fig. 8.9) yields

$$P_s = (0.98)(0.88)[1 - (1 - 0.65)(1 - 0.65)](0.98)$$
$$= \cdots [1 - (0.35)(0.35)] \cdots$$
$$[1 - 0.123],$$

$$(0.98)(0.88)[0.88](0.98) = 0.744. \tag{8.20}$$

8.6   RELIABILITY[7]

Reliability of systems and components depends on the **reliability** of individual components and the manner in which they are assembled into a system. Broadly, these include considerations of determining the reliability of individual components and assuring that they are engineered so as to interact to minimize the effect of failure.

The generalized curve of component (and system) failure is shown in Fig. 8.10. It consists of three phases. The relatively short initial phase, with a rapidly decreasing

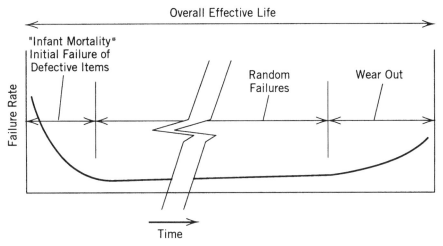

Fig. 8.10 Failure rate curve.

rate of failure often called **infant mortality,** covers failures due to poor materials, improper application, and similar factors. This rarely reaches $\frac{1}{2}\%$ and is more typically $\frac{1}{10}\%$ or less of the overall effective life. Frequently, this is covered by the warranty period for the component. The second phase is characterized by a level or nearly level rate of failure caused primarily by random and external factors; and the final phase is characterized by a gradually increasing **failure rate** due to wear out. It is important to note that in the case of **electronic** components, the rate of technological obsolescence is normally far shorter than the time to the beginning of the period of wear out. Thus this is often not a factor when considering these applications.

To minimize the impact of these effects, significant testing is often employed to assure quality and thus reliability. Typical of these is the **burn-in** often used with electronic components, together with simulations of environmental factors such as temperature, humidity, radiation, and gravitational forces. Testing methods are covered later in this section.

The terms *mean time to failure* (MTTF) and *mean time between failure* (MTBF) have slightly different meanings. The distinction is somewhat arcane, and most practicing engineers use the term **mean time between failure** to describe the likely performance life of a component. Broadly, it can be defined as the duration between placing a component into service and the failure of that component.

Specifications today include more data on the requirements for both testing and analytical data demonstrating MTBF. One example is the B10 life often specified for antifriction bearings. This requires that no more than 10% of the bearings fail in less than the number of operating hours specified. The reliability of systems containing multiple components can be calculated as described in Sections 8.5 and 8.6.

8.6.1 Life Cycle Costs

Of increasing importance is the estimation of **life cycle costs.** In its ultimate form this includes not only the cost of acquisition (including design and development), direct operational costs, and replacement costs and salvage value, but in addition, indirect costs such as income loss for funds tied up in inventory. A typical list of these costs is shown in Table 8.1.

These are direct costs only; items such as the cost of money or lost profit due to investment are not included, but are added in some analyses.

8.6.2 Failure Analysis

Failure analysis[4] requires the acquisition or development of failure rate data and their reduction. The acquisition of data typically yields numerical information on failure rates. These data can then be classified by type, as shown on Pareto diagrams (histograms) (Fig. 8.2), and the most important causes corrected. Cause identification can be facilitated by several graphic methods which are in wide use. These are sometimes called **Ishikawa diagrams** after Kaoru Ishikawa, a noted authority on quality. Examples of the diagrams are:

Table 8.1   Life Cycle Cost Elements

1. Design and development
2. Acquisition and installation of equipment
3. Purchase of initial stock of spare parts
4. Establishment of operating and maintenance support facilities
5. Initial training of operating and maintenance personnel
6. Operating costs
7. Purchase and maintenance of ongoing spare parts
8. Unplanned maintenance
9. Damaged materials and equipment due to equipment failure
10. Secondary costs such as claims, legal, and environmental costs due to item 9
11. Personnel training on a continuing basis
12. Engineering, technical, and management support
13. Cost to decommission
14. Salvage or scrap value at end of life

- Cause enumeration diagram (Fig. 8.11)
- Dispersion analysis diagram (Fig. 8.12)
- Process analysis diagram (Fig. 8.13)

8.6.3   Testing

Testing methods fall into two general classifications, *nondestructive* and *destructive*.[8,9] As the names imply, one does not damage the item, while the other subjects it to forces or effects that effectively damage it beyond use. Both methods normally

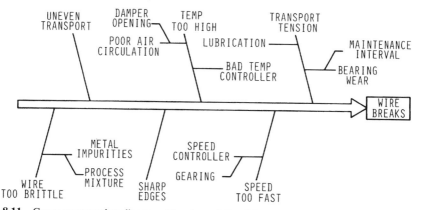

Fig. 8.11 Cause enumeration diagram. (Courtesy of DataMyte Business of Rockwell Automation. Copyright Allen-Bradley Company, Inc., 1995. Used with permission.)

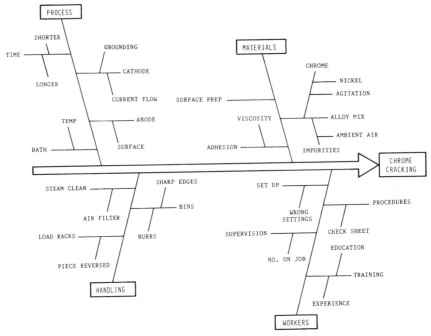

Fig. 8.12 Dispersion analysis diagram. (Courtesy of DataMyte Business of Rockwell Automation. Copyright Allen-Bradley Company, Inc., 1995. Used with permission.)

use some form of sampling and statistical analysis to determine the quantity of the product to be tested. For a particular level of confidence the size of the sample to be tested depends on the **standard deviation** of the population of components from which the sample is drawn. The better the control of the process that produces the part, the smaller the standard deviation and the smaller the sample that will be needed. Sampling methods and precalculated tables for acceptance or rejection of various lot sizes are available in **ANSI/ASQC Standard Z1.4.**

Less sophisticated methods may arbitrarily specify a percentage of the material to be tested based on experience or in some cases, upon code requirements. In general, the size of the samples to be tested from lots of 1000 or more are in the range 0.25

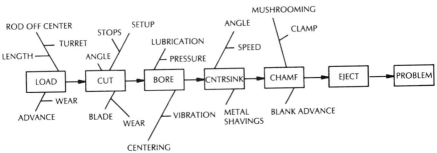

Fig. 8.13 Process analysis diagram. (Courtesy of DataMyte Business of Rockwell Automation. Copyright Allen-Bradley Company, Inc., 1995. Used with permission.

Table 8.2 Nondestructive Testing Methods

Magnetic particle testing—Utilizes magnetic powder, which gathers at discontinuities; rapid. Accurate for surface and near-surface defects; ability to detect subsurface defects decreases rapidly with depth, a typical depth limit being 1 mm. Used on ferromagnetic materials; requires some operator training. Variations of this method include magnetic field perturbation, electric current injection, magnetic resonance, and eddy current testing.

Die penetrant—Utilizes a penetrating liquid, followed by a developer fluid, creating a high-contrast indication. Discloses surface irregularities; rapid; good for fatigue cracking such as aircraft wing roots. Requires some operator training.

Ultrasonic testing—Utilizes high-frequency sound (10^6 to 10^7 Hz) and a coupling fluid. Requires interpretation of results; significant training for operator is critical. Scanning speed only modest; can detect internal irregularities. Large-grain-size materials (e.g. cast stainless steel) can give false readings.

Radiographic testing (x-ray, etc.)—Utilizes an x-ray source (10^{-10} to 10^{-7} cm) and special photographic film. Requires careful setup, personnel control, and penetration standards for **film interpretation.** Significant special training required to perform testing or interpret film images. Slow; accurate with qualified personnel. Variations of this basic technique utilize gamma rays and neutron radiography, Mossbauer analysis, and neutron activation analysis.

to 1.0%. Where parts are produced in small lots, as low as 1, it may be necessary to perform 100% examination of certain characteristics and processes, particularly where performance is critical, such as integrated electronic circuit chips or where the health and safety of the public may be involved. Often, codes will establish required examination methods and sample sizes. Selected widely used **nondestructive testing methods,** often called **nondestructive examination** (NDE), are shown in Table 8.2.

There are a host of **destructive testing methods,** depending on the type of use the component or material is expected to receive. In general, they all depend on exceeding the ultimate or yield strength of the material, its electrical or physical properties, or ambient conditions (often including acceleration or impact) Tests include tensile, compressive, impact, resistance to solvents or reagents, thermal resistance, and similar tests carried to yield or failure of the material or component. The test method is dependent on the characteristics being investigated. In general, these tests requires instrumentation to obtain accurate results. With significant numbers to be tested, they can be rapid, with results normally accurate, but the cost may be significant for small lot sizes. **Prototype** testing normally takes the form of a functional test rather than a test of specific characteristics. Usually, they are conducted to establish, **form, fit,** or **function** of the component. The cost can be very high with small lot sizes. Prototype testing usually combines several of the type tests for the test article.

8.7 ECONOMIC CHOICE

8.7.1 Minimum-Cost Point

The **minimum-cost point** is time independent. When only immediate expenditures are relevant to the choice among alternatives, the time value of money is not involved. The problem then is to obtain the design with the minimum first cost. These problems are solved graphically or analytically by expressing the costs as functions of the design variable, adding the separate costs, and locating the point of minimum total cost.

Economic Balance. When increase in annual capital charges results in decreased annual operating charges, both expressed in terms of a common design variable, the solution for minimum total annual cost is called **economic balance.** Although graphical or tabular constructions are generally applicable, many of these problems may be solved analytically. For example, if x is the common design variable; a, b, c, and d are constants related to the specific problem; and

$$\text{annual capital cost} = ax + c, \tag{8.21}$$

$$\text{annual operating cost} = \frac{b}{x} + d, \tag{8.22}$$

$$\text{total annual cost} = ax + \frac{b}{x} + (c + d), \tag{8.23}$$

differentiating total cost and equating to zero gives the optimum solution as

$$x = \sqrt{\frac{b}{a}}. \tag{8.24}$$

When there are two common design variables x and y,

$$\text{total annual cost} = ax + \frac{b}{xy} + cy + d, \tag{8.25}$$

from which a plot of total annual cost versus x at various values of y gives a cost curve with the minimum point as the optimum.

8.7.2 Measures of Profitability

In the choice among alternative investments, the objective is to make the best possible use of a limited resource, capital. Decision criteria are thus required, among which the principal ones are described here using the convention of **continuous interest** r with periodic receipts and disbursements. To convert these equations to periodic interest, substitute $(1 + i)^n$ for e^{rn}. Periodic and continuous interest should not be intermixed in economic comparisons. All results should be converted to one form of interest before comparing profitabilities.

Minimum Acceptable Return

In the array of prospective investments, some more profitable than others, a minimum standard of comparison is needed. **Minimum acceptable return** p may be looked upon as the interest rate of the marginal investment that is generally available to the investor. It is commonly taken as the average rate of earnings on the total assets of the company, often called the **pool rate.** Rates of return are usually computed after income tax.

Internal Rate Method

The well-known **internal rate method,** recently referred to as **discounted cash flow** and **interest rate of return,** serves adequately for the great majority of investment and budgetary decisions. It may be applied to incremental cash flow from an incremental investment in a project as well as to an array of unrelated projects. The internal rate method postulates that the algebraic sum of the compound amounts of all cash flows for a project is zero at some internal rate of return found by trial-and-error solution of Equation (8.27).

$$\sum_{n=1}^{n=L} R_n e^{r(L-n)} - \sum_{n=0}^{n=L} D_n e^{r(L-n)} = 0 \tag{8.26}$$

or

$$\sum_{n=1}^{n=L} R_n e^{-rn} - \sum_{n=0}^{n=L} D_n e^{-rn} = 0. \tag{8.27}$$

The decision criterion is the internal rate r, larger r being preferred.

This method is not entirely satisfactory because of two inherent defects: (1) it is based on the questionable assumption that the receipts from a project will be reinvested in an equally profitable investment, and (2) the solution for r may be indeterminate (imaginary or multiple roots) when there is more than one reversal in the direction of annual net cash flow.

Proportional Gain Method

The **proportional gain method,** attributed to Bernoulli, avoids trial-and-error solutions and is suitable for choosing between mutually exclusive alternates or for ranking an array of investment opportunities. This formulation postulates that the net receipts are accumulated in one account and the net investments in another account, both at interest rate p. When the project is terminated at time L, the relative gain G is the ratio of the two accounts or the ratio of their present worths:

$$\frac{\sum\limits_{n=1}^{n=L} R_n e^{p(L-n)}}{\sum\limits_{n=0}^{n=L} D_n e^{p(L-n)}} = \frac{\sum\limits_{n=1}^{n=L} R_n e^{-pn}}{\sum\limits_{n=0}^{n=L} D_n e^{-pn}} = G = e^{kL}. \tag{8.28}$$

The proper decision criterion is G, not k as erroneously used by some authors. Since this method is biased in favor of long-term investments, it is generally reliable only for comparing investments with nearly equal lives as originally proposed by Bernoulli.

Present Worth Method

The **present worth method,** also referred to as **venture worth** and **incremental present worth,** is restricted to comparison of projects that have identical lives L or that cover the same total time span. If it can be assumed that the costs and returns of replacements will repeat those of the original asset, multiple cycles of a short-term project may be compared with a single long-term project covering the same time span (e.g., three 5-year lives considered equivalent to one 15-year life).

With this method, present worth is the decision criterion, larger present worth being preferred. The present worth P of each project is computed, with r equal to the minimum acceptable return, as the algebraic sum of the present worths of the annual net cash flows, with salvage value taken as a receipt at the end of the project life.

$$P = \sum_{n=1}^{n=L} R_n e^{-rn} - \sum_{n=0}^{n=L} D_n e^{-rn} \qquad (8.29)$$

This method is suitable for projects that have no positive receipts, a situation in which the other two methods are indeterminate.

8.8   QUALITY ASSURANCE

Quality assurance practices are based on organization and control of activities and requires organization, planning of activities, control of those activities, cause determination and loop closing of deviations, and independent reviews of effectiveness. The villain in quality is **variability**—the normal variations in materials and fabrication, tooling, wear, human error, and other similar factors that contribute to unpredictable changes in the output of the process. Since the main function of quality systems is to remove the variability from the production processes, quality systems measure these variations and put into place methods that avoid or identify nonconformance before they produce out-of-specification work. When nonconformance occur, these systems identify the deviation and set into action both **remedial** (cause) and **corrective** (output) measures.

Quality systems require not only a statement of the managerial goals, including the importance of quality and the overall approach that will be taken, but also the more detailed methods by which quality will be achieved and ensured. There are in effect two levels at which the subject is addressed: the **program** level and the **procedural** level.

Typically, quality programs focus on the managerial aspects of the programs with the details of implementation found in departmental procedural documents. The principles set forth in the programs are sufficiently broad to be applicable to most situations and thus can be applied with a high degree of confidence. As would be expected, the implementation (procedural) aspects will vary widely from company to company, product to product, and industry to industry.

8.8.1   Quality Programs

There are several **quality programs** used in the United States and abroad that provide overall direction to the content and requirements and are suitable for a variety of industries and products.[10–12] The ANSI/ASQC Q90-1987 series of quality standards have been widely adopted for use in the United States and include generally the standards used in other countries. (In particular, these standards are considered to be the equivalent of International Standards Organization's ISO 9000, described further below.) Q90 consists of an **overall standard** that is supplemented by a series of **daughter standards** which are to be applied to the type and scope of work undertaken. This family of standards is as follows:

- ANSI/ASQC Q90-1987 Quality Management and Quality Assurance Standards—Guidelines for Selection and Use
- ANSI/ASQC Q91-1987 Quality Systems—Model for Quality Assurance in Design Development, Production, and Installation
- ANSI/ASQC Q92-1987 Quality Systems—Model for Quality Assurance in Production and Installation
- ANSI/ASQC Q93-1987 Quality Systems—Model for Quality Assurance in Final Inspection and Test
- ANSI/ASQC Q94-1987 Quality Management and Quality Systems Elements—Guidelines

Another quality systems standard that has been widely adopted and is considered essential for selling in the European market is ISO (International Standards Organization) 9000-1987. This standard is identical to BS (British Standard) 5750. ISO 9000 again uses an overall standard that establishes concepts and applications and a series of daughter standards as follows:

- ISO 9000, Quality systems—Principal concepts and applications.
- ISO 9001, Quality systems—Model for quality assurance in design/development, production, installation, and servicing.
- ISO 9002, Quality systems—Model for quality assurance in production and installation.
- ISO 9003, Quality systems—Model for quality assurance in final inspection and test.
- ISO 9004, Quality management and quality system elements—Guidelines.

Similarly, ISO Standard 14000 (and its daughter standards) is being promulgated to provide a framework for the development of environmental management systems. While broadly following the pattern of the ISO 9000 series, the thrust of 14000 is to establish standards for environmental compatability, impact assessment, certification, and labeling. With increasing worldwide concern being placed on these issues, it is likely to have increasing importance.

With these U.S. and foreign standards, the intent is to apply a particular standard to the activities performed, and not all portions of a standard need to be implemented. Thus a design organization would utilize those portions of 9001 that deal with design activities, and requirements for production control, installation activities, and such would not be utilized.

At the operational level numerous organizations have established more detailed and specific requirements for the scope and content of quality programs. Usually these quality programs make a distinction between quality assurance and quality control. Here **quality assurance** is considered the overall program and the activities (usually of a quality assurance organization) that ensure the quality of the product. The term **quality control** is generally applied to production activities, where the interest is in controlling accuracy of output. Production as used in this sense can apply to an engineering design operation. Further distinctions are usually made between quality assurance and reliability. Thus quality assurance practices are based on organization and control of activities whereas reliability deals with predictability of results.

For example, U.S. federal regulations set forth 18 quality criteria to which nuclear projects must conform (see 10 Code of Federal Regulations, Part 50, Appendix B). These regulations are typical of the philosophy and scope of formal quality assurance programs applied throughout industry, although all elements may not be appropriate for a particular activity or component. In summary form, these elements are:

I.   An organization with sufficient independence shall be established to control and verify the performance of functions affecting quality.

II. A documented program defining scope and responsibilities shall be established.

III. Measures to assure proper control of design activities, including verification or testing, shall be applied.

IV. Documents that control procurements shall include the applicable quality assurance program requirements.

V.   Activities affecting quality shall be documented, accomplished in accordance with these documents (procedures), and include acceptance criteria.

VI. The review, issuance, and change of documents affecting quality is controlled.

VII. Conformance to procurement documents of purchased materials, equipment, and services shall be established, including both control and documentation.

VIII. Identification and control of material, parts, and components shall be established.

IX. Welding, heat treating, nondestructive testing, and other special processes are to be controlled.

X.   A program for the inspections of activities affecting quality shall be established.

XI. A controlled testing program shall be established and its results shall be documented.

   XII.   Measuring and test equipment shall be controlled and calibrated.

   XIII.   Measures to control handling, shipping, and storage to prevent damage or deterioration shall be established.

   XIV.   Measures shall be established to control and indicated inspection, test, and operating status.

   XV.   Nonconforming materials shall be identified and controlled to prevent their installation or inadvertent use.

   XVI.   Conditions adverse to quality shall be identified, corrected, and for significant items reported to management.

   XVII.   Sufficient records shall be maintained to furnish evidence of activities affecting quality.

   XVIII.   A system of planned, periodic audits shall be established to verify compliance with and the effectiveness of the quality assurance program.

To avoid excessive cost and waste, it is important that the requirements for quality assurance be applied selectively only to necessary components and/or characteristics. In the defense sector similar quality assurance programs are used but with somewhat more specific requirements. Typical of this are MIL-Q-9858A and MIL-STD-1535A (USAF).

Certain industries have developed quality assurance programs, such as API Spec. Q1 for work in the petroleum industries and ANSI/ASME NQA-1 for work in the nuclear field. In addition, some standards organizations have developed **generic standards,** such as ANSI/ASQC Z1.15. For the most part these standards are similar to the general standards described earlier, and overall compliance involves the same type of organization and approach. Because there are some differences, it is important to establish early in the work which quality program will be used. The standard should then be compared carefully with the program in place, and reviewed in terms of the appropriate departmental procedures and corrections or additions made to ensure compliance.

8.8.2 Good Manufacturing Practices (for Medical Devices)

Generally referred to as **good manufacturing practices** (GMPs), 21 Code of Federal Regulations Ch. 1, Part 820,[13] establishes regulations governing current good manufacturing practices for "methods used in, and the facilities and controls used for, the design, manufacture, packing, storage, and installation of all finished devices intended for human use." The regulation is intended to assure that the devices are safe and effective. Its sister section, 821, "Medical Devices Tracking Requirements," establishes requirements for tracking of medical devices to permit manufacturers to take suitable action to correct problems arising with their devices. An index to Part 820 is given in Table 8.3.

The act is sufficiently broad that it affects all manufacturers of "finished" (as distinct from components of) medical devices and has become the standard to be followed in their activities. It is similar to other quality assurance standards and prescribes a series of practices to be followed in activities involving medical equipment. It is intended to cover design, procurement of components, fabrication, instal-

Table 8.3   Good Manufacturing Practice for Medical Devices

Subpart A—General Provisions	*Subpart G—Packing and Labelling Control*
820.1 Scope	820.120 Device labeling
820.3 Definitions	820.121 Critical devices, device labeling
820.5 Quality assurance program	820.130 Device packaging
Subpart B—Organization and Personnel	*Subpart H—Holding, Distribution, and Installation*
820.20 Organization	820.150 Distribution
820.25 Personnel	820.151 Critical devices, distribution, records
Subpart C—Buildings	820.152 Installation
820.40 Buildings	*Subpart I—Device Evaluation*
820.46 Environmental control	820.160 Finished device inspection
820.56 Cleaning and sanitation	820.161 Critical devices, finished device inspection
Subpart D—Equipment	820.162 Failure investigation
820.60 Equipment	*Subpart J—Records*
820.61 Measurement equipment	820.180 General requirements
Subpart E—Control of Components	820.181 Device master record
820.80 Components	820.182 Critical devices, device master record
820.81 Critical devices, components	820.184 Device history record
Subpart F—Production and Process Controls	820.185 Critical devices, device history record
820.100 Manufacturing specifications and processes	820.195 Critical devices, automated data processing
820.101 Critical devices, manufacturing specifications, and processes	820.198 Complaint files
820.115 Reprocessing of devices or components	
820.116 Critical devices, reprocessing of devices or components	

Source: 43 FR 31508, July 21, 1978, unless otherwise noted. Authority: Secs. 501, 502, 515, 518, 519, 520, 701, 704 of the Federal Food, Drug, and Cosmetic Act (21 U.S.C. 351, 352, 360e, 360h, 360i, 360j, 371, 374).

lation, and testing of these devices. As with other quality programs, a rigorous system of audits (for these products by the FDA) are conducted, sometimes on an unscheduled basis. The audits may be triggered by complaints or take place on a periodic basis. Satisfactory passing of audits permits a company to certify and label its product(s) as conforming and is essential for sales in the medical field.

8.8.3   Total Quality Management

Another management level approach, and one that is finding increasing acceptance, is the concept of **total quality management** (TQM). This methodology is applicable to all types of activities regardless of whether product or service oriented. TQM includes such different industries as hospitals and electronic manufacturing firms. Basically, the TQM programs focus on control of all the processes of the company.

This includes the orientation and motivation of the workforce, starting with and including the senior management, analysis of current operating practices and the development of corrective and remedial approaches, and finally, implementation of the improvements identified. The concept of TQM is to involve the entire organization at all levels in a continuing process of improvement in the activities of the business. A proper TQM program will involve such diverse groups as sales, legal, and accounting, as well as the technical and production personnel. It may in some cases also include customers both within and outside the organization. The methodology also may include techniques such as **partnering** with suppliers, the use of **just-in-time** practices to reduce in-process inventories in manufacturing firms, and an overall approach that stresses "do it right, once, the first time." The TQM programs are tailored to each firm and are highly individualized—they cannot merely be adapted from an existing program. As with other quality programs, when properly introduced and supported by top management, dramatic improvements have often been seen within a very short time. Some clients now require that bidders submit their TQM programs for evaluation before being permitted to offer proposals.

As stated earlier, the quality program for a typical firm establishes requirements that cover the procedures and practices used to **control** the work, as distinct from the work itself. The actual control of the work product is left in the hands of the operational personnel and appears in their procedures. In an engineering office, for example, the design personnel would use some form of quality control (QC) system to control their day-to-day work. The quality assurance group will normally concern itself with checking the procedures and methods and their **compliance** and will not normally perform reviews of the technical adequacy of the work itself. In many cases, because of overriding concern for accuracy, **technical audits** are conducted on the work product itself to ensure that it is correct and in accordance with required criteria.

8.8.4 Less Complex Programs

For many smaller engineering offices, such far-reaching and formal programs are not necessary, and there is no need to establish detailed systems such as these used in nuclear or very high-tech work. For such cases common sense should be the rule with a few principles for overall guidance:

- Quality starts at the top.
- Keep it simple.
- Check it periodically—independently.
- The doers are responsible and create quality—not outsiders.
- Establish and maintain a desk book of how a job is performed. Make sure that it is both accurate and realistic, and revise it when practices change.
- Get input from the doers, and permit them to make improvements.
- Fit the practices to the goals and risks of the design.
- If there is an error, fix it and the cause as well.
- Quality is a continual process, and improvements in methods, tooling, activities, and systems must be continual over time.

Table 8.4   Amount at Compound Interest $(1 + i)^n$

n	1¼%	2%	2½%	3%	3½%	4%	4½%	5%
1	1.01500	1.02000	1.02500	1.03000	1.03500	1.04000	1.04500	1.05000
2	1.03023	1.04040	1.05062	1.06090	1.07122	1.08160	1.09203	1.10250
3	1.04568	1.06121	1.07689	1.09273	1.10872	1.12486	1.14117	1.15763
4	1.06136	1.08243	1.10381	1.12551	1.14752	1.16986	1.19252	1.21551
5	1.07728	1.10408	1.13141	1.15927	1.18769	1.21665	1.24618	1.27628
6	1.09344	1.12616	1.15969	1.19405	1.22926	1.26532	1.30226	1.34010
7	1.10984	1.14869	1.18869	1.22987	1.27228	1.31593	1.36086	1.40710
8	1.12649	1.17166	1.21840	1.26677	1.31681	1.36857	1.42210	1.47746
9	1.14339	1.19509	1.24886	1.30477	1.36290	1.42331	1.48610	1.55133
10	1.16054	1.21899	1.28008	1.34392	1.41060	1.48024	1.55297	1.62889
11	1.17795	1.24337	1.31209	1.38423	1.45997	1.53945	1.62285	1.71034
12	1.19362	1.26824	1.34489	1.42576	1.51107	1.60103	1.69588	1.79586
13	1.21355	1.29361	1.37851	1.46853	1.56396	1.66507	1.77220	1.88565
14	1.23176	1.31948	1.41297	1.51259	1.61869	1.73168	1.85194	1.97993
15	1.25023	1.34587	1.44830	1.55797	1.67535	1.80094	1.93528	2.07893
16	1.26899	1.37279	1.48451	1.60471	1.73399	1.87298	2.02237	2.18287
17	1.28802	1.40024	1.52162	1.65285	1.79468	1.94790	2.11338	2.29202
18	1.30734	1.42825	1.55966	1.70243	1.85749	2.02582	2.20848	2.40662
19	1.32695	1.45681	1.59865	1.75351	1.92250	2.10685	2.30786	2.52695
20	1.34685	1.48595	1.63862	1.80611	1.98979	2.19112	2.41171	2.65330
21	1.36706	1.51567	1.67958	1.86029	2.05943	2.27877	2.52024	2.78596
22	1.38756	1.54598	1.72157	1.91610	2.13151	2.36992	2.63365	2.92526
23	1.40838	1.57690	1.76461	1.97359	2.20611	2.46472	2.75217	3.07152
24	1.42950	1.60844	1.80873	2.03279	2.28333	2.56330	2.87601	3.22510
25	1.45095	1.64061	1.85394	2.09378	2.36324	2.66584	3.00543	3.38635

n	5½%	6%	7%	8%	10%	12%	15%	20%
1	1.055	1.060	1.070	1.080	1.100	1.120	1.150	1.200
2	1.113	1.124	1.145	1.166	1.210	1.254	1.322	1.440
3	1.174	1.191	1.225	1.260	1.331	1.405	1.521	1.728
4	1.239	1.262	1.311	1.360	1.464	1.574	1.749	2.074
5	1.307	1.338	1.403	1.469	1.611	1.762	2.011	2.488
6	1.379	1.419	1.501	1.587	1.772	1.974	2.313	2.986
7	1.455	1.504	1.606	1.714	1.949	2.211	2.660	3.583
8	1.535	1.594	1.718	1.851	2.144	2.476	3.059	4.300
9	1.619	1.689	1.838	1.999	2.358	2.773	3.518	5.160
10	1.708	1.791	1.967	2.159	2.594	3.106	4.046	6.192
11	1.802	1.898	2.105	2.332	2.853	3.479	4.652	7.430
12	1.901	2.012	2.252	2.518	3.138	3.896	5.350	8.916
13	2.006	2.133	2.410	2.720	3.452	4.363	6.153	10.699
14	2.116	2.261	2.579	2.937	3.797	4.887	7.076	12.839
15	2.232	2.397	2.759	3.172	4.177	5.474	8.137	15.407
16	2.355	2.540	2.952	3.426	4.595	6.130	9.358	18.488
17	2.485	2.693	3.159	3.700	5.054	6.866	10.761	22.186
18	2.621	2.854	3.380	3.996	5.560	7.690	12.375	26.623
19	2.766	3.026	3.617	4.316	6.116	8.613	14.232	31.948
20	2.918	3.207	3.870	4.661	6.727	9.646	16.367	38.338
21	3.078	3.400	4.141	5.034	7.400	10.804	18.821	46.005
22	3.248	3.604	4.430	5.437	8.140	12.100	21.645	55.206
23	3.426	3.820	4.741	5.871	8.954	13.552	24.891	66.247
24	3.615	4.049	5.072	6.341	9.850	15.179	28.625	79.497
25	3.813	4.292	5.427	6.848	10.835	17.000	32.919	95.396

Table 8.5 Present Worth $(1 + i)^{-n}$

n	2%	3%	4%	5%	6%	7%	8%	10%
1	0.98039	0.97087	0.96154	0.95238	0.94340	0.9346	0.9259	0.9091
2	0.96117	0.94260	0.92456	0.90703	0.89000	0.8734	0.8573	0.8264
3	0.94232	0.91514	0.88900	0.86384	0.83962	0.8163	0.7938	0.7513
4	0.92385	0.88849	0.85480	0.82270	0.79209	0.7629	0.7350	0.6830
5	0.90573	0.86261	0.82193	0.78353	0.74726	0.7130	0.6806	0.6209
6	0.88797	0.83748	0.79031	0.74622	0.70496	0.6663	0.6302	0.5645
7	0.87056	0.81309	0.75992	0.71068	0.66506	0.6227	0.5835	0.5132
8	0.85349	0.78941	0.73069	0.67684	0.62741	0.5820	0.5403	0.4665
9	0.83676	0.76642	0.70259	0.64461	0.59190	0.5439	0.5002	0.4241
10	0.82035	0.74409	0.67556	0.61391	0.55839	0.5083	0.4632	0.3855
11	0.80426	0.72242	0.64958	0.58468	0.52679	0.4751	0.4289	0.3505
12	0.78849	0.70138	0.62460	0.55684	0.49697	0.4440	0.3971	0.3186
13	0.77303	0.68095	0.60057	0.53032	0.46884	0.4150	0.3677	0.2897
14	0.75788	0.66112	0.57748	0.50507	0.44230	0.3878	0.3405	0.2633
15	0.74301	0.64186	0.55526	0.48102	0.41727	0.3624	0.3152	0.2394
16	0.72845	0.62317	0.53391	0.45811	0.39365	0.3387	0.2919	0.2176
17	0.71416	0.60502	0.51337	0.43630	0.37136	0.3166	0.2703	0.1978
18	0.70016	0.58739	0.49363	0.41552	0.35034	0.2959	0.2502	0.1799
19	0.68643	0.57029	0.47464	0.39573	0.33051	0.2765	0.2317	0.1635
20	0.67297	0.55368	0.45639	0.37689	0.31180	0.2584	0.2145	0.1486
21	0.65978	0.53755	0.43883	0.35894	0.29416	0.2415	0.1987	0.1351
22	0.64684	0.52189	0.42196	0.34185	0.27751	0.2257	0.1839	0.1228
23	0.63416	0.50669	0.40573	0.32557	0.26180	0.2109	0.1703	0.1117
24	0.62172	0.49193	0.39012	0.31007	0.24698	0.1971	0.1577	0.1015
25	0.60953	0.47761	0.37512	0.29530	0.23300	0.1842	0.1460	0.0923

n	12%	15%	20%	25%	30%	35%	40%	45%
1	0.8929	0.8696	0.8333	0.8000	0.7692	0.7407	0.7143	0.6897
2	0.7972	0.7561	0.6944	0.6400	0.5917	0.5487	0.5102	0.4756
3	0.7118	0.6575	0.5787	0.5120	0.4552	0.4064	0.3644	0.3280
4	0.6355	0.5718	0.4823	0.4096	0.3501	0.3011	0.2603	0.2262
5	0.5674	0.4972	0.4019	0.3277	0.2693	0.2230	0.1859	0.1560
6	0.5066	0.4323	0.3349	0.2621	0.2072	0.1652	0.1328	0.1076
7	0.4523	0.3759	0.2791	0.2097	0.1594	0.1224	0.0949	0.0742
8	0.4039	0.3269	0.2326	0.1678	0.1226	0.0906	0.0678	0.0512
9	0.3606	0.2843	0.1938	0.1342	0.0943	0.0671	0.0484	0.0353
10	0.3220	0.2472	0.1615	0.1074	0.0725	0.0497	0.0346	0.0243
11	0.2875	0.2149	0.1346	0.0859	0.0558	0.0368	0.0247	0.0168
12	0.2567	0.1869	0.1122	0.0687	0.0429	0.0273	0.0176	0.0116
13	0.2292	0.1625	0.0935	0.0550	0.0330	0.0202	0.0126	0.0080
14	0.2046	0.1413	0.0779	0.0440	0.0254	0.0150	0.0090	0.0055
15	0.1827	0.1229	0.0649	0.0352	0.0195	0.0111	0.0064	0.0038
16	0.1631	0.1069	0.0541	0.0281	0.0150	0.0082	0.0046	0.0026
17	0.1456	0.0929	0.0451	0.0225	0.0116	0.0061	0.0033	0.0018
18	0.1300	0.0808	0.0376	0.0180	0.0089	0.0045	0.0023	0.0012
19	0.1161	0.0703	0.0313	0.0144	0.0068	0.0033	0.0017	0.0009
20	0.1037	0.0611	0.0261	0.0115	0.0053	0.0025	0.0012	0.0006
21	0.0926	0.0531	0.0217	0.0092	0.0040	0.0018	0.0009	0.0004
22	0.0826	0.0462	0.0181	0.0074	0.0031	0.0014	0.0006	0.0003
23	0.0738	0.0402	0.0151	0.0059	0.0024	0.0010	0.0004	0.0002
24	0.0659	0.0349	0.0126	0.0047	0.0018	0.0007	0.0003	0.0001
25	0.0588	0.0304	0.0105	0.0038	0.0014	0.0006	0.0002	0.0001

Table 8.6   Amount of Annuity $[(1 + i)^n - 1]/i$

n	$1\frac{1}{2}$%	2%	$2\frac{1}{2}$%	3%	$3\frac{1}{2}$%	4%	$4\frac{1}{2}$%	5%
1	1.00000	1.00000	1.00000	1.00000	1.00000	1.00000	1.00000	1.00000
2	2.01500	2.02000	2.02500	2.03000	2.03500	2.04000	2.04500	2.05000
3	3.04522	3.06040	3.07562	3.09090	3.10623	3.12160	3.13702	3.15250
4	4.09090	4.12161	4.15252	4.18363	4.21494	4.24646	4.27819	4.31013
5	5.15227	5.20404	5.25633	5.30914	5.36247	5.41632	5.47071	5.52563
6	6.22955	6.30812	6.38774	6.46841	6.55015	6.63298	6.71689	6.80191
7	7.32299	7.43428	7.54743	7.66246	7.77941	7.89829	8.01915	8.14201
8	8.43284	8.58297	8.73612	8.89234	9.05169	9.21423	9.38001	9.54911
9	9.55933	9.75463	9.95452	10.1591	10.3685	10.5828	10.8021	11.0266
10	10.70272	10.94972	11.2034	11.4639	11.7314	12.0061	12.2882	12.5779
11	11.86326	12.16872	12.4835	12.8078	13.1420	13.4864	13.8412	14.2068
12	13.04121	13.41209	13.7956	14.1920	14.6020	15.0258	15.4640	15.9171
13	14.23683	14.68033	15.1404	15.6178	16.1130	16.6268	17.1599	17.7130
14	15.45038	15.97394	16.5190	17.0863	17.6770	18.2919	18.9321	19.5986
15	16.68214	17.29342	17.9319	18.5989	19.2957	20.0236	20.7841	21.5786
16	17.93237	18.63929	19.3802	20.1569	20.9710	21.8245	22.7193	23.6575
17	19.20136	20.01207	20.8647	21.7616	22.7050	23.6975	24.7417	25.8404
18	20.48938	21.41231	22.3863	23.4144	24.4997	25.6454	26.8551	28.1324
19	21.79672	22.84056	23.9460	25.1169	26.3572	27.6712	29.0636	30.5390
20	23.12367	24.29737	25.5447	26.8704	28.2797	29.7781	31.3714	33.0660
21	24.47052	25.78332	27.1833	28.6765	30.2695	31.9692	33.7831	35.7193
22	25.83758	27.29899	28.8629	30.5368	32.3289	34.2480	36.3034	38.5052
23	27.22514	28.84496	30.5844	32.4529	34.4604	36.6179	38.9370	41.4305
24	28.63352	30.42186	32.3490	34.4265	36.6665	39.0826	41.6892	44.5020
25	30.06302	32.03030	34.1578	36.4593	38.9499	41.6459	44.5652	47.7271

n	$5\frac{1}{2}$%	6%	7%	8%	10%	12%	15%	20%
1	1.00000	1.00000	1.000	1.000	1.000	1.000	1.000	1.000
2	2.05500	2.06000	2.070	2.080	2.100	2.120	2.150	2.200
3	3.16803	3.18360	3.215	3.246	3.310	3.374	3.472	3.640
4	4.34227	4.37462	4.440	4.506	4.641	4.779	4.993	5.368
5	5.58109	5.63709	5.751	5.867	6.105	6.353	6.742	7.442
6	6.88805	6.97532	7.153	7.336	7.716	8.115	8.754	9.930
7	8.26689	8.39384	8.654	8.923	9.487	10.089	11.067	12.916
8	9.72157	9.89747	10.260	10.637	11.436	12.300	13.727	16.499
9	11.2563	11.4913	11.978	12.488	13.579	14.776	16.786	20.799
10	12.8754	13.1808	13.816	14.487	15.937	17.549	20.304	25.959
11	14.5835	14.9716	15.784	16.645	18.531	20.655	24.349	32.150
12	16.3856	16.8699	17.888	18.977	21.384	24.133	29.002	39.580
13	18.2868	18.8821	20.141	21.495	24.523	28.029	34.352	48.497
14	20.2926	21.0151	22.550	24.215	27.975	32.393	40.505	59.196
15	22.4087	23.2760	25.129	27.152	31.772	37.280	47.580	72.035
16	24.6411	25.6725	27.888	30.324	35.950	42.753	55.717	87.442
17	26.9964	28.2129	30.840	33.750	40.545	48.884	65.075	105.931
18	29.4812	30.9057	33.999	37.450	45.599	55.750	75.836	128.117
19	32.1027	33.7600	37.379	41.446	51.159	63.440	88.212	154.740
20	34.8683	36.7856	40.995	45.762	57.275	72.052	102.443	186.688
21	37.7861	39.9927	44.865	50.423	64.002	81.699	118.810	225.025
22	40.8643	43.3923	49.006	55.457	71.403	92.502	137.631	271.031
23	44.1118	46.9958	53.436	60.893	79.543	104.603	159.276	326.237
24	47.5380	50.8156	58.177	66.765	88.497	118.155	184.167	392.484
25	51.1526	54.8645	63.249	73.106	98.347	133.334	212.793	471.981

Table 8.7 Present Worth of Annuity $[1 - (1 + i)^{-n}]/i$

n	2%	3%	4%	5%	6%	7%	8%	10%
1	0.98039	0.97087	0.96154	0.95238	0.94340	0.935	0.926	0.909
2	1.94156	1.91347	1.88609	1.85941	1.83339	1.808	1.783	1.736
3	2.88388	2.82861	2.77509	2.72325	2.67301	2.624	2.577	2.487
4	3.80773	3.71710	3.62990	3.54595	3.46511	3.387	3.312	3.170
5	4.71346	4.57971	4.45182	4.32948	4.21236	4.100	3.993	3.791
6	5.60143	5.41719	5.24214	5.07569	4.91732	4.767	4.623	4.355
7	6.47199	6.23028	6.00205	5.78637	5.58238	5.389	5.206	4.868
8	7.32548	7.01969	6.73274	6.46321	6.20979	5.971	5.747	5.335
9	8.16224	7.78611	7.43533	7.10782	6.80169	6.515	6.247	5.759
10	8.98258	8.53020	8.11090	7.72173	7.36009	7.024	6.710	6.144
11	9.78685	9.25262	8.76048	8.30641	7.88687	7.499	7.139	6.495
12	10.57534	9.95400	9.38507	8.86325	8.38384	7.943	7.536	6.814
13	11.34837	10.6350	9.98565	9.39357	8.85268	8.358	7.904	7.103
14	12.10625	11.2961	10.5631	9.89864	9.29498	8.745	8.244	7.367
15	12.84926	11.9379	11.1184	10.3797	9.71225	9.108	8.559	7.606
16	13.57771	12.5611	11.6523	10.8378	10.1059	9.447	8.851	7.824
17	14.29187	13.1661	12.1657	11.2741	10.4773	9.763	9.122	8.022
18	14.99203	13.7535	12.6593	11.6896	10.8276	10.059	9.372	8.201
19	15.67846	14.3238	13.1339	12.0853	11.1581	10.336	9.604	8.365
20	16.35143	14.8775	13.5903	12.4622	11.4699	10.594	9.818	8.514
21	17.01121	15.4150	14.0292	12.8212	11.7641	10.836	10.017	8.649
22	17.65805	15.9369	14.4511	13.1630	12.0416	11.061	10.201	8.772
23	18.29220	16.4436	14.8568	13.4886	12.3034	11.272	10.371	8.883
24	18.91393	16.9355	15.2470	13.7986	12.5504	11.469	10.529	8.985
25	19.52346	17.4131	15.6221	14.0939	12.7834	11.654	10.675	9.077

n	12%	15%	20%	25%	30%	35%	40%	45%
1	0.893	0.870	0.833	0.800	0.769	0.741	0.714	0.690
2	1.690	1.626	1.528	1.440	1.361	1.289	1.224	1.165
3	2.402	2.283	2.106	1.952	1.816	1.696	1.589	1.493
4	3.037	2.855	2.589	2.362	2.166	1.997	1.849	1.720
5	3.605	3.352	2.991	2.689	2.436	2.220	2.035	1.876
6	4.111	3.784	3.326	2.951	2.643	2.385	2.168	1.983
7	4.564	4.160	3.605	3.161	2.802	2.507	2.263	2.057
8	4.968	4.487	3.837	3.329	2.925	2.598	2.331	2.109
9	5.328	4.772	4.031	3.463	3.019	2.665	2.379	2.144
10	5.650	5.019	4.192	3.571	3.092	2.715	2.414	2.168
11	5.938	5.234	4.327	3.656	3.147	2.752	2.438	2.185
12	6.194	5.421	4.439	3.725	3.190	2.779	2.456	2.196
13	6.424	5.583	4.533	3.780	3.223	2.799	2.469	2.204
14	6.628	5.724	4.611	3.824	3.249	2.814	2.478	2.210
15	6.811	5.847	4.675	3.859	3.268	2.825	2.484	2.214
16	6.974	5.945	4.730	3.887	3.283	2.834	2.489	2.216
17	7.120	6.047	4.775	3.910	3.295	2.840	2.492	2.218
18	7.250	6.128	4.812	3.928	3.304	2.844	2.494	2.219
19	7.366	6.198	4.844	3.942	3.311	2.848	2.496	2.220
20	7.469	6.259	4.870	3.954	3.316	2.850	2.497	2.221
21	7.562	6.312	4.891	3.963	3.320	2.852	2.498	2.221
22	7.645	6.359	4.909	3.970	3.323	2.853	2.498	2.222
23	7.718	6.399	4.925	3.976	3.325	2.854	2.499	2.222
24	7.784	6.434	4.937	3.981	3.327	2.855	2.499	2.222
25	7.843	6.464	4.948	3.985	3.329	2.856	2.499	2.222

REFERENCES

1. T. Gonen, *Engineering Economy for Engineering Mangers,* John Wiley & Sons, Inc., New York, 1990.
2. J. A. White, M. H. Agee, and K. E. Case, *Principles of Engineering Economic Analysis,* John Wiley & Sons, Inc., New York, 1977.
3. E. L. Grant and W. G. Ireson, *Principles of Engineering Economy,* 7th ed., The Ronald Press Company, New York, 1982.
4. *Data Myte Handbook,* 6th ed., Data Myte Division, Allen-Bradley Co., Inc., Minnetonka, Minn., 1995.
5. W. S. Messina, *Statistical Quality Control for Manufacturing Managers,* John Wiley & Sons, Inc., New York, 1987.
6. *Statistical Quality Control Handbook,* Western Electric Co., Inc., Indianapolis, Ind. 1984.
7. H. S. Blanks, *Reliability in Procurement Use,* John Wiley & Sons, Inc., New York, 1992.
8. D. E. Bray and D. McBride, *Nondestructive Testing Techniques,* John Wiley & Sons, Inc., New York, 1992.
9. J. M. Juran, F. M. Gryna, and R. S. Bingham, Jr., *Quality Control Handbook,* 4th ed., McGraw-Hill Book Company, New York, 1988.
10. B. Margulio, *Quality Systems in the Nuclear Industry,* American Society for Testing and Materials, Philadelphia, 1977.
11. 10 Code of Federal Regulations, Part 50, App. B.
12. S. L. Jackson, *ISO 14001 Implementation Guide: Creating an Integrated Management System,* John Wiley & Sons, Inc., New York, 1996.
13. 21 Code of Federal Regulations, Ch. 1, Part 820.

CHAPTER 9
ENERGY SOURCES

9.1 GENERAL[1]

Energy sources can be classified broadly depending on fuel types. A typical classi-fication identifies them as solid, liquid, gaseous, nuclear, geothermal, solar, wind, and advanced (still largely experimental) forms such as fuel cells and magnetohydrody-namics. This ranking is relatively widely used and broadly follows the common applications of energy sources around the world. It is possible also to categorize fuels and energy sources as to whether they are renewable or nonrenewable; thus wind and solar would be considered renewable sources, whereas coal, lignite, geothermal in some cases, petroleum, and natural and liquefied petroleum gases would be consid-ered nonrenewable.

Historical data on production of energy in the United States is shown in Fig. 9.1. There are numerous projections as to the kind of fuel that will be in use in the various nations around the world and within the United States in particular. Broadly, there is a projected increase in fuel costs, as has been the case in the United States in the last 6 or 8 years, although the forecast annual increases are smaller. The use of smaller automobiles (a major consumer of liquid fuels) and the transition from petroleum to coal, with its larger reserves available in the United States, will be accelerated. This will tend to shift the use of fuels away from those that are imported to those that are indigenous. Further, the higher costs will tend to shift the use of fuels away from those that are imported to those that are indigenous. Further, the higher costs will tend to provide increased economic incentives for improvements in the utilization of fuels through use of small automobiles, improved insulation in homes, more wide-spread use of heat-recovery units in industrial processes, and improved efficiency in heat utilization equipment. Concurrently, there will be a movement toward increased use of the renewable sources of energy; windpower and solar are the two principal sources given increased emphasis in the near term. These sources are currently un-dergoing fairly extensive development in the United States.

9.2 HEATING VALUE AND COMBUSTION

Higher Heating Value for a Solid Fuel. In the case of a solid fuel that contains C percent* of fixed and volatile carbon, H percent of hydrogen, O percent of oxygen, and S percent of sulfur, the heating value (Q) per pound of fuel-as-fired is

*Percentages by weight are expressed as a decimal fraction.

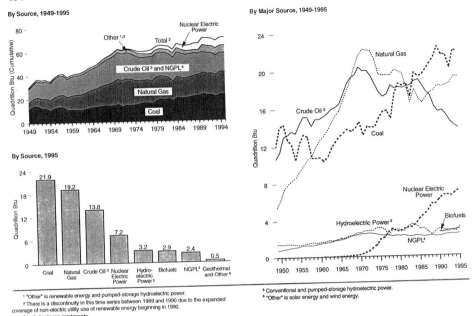

Fig. 9.1   Production of energy by type. (From Ref. 1.)

$$Q = 14{,}540C + 62{,}030\left(H - \frac{O}{8}\right) + 4050S \quad \text{Btu.} \tag{9.1}$$

The **higher heating value** (HHV) of a fuel is closely produced in a bomb-type calorimeter when the condensation and cooling of the hydrogen products (water) of combustion occurs. This adds approximately 1050 Btu/lb (710 kcal/kg) of H_2O formed (heat of condensation) to the **lower heating value** (LHV) of the fuel. For processes where vapor is not condensed and cooled, LHV is more indicative. Because of the difficulties of measuring LHV, the prevailing practice in the United States is to use HHV, whereas in Europe, LHV is frequently used.

Heating Value for Liquid Fuel.   In the case of a liquid fuel that contains C percent of carbon and H percent of hydrogen, the heating value Q per pound of fuel is

$$Q = 13{,}500C + 60{,}890H \quad \text{Btu.} \tag{9.2}$$

If the Baumé reading is known, the heating value Q per pound of fuel is

$$Q = 18{,}650 + 40(\text{Baumé reading} - 10) \quad \text{Btu.} \tag{9.3}$$

Weight of Dry Flue Gases per Pound of Carbon.   The flue gas analysis indicates the percentage CO_2, CO, O_2, and N_2 by volume. The weight (G_1) of dry flue gas per pound of carbon is given by

$$G_1 = \frac{11CO_2 + 8O_2 + 7(CO + N_2)}{3(CO_2 + CO)} \quad \text{pounds}$$

(9.4)

$$= \frac{4CO_2 + O_2 + 700}{3(CO_2 + CO)} \quad \text{pounds.}$$

Weight of Dry Flue Gases per Pound of Coal-as-Fired. The percentage by weight of carbon in the coal (C_c) and in the ash (C_a) can be determined from the coal and ash analyses. If the pounds of coal-as-fired (M_c), the pounds of ash (M_a), and the gas analysis are known, then the weight (G_2) of dry flue gas per pound of coal-as-fired is given by

$$G_2 = \frac{M_c C_c - M_a C_a}{M_c} \left[\frac{4CO_2 + O_2 + 700}{3(CO_2 + CO)} \right] \quad \text{pounds.}$$

(9.5)

Actual Weight of Dry Air per Pound of Coal-as-Fired. If air is assumed to be 77% nitrogen (N_2) by weight, the weight (G_3) of dry air per pound of coal-as-fired is given by

$$G_3 = \frac{M_c C_c - M_a C_a}{M_c} \left(\frac{3.032 N_2}{CO_2 + CO} \right) \quad \text{pounds.}$$

(9.6)

Theoretical Weight of Dry Air per Pound of Coal-as-Fired. In the case of a coal-as-fired that contains C percent by weight of fixed and volatile carbon, H percent of hydrogen, O percent of oxygen, and S percent of sulfur, the theoretical weight (G_4) of dry air per pound of coal-as-fired is given by

$$G_4 = 11.57C + 34.8\left(H - \frac{O}{8} \right) + 4.35S \quad \text{pounds.}$$

(9.7)

Percentage of Excess Air. Since G_3 gives the actual weight and G_4 the theoretical weight of air required for combustion, it follows that

$$\text{excess air} = \left(\frac{G_3 - G_4}{G_4} \right) 100 \quad \text{percent.}$$

(9.8)

Some common chemical reactions of combustion are shown in Table 9.1.

9.3 SOLID FUELS

Solid fuels are widely distributed and have formed the basis for most energy development until the 1930s. For large stationary energy applications, they are still considered a primary candidate.

Table 9.1   Common Chemical Reactions of
Combustion

Combustible	Reaction
Carbon (to CO)	$2C + O_2 = 2CO$
Carbon (to CO_2)	$C + O_2 = CO_2$
Carbon monoxide	$2CO + O_2 = 2CO_2$
Hydrogen	$2H_2 + O_2 = 2H_2O$
Sulfur (to SO_2)	$S + O_2 = SO_2$
Sulfur (to SO_3)	$2S + 3O_2 = 2SO_3$
Methane	$CH_4 + 2O_2 = CO_2 + 2H_2O$
Acetylene	$2C_2H_2 + 5O_2 = 4CO_2 + 2H_2O$
Ethylene	$C_2H_4 + 3O_2 = 2CO_2 + 2H_2O$
Ethane	$2C_2H_6 + 7O_2 = 4CO_2 + 6H_2O$
Hydrogen sulfide	$2H_2S + 3O_2 = 2SO_2 + 2H_2O$

Source: Ref. 2.

9.3.1   Characteristics of Common Solid Fuels

The more common solid fuels, in descending order from coal, are shown in Table
9.2.

9.3.2   Coal

Coal is an extremely widely used fuel not only in the United States but throughout
the world. It is the most generally disbursed, commercially valuable fuel and it is
available in a variety of types, sizes, and heating values. The coals are typically
ranked as anthracite, bituminous and lignites, from the very hard to the very soft.

Table 9.2   Typical Characteristics of Solid Fuels

Type	HHV (Btu/lb)	LHV (Btu/lb)	Fixed Carbon (%)	Moisture (%)	Volatile Matter (%)	Ash (%)
Anthracite coal	14,600	12,900	92–98	5	2–8	11
Bituminous coal	15,200	13,800	69–78	3–5	22–31	7
Subbituminous coal	9,500	9,000	52	17	39	8
Lignite	7,500	6,900	40	35	45	12
Peat	10,000[a]	3,500	23	70	67	2–70
Wood	8,800	8,300[b]	20	45	78	2
Bagasse	8,300	3,000	45	50	—	3
Hogged fuel	8,800	8,300	20	50	78	2
Municipal garbage	9,500[c]	—	—	35–50	—	8–12
Manure	7,400	—	—	45–65	—	16

[a] Moisture and ash free.
[b] Can be as low as 4300 Btu with 50% moisture.
[c] Varies widely.

Characteristics of coals vary widely and each deposit should be properly sampled to provide firm data for design.

The coals are solid materials found principally underground in seams as narrow as a few inches to very thick beds of coal up to 50 ft (15 m) thick that can be handled with large-scale excavating equipment. Thick seams are frequently found on or near the surfaces in the United States, Germany, and elsewhere, where they are mined with very large excavating equipment such as drag lines and mining wheels. The coal must be handled as a solid and conveyed either as a solid or (with newer coal technology currently being developed) as particles of solid in suspension in a liquid.

Care must be taken in stockpiling the coal at the point of use, because spontaneous combustion may begin if the coal pile is not well sealed. Such a fire typically extends into the interior of the coal pile itself where it is most difficult to extinguish. Thus, for any significant application of coal, it is essential that the coal pile be properly sealed (airtight) by compacting and rolling to avoid supporting combustion.

The burning of coal involves combustion of the material either as a solid on a grate, which permits air to pass through the burning fuel, or as a finely divided suspension in an airstream (as a pulverized material). The use of pulverizing equipment is typically limited to rather large installations, such as central power plants or large industrial boilers, and is not widely found in small installations. Small installations typically use a grate with a series of slots through which combustion air is introduced sufficient for the rapid combustion of the coal. In the case of pulverized coal, some air is introduced with the coal while other air is introduced around the coal and provides the added oxygen needed for combustion. In the burning of pulverized coal, particularly where the coal tends to be low in volatiles or of the lower rankings, it is frequently necessary to burn fuel oil as a stabilizing agent. This assures that the combustion is stable and avoids puffing or other types of intermittent combustion. Furnaces have been known to experience severe flashbacks and blowbacks where a flame has been lost and fuel had continued to be fed into the furnace. With their low volatility, this is not so much a problem with the coals and is more usually a concern with gaseous or liquid fuels.

When burning noncleaned coals as pulverized coal, the coal may very often contain significant amounts of sulfur, perhaps as much as 2 to 3%. Current technology is developing methods to introduce desulfurizing agents into the coal at the time of burning on the grate. One method is the **fluidized bed method,** wherein the chemical reaction between the sulfur and the desulfurization agent (typically dolomite, $Ca,MgCO_3$) takes place on the grate surface itself. This requires the introduction of large amounts of air and a rather open grate structure, but has the distinct advantage that virtually the entire desulfurization takes place as a part of the combustion process rather than farther back in the gas stream flow path, where the gases are cooler. This contrasts with other sulfur removal systems when the gas stream is scrubbed (i.e., desulfurized), usually involving both a cooling and chemical combination and thus creating a significant efficiency penalty for the cycle.

Coal contains a substantial amount of ash or nonburnable material. The ash can be broadly divided into the categories of fly ash and bottom ash. Fly ash is the very light ash that finds its way out of the furnace combustion zone and is carried by the air stream up through the furnace and on out toward the stack. Bottom ash is a heavy, more granular type of material that may stick to the lower walls of the furnace or fall through the grate structure itself and be discharged from the actual combustion zone in that fashion. An important part of the concern for handling of both the fly ash and the bottom ash is their temperature; that is, they are at roughly the combustion temperature of the furnace and must be cooled, removed from the furnace or the gas stream, and disposed of. Furnaces in the larger sizes are frequently built with a wet bottom in which the bottom ash is discharged into a sealed wet tank integral with the bottom of the furnace where the ash is typically sluiced away by a water stream. Alternatively, a dry-bottom furnace may be employed where the ash is dumped into a pit below the furnace, mixed with water and sluiced or pumped away as a slurry. Fly ash, being very light, is carried with the gas stream and will be discharged up the stack unless fly ash removal equipment is installed. In many large installations today, such as power utility boilers, the cost of fly ash equipment may cost in the tens of millions of dollars. The removal of fly ash is frequently handled by an electrostatic system. The electrostatic generator places a charge on a series of wires or plates, which selectively attract and hold the fly ash. The electrostatic system then bypasses one or more of the sections of the fly ash collector and the electric charge is shut off. The plates or wires are then rapped or shaken, the fly ash then dumping into a hopper system beneath. Alternatively, many furnaces employ a passive system of (filter) bag collectors; the bag collectors have proven to be simple to operate and require no power consumption. Although the bag systems have the disadvantage of being unable to operate at very high temperatures, they have proven to be extremely useful for many installations. The bags act as a filtering agent to remove the fly ash from the flue gas stream, and again, periodically, the gas stream is diverted and the bags are shaken or moved to break loose the accumulated ash from the bag surface. The ash is collected in a series of hoppers from which it is disposed.

Of particular concern with the design of furnaces burning coal is the ash fusion temperature. The bottom ash has a tendency to fuse on the walls of the furnace and can agglomerate into extremely large nodules of several hundred pounds if the ash fusion temperature is substantially exceeded. Thus, the ash fusion temperature for the coals to be burned must be established rather carefully to assure that furnace design does not create a situation in which large masses of ash accumulate on the lower walls of the furnaces. Where this does happen, it may be necessary to shut down the furnace, enter it, and break up the agglomerate with jackhammers, pinch bars, or other mechanical means.

One of the more significant areas of potential savings in a furnace is the use of air preheaters. Typically, for large furnaces, the combustion air, which will be introduced with the fuel, flows through an air preheater where its temperature is increased. The air preheater is an air heat exchanger, exchanging the heat from the discharge gas stream with the incoming cold air, thus improving cycle efficiency.

9.3.3 Lignite and Peat

Both **lignite** and **peat** are solid fuels having a high moisture content and are low-ranked coals in terms of the coal formation geological process. They are the predecessors of the bituminous and the anthracite coals. Because of their high moisture content, large amounts of combustion air are needed to carry off the steam resulting from the water given off during combustion; and, as a consequence, fuels generally have a reduced heating value since their net heating value or lower heating value includes the energy required to drive the moisture out of the fuel itself.

Because of the weight of water in the fuel and its relatively low heating value, they are not considered economic to transport over long distances. Where these deposits occur close to the point of use, they can be economically exploited; for example, in the brown coal region of Germany, where a very low-grade coal similar to a lignite or peak is burned, much local industry has been based on this source of fuel.

9.3.4 Wood, Hogged Fuel, and Bagasse

Several **solid fuels** come from the wood family. **Bagasse** is the residual from sugarcane processing (stalks, etc.). In more recent years, wood and its by-product, **hogged fuel,** which is a waste product of wood (chips, branches, slashings, etc.), have been more valuable as building and furniture products in themselves than as a fuel. In some areas of the world, where they exist in surplus, they are used for fuels. Again, the economics of transportation apply and tend to limit their usable geographic area, although it typically is wider than the peat/lignite area described previously. Wood, hogged fuel, and bagasses have a high volume/weight ratio since they are in the form of chips or random sizes, and rail cars that carry these normally have high extended sides to permit loading large volumes of this material in a car.

Normally, the combustion air requirements for these fuels are not excessive since large amounts of surface area are available when the fuel is burned and there is a relatively free flow of air through the burning material itself. When sawdust is burned, however, particularly if fresh and fairly damp, it tends to compact, requiring more combustion air and occasional agitation. Usually, these materials are burned on a grate or in a stoker furnace. If the sawdust or a finely divided wood material is burned in suspension, it is important that there be sufficient residence time within the furnace to dry the material prior to combustion.

The moisture in these woody or lignocellulose-type fuels varies widely depending on how recently the fuel was cut, its form, matter of transport, and so on. Of the several fuels, hogged fuel is more predictable, because it is typically produced in a chipper, which yields flakes or chips of a relatively uniform size. Bagasse tends to have a high content of dirt, silica, ash, and other noncombustibles, and extra provisions must be provided to handle the large quantities of ash that occur with its combustion.

9.3.5 Municipal Garbage

Municipal garbage is a unique and highly variable fuel. Typically, it contains substantial amounts of inert or metallic materials. These range from small metallic materials, such as nails and metal fasteners, to objects as large as entire bed springs. Other typical materials in the as-received condition include glass bottles, organic materials, plastics, aluminum containers, and utensils.

Normally, in handling municipal garbage there is a system for removal of the metals from the incoming material flow; magnets are used to remove the ferrous materials and gravity separation is used for dropping out materials such as aluminum or other nonferrous or glass materials. Hammer mills are often used to reduce the size of the materials prior to charging into the furnace.

Fairly high quantities of combustion air are required because of typically high amounts of moisture in the fuel. With the widely varying mix of materials present in the municipal garbage, grate-type furnaces producing extended residence time are normally employed to burn this material. Because of transportation costs, availability of this fuel is limited to the garbage produced in the immediate area. Its composition will vary over the years as increased conservation measures reduce the amounts of recoverable materials in the garbage stream such as aluminum, steel, and plastics. The municipal garbage installations to date are able to operate on a self-sustaining basis as regards the energy available versus the energy used and in fact have a slight net excess operating energy. It is relatively small, however, and has been used more widely primarily as a disposal method rather than a system for power generation.

9.4   LIQUID FUELS

Liquid fuels principally used are hydrocarbons of petroleum base and are usually ranked as shown in Table 9.3.

9.4.1   Characteristics of Liquid Fuels

Liquid fuels can be conveniently ranked as shown in Table 9.3 (see also Table 12.6). Liquid hydrocarbon fuels rank in decreasing volatility as follows[3]:

- **Volatile products:** liquefied gases and natural gasoline
- **Light oils:** gasolines, jet and tractor fuels, kerosene
- **Heavy distillates:** burner oil, furnace distillates, diesel fuel, gas oil
- **Residues:** fuel oil, asphalt, coke

These reflect the results of the initial separation of crude petroleum into several *cuts* of average boiling point. The *lighter* fractions (low boiling point) become gasoline directly or through intermediate processing. *Heavier* fractions become distillates, or residual fuel oil.

Gasoline is the major objective of the refinery and is the end product of a variety of alternative processes. Light fractions from the crude are *hydrotreated* to remove contaminants and *re-formed* to raise octane rating. Heavier fractions are *cracked,* to break the high-molecular-weight hydrocarbon into lower-weight components either by purely thermal treatment or by catalytic cracking or hydrocracking in the presence of hydrogen.

Other products, of lesser production importance, are lubricating oils, grease or wax, petroleum coke, and residue. Feedstock for conversion to petrochemicals is a potential refinery product as well. All of the crude petroleum feedstock may be converted to useful forms.

Table 9.3 Typical Characteristics of Liquid Fuels[a]

Type	HHV (Btu/gal)	lb/gal at 60°F	Specific Gravity	Pour Point (°F)	Viscosity (cStokes 100°F)	API Gravity (60°F)	Sulfur (max. %)	Sediment and Water (%)	Recommended Pumping Temperature (°F)
No. 1 fuel oil (light distillate)	137,000	6.87	0.825	<0	1.6[b]	40	0.1	Trace	Ambient
No. 2 fuel oil (distillate)	141,000	7.21	0.865	<0	2.7	32	0.5	Trace	Ambient
No. 4 fuel oil (very light residual)	146,000	7.73	0.928	10	15.0	21	1.0	0.5 (max.)	15 (min.)
No. 5 fuel oil (light residual)	148,000	7.94	0.953	30	50.0	17	2.0 (max.)	1.0 (max.)	35 (min.)
No. 6 fuel oil (residual)	150,000	8.21	0.986	65	360.0	12	2.8 (max.)	2.0 (max.)	100 (min.)
Gasoline (90 octane)	127,000	6.25	0.750	<0	b	—	—	—	Ambient
Methanol (106 octane)	66,700	6.60	0.790	<0	b	—	Nil	Nil	Ambient

[a] See also Table 12.6.
[b] Viscosity not usually a consideration.

Depending on geographic source, **crude petroleum** will reflect a predominance of one or more of the following series, that is, the "base."*

- **Paraffin** series, saturated chain compounds
- **Olefin** series, unsaturated chain compounds
- **Naphthene** series, saturated ring compounds
- **Aromatic** series, unsaturated ring compounds

Fuel and lubrication products[4] are produced by splitting the crude into fractions, mainly by distillation. Each fraction is a mixture whose average molecular weight is related to its boiling range, the heavier fractions are those having higher boiling points. For most purposes, these stocks can be defined in terms of simple physical properties.

9.4.2 Tests and Measures[3,4]

- **API gravity:** a measure of specific gravity, according to

$$\text{degrees API} = \frac{141.5}{\sigma} - 131.5, \qquad (9.9)$$

  where σ is the specific gravity, referred to water at 60°F.
- **Baumé gravity:** a measure of specific gravity, according to

$$\text{degrees Baumé} = \frac{140}{\sigma} - 130. \qquad (9.10)$$

  [For liquids heavier than water $= 145 - (145/\sigma)$.]
- **Saybolt universal viscosity:** a measure of the kinematic viscosity of a liquid; specifically the time, in seconds for 60 cm^3 of liquid to drain through a standardized orifice at a specified temperature.
- **Flash point:** the minimum temperature of a liquid at which vapor issuing from liquid exposed to air will momentarily ignite.
- **Fire point:** the minimum temperature, as above, at which the vapor from the liquid will remain ignited.
- **Pour point:** the maximum oil temperature plus 5°F (2.8°C) at which no discernible movement is apparent in a jar of oil within 5 sec after being tipped 90°.
- **Reid vapor pressure:** a standardized measure of the volatility of liquid; approximately equal to the true vapor pressure of the test liquid.

---

*Naphthene base is sometimes termed *asphalt base.*

- **Fractionation:** the process of distillation by which raw petroleum liquid is separated into mixtures (cuts) of hydrocarbons of relatively narrow boiling ranges.
- **ASTM distillation:** an ASTM nonrectifying test in which a liquid sample is distilled at uniform rate, yielding percent vaporized as a function of liquid temperature, the 10, 50, and 95% points being of particular interest (ASTM D86). Analogous results can be obtained by the Hempel, Saybolt, Engler, and flash vaporization tests.
- **Cetane number:** a measure of ignition quality of a liquid fuel; specifically, the volume percent of cetane in a mixture with α-methyl naphthalene showing the same ignition characteristics as the fuel.
- **Octane number:** a measurement of the knock characteristics in a spark-inputed engine, the percentage by volume of iso-octane which must be mixed with normal heptane to match the knock intensity of the fuel under test. The "motor" and "research" octane numbers are measured at different engine speeds.

9.4.3 Viscosity Conversion

Viscosity standards have been developed using a variety of instruments. Their conversion is shown in Fig. 9.2.

9.4.4 Liquefied Petroleum Gases[5]
Liquefied petroleum gases can be classified as:

- **Commercial propane:** a hydrocarbon product for use where high volatility is required.
- **Commercial butane:** a hydrocarbon product for use where low volatility is required.
- **Commercial PB mixtures:** mixture of propane and butane for use where intermediate volatility is required.
- **Special-duty propane:** a high-quality product composed chiefly of propane, which exhibits superior antiknock characteristics when used as an internal combustion engine fuel.

ASTM Standard Specification D1835-76 establishes detail requirements as shown in Table 9.4.

9.4.5 Transportation Fuels[5,7–9]

Automotive gasolines are defined by the SAE as the fuels for internal combustion engines in which ignition is spark induced and which are used primarily for passenger car and truck service. They are blends of petroleum hydrocarbons, with additives to suppress bad properties or promote good ones. Their properties are a function of geographic location in the United States. The *SAE Handbook* provides recommen-

$$\mu = \nu\rho' = \nu S$$

The empirical relation between Saybolt Universal Viscosity and Saybolt Furol Viscosity at 100 F and 122 F, respectively, and Kinematic Viscosity is taken from A.S.T.M. D2161-63T. At other temperatures, the Saybolt Viscosities vary only slightly.

Saybolt Viscosities above those shown are given by the relationships:

Saybolt Universal Seconds = Centistokes x 4.6347

Saybolt Furol Seconds = Centistokes x 0.4717

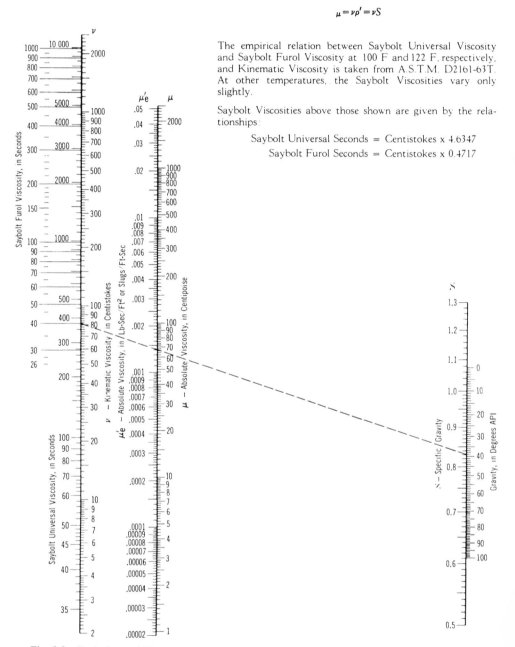

Fig. 9.2 Equivalents of kinematic, Saybolt universal, Saybolt furol, and absolute viscosity. (From Ref. 6. Courtesy of the Crane Co.)

Table 9.4 Detail Requirements for Liquefied Petroleum Gases

Characteristic	Product Designation			
	Commercial Propane	Commercial Butane	Commercial PB Mixtures	Special-Duty Propane[a]
Vapor pressure at 100°F (37.8°C), max., psig (kPa)	208 (1430)	70 (485)	[b]	208 (1430)
Volatile residue Evaporated temperature, 95%, max., °F (°C)	−37 (−38.3)	36 (2.2)	36 (2.2)	−37 (−38.3)
or				
Butane and heavier, max., vol. %	2.5	—	—	2.5
Pentane and heavier, max., vol. %	—	2.0	2.0	—
Propylene content, max., vol. %	—	—	—	5.0
Residual matter				
Residue on evaporation 100 mL, max., mL	0.05	0.05	0.05	0.05
Oil stain observation	Pass[c]	Pass[c]	Pass[c]	Pass[c]
Relative density (specific gravity) at 60/60°F (15.6/15.6°C)	[d]	[d]	[d]	—
Corrosion, copper, strip, max.	No. 1	No. 1	No. 1	No. 1
Sulfur, grains/100 ft³ max. at 60°F and 14.92 psia mg/m³ (15.6°C and 101 kPa)	15 (343)	15 (343)	15 (343)	10 (229)
Hydrogen sulfide content	Pass	—	—	Pass[e]
Moisture content	Pass	—	—	Pass
Free water content	—	None[f]	None[f]	—

[a] Equivalent to propane HD-5 of GPA Publication 2140.

[b] The permissible vapor pressures of products classified as PB mixtures must not exceed 20 psig (1380 kPa) and additionally must not exceed that calculated from the following relationship between the observed vapor pressure and the observed specific gravity:

$$\text{Vapor pressure, max.} = 1167 - 1880 \ (\text{sp. gr. } 60/60°F) \text{ or } 1167 - 1880 \ (\text{density at } 15°C).$$

A specific mixture shall be designated by the vapor pressure at 100°F in pounds per square inch gauge. To comply with the designation, the vapor pressure of the mixture shall be within +0 to −10 psi of the vapor pressure specified.

[c] An acceptable product shall not yield a persistent oil ring when 0.3 mL of solvent residue mixture is added to a filter paper in 0.1-mL increments and examined in daylight after 2 min as described in Method D2158.

[d] Although not a specific requirement, the specific gravity must be determined for other purposes and should be reported. Additionally, the specific gravity of PB mixture is needed to establish the permissible maximum vapor pressure (see footnote b).

[e] An acceptable product shall not show a distinct coloration.

[f] The presence or absence of water shall be determined by visual inspection of the samples on which the gravity is determined.

dations of volatility class for all states in the United States. ASTM Standard Specification D439-79 provides detailed requirements.

Flash point	$-45°F$ ($-43°C$) (approx.)
Autoignition temperature	$500°F$ ($260°C$) (approx.)
Flammability limits, volume in air	1.4 to 7.6%
Vapor pressure at $70°F$ ($21°C$)	4 to 8 psi (28 to 55 kPa)
Concentration in saturated air at $68°F$ ($20°C$), volume	25 to 50% (SAE)

Gasoline flash point is so low that a flammable mixture is always presumed to exist over the liquid in the open. In covered tanks the saturated vapor is likely to be too rich to ignite, but explosive conditions may exist at extremely low temperatures or during transfer operations.

Diesel fuel oils[7] are evaluated by Standard Specification ASTM D975 for Grades 1D, 2D, and 4D. Ignition quality (ASTM D613) is determined by the cetane numbers, or from API gravity and midboiling point. Heating value (ASTM D240) can be either measured in the calorimeter or estimated from the aniline point (ASTM D611) and the API gravity of the fuel.

Jet fuels[7] are kerosene or a kerosene/gasoline mixture having low Reid vapor pressure, 0.2 to 3 psi (1.4 to 2.1 kPa) and very low freezing points, -58 to $-80°F$ (-50 to $-62°C$). Although similar to motor gasolines, quality control for this fuel must be more rigorous for reasons of safety. Tetraethyllead is used to control premature ignition or knocking, which is usually inaudible and can damage or destroy an engine. Substitution of motor gasoline for aviation use introduces additional hazards of vapor lock and carburetor icing.

9.4.6   Fuel Oils[7]

Fuel oils are categorized as either distillates or residual oils. ASTM Standard Specification D396 specifies ranges of properties conducive to good burner performance. The oil grades are:

- **Grade 1:** a light distillate of high volatility for burners of the vaporizing type.
- **Grade 2:** a distillate heavier than Grade 1 oil, for use in burners of the atomizing type to meet domestic or medium commercial duty.
- **Grades 4 and 4 (light):** light residual oil or heavy distillate capable of atomization at relatively low storage temperatures.
- **Grades 5 and 5 (light):** residual fuels of viscosity intermediate between those of Grades 4 and 6, occasionally requiring preheating in cold climates.
- **Grade 6:** sometimes referred to as Bunker C, a high-viscosity residual oil used for commercial and industrial heating, requiring preheating at the storage tanks to aid pumping, and preheating at the burners to support atomization.

Petroleum products are widely available around the world and vary from the light crude oils to residual fuels or the extremely heavy petroleum cokes, the residues from the refining processes. Typically, the lighter grades of products such as the gasolines, No. 2 oils, and so on, are used for transportation purposes and chemical feedstocks. These grades have low flash points and ventilation requirements become important because combustion may occur if insufficient care is given during the transportation, processing, handling, and storage of them. In most regions, lightning protection must be provided as well as explosion-proof electrical systems, and so on, to avoid initiating combustion explosions.

Heavier fuels are used in stationary engines or large energy users such as central station power boilers, refinery boilers, and other major installations that can make an investment in the heating and other equipment necessary to handle the high-viscosity fuels together with their contaminants such as sulfur and vanadium. The heavier fuels require significant heating to reduce the viscosity and may require systems such as steam-traced piping and equipment to maintain the viscosity sufficiently low to permit ready pumping and firing.

The heavy fuels in some cases have a substantial amount of vanadium (as much as 2%), which creates problems of erosion in large boilers or gas turbines using these fuels. For these cases, water washing and centrifuging is sometimes used to remove the vanadium, or chemical agents such as magnesium and zinc may be added to the fuel to reduce the effect of vanadium. The lighter fuels, such as gasoline and naphthas, are liquid at normal temperatures and despite their volatility, their handling and transport are relatively easy. Typically, filtration of these lighter fuels will be necessary because of end-use equipment. In the case of diesel engine unit injectors, filtration systems removing very small particles in the range of 1 micron are necessary.

A typical system for serving a burner with Bunker C grade oil is illustrated in Fig. 9.3. Provision for heating the fuel and recirculating it with positive displacement pumps to the burners is shown. The heating medium—steam—can also be used to smother flames at the holding tank if necessary. The lines are flushed with light oil at system shutdown to prevent freezeup.

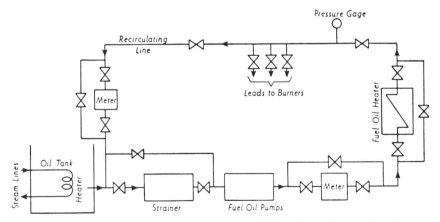

Fig. 9.3 Schematic diagram of fuel-oil system. (From Philip J. Potter, *Power Plant Theory and Design.* Copyright © 1959 by The Ronald Press Co. Reprinted with permission.)

9.4.7   Methanol and Ethanol[10,11]

Methanol is a versatile liquid fuel produced by the distillation of biomass materials, such as wood, peat, coal, and farm products, or by chemical conversion from natural gas. It is relatively low in cost, less flammable than gasoline, and is extremely clean burning, yielding no particulates or sulfur dioxides (Table 9.5). Because of its relatively low combustion temperature, fewer nitrogen oxides are produced than with petroleum-based fuels. Methanol dissolves in water and also has important industrial uses as a solvent. It has a very high octane rating (from 106 to 120) and up to 10% can be added to unlead gasoline to improve its performance. It can be burned at a nominal 100% concentration as a fuel, and small amounts of water (1 to 3%) may be added to assist combustion.

Methanol has been used as an automotive fuel since the 1900s, particularly for high-performance engines where cost is not a significant factor. At the present time (1998), it's being produced and sold commercially at a cost roughly three-fourths that of gasoline. Large-scale production plants coming on line are expected to further reduce its cost. Recently, it has also become the feedstock in a developing process for synthesizing gasoline.

Internal combustion engines can utilize methanol at compression ratios of 12 to 16 or higher to 1, with relatively high combustion efficiencies (roughly twice those utilized for gasoline). To a large extent, this compensates for its lower heating value per unit volume. Diesel engines can use regular fuel for starting and warmup and then switch over to pure methanol so that of the total fuel used approximately 90% is methanol and 10% diesel fuel. For combustion turbines or for boilers, only a minor change of burner tips or combustors is necessary to adapt to this fuel with significantly higher combined cycle efficiency.

Table 9.5   Methanol and Gasoline Characteristics

Characteristic	Methanol	Gasoline
Specific gravity, 60°F (15.6°C)	0.796	0.72
Boiling point	148°F (64°C)	100–400°F (38–204°C)
Freezing point	−143.7°F (−97.6°C)	−40°F (−40°C)
Viscosity, 77°F (25°C)	0.5 cps	0.5 cps
Vapor pressure 100°F (38°C)	0.32 kg/cm^2	0.6–0.84 kg/cm^2
Heat of vaporization at 68°F (20°C)	506 Btu/lb (342 kcal/kg)	150 Btu/lb (101 kcal/kg)
Heat of combustion	8640 Btu/lb (5830 kcal/kg)	18,650 Btu/lb (12,590 kcal/kg)
Flashpoing, closed cup	52°F (11°C)	−40°F (−40°C)
Ignition temperature	878°F (470°C)	800–950°F (427–510°C)
Flammable limits	6–37%	1–8%
Flame speed	1.6 ft/sec (0.48 m/s)	1.1 ft/sec (0.34 m/s)
Octane number	110–112	90–100

Source: Ref. 12.

Ethanol can be produced by fermentation of any carbohydrate, such as saccharin (sugarcane, sugar beets, molasses, and other juices), starch (grains and potatoes), or cellulose (wood, bagasse, and straw). The thermal properties are similar to those of methanol, although chemically it is quite different and much less stable than methanol. Its production cost today (1998) is substantially higher than methanol.

9.4.8 Oil Sands[13]

The **oil sands** of Canada and the United States are a large-scale source of *bitumen,* a crude hydrocarbon exhibiting the properties of an asphalt or tar but capable of being upgraded to fuels. The oil sands contain bitumen, which acts as a matrix holding together particles of sand. When treated with hot water, the agglomerate releases the hydrocarbon, which can be separated as a heavy viscous liquid (8 to 12° API), suitable as a refinery feedstock. The raw material is readily mined as a solid using bucket wheel excavators or draglines from surface deposits.

The hydrocarbon is easily displaced from the mineral by the strong wetting action of water, aided if necessary by surfactants. Commercial ventures use the hot-water process to effect this treatment. A family of methods for development of synthetic oils and extraction of oils from nonconventional sources is under way. The processes are still largely either experimental or developmental at the pilot plant stage and have varying degrees of potential for large-scale economic development. Typically, these processes for shale oil, coal liquefaction, and coal-derived oils require extensive investment in processes and facilities and their development is dependent on and heavily influenced by world market crude oil pricing, with which they directly compete.

9.4.9 Shale Oil

Shale oil is widely found in the United States and elsewhere. In the United States, significant deposits are located in the Rocky Mountains [estimated at 2.2 trillion barrels (bbl)] in thick deposits and in a belt including the Western Appalachian Mountains, Indiana, Illinois, and Michigan (estimated at 400 billion bbl).

The shale contains a form of bitumen that when heated to 900°F (480°C) yields 30 to 40 gal of kerogen/ton (0.12 to 0.17 gal/kg) for western shale and 15 to 18 gal/ton (0.06 to 1.075 gal/kg) for eastern shale. **Kerogen** is a heavy synthetic oil whose release is accompanied by high-quality synthetic gas. Eastern shales tend to produce coke when heated; thus some current processes use hydrogen to reduce this tendency. Other processes utilize the coke directly as a process fuel, yielding oil as well as significant amounts of waste heat.

Current extraction methods utilize mining, crushing, and retorting the shale in large "mine-mouth" plants. These have the disadvantages of the mining and transportation costs as well as the disposal of spent shales. Various *in-situ* processes are under development to liberate the oil by burning a fraction of it underground in place, the resultant heat liberating the kerogen, which flows to sumps where the liquid is pumped to the surface. This process avoids most of the surface mining problems, but presents other technical problems such as methodology and control of underground shale fracturing, maintenance and control of combustion air, rate of advance, and temperature of flame front.

9.4.10   Coal Liquefaction

A variety of processes are available for the production of liquid fuels from coal. Unless the resultant liquids are burned directly as a fuel, these generally require further refining to produce other hydrocarbons of commercial value. At the present time (1998), with the price of crude oil below $25/bbl, none of the processes are sufficiently (economically) feasible to be a significant source of fuel. It appears crude oil prices in excess of $35/bbl or low interest rates (e.g., 5 to 7%) are necessary to stimulate the large investment necessary for full-scale production facilities.

9.5   GASEOUS FUELS

Gaseous fuels are considered by many to be the ideal fuel since they are extremely easy to handle, can be readily controlled, are economically competitive with other fuels, and are clean burning, leaving little or no ash or residue. Their disadvantage lies in their relatively low heating value per unit volume, which requires either a continuous piped supply or a compressed-gas storage system, rendering them some-what unattractive for portable uses.

9.5.1   Characteristics of Gaseous Fuels

Table 9.6 lists the characteristics of the more commonly used gaseous fuels. See also Table 12.6.

9.5.2   Natural Gas

Natural gas is found compressed in porous rock, shale formations, or overlying oil deposits at pressures up to several thousand psi (20×10^3 kPa) or more. As the gas is removed from the field, the pressure drops until the gas in the field can no longer be withdrawn. In many areas, however, where the gas field is large and rock porosity

Table 9.6   Typical Characteristics of Gaseous Fuels

Type	HHV (Btu/ft^3, dry)	Specific Gravity (Air = 1.0)	Nitrogen (%)
Natural gas	1100	0.62	0.5
Blast furnace gas	110	0.97	58
Producer gas	160	0.84	53
Propane (C$_3$H$_8$)	2520	1.56	—
Butane (C$_4$H$_{10}$)	3390	2.01	—
Hydrogen	324	0.07	—

sufficient, pressure reduction is not a problem and production pressure can be considered a constant over substantial periods of time.

Natural gas consists principally of methane with smaller quantities of other hydrocarbons, particularly ethane, although carbon dioxide and nitrogen are usually present in small amounts. Appreciable amounts of hydrogen sulfide are also sometimes present; these are usually removed at the field before transmission.

Oxygen is present only where there has been atmospheric air infiltration, although in some fields nitrogen (in the range of 5 to 30%) or carbon dioxide (to 6.0%) may be found. In some fields in Texas as much as 1% helium is also found.

Natural gas characteristics are determined by underground conditions. Where the field contains oil deposits, the gas may contain heavy saturated hydrocarbons, which are liquid at ordinary temperatures and pressures. This *wet gas* is dried by stripping out the liquids at the wellhead. Where sulfur is present in the oil, the gas will contain hydrogen sulfide. *Dry gas* contains less than 0.1 gal of gasoline vapor per 1000 ft^3 (0.013 L/10^3 L^3) and is produced by wells generally remote from oil-bearing areas. Natural gases are also classified as *sweet* or *sour*. Sour gases contain a high percentage, frequently to 7% or more, of hydrogen sulfide as well as some mercaptans.

The higher heating value of natural gas is usually around 1000 Btu/ft^3 (8.9 kcal/L), and it can be computed by adding together the heat contributed by the volumetric percentage of the various component gases. Natural gas normally is odorless, and methyl mercaptan, a distinctive aromatic, is added to make leak detection easier.

9.5.3   Biomass[14]

The main process by which energy may be obtained from **biomass** includes direct combustion, pyrolysis, hydrogasification, anaerobic digestion, alcoholic fermentation, and biophotolysis. Each technology has advantages and these depend on the biomass source and the type of energy needed.

Among the advantages of utilizing biomass materials as an energy source are: (1) biomass provides an effective low-sulfur fuel; (2) it can provide an inexpensive source of energy (e.g., fuel-wood, dry animal manure, methane gas), and (3) in some cases, processing biomass materials for fuel reduces the environmental impact for these materials (e.g., biomass from sewage and processing wastes).

The major difficulties in utilizing biomass materials from solar energy conversion are: (1) the relatively small percentage (less than 0.1%) of radiant energy converted into biomass by plants; (2) the relatively sparse and low concentration of biomass per unit area of land and water (causing high labor costs for collection and transport); (3) the scarcity of additional land suitable for growing plants; and (4) the high moisture content (50 to 95%) of biomass that makes collection and transport expensive and energy conversion relatively inefficient. All of these factors make biomass energy costly in terms of energy and other activities expended in the conversion process and reduce the net energy yield. In general, the use of wastes such as livestock manure and municipal sewage for biomass generation would help reduce environmental problems associated with the management of these wastes. In addition, the residual materials remaining after processing can be used as fertilizers.

Some environment benefits occur in the conversion of urban and industrial wastes for heat and electricity. These organic wastes generally have a low sulfur content, their combustion under carefully managed conditions would cause minimal air pollution problems and reduce the need for landfills, and the residue ash can (dependent on chemical makeup) be used as a fertilizer for crops and lawns. Urban and industrial wastes can also be converted into liquid and solid fuels by pyrolysis. The pyrolysis of these materials requires the majority of the combustion energy of the original wastes and, therefore, is less attractive as an energy source. Production of oil, char, and gas gives more freedom to the energy needs with the fuels produces.

The use of surplus sugar crops for the production of ethanol is feasible in areas where availability of fertile land is not a constraint and when plentiful solar radiation and manpower are available. The lack of enough land for sugarcane raising and the comparatively lower solar radiation are among the causes for the low potential contribution of sugar crops to provide energy in the United States. The potential of biomass energy conversion from various biological materials can be categorized as shown in Table 9.7.

9.6 NUCLEAR ENERGY

9.6.1 Equivalence of Energy and Mass

In nuclear physics it is the composite of mass and energy that is conserved. From *Einstein's law,* when velocity is small relative to c,

$$E = mc^2. \tag{9.11}$$

For example, complete conversion of 1 gram of matter into energy yields 9×10^{20} ergs or 2.5×10^7 kilowatthours of energy. In the practical development of nuclear power, the yield of energy is very small, only about 0.1% of the mass of U-235 being converted to energy in the fission process, for example.

Table 9.7 Production of Biofuels from Various Biomass Sources

Biomass Substrate	Conversion Technology	Biofuels Produced
Livestock manure	Anaerobic digestion	CH_4
Urban refuse	Pyrolysis	Fuel oil
		Gas
		Char
Urban refuse	Incineration	Heat/electricity
Food processing waste	Anaerobic digestion	CH_4
Corn	Ethanol fermentation	C_2H_5OH
Sugar crops	Ethanol fermentation	C_2H_5OH
Forest biomass	Incineration	Heat/electricity
Municipal sewage	Anaerobic digestion	CH_4

Source: P. Auer, *Advances in Energy Systems and Technology,* Vol. 1, © 1978, Academic Press. Used with permission.

Mass Defect. When a compound nucleus has less mass than the sum of the masses of its particles or nucleons, it is said to have a **mass defect.** The mass defect is equivalent to the energy radiated when the particles combined or to the energy required to separate the nucleus into its particles, which is called the **binding energy.** The binding energy per nucleon increases up to mass number of about 50 and then decreases gradually with increasing mass number (see Fig. 9.4).

9.6.2 Nuclear Reactions

The conventional notation for a nuclear reaction is

$$(_z C^A)_1 + (_z C^A)_2 \longrightarrow (_z C^A)_3 + (_z C^A)_4 + Q, \tag{9.12}$$

in which z is the number of protons, A is the mass number, C is the chemical symbol for the atom, electron, or nucleon, and Q is the energy released.

The conservation equations are

$$\sum Z_i = 0 \quad \text{and} \quad \sum A_i = 0. \tag{9.13}$$

$$\text{initial mass} - \text{final mass} = Q. \tag{9.14}$$

An example is the **fission reaction:**

$$_{92}U^{235} + {}_0n^1 \longrightarrow {}_{92}U^{236} \longrightarrow {}_{38}Sr^{94} + {}_{54}Xe^{140} + 2{}_0n^1 + Q. \tag{9.15}$$

The strontium and xenon products are highly radioactive and decay further to other products. The final result is a spectrum of products. Table 9.8 gives frequently encountered constants.

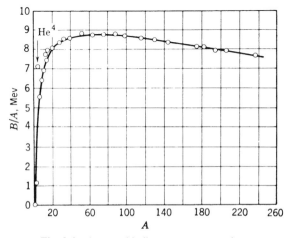

Fig. 9.4 Average binding energy per nucleon.

Table 9.8   Some Nuclear Constants

	amu	grams $\times 10^{-24}$	Rest Energy (MEV)
Unit mass, m	1	1.65990	931.16
Electron, $_1e^0$ or β^-	0.00054862	0.00091091	0.51083
Proton, p or $_1p^1$	1.007595	1.67247	938.17
Neutron, n or $_0n^1$	1.008983	1.67472	939.43
Hydrogen atom, $_1H^1$	1.00812	1.67338	938.68
Alpha particle, α or $_2He^4$	4.00280	6.64424	3727.07

Charge on electron $(1.60206 \pm .00007) \times 10^{-19}$ coulomb.
Radius of electron $(2.81784 \pm .00010) \times 10^{-13}$ cm.
Radius of nucleus $(1.5 \pm 0.15) \sqrt{A} \times 10^{-13}$ cm.

Modes of Radioactive Decay

Negative beta (electron) emission:

$$_0n^1 \longrightarrow \;_{-1}e^0 + \;_1H^1 + \text{neutrino}. \tag{9.16}$$

$$_{38}Sr^{94} \xrightarrow{\beta-} \;_{39}Y^{94} \xrightarrow{\beta-} \;_{40}Zr^{94}. \tag{9.17}$$

$$_{54}Xe^{140} \longrightarrow 4(_{-1}e^0) + \;_{58}Ce^{140}. \tag{9.18}$$

Positive beta (positron) emission:

$$_7N^{13} \longrightarrow \;_{+1}e^0 + \;_6C^{13} + \text{neutrino}. \tag{9.19}$$

$$_{+1}e^0 + \;_{-1}e^0 \longrightarrow 2 \text{ gammas of 0.51 MeV each}. \tag{9.20}$$

Alpha emission:

$$_{94}Pu^{239} \longrightarrow \;_2He^4 + \;_{92}U^{235}. \tag{9.21}$$

Neutron emission:

$$_{53}I^{137} \xrightarrow{\beta-} \;_{54}Xe^{137} \longrightarrow \;_0n^1 + \;_{54}Xe^{136}. \tag{9.22}$$

Orbital electron (K) capture:

$$_{29}Cu^{64} + \;_{-1}e^0 \longrightarrow \;_{28}Ni^{64}. \tag{9.23}$$

Gamma emission by:

1. Ejection of a gamma photon from an excited nucleus
2. Isomeric transition of a nucleus from one energy level to another
3. Annihilation of an electron following positive beta emission

Nuclei with an excess of neutrons are usually electron emitters. Among nuclei with a deficiency of neutrons, the heavy ones usually decay by alpha emission, and the light ones by positron emission or orbital electron capture. Gamma emission often accompanies other types of decay.

9.6.3 Decay with Time

From statistical considerations, the rate of decay (alpha emission) is proportional to the number N of radioactive nuclei present,

$$\frac{dN}{dt} = -\lambda N, \tag{9.24}$$

where the proportionality constant λ is called the **disintegration constant** or the **radioactive decay constant.** Integration of Equation (9.24) gives

$$N = N_0 e^{-\lambda t}, \tag{9.25}$$

where N_0 is the initial number of radioactive nuclei. The **half-life** ($t_{1/2}$) the time required for one-half of the original atoms to decay, by substitution in Equation (9.25), is

$$t_{1/2} = \frac{0.693}{\lambda}. \tag{9.26}$$

When there is more than one radioisotope in the decay chain,

$$P \xrightarrow{\lambda_1} Q \xrightarrow{\lambda_2} R \xrightarrow{\lambda_3} \tag{9.27}$$

For the second member or daughter Q,

$$\frac{N_Q}{N_{P_1}} = \frac{\lambda_1}{\lambda_2 - \lambda_1} e^{-\lambda_1 t} + \frac{\lambda_1}{\lambda_1 - \lambda_2} e^{-\lambda_2 t}. \tag{9.28}$$

If the half-life of the parent P is longer than the half-life of the daughter $Q(\lambda_2 > \lambda_1)$, after a lapse of time $e^{-\lambda_2 t}$ becomes negligible and

$$\frac{N_Q}{N_P} = \frac{\lambda_1}{\lambda_2 - \lambda_1} e^{-\lambda_1 t}. \tag{9.29}$$

With naturally occurring radioisotopes, the half-life of the parent is very long compared to that of the daughter, so that Equation (9.29) reduces to

$$\frac{N_Q}{N_P} = \frac{\lambda_1}{\lambda_2} e^{-\lambda_1 t}. \tag{9.30}$$

9.6.4 Characteristics of Selected Radioisotopes

Half-life and emitted particle energy for selected radioisotopes are shown in Table 9.9.

9.6.5 Biological Effects

Relative biological effectiveness (RBE) is the ratio of gamma radiation to another type of radiation that produces the same biological effect, both expressed in rads. The **roentgen equivalent man** (rem) is the unit of absorbed radiation dose for biological effects. It is defined as

$$\text{rem} = \text{RBE} \times (\text{number of rad}). \tag{9.31}$$

The maximum permissible external radiation dose for adult workers in the radiation industries is 1.25 R per quarter. A single chest x-ray is about (0.2 rem) and continuous exposure to natural background radiation at sea level and $50°$ latitude is about one-thirtieth of this rate (0.0033 rem per week) (Fig. 9.5).

   The ingestion of radioactive materials in air, water, or food presents hazards that depend not only on the concentration and absorption but also on the rate of elimination. For example, Sr-90 with a long half-life is absorbed in bone structure and persists for years.

9.6.6 Shielding

Since alpha and beta particles have relatively short range, **shielding** is provided mainly for neutron and gamma radiation. Shielding for neutrons usually involves thermalization of the neutrons followed by their absorption in a material with high-absorption cross section. Neutron shielding also involves provision for photons from neutron–gamma reactions in the shield. Gamma shielding normally is provided by a material with a high absorption coefficient or by large thickness. Design considerations include geometrical arrangement and balance of thickness and effectiveness of materials against cost. Effectiveness depends both on the material and on the energy of radiation, as shown in Table 9.9.

   The attenuation with distance x is the intensity I of gamma radiation from a plane source is given by

$$\frac{I_x}{I_0} = be^{-\mu x}, \tag{9.32}$$

and from a point source surrounded by a spherical shield by

$$J = \frac{bI_0 e^{-\mu x}}{4\pi x^2}, \tag{9.33}$$

where I_0 is the intensity at the source (MeV/cm$^2 \cdot$ s), μ is the attenuation coefficient

Table 9.9 Characteristics of Selected Radioisotopes

Element	Mass Number	β (MeV)	γ (MeV)	Half-life
Aluminum	29	2.5	—	6.7 m
Antimony	122	1.36, 1.94	0.57	2.8 d
	124	0.74, 2.45	1.72	60 d
	125	0.3, 0.7	0.55	2.7 y
Arsenic	77	0.8	None	40 h
Beryllium	10	0.56	None	3×10^6 y
Bismuth	210	1.17	None	5 d
Bromine	82	0.465	0.547, 0.787, 1.35	34 h
Cadmium	115	0.6, 1.11	0.65	2.8 d
Calcium	45	0.2, 0.9	None	180 d
	49	2.3	0.8	2.5 h
Carbon	11	$0.97e^+$	—	20.5 m
	14	0.145	None	5100 y
Cerium	141	0.6	0.21	28 d
Chlorine	36	0.66	None	10^6 y
	38	1.1, 2.8, 5.0	1.65, 2.15	37 m
Cobalt	60	0.31	1.10, 1.30	5.3 y
Columbium	95	0.15	0.75	35 d
Copper	61	0.9, 1.23	None	3.4 h
Europium	154	0.9	1.4	5.4 y
	155	0.23	0.084	2 y
Fluorine	18	0.7	—	1.86 h
Gallium	72	0.8, 3.1	0.84, 2.25	14.1 h
Germanium	71	1.2	—	40 h
	77	1.9	—	12 h
Gold	198	0.98	0.12, 0.41	2.7 d
	199	1.01	0.45	3.3 d
Hafnium	181	0.8	0.5	46 d
Hydrogen	3	0.011, 0.015	None	12 y
Iodine	131	0.315, 0.600	0.367, 0.080, 0.284, 0.638	8.0 d
Iridium	192	0.67	$0.137 \rightarrow 0.651$ (12γ)	74.7 d
	194	0.48, 2.18	0.38, 1.43	19 h
Iron	59	0.26, 0.46	1.1, 1.3	46.3 d
Krypton	85	1.0	0.17, 0.37	4.5 h
Lanthanum	140	1.32, 1.67, 2.26	$0.093 \rightarrow 2.5$	40 h
Magnesium	27	0.9, 1.80	0.64, 0.84, 1.02	9.58 m
Mercury	203	0.205	0.286	43.5 d
Molybdenum	99	0.445, 1.23	0.04, 0.741, 0.780	68.3 h
Neodymium	147	0.17, 0.78	0.035, 0.58	11 d
Nitrogen	13	$1.23e^+$	None	10.1 m
Osmium	191	0.142	0.039, 0.127	15.0 d
	193	1.15	1.58	32 h
Phosphorus	32	1.718	None	14.3 d
Platinum	197	0.65	None	18 h
Potassium	40	1.40	1.45, K	9.9×10^8 y
	42	3.5	None	12.4 h
Praeseodymium	142	0.636, 2.154	1.57	19.1 h
	143	0.932	None	13.8 d
Promethium	147	0.229	None	2.26 y
Rhenium	186	0.64, 0.95, 1.09	0.132, 0.275, 1.70	92.8 h
Rhodium	105	0.57	0.33	36.2 h

Table 9.9 *(Continued)*

Element	Mass Number	β (MeV)	γ (MeV)	Half-life
Rubidium	86	0.72, 1.80	1.08	19.5 d
Ruthenium	103	0.205, 0.670	0.494	42 d
Samarium	153	0.68, 0.80	0.070, 0.103, 0.61	47 h
Scandium	46	0.36, 1.49	0.89, 1.12	85 d
Silicon	31	1.5	None	2.59 h
Silver	110	0.087, 0.53	0.885, 0.935, 0.1389, 1.516	270 d
	111	1.06	None	7.5 d
Sodium	22	$0.58e^+$	1.3	2.6 y
	24	1.390	1.380, 2.758	15.0 h
Strontium	89	1.50	None	53 d
	90	0.54	None	19.9 y
Sulfur	35	0.167	None	87.1 d
Tantalum	182	0.52 m, 1.1	$0.05 \rightarrow 1.24(33\gamma)$	115 d
Technetium	97	IT	0.097	90 d > 10^3 y
	99	0.30	0.140	2.1×10^5 y
Tellurium	127	IT ~ 0.8	0.089	90 d, 9.3 h
	129	IT 1.46	0.102, 0.3, 0.8	32 d, 72 m
	131	IT > 1.8	0.177	30 h, 25 m
Thallium	206	0.58	None	2.7 y
Titanium	51	0.36	1.0	72 d
Tungsten	185	0.428	None	73.2 d
	187	0.627, 1.318	0.086, 0.70	24.1 h
Yttrium	90	2.24	None	61 h

(see Table 9.10), b is the buildup factor for shields of appreciable thickness, and J is the current density (MeV/cm^2 · s). The factor b, determined experimentally, is found to depend on μx and I_0, as well as on the shield material. Equations (9.32) and (9.33) can be applied to the attenuation of thermal neutrons where $\Sigma = \mu$. Attenuation of fast neutrons is a much more complex problem and requires specialized knowledge.

The energy of radiation attenuated by a shield is nearly all converted to heat energy. With the assumption that all the energy is absorbed within one free path, the maximum temperature rise in a plane shield is given by

$$\Delta T = \frac{I_0}{K\mu}, \tag{9.34}$$

and in a spherical shield by

$$\Delta T = \frac{I_0}{4\pi K}\left(\frac{1}{r_1} - \frac{1}{r_2}\right), \tag{9.35}$$

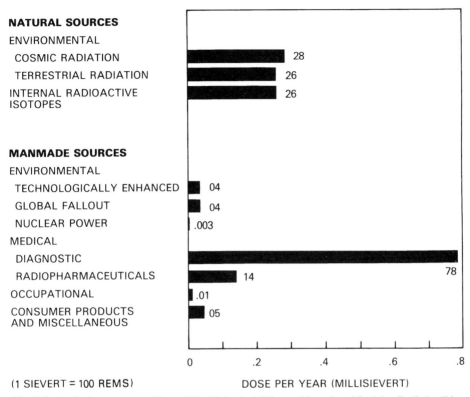

NATURAL SOURCES

ENVIRONMENTAL

COSMIC RADIATION — 28

TERRESTRIAL RADIATION — 26

INTERNAL RADIOACTIVE ISOTOPES — 26

MANMADE SOURCES

ENVIRONMENTAL

TECHNOLOGICALLY ENHANCED — 04

GLOBAL FALLOUT — 04

NUCLEAR POWER — .003

MEDICAL

DIAGNOSTIC — 78

RADIOPHARMACEUTICALS — 14

OCCUPATIONAL — .01

CONSUMER PRODUCTS AND MISCELLANEOUS — 05

0   .2   .4   .6   .8

(1 SIEVERT = 100 REMS)   DOSE PER YEAR (MILLISIEVERT)

Fig. 9.5 Radiation exposure. (From "The Biological Effects of Low-Level Ionizing Radiation," by A. C. Upton. Copyright © 1982 by Scientific American, Inc. All rights reserved.)

Table 9.10   Shielding Properties of Materials

	Water	Iron	Lead	Portland Concrete	Barite Concrete
Density, g/cm^3	1.00	7.78	11.3	2.37	3.49
Thermal neutrons Σ, cm^{-1}	0.100	0.156	0.113	0.094	0.094
γ at 0.5 MeV					
μ, cm^{-1}	0.096	0.653	1.72	0.20	0.30
Build-up b					
at $\mu x = 1$	2.46	2.80	1.51		
At $\mu x = 10$	71.5	34.2	3.01		
γ at 6 MeV					
μ, cm^{-1}	0.027	0.237	0.503	0.065	0.109
Build-up b					
At $\mu x = 1$	1.46	1.30	1.14		
At $\mu x = 10$	5.18	7.10	4.20		

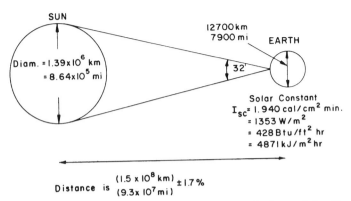

Fig. 9.6 Sun–earth relationships. (From J. A. Duffie and W. A. Beckman, *Solar Thermal Energy Process.* Copyright © 1974 by John Wiley & Sons, Inc. Reprinted with permission.)

where K is the thermal conductivity of the shield (in MeV/s · cm · °C). Shielding properties of some selected materials are shown in Table 9.10.

9.7 SOLAR ENERGY[15–17]

9.7.1 Solar Characteristics

The relationship of the earth and sun with respect to the incident solar energy is shown in Fig. 9.6. Variation of earth–sun distances during the year affect the amount of radiation received (Fig. 9.7). Note that the radiation theoretically received is actually lower in the summer because the earth–sun distance is greater.

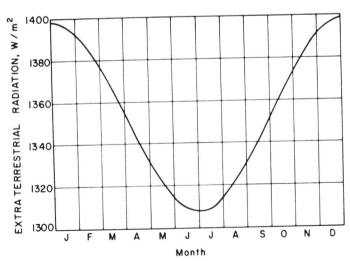

Fig. 9.7 Variation of the extraterrestrial solar radiation with time of the year. (From J. A. Duffie and W. A. Beckman, *Solar Thermal Energy Process.* Copyright © 1974 by John Wiley & Sons, Inc. Reprinted with permission.)

Solar radiation received can be estimated:

$$H_{av} = H'_o \left(a + b \frac{n}{N} \right) \tag{9.36}$$

where H_{av} is the average horizontal radiation for the period in question (e.g., month), H'_o is the clear-day horizontal radiation for the same period, n is the average daily hours of bright sunshine for same period, N is the maximum daily hours of bright sunshine for same period, and a and b are constants. Values frequently used are $a = 0.35$ and $b = 0.61$. Values of H'_o for use in Equation (9.36) can be obtained from Fig. 9.8.

9.7.2 Solar Utilization Systems

Solar utilization systems can be defined as either concentration or collector systems, with direct conversion a further variation. The use of systems with reflectors or concentrators provide ways to increase the energy level to the point that steam can be generated from solar rays and the steam used to generate electric power. This development appears most promising, although it does have the disadvantage of high first cost and the requirement for a large amount of land to gather the relatively low-energy incident sunlight. A potential problem with this is that little data is available on maintenance requirements for the solar collector arrays and the problems of maintaining a sufficiently high level of reflectivity of the mirror reflector. High concentration ratios that yield very high operating fluid temperatures are readily achievable with some systems (Fig. 9.9). These tend to fall into a lower range group, 600°C and below, which is the basis for most of the pilot applications under way and very high temperature applications still largely considered experimental.

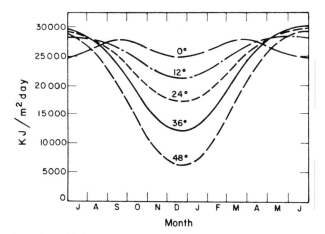

Fig. 9.8 Clear-day solar radiation on a horizontal plane for various latitudes. (From J. A. Duffie and W. A. Beckman, *Solar Thermal Energy Process.* Copyright © 1974 by John Wiley & Sons, Inc. Reprinted with permission.)

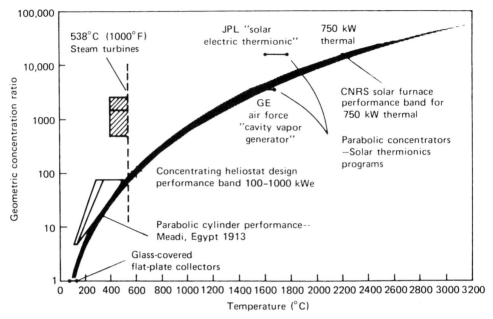

Fig. 9.9 Typical temperature performance of solar collectors. (From F. Kreith and J. F. Kreider, *Principles of Solar Engineering.* Copyright © 1978 by Hemisphere Publishing Corporation. Reprinted with permission.)

9.7.3 Self-Contained Collectors

Self-contained collectors typically are of the parabolic cross-section type with the working fluid circulating through a pipe or pipes at the focal line of the mirror system (Fig. 9.10). A tracking system is needed, but the system need not focus as accurately as in a concentrator-type system because the solar energy is only reflected a short

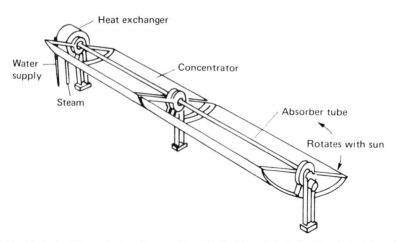

Fig. 9.10 Typical self-contained collector. (From F. Kreith and J. F. Kreider, *Principles of Solar Engineering.* Copyright © 1978 by Hemisphere Publishing Corporation. Reprinted with permission.)

distance to the focus of the reflector. The system must be arranged to provide for insulation of the working fluid as well as a positive head for movement of the fluid (generally through pumping). (The use of the thermal siphon can be considered for those installations where the utilization equipment is positioned significantly above the collector.)

Flat-plate collectors are also suitable for large arrays and can be made relatively efficient. The flat-plate mirror systems have the advantage of low-cost fabrication and low maintenance, but require careful focusing (as with the single- and double-curvature devices) to achieve reasonable concentration ratios.

9.7.4 Mirror and Tracking Systems[15]

Mirror and tracking systems have high efficiency factors and can achieve temperatures above 3000°C when grouped in multiple arrays. Arrays can be furnished as either single- or double-curvature devices; the multiple-curvature devices provide higher concentration ratios (Fig. 9.11). Tracking types include either intermittent tilt change types or continuously tracking reflectors, refractors, or receivers. If oriented east–west, they require an approximate ±30°/day motion; if north–south, a 15°/hr motion. Both must accommodate to a ±23°/yr declination excursion. Single-curvature devices may be of either type but double-curvature devices are usually of the continuously tracking, high-concentration ratio type.

One company has developed a concentrating solar collector that can yield temperatures up to about 530°F (277°C). The system uses a parabolic trough reflector with special heat-transfer fluid flowing through tubes placed in transparent envelopes at the parabola's focus; air having been exhausted from the envelopes. The reflector tracks the sun; its movement around a single axis is controlled by a microcomputer that draws relevant information from a series of sensors.

The reflectors and collectors are built in standard modules with an area of 135 ft³ (12.5 m²) each. There is an automatic flushing device and wiper to remove accu-

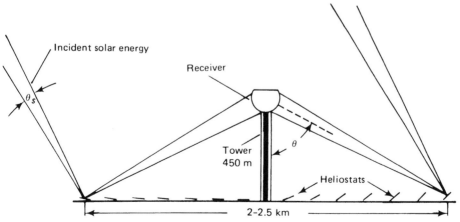

Fig. 9.11 Cross-sectional view of solar power tower system. (From F. Kreith and J. F. Kreider, *Principles of Solar Engineering.* Copyright © 1978 by Hemisphere Publishing Corporation. Reprinted with permission.)

mulating dust. As many modules as desired can be assembled, which are controlled by a single computer unit that tracks the sun, activates the cleaning equipment, operates the heat-transfer fluid pump, and otherwise manages the system for optimum productivity. The units have a design capability to collect an average of 2.7 kWh/day/m^2 of collector area in sunny locations.

9.7.5 Direct Conversion[15]

Thermal conversion utilizes the solar absorption phenomenon with its accompanying heating. Solar energy collectors working on this principle consist of a dark surface facing the sun, which transfers part of the energy it absorbs to a working fluid in contact with it. To reduce atmospheric heat losses, one or two sheets of glass are usually placed over the absorber surface to improve its efficiency. These types of thermal collectors suffer from heat losses due to radiation and convection, which increase rapidly as the temperature of the working fluid increases. Improvements such as the use of selective surfaces, insulation, evacuation of the collector to reduce heat losses, and special kinds of glass are used to increase the efficiency of these devices. These simple thermal-conversion devices are called *flat-plate collectors.* They are available today for operation over a range of temperatures up to approximately 200°F (93°C). These collectors are suitable mainly for providing hot service water and space heating and may operate absorption-type air-conditioning systems.

The thermal utilization of solar energy for the purpose of generating low-temperature heat is at the present time (1998) technically feasible and economically viable for producing hot water and heating swimming pools. In some parts of the world thermal low-temperature utilization is also economically attractive for heating and cooling buildings. Other applications, such as the production of low-temperature steam are under development.

Production of higher working temperatures requires the use of focusing devices in connection with a basic absorber–receiver. Operating temperatures as high as 3700°C (6740°F) have been achieved in the Odeillo solar furnace in France, and the generation of steam to operate pumps for irrigation purposes has also proved technologically feasible. At the present time a number of focusing devices for the generation of steam to produce electric power are under construction in different regions of the world, and estimates suggest that the cost of solar power equipment in favorable locations may be reduced to the range of $1500/per kilowatt of capacity. To compare with costs for conventional power sources, the percent availability of sunshine must be considered (i.e., number of hours/day of sunshine, number of days with significant cloud cover, rain, etc.).

9.7.6 Photovoltaic Conversion[15]

Photovoltaic solar cells utilize energetic photons of the incident solar radiation directly to produce electricity. This technique is often referred to as **direct solar conversion.** Conversion efficiencies of thermal systems are limited by collector temperatures, whereas the conversion efficiency of photo cells is limited by other factors.

The conversion of solar radiation into electrical energy by means of solar cells has been developed as a part of satellite and space-travel technology. The theoretical efficiency of solar cells is about 24%, and, in practice, efficiencies as high as 15% have been achieved with silicon photovoltaic devices. The technology of photovoltaic conversion is well developed, but large-scale application is hampered by the high price of photo cells. A cost reduction of the order of 10:1 will be necessary before photovoltaic conversion can be economically viable.

Efforts are currently under way to develop new technologies for producing solar cells in the hope that such methods might lead to an appreciable lowering in price. Silicon cells, which have performed well for spacecraft application, are still expensive. Efforts are directed toward producing thin monocrystalline foil-bands because such a method might lead to appreciable lowering of the production cost. Also, cell materials are being considered, but unfortunately, the lower price is somewhat offset by lower efficiency of these polycrystalline cells. Copper–indium–gallium selenide cells (one of the candidates) have an efficiency of about 16%, with a potential of 20%.

9.8   GEOTHERMAL ENERGY[18,19]

Geothermal energy utilizes the heat within the earth, either from steam venting through human-made wells or by heating water injected into the earth through wells. In several locations (Larderello, Italy; Mexico, the Philippines, Indonesia, New Zealand; and California, Nevada, and Utah in the United States) geothermal energy has been developed sufficiently to yield significant amounts of energy.

Geothermal steam is utilized most widely for **electric power generation.** The cycles most frequently employed are *direct,* sometimes called *flash* (often using a steam–liquid–solids separator); *binary,* where a heat exchanger transfers heat to a working hydrocarbon fluid, such as fluorocarbon, isobutane, or pentane, which is then used in a closed-loop **turbine** generator cycle; and *hybrid,* where combustible gases may be present and which are separated and burned for power generation separately from a turbine generator cycle used for the balance of the steam energy. Further, wells are broadly classified as either *water dominated* or *steam dominated,* depending on the percentage of liquid entrained.

Geothermal steam generally includes noncondensibles (gases) as well as hydrogen sulfide and carbonates. When condensed, the **hydrogen sulfide** converts to weak sulfuric acid, and combined with the aerated water of the cooling side, requires widespread use of corrosion-resistant materials (typically type 316 stainless steel or titanium for wetted portions). For power recovery applications utilizing turbines, minimal steam treatment is required. Equipment to provide water and suspended solids removal is necessary. Very large air ejector capacity is necessary to remove the non-condensibles.

Wellhead steam conditions for steam-dominated systems are typically near saturation, with perhaps 10 to 20°F (6 to 11°C) of superheat, flows of 50,000 lb/hr (22,700 kg/hr) per well, and flowing at 120 psig (8.4 kg/cm²). Well bores are completed generally with 8 to 12 in. (20 to 30 cm)-diameter casing. Reservoir pressures are initially 500 psig, but mature systems often have reservoir pressures down to 150 psig (10.5 kg/cm²). Most systems utilize some form of reinjection of the geothermal liquid to extend the life of the steam fields.

In some water-dominated systems, 2 to 3 lb (0.9 to 1.4 kg) of water may be produced per pound of steam flow. Disposal of spent geothermal fluids needs to be considered carefully with disposition of solids, pH control, and emissions the most prevalent problems. Reservoir pressure and volumetric flow declines have to be planned for in any geothermal project's economics. The characteristics of any geothermal field will vary widely; however, Table 9.11 indicates some typical geothermal data.

9.9 WIND ENERGY[20]

Although in use for centuries as distributed single-purpose installations, **wind energy** has not until recently been considered seriously for electric power generation or central power grid use. Previous uses included grain milling, water pumping, and some general manufacturing applications. Recently, with the development of inexpensive voltage regulators, limited windmill output can be compatible with the voltages used by public utility systems. Installed in clusters, as wind farms, at favorable locations, they are becoming more common and provide for significant additions to the power generation mix.

Designs principally use three or four bladed horizontal-axis machines mounted on towers as well as vertical-axis Darrius-type rotors. While the vertical-axis machines

Table 9.11 Typical Geothermal Data

Place	Activity	Description
Italy Larderello	Power generation	420,000 kW, 27 units
	Average well 37,500 lb/hr (17,000 kg/hr) at 30–100 psig (210–700 kPa), 285–430°F (140–204°C), 1–20% noncondensibles	
Iceland Reykjavik	Primary space heating plus power generation	65% of total island usage for 214,000 people, 100% of Reykjavik, 30,000 kW
	Average steam content, 1000 ppm dis. solids, 300–350°F (149–177°C), 1% noncondensibles	
New Zealand Wairakei, Kawerau, Rotorua	Power generation plus lumber drying, space heating	200,000 kW
	Average well 20,000–35,000 lb/hr (9–16,000 kg/hr) at 90–125 psig (628–873 kPa), 350°F ± (177°C), 0.5–2.5% noncondensibles	
California The Geysers	Power generation	2,000,000 kW, 22 units
	Average well 40,000 lb/hr (18,000 kg/hr) at 130 psig (9.1 kg/cm²), 350°F ± (177°C), 0.3–1% noncondensibles	

have their machinery located at ground level, a maintenance advantage, they are not usually tall enough to receive higher-velocity wind, away from ground effects. Large machines with two blades are still considered developmental but are becoming more common.

Power available is proportional to the cube of the wind speed, so even small increases above the nominal 12 mph (19 km/hr), considered the baseline for power generation, can increase output significantly. Variable-speed horizontal-shaft machines typically use tower-mounted synchronous generators and gearing, as well as solid-state electronic inverters to provide constant output voltage. Machine sizes in the range 300 to 500 kW are commonly available, with larger sizes evolving continuously.

Blade contamination from dust, insects, and such, together with surface pitting, can substantially reduce output, by 30% or more; hence blade washing programs are widely used. Blade design is critical to satisfactory performance, and the development and testing of blades with improved performance across a wide spectrum of wind velocities and with extended lives is receiving great attention. Generally designed to start generating in winds of 12 to 14 mph (19 to 23 km/hr), horizontal-shaft designs often have variable-pitch blading, and all types are typically designed to withstand winds up to 125 mph (200 km/hr).

9.10 FUEL CELLS[21-24]

Fuel cells are direct energy conversion devices in which the conversion from chemical to electrical energy occurs without the intermediate steps of combustion, conversion to mechanical energy, and (by a generator) conversion to electrical energy. The fuel cell process is a simple step where the chemical energy from a methanol or a hydrocarbon fuel such as coal, oil, or gas and an oxidizing agent such as oxygen or air are converted directly to electrical energy (Fig. 9.12). Since the conversion is direct, it is extremely efficient and fuel cells used in aerospace applications have achieved efficiencies greater than 75%.

Basically, the fuel cell consists of three principal elements: an anode, an electrolyte, and a cathode. Hydrogen-rich fuel is fed to the anode, where the hydrogen loses its electrons. On the other side of the cell, oxygen is fed to the cathode, which picks up electrons and develops a positive charge on the cathode. The flow of excess electrons proceeds from the anode toward the cathode and creates the electric power, which is the useful output of the cell. The waste product of the excess hydrogen ions from the anode and the oxygen ions from the cathode pass into the electrolyte where they combine to form water, which is exhausted from the cell. Because of its elevated operating temperature, the water is exhausted as steam. The electron flow produces dc energy which for many applications must be converted to ac, requiring the use of a converter in the system. Typically, ac-to-dc converters have efficiencies on the order of 90%, so that the conversion is highly efficient and little energy is lost in this step.

The flexibility of fuel cells when arranged in systems is a major advantage because the fuel cells can be stacked and arranged in a modular fashion and connected in parallel to provide sufficient power. Potentially, there are no major restrictions as to size or capacity, only those limited by the mechanical strength of the structural array.

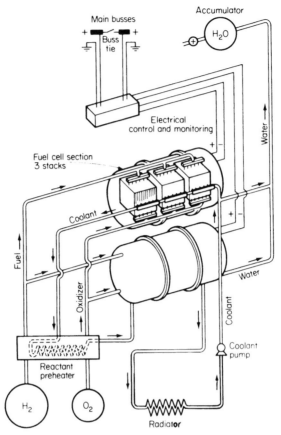

Fig. 9.12 Gemini fuel cell schematic diagram. (From A. McDougall, *Fuel Cells.* © 1976. Reprinted by permission of Macmillan, London and Basingstoke.)

Fuel cells operate efficiently at part load as well as full load and provide very rapid response to variations to electrical demand. For this reason they have a high potential for use in systems that have widely varying demands.

A major advantage of the fuel cells is the flexibility to accept a variety of fuels, including methanol and hydrocarbons such as light distillates, natural gas, and high-, medium- or low-Btu gases, although for some fuels it may be necessary to process the fuel to assure that it is clean enough to avoid contamination of the anode. Other advantages of the fuel cell plants include low water requirements, quiet operation, and limited emissions that are environmentally compatible. These help to make fuel cell plants an attractive concept for distributed power generation. It is possible to cool these plants by low-speed fans and, since combustion processes are not involved, the emissions from these plants are low temperature and do not yield the nitrous oxides, unburned hydrocarbons, and other typical products of combustion that we find with combustion engines or the particulate discharge and the sulfur compounds in the flue gases released with larger fossil fuel burning units.

The 200°C **phosphoric acid fuel cell** (PAFC) is the only fuel cell available commercially today (1998) and one standardized size will produce 200 kW of electricity and approximately 200 kW of heat at 120°C per hour when consuming 50 m³ of natural gas. The power output is readily connected to the local utility grid, while the thermal output can be used for space or hot water heating. The cost of this unit is approximately $3000 per kW, with significant reductions of up to 40 to 50% expected over the next several years.

Other types under development include the 200°C *proton exchange membrane* (PEM) fuel cell, the 650°C *molten carbonate fuel cell* (MCFC), and the 100°C *solid oxide fuel cell* (SOFC). Development work on these types is intensive, and it is likely that the near term will see significant improvements and the beginnings of their commercialization.

REFERENCES

1. *Annual Energy Review, 1995,* U.S. Department of Energy, Washington, D.C., 1996.

2. J. G. Singer, *Combustion-Fossil Power Systems,* Combustion Engineering, Inc., Windsor, Conn., 1981.

3. W. L. Nelson, *Petroleum Refinery Engineering,* 3rd ed., McGraw-Hill Book Company, New York, 1949.

4. J. H. Perry, *Chemical Engineer's Handbook,* 5th ed., McGraw-Hill Book Company, New York, 1973.

5. *Annual Book of ASTM Standards,* American Society for Testing and Materials, Philadelphia, Pa., 1981.

6. *Flow of Fluids Through Valves, Fittings and Pipes,* Technical Paper 410, Crane Co., New York, 1980.

7. E. M. Shelton, *Aviation Turbine Fuels, 1980,* DOE/BETC/PPS-81/2; *Diesel Fuel Oils, 1981,* DOE/BETC/PPS-81/5; *Heating Oils, 1981,* DOE/BETC/PPS-81/4; *Motor Gasolines, Winter 1980–1981,* DOE/BETC/PPS-81/4; Department of Energy, Bartlesville, Okla.

8. *SAE Handbook 1981,* Society of Automotive Engineers, Warrendale, Pa., 1981.

9. A. C. Upton, "The Biological Effects of Low-Level Ionizing Radiation," *Scientific American,* Feb. 1982.

10. D. M. Considine, *Chemical and Process Technology Encyclopedia,* McGraw-Hill Book Company, New York, 1974.

11. D. F. Othmer, *Methanol as a Fuel: Methanol Production and Use as a Low Cost Fuel,* and *Methanol: The Efficient Conversion of Valueless Fuels into Versatile Fuel and Chemical Feedstock,* Polytechnic Institute of New York, New York, 1981 and 1982, respectively.

12. R. E. Kirk and D. F. Othmer, *Encyclopedia of Chemical Technology,* 3rd ed., Wiley-Interscience, New York, 1981.

13. *Alternate Liquid Fuels,* Report 652-1981, Stanford Research Institute, Menlo Park, Calif., 1981.

14. P. Auer, *Advances in Energy Systems and Technology,* Vol. 1, Academic Press, Inc., San Diego, Calif., 1978.

15. F. Kreith and J. F. Kreider, *Principles of Solar Engineering,* Hemisphere Publishing Corp., New York, 1978.

16. "Renewable Energy on Line," *Compressed Air Magazine,* Sept. 1996.

17. J. A. Duffie and W. A. Beckman, *Solar Energy Thermal Processes,* Wiley-Interscience, New York, 1974.

18. J. Rudisill, "Geothermal Energy," Calpine Corp., Santa Rosa, Calif., private communication, 1996.

19. R. Benoit, *Review of Geothermal Power Generation in the Basin and Range Province in 1995,* Oxbow Power Services, Reno, Nev.

20. S. Ashley, "Turbines Catch Their Second Wind," *Mechanical Engineering,* Nov. 1992.

21. A. McDougall, *Fuel Cells,* Halsted Press, New York, 1976.

22. *Fuel Cells,* Publ. 0024-1978, U.S. Department of Energy, Washington, D.C., 1978.

23. L. O'Conner, "Fuel Cells Turn Up the Heat," *Mechanical Engineering,* Dec. 1994.

24. J. H. Hirschenhofer and R. H. McClelland, "The Coming of Age of Fuel Cells," *Mechanical Engineering,* Oct. 1995.

CHAPTER 10
THE DESIGN PROCESS

10.1 INFORMATION BASES

With the enormous growth in the body of knowledge, and given the current state of the art, theoretical bases are available for most all design problems facing the engineer. However, in developing a basis for solution of a **design problem,** it is useful to bear in mind the general way in which design solutions have evolved.

With any of the technological branches, the initial data acquired were **empirical.** Subsequently, **hypotheses** were developed and tested experimentally. After a number of revisions of the hypothesis, improved theories were developed which more closely matched the actual phenomena observed. Eventually, as is the case today, with more powerful and widely used analytical tools and with a better understanding of the composition and behavior of materials, often at the atomic level, the theory and predictions of performance have vastly improved.

These theoretical bases are often arcane and not necessarily well understood. As a result, the use of **application data** is still widely practiced and is extremely useful to most engineers for the following reasons:

- It reduces the time to arrive at a solution that is suitable both functionally and economically.
- It avoids the need for the engineer to be totally knowledgeable of all aspects of the formulation of the solution.
- It normally includes sufficient **margins** to provide suitable protection for the health and safety of the public.
- It normally provides a solution that matches the availability of materials and products.

This does not relieve the engineer of the responsibility for selecting the correct methodology for the solution, knowledge of the general theory behind the solutional method, and for suitable checking to assure accuracy. Indeed, these requirements are fundamental and cannot be waived.

10.2 DESIGN APPROACHES

In the past, design of complex systems utilized fairly precise methodology for design of specific components in a system. However, the interaction of these components and the dynamic response of the system overall used **rules of thumb,** often not necessarily well understood but recognized as providing the type of conservatism to provide a safe and reasonable economic **solution;** many of these rules are found in codes to this day. The availability of more powerful computational equipment has permitted a more accurate and dependable solution to the problem of the linkage and interaction of complex systems. In addition, dynamic systems are now analyzed and solved more readily and accurately. Despite these powerful tools, the overall approach and development of a solution for many designs is as much an art as a science. Often there are alternative methodologies with advantages of accuracy or timeliness which may be employed, and their selection favors an experienced approach.

Design solutions can take either an **active** or a **passive** approach. For dynamic systems such as fluid flow, heat transfer, aerodynamic, and similar problems, active solutions are favored, while for static systems, such as foundations and structures, passive solutions have the edge. It is often tempting to develop a solution that is technologically elegant, but such approaches often are too costly or unproven to be utilized, and indeed, this is often the plight facing the young engineer with limited experience. While it is normal to seek to improve design solutions, this work takes place within the context of a specific industry. Thus the philosophy of the industry with respect to the degree of conservatism must be borne in mind. Broadly speaking, older, more basic industries have extensive operating experience and are more conservative than evolving industries. For the basic industries, improvements and developments are highly controlled often in the form of extensive experimentation before a major commitment is made.

The design approach chosen for a specific problem will vary widely, but all will contain consideration of the following factors:

- Do actual **data** exist to predict the requirements of the problem (system)? Are they adequate? Loadings, dynamics, response to various modes, material characteristics, and so on.
- Are the **mathematical models** available for solution of the design problem adequate? Are they academic (too complex), amateurish (too simple), or at the correct level of sophistication for the problem? Are they well proven?
- What is the **accuracy** of these models?
- Should a passive or an active solution be pursued?
- Does the software or hardware exist for implementing the problem solution?
- Should a **"robust" design** be considered?
- What are the consequences of failure, and how do these relate to the end use?
- What is the **cost** of the solution? It is the most economical reasonably attainable?
- What trade-offs are necessary in solving the design problem?
- How much time and effort are available for developing the solution?

10.3 SCOPING THE DESIGN PROBLEM

In scoping the design problem, it is necessary to consider a number of factors:

- What are the **operating modes** of the system? Three to ten are not uncommon, and for electronic systems and analyses, multiples of these are possible.

- Are all the modes of the same type or are some of different types (e.g., static loading, impact, energy flow, environmental, hydraulic, electro-optic)?
- What **operational factors** are to be considered? Is this a use-once device? Is it to be fail-safe? Does it have salvage value? Is it to be a modular device?
- Are specific **environmental factors** to be considered? Typical of these are aging, vibration, radiation, seismic, thermal, and chemical factors.
- What is the level of technical sophistication of the persons using the solution /system/device?
- Will the system users require special technical training?

10.4 DESIGN MARGINS

The establishment of **design margins** will vary with the ability to predict the loadings or performance requirements for the problem. Wherever possible, design margins should be based on a firm mathematical basis, as is widely done in manufacturing. Each industry has standards that vary widely; for instance, the design margin for a foundation design might be a factor of 3 or 4, while the design margin for an aircraft missile might be only 50%. For still other uses, where the consequences of failure are low, the design margin may be only 20 or 30%. To establish these margins it is necessary to consider the ability to accurately predict the performance loadings or requirements, the criticality of function of the device, its cost, and the cost of the design, manufacturing, and test effort for the incremental gain, and to compare them to the value of improvement in reliability: to, in effect, perform a cost/benefit analysis (see Sections 8.4 and 8.6).

Regardless of the methodology or database chosen, the design margins must always include some allowance for unknown or unforeseen events. Thus normal variations in materials and workmanship will require that for certain types of devices or activities, these margins be substantial, while for others they may be significantly lower.

Another way to view this is from the overall question of reliability of performance. Where a body exists of both event requirements and performance or test data for the component or system, it is possible to construct curves of the likely requirements of the event(s) which establish the performance of the device. This is then compared to the performance or capability of the device or structure to determine whether a shortfall exists and if appropriate, its magnitude. Although it is often difficult to determine the frequency and severity of events affecting performance requirements, this approach is useful even in a qualitative way. It also improves understanding not only of the relationship between design and performance but also the importance of controlling the quality of the device.

Typically, a device or structure is called upon to resist only modest stresses or loadings. The normal design approach calculates a higher design value based on codes, previous performance requirements, or in some cases an engineering judgment. The design value is then rounded up because of geometry, standard sizes available, and so on. This is specified to the supplier (or producer), who adds a **margin** to that to assure production of an acceptable product. This margin, together with the production process, yields a process output which has some degree of variation around a mean value. The actual capability of the product (i.e., variation about the mean) will vary depending on the quality of control of the production process. Good **process control** will yield large percentages of **conforming** material, whereas poor process control may yield a significant percentage of materials where the strength or other characteristic is insufficient. Applicable to many forms of engineering work, Fig. 10.1 illustrates this concept using an example of concrete compressive strength.

10.5   DESIGN CHECKLIST

As a part of the design process, and independent of organization size, it is necessary to have an orderly way to approach the design problem. It is not enough merely to rely upon experience or a code requirement. Design by its very nature is an **iterative** process, and thus reliance on a cookbook or checklist alone is not sufficient. Often, the functional or environmental requirements imposed on the design differ from previous ones, different materials are proposed, and the circumstances of use of the design are different. In some cases, because of the complicated nature of the design, it is necessary to diagram the approaches considered and where time may be important, to develop and perform different approaches to the design in parallel. (This approach is expensive and is only rarely justified.) More commonly, in complex cases where scheduling is important, a simplified **precedence schedule** is prepared to indicate the major activities and their interrelationships. Then based on this work, priorities can more accurately be established. When these factors are borne in mind, the use of design checklists is helpful and assures that important factors are not overlooked.

An excellent checklist of design input considerations has been promulgated by the American National Standards Institute (ANSI) and the American Society of Mechanical Engineers (ASME) in standards N45.2.11 and NQA 1. Although developed primarily from a quality assurance point of view, because of their value and applicability, these standards have been widely adopted for general use. Figure 10.2 abstracts this material.

Of all the design objectives the most important is **technical adequacy.** If the design is not technically correct and cannot meet its performance requirements, the design is unsuccessful. For this reason at the beginning of the work the technical or performance criteria must be clearly established. These **criteria** need to be stated in specific terms with ranges of accuracy given where appropriate.

For example, a pump design project might begin with the criteria that it should have a capacity to pump a minimum of 125 U.S. gal of clear water per minute, ranging from 40 to 120°F at a discharge pressure of 80 ft with a maximum suction lift of 10 ft. The pump shall be self-priming, powered by a gasoline engine with the entire unit mounted on a portable skid, able to be moved by two workers. It should have sufficient fuel capacity to run for one hour without refueling and must be able to operate outdoors in weather ranging from 20 to 140°F in direct sun as well as in heavy rain.

The specification above, while brief, provides the design team with a good bit of detail and answers many of the questions that may arise during the course of design. Although a statement of the criteria cannot and should not take the place of proper design decisions, it can answer many of the questions that arise early in the effort and thus avoid false starts and misdirection of the work. It offers a further advantage,

(*text continues on page 527*)

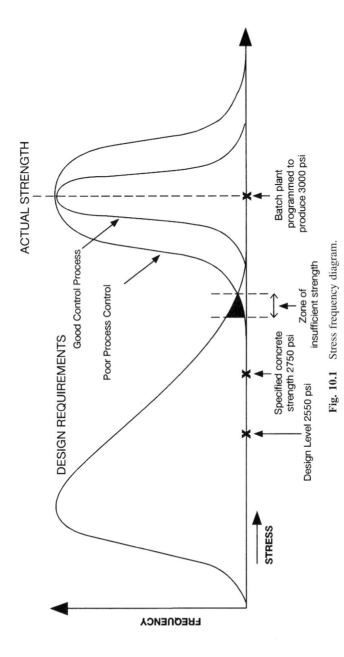

Fig. 10.1 Stress frequency diagram.

ACTUAL STRENGTH

DESIGN REQUIREMENTS

Good Control Process

Poor Process Control

Batch plant programmed to produce 3000 psi

Zone of insufficient strength

Specified concrete strength 2750 psi

Design Level 2550 psi

STRESS

FREQUENCY

1. Basic functions of each structure, system and component.

2. Performance requirements such as capacity, rating and system output.

3. Codes, standards and regulatory requirements including the applicable issue and/or addenda.

4. Design conditions such as pressure, temperature, fluid chemistry and voltage.

5. Loads such as seismic, wind, thermal and dynamic.

6. Environmental conditions anticipated during storage, construction, and operation such as pressure, temperature, humidity, corrosiveness, site elevation, wind direction, nuclear radiation, electromagnetic radiation and duration of exposure.

7. Interface requirements including definition of the functional and physical interfaces and the effects of cumulative tolerances involving structures, systems and components.

8. Material requirements including such items as compatibility, electrical insulation properties, protective coating and corrosion resistance.

9. Mechanical requirements such as vibration, stress, shock and reaction forces.

10. Structural requirements covering such items as equipment foundations and pipe supports.

11. Hydraulic requirements such as pump net positive suction heads (NPSH), allowable pressure drops and allowable fluid velocities.

12. Chemistry requirements such as provisions for sampling and limitations on water chemistry.

13. Electrical requirements such as source of power, voltage, raceway requirements, electrical insulation and motor requirements.

14. Layout and arrangement requirements.

15. Operational requirements under various conditions such as plant startup, normal plant operation, plant shutdown, plant emergency operation, special or infrequent operation and system abnormal or emergency operation.

16. Instrumentation and control requirements including indicating instruments, controls and alarms required for operation, testing and maintenance. Other requirements such as the type of instrument, installed spares, range of measurement and location of indication are included.

17. Access and administrative control requirements for plant security.

18. Redundancy, diversity and separation requirements of structures, systems and components.

19. Failure effects requirements of structures, systems and components including a definition of those events and accidents which they must be designed to withstand.

20. Test requirements including pre-operational and subsequent periodic in-plant tests and the conditions under which they will be performed.

21. Accessibility, maintenance, repair and inservice inspection requirements for the plant including the conditions under which these will be performed.

22. Personnel requirements and limitations including the qualifications and number of personnel available for plant operation, maintenance, testing and inspection, and radiation exposures to the public and plant personnel.

23. Transportability requirements such as size and shipping weight, limitation and I.C.C. regulations.

24. Fire protection or resistance requirements.

25. Handling, storage, cleaning and shipping requirements.

26. Other requirements to prevent undue risk to the health and safety of the public.

27. Materials, processes, parts and equipment suitable for application.

28. Safety requirements for preventing personnel injury including such items as radiation safety, criticality safety, restricting the use of dangerous materials, escape provisions from enclosures and grounding of electrical systems.

29. Quality and quality assurance requirements.

30. Reliability requirements of structures, systems and components including their interaction, which may impair functions important to safety.

31. Interface requirements between plant equipment and operation and maintenance personnel.

32. Requirements for criticality control and accountability of nuclear materials.

Fig. 10.2 Design input considerations. (Reprinted from ASME/ANSI N45.2.11 and ASME NQA-1, courtesy of the American Society of Mechanical Engineers.)

since the important performance requirements are clearly stated at the beginning of the work, that decisions will not be taken up piecemeal during the work causing dislocations. Further, since the design process is often very hectic, if important criteria are left to evolve during the design work, it is difficult, if not impossible, to later determine when they were introduced, by whom, and the rationale behind them. Also, important criteria, on many large or complex projects are subject to review and approval by **boards of review,** or similar bodies, and the value of these review steps is lost if they occur in a developmental fashion rather than given the consideration they merit.

10.6   DESIGN CONTROL

Design control includes the following elements:

- A system of **procedures** to control the design process
- A technical **scope description,** or **scope book**
- A system of **design reviews**
- A credible design **schedule**
- A schedule that lists the **responsibilities** of the various parties for activities such as drawing preparation, requisitioning, reviews of supplier drawings and data, material acquisition, and such
- A system of **drawing progress control**
- A system of control for **supplier** drawings and data
- Systems to control development of equipment lists, motor lists, instrument I/ O lists, control valve and instrument lists, and so on
- A **numbering** and **checking system** for engineering calculations and design drawings

and most important,

- A system for controlling **changes** to both drawings and calculations

For small projects the control documents can be prepared and maintained manually; for more complex or longer projects, some form of computerized documentation is normally used, and timely status and exception reports can readily be produced.

Some care needs to be given to the content and level of detail of the control documents to assure they do not become an end in themselves and that they provide timely and useful information. In general, the simpler the documents, the better. It is better to err on the side of too little information rather than create documents with exhaustive information, little of which will be used. Numerous versions of control documents are found in commercially available computer programs.

For more simple work, Fig. 10.3 depicts a drawing control form of the manual type. Similarly, scheduling of the work can be depicted using either one of the many scheduling programs, or manually, using simple bar charts or precedence schedules as appropriate. Because of the dependence of the design on outside information, and the sequential development of data, it is customary to utilize precedence schedules for all but the most simple design efforts.

10.7 CODES AND STANDARDS

Codes and **standards** are a form of design criteria shortcut instituted to assure that components, systems, and structures perform as intended and that the health and safety of the public is not compromised. In the United States, the principal standards agency is the **American National Standards Institute** (ANSI). They act as the clearinghouse and coordinator for standards development across the various technical societies and user groups. The principles of due process and development of consensus are critical to the development of an ANSI standard. A standard will often carry an identification such as ANSI/ASME NQA-1. Other standards will carry only the designation of the originating society.

Standards will occasionally be revised, and often the title of a standard will also carry the year of original adoption or last revision, such as ANSI C78.1500-1993. Most frequently the year is omitted and the standard merely referred to without including the year designation. As a result, some care needs to be taken to assure that you are working to the latest standard.

The **American Society for Testing and Materials** (ASTM) is the principal organization for issuing standards covering the chemical and physical properties of materials, including testing methods. The ASTM standards cover virtually the entire range of materials, including, for example, metallic and nonmetallic products, fabrics, and paint, and are considered definitive.

International standards are becoming increasingly important and the **International Standards Organization** (ISO) has issued numerous standards, two in particular having a major effect on export activities: ISO 9000 (Quality System) and ISO 14000 (Environmental Management Systems). It is likely that in the near term additional ISO standards will be issued having a significant effect upon export activities and design as well. **Technical advisory groups** (TAGs) represent ANSI in the development or application of most of the significant international standards.

Technical societies such as ASME, ASCE, IEEE, and ISA issue technical standards for work in their own fields. Although their methods of designation vary slightly, they all employ a system that identifies the standard with the society designation (e.g., IEEE xx.x). In a similar way, industry groups promulgate standards. Virtually every industrial group, such as the American Petroleum Industry (API), American Water Works Association (AWWA), National Electrical Manufacturers Association

LEGEND

— DRAWING STARTED
╲ ISSUED FOR APPROVAL
▽ ISSUED FOR CONST.

PROGRESS LAST PERIOD
PROGRESS THIS PERIOD
ISSUED FOR A - APPROVAL Q - QUOTATION
C - CONSTRUCTION P - PURCHASE

DRAWING CONTROL

JOB NO. 3OCC PERIOD ENDING 3-1-90
JOB SAMPLE
SHEET ___ OF ___ SHEETS

CODE	DRAWING NUMBER	TITLE	START DATE SCHED.	START DATE FOR	START DATE ACTUAL	COMP DATE SCHED.	COMP DATE FOR	COMP DATE ACTUAL	APP'VD DATE	DATE ISSUED REV. 0	REV. 1	REV. 2	REV. 3	REV. 4	REV. 5	REV. 6	REV. 7	REV. 8	REMARKS * DESIGNATES PRINCIPAL DRAWING
		EQUIPMENT LOCATION DRAWINGS																	
	M-1	Ground Floor Plan	4-15-89		4-13-89	8-1-89	9/1	9-15-89	9/17	A 9/15	C 9/20								
	M-2	Plan at Turbine Floor	5-3-89		5-1-89	8-1-89	10/1	9-26-89		A 9/27									
	M-3	Plan at Control Floor	7-20-89	8/15	8-10-89	2-1-90	3/4												
		F. O. HEATER AND PUMPS																	
	M-20	F. O. Heater Assembly	5-15-89		5-9-89	6-30-89		6-14-89	6/16	Q 6/9	A 6/14	P 6/23	C 6/25	C 7/13					
	M-21																		
		AREA NO. 2																	
	M-100	Plan Below Elev. 30'	8-1-89		7-21-	10-1-89		9-24-		A 9/25									
	M-101	Misc. Turbine Piping	2-1-90	3/10		3-1-90	4/10												

Fig. 10.3 Drawing control form (manual).

(NEMA), and the Underwriters' Laboratories (UL), issue standards to supplement ANSI, ASTM, and the technical society standards.

Governmental standards are frequently used in lieu of or in conjunction with the various standards types noted above. Generally, they are in either the federal or military family. In addition, jurisdictions will often impose standards or mandate them into law. For example, the **ASME Boiler and Pressure Vessel Code** is the required code for this type of work in most jurisdictions within the United States and in some foreign countries as well.

Unless the designer is thoroughly familiar with a standard, he or she should review the standard to be sure that it applies to the work. As codes and standards are continuously evolving and being revised, it is important to assure that the version of the standard imposed is the latest. Recently in the United States, many military (MIL) standards have been withdrawn and may no longer apply to the work. Some of these standards have had wide use and have been replaced with other standards (e.g., **MIL-STD-105E** has been replaced by **ASQC Z1.4**). When specifying a MIL standard, it is wise to check to assure that it is still relevant and in force. Where there is a doubt, it is better to omit a standard, particularly an obscure one. Specifying unneeded standards merely delays and drives up the cost of the work with no real benefit. Similarly, it is essential that the jurisdiction in which the work will take place be contacted to assure that the designer has full knowledge of any additional local, state, or regional requirements.

10.8   SPECIFICATIONS

Together with drawings, **specifications** form the principal method by which the design is described and transmitted to others. Specifications are of several types: those necessary for procurement activities, those used for construction or subcontracting work, and those used for test and acceptance work. For replacement parts, or for simple or inexpensive items, a formal specification may not be necessary; merely a statement setting forth the model number or the basic performance requirements may be adequate. Where, however, a formal specification is necessary, certain principles are common to the preparation of them regardless of the purpose. These include clarity, simplicity, precision, quantification, use of tolerances, and brevity.

Clarity is certainly a fundamental requirement. The specification needs to be written in a direct way using simple unambiguous language to define the requirements. Often, specifications are used by personnel who have little or no formal technical training, and the writer must keep this in mind. Technical terms should be used where necessary, but the specification should never be used as an exercise to demonstrate the deep technical knowledge of the writer. Rather, it is a way to convey meaning, intent, and requirements and as such should, if possible, be written at the educational level of those who will use it.

Where there is any question of clarity, that portion of the specification should be rewritten to ensure improved understanding. Often having someone unfamiliar with the work read it and ask questions will highlight areas where rewriting is necessary. Precision of language tends to improve the clarity and understanding also. Where there is a choice of words, the most definitive and precise ones should be used. Certain phrases in common use add little to a specification and merely become a basis for argument. Examples of these are "of the highest commercial quality," or "suitable for the service intended." Although desirable goals, such highly subjective language does little to help the vendor or constructor. It is far better to include **quantifiable** values such as specifying a "B-10 bearing life of 20,000 hours," requiring "that the minimum safety factor for rotating parts be 3.5," and so on. This approach removes the subjectiveness in determining whether the criteria is met and provides an objective measurable standard against which to judge **specification compliance.**

Related to clarity is the question of avoidance of **conflict** in the specification documents. The only real way to guard against this is to state the requirement only once. If the requirement is stated in more than one place, it can almost be guaranteed that in the course of design evolution, the value will change, and it will only be corrected in one place thus leading to uncertainty and conflict.

Standard specifications are extremely useful if developed early in the project. They can then be included as attachments in specifications where the component is incorporated. For example, standard specifications for motor starters and finish painting could be included in a procurement specification for a conveyor system, where motor starters and finish painting are included. This has the advantage of permitting the designer of the system—in this case, a mechanical engineer—to concentrate on the area of his or her expertise and incorporate well-thought-out standards, developed by others, that apply to the entire project. There is an added advantage in that standard specifications can identify particular brands and models of components. There is a great advantage to simplifying the stock of spare parts and the later maintenance of the equipment or facility.

National codes and **standards** as well as government specifications, very often the military (MIL) series, provide a convenient source for standard specifications. They are available for all the disciplines and most common products. In addition, technical professional societies and manufacturers' groups have prepared standard specifications that can be a great help. It is not necessary to incorporate all portions of these standard specifications; to do so would in most cases be excessive. They can, however, be used effectively to generate a specification and often serve as a checklist to ensure that no major features or requirements are overlooked. **Manufacturers' specifications** can be the starting point, but particular care should be taken that they do not result in a specification that limits competition, for they will sometimes include clauses and requirements that are proprietary or met only by their particular size or features.

An important consideration that needs to be decided upon early is the format of the specifications to be used. It will probably be different from the guide specs used and should be followed for all specifications prepared, although not all sections are necessary in all specifications. One commonly used format has the following sections; the example shown is the technical specification for procurement of a centrifugal pump*:

1. *Scope of work:* pump, baseplate, driver (if included), interconnecting piping, lubricating oil pump and piping, spare parts, instrumentation (pump-mounted), erection supervision.

---

*Reprinted from I. J. Karassik and William C. Krutzch, *The Pump Handbook,* by permission of McGraw-Hill, Inc. Copyright © 1986 by McGraw-Hill, Inc.

2. *Work not included:* foundations, installation labor, anchor bolts, external piping, external wiring, motor starter.

3. *Rating and service conditions:* fluid pumped, chemical composition, temperature, flow, head, speed range preference, load conditions, overpressure, runout, off-standard operating requirements, transients.

4. *Design and construction* (Care should be taken to provide latitude in this section, as this borders on dictating construction requirements): codes, standards, materials, type of casing, stage arrangement, balancing, nozzle orientation, special requirements for nozzle forces and moments (if known), weld-end standards, supports, vents and drains, bearing type, shaft seals, baseplates, interconnecting piping, resistance temperature detectors, instruments, insulation, appearance jacket.

5. *Lubricating oil system* (if applicable): system type, components, piping, mode of operation, interlocks, instrumentation.

6. *Driver:* motor voltage standards, power supply and regulation, local panel requirements, wiring standards, terminal boxes, electric devices, for internal combustion drivers, fuel type preferred (or required), number of cylinders, cooling system, speed governing, self-starting or manual, couplings or clutches, exhaust muffler.

7. *Cleaning:* cleaning, painting, preparation for shipment, allowable primers and finish coats, flange and nozzle protection, integral piping protection, storage requirements.

8. *Performance testing:* satisfactory for the service, smooth-running, free of cavitation and vibration, shop tests (Hydraulic Institute standards) for pump and spare rotating elements, overspeed tests, hydrostatic tests, test curves, field testing.

9. *Drawing and data:* drawings and data to be furnished, outline, speed versus torque curves, Wk^2 data, instruction manuals, completed data sheets, recommended spare parts.

10. *Tools:* one set of any special tools, including wheeled carriage for rotor if needed for servicing and maintenance.

11. *Evaluation basis:* power, efficiency, proven design.

Supplementing these may be technical specifications relating to other requirements of the order, such as specifications for the electric motor, steam turbine, or other type of driver; a specification on marking for shipment; a specification on painting; and requirements for any supplementary quality control testing.

In addition it is important that any unusual requirements be listed in the technical specification so that the manufacturers are aware of them. Examples of these are special requirements for repair of defects in pump castings, a sketch of the intake arrangement for wet-pit applications, and special requirements regarding unique testing, for example, metallurgical testing that may be required during manufacture apart from performance testing.

It is helpful to the pump supplier to provide system-head curves, sketches of the piping system (dimensioned, if this is significant), listings of piping and accessories required, etc.

Pump data sheets are extremely useful in providing a summary of information to the bidder and also in allowing the ready comparison of bids by various manufacturers. Some of the items on these sheets are filled in by the purchaser and the balance by the bidder to provide a complete summary of the characteristics of the pump, the materials to be furnished, accessories, weight, etc. The data sheets should be included with the technical specification.

One common way to develop a specification is to rework previous specifications. The **cut-and-paste method** has the advantage of speed and in general ensures that the specification format and structure fit the standard previously used. Its main weakness is that less-experienced personnel will often include items not required, and the specification may become much more complicated and far-reaching than needed. Perhaps items needed may be omitted. Often when this method is used, old requirements are carried forward, and in many instances specifications contain requirements or references that had ceased to exist many years earlier. This has the effect not only of increasing the price for the item but forgoes the benefit of technological development. It further often has the effect of excluding bids that may be favorable in both technology and price.

When using this method, then, it is important that the writer think critically about each clause and requirement in the specification to ensure that it is truly necessary. Besides improving precision, this approach has the further advantage of making the specification shorter and more understandable. Where there is truly a doubt whether or not to include a requirement, and where advice is not readily available, it is usually satisfactory to omit the requirement, since the bidders will normally have their standards that cover such requirements.

To summarize, there are a number of sources for standard or guide specifications that can be used as a starting point for specification preparation:

- Previous specifications
- Industry standard specifications
- Government specifications
- National codes and standards
- Manufacturers' "guide" specifications

Specifications fall into two distinct categories: performance or design. **Performance specifications** state input and output parameters but do not normally restrict or establish the way in which they are met. They may also in a nondetailed way establish the features required. They have the advantage of permitting the supplier to offer a standard unit (or alternative units) most nearly matching the overall required performance and place no responsibility on the purchaser. They are most applicable to standard items for which some design, manufacturing, and operating experience is available.

Design specifications normally state not only input and output parameters but also specific design requirements, materials of construction, and so on. They establish not only what must be achieved, but to some extent how to do it as well. As a consequence, standard designs rarely apply, alternates are limited, and the purchaser may bear some degree of responsibility if performance is not met. They are most appropriate for innovative designs or developmental work, particularly where new technologies are involved. Frequently the number of sources of supply is limited, and prequalification of bidders may be necessary to ensure technical competence.

When preparing specifications, preciseness is essential. Terms such as *fitness for use,* and *good commercial quality* are not sufficient to define requirements for complex, costly, or high-impact items. For these cases specific **acceptance criteria** should be stated, and, where codes are involved, they should be referenced or preferably included in the body of the specification.

Often specifications will take a **black box** approach, where there is little or no interest in the internals of a supplied package and the only parameters spelled out in the specification are the inputs and outputs with perhaps limitations of space or weight if they are important. This can be a convenient way to avoid getting embroiled in a supplier's offering, but some care must be exercised to ensure that there are no hidden problems which arise because of integration requirements.

Each industry or trade has standards of **tolerance.** It is good practice to list or refer to these standards in the specification. Where the standards do not exist, or where the writer wishes something tighter, the tolerances required should be stated. This can be extremely important to the cost of the item specified because tightened tolerancing can increase costs dramatically, particularly if only a few are to be produced or purchased. The caution here, is to never impose tolerances tighter than actually needed for the item.

Allowing for **alternates** in procurement specifications usually helps the purchaser. Equipment is generally produced to preengineered designs in standard sizes, and bidders are more familiar with their equipment capability, features, and size breaks than are purchasers. As a result, it is often possible for a supplier to offer equipment that may differ slightly from some details of the specification but that complies with the performance requirements and may have significant advantages in terms of cost or other considerations.

Specifications will normally take precedence over drawings where conflicts exist, and contracts, if properly drawn, will state the precedence of the documents. There are many rules of interpretation that a court might apply in determining the intention of the parties and the precedence of documents. The best course is that the contract be clear and unambiguous on these matters. Such contract provisions will govern, notwithstanding that the parties may have selected an order of precedence different from that which is usual or customary.

A typical tier of documents in descending order of precedence is:

- Equipment- or component-unique specifications
- Basic design or data sheets
- Process requirements
- Project, equipment, or component drawings
- Standard specifications
- Standard drawings
- Reference drawings
- Industrial standards

10.8.1   Developmental Work*

Where specifications are prepared to define a work product for a customer, as for a costly, complex semicustom device, it is useful to agree on a format and the generic contents of the document. The specifications may not only serve as a contract between the supplier and the purchaser, but also as a basis for detailed development of the work packages that make up the product. It is in both parties' interest to make the document as accurate and understandable as possible. Responsibility for developing and controlling the information, its schedule, and most important, the manner of handling changes, are all of prime concern.

Since the specification is normally used as a basis for scheduling of **deliverables,** it is important that changes to it be carefully controlled and tracked. It is inevitable that as work progresses, the purchaser will have changes to the specifications; it is the responsibility of the supplier to manage those changes effectively.

Early in the process, as a part of the initial procedural steps and before design begins, it is essential to establish a method for **change control.** Where data flow occurs between the supplier and the purchaser, responsibility for developing and controlling the information and its schedule are of prime concern. Specifications are best developed by the purchaser (who has knowledge of what is needed) with guidance from the supplier (who has the most knowledge of how the need can best be filled). The purchaser and the supplier agree on the specifications, and the purchaser signs to indicate that the specifications represent the need. The supplier must identify and pursue any clarifications needed from the purchaser during the work.

10.8.2   Construction Industry Specifications

The **Construction Specifications Institute** has established a structure and classification system for specifications, which are used almost universally for major construction projects:

Division 1	General Requirements
Division 2	Site Work
Division 3	Concrete
Division 4	Masonry
Division 5	Metals
Division 6	Carpentry
Division 7	Thermal and Moisture
Division 8	Doors and Windows
Division 9	Finishes
Division 10	Specialties
Division 11	Equipment
Division 12	Furnishings

*From Helen Belcastro, copyright 1996.

Division 13　　Special Construction

Division 14　　Conveying Systems

Division 15　　Mechanical

Division 16　　Electrical

Within each of these divisions, there may be several specifications covering the major categories of work from which those that apply are selected. For example, Division 8, Doors and Windows, includes the following sections:

081　　Metal Doors and Frames

082　　Wood and Plastic Doors

083　　Special Doors

084　　Entrances and Storefronts

085　　Metal Windows

086　　Wood and Plastic Windows

087　　Hardware

088　　Glazing

089　　Glazed Curtain Walls

Each of these section specifications will have four principal parts. The exact content of the parts will vary, depending on the work involved, but all include at least the following categories:

Part 1—General covers a description of the work, **quality control** requirements, submittals, applicable standards, and product delivery, storage, and handling.

Part 2—Products lists acceptable manufactures, material specifications, details of the fabrication and workmanship requirements for each component or aspect of the work covered, and painting.

Part 3—Execution establishes the manner of work performance.

Part 4—Measurement and Payment establishes the method of measurement and payment for the work.

10.9 CALCULATIONS

It is normal practice that the basis for the design be clearly stated at the beginning of the calculation. Typically, the calculation includes statements for items that are **given,** items that are **to be found** (or calculated), and the necessary **assumptions** used in the calculations. It is important that all calculations be **signed** and **dated** by the person performing them. Calculations must be legible and arranged in a logical sequence so that someone unfamiliar with them but of experience equal to the originator, performing an independent check, can readily follow them and verify their accuracy. Thus they should call out codes, standards, references used, and so on, so that both the originator and the checker can return to them at a later date, if necessary, and reestablish both their applicability and their accuracy.

All calculations should be checked. The checking may include any of these methods:

- A rough rule-of-thumb check
- An independent separate check using different but equivalent formulas
- An estimate by a highly experienced person
- A detailed, meticulous check of each of the steps in the calculation

When checking calculations, an experienced professional will always rely to some extent on an intuitive understanding of what the calculation should indicate. Where results are counterintuitive, the entire calculation, including any simplifying assumptions, methodology, formulas used, and so on, should be reviewed carefully.

Calculations will have different levels of accuracy depending on their end use. A calculation for a precision machine part might have a very high level of accuracy, while a calculation used in a gross analysis of earth to be moved for a large construction project might be suitable at a level of rough approximation only. It is important to state the level of accuracy desired in the calculations so that the use is clear and the purpose of the calculation is clearly understood.

Calculations should be oriented to using standard sizes of materials and equipment wherever possible. In general, the cost to produce a special off-standard size or type of material is prohibitive and, even if justified initially, often creates maintenance problems later.

Calculations are often prepared on quadrille paper, usually of the vellum variety. This provides ease of reproduction by the ozalid as well as the photocopy processes. The quadrille ruling makes it convenient to prepare those sketches that are a part of the calculation.

With the development and widespread use of **personal computers** the vast majority of all the calculations performed in today's engineering offices are computerized. Three basic principles must be followed:

1. The calculation should be appropriate to the complexity and risk of the design.
2. Input data must be accurate and data entry must be error free.
3. The program used must be accurate and of proven applicability.

It matters little whether the calculation is performed on a large mainframe or on a PC because the major concern is the **suitability** of the program used to perform the calculation. Since the steps of the calculation are internal to the program and hence hidden from the view of the engineer, it is imperative to provide some form of control and overview of the programs used.

The normal program concerns include **validation** of the accuracy of the program, obtaining or establishing **documentation** on the program, and controlling the **distribution** and **use** of the program, including revisions or changes to it. To provide for this, larger firms usually establish a sponsor who controls and approves the program, approves all modifications to it, including options, and maintains its documentation, including establishing limits for use. Since these programs are developed on a design discipline basis, it is usual to place this responsibility on the chief engineers of each of the design disciplines.

Most programs used for engineering calculations today (1998) are commercially available rather than being developed in house. Because of their development and widespread use, establishing their documentation is not usually necessary, and simple checking to determine their accuracy and applicability is all that is required. Prior to use the confirmation consists of running the program for a series of calculations both for normal and limiting conditions and comparing the results to the calculation run by another means, usually by hand. This permits determination of the accuracy and limits of the program and provides the basis for its use within the firm. Following assurance that the program is satisfactory, it is authorized for use.

Despite the controls described above, errors may occur or be detected either in the programs or in their use. To correct this problem, a system of computer error reporting is usually established. The system provides for the classification of errors and their correction. **Error classification** normally uses three levels:

1. Errors that have no effect on the validity of the results.
2. Errors that give obvious meaningless or absurd results.
3. Errors that can be interpreted as being valid.

When an error is detected, it is, until proved otherwise, classified as being in the third category. The program sponsor is immediately notified and normally requires that users suspend use of the program pending resolution of the problem. Where the program is shown to be incorrect, the program is corrected, and copies are distributed to the users. In addition, and most important, all **previous** calculations performed using the program are reviewed to determine the error's impact and fixed where necessary. Because of the far-reaching effect of an error, it is imperative that the program sponsor exercise extreme care when authorizing programs for use.

All calculations should be checked independently and the date of the check and the checkers initials added to the calculation sheet. Where the checker has used a method different from the originator, it is helpful to briefly indicate the method used to verify that the calculation is correct.

Since calculations are performed for the benefit of and paid for by the owner, requests to provide them, or copies, to the owner or other outside parties should be complied with. Where calculations are provided to owners during the design process, frequent questions may arise with necessitate considerable discussion to resolve. This of course has an adverse impact upon the time of the senior project personnel and as a result may delay completion of the design. The best course of action is to provide the calculations at the conclusion of the project as a part of the close out activities.

Important to the proper performance of the work is maintenance of a rigorous system of control and custody of the **design calculations** as they are prepared. Normally, a calculation binder is setup in which all calculations are filed. For large projects this might comprise one or more binders for each of the engineering disciplines or interest areas. Hard copies of computerized calculations are included as well as manually prepared calculations. All calculations, including their current revisions, must be maintained in the binder(s) so that they are available for review or revision. It is important to designate one person as responsible for maintenance of the binder(s). For convenience on medium- and large-size jobs, indexes of the calculations prepared should be maintained with the binders. Each of the design disciplines maintains the calculation binder for its own work.

10.10   PROCESS ENGINEERING

Process engineering is broadly defined as the engineering of a system of components in which materials are extracted, treated, processed, or where dynamic interaction between components occurs. Examples include block diagrams, material and energy balances, flow diagrams, logic diagrams, piping and instrumentation diagrams, and hydraulic (power) diagrams. These diagrams are used initially to establish the broad relationship between elements in a dynamic system. As the system parameters and requirements develop during the design cycle, detail is added as appropriate. They are particularly useful where the system may operate at various levels, as they can reflect efficiency, throughput, and other considerations necessary for execution of the detailed design. In most cases they guide the overall design, are continuously updated, and at the conclusion of the design process represent the final design relationships. They do not necessarily represent physical relationships, as this is normally shown on the design drawings, although many diagrams will represent in an approximate way the physical orientation or arrangement of the process elements. Because of the common use of nonstandard symbols for diagrams, it is wise, for all but the most straightforward block diagrams, to include a legend for each set.

Because of its simplicity, ease of understanding, and versatility, the **block diagram** is used almost universally for process work. Examples are shown as Figs. 10.4 and 10.5. These diagrams represents two levels of development and cover the design, fabrication, marketing, and use of a medical device.

As their name implies, **material and energy balances** are of great value in establishing the overall design requirements. For classical process plant work, such as chemical or metallurgical plants they are fundamental to establishing the basic design parameters and particularly capacities, for work in other industries where energy distribution and utilization is critical, a **heat balance** is typically used.

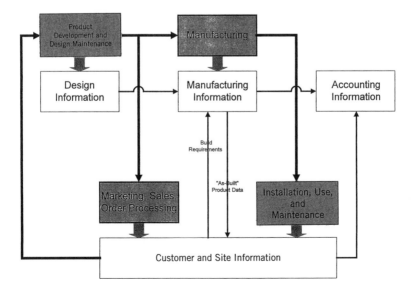

Operational Flow: Activity and Information

Fig. 10.4   Operational flow: activity and information. (Reprinted with permission from Varian Associates.)

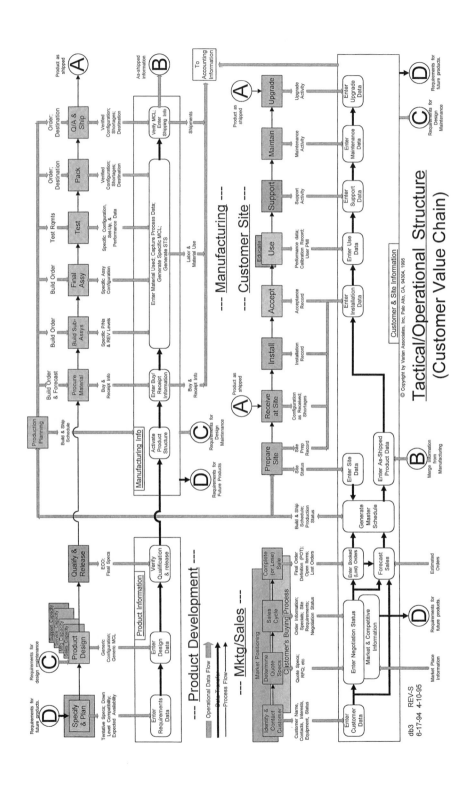

Tactical/Operational Structure
(Customer Value Chain)

© Copyright by Varian Associates, Inc, Palo Alto, CA, 94304, 1995

db3   REV-S
6-17-94 4-10-95

Other forms of **process diagrams** are discussed in Section 10.13.

10.11   DESIGN FOR MANUFACTURE

The trade-offs for the design process include considerations of product performance, product cost, development program expense, and development speed. While each must be considered the primary concerns must be for product **performance** and product **cost.** Even though early entry into the market is often critical to capturing market share, development speed and expense must be considered secondary to the basic functional requirement for product performance and quality and the general goal of optimum product cost. Great care should be taken to ensure that in the rush to bring a product to market, deficiencies and defects are not introduced.

Designing for manufacture requires above all a commonsense approach to the goal of minimum product cost, without sacrificing product quality or performance. To achieve this effectively, it is necessary to establish specific, measurable product requirements and to limit expansion or enhancement. The design team has several "customers" apart from the actual purchaser. They include manufacturing, installation, service, test, and maintenance. In a sense, even marketing and sales are customers because they will sell the product. Of course, the restraints and requirements of the manufacturing organization must be considered and incorporated by the design organization, since these are fundamental to being able to produce the product.

The principles that should be utilized are well known and follow what experience normally dictates. The total number of parts should be minimized. In that way not only does it avoid the cost of producing unnecessary parts, but also it cuts down on the time and effort for the design, administration, material handling, and manufacturing effort. Where one part can be designed to include the functions of two or more parts, the savings may be significant, even though the cost to produce the single more complex part may be higher. Since the potential for interaction between components roughly doubles with the addition of each new element, the effort to reduce the number of elements, and thus reduce the **interactions** and improve **reliability,** can be easily justified. Care needs to be taken to maintain an optimum design and not merely to reduce the number of parts for its own sake. Figure 10.6 shows how these interactions increase with added elements, while Fig. 10.7 indicates the difficulty of maintaining a 99% reliability with increasing numbers of elements in the product.

In all systems some components have a significantly higher failure rate than others. One way to deal with this is to group developmental functions that have a potential for higher failure rates in a single component. This will greatly improve the reliability of the entire assembly. It also will permit isolation and remedy of problems more quickly. Thus the design will also be more easily maintained in service. **Modular designs** favor this risk concentration approach and should be used where possible. Figure 10.8 indicates how risk concentration and risk distribution among the design components affects the overall reliability of the design.

Standard components should be used wherever possible. Their reliability is known, and developmental problems can be avoided, they are readily available, often provided by several sources at competitive cost and offer assurance against loss of supply. Use of standard components also frees up in-house resources for more creative work.

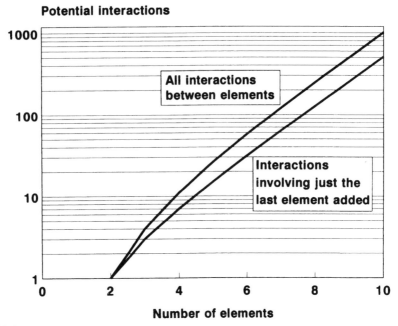

Fig. 10.6 Interactions between design elements. (From P. G. Smith and D. G. Reinertsen, *Developing Products in Half the Time,* Van Nostrand Reinhold, New York, 1991.)

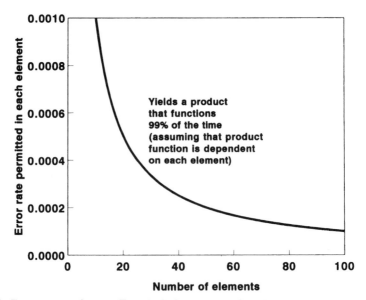

Fig. 10.7 Error rate per element. (From P. G. Smith and D. G. Reinertsen, *Developing Products in Half the Time,* Van Nostrand Reinhold, New York, 1991.)

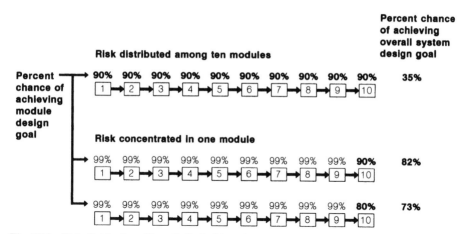

Fig. 10.8 Risk distribution. (From P. G. Smith and D. G. Reinertsen, *Developing Products in Half the Time,* Van Nostrand Reinhold, New York, 1991.)

Parts must be designed for ease of **assembly.** The use of suitable tolerances, proper radii, generous tapers, self-guiding features, and similar considerations will do much to prevent assembly difficulties and high cost. The use of separate fasteners should be minimized, and integral clips or similar features should be provided. Where fasteners are necessary, they should be of the captive or self-guiding type designed to provide maximum flexibility and provision for ease of alignment during assembly.

Where possible, parts should be designed to perform several functions. As mentioned above, a more expensive part may in the long run offer a significant advantage in enabling simplification of the production process. A corollary to this is for a part to be designed to have multiple uses. For example, a small shaft could also be a guide pin, a hinge, or a similar component on the same assembly, thus simplifying handling, storage, stocking, and, more important, production of the part. However, care must be taken to avoid the temptation to use custom-designed components where the functional gain is only modest, or the cost impact negative.

Tolerances established for the part should be based on rational **statistical** decisions. Often, designers will decrease tolerances to ensure that there is sufficient margin to accept normal variations in the manufacturing or fabrication process. The manufacturing planners then may further tighten them to ensure a lower rejection rate after production. The result of this is a cutting of tolerances, often to one-fourth or less of that necessary, with a resultant large increase in cost. It further classifies items that may be usable as out of tolerance. This requires that they be reviewed and dispositioned by a Material Review Board or similar technical group. Out-of-tolerance parts not only generate unnecessary paperwork and effort, but they undermine the integrity of the design, since many of the deviations then are dispositioned as **"use as is."** The loss of confidence in the design can become a real problem and a point of controversy, particularly when tolerances are not maintained for critical items. By far the best and most economic policy is to determine what tolerances are necessary, state them, and most importantly, enforce them.

The design of individual parts should minimize manufacturing cost by finding the lowest-cost method of production considering production volume, available tooling, and so forth. Sometimes what may seem to be the cheapest manufacturing process may not be the lowest cost after all, as assembly, fastening, finish, and other processes add to the cost of the completed assembly.

Design should ideally provide for assembly from **one direction** only. This will avoid additional handling and positioning of either the assembly itself or positioning and assembling parts and fasteners in other planes and axes. For **robotic**-assisted manufacturing, this can often avoid the need for robots with five or six degrees of freedom of motion. As a corollary, all handling of the parts and assembly should be minimized. There should be as little repositioning as possible, and to the greatest extent handling should be linear in direction.

Manufacturing tolerances should be established statistically where possible, and if approaches such as the 6-sigma philosophy (described below) are used, inspection can be reduced and even avoided.

For large production runs, a methodology of 6-sigma quality has been adopted by some firms. This approach uses a plus or minus 6-sigma band (12 sigma) between the upper and lower specification limits and permits a 1.5-sigma drift of the mean in response to normal manufacturing variations, while limiting the defect rate of 3.4 parts per million. The use of sigma, the standard deviation, to define the quality band permits the linking of the degree of process variation to specification limits and allows the derivation of correlation and capability ratios that simplify application. Where less rigorous approaches are adopted, the possibility of nonconforming parts will increase and automatic or self-inspection will be required to detect and eliminate them.

Improvements in product design should be made in **small increments** and be continual. This permits a shorter time to market and reduces the risk of significant problems being discovered at a later date.

In some cases it may be prudent to consider the use of nonstandard sizes and components to foil attempts at counterfeiting or the unauthorized production of re-placement parts. But this extreme step, though sometimes justified, should be care-fully weighed before implementation.

In general, manufacturing systems should be designed to minimize production time without **value being added** to the product. The timing of operations involving order entry, queueing time, in-process storage, material movement, inspections, time await-ing management decisions, and Material Review Boards should be minimized, and eliminated, to cut down on manufacturing costs. Just-in-time inventory and produc-tion control measures might need to be instituted to reduce such costs.

The use of robotic equipment to assist in the manufacturing process has been long recognized. In general, robots can be separated into two types: those that transfer or move materials and those that manipulate a component and perform some manufac-turing operation. The decision to design for and install robotic equipment involves both design and manufacturing considerations. The mere act of designing for robotics, but using people to perform the work actually yields about 85% of the benefit. To gain the 15% additional benefit yield, high-volume production is normally required. Although if elements of the work are performed on NC tooling where the program-ming is convenient, the break-even size of the lot can be significantly reduced.

The great advantage of using robotics is due not so much to production volume as to **accuracy.** With predictability of production rates and increased amortization costs, firmer economic data are available for pricing. Economic analyses of robotics manufacturing need to quantify all of these factors to properly reflect their impact.

The economic justification for robotics should additionally include criteria such as **return on investment** (ROI). Suppose that the ROI analysis for automatic test equipment does not indicate a clear cost advantage. Since robotic equipment can reduce the need for highly trained test technicians (where there is a potential for increasing production levels), the time needed for ramp up to a second or third shift would be reduced and thus its selection, while higher in cost may be the correct decision.

Yet another critical factor in the design process is the time it takes to complete the design, implement it, and produce and distribute the product to the market. Throughout the last quarter century the effective life of products has been continually decreasing. As a result the time required to develop and produce a new product, that is, **time to market,** has become increasingly important. Clearly the company reaching the market first can command higher prices, capture a significant share of the market, and establish name recognition and perhaps customer loyalty. In addition a firm that establishes a reputation for innovation ensures public acceptance of successive models and products. Later, as products of other firms enter the market, there is less differentiation between competing products, so profit margins are reduced, and customer (brand) loyalty becomes increasingly important. These effects are shown graphically in Figs. 10.9 and 10.10.

Break-even time is a recent development that promises to be an effective alternative to the "time to market" approach. Break-even time is the time is takes to develop, market, and profit from a new product. This includes the time to recover not only the production costs but also the development costs. It affords the advantage

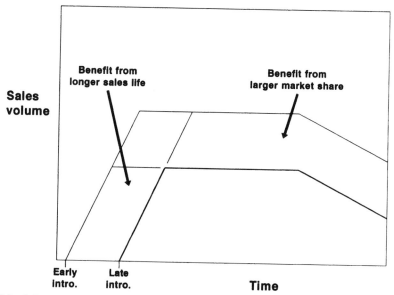

Fig. 10.9 Sales volume versus time. (From P. G. Smith and D. G. Reinertsen, *Developing Products in Half the Time,* Van Nostrand Reinhold, New York, 1991.)

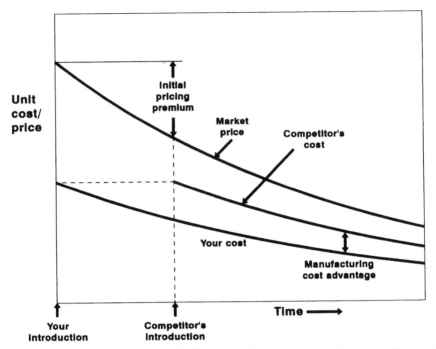

Fig. 10.10   Pricing versus time. (From P. G. Smith and D. G. Reinertsen, *Developing Products in Half the Time,* Van Nostrand Reinhold, New York, 1991.)

of establishing the true cost of the product rather than permitting development costs to be spread across other product costs or carried as an overhead charge. This is based in part on the concept of charging individual production operations with all identifiable charges, such as the appropriate tooling and overhead, to ensure that all the costs attributable to a specific operation appear in the economic analyses.

Concurrent design development is used widely to achieve many of the reductions discussed above. In this approach design teams are established to pursue different components of the design. Each team is autonomous and has sufficient authority to obtain priority support from the organization. Each team can call upon groups as diverse as sales, maintenance, and shop personnel, in addition to the manufacturing engineering personnel. The product design is developed in parallel rather than sequentially, and this shortens the schedule. For example, a design including a power supply, electronic components, structural and mechanical systems, and plastic appearance housings might have each of these elements designed at the same time, relatively independently of each other. To minimize the risk of major problems, a preliminary space envelope, configuration, and performance characteristics of each component is agreed upon very early in the work and as the design of each evolves, frequent coordination meetings take place to update the groups on the development of the design, particularly as it may affect their interfaces. If the group is housed in a common area, the coordination may be ongoing, with the need for formal meetings reduced. With this system some rework will be needed but it will normally be offset many times over by the saving in time to complete the design. As the design teams work on additional projects, their initial estimates become more accurate, and less rework is required.

One of the important elements in the development of a design is liason with the **shop personnel** who will produce the product. The liason with the shop floor is essential if the design is to be readily producible. The design engineers must maintain a close and continuing relationship with the shop floor personnel who will actually produce the product to ensure that the design is practical and suited to the capability of both the personnel and the production facility. The attitude of the design personnel toward the shop personnel is also important. It is easy to forget that the shop work force that will produce the item have accumulated experience and skill that often spans decades. They should therefore be treated with respect and their views given proper weight. At times shop personnel may be less articulate and convincing than designers, but that does not lessen their contribution nor the importance of their views.

In some instances it is useful to form an ad hoc committee of both design and shop personnel to work on development of a new or improved design. Such a group meets periodically to review the development of the design and to suggest improvements and simplifications to facilitate production. The tendency to abdicate design responsibility to the shop personnel should be guarded against, but these people can make a valuable and very practical contribution to the design process.

Where new tooling is to be introduced, the shop personnel should be made a part of the decision process. This is particularly the case if automation, robotics, or other such improvements are contemplated. In those cases where there is a significant impact on the workforce, it is better to deal with it early than to attempt to do it later with an unhappy workforce. It is good practice to maintain the trust of the shop workforce by letting them know early what is happening and what will be expected of them. This goes a long way to easing the developmental problems and implementation difficulties associated with a changed design or a new product.

Another form of liason with the shop floor occurs where specially designed products are manufactured outside the organization. Wherever possible it is wise to consult with the outside vendors on the **manufacturability** of the item. Often, design engineers will not be fully knowledgeable of manufacturing techniques and limitations, standard sizes of materials available, secondary effects of processes, and such, and will produce a design that, though functional, may be expensive to manufacture. It is not unusual for outside vendors to recommend nonfunctional design changes that can reduce cost and hence the purchase price by factors of 60 to 90% or more. An example might be a thin plate whose design calls for a series of small holes to be punched in it. A clever fabricator may wish to produce them by chemical milling at a fraction of the cost per hole, significantly cutting down on the cost of the part. Other examples could include part size selection that creates large quantities of scrap, surface finish, or appearance problems.

Economic lot size depends primarily on the complexity of the product, the degree of precision needed in replicated parts, the tooling in use, and the cost target for the part produced. In most cases there is a trade-off between these variables. An individual case can only be analyzed by costing the various alternates. With labor costs becoming a smaller portion of the cost of the product, other production costs are becoming important. As a result, in estimating the costs of manufacturing parts, companies are increasingly turning to **activity-based financial management** (ABM).

This method involves breaking down in some detail costs formerly charged to overhead or other general accounts and charging them to the part under consideration. Included are charges such as rent, utilities, depreciation, and interest on costs of special tooling, operator training, and so on. This approach provides a more accurate and useful model upon which to base decisions.

For simple parts the economic lot size may be fairly small because replication is fairly easy. Similarly, for parts requiring low precision, relatively coarse tooling may be used, and it may be possible to produce the part satisfactorily with small lot sizes. Where high precision is required, the setup necessary to produce the part may require semi- or fully automatic machines, together with sophisticated inspection and gaging equipment. Where **numerically controlled** (NC) machines are not available, this requirement would involve a significant investment in either special machinery or jigs and fixtures, and thus a fairly large lot size to amortize the equipment and setup costs. With NC tools, once the programming is done, successive lots can be produced with virtually no special setup required. Often, NC tools can lower the economic lot size to one unit. One example is the NC-directed laser cutting of light thickness parts.

Ultimately, the **cost target** for the part will dictate the final production method employed. Very low cost targets will favor larger lot sizes and will often incur a higher degree of risk because of the difficulty of accurately predicting final costs, particularly with complex or new manufacturing systems. Further, where special tooling or processes are required, the machine tools utilized are often costly, and insufficient work may be in hand or confidently projected to amortize their cost. Faced with this circumstance, some manufacturers are contracting work out to other more specialized manufacturers. The advantages to this are lower market risk, simplified operations, and predictability of final costs (through the contract).

Steady improvements in the manufacturing processes will reduce the cost of successive units. The result is called the **learning curve,** and it applies to virtually any type of manufacture with more complex products having longer improvement curves. A rough rule of thumb that is used is that doubling the quantity produced causes a decline of 40 to 50% of the unit cost. This depends to a great extent upon the product, work force, duration of the work, and so on.

Bar codes are used almost universally to control manufacturing operations. They are found on shop work orders, travelers, drawings, and in many cases on the material itself. In some instances, tooling is installed which reads the bar code on the part and performs the appropriate manufacturing operations, such as machining and assembly.

10.12 GRAPHIC STANDARDS

10.12.1 Drawing Sizes

Standard sizes for drawings are $8\frac{1}{2} \times 11$ in. (outside sheet dimensions) or multiplies thereof:

A $8\frac{1}{2} \times 11$ in.   D 22×34 in.
B 11×17 in.   E 34×44 in.
C 17×22 in.

This arrangement of sizes permits folding into letter size (8½ × 11 in.) for convenient mailing or filing in standard office files.

Drawing sheets using preprinted borders are preferred to hand-drawn borders and are typically available from engineering and drafting supply firms in the sizes noted; in many cases sheets with preprinted title blocks bearing the company name, standard numbering systems, and so on, are stocked.

10.12.2 Scales and Materials

Scale requirements vary widely depending on the type and purpose of the drawing. Scales of ³⁄₃₂ in. and ³⁄₁₆ in. equal 1 ft should be avoided where possible.

Since most drawings are reproduced, transparent drawing sheet material is typically used. Vellum has been the standard material used very satisfactorily for several decades. More recently, Mylar drawing sheets have come into wide use. Mylar has the advantage of permitting more repeated erasures without damaging the transparent characteristics of the sheets, permits cleaner erasures without "ghosts," and has greater resistance to tearing or creasing than vellum. The disadvantages of Mylar are its higher cost and the requirements to use special leads and erasers.

10.12.3 Drafting Practice

Some drafting practices that are economic and minimize error or confusion include:

- Draw repetitive details only once and reference on other drawings.
- Eliminate unnecessary views.
- Make full use of industry standard details.
- Use templates for drawing symbols and common shapes.
- Cross-referencing must be accurate.
- Independent checking is essential and should be indicated in the sign-off block.
- Details of backgrounds should be omitted unless essential to a clear understanding and use of the drawing.

10.12.4 Revision Control and Check Prints

Absolute control of revisions is essential to any drafting operation. A convention of using letter designations for revisions to "in house" issues and number designations for revisions issued for external use is widely followed. Regardless of the convention adopted, total control using a formal register or control record is essential to avoid reuse of a revision with different dates and, more importantly, different information.

Check prints used for internal review or coordination need not in most cases be controlled as closely as formal revisions. One system having wide acceptance is to indicate a check mark and date in the revision block of the drawing, thus clearly identifying the purpose and timing of the print.

10.12.5   Canceled Numbers

Canceled drawing numbers should be withdrawn from use and not reused. The register should list the drawing number and a notation such as "Superseded by Drawing XXX," or "Canceled." The same practice should be followed with specifications or other types of unique documentation.

10.13   PROCESS DIAGRAMS

Diagrams of several different types are used to depict and develop process designs. **Flow diagrams** may range from simple block diagrams to those carrying complete and detailed flow or material balance data. **Piping and instrumentation diagrams** permit line sizing, valve selection, and instrument and control function determination/inclusion. **Logic diagrams,** for more complex control circuits, indicate functions such as "and," "or," and "not" required for control purposes.

10.13.1   Piping and Instrumentation Diagrams

Piping and instrumentation diagrams (P&IDs) utilize a wide variety of symbols and alphabetic codes to describe functions and are relatively standardized. A selection of the more common standard symbols and a portion of a typical P&ID are shown in Figs. 10.11 to 10.13.

10.13.2   Logic Diagrams

Logic diagrams establish the requirements for signals (generally electrical) to cause resultant actions. They utilize standardized symbolism and frequently are used in conjunction with P&IDs to define the control logic required. Examples of some logic diagrams are shown in Figs. 10.14a to c.

10.13.3   Hydraulic Schematic Diagrams

Hydraulic schematic diagrams are the equivalent of P&IDs (Section 10.13.1) but use a different symbology. Widely used for high-pressure systems, both air and hydraulic, a selection of the more common symbols is shown in Fig. 10.15.

10.14   ELECTRICAL DIAGRAMS

10.14.1   One-Line Diagrams

Power generation, transmission, and distribution systems are frequently presented in a one-line format sometimes called **single-line diagrams.** The term **riser diagram** is sometimes used for power distribution systems in commercial or industrial buildings. These diagrams are analogous to a mechanical flow or P&ID and, by use of symbols and single lines, show the generators, transformers, buses, switchgear, circuit

	FIRST LETTER	SENSING DEVICE	TRANSMITTER		DISPLAY DEVICE		ALARM (NOTE 5)			CONTROL DEVICE (NOTE 3)					
										CONTROLLER (NOTE 3)					
SYMBOL	MEASURED OR INITIATING VARIABLE	PRIMARY ELEMENT	BLIND	INDICATING	RECORDER	INDICATOR	LOW	HIGH	HIGH & LOW	BLIND	INDICATING	RECORDING	CONTROL VALVE (NOTE 6)	SELF-ACTUATED VALVE	SWITCH (NOTE 3)
( )	TYPICAL SYMBOL	( )E	( )T	( )IT	( )R	( )I	( )AL	( )AH	( )AHL	( )C	( )IC	( )RC	( )V	( )CV	( )S
A	ANALYSIS (NOTE 2)	AE	AT	AIT	AR	AI	AAL	AAH	AAHL	AC	AIC	ARC	AV		AS
B	BURNER, FIRE & FLAME	BE						BAH							
C	CONDUCTIVITY	CE	CT	CIT	CR	CI	CAL	CAH	CAHL				CV		CS
D	DENSITY	DE	DT	DIT	DR	DI	DAL	DAH	DAHL	DC	DIC	DRC	DV		DS
E	VOLTAGE (EMF)	EE	ET	EIT	ER	EI	EAL	EAH	EAHL	EC	EIC	ERC			ES
F	FLOW	FE	FT	FIT	FR	FI	FAL	FAH	FAHL	FC	FIC	FRC	FV	FCV	FS
G	GAGING	GE	GT	GIT	GR	GI	GAL	GAH	GAHL	GC	GIC	GRC			GS
H	HAND (MANUAL)									HC	HIC	HRC	HV	HCV	HS (NOTE 9)
I	CURRENT	IE	IT	IIT	IR	II	IAL	IAH	IAHL	IC	IIC	IRC			IS
J	POWER	JE	JT	JIT	JR	JI	JAL	JAH	JAHL	JC	JIC	JRC			JS
K	TIME		KT	KIT	KR	KI	KAL	KAH	KAHL	KC	KIC	KRC	KV		KS
L	LEVEL	LE	LT	LIT	LR	LI	LAL	LAH	LAHL	LC	LIC	LRC	LV	LCV	LS
M	MOISTURE	ME	MT	MIT	MR	MI	MAL	MAH	MAHL	MC	MIC	MRC	MV		MS
N	UNCLASSIFIED (NOTE 4)														
O	TORQUE	OE	OT	OIT	OR	OI	OAL	OAH	OAHL	OC	OIC	ORC			OS
P	PRESSURE	PE	PT	PIT	PR	PI	PAL	PAH	PAHL	PC	PIC	PRC	PV	PSV (NOTE 7)	PS
PD	PRESSURE DIFFERENTIAL	PDE	PDT	PDIT	PDR	PDI	PDAL	PDAH	PDAHL	PDC	PDIC	PDRC	PDV	PDCV	PDS
Q	QUANTITY OR EVENT	QE	QT	QIT	QR	QI	QAL	QAH	QAHL	QC	QIC	QRC	QV		QS
R	RADIOACTIVITY	RE	RT	RIT	RR	RI	RAL	RAH	RAHL	RC	RIC	RRC	RV		RS
S	SPEED OR FREQUENCY	SE	ST	SIT	SR	SI	SAL	SAH	SAHL	SC	SIC	SRC			SS
T	TEMPERATURE	TE	TT	TIT	TR	TI	TAL	TAH	TAHL	TC	TIC	TRC		TCV	TS
TV	TELEVISION	TVE	TVT	TVIT	TVR	TVI				TVC	TVIC	TVRC			TVS
U	MULTIVARIABLE				UR	UI	UAL	UAH	UAHL	UC	UIC	URC	UV		US
V	VISCOSITY	VE	VT	VIT	VR	VI	VAL	VAH	VAHL	VC	VIC	VRC	VV		VS
W	WEIGHT (NOTE 10)	WE	WT	WIT	WR	WI	WAL	WAH	WAHL	WC	WIC	WRC	WV		WS
X	UNCLASSIFIED (NOTE 4)														
Y	OBJECT OR MOTION SENSOR	YE	YT	YIT	YR	YI	YAL	YAH	YAHL	YC	YIC	YRC	YV	YCV	YS
Z	POSITION	ZE (NOTE 11)	ZT	ZIT	ZR	ZI	ZAL	ZAH	ZAHL	ZC	ZIC	ZRC			ZS

NOTE 1: THE INSTRUMENT LEGEND IS BASED ON ISA STANDARD S5.1 1975

2. THE LETTER "A" IS USED FOR ALL ANALYSIS VARIABLES. TERMS ARE PLACED OUTSIDE THE INSTRUMENT CIRCLE OF A LOOP TO DENOTE THE SPECIFIC VARIABLES. SOME EXAMPLES ARE

CO – CARBON MONOXIDE
COMB – COMBUSTIBLES
DH – DISSOLVED HYDROGEN
H_2 – GASEOUS HYDROGEN
M_4 – METHANE
NOx – NITROGEN OXIDES

DO – DISSOLVED OXYGEN
O_2 – GASEOUS OXYGEN
pH – PERCENT HYDROGEN
Cl_2 – CHLORINE
SMOKE – SMOKE DENSITY
SO_2 – SULPHUR DIOXIDE
TRB – TURBIDITY
TSP – TOTAL SUSPENDED PARTICULATE

3. A DEVICE THAT CONNECTS, DISCONNECTS, OR TRANSFERS ONE OR MORE CIRCUITS MAY BE EITHER A SWITCH, A RELAY, OR AN ON-OFF CONTROLLER, DEPENDING ON THE APPLICATION

• A SWITCH, IF IT IS ACTUATED BY HAND OR THE DEVICE IS USED FOR ALARM, PILOT LIGHT, SELECTION, INTERLOCK, OR SAFETY.
• A CONTROLLER, IF THE DEVICE IS USED FOR NORMAL ON/OFF OPERATING CONTROL, SUCH AS A SAMPLE HEATING THERMOSTAT.
• THE LETTERS H AND L ARE ADDED TO THE MEASURED VARIABLES FOR HIGH AND LOW RESPECTIVELY, LETTER AS FOR ALARMS (LSH, PSL, ETC.)
• A CONTROL OR SENSING DEVICE HAVING A DISPLAY FUNCTION SHOULD HAVE THE APPROPRIATE DISPLAY LETTERS ADDED AFTER THE MEASURED VARIABLE DESIGNATION, E.G. A/C DESIGNATES ANALYSIS INDICATING CONTROL STATION.

4. AN UNCLASSIFIED LETTER MAY BE USED FOR UNLISTED MEANINGS THAT WILL BE USED REPETITIVELY ON A PARTICULAR PROJECT. THE MEANING(S) WILL BE DEFINED ONLY ONCE FOR THAT PROJECT AND HAVE ONE MEANING AS THE FIRST LETTER AND ANOTHER SINGLE MEANING AS THE SUCCEEDING LETTER

5. HIGH-HIGH ALARMS HAVE ( ) AHH AND LOW-LOW ALARMS HAVE ( ) ALL IN INSTRUMENT CIRCLE E.G. LAHH – DESIGNATES HIGH-HIGH LEVEL ALARM LALL DESIGNATES LOW-LOW LEVEL ALARM

6. VALVES
• IF A DEVICE MANIPULATES A FLUID PROCESS STREAM AND IS NOT A MANUAL LY ACTUATED ON-OFF BLOCK VALVE, IT SHALL BE DESIGNATED AS A CONTROL VALVE.
• A HAND CONTROL VALVE HCV IS A MANUALLY ACTUATED VALVE THAT EITHER MODULATES (THROTTLES) A PROCESS STREAM OR IS USED AS AN INSTRUMENT DEVICE
• MOTORIZED VALVES ARE DESIGNATED THE SAME AS OTHER CONTROL VALVES, E.G. FV, PV, HCV, HV, ETC.
• AN ON-OFF VALVE REMOTELY CONTROLLED BY A HAND SWITCH IS DESIGNATED AS A HAND VALVE HV.

7. THE DESIGNATION PSV APPLIES TO ALL VALVES INTENDED TO PROTECT AGAINST EMERGENCY PRESSURE CONDITIONS. RUPTURE DISCS SHALL BE DESIGNATED PSE

8. USE OF MODIFYING TERMS HIGH, LOW, AND MIDDLE OR INTERMEDIATE SHALL CORRESPOND TO VALUES OF THE MEASURED VARIABLE, NOT OF THE SIGNAL, UNLESS OTHERWISE NOTED

9. SPECIAL SWITCH CONTROLS SHALL BE DENOTED IN POSITION C OF SWITCH SYMBOLS.
JOG – JOGGLE CONTROLS
KEY – KEYLOCKED SWITCHES

10. LOAD AND PRESSURE CELLS ARE USUALLY DENOTED "WE"

11. DENOTES STRAIN GAUGE

Fig. 10.11 Selected piping and instrumentation diagram legend.

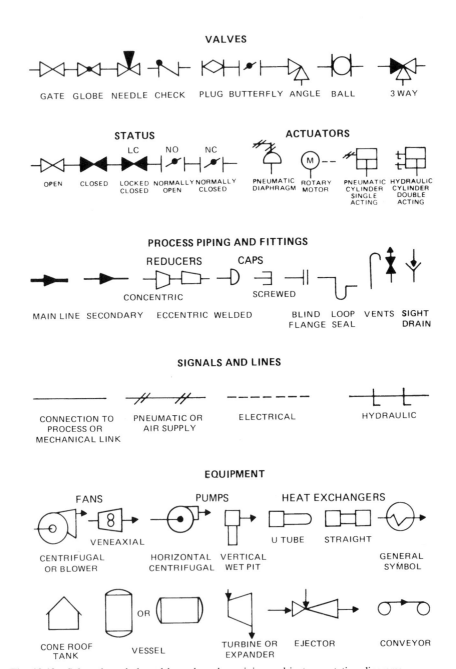

Fig. 10.12 Selected symbols and legend used on piping and instrumentation diagrams.

GENERAL INSTRUMENTS

INSTRUMENT FOR SINGLE
MEASURED VARIABLE OR
FUNCTIONAL INSTRUMENT
LOCALLY MOUNTED

INSTRUMENT FOR MORE
THAN ONE FUNCTION

TYPICAL CONNECTION
ANY VARIABLE

DIRECT CONNECTION

CAPILLARY FILLED SYSTEM
WITH CHEMICAL SEAL

IN LINE DEVICE

OPTICAL, SONIC, OR
RADIATION

SELF-ACTUATED DEVICES —FLOW

FLOW REGULATOR
SELF-CONTAINED

SELF-ACTUATED DEVICES—LEVEL

LEVEL REGULATOR WITH
MECHANICAL LINKAGE

SELF-ACTUATED DEVICES—
PRESSURE

PRESSURE REDUCING
REGULATOR, SELF-
CONTAINED

PRESSURE REDUCING
REGULATOR WITH
EXTERNAL PRESSURE
TAP

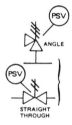

ANGLE

PSV

STRAIGHT
THROUGH

PRESSURE RELIEF OR
SAFETY VALVE, SPRING
OR WEIGHT LOADED, OR
WITH INTEGRAL PILOT

PRESSURE RELIEF OR
SAFETY VALVE, ANGLE
PATTERN, TRIPPED BY
INTEGRAL SOLENOID

PSE

RUPTURE DISK OR SAFETY
HEAD FOR PRESSURE RELIEF

SELF-ACTUATED DEVICES—
TEMPERATURE

TEMPERATURE REGULATOR,
FILLED-SYSTEM TYPE

SELF ACTUATED
TEMPERATURE REGULATOR

Fig. 10.12   (*Continued*)

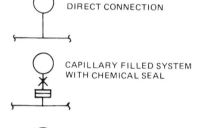
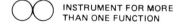
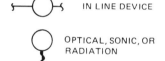

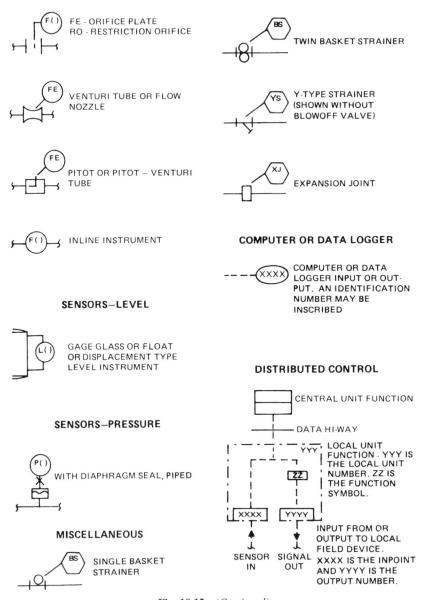

Fig. 10.12   (*Continued*)

breakers, disconnect switches, and so on, and their interconnections for all or a portion of an electrical power system. Equipment ratings are usually also given. Instrument transformers, meters, and protective relays are frequently shown also. Standard device numbers are used where appropriate. Figure 10.16 shows a typical single line diagram.

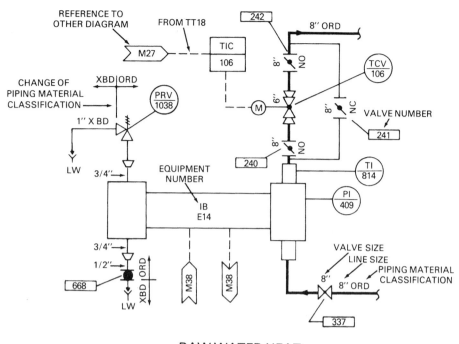

Fig. 10.13 Typical piping and instrumentation diagram.

10.14.2 Schematic Diagrams

Schematic diagrams, sometimes called **elementary diagrams,** are prepared for power systems, or portions thereof, and for control and instrumentation circuits. These diagrams are in three-line or two-line format and show individual items of equipment, devices within equipment, their coils, contacts, windows, terminals, and so on, and each connection (wire, cable, or bus) between equipment or devices. Schematics can be interpreted to indicate the function of the circuits. They are used in the preparation of wiring diagrams for the internals of individual items of equipment and connection diagrams for field wiring to equipment. Figure 10.17 shows a typical schematic diagram.

10.14.3 Wiring Diagrams

Wiring diagrams are based on the schematic diagrams and are prepared to permit shop wiring of individual items of equipment. They show each device terminal, terminal point for external connection, and the wires interconnecting these terminals. **Wire lists** are sometimes used in lieu of these diagrams, showing individual wires in line (tabular) form. These devices and terminals are usually shown in their relative

GENERAL NOTES

1. Logic symbols represent system functions and do not necessarily duplicate circuit arrangement or devices. System control logic diagrams do not inherently imply energized, de-energized, or other circuit operation states.

2. Process equipment will change state when a change is initiated, and will remain in that state until a change to another state is initiated.

3. Process equipment will remain in, or return to, the original state after a loss and restoration of power, unless otherwise noted.

4. Inherent equipment interlocks such as circuit breaker trip free and reversing starter cross interlocks are not shown.

5. Some protection actions are shown also as start permissives. Trip free design prevents equipment operation when a protection action exists, even if a start permissive is not provided.

6. Final instrument set points are shown elsewhere. Set points shown on system control logic diagrams are approximate.

7. See electrical drawings for details of equipment electrical overcurrent, short circuit, and differential protection and space heaters.

8. The memory, reset, and start permissive logic associated with the operation of electrical protection devices is not shown. Electrical auxiliary system breakers are reset by operation of the control room switch to trip. Mechanical auxiliary system circuits are reset by operation of a switch at the switchgear or motor control center.

9. The test control switches at the switchgear which function only when a circuit breaker is in the test position are not shown.

10. All circuit controls, except interlocks with other equipment, function when a circuit breaker is in the test position to allow circuit testing.

11. The logic to show that valve and damper position lights are both on when the equipment is in an intermediate position is not shown.

12. Limit and torque switches to stop valve and damper motor actuators at the end of travel are not shown in the logic. The valve type and required actions will be noted on the diagram when available.

ABBREVIATIONS

C01	— Unit control panel
C02	— Auxiliary control panel
C03	— Hot shutdown panel
L	— Local to controlled equipment
MCC	— Motor control center
SWGR	— Switchgear

(a)

Function	Symbol	Definition
MANUAL INPUT		Momentary hand switch input to logic
		Maintained hand switch input to logic
AND		Output exists only when all inputs are present.
OR		Output exists only when one or more inputs are present.
NOT		Output exists only when input is not present.
ON DELAY		Output exists only when input has been continuously present for a preset time and remains present.
OFF DELAY (TIMED MEMORY)		Output exists only when input is present and for a preset time after the input is not present.
MEMORY	S R	Set output exists when set input is present and continues until the reset input is present. Reset output exists only when set output is not present.
COINCIDENCE MATRIX	A/B	Output exists only when at least A out of B inputs are present.
LOW BISTABLE	L S.P.	Digital output exists only when analog input is lower than set point.
HIGH BISTABLE	H S.P.	Digital output exists only when analog input is higher than set point.
ISOLATION	ISO	Output is electrically isolated from input.
TEST DEVICE	T TEST R	Test signal can be inserted manually in place of normal signal.
LIGHT		RED—Operating, flowing, or increasing GREEN—Not operating, not flowing, or decreasing AMBER—Automatic, standby, or intermediate WHITE—Manual or protective trip BLUE—
ANNUNCIATOR		Input to annunciator
COMPUTER		Input to computer
CONTINUATION		Logic continuation

Fig. 10.11. Logic diagram legend.

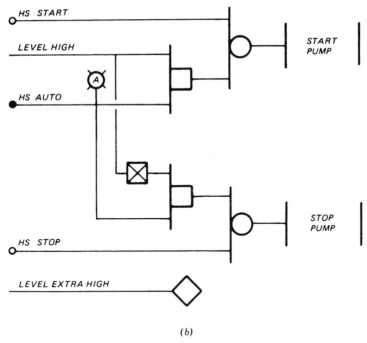

(b)

Fig. 10.14b (*Continued*)

physical location (e.g., top-to-bottom, left-to-right). Figure 10.18 shows a typical wiring diagram.

10.14.4 Connection Diagrams

Connection diagrams show the field or external connections to individual items of equipment. Each terminal is identified and usually shown in its relative physical

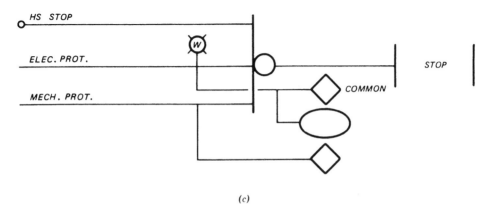

(c)

Fig. 10.14c (*Continued*)

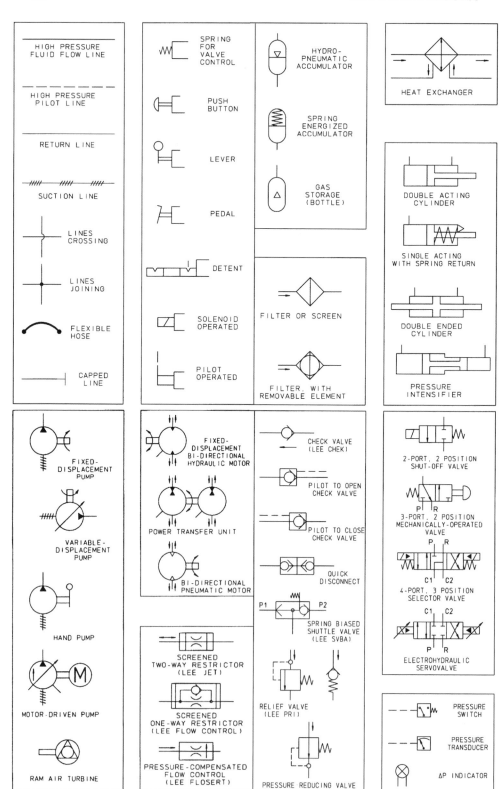

Fig. 10.15 Hydraulic schematic symbols. (Courtesy of the Lee Company.)

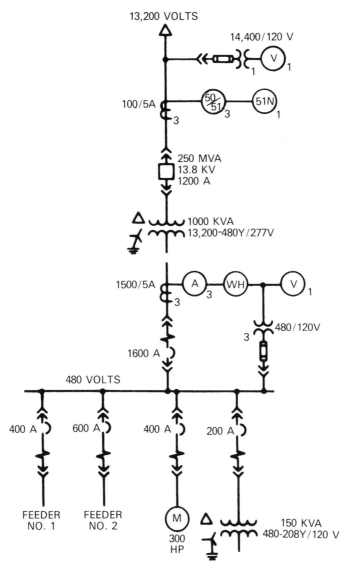

Fig. 10.16 Typical single-line diagram.

location. Connection lists are sometimes used in lieu of showing individual external wires in line form. Figure 10.19 shows a typical connection diagram.

10.14.5 Electrical Devices

Various electrical devices, including circuit breakers, relays, and switches have been given standardized device numbers for use on electrical single-line, schematic, and wiring diagrams. These devices are listed in ANSI C37.2-1979, and some of the more commonly used ones are shown in Table 10.1. A more complete alphabetical listing

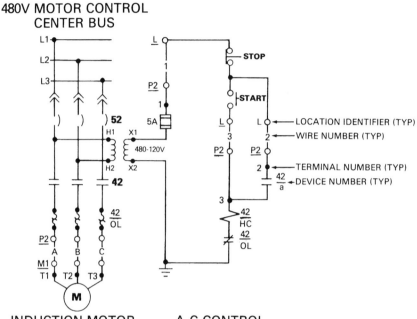

Fig. 10.17 Typical schematic diagram.

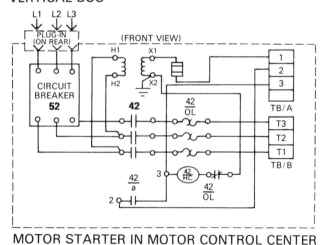

Fig. 10.18 Typical wiring diagram.

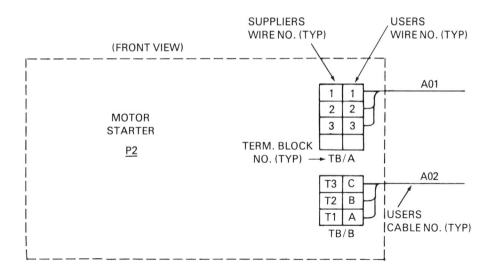

MOTOR STARTER IN MOTOR CONTROL CENTER

Fig. 10.19   Typical connection diagram.

is given in Table 6.16. These device numbers are typically used with suffixes to indicate auxiliary devices, condition, location, and so on, and may be given a unique identification number as well. Because of this, electrical drawings typically carry legends or notes to assist the user in understanding the specific meaning intended.

10.14.6   Selected ANSI Device Numbers

Table 10.1 shows some of the more commonly used ANSI device numbers.

10.14.7   Selected Suffix Letters

Suffix letters are used with device numbers for various purposes. In order to prevent possible conflict, any suffix letter used singly, or any combination of letters, denotes only one word or meaning in individual equipment. Suffix letters generally form part of the device function designation and thus are written directly behind the device number, such as 23X and 90V. Some selected suffix letters are shown in Tables 10.2 and 10.3.

Location of the main device in the circuit, or the type of circuit in which the device is used, or the type of circuit or apparatus with which it is associated, can be shown by the suffixes in Table 10.4.

In addition to the suffix letters, other letters are used to denote parts of the main device or other features, characteristics, or conditions. These are generally shown directly below the device number, such as $\frac{20}{2S}$ or $\frac{52}{a}$. A partial listing of these letters is shown in Table 10.5.

Table 10.1 **Selected ANSI Device Numbers**

Device No.	Function
1	Master element (initiating device)
2	Time-delay starting or closing relay
3	Checking or interlocking relay
20	Electrically operated valve
21	Distance relay
23	Temperature control device
25	Synchronizing or synchronism-check device
27	Undervoltage relay
30	Annunciator relay
32	Directional power relay
33	Position switch
37	Undercurrent or underpower relay
40	Field relay
41	Field circuit breaker
42	Running circuit breaker (motor starter contactors)
43	Manual transfer or selective device
44	Unit sequence starting relay
46	Reverse-phase or phase balance current relay
47	Phase sequence voltage relay
48	Incomplete sequence relay
49	Machine or transformer thermal relay
50	Instantaneous overcurrent or rate-of-rise relay
51	Ac time overcurrent relay
52	Ac circuit breaker
53	Exciter or dc generator relay
56	Field application relay
59	Overvoltage relay
62	Time-delay stopping or opening relay
63	Pressure switch
64	Ground detector relay
67	Ac directional overcurrent relay
69	Permissive control
71	Level switch
72	Dc circuit breaker
74	Alarm relay
76	Dc overcurrent relay
78	Phase-angle measuring or out-of-step protective relay
79	Ac reclosing relay
80	Flow switch
81	Frequency relay
82	Dc reclosing relay
83	Automatic selective control or transfer relay
85	Carrier or pilot-wire receiver relay
86	Lockout relay
87	Differential protective relay
90	Regulating device
92	Voltage and power directional relay

Table 10.2   Separate Auxiliary Devices

Suffix	Device
C	Closing relay or contactor
CL	Auxiliary relay, closed (energized when main device is in closed position)
CS	Control switch
D	"Down"-position switch relay
L	Lowering relay
O	Opening relay or contactor
OP	Auxiliary relay, open (energized when main device is in open position)
PB	Pushbutton
R	Raising relay
U	"Up"-position switch relay
X	Auxiliary relay
Y	Auxiliary relay
Z	Auxiliary relay

With the expanding emphasis on electronic and electrical circuits, the preparation and use of logic diagrams is of increasing importance.

10.15   DESIGN COMPLETION

The **degree of completion** of design is often a point of concern. Typically, several tiers of drawings are required to execute a design and the responsibility for preparation of these and their level of detail may vary from one customer to another and from one project to another. Each industry has its own standards for **documentation, qualification,** and **testing,** which may differ widely from one another. In the construction industry it is common practice to require that the contractor (and its subcontractors) provide shop detail drawings, and in some cases where the design, as bid, is not fully detailed, to complete the design as well. For manufacturing industries, machine designs, as bid, may not fully detail the components in an assembly but may merely establish critical working points, controlling dimensions, and so on.

Table 10.3   Condition or Electrical Quantity

Suffix	Denotes
A	Air, or amperes
C	Current
F	Frequency, or flow
L	Level, or liquid
P	Power, or pressure
PF	Power factor
S	Speed
T	Temperature
V	Voltage, volts, or vacuum
VAR	Reactive power
W	Water, or watts

Table 10.4 Typical Location or Circuit Identification

Suffix	Denotes
A	Alarm or auxiliary power
AC	Alternating current
BK	Brake
C	Capacitor, or condenser, or compensator, or carrier current
CA	Cathode
DC	Direct current
E	Exciter
F	Feeder, or field, or filament
G	Generator, or ground
M	Motor, or metering
N	Network, or neutral
P	Pump
R	Reactor, or rectifier
S	Synchronizing
T	Transformer, or test, or thyratron
TM	Telemeter
U	Unit

For **electronic** work, *design completion* (also called *design qualification*) commonly requires first a demonstration that the design meets specifications, and then a **documentation** package, which includes electrical schematics, bills of materials along with qualified suppliers, fabrication and assembly drawings for printed circuit boards, assembly drawings at the chassis level, and circuit descriptions and test instructions for all levels of assembly. The degree of adherence to these standards and the specific format requirements for each document will vary for different companies and with whether or not there are applicable regulatory requirements (e.g., military, medical device, etc.) Where any of the procurement, assembly, or test information is omitted, outcome of that process is *de facto* left to the discretion of whoever is doing the work, often leading to inconsistent results.

Where manufacturing considerations are important, **design qualification** is often linked to repeatability of the manufacturing process. Sufficient pilot production must occur to validate the process and to provide **quality control metrics,** typically the mean and **standard deviation** of critical parameters. Design completion can be specifically linked to having manufacturing processes that can repeatedly (or with a predictable defect rate) produce products that meet specifications. For such products, design, qualification testing, and documentation of the manufacturing process have importance equal to product definition. Sophisticated design teams will extend these concepts and ensure that the design effort accounts for the needs of all steps in a product's life cycle, from material procurement through use and maintenance and even to end-of-life disposition. For mission critical products (such as for military, medical or NASA applications) design qualification must satisfy well-defined standards throughout the design cycle. The contents of the design package, and often the format, is carefully controlled to assure this result.

In summary, the degree of completion for the design depends on several factors, including:

- What does the contract between the parties require?

Table 10.5 Typical Device Parts Identification[a]

Letter	Denotes
C	Coil, or condenser, or capacitor
CC	Closing coil
HC	Holding coil
LS	Limit switch
M	Operating motor
S	Solenoid
TC	Trip coil
a	Auxiliary switch, open when the main device is in the de-energized or nonoperated position
b	Auxiliary switch, closed when the main device is in the de-energized or nonoperated position
A	Accelerating or automatic
B	Blocking, or backup
C	Close, or cold
D	Decelerating, or detonate, or down
E	Emergency
F	Failure, or forward
H	Hot, or high
HR	Hand reset
HS	High speed
IT	Inverse time
L	Left, or local, or low, or lower, or leading
M	Manual
OFF	Off
ON	On
O	Open
P	Polarizing
R	Right, or raise, or reclosing, or receiving, or remote, or reverse
S	Sending, or swing
T	Test, or trip, or trailing
TDC	Time-delay closing
TDO	Time-delay opening
U	Up

[a]Typical examples of the use of device numbers, suffixes, and other identifiers are:

$\dfrac{52G}{a}$ = auxiliary contact, normally open, on ac generator circuit breaker.

86X = auxiliary relay to lockout relay.

87G = differential protective relay for generator.

$\dfrac{74T}{TDC}$ = time-delay closing contact on temperature alarm relay.

- What is the purpose of the design? Is it for overall manufacturing guidance or for actual use on the shop floor as a shop drawing?
- Is the function of the device critical?
- What is the cost to prepare each of the tiers of the design (i.e., overall, definitive, shop details, etc.)?
- What is the normal distribution of work within the industry (e.g., what are the contractor's and subcontractor's or supplier's normal responsibilities)?
- Will the design and manufacture be outsourced? What does the supplier need to know to produce the design?

- Are there legal, jurisdictional, or code requirements or liability considerations that favor or require one party or another to prepare specific portions or levels of the design?

10.16   TOLERANCING

Increasingly, the requirements of ASME standard **Y14.5M-1994** for geometric dimensioning and tolerancing are being imposed. This standard establishes a uniform method of stating dimensional and geometric requirements and avoids the problem of self-interpretation by production personnel. An important advantage of the use of this standard is that it provides a clearer display of tolerances and their interrelationship with other geometric characteristics. This has the effect of increasing the tolerance permitted, thus facilitating production and reducing cost. Further, since geometric dimensioning and tolerancing are widely used in Europe, this methodology is becoming more important for companies engaged in international trade. Some of the more commonly used symbols and an example are shown in Fig. 10.20.

10.17   COMPUTER-AIDED DESIGN

While traditional hand drafting continues to be available, the continuing expansion of **computer-aided design** (CAD) equipment capability has relegated it to either small projects or quick sketches. As a result, most drawings today are prepared on CAD equipment.

A major advantage of the CAD drawings is their uniformity, legibility, and the ease of making additions and changes. CAD drawings do not suffer from handling or staining and do not become torn, dog-eared, or faded. Erasures are clean, and there is no problem with repeated handling. Where engineering is pursued concurrently with manufacturing or other external activities the ease of revision of drawings and preparation of bills of material greatly facilitate the work. In addition these drawings can be transmitted electronically, thus speeding up the comment and distribution processes. CAD drawings can be manipulated electronically to produce bills of material and in some cases have **extraction** capability that permits the automatic preparation of specialized discipline drawings. Many CAD programs have a **reference file** feature that ensures that the different workstations using the system are using the same data base and that the drawing information is current. This avoids the problem of two or more engineering disciplines revising the same drawing and ending up with two different versions of it.

Experience to data indicates that for **two-dimensional** (2D; conventional drafting) CAD systems using a common database with two or more workstations and with dedicated CAD operators, when the cost to amortize the computer equipment and programs are added to the actual cost of the time of the CAD operator, it is about 6 to 10% cheaper in cost then traditionally prepared hand-drafted drawings.

The initial investment in a central database-type 2D CAD workstation, having capacity to handle say 500 E-size drawings, is significant. While the cost of the workstation hardware is declining constantly, the cost of the software to drive them is tending to increase due to continuing expansion of capability. Today this increase

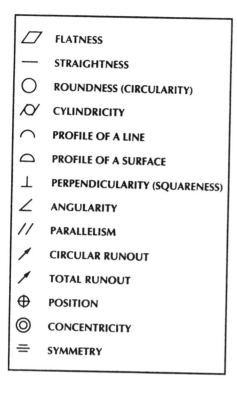

FLATNESS

STRAIGHTNESS

ROUNDNESS (CIRCULARITY)

CYLINDRICITY

PROFILE OF A LINE

PROFILE OF A SURFACE

PERPENDICULARITY (SQUARENESS)

ANGULARITY

PARALLELISM

CIRCULAR RUNOUT

TOTAL RUNOUT

POSITION

CONCENTRICITY

SYMMETRY

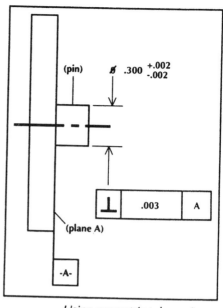

Using geometric tolerancing, a manufacturer can specify tolerance and location by referencing a dimension, a plane, and a perpendicularity callout to that plane. In this example, the pin diameter must be within the specified limits (±.002). The specified tolerance of perfect form (.003 max), which actually specifies the pin form, will be perpendicular to plane A. The pin must be the correct size and perpendicular to plane A within .003.

Fig. 10.20 Selected geometric and tolerancing symbols. (Courtesy of DataMyte Business of Rockwell Automation. Copyright Allen-Bradley Company, Inc., 1995. Used with permission.)

is still less than the equipment decrease, resulting in a net reduction in cost. Formerly these systems were often operated in a two-shift mode: 16 hours per day and sometimes six or seven days a week to reduce the unit cost that must be charged to the drawing(s) produced. However, with the continuing reduction in overall cost and the increased productivity from advanced software, this is the exception rather than the rule.

To overcome the high cost associated with these types of systems, **stand-alone workstations** are available that range in price from $7000 to over $100,000, depending on CPU, processing speed, memory, disk drive capacity, size and resolution of the monitor, number of monitors, and so forth. Centralized CPU-based systems are little used, and the current approach is to connect combinations of stand-alone workstations through a Local Area Network (LAN) by a file server. The file server normally has additional disk drives as well as plotters attached to it. This approach is useful on smaller installations, as small as two or three workstations where one of the workstations acts as the **file server.** The state of the art today is to utilize stand-alone workstations, with development pointed toward higher-capacity workstations in the future.

While most CAD work is performed as a replacement for conventional drafting, so-called 2D CAD, a system of using **three-dimensional** (3D) CAD is becoming more common. The 3D CAD system is essentially an evolution of the 2D CAD systems, and much of the basic CAD technology and hardware is used. However, instead of developing drawings directly, a three-dimensional **mathematical model** of the item, or facility, is developed, and drawings are prepared by the computer based upon data resident in the mathematical model.

Both 2D and 3D systems begin with intelligent flow diagrams and P&IDs prepared using equipment symbol cells. A common data base that includes all information related to the specific piece of equipment is used, or developed. This information is used on the P&IDs, flow diagrams, and electrical schematics.

In 3D CAD many of the drawings are prepared by cutting sections through the three-dimensional model, the sections becoming the drawings. In some cases, for example, structural drawings, **automatic drawing extraction** is available that prepares structural steel plans and elevations from the data base, without showing other features of the design, such as piping and equipment. In addition to the preparation of orthographic and extracted drawings, 3D CAD permits the preparation of perspective and walk-thru-type drawings as well.

To construct the model requires that dimensional and other data on components, piping, structural systems, and similar elements be entered into the computer. In the case of equipment and many major components, a **library** of them is established and drawn upon for principal characteristics such as the space envelope, weight, connection size, and location, with the engineer, designer, or drafter providing orientation and centerline location data. Some equipment vendors are beginning to provide equipment data in **electronic format** that can be input directly into the electronic library for use by these 3D systems. For piping and wiring components, entering centerline location, routing, system identifiers, locations of valves, junction and pull boxes, and similar data permit the computer to establish all the significant parameters and incorporate their configuration into the design. The same approach is used with the structural systems and other elements of the design and creates the three-dimensional model in the computer. The preparation of **orthographic** design drawings from this model involves the cutting of sections through the model and then printing them on **high-resolution plotters** along with necessary dimensions and notes. The systems can produce ghost backgrounds to permit the highlighting of a particular type of work, such as piping. In addition it is possible to omit certain types of data such as structural systems from drawings to permit easier reading.

While the system only requires from 50 to 60% of the time of conventional drafting, there is a considerable investment in establishing the data base or library of standard equipment, piping, electrical equipment, structural members, and so on. One major advantage to 3D CAD is the reduction not only in the physical drafting effort but also the reduction in **interferences** in the field. The combined saving in these two items often equals the total cost of the engineering effort.

Figure 10.21 gives a cost comparison for different methods of drawing preparation and is reasonably representative of the economic factors present.

With either form of CAD, automated **checking** and **interference** identification are built into the system, and bills of material can be automatically produced. This is

EXAMPLE PROJECT DOCUMENT PRODUCTION			
	MANUAL NO CAE	BASIC CAE 2D DRAFTING	FULL CAE 3D PDS
NUMBER OF DRAWINGS (1)	909	909	875
TOTAL MANHOURS (1)	101,324	88,861	65,073
HOURS/DRAWING	111	98	74
MANUAL DRAFTING HOURS (1)	24,925		
CAE HOURS (1)		12,463	21,546
TOTAL LABOR	$5,319,491	$4,976,771	$3,954,983
$/HOUR	$52.50	$56.01	$60.78
MULTIPLIER (2)	2.10	2.24	2.43
DIRECT LABOR $/HOUR	$25.00	$25.00	$25.00
SAVINGS FROM MANUAL		$342,720	$1,364,509
% SAVINGS FROM MANUAL		6.44%	25.65%
TOTAL LABOR (A)	$5,319,491	$4,976,771	$3,954,983
EQUIPMENT COST	$45,500,000	$45,500,000	$45,500,000
CONSTRUCTION COST (C)	$84,500,000	$84,500,000	$84,500,000
TOTAL PROJECT COST (B)	$130,000,000	$130,000,000	$130,000,000
% A OF B	4.09%	3.83%	3.04%
% A OF C	6.30%	5.89%	4.68%
SAVINGS FROM MANUAL DURING ENGINEERING DURING CONSTRUCTION		$342,720	$1,364,509 $1,690,000
TOTAL SAVINGS		$342,720	$3,054,509
% SAVINGS OF B	0.00%	0.26%	2.35%
% SAVINGS OF C	0.00%	0.41%	3.61%

NOTES:
1. The preparation of the following drawings have been excluded from the number of drawings, total manhours, manual drafting hours and CAE hours, above:
   o 470 Piping Isometrics
   o 130 Instrument Loop Diagrams
   o 20 Perspectives (3D General Arrangements) in FULL CAE 3D PDS, only
2. The cost of CAE has been included in the multiplier. The multiplier on labor, only, is 2.10. The cost of CAE is $25 per hour.

Fig. 10.21 Drawing preparation cost comparison. (Courtesy of ICF Kaiser Engineers.)

more extensive and accurate in the 3D than the 2D systems. In 2D CAD there is no physical tie between plans and sections or plans and elevations. Plans are only 2D, and no third dimension exists. As a result the quantity of material in the third dimension is an approximate allowance or has to be manually input. Similarly 2D CAD provides interference detection by overlay detection but, with different elements appearing at different elevations, may not be entirely accurate. Thus computerized data cannot be totally relied upon, and some manual checking is required. With either form of CAD, **bills of material** can be automatically produced and then electronically input to requisitions to begin the purchasing process. Because of the greater memory and processing capacity, there is an exponential increase in this capability in the 3D systems. Phased drawings can be produced to indicate both sequence of operations and completion with time, enabling the preparation of drawings and diagrams to assist in scheduling and estimating. **Walk-throughs,** which simulate the appearance of the facility, can be produced to permit review of clearances, access, and similar considerations. Architectural and other type presentation drawings can be prepared, and they are often of great value in dealing with licensing or permitting bodies.

The term **computer-aided design and drafting** (CADD) is often used interchangeably with the term CAD, described above. With increased emphasis on the design aspect of the work, CADD employs one or a series of computer programs, such as finite element analysis programs, to solve the specific design problem at hand. Usually linked to a drafting capability, the solution of the design program is readily applied to produce the design or design modifications indicated by the calculation model.

10.18 COMPONENT AND SYSTEM TESTING

This testing is performed either as in-process or final testing. As its name implies, **in-process testing** provides assurance of the correctness of manufacture by tests performed while the work is in process. Typically, **statistical process control** (SPC) is most adaptable to medium-length and longer production runs. The larger the run, the greater the population considered and the more accurate the SPC calculations, inferences, and decisions. By performing SPC while work is in process, **deviations** can be detected and the process corrected to minimize effects on later processes or the finished goods themselves. Further, where SPC is used on early processes in a multistep fabrication system, **nonconforming** work is detected before later value-added steps have been performed, and the economic impact of nonconformances is minimized. Testing is usually **nondestructive.** For prototypes, and often for short production runs, 100% testing is universally used.

Where a high degree of confidence exists that the individual fabrication steps are being performed correctly, only **final testing** is employed. With most manufactured goods, final testing is employed either on a 100% basis for smaller lots of goods or on a statistical basis for larger lots. Despite this, the trend is to use 100% final testing for all high-value goods of a critical nature or those that are relatively complex. Often with complex items intended to operate in various modes, such as computer chips, control equipment, and defense electronic equipment, 100% testing is performed at steps in the manufacturing process with computer-driven programs used to simulate the various operating conditions.

10.19 REFERENCE SOURCES

Although the specific requirements for **references** will vary depending on the type of engineering being performed, certain volumes are useful as a core library of ready reference data. The following titles have proven particularly useful to the author.

- *Webster's New Collegiate Dictionary,* Merriam-Webster
- *Roget's International Thesaurus,* Crowell
- *Harbrace College Handbook,* Hodges and Whitten, Harcourt Brace Jovanovich
- *World Almanac,* Newspaper Enterprise Assoc., Inc.
- *The Wiley Engineer's Desk Reference,* Heisler, Wiley
- *The Wiley Project Engineer's Desk Reference,* Heisler, Wiley
- *Handbook of Engineering Fundamentals,* Eshbach, Wiley

- *The Way Things Work,* Simon & Schuster
- *Engineering Economy,* Grant, Ireson, and Leavenworth; Ronald Press
- *The Procedure Handbook of Arc Welding,* Lincoln Electric Co.
- *Materials and Processes in Manufacturing,* DeGarmo, Macmillian
- *Manual of Steel Construction,* American Institute of Steel Construction

Some sources of **engineering information** include the following, as listed in *Reference Sources in Science, Engineering, Medicine, and Agriculture,** a master list of reference sources arranged by type and fields of engineering and indexed by title, author, and subject.

Applied Science and Technology Index. New York: Wilson, 1958– , monthly with quarterly cumulations. v. 1– , index. service basis. ISSN: 0003-6986. This subject index lists the articles in the more popular periodicals in science and technology. Citations to book reviews are listed separately. It covers aeronautics, space science, chemistry, computer technology, construction, engineering, petroleum and gas, robotics, telecommunications, and transportation. It is also available online and on CD-ROM.

Bibliographical Guide to Technology. Boston: G. K. Hall, 1974– , annual. v. 1– , index. ISSN: 0360-2761. Previously called *Technology Book Guide,* this guide gives full bibliographic information of books cataloged by the New York Public Library and the Library of Congress in the fields of industrial, structural, civil, transportation, hydraulic, sanitary, highway, mechanical, electrical, nuclear, and mining engineering.

Engineering Index Monthly. New York: Engineering Index, Inc, 1906– , monthly with annual cumulations. v. 1– , index. ISSN: 0742-1974. This monthly indexing/abstracting service covers all areas of engineering technology. Access to information can be made by numerous special indexes. It is also available online and on CD-ROM. *Bioengineering and Biotechnology Abstracts* is a spin-off from this database.

Comprehensive Dictionary of Engineering and Technology, with Extensive Treatment of the Most Modern Techniques and Processes. R. Ernst. New York: Cambridge University Press, 1985. 2 v., illus., index. ISBN: 0-521-30377-Xv.1; 0-521-30378-8v.2. This very comprehensive bilingual dictionary covers engineering terminology in French and English. Richard Ernst also has the same dictionary in German and English called *Dictionary of Engineering Technology* (Oxford University Press, 1985–1989).

Annual Book of ASTM Standards. Philadelphia, PA: American Society for Testing and Materials, 1939– , annual. v. 1– , illus., bibliog., index (59v. per year). This book is the world's most comprehensive compilation of standards, tests, and specifications of one society, the American Society for Testing and Materials. Each of the 59 annual volumes is published at a different time during the year and

*Modified from H. Robert Malinowsky, *Reference Sources in Science, Engineering, Medicine, and Agriculture.* Used with permission of The Oryx Press, 4041 North Central Avenue, Suite 700, Phoenix, AZ 85012. (800) 279-6799.

covers a specific topic such as coated steel or electrical conductors. The standards are accepted industry-wide and written into specifications.

Index and Directory of Industry Standards. Englewood, CO: Information Handling Services, 1983– , annual. v. 1– , index (several volumes per year). Previously called *Index and Directory of U.S. Industry Standards,* this standard source helps locate over 113,000 international and domestic standards of some 373 professional societies by subject and numerical designation. For just those standards from the American National Standards Institute, consult the *Catalog of American National Standards, 1990–1991,* ANSI, 1990).

Index of Federal Specifications Standards and Commercial Item Descriptions. Washington, DC: U.S. Federal Supply Service, 1952– , annual. v. 1– . This source provides alphabetical, numerical, and Federal Supply classification indexing to government specifications in general use.

In addition to the writings in engineering literature, **manufacturers' data** often provide highly useful and practical information. This material is typically written with an engineering user in mind and is application oriented. It is frequently more current than either magazine articles or books, but may not be as current as electronic database information. It should be used with some care to assure that data are neither proprietary nor excludes others from consideration. With these caveats, however, they serve as a convenient and useful source.

CHAPTER 11
ENGINEERING OPERATIONS

11.1 COMMERCIAL CONSIDERATIONS

Of all the commercial considerations the content and type of the contract are the most critical. For all engineering work, even the most simple, some form of **contract** is essential. The contract must be in writing and can be as brief as a letter of agreement or a complex and lengthy document. The contract must spell out the specifics of the relationship between the parties, including the overall scope of the work, liabilities, warranties, contingencies, payment, schedules, unit prices, and such. Detailed considerations, such as the plans and specifications, are usually made a part of the contract by reference. Each contract is unique and should be drawn carefully to cover the particulars of the specific project. Naturally, legal help should be obtained when drawing a contract or when reviewing one prepared by others. There are some key points that should be covered in any contract, no matter how abbreviated.

- A brief description of the scope of the project
- A description of the services to be provided
- Basis for the service—reimbursability and fee arrangement
- Overall schedule
- Delays
- Changes
- Force majeure
- Liability limitations
- Termination
- Payment
- Completion

11.2 CONTRACT TYPES

A prime concern in operating an engineering practice is the basis on which the engineering services will be rendered. Broadly, the contractual bases available are lump sum, cost plus fee (either fixed or percentage), and unit priced.

11.2.1 Lump Sum

Lump-sum contractual arrangements are preferred by clients because the client assumes no risks for overruns on the part of the engineer. Since the engineer assumes all risk, he or she therefore must take great pains to carefully describe the **scope** of his or her activities such that the lump-sum offering can be properly defined and **limited.** With a lump-sum arrangement, **scope changes** are the basis for "extras," which require negotiation and can be acrimonious. Nevertheless, where indicated, scope changes should be pursued since their cost is directly a bottom-line cost to the engineer. Particular care should be given to time delays, as these can have serious effects on lump-sum work, usually causing cost overruns. Time delays affect the work in two principal ways. They always tend to extent supervisory and overhead costs and, in addition, often cause a reengineering of work already completed. Although each of these can be significant, usually the rework has an effect several times larger than the supervisory and overhead cost increases.

11.2.2 Cost Plus Fee

Cost-plus-fee arrangements are preferred by the engineer when scope cannot be well defined or subject to changes, when schedules are indeterminant, or when new technologies are being utilized for the development of the engineering work. In a cost-plus-fee arrangement the purchaser, that is, the client, assumes the risk for overruns but gains the benefit of underruns. Since it is a separate item, the fee represents a secure form of profit for the engineer. However, the fee may not be totally net profit since entertainment and other costs may be charged against it. Obviously, for those cases where some form of reimbursable contract is entered into, the client will typically expect to involve himself or herself to a much greater extent in those engineering decisions made as the work proceeds on a day-by-day basis. This has the potential disadvantage of causing a loss of much of the engineer's freedom of action since the client participation is so great. This involvement can also adversely affect schedule, since clients have difficulty understanding the effects of even small changes that may be requested. In the lump sum contract, for example, the engineer can ask for total freedom with no restraints or involvement by the client until the finished product is delivered, because the engineer is furnishing the entire package and assuming the financial risk if the lump sum work overruns its contractual amount. It is not possible to take this position with cost-plus-fee work.

The fee determination for either **fixed** or **percentage fee** can be arranged as necessary to suit the preferences of the owner. Typically, fixed fees are used for those cases where it is possible to reasonably establish the scope in a very broad way, and the fixed fee is not normally subject to renegotiation except in those cases where the scope changes have been significant, say on the order of 20% or more. The use of percentage fee is not usually favored because of the possibility of expanding the basic scope to automatically earn additional fee.

11.2.3 Unit Pricing

For some work where the scope and the extent of the engineering is either totally unpredictable or difficult to anticipate, unit pricing may instead be used. Typically, the unit pricing includes both **direct** and **indirect payroll costs, overhead (burden),** and **fee,** all normally structured on the basis of dollars per manhour. **Unit pricing** will usually establish different unit prices for different categories of work. Thus, the unit price in dollars per hour for a drafter will differ from that for a checker, a design engineer, a supervising engineer, and so on.

11.2.4 Cost Structure

Pricing for services typically includes (1) direct payroll (gross amounts paid employees); (2) an allowance for indirect payroll costs, which is a direct multiplier on payroll costs, ranging from 25 to 40% to include the costs for employer payroll taxes, vacation, sick leave, retirement, and so on; (3) an overhead or burden percentage on the order of 75 to 100% to cover a large variety of office expenses (e.g., office rent, furniture and fixtures, insurance, personnel recruiting, training, development and management, legal services, technical society participation, and sales costs including proposal preparation); and (4) an allowance for fee (profit) often arbitrarily set at 10 or 15%.

Typically, the overall multiplier of direct payroll (including indirect costs) will range from $2\frac{1}{2}$ to $3\frac{1}{2}$; this includes fee (profit) plus overhead costs as well. As a part of providing engineering services, there are frequently costs incurred for laboratory testing, transportation and travel, consultants' fees, and so on; all accrued for the benefit of the client. Typically, these costs are passed along to the client with no or a modest 10 to 15% mark-up only. Normal engineering practice does not include the addition of fees to these types of costs.

11.2.5 Fee Curves[1]

The ASCE has published fee curves which, although intended for civil engineering, are widely used as guides in engineering consulting practices more generally (Figs. 11.1 and 11.2). The fee curves represent total remuneration, as in the case of lump sum work, and not merely that portion of fee (profit) described earlier.

11.3 PERFORMANCE STANDARDS

Performance standards include considerations of accuracy, timeliness, economy, and ethics. While the relative weighting of the importance of the factors will vary from one assignment to another, it is of paramount importance to assure first, that the suitability of the design is not compromised.

11.3.1 Accuracy

All calculations do not require the same degree of precision, particularly when given the less than rigorous **accuracy** of basic assumptions or certain input data. Thus, although not all designs need to be refined to an ultimate degree, it is necessary that the accuracy of the design be appropriate to the accuracy of the basic input data, suited for the purpose and utility intended, and that the health and safety of the public are not jeopardized. A degree of reasonableness and prudence is required when performing design work and it may be necessary to provide additional margins to assure

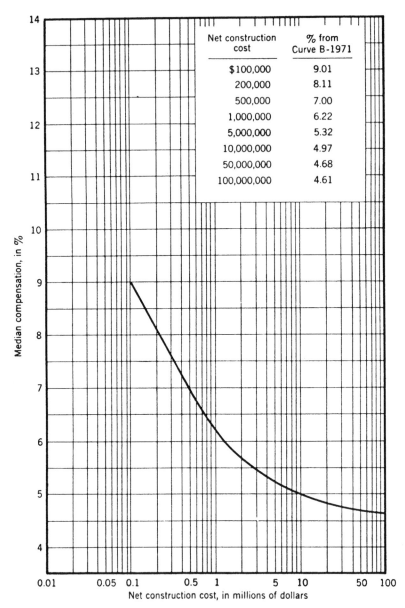

Net construction cost	% from Curve B-1971
$100,000	9.01
200,000	8.11
500,000	7.00
1,000,000	6.22
5,000,000	5.32
10,000,000	4.97
50,000,000	4.68
100,000,000	4.61

Fig. 11.1 Fee curve (average complexity). (From Ref. 1. Copyright American Society of Civil Engineers, New York, Oct. 1981.)

adequacy. Fortunately, industry codes and standards provide guidance and assistance, which is particularly useful to the less experienced engineer, and these together with mandated standards and criteria are usually enough to avoid significant problems. Where unusual questions arise, it is helpful to obtain guidance from peers, technical society headquarters personnel, and personnel on code and standards committees.

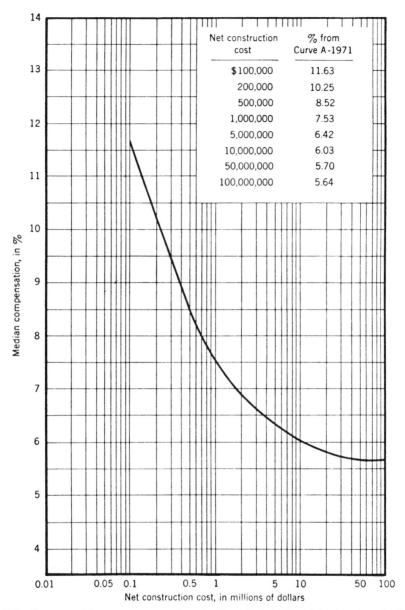

Net construction cost	% from Curve A-1971
$100,000	11.63
200,000	10.25
500,000	8.52
1,000,000	7.53
5,000,000	6.42
10,000,000	6.03
50,000,000	5.70
100,000,000	5.64

Median compensation, in %

Net construction cost, in millions of dollars

Fig. 11.2 Fee curve (above-average complexity). (From Ref. 1. Copyright American Society of Civil Engineers, New York, Oct. 1981.)

11.3.2 Timeliness

Timeliness is of increasing importance as the time from conception to implementation decreases and can become particularly critical where interest rates are high or where time to market for competing products is a concern. It is almost always better to have a 95% solution that is on time than a 99% one that is late. It is useful to bear in mind that not too long ago, before the widespread use of computers, slide rules were the traditional calculational tool and that they provided accuracy only to two, and on occasion, three significant figures. This does not mean that designs can be produced in a slapdash manner, but rather, that absolute precision may not be appropriate, particularly where it delays completing the design on time. Overriding all this is, of course, the requirement that the design be adequate. There is an old saying which nicely summarizes the problem: "If you don't have time to do it right, when are you going to have time to do it over?" While the difficulty is to define *right,* which necessarily varies depending on the purpose of the design, the principle remains that sufficient time must be devoted to the effort to assure correctness and accuracy.

11.3.3   Economy

Economy is always a goal of design. It can be economy of materials, fabrication time or effort, economy of the design effort itself, or economy of use for the end product. Each assignment has a different requirement for weighting the importance of economy, but in no case can it be ignored. Frequently, design criteria will establish the value of elements of design, such as heat or electrical energy saved, or weight saved.

11.3.4   Ethics

The engineering profession has over the years developed a code of ethics intended to assure integrity and competence in the conduct of the engineering profession. A copy of this standard and its detailed guidelines is shown in Fig. 11.3.

At the 1992 commencement ceremony for the College of Engineering of the University of California at Berkeley, civil engineering professor emeritus Alex C. Scordelis put forth a number of words of wisdom for new graduates. Some of them, reprinted below,* apply as well to more experienced practitioners and are worth bearing in mind:

- You got a good education. Use it, and remember your fundamentals.
- Your reputation follows you wherever you go. Do a good job.
- Learn from your mistakes—but don't repeat them.
- If it doesn't look right, it probably isn't right.
- Try to understand the big picture and the costs, as well as your own part, of each project.
- Make sure that whatever you design can be built or produced.
- At the end of a project, always ask yourself, "How could I do it differently and better next time?"

*Courtesy of Alex C. Scordelis. Professor Emeritus, University of California at Berkeley.

- Don't make an engineering decision on something you don't understand or don't know anything about.
- Use your common sense to check results. If you don't have common sense, you'd better develop some.

CODE OF ETHICS OF ENGINEERS

THE FUNDAMENTAL PRINCIPLES

Engineers uphold and advance the integrity, honor and dignity of the engineering profession by:

I.   using their knowledge and skill for the enhancement of human welfare;

II. being honest and impartial, and serving with fidelity the public, their employers and clients;

III. striving to increase the competence and prestige of the engineering profession; and

IV. supporting the professional and technical societies of their disciplines.

THE FUNDAMENTAL CANONS

1. Engineers shall hold paramount the safety, health and welfare of the public in the performance of their professional duties.

2. Engineers shall perform services only in the areas of their competence.

3. Engineers shall issue public statements only in an objective and truthful manner.

4. Engineers shall act in professional matters for each employer or client as faithful agents or trustees, and shall avoid conflicts of interest.

5. Engineers shall build their professional reputation on the merit of their services and shall not compete unfairly with others.

6. Engineers shall act in such a manner as to uphold and enhance the honor, integrity and dignity of the profession.

7. Engineers shall continue their professional development throughout their careers and shall provide opportunities for the professional development of those engineers under their supervision.

Approved by the Board of Directors, October 5, 1977

Fig. 11.3 Code of ethics.

Engineers' Council for Professional Development

SUGGESTED
GUIDELINES FOR USE WITH
THE FUNDAMENTAL CANONS OF ETHICS

1. Engineers shall hold paramount the safety, health and welfare of the public in the performance of their professional duties.

a. Engineers shall recognize that the lives, safety, health and welfare of the general public are dependent upon engineering judgments, decisions and practices incorporated into structures, machines, products, processes and devices.

b. Engineers shall not approve nor seal plans and/or specifications that are not of a design safe to the public health and welfare and in conformity with accepted engineering standards.

c. Should the Engineers' professional judgment be overruled under circumstances where the safety, health, and welfare of the public are endangered, the Engineers shall inform their clients or employers of the possible consequences and notify other proper authority of the situation, as may be appropriate.

   (c.1) Engineers shall do whatever possible to provide published standards, test codes and quality control procedures that will enable the public to understand the degree of safety or life expectancy associated with the use of the design, products and systems for which they are responsible.

   (c.2) Engineers will conduct reviews of the safety and reliability of the design, products or systems for which they are responsible before giving their approval to the plans for the design.

   (c.3) Should Engineers observe conditions which they believe will endanger public safety or health, they shall inform the proper authority of the situation.

d. Should Engineers have knowledge or reason to believe that another person or firm may be in violation of any of the provisions of these Guidelines, they shall present such information to the proper authority in writing and shall cooperate with the proper authority in furnishing such further information or assistance as may be required.

   (d.1) They shall advise proper authority if an adequate review of the safety and reliability of the products or systems has not been made or when the design imposes hazards to the public through its use.

   (d.2) They shall withhold approval of products or systems when changes or modifications are made which would affect adversely its performance insofar as safety and reliability are concerned.

e. Engineers should seek opportunities to be of constructive service in civic affairs and work for the advancement of the safety, health and well-being of their communities.

f. Engineers should be commited to improving the environment to enhance the quality of life.

2. Engineers shall perform services only in areas of their competence.

a. Engineers shall undertake to perform engineering assignments only when qualified by education or experience in the specific technical field of engineering involved.

b. Engineers may accept an assignment requiring education or experience outside of their own fields of competence, but only to the extent that their services are restricted to those phases of the project in which they are qualified. All other phases of such project shall be performed by qualified associates, consultants, or employees.

c. Engineers shall not affix their signatures and/or seals to any engineering plan or document dealing with subject matter in which they lack competence by virtue of education or experience, nor to any such plan or document not prepared under their direct supervisory control.

3. Engineers shall issue public statements only in an objective and truthful manner.

a. Engineers shall endeavor to extend public knowledge, and to prevent misunderstandings of the achievements of engineering.

b. Engineers shall be completely objective and truthful in all professional reports, statements, or testimony. They shall include all relevant and pertinent information in such reports, statements, or testimony.

c. Engineers, when serving as expert or technical witnesses before any court, commission, or other tribunal, shall express an engineering opinion only when it is founded upon adequate knowledge of the facts in issue, upon a background of technical competence in the subject matter, and upon honest conviction of the accuracy and propriety of their testimony.

d. Engineers shall issue no statements, criticisms, nor arguments on engineering matters which are inspired or paid for by an interested party, or parties, unless they have prefaced their comments by explicitly identifying themselves, by disclosing the identities of the party or parties on whose behalf they are speaking, and by revealing the existence of any pecuniary interest they may have in the instant matters.

e. Engineers shall be dignified and modest in explaining their work and merit, and will avoid any act tending to promote their own interests at the expense of the integrity, honor and dignity of the profession.

4. Engineers shall act in professional matters for each employer or client as faithful agents or trustees, and

Fig. 11.3   *(Continued)*

shall avoid conflicts of interest.

a. Engineers shall avoid all known conflicts of interest with their employers or clients and shall promptly inform their employers or clients of any business association, interests, or circumstances which could influence their judgment or the quality of their services.

b. Engineers shall not knowingly undertake any assignments which would knowingly create a potential conflict of interest between themselves and their clients or their employers.

c. Engineers shall not accept compensation, financial or otherwise, from more than one party for services on the same project, nor for services pertaining to the same project, unless the circumstances are fully disclosed to, and agreed to, by all interested parties.

d. Engineers shall not solicit nor accept financial or other valuable considerations, including free engineering designs, from material or equipment suppliers for specifying their products.

e. Engineers shall not solicit nor accept gratuities, directly or indirectly, from contractors, their agents, or other parties dealing with their clients or employers in connection with work for which they are responsible.

f. When in public service as members, advisors, or employees of a governmental body or department, Engineers shall not participate in considerations or actions with respect to services provided by them or their organization in private or product engineering practice.

g. Engineers shall not solicit nor accept an engineering contract from a governmental body on which a principal, officer or employee of their organization serves as a member.

h. When, as a result of their studies, Engineers believe a project will not be successful, they shall so advise their employer or client.

i. Engineers shall treat information coming to them in the course of their assignments as confidential, and shall not use such information as a means of making personal profit if such action is adverse to the interests of their clients, their employers, or the public.

(i.1) They will not disclose confidential information concerning the business affairs or technical processes of any present or former employer or client or bidder under evaluation, without his consent.

(i.2) They shall not reveal confidential information nor findings of any commission or board of which they are members.

(i.3) When they use designs supplied to them by clients, these designs shall not be duplicated by the Engineers for others without express permission.

(i.4) While in the employ of others, Engineers will not enter promotional efforts or negotiations for work or make arrangements for other employment as principals or to practice in connection with specific projects for which they have gained particular and specialized knowledge without the consent of all interested parties.

j. The Engineer shall act with fairness and justice to all parties when administering a construction (or other) contract.

k. Before undertaking work for others in which Engineers may make improvements, plans, designs, inventions, or other records which may justify copyrights or patents, they shall enter into a positive agreement regarding ownership.

l. Engineers shall admit and accept their own errors when proven wrong and refrain from distorting or altering the facts to justify their decisions.

m. Engineers shall not accept professional employment outside of their regular work or interest without the knowledge of their employers.

n. Engineers shall not attempt to attract an employee from another employer by false or misleading representations.

o. Engineers shall not review the work of other Engineers except with the knowledge of such Engineers, or unless the assignments/or contractual agreements for the work have been terminated.

(o.1) Engineers in governmental, industrial or educational employment are entitled to review and evaluate the work of other engineers when so required by their duties.

(o.2) Engineers in sales or industrial employment are entitled to make engineering comparisons of their products with products of other suppliers.

(o.3) Engineers in sales employment shall not offer nor give engineering consultation or designs or advice other than specifically applying to equipment, materials or systems being sold or offered for sale by them.

5. Engineers shall build their professional reputation on the merit of their services and shall not compete unfairly with others.

a. Engineers shall not pay nor offer to pay, either directly or indirectly, any commission, political contribution, or a gift, or other consideration in order to secure work, exclusive of securing salaried positions through employment agencies.

b. Engineers should negotiate contracts for professional services fairly and only on the basis of demonstrated competence and qualifications for the type of professional service required.

c. Engineers should negotiate a method and rate of compensation commensurate with the agreed upon scope of services. A meeting of the minds of the parties to the contract is essential to mutual confidence. The public interest requires that the cost of engineering services be fair and reasonable, but not the controlling consideration in selection of individuals or firms to provide these services.

(c.1) These principles shall be applied by Engineers

Fig. 11.3 (*Continued*)

in obtaining the services of other professionals.

d. Engineers shall not attempt to supplant other Engineers in a particular employment after becoming aware that definite steps have been taken toward the others' employment or after they have been employed.

(d.1) They shall not solicit employment from clients who already have Engineers under contract for the same work.

(d.2) They shall not accept employment from clients who already have Engineers for the same work not yet completed or not yet paid for unless the performance or payment requirements in the contract are being litigated or the contracted Engineers' services have been terminated in writing by either party.

(d.3) In case of termination of litigation, the prospective Engineers before accepting the assignment shall advise the Engineers being terminated or involved in litigation.

e. Engineers shall not request, propose nor accept professional commissions on a contingent basis under circumstances under which their professional judgments may be compromised, or when a contingency provision is used as a device for promoting or securing a professional commission.

f. Engineers shall not falsify nor permit misrepresentation of their, or their associates', academic or professional qualifications. They shall not misrepresent nor exaggerate their degree of responsibility in or for the subject matter of prior assignments. Brochures or other presentations incident to the solicitation of employment shall not misrepresent pertinent facts concerning employers, employees, associates, joint ventures, or their past accomplishments with the intent and purpose of enhancing their qualifications and work.

g. Engineers may advertise professional services only as a means of identification and limited to the following:

(g.1) Professional cards and listings in recognized and dignified publications, provided they are consistent in size and are in a section of the publication regularly devoted to such professional cards and listings. The information displayed must be restricted to firm name, address, telephone number, appropriate symbol, names of principal participants and the fields of practice in which the firm is qualified.

(g.2) Signs on equipment, offices and at the site of projects for which they render services, limited to firm name, address, telephone number and type of services, as appropriate.

(g.3) Brochures, business cards, letterheads and other factual representations of experience, facilities, personnel and capacity to render service, providing the same are not misleading relative to the extent of participation in the projects cited and are not indiscriminately distributed.

(g.4) Listings in the classified section of telephone directories, limited to name, address, telephone number and specialties in which the firm is qualified without resorting to special or bold type.

h. Engineers may use display advertising in recognized dignified business and professional publications, providing it is factual, and relates only to engineering, is free from ostentation, contains no laudatory expressions or implication, is not misleading with respect to the Engineers' extent of participation in the services or projects described.

i. Engineers may prepare articles for the lay or technical press which are factual, dignified and free from ostentatious or laudatory implications. Such articles shall not imply other than their direct participation in the work described unless credit is given to others for their share of the work.

j. Engineers may extend permission for their names to be used in commercial advertisements, such as may be published by manufacturers, contractors, material suppliers, etc., only by means of a modest dignified notation acknowledging their participation and the scope thereof in the project or product described. Such permission shall not include public endorsement of proprietary products.

k. Engineers may advertise for recruitment of personnel in appropriate publications or by special distribution. The information presented must be displayed in a dignified manner, restricted to firm name, address, telephone number, appropriate symbol, names of principal participants, the fields of practice in which the firm is qualified and factual descriptions of positions available, qualifications required and benefits available.

l. Engineers shall not enter competitions for designs for the purpose of obtaining commissions for specific projects, unless provision is made for reasonable compensation for all designs submitted.

m. Engineers shall not maliciously or falsely, directly or indirectly, injure the professional reputation, prospects, practice or employment of another engineer, nor shall they indiscriminately criticize another's work.

n. Engineers shall not undertake nor agree to perform any engineering service on a free basis, except professional services which are advisory in nature for civic, charitable, religious or non-profit organizations. When serving as members of such organizations, engineers are entitled to utilize their personal engineering knowledge in the service of these organizations.

o. Engineers shall not use equipment, supplies, laboratory nor office facilities of their employers to carry on outside private practice without consent.

p. In case of tax-free or tax-aided facilities, engineers should not use student services at less than rates of other employees of comparable competence, including fringe benefits.

Fig. 11.3 (*Continued*)

6   Engineers shall act in such a manner as to uphold and enhance the honor, integrity and dignity of the profession.

  a. Engineers shall not knowingly associate with nor permit the use of their names nor firm names in business ventures by any person or firm which they know, or have reason to believe, are engaging in business or professional practices of a fraudulent or dishonest nature.

  b. Engineers shall not use association with non-engineers, corporations, nor partnerships as 'cloaks' for unethical acts.

7. Engineers shall continue their professional development throughout their careers, and shall provide opportunities for the professional development of those engineers under their supervision.

  a. Engineers shall encourage their engineering employees to further their education.

  b. Engineers should encourage their engineering employees to become registered at the earliest possible date.

  c. Engineers should encourage engineering employees to attend and present papers at professional and technical society meetings.

d. Engineers should support the professional and technical societies of their disciplines.

e. Engineers shall give proper credit for engineering work to those to whom credit is due, and recognize the proprietary interests of others. Whenever possible, they shall name the person or persons who may be responsible for designs, inventions, writings or other accomplishments.

f. Engineers shall endeavor to extend the public knowledge of engineering, and shall not participate in the dissemination of untrue, unfair or exaggerated statements regarding engineering.

g. Engineers shall uphold the principle of appropriate and adequate compensation for those engaged in engineering work.

h. Engineers should assign professional engineers duties of a nature which will utilize their full training and experience insofar as possible, and delegate lesser functions to subprofessionals or to technicians.

i. Engineers shall provide prospective engineering employees with complete information on working conditions and their proposed status of employment, and after employment shall keep them informed of any changes.

Engineers' Council for Professional Development
345 East 47th Street
New York, N.Y.

Fig. 11.3 (*Continued*)

11.4 COST CONTROL

Cost control for the work in progress is usually handled by carefully estimating manhours per unit of work (e.g., per specifications, per drawing). The value of the completed component parts of the work is then established, and as the percentages are worked off, the amount of manhours "earned" is established. Particular care should be taken to realistically evaluate the value of the work completed and of that remaining. It is typical to substantially underestimate the remaining work and thus indicate higher percentage of completion than is actual. By comparing the amount of manhours earned to the amount of manhours estimated, it is possible to determine broadly whether the work is on schedule and whether the revenue is being realized at the anticipated rate. For deviations from the plan, it is essential to analyze and determine their causes and establish a plan to get back on schedule and within budget. Alternatively, where the budget has been substantially overrun or underrun it may be necessary to obtain suitable agreement on changes of scope and possibly schedule. In estimating the effort remaining to complete the work, it is useful to remember that the final 10% of the work can require up to 30% of the total time and effort. Adequate time and budget amounts should be allowed for this as early in the design period as possible.

    The question of **scope changes** is frequently the most significant in administration of an engineering contract. It is absolutely essential to maintain accurate and complete records of changes in scope, which may occur by virtue of conversations, phone calls, a brief note, or informal requests. Each of these changes in effect redefines the contract and modifies the initial agreement for the work. Without a suitable scope change procedure and rigorous administration, it is virtually impossible to maintain proper cost and schedule control of the work.

Prompt submittal of **billings** for work is essential. Wherever possible they should be submitted monthly with sufficient backup so that charges are justified. To improve collections, some consultants provide discounts for prompt payment, such as 2% discount for payment within 10 days and with interest to run (usually at a stated amount) if payments are delayed more than 60 days.

11.5 SCHEDULE CONTROL

Schedule control for engineering work in progress is of critical importance regardless of the contractual basis for the work. Except for the most simple work, some form of **critical path** or **arrow diagram** is necessary to assure consideration of the interfaces and the restraints between the development and utilization of data. For most engineering activities, the ability to perform a calculation and prepare a drawing, for example, is dependent on a receipt of information from manufacturers, data from outside sources, field survey data, and so on. It is essential to reflect the interrelation of these activities in some form of a diagram, usually of the critical path type. The diagram should be arranged with a calendar scale, either an absolute calendar indicating days, weeks, and months, or a calendar indicating time from award of work or release of engineering measured in days or weeks.

The diagram, if properly constructed, will indicate (1) significant milestones, (2) those activities that are restraining on the completion of the work, as well as (3) the several critical paths for completion of the work. (These paths are those that have no or minimal **float,** i.e., free time not required for the work.) Based on such a diagram, the early required critical decisions can be identified and those activities monitored particularly closely to assure that the overall schedule does not slip. Noncritical activities, that is, activities not on the critical path, should also be followed, but need not be monitored as closely as those on critical path since their impact is less. If, however, items not on the critical path are delayed beyond their allowable float (i.e., started or completed after their "late start" or "late completion" dates), they will become critical and may in fact govern completion of the work. Thus, a reasonable degree of surveillance must be applied to these noncritical activities as well.

For extensive and complex activities where the volume of work exceeds the capabilities of the individual design office, it may be necessary to place blocks of work with other offices or companies. For such activities, milestone schedules establishing major events and data exchanges taken from the overall critical path schedule are a highly important tool for monitoring this work.

The level of detail to be indicated on the critical path diagrams would depend on the degree of monitoring and control that is desired. It is possible to monitor modest activities with a critical path diagram having perhaps 20 to 30 activities, whereas on very large, multimillion dollar engineering contracts, the critical path diagram may include several hundred activities and may be converted into a computerized schedule that computes float and critical path(s) and generates exception reports, critical item lists, and so forth. Figure 11.4 shows a typical critical path diagram.

Extremely simple work may be controlled using **bar charts** or other less sophisticated tools. Considerable caution should be exercised in their use, since they in-

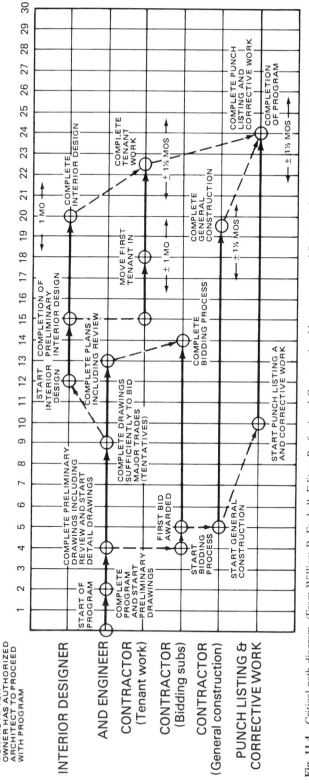

Fig. 11.4 Critical path diagram. (From William B. Foxhall, Editor, *Professional Construction Management and Project Administration*, 2nd ed. Copyright 1972, 1976 by the American Institute of Architects and Architectural Record. Reproduced with permission of the American Institute of Architectures under permission number 82060. Further reproduction is prohibited. Also, copyright © 1972 by McGraw-Hill Book Company. Used with the permission of McGraw-Hill Book Company.)

herently require that the user maintain continuously in mind the interrelation of the events, and if even modest complications present themselves, control of the work can be lost. In this circumstance, a critical path diagram should be constructed and utilized as the control device for the work.

11.6   REPORT WRITING

Report writing is an essential skill of the engineer and must be performed with a view toward maximizing the amount of information for the reader or user of the report. Typically, reports are structured with sections titled:

- Purpose
- Scope
- Conclusions and Recommendations
- Discussion
- Appendices

Further, a portion of the report (typically the introduction) would indicate who participated in the preparation of the report, when the report was prepared, and so forth. This provides a source for clarifications and also permits the reader to determine the validity of the report based on the credentials or reputation of the author(s) and contributors.

The **purpose** sets forth why the report was written, the factors that led to its writing, and so on. The **scope** describes the extent of the report, limitations on its coverage, and so on. This serves to put the background of the report in clear perspective so that the reader will understand fully not only the purpose of the report but any limitations imposed.

Conclusions and **recommendations** may be one section or separate sections, but should be placed toward the front of the report, so that a busy person can easily locate this section and readily determine the findings of the report. Each conclusion must be supported by material elsewhere in the report. Conclusions are rarely useful without recommendations. Therefore, typically, recommendations will be included and are essential to a full use of the report.

Sections on **discussion** and **appendices** are self-evident and become a convenient way of dividing the material between that found in the course of the investigation (Discussion) and Appendices, which typically are secondary or reference material of limited interest to most users of the report, although of possible interest to some.

For extensive or lengthy reports a separate **executive summary** is often prepared which, although placing major emphasis on the conclusions and recommendations, provides an overview of the entire report.

Reports should be written in clear concise language with emphasis on short sentences and clarity of thought. Ponderous or obscure language should be avoided and technical or complex material should be put into lay terms to the maximum extent possible. Wherever possible, abstractions should be avoided and the report made as concrete and specific as possible.

For complex reports it is often convenient to separate the reference material from the report itself and to assign citation numbers to important information. The (reference) documents that contain this information are often tabbed by citation number and grouped together, sometimes in one or more appended volumes to provide ready reference. Graphical material is preferred over tabular material, although tabular material can be placed in the discussion or appendices if necessary.

11.7 MATERIALS ACQUISTION

Materials acquisition includes material identification and requisitioning, bid solicitation and evaluation, purchase order or supply contract award, fabrication, inspection and test, shipping, receiving, and issuance to user. The most common areas involving engineering personnel are identification and requisitioning, technical bid evaluation, and inspection and test.

Materials that are procured by purchase order usually are **requisitioned** by the specifying or design engineer from bills of material developed as a part of the design drawings. Often, these are prepared automatically by CAD programs. Typically, the **requisition** establishes the material by its specification or other description, length or size, and calls out the quantity required. Frequently for bulk or raw materials, to allow for fabrication and scrap losses, an arbitrary amount, often 10%, is added.

Where equipment or engineered materials are being purchased, a detailed **specification** is typically prepared. This, together with standardized purchasing documents such as **general conditions** (which establish the contractual conditions of the purchase), drawings, drawings and data requirements, painting standards, marking and shipping instructions, and other data to describe the equipment and its appurtenances are issued to bidders or prospective suppliers.

After receipt of bids, the bid analysis is performed. Where performance concerns are significant, a technical **bid analysis** is used, and the engineer is often required to perform an analysis of efficiency, size, weight, or other significant parameters. Frequently, these consider life cycle costs, including initial cost, cost of spares, maintenance, downtime, overall expected life, and similar factors (Section 8.6.1). Even for simple pieces of equipment, differences in efficiency or performance may exist and a simple comparison of cost, including the value of the efficiency difference, can be important to the purchasing decision.

Testing and inspection of materials, particularly fabrication methods and qualification testing for special operations such as welding and heat treating, may be required. In many cases the procurement documents provide for testing to be witnessed by (often the specifying engineer for) the purchaser. For special materials and equipment, prototype testing, which may include extensive simulations or actual performance tests, is frequently required.

11.8 RESOURCE AND LABOR ALLOCATION

With the increasing stress placed on budget discipline, **resource** and **labor allocation** is of increasing importance. Leveling techniques to reduce large swings in personnel or resource use, through adjustments to the activities internal to the schedule, are widely used. In addition, temporary personnel or personnel working under contract to supplement the normal workforce are often used. One advantage to these methods is to permit the firm to look over the temporary or contract personnel with the possibility of permanent employment while reducing the risk of a poor hiring decision. A second advantage is that the terms of use of these personnel usually exclude benefits, thus being more economical than a permanent employee. In addition, the most overriding advantage to these arrangements is the ability to terminate these personnel on short notice, without incurring exposure for unlawful termination. This provides maximum flexibility in changing staffing levels to meet varying workloads.

Where workload swings are too extreme to be handled by these techniques, it is possible to outsource work to contract firms. If done for design work, great care needs to be exercised to assure proper quality, and in some cases confidentiality, of the work. This is normally done by a well-defined scope and criteria document, supplemented by frequent in-process reviews, occasionally placing one or more personnel in the outsourced location, and nondisclosure agreements.

11.9   TECHNICAL AND INDUSTRIAL SOCIETY PARTICIPATION

Technical and industrial society participation is an important aspect of professional development. Not only does it provide an opportunity to establish contacts in the technical community, but it also provides a continuing source of information for expanding knowledge and technical information. Participation on technical committees is particularly useful, as it develops a network of colleagues who have often faced and solved the same problems in the past and who can provide suggestions and guidance when troublesome issues arise. The writing and publication of professional papers offers a way to present new ideas or considerations to a society and facilitates the spread of technical information.

Further, any of these activities offers opportunities for work should there be a wish to relocate or change employers. For those in the consulting field it is a useful way to disseminate information on your capabilities to potential users of your services.

11.10   REGISTRATION

Registration, although not a legal requirement for all engineers working for corporations, is nevertheless considered essential for professional growth. In the case of engineers in consulting practice (as distinguished from corporate practice), registration is virtually mandatory. In many jurisdictions the engineer in responsible charge of the work must be registered in that jurisdiction (normally state), and the drawings and sometimes calculations supporting them must be sealed (have the engineer's registration stamp affixed or drawings embossed).

Registration authorities are normally concerned with both technical competence and moral character of applicants. The technical competence is usually established by a two-part examination system covering (1) engineering theory, mathematics, physics, and so on; and (2) a second portion, typically, a few years later, covering application of engineering knowledge to practical problems. Usually the law requires that during this period actual engineering work be performed under the supervision of a registered engineer. In addition, references active in the engineering field are required to attest to the moral character of the applicant.

In general, application for registration should be made as early in one's career as possible, while the theory remains fresh. Preparatory courses are of great help in preparing for the examinations and should be taken where possible. Registration supports the status as expert witness, an increasingly important aspect of professional engineering.

11.11   WRITING FOR PUBLICATION

Writing for **publication** can be both the most rewarding and the most frustrating of engineering activities. Although a source of considerable personal satisfaction, great care must be taken not only to treat the topic properly, but also to assure accuracy, give credit, make certain that figures are clear and appropriate, meet publication deadlines, obtain **peer reviews** and permissions to reprint material from outside sources—the list seems to go on and on. Although appearing difficult, it can be made simpler; and the following steps have proven helpful to the author.

- Start the work early. There will be surprises and developments during the work and it is essential that plenty of time be available to avoid hasty or ill-prepared work completed to meet publication deadlines. Often, outside information is needed for the work and it may not be readily available, be delayed, or require development, all of which use up schedule time.
- A suitable **outline** must be prepared for the work at the earliest date—it should precede all other work on the writing. The outline incorporates a line for each paragraph using a word or two or a short phrase to list the important idea in that paragraph. After preparing the outline, it should go through a series of revisions as additional thought is given the content of the writing. Only after several revisions should the actual writing begin. The revisions may be as few as three or four for brief writing or as many as 10 to 20 for longer or more widely disseminated work. The revisions to the outline continue during the writing process, and often a final version is prepared when the work is completed. With word processing equipment, it is easy to reorganize the work and move paragraphs and sections about. It is important not to be constrained, and thus the outline should be considered a living document.
- Begin the writing as soon as it is felt that the outline is sufficiently mature so that the overall thrust of the work will not be changed significantly. Allow plenty of time for rewriting and revisions. The more time that can be allowed for rewrite and revision, the better the work will be. Do not be afraid to edit and rework the material heavily. As the writing is reworked it will improve in precision. Accuracy is paramount, and even a small, seemingly inconsequential error can impeach the value of the entire work. If there is a question, either state it or don't use the material. Be ruthless in editing out superfluous phrases and material that does not add to the writing. Keep the work crisp and to the point. Allow enough time so that the work can be set aside for a short time

near the end to allow some reflection and final editing; the work is often improved significantly by this simple step.

- Arrange for one or more **peer reviews.** These can be as simple as review by a co-worker or friend, or more formal, where the work is submitted to a review committee. Be scrupulous in giving credit for ideas or the work of others. Often, it is necessary to draw on the work of others and incorporate it into the writing. Be certain to obtain their **permission to reprint** this material, to avoid plagiarism and potential legal complications. Reprinted material can take the form of extracts from writings, figures, tables, computer programs, or unique formulae—all of these require permission to use. The use of footnotes or citations is the normal way to assure that credit is properly given.

- For scholarly works, the use of footnotes and references is essential. It is better to include a reference of potentially little value than to omit one that may be significant.

- Writing style will vary depending on the purpose of the writing. Absolute accuracy is fundamental and care should be taken to assure correct spelling, punctuation, and proper use of the English language. When using spell-checking computer programs, many words having different meanings have similar but slightly different spellings which the program may not pick up. After the work is in final draft form, it should be proofread for a series of concerns: once for spelling, once for content, once for format, once for numbering of paragraphs or sections, and once for footnotes and references. It is then wise to have the work proofread by at least two other persons.

11.12   DISPUTE RESOLUTION

Despite the greatest care, disputes often arise, particularly if the client is unsophisticated and not accustomed to dealing with design issues. Resolution is critical and should take place at the earliest possible date, before the positions of the parties have hardened. Resolution of these disputes normally follows one of several paths, the most extreme being **litigation** in a local, state, or federal court.

Alternative dispute resolution systems are being used increasingly to expedite resolution and reduce costs by avoiding litigation. These methods include mediation, minitrials, dispute resolution boards or boards of contract review or appeal, and arbitration. In addition, some recent contracts include provisions for **partnering,** which often provides for the use of a neutral to evaluate and recommend resolution to the parties.

Mediation seeks to develop resolution of the dispute by the voluntary agreement of the parties. Similarly, **minitrials** simulate the actual judicial trial of the dispute in an attempt to demonstrate to the parties the strengths and weaknesses of their positions and thus facilitate voluntary settlement. Large contracts often provide for the establishment and use of **dispute resolution boards** over the active life of the contract to hear and recommend (in come cases binding) resolution. For federal government work, **boards of contract review and appeals** are often available, which provide for a binding resolution of the dispute.

Of all the methods used, **arbitration** is the most widespread. It provides for a binding judgment of the dispute and is favored by most parties involved. The dispute is heard by an arbitrator or, in the case of larger disputes, a panel of three arbitrators, who will reach a decision based on the material presented. The arbitrator(s) are attorneys, judges, and trained executives who are experienced in the industry. Their decision is binding and can be enforced by the courts. This requirement is incorporated in the contract by a clause that includes the recommended language of the **American Arbitration Association:** "Any controversy, or claim arising out of or relating to this contract, or the breach thereof, shall be settled by arbitration in accordance with the Rules of the American Arbitration Association, and judgment upon the award rendered by the arbitrator(s) may be entered in any court having jurisdiction thereof."*

Finally, litigation is available to resolve disputes. In general, it is avoided because of the high cost, the extended time required, and the uncertainty of the outcome.

Apart from participating as a principal in a dispute, the engineer is frequently called upon to testify or give depositions as an **expert witness.** Testimony occurs in a courtroom with the usual complement of a judge, perhaps a jury, counsel, the defendant, plaintiff, and so on. **Depositions** are typically sworn statements with legal counsel present, taken by a legal stenographer at a location distant from the court. Depositions are given at a location generally of convenience to the deposer, that is, the person testifying. The deposition is given under oath and counsel for both parties are usually present. In every aspect the deposition is identical to testimony given in the court and constitutes admissible testimony.

Typically, the person testifying will initially be asked to establish his credentials. This takes the form of a series of questions that establish the educational background, experience levels, and other aspects of the training and experience of the expert—permitting the testimony to be accepted as unique and expert and outside the sphere of the layperson. In most cases, the testimony of the expert witness is not contested and the testimony of the witness is used to provide an improved understanding of an engineering matter for the benefit of the court. For example, a witness might testify, in the case of a failure of a concrete structure, on the rate at which concrete acquires strength during the curing process; present test data indicating a strength of a certain member for a vehicular component where a case of product liability was involved and so forth.

The testimony of the witness needs to be clear and concise and in lay terms as much as possible. Although the expert witness is engaged because of his or her ability to understand and evaluate the problem, the testimony will be used either by a judge who may not be familiar with the technological details, or a jury who are themselves not experts in the field. The engineer's testimony, therefore, needs to be in a language that the layperson can readily understand. The use of similes, parallels, and the drawing of comparisons are very useful ways in which the testimony can be made realistic and understandable to the jury. The use of tabular data should be minimized, but charts, visual aids, and models should be maximized, as these provide a ready way in which the jury can understand the item being presented.

---

*Quoted with permission from the American Arbitration Association.

The testimony of the expert witness must be conducted with suitable decorum, and absolute integrity is essential in the presentation of material to the court. Since the layperson does not, in many cases, understand concepts such as design margins and statistical analysis, simplifications that may be necessary should be carefully chosen to avoid distorting the basic information. The maintenance of integrity is absolutely essential for the purpose of establishing and maintaining the credibility of the testimony. Typically, the expert witness will be counseled not to volunteer information, but merely to answer the questions asked of him or her by the attorneys. Normally, the testimony of the witness is carefully reviewed before the actual deposition is taken to assure the attorney understands fully the scope and implications of the material and thus can properly explore all aspects through his or her line of questioning.

11.13   INTELLECTUAL PROPERTY

The entire subject of **intellectual property,** including patents, copyrights, trademarks, and trade secrets, is complex and subject to rigorous legal and procedural requirements. The ownership of intellectual property, sometimes called a **proprietary interest,** is defined as ownership of a process, procedure, method, or identification which is developed or used in a way to permit a specific advantage in the conduct of business. Normally, this is applied to processes or methodology, or in some cases to design of equipment. As such, it is often of great value to an enterprise, and maintaining control over its secrecy and use is of great importance.

Where specific issues or questions arise, suitable legal advice should be obtained. Nevertheless, a brief overview of protections and procedures for two of the most common ownership forms, patents and copyrights for the United States and its possession, is useful to the practitioner. **Copyrights** deal with items of creative expression such as writings, phonograph records, and CD ROMs; **patents** deal with machinery and processes. Extensive written guidance in the form of pamphlets and circulars is available from the U.S. government office having jurisdiction.

11.13.1   Copyrights[2]

Specific copyright rules apply to works created prior to 1978, while for works created later, the following general guidelines are useful. The work must be original with yourself and cannot merely be an idea. It must be an "expression" that is fixed in a tangible medium, such as a writing, recording, or other representation. The protections afforded by copyright coverage are very broad. In the specific case of computer works, the coverage may include both the source (programming language) and the machine language (program) that implement it. Where writing is prepared by an employee within the scope of his or her employment, the employer is normally considered the author and the work considered a "work made for hire."

The duration of a copyright is the life of the author plus 50 years. A work for hire has a life of 75 years from first publication or 100 years from the date of creation, whichever is shorter. A major exception to this is the 10-year copyright life for masks used for computer chip manufacture. Copyrights can be sold, transferred, donated, and otherwise treated as a piece of tangible personal property. Infringement on a copyright is a very serious matter and can carry substantial legal penalties. In working with material prepared by others, it is likely that a copyright issue will arise. Wherever there is a question regarding whether an item is copyrighted, seek permission to reprint or use it in your work. Permissions are generally given by copyright holders but are often limited to a specific publication, edition, or purpose.

Copyrights are commonly designated by the encircled letter c, p, or m (for writings, sound recordings, or masks) together with the year of the copyright and the name of the owner. Copyrights may be registered with the U.S. Copyright Office, Library of Congress, Washington, DC 20559. Depending on the form of the work, one or more copies may or may not be required to be deposited. For example, written works, recordings, and the like, are deposited, whereas scientific or technical diagrams and models are not.

11.13.2 Patents[3,4]

As with copyrights, patents grant an exclusive ownership right to a holder. The **GATT treaty,** signed in December 1994, changed the ownership period from 17 to 20 years after the filing date. It also broadened the definition of patent infringement. In addition, a patent process was established on a trial basis and because of the international nature of the agreement gave credit in the United States for certain activities conducted abroad.

Broadly, a patent grants exclusive rights to the holder. This permits the holder to prevent others without authorization from manufacturing, using, or selling the device (invention) for the term of the patent. A patent application is a set of standardized papers that describes the device. To obtain a patent, the device must be "novel" and "unobvious." Patents are issued in three types: the **utility patent,** the most common type, covering a device that provides a utilitarian result; a **design patent,** which covers a unique shape or configuration (there is some overlap with copyrights for this category); and a **plant patent** (horticultural) (e.g., a patented rose).

When developing an **invention,** it is important that suitable documentation be prepared and maintained. This should be in an orderly form, and dated, signed, and witnessed. The typical patent application contains sections dealing with description, summary, main and dependent claims, a detailed specification, and drawing(s) with all parts labeled. A special style has been developed for patent drawings and there are specific rules on the details of drafting and drafting standards to be used. Only rarely must models be submitted. There are numerous books on the market that assist the inventor in documenting the invention and preparing a patent application.

In general, it is better to file a patent application before attempting to commercialize the device, as proving the date of conception of the idea is then less difficult. Ownership of a patent means that the owner has the ability to prevent others from using the idea or device. It does not automatically mean that the idea or device is commercially feasible or marketable; production and marketing of a patented device is a critical step in bringing it to economic fruition. Patent **infringement** is a serious matter and severe economic and legal penalties may result if infringement litigation is pursued successfully.

As with other personal property, patents can be sold, transferred, licensed, and so on. To maintain a patent in force, several maintenance payments are required over its life. U.S. patents are issued by the U.S. Patent and Trademark Office, Washington, DC 20231.

REFERENCES

1. *A Guide for the Engagement of Engineering Services,* Manual 45, American Society of Civil Engineers, New York, 1981.
2. *Copyright Basics,* Circular 1, Copyright Office, Library of Congress, Washington, D.C., 1992.
3. *General Information Concerning Patents,* U.S. Department of Commerce, Patent and Trademark Office, Washington, D.C.
4. D. Pressman, *Patent It Yourself,* 4th ed., Nolo Press, Berkeley, Calif., 1995.
5. *Professional Construction Management and Project Administration,* Wm. B. Foxhall, Architectural Record, New York, and The American Institute of Architects; 1972 and 1976.
6. *The Copyright Book,* 4th Edition, Wm. S. Strong 1993, MIT Press, Cambridge Massachusetts.

CHAPTER 12
TABLES

Table 12.1 Weights of Materials

Material	Lbs. per cu. ft.	Material	Lbs. per cu. ft.
Air *	0.0809	copper, pure	554
acetylene gas *	0.0733	" cast	549–558
alabaster	168	" wrought	552–558
alcohol	49–57	" wire	555–558
aluminum, pure	168	cork	15.6
" cast	160		
" wire	168	Erbium	297
amber	67	emery	250
ammonia *	0.0482		
antimony	414	Feldspar	158–162
argon *	0.113	flint	162
arsenic	357	fluorine *	0.0920
asbestos	125–175		
asphaltum	69–94	Germanium	341
		german silver	515–535
Barium	234	glass, common	150–175
basalt	180	" flint	180–280
bismuth	609	glucinum	122
boron	159	glycerine	78.6
brass	510–542	gold	1203
brick	100–150	granite	125–187
bromine	196	gravel	90–147
bronze	545–555	gum arabic	90
		gun metal	533
Cadmium	540	gutta percha	61
caesium	117	gypsum	144
calcium	98.6		
carbon	125–144	Hydrogen *	0.00562
" bisulphide	80.6		
" dioxide *	0.124	Ice	55–57
" monoxide *	0.0782	iodine	300
celluloid	90	iridium	1399
cement, loose	72–105	iron, pure	491
" set	168–187	" gray cast	439–445
cerium	437	" white cast	473–482
chalk	119–175	" wrought	487–492
charcoal	17–35	" steel	474–494
chlorine *	0.196	ivory	114
chromium	368		
clay, hard	129–133	Lead	710
" soft	118	leather, dry	54
coal, anthracite	81–106	" greased	64
" " loose	47–58	lime	53–75
" bituminous	78–88	limestone	156–162
" " loose	44–54	lithium	39
" lignite	52	loam	65–88
cobalt	530–563		
coke	62–105	Magnesium	107
" loose	23–32	" carbonate	150
columbium	452	manganese	462
concrete (1 : 2 : 4)	146	marble	157–177
" (1 : 1½ : 3)	139	masonry	100–165
" (1 : 3 : 6)	156	mercury *	849
		mica	165–200
		molybdenum	529

*At 0°C and atmospheric pressure.

Table 12.1 *(Continued)*

Material	Lbs. per cu. ft.	Material	Lbs. per cu. ft.
mortar, hard............	103	steel....................	474–494
muck....................	40–74	strontium..............	158
mud....................	80–130	sulphur.................	120–130
Naptha.................	53	Talc....................	168
nickel..................	540–550	tantalum...............	1040
nitrogen *..............	0.0782	tar.....................	62.4
nitrous oxide *..........	0.0838	tellurium...............	389
Oil, cotton-seed.........	60.2	thallium................	739
" lard................	57.4	thorium................	686
" linseed.............	58.8	tile....................	113
" lubricating..........	56.2–57.7	" hollow.............	26–45
" petroleum..........	54.8	tin.....................	455
" transformer.........	52.6–54.2	titanium................	218
" turpentine..........	54.2	trap rock...............	187–190
" whale..............	57.3	tungsten...............	1174
osmium.................	1400	turf....................	20–30
oxygen *................	0.0895	Uranium................	1165
Palladium...............	711		
paper...................	44–72	Vanadium..............	343
paraffin.................	54–57	Water, max. dens.......	62.4
peat....................	20–30	" sea.............	64.0–64.3
phosphorus..............	146	wax, bees...............	60.5
pitch...................	67	wood, ash..............	45–47
plaster of Paris.........	144	" bamboo..........	22–25
platinum................	1336	" beech............	43–56
porcelain................	143–156	" birch............	32–48
potassium...............	53.7	" butternut........	24–28
pumice stone............	23–56	" cedar............	37–38
Quartz..................	165	" cherry...........	43–56
		" chestnut..........	38–40
Resin...................	67	" cypress...........	32–37
rhodium................	773	" ebony............	69–83
rubber, pure............	58.0–60.5	" elm..............	35–36
" compound........	106–124	" fir...............	34–35
" ebonite...........	74.9–78.0	" hemlock..........	25–29
rubidium................	955	" hickory..........	53–58
ruthenium...............	767	" lig. vitæ.........	78–83
		" mahogany........	32–53
Salt....................	129–131	" maple............	49–50
sand....................	90–120	" oak..............	37–56
sandstone...............	124–200	" pine.............	24–45
selenium................	300	" poplar...........	24–27
shale...................	162	" red wood........	30–32
silicon..................	131	" spruce...........	25–32
silver...................	660	" walnut...........	38–45
slate...................	162–205	" willow...........	24–37
snow, fresh fallen........	5–12	Xenon *.................	0.284
" wet compact.......	15–50		
soapstone...............	162–175	Zinc....................	448
sodium.................	60.5	zirconium...............	258
spermaceti..............	59		

*At 0°C and atmospheric pressure.

Table 12.2 Specific Heats

Average values (0 to 100°C unless otherwise stated) of c in the formula $Q = Mkc(t_2 - t_1)$, c being measured in gram-calories per gram per °C or British thermal units per pound per °F.

Acetylene * (15)	0.383	Ice (−20 to 0)	0.505
air * (−30 to +10)	0.238	iridium	0.0323
air † (−30 to +10)	0.169	iron, cast	0.119
alcohol, ethyl (30)	0.615	iron, wrought	0.115
aluminum	0.226		
ammonia (liq. 0)	1.098	Lead	0.0297
ammonia *	0.520	leather, dry	0.360
ammonia †	0.391		
antimony	0.0504	Marble	0.206
asbestos	0.195	mercury	0.0331
		mica	0.208
Beryllium	0.425	Nickel	0.109
bismuth	0.0297	nitrogen *	0.244
brass (60 Cu, 40 Zn)	0.0917	nitrogen †	0.173
bronze (80 Cu, 20 Sn)	0.0860		
		Oxygen *	0.224
Calcium	0.149	oxygen †	0.155
carbon, gas	0.315	osmium	0.0311
carbon, graphite	0.310		
carbon dioxide * (15 to 100)	0.202	Paraffin	0.589
carbon dioxide † (15 to 100)	0.168	petroleum	0.504
carbon monoxide *	0.243	platinum	0.0319
carbon monoxide †	0.173	porcelain (15 to 950)	0.260
cement, Portland	0.271		
chalk	0.220	Quartz (12 to 100)	0.188
chloroform (liq., 30)	0.235		
chloroform (gas, 100 to 200)	0.147		
chromium	0.111	rubber, hard	0.339
clay, dry (20 to 100)	0.220		
coal	0.201	Selenium (−188 to +18)	0.0680
cobalt	0.103	silicon	0.175
copper	0.0928	silver	0.0560
cork	0.485	steam (100 to 200)	0.480
cotton	0.362	steel	0.118
		sulphur (−188 to +18)	0.137
Gasoline	0.500	Tantalum (58)	0.0360
german silver	0.0945	tin	0.0556
glass	0.180	tungsten	0.0340
glycerine (15 to 50)	0.576	turpentine (0)	0.411
gold	0.0312		
granite (12 to 100)	0.192	Water (15)	1.000
		wood	0.420
		wool	0.393
Hydrogen *	3.41		
hydrogen †	2.42	Zinc	0.0950

*Constant pressure of 1 atm.

†Constant volume.

Table 12.3 Coefficients of Linear Expansion

Average values (0 to 100°C unless otherwise stated) of a in the formula $l = l_0(1 + at)$, t being measured in °C.

Substance	$a \times 10^6$	Substance	$a \times 10^6$
Aluminum (20 to 100)...	23.8	Marble, Rutland blue (15 to 100)..............	15.0
antimony (15 to 101)....	10.9	marble, Georgia gray (20 to 65)................	1.00
Beryllium (20).........	12.2	mercury (− 78 to − 38)..	41.0
bismuth (19 to 101).....	13.4	mica..................	7.60
brass.................	18.7	monel metal (25 to 100)..	14.1
brick.................	9.50		
bronze (80 Cu, 20 Sn) (0 to 800)............	27.0	Nickel (25 to 100).......	12.9
Cadmium..............	31.6	Osmium (40)...........	6.57
calcium (0 to 21)........	25.0		
carbon, diamond (40)...	1.18	Paraffin (0 to 16)	107.
" gas (40).........	5.40	paraffin (16 to 38)	130.
" graphite (40)....	7.86	phosphorous (6 to 44)...	124.
celluloid (20 to 70)	109.	platinum (20)..........	8.93
cobalt (20).............	12.3	porcelai., average......	3.50
copper (25 to 100).......	16.8		
		Quartz, fused..........	0.500
Duralumin, cast (20 to 100)	23.6	Rubber, hard (20 to 60) .	80.0
duralumin, cold rolled (20 to 100)............	23.7	Selenium (40)..........	36.8
		silicon (40)............	7.63
German silver.........	18.4	silver (20).............	18.8
glass, crown...........	8.97	slate (20).............	8.00
" flint (50 to 60).....	7.88	solder................	25.1
gold (16 to 100).........	14.3	sodium (−188 to +17)..	62.2
granite................	8.30	steel, cast.............	13.6
gutta percha...........	198.	sulphur (40)...........	64.1
Ice (− 20 to − 1).......	51.0	Tin (18 to 100)..........	26.9
iridium (− 183 to + 19)..	5.71	tungsten (0 to 500)......	4.60
iron, pure.............	11.9	tungsten (1000 to 2000)..	6.10
" cast (40)	10.6		
" wrought (− 18 to + 100)..........	11.4	Wood, beech (2 to 34)...	2.57
		wood, walnut (2 to 34) ..	6.58
Lead (18 to 100)........	29.4	Zinc (10 to 100)	26.3

Table 12.4 Melting and Boiling Points (at Atmospheric Pressure)

Substance	Melts (°C)	Boils (°C)	Substance	Melts (°C)	Boils (°C)
Acetylene	−81.3	−72.2	Magnesium	651	1110
Alcohol, ethyl	−115	78.3	Manganese	1260	1900
Alcohol, methyl	−97.8	64.7	Mercury	−38.87	356.9
Aluminum	659.7	1800	Molybdenum	2620	3700
Ammonia	−75	−33.5	Neon	−248.7	−245.9
Antimony	630.5	1380	Nickel	1455	2900
Argon	−189.2	−185.7	Nitric oxide	−160.6	−153
Barium	850	1140	Nitrogen	−209.9	−195.8
Beryllium	1350	1500	Oxygen	−218.4	−183
Bismuth	271.3	1450	Ozone	−251.4	−112
Borax	561	—	Palladium	1553	2200
Boron	2300	2550	Paraffin	52.4	—
Brass	950±	—	Phosphorus	44.1	280
Bromine	−7.2	58.8	Platinum	1773.5	4300
Bronze	1000±	—	Potassium	62.3	760
Cadmium	320.9	767	Radium	960	1140
Calcium	810	1170	Radon	−110	—
Carbon	>3500	4200	Rhenium	3000	—
Carbon dioxide	−57	−80	Rhodium	1985	>2500
Carbon monoxide	−207	−191.5	Rubber	100	—
Cerium	640	1400	Rubidium	38.5	700
Cesium	28.5	670	Ruthenium	2450	>2700
Chlorine	−101.6	−34.6	Selenium	220	688
Chromium	1615	2200	Silicon	1420	2600
Cobalt	1480	3000	Silver	960.5	1950
Columbium	1950	2900	Sodium	97.5	880
Copper	1083	2300	Sodium chloride	772	—
Fluorine	−223	−187	Steel, carbon	1400	—
Gallium	29.75	>1600	Steel, stainless	1450	—
German silver	1100±	—	Strontium	800	1150
Germanium	958.5	2700	Sugar	160	—
Glass, flint	1300	—	Sulfur	112.8	444.6
Gold	1063	2600	Tantalum	2850	>4100
Hafnium	1700	>3200	Tellurium	452	1390
Helium	≤272.2	−268.9	Thallium	303.5	1650
Hydrogen	−259.1	−252.7	Thorium	1845	>3000
Indium	155	1450	Tin	231.9	2260
Iodine	113.5	184.3	Titanium	1800	>3000
Iridium	2350	>4800	Tungsten	3370	5900
Iron, pure	1535	3000	Turpentine	—	161
Iron, gray pig	1200	—	Uranium	<1850	—
Iron, white pig	1050	—	Vanadium	1710	3000
Krypton	−169	−151.8	Xenon	−140	−109
Lanthanum	826	1800	Yttrium	1490	2500
Lead	327.4	1620	Zinc	419.5	907
Lithium	186	>1220	Zirconium	1900	>2900

Table 12.5 Physical Constants of Various Fluids

Fluid	Formula	Molecular Weight	Boiling Point (°F at 14.696 psia)	Vapor Pressure at 70°F (psig)	Critical Temp. (°F)	Critical Pressure (psia)	Specific Gravity Liquid 60/60°F	Specific Gravity Gas
Acetic acid	$HC_2H_3O_2$	60.05	245		455	691	1.05	2.01
Acetone	C_3H_6O	58.08	133				0.79	1.0
Air	N_2O_2	28.97	-317		-221	547	0.86^a	
Alcohol								
Ethyl	C_2H_6O	46.07	173	2.3^b	470	925	0.794	1.59
Methyl	CH_4O	32.04	148	4.63^b	463	1174	0.796	1.11
Ammonia	NH_3	17.03	-28	114	270	1636	0.62	0.59
Aniline	C_6H_7N	93.12	365		798	770	1.02	
Argon	A	39.94	-302		-188	705	1.65	1.38
Bromine	Br_2	159.84	138		575		2.93	5.52
Carbon dioxide	CO_2	44.01	-109	839	88	1072	0.801^a	1.52
Carbon disulfide	CS_2	76.1	115				1.29	2.63
Carbon monoxide	CO	28.01	-314		-220	507	0.80	0.97
Carbon tetrachloride	CCl_4	153.84	170		542	661	1.59	5.31
Chlorine	Cl_2	70.91	-30	85	291	1119	1.42	2.45
Ether	$(C_2H_5)_2O$	74.12	34				0.74	2.55
Fluorine	F_2	38.00	-305	300	-200	809	1.11	1.31
Formaldehyde	H_2CO	30.03	-6				0.82	1.08
Formic acid	HCO_2H	46.03	214				1.23	
Furfural	$C_5H_4O_2$	96.08	324				1.16	
Glycerine	$C_3H_8O_3$	92.09	554				1.26	
Glycol	$C_2H_6O_2$	62.07	387				1.11	
Helium	He	4.003	-454		-450	33	0.18	0.14
Hydrochloric acid	HCl	36.47	-115				1.64	
Hydrofluoric acid	HF	20.01	66	0.9	446		0.92	

Hydrogen	H_2	2.016	−422		−400	188	0.07[a]	0.07
Hyrogen chloride	HCl	36.47	−115	613	125	1198	0.86	1.26
Hydrogen sulfide	H_2S	34.07	−76	252	213	1307	0.79	1.17
Isopropyl alcohol	C_3H_8O	60.09	180				0.78	2.08
Linseed oil			538				0.93	
Magnesium chloride[a]	$MgCl_2$						1.22	
Mercury	Hg	200.61	670				13.6	6.93
Methyl bromide	CH_3Br	94.95	38	13	376		1.73	3.27
Methyl chloride	CH_3Cl	50.49	−11	59	290	969	0.99	1.74
Naphthalene	$C_{10}H_8$	128.16	424				1.14	4.43
Nitric acid	HNO_3	63.02	187				1.5	
Nitrogen	N_2	28.02	−320		−233	493	0.81[a]	0.97
Oxygen	O_2	32	−297		−181	737	1.14[a]	1.105
Phosgene	$COCl_2$	98.92	47	10.7	360	823	1.39	3.42
Phosphoric acid	H_3PO_4	98.00	415				1.83	
Refrigerant 11	CCl_3F	137.38	75	13.4	388	635		5.04
Refrigerant 12	CCl_2F_2	120.93	−22	70.2	234	597		4.2
Refrigerant 13	$CClF_3$	104.47	−115	458.7	84	561		
Refrigerant 21	$CHCl_2F$	102.93	48	8.4	353	750		3.82
Refrigerant 22	$CHClF_2$	86.48	−41	122.5	205	716		
Refrigerant 23	CHF_3	70.02	−119	635	91	691		
Sulfuric acid	H_2SO_4	98.08	626					1.83
Sulfur dioxide	SO_2	64.6	14	34.4	316	1145	1.39	2.21
Turpentine			320				0.87	
Water	H_2O	18.016	212	0.9492[b]	706	3208	1.00	0.62

Source: Fisher Controls, Control Valve Handbook, 2nd ed. © 1977. Reproduced with permission of Fisher Controls International, Inc.

[a] Density of liquid, g/mL at normal boiling point.

[b] Vapor pressure in psia at 100°F.

Table 12.6 Physical Constants of Hydrocarbons

Compound	Formula	Molecular Weight	Boiling Point at 14.696 psia (°F)	Vapor Pressure at 100°F (psia)	Freezing Point at 14.696 psia (°F)	Critical Constants		Specific Gravity at 14.696 psia	
						Critical Temperature (°F)	Critical Pressure (psia)	Liquid,[a,b] 60°F/60°F	Gas at 60°F (Air = 1)[c]
Methane	CH_4	16.043	−258.69	(5000)[d]	−296.46[e]	−116.63	667.8	0.3[f]	0.5539
Ethane	C_2H_6	30.070	−127.48	(800)[d]	−297.89[e]	90.09	707.8	0.3564[g]	1.0382
Propane	C_3H_8	44.097	−43.67	190.	−305.84[e]	206.01	616.3	0.5077[g]	1.5225
n-Butane	C_4H_{10}	58.124	31.10	51.6	−217.05	305.65	550.7	0.5844[g]	2.0068
Isobutane	C_4H_{10}	58.124	10.90	72.2	−255.29	274.98	529.1	0.5631[g]	2.0068
n-Pentane	C_5H_{12}	72.151	96.92	15.570	−201.51	385.7	488.6	0.6310	2.4911
Isopentane	C_5H_{12}	72.151	82.12	20.44	−255.83	369.10	490.4	0.6247	2.4911
Neopentane	C_5H_{12}	72.151	49.10	35.9	2.17	321.13	464.0	0.5967[g]	2.4911
n-Hexane	C_6H_{14}	86.178	155.72	4.956	−139.58	453.7	436.9	0.6640	2.9753
2-Methylpentane	C_6H_{14}	86.178	140.47	6.767	−244.63	435.83	436.6	0.6579	2.9753
3-Methylpentane	C_6H_{14}	86.178	145.89	6.098	—	448.3	453.1	0.6689	2.9753
Neohexane	C_6H_{14}	86.178	121.52	9.856	−147.72	420.13	446.8	0.6540	2.9753
2,3-Dimethylbutane	C_6H_{14}	86.178	136.36	7.404	−199.38	440.29	453.5	0.6664	2.9753
n-Heptane	C_7H_{16}	100.205	209.17	1.620	−131.05	512.8	396.8	0.6882	3.4596
2-Methylhexane	C_7H_{16}	100.205	194.09	2.271	−180.89	495.00	396.5	0.6830	3.4596
3-Methylhexane	C_7H_{16}	100.205	197.32	2.130	—	503.78	408.1	0.6917	3.4596
3-Ethylpentane	C_7H_{16}	100.205	200.25	2.012	−181.48	513.48	419.3	0.7028	3.4596
2,2-Dimethylpentane	C_7H_{16}	100.205	174.54	3.492	−190.86	477.23	402.2	0.6782	3.4596
2,4-Dimethylpentane	C_7H_{16}	100.205	176.89	3.292	−182.63	475.95	396.9	0.6773	3.4596
3,3-Dimethylpentane	C_7H_{16}	100.205	186.91	2.773	−210.01	505.85	427.2	0.6976	3.4596
Triptane	C_7H_{16}	100.205	177.58	3.374	−12.82	496.44	428.4	0.6946	3.4596

n-Octane	C$_8$H$_{18}$	114.232	0.537	258.22	−70.18	564.22	360.6	0.7068	3.9439
Diisobutyl	C$_8$H$_{18}$	114.232	1.101	228.39	−132.07	530.44	360.6	0.6979	3.9439
Isooctane	C$_8$H$_{18}$	114.232	1.708	210.63	−161.27	519.46	372.4	0.6962	3.9439
n-Nonane	C$_9$H$_{20}$	128.259	0.179	303.47	−64.28	610.68	332.	0.7217	4.4282
n-Decane	C$_{10}$H$_{22}$	142.286	0.0597	345.48	−21.36	652.1	304.	0.7342	4.9125
Cyclopentane	C$_5$H$_{10}$	70.135	9.914	120.65	−136.91	461.5	653.8	0.7504	2.4215
Methylcyclopentane	C$_6$H$_{12}$	84.162	4.503	161.25	−224.44	499.35	548.9	0.7536	2.9057
Cyclohexane	C$_6$H$_{12}$	84.162	3.264	177.29	43.77	536.7	591.	0.7834	2.9057
Methylcyclohexane	C$_7$H$_{14}$	98.189	1.609	213.68	−195.87	570.27	503.5	0.7740	3.3900
Ethylene	C$_2$H$_4$	28.054	—	−154.62	−272.45^e	48.58	729.8	—	0.9686
Propene	C$_3$H$_6$	42.081	226.4	−53.90	−301.45^e	196.9	669.	0.5220^g	1.4529
1-Butene	C$_4$H$_8$	56.108	63.05	20.75	−301.63^e	295.6	583.	0.6013^g	1.9372
cis-2-Butene	C$_4$H$_8$	56.108	45.54	38.69	−218.06	324.37	610.	0.6271^g	1.9372
trans-2-Butene	C$_4$H$_8$	56.108	49.80	33.58	−157.96	311.86	595.	0.6100^g	1.9372
Isobutene	C$_4$H$_8$	56.108	63.40	19.59	−220.61	292.55	580.	0.6004^g	1.9372
1-Pentene	C$_5$H$_{10}$	70.135	19.115	85.93	−265.39	376.93	590.	0.6457	2.4215
1,2-Butadiene	C$_4$H$_6$	54.092	(20.)^d	51.53	−213.16	(339.)^d	(653.)^d	0.658^g	1.8676
1,3-Butadiene	C$_4$H$_6$	54.092	(60.)^d	24.06	−164.02	306.	628.	0.6272^g	1.8676
Isoprene	C$_5$H$_8$	68.119	16.672	93.30	−230.74	(412.)^d	(558.4)^d	0.6861	2.3519
Acetylene	C$_2$H$_2$	26.038	—	119.^{h}	−114.^{e}	95.31	890.4	0.615^i	0.8990
Benzene	C$_6$H$_6$	78.114	3.224	176.17	41.96	552.22	710.4	0.8844	2.6969
Toluene	C$_7$H$_8$	92.141	1.032	231.13	−138.94	605.55	595.9	0.8718	3.1812
Ethylbenzene	C$_8$H$_{10}$	106.168	0.371	277.16	−138.91	651.24	523.5	0.8718	3.6655
o-Xylene	C$_8$H$_{10}$	106.168	0.264	291.97	−13.30	675.0	541.4	0.8848	3.6655
m-Xylene	C$_8$H$_{10}$	106.168	0.326	282.41	−54.12	651.02	513.6	0.8687	3.6655
p-Xylene	C$_8$H$_{10}$	106.168	0.342	281.05	55.86	649.6	509.2	0.8657	3.6655
Styrene	C$_8$H$_8$	104.152	(0.24)^d	293.29	−23.10	706.0	580.	0.9110	3.5959
Isopropylbenzene	C$_9$H$_{12}$	120.195	0.188	306.34	−140.82	676.4	465.4	0.8663	4.1498

Source: Fisher Controls, *Control Valve Handbook*, 2nd ed. © 1977. Reproduced with permission of Fisher Controls International, Inc.

^{a} Air-saturated hydrocarbons.

^{b} Absolute values from weights in vacuum.

^{c} Calculated values.

^{d} ( )-Estimated values.

^{e} At saturation pressure (triple point).

^{f} Apparent value for methane at 60°F.

^{g} At saturation pressure and 60°F.

^{h} Sublimation point.

^{i} Specific gravity, 119°F/60°F (sublimation point).

Table 12.7 Specific Gravity of Hydrocarbons

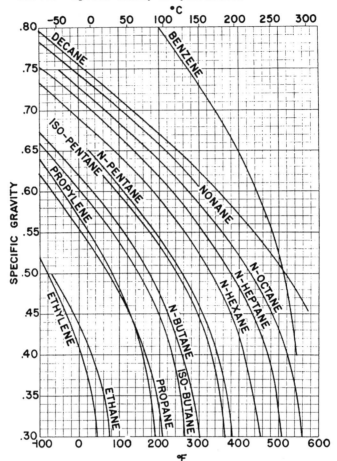

Source: Courtesy of Ingersoll-Dresser Pumps Co., based on data from Gas Processors & Suppliers Association.

Table 12.8   Specific Gravity of Miscellaneous Liquids

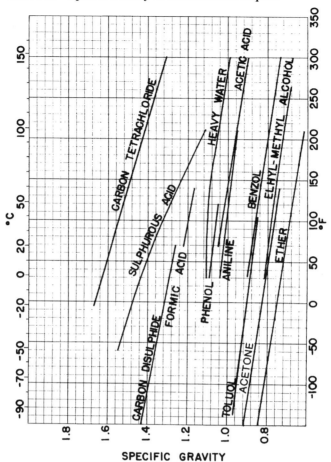

SPECIFIC GRAVITY

Source: Courtesy of Ingersoll-Dresser Pump Co., based on data from various chemical handbooks.

607

Table 12.9 Higher Heating Values

Substance[a]	Btu		
	Per Pound	Per Gallon	Per Foot[3b]
Acetylene	21,500	—	1,480
Alcohol, ethyl, denatured	11,600	78,900	—
Alcohol, ethyl, pure (0.816)	12,400	84,300	—
Alcohol, methyl (0.798)	9,540	63,700	—
Benzene C_6H_6	18,200	—	3,740
Butane C_4H_{10}	21,300	—	3,390
Carbon, to CO	4,000	—	—
Carbon, to CO_2	14,100	—	—
Carbon disulfide	5,820	62,700	—
Carbon monoxide, to CO_2	4,370	—	323
Charcoal, peat	11,600	—	—
Charcoal, wood	13,500	—	—
Ethane C_2H_6	22,400	—	1,770
Ethylene C_2H_4	21,600	—	1,600
Gas, blast furnace	—	—	90–110
Gas, coal	—	—	630–680
Gas, coke oven	—	—	430–600
Gas, illuminating	—	—	550–600
Gas, natural	—	—	700–2,470
Gas, oil	—	—	450–950
Gas, producer	—	—	110–185
Gas, water, blue	—	—	290–320
Gas, water, carburetted	—	—	400–680
Hexane (liq.) C_6H_{14}	20,700	—	—
Kerosene (0.783)	20,000	131,000	—
Kerosene (0.800)	20,160	136,000	—
Methane CH_4	23,900	—	1,010
Octane (liq.) C_8H_{18}	20,500	—	—
Pentane C_5H_{12}	21,100	—	4,010
Propane C_3H_8	21,700	—	2,520
Straw	5,100–6,700	—	—
Sulfur	4,020	—	—

[a]Numbers in parentheses indicate specify gravity.
[b]At 60°F and atmospheric pressure.

Table 12.10 Thermal Conductivity[a]

Substance	Temp. Range (°C)	$k \times 10^3$	Substance	Temp. Range (°C)	$k \times 10^3$
Air	0	0.0568	Fiberglass		
Aluminum	18	480	Blanket	0–120	0.25–0.31[b]
Antimony	0	44.2	Semirigid	0–250	0.23–0.24[b]
Argon	0	0.0389		0–530	0.30–0.23[b,c]
Asbestos, paper	—	0.6	Flannel	50	0.035
Bismuth	0	17.7	German silver	0–100	80
Brass	0	204	Glass		
Brick			Crown	—	2.5
Alumina	0–700	2.0	Flint	—	2.0
Building	15–30	1.5	Gold	18	700
Carborundum	100–1000	23	Granite	100	4.5
Fire	0–1300	3.1	Graphite	—	12
Graphite	100–1000	25	Gypsum	—	3.1
Magnesia	100–1000	7.1	Hair	20–155	0.15
Silica	100–1000	2.0	Hair cloth, felt	—	0.042
Cadmium	18	222	Helium	0	0.339
Cambric, varn	—	0.60	Hydrogen	0	0.327
Carbon			Ice	—	3.9
Gas	100–942	130	Iron		
Graphite	100–914	290	Cast	18	109
Carbon dioxide	0	0.0307	Pure	18	161
Carbon monoxide	0	0.0499	Wrought	18	144
Carborundum	20–100	0.50	Lampblack	100	0.07
Cardboard	—	0.50	Lead	18	83
Cement, portland	0–700	0.17	Leather		
Chalk	0–100	0.28	Chamois	—	0.15
Charcoal, powd'd	0–100	0.22	Cowhide	—	0.42
Clinkers, small	0–700	1.1	Lime	—	0.29
Coal	—	0.30	Linen	—	0.21
Coke, powdered	0–100	0.44	Magnesia	—	0.3
Concrete, cinder	—	0.81	Magnesium	100	0.23
Concrete	—	2.2	carbonate		
Copper	18	918	Marble	15–30	8.4
Cotton batting			Mercury	17	19.7
Loose	—	0.11	Mica	—	0.86
Packed	—	0.072	Nickel	18	142
Cotton wool	—	0.043	Nitrogen	0	0.0524
Earth, average	—	4.0	Oxygen	0	0.0563
Eiderdown			Paper	—	0.31
Loose	—	0.108	Paraffin	—	0.62
Packed	—	0.045	Pasteboard	—	0.45
Feathers	20–155	0.16	Plaster of paris	20–155	0.42
Felt	21–175	0.22	Plaster, mortar	—	1.3
Fiber, red	—	1.1	Platinum	18–100	170

Table 12.10 *(Continued)*

Substance	Temp. Range (°C)	$k \times 10^3$	Substance	Temp. Range (°C)	$k \times 10^3$
Petroleum	23	0.39	Terra-cotta	100–1000	2.3
Porcelain	165–1055	4.3	Tin	18	155
Pumice stone	20–155	0.43	Water	0	1.4
Quartz				30	1.6
‖ to axis	—	30	Wood		
⊥ to axis	—	160	Fir, with grain	—	0.30
Rubber, hard	—	0.43	Fir, cross-grain	—	0.09
Rubber	—	0.38	Wool		
Sand, dry	20–155	0.86	Mineral	0–175	0.11
Sandstone	—	5.5	Sheep's	20–100	0.14
Sawdust	—	0.14	Steel	100	0.20
Silica, fused	100	2.55	Woolen		
Silk	50–100	0.13	Loose	—	0.12
Silver	18	974	wadding		
Slate	94	4.8	Packed	—	0.055
Snow	—	0.60	wadding		
Steel	18	115	Zinc	18	265

[a]Average values of k in the formula $Q = ckS\theta t/x$. See Section 5.7 for descriptions of units.
[b]Btu/hr/ft²/in./°F.
[c]Values descrease with increasing temperature.

Property		Natural Rubber	Buna-S	Nitrile	Neo-prene	Butyl	Thiokol[1]	Silicone[1]	Hypalon[2]	Viton[2,3,4]	Poly-urethane[5]	Poly-acrylic[3]	Ethylene Propylene[5]
Tensile Strength, Psi (Bar)	Pure Gum	3000 (207)	400 (28)	600 (41)	3500 (241)	3000 (207)	300 (21)	200-450 (14-31)	4000 (276)	...	...	100 (7)	...
	Reinforced	4500 (310)	3000 (207)	4000 (276)	3500 (241)	3000 (207)	1500 (103)	1100 (76)	4400 (303)	2300 (159)	6500 (448)	1800 (124)	2500 (172)
Tear Resistance		Excellent	Poor-Fair	Fair	Good	Good	Fair	Poor-Fair	Excellent	Very Good	Excellent	Good	Good
Abrasion Resistance		Excellent	Good	Good	Excellent	Fair	Poor	Poor	Excellent	Good	Excellent	Good	Good
Aging: Sunlight		Poor	Poor	Poor	Excellent	Excellent	Good	Good	Excellent	Excellent	Excellent	Excellent	Excellent
Oxidation		Good	Fair	Fair	Good	Good	Good	Very Good	Very Good	Excellent	Excellent	Excellent	Good
Heat (Max. Temp.)		200°F (93°C)	200°F (93°C)	250°F (121°C)	200°F (93°C)	200°F (93°C)	140°F (60°C)	450°F (232°C)	300°F (149°C)	400°F (204°C)	200°F (93°C)	350°F (177°C)	350°F (177°C)
Static (Shelf)		Good	Good	Good	Very Good	Good	Fair	Good	Good	...	...	Good	Good
Flex Cracking Resistance		Excellent	Good	Good	Excellent	Good	Fair	Fair	Excellent	...	Excellent	Good	...
Compression Set Resistance		Good	Good	Very Good	Excellent	Fair	Poor	Good	Poor	Poor	Good	Good	Fair
Solvent Resistance:													
Aliphatic Hydrocarbon		Very Poor	Very Poor	Good	Fair	Poor	Excellent	Poor	Fair	Excellent	Very Good	Good	Poor
Aromatic Hydrocarbon		Very Poor	Very Poor	Fair	Poor	Very Poor	Good	Very Poor	Poor	Very Good	Fair	Poor	Fair
Oxygenated Solvent		Good	Good	Poor	Fair	Good	Fair	Poor	Poor	Good	Poor	Poor	Poor
Halogenated Solvent		Very Poor	Very Poor	Very Poor	Poor	Poor	Poor	Very Poor	Very Poor	...	...	Poor	Poor
Oil Resistance:													
Low Aniline Mineral Oil		Very Poor	Very Poor	Excellent	Fair	Very Poor	Excellent	Poor	Fair	Excellent	...	Excellent	Poor
High Aniline Mineral Oil		Very Poor	Very Poor	Excellent	Good	Very Poor	Excellent	Good	Good	Excellent	...	Excellent	Poor
Synthetic Lubricants		Very Poor	Very Poor	Fair	Very Poor	Poor	Poor	Fair	Poor	...	...	Fair	Poor
Organic Phosphates		Very Poor	Very Poor	Very Poor	Very Poor	Good	Poor	Poor	Poor	Poor	Poor	Poor	Very Good
Gasoline Resistance:													
Aromatic		Very Poor	Very Poor	Good	Poor	Very Poor	Excellent	Poor	Poor	Good	Fair	Fair	Fair
Non-Aromatic		Very Poor	Very Poor	Excellent	Good	Very Poor	Excellent	Good	Fair	Very Good	Good	Poor	Poor
Acid Resistance:													
Diluted (Under 10%)		Good	Good	Good	Fair	Good	Poor	Fair	Good	Excellent	Fair	Poor	Very Good
Concentrated[6]		Fair	Fair	Fair	Fair	Fair	Very Poor	Poor	Good	Very Good	Poor	Poor	Good
Low Temperature Flexibility (Max.)		-65°F (-54°C)	-50°F (-46°C)	-40°F (-40°C)	-40°F (-40°C)	-40°F (-40°C)	-40°F (-40°C)	-100°F (-73°C)	-20°F (-29°C)	-30°F (-34°C)	-40°F (-40°C)	-10°F (-23°C)	-50°F (-45°C)
Permeability to Gases		Fair	Fair	Fair	Very Good	Very Good	Good	Fair	Very Good	Good	Good	Good	Good
Water Resistance		Good	Very Good	Very Good	Fair	Very Good	Fair	Fair	Fair	Excellent	Fair	Fair	Very Good
Alkali Resistance:													
Diluted (under 10%)		Good	Good	Good	Good	Good	Fair	Fair	Good	Excellent	Fair	Poor	Excellent
Concentrated		Fair	Fair	Fair	Good	Good	Poor	Poor	Good	Very Good	Poor	Poor	Good
Resilience		Very Good	Fair	Fair	Very Good	Very Good	Good	Good	Good	Good	Fair	Very Poor	Very Good
Elongation (Max.)		700%	500%	500%	500%	700%	400%	300%	300%	425%	625%	200%	500%

Source: Control valve Handbook, 2nd ed. © 1977. Reproduced with permission of Fisher Controls International, Inc.

[1] Trademark of Thiokol Chemical Co. [2] Trademark of E.I. DuPont Co.
[3] Do not use with steam. [4] Do not use with ammonia.
[5] Do not use with petroleum base fluids. Use with ester base non-flammable hydraulic oils and low pressure steam applications to 300°F (149°C).
[6] Except for nitric acid and sulfuric acid.

Table 12.12 Density of Gases (at 60°F and 30 in. Hg)[a]

Gas	Molecular Formula	Molecular Weight	Specific Gravity Air = 1.0	Weight lb per cu ft	Volume cu ft per lb
Air	—	28.9	1.000	0.07655	13.063
Oxygen	O_2	32.00	1.105	0.08461	11.819
Hydrogen	H_2	2.02	0.070	0.00533	187.723
Nitrogen (atmospheric)	N_2	28.02	0.972	0.07439	13.443
Carbon Monoxide	CO	28.01	0.967	0.07404	13.506
Carbon Dioxide	CO_2	44.01	1.528	0.1170	8.548
Methane	CH_4	16.04	0.554	0.04243	23.565
Acetylene	C_2H_2	26.04	0.911	0.06971	14.344
Ethylene	C_2H_4	28.05	0.974	0.07456	13.412
Ethane	C_2H_6	30.07	1.049	0.08029	12.455
Sulphur Dioxide	SO_2	64.06	2.264	0.1733	5.770
Hydrogen Sulphide	H_2S	34.08	1.190	0.09109	10.979

Source: *Fuel Flue Gases,* 1941, courtesy of the American Gas Association.

[a]Approximate percentage composition of air:

	By Weight	By Volume
Nitrogen	76.8	79.0
Oxygen	23.2	21.0

Table 12.13 Atmospheric Pressures and Barometer Readings at Different Altitudes (Approximate Values)

Altitude Below or Above Sea Level Feet	Barometer Reading Inches Merc at 32°F	Atmospheric Pressure Lb-Sq In	Equivalent Head of Water (75°) Feet	Boiling Point of Water °F
−1000	31.02	15.2	35.2	213.8
− 500	30.47	15.0	34.7	212.9
0	29.921	14.7	34.0	212.0
+ 500	29.38	14.4	33.4	211.1
+1000	28.86	14.2	32.8	210.2
1500	28.33	13.9	32.2	209.3
2000	27.82	13.7	31.6	208.4
2500	27.31	13.4	31.0	207.4
3000	26.81	13.2	30.5	206.5
3500	26.32	12.9	29.9	205.6
4000	25.84	12.7	29.4	204.7
4500	25.36	12.4	28.8	203.8
5000	24.89	12.2	28.3	202.9
5500	24.43	12.0	27.8	201.9
6000	23.98	11.8	27.3	201.0
6500	23.53	11.5	26.7	200.1
7000	23.09	11.3	26.2	199.2
7500	22.65	11.1	25.7	198.3
8000	22.22	10.9	25.2	197.4
8500	21.80	10.7	24.8	196.5
9000	21.38	10.5	24.3	195.5
9500	20.98	10.3	23.8	194.6
10,000	20.58	10.1	23.4	193.7
15,000	16.88	8.3	19.1	184
20,000	13.75	6.7	15.2	—
30,000	8.88	4.4	10.2	—
40,000	5.54	2.7	6.3	—
50,000	3.44	1.7	3.9	—

Source: Used with permission of Ingersoll-Rand Co. © 1962.

Table 12.14 Corrosion Information

The following table is intended to give only a general indication of how various materials will react when in contact with certain fluids. The recommendations cannot be absolute because concentration, temperature, pressure, and other conditions may alter the suitability of a particular material. There are also economic considerations that may influence material selection. Use this table as a guide only.

FLUID	Carbon Steel	Cast Iron	302 or 304 Stainless Steel	316 Stainless Steel	Bronze	Monel*	Hastelloy† B	Hastelloy† C	Durimet‡ 20	Titanium	Cobalt-Base Alloy 6	416 Stainless Steel	440C Hard Stainless Steel	17-4PH Hard Stainless Steel
Acetaldehyde	A	A	A	A	A	A	I.L.	A	A	I.L.	I.L.	A	A	A
Acetic Acid, Air Free	C	C	B	B	B	B	A	A	A	A	A	C	C	B
Acetic Acid, Aerated	C	C	A	A	A	A	A	A	A	A	A	C	C	B
Acetic Acid Vapors	C	C	A	A	B	B	I.L.	A	B	A	A	C	C	B
Acetone	A	A	A	A	A	A	A	A	A	A	A	A	A	A
Acetylene	A	A	A	A	I.L.	A	A	A	A	I.L.	A	A	A	A
Alcohols	A	A	A	A	A	A	A	A	A	A	A	A	A	A
Aluminum Sulfate	C	C	A	A	B	B	A	A	A	A	I.L.	C	C	I.L.
Ammonia	A	A	A	A	C	A	A	A	A	A	A	A	A	I.L.
Ammonium Chloride	C	C	B	B	B	B	A	A	A	A	B	C	C	I.L.
Ammonium Nitrate	A	C	A	A	C	C	A	A	A	A	A	C	B	I.L.
Ammonium Phosphate (Mono-Basic)	C	C	A	A	B	B	A	A	B	A	A	B	B	I.L.
Ammonium Sulfate	C	C	B	A	B	A	A	A	A	A	A	C	C	I.L.
Ammonium Sulfite	C	C	A	A	C	C	I.L.	A	A	A	A	B	B	I.L.
Aniline	C	C	A	A	C	B	A	A	A	A	A	C	C	I.L.
Asphalt	A	A	A	A	A	A	A	A	A	I.L.	A	A	A	A
Beer	B	B	A	A	B	A	A	A	A	A	A	B	B	A
Benzene (Benzol)	A	A	A	A	A	A	A	A	A	A	A	A	A	A
Benzoic Acid	C	C	A	A	A	A	I.L.	A	A	A	I.L.	A	A	A
Boric Acid	C	C	A	A	A	A	A	A	A	A	A	B	B	I.L.
Butane	A	A	A	A	A	A	A	A	A	I.L.	A	A	A	A
Calcium Chloride (Alkaline)	B	B	C	B	C	A	A	A	A	A	I.L.	C	C	I.L.
Calcium Hypochlorite	C	C	B	B	B	B	C	A	A	A	I.L.	C	C	I.L.
Carbolic Acid	B	B	A	A	A	A	A	A	A	A	A	I.L.	I.L.	I.L.
Carbon Dioxide, Dry	A	A·	A	A	A	A	A	A	A	A	A	A	A	A
Carbon Dioxide, Wet	C	C	A	A	B	A	A	A	A	A	A	A	A	A
Carbon Disulfide	A	A	A	A	C	B	A	A	A	A	A	B	B	I.L.
Carbon Tetrachloride	B	B	B	B	A	A	B	A	A	A	A	I.L.	C	I.L.
Carbonic Acid	C	C	B	B	B	A	A	A	A	I.L.	I.L.	A	A	A
Chlorine Gas, Dry	A	A	B	B	B	A	A	A	A	C	B	C	C	C
Chlorine Gas, Wet	C	C	C	C	C	C	C	B	C	A	B	C	C	C
Chlorine, Liquid	C	C	C	C	B	C	C	A	B	C	B	C	C	C
Chromic Acid	C	C	C	B	C	A	C	A	C	A	B	C	C	C
Citric Acid	I.L.	C	B	A	A	B	A	A	A	A	I.L.	B	B	B
Coke Oven Gas	A	A	A	A	B	B	A	A	A	A	A	A	A	A
Copper Sulfate	C	C	B	B	B	C	I.L.	A	A	A	I.L.	A	A	A
Cottonseed Oil	A	A	A	A	A	A	A	A	A	A	A	A	A	A
Creosote	A	A	A	A	C	A	A	A	A	I.L.	A	A	A	A
Ethane	A	A	A	A	A	A	A	A	A	A	A	A	A	A
Ether	B	B	A	A	A	A	A	A	A	A	A	A	A	A
Ethyl Chloride	C	C	A	A	A	A	A	A	A	A	A	B	B	I.L.
Ethylene	A	A	A	A	A	A	A	A	A	A	A	A	A	A
Ethylene Glycol	A	A	A	A	A	A	I.L.	I.L.	A	I.L.	A	A	A	A
Ferric Chloride	C	C	C	C	C	C	C	B	C	A	B	C	C	I.L.
Formaldehyde	B	B	A	A	A	A	A	A	A	A	A	A	A	A
Formic Acid	I.L.	C	B	B	A	A	A	A	A	C	B	C	B	B
Freon, Wet	B	B	B	A	A	A	A	A	A	A	A	I.L.	I.L.	I.L.
Freon, Dry	B	B	A	A	A	A	A	A	A	A	A	I.L.	I.L.	I.L.
Furfural	A	A	A	A	A	A	A	A	A	A	A	B	B	I.L.
Gasoline, Refined	A	A	A	A	A	A	A	A	A	A	A	A	A	A
Glucose	A	A	A	A	A	A	A	A	A	A	A	A	A	A
Hydrochloric Acid (Aerated)	C	C	C	C	C	C	A	B	C	C	B	C	C	C
Hydrochloric Acid (Air Free)	C	C	C	C	C	C	A	B	C	C	B	C	C	C
Hydrofluoric Acid (Aerated)	B	C	C	B	C	C	A	A	B	C	B	C	C	I.L.
Hydrofluoric Acid (Air Free)	A	C	C	B	C	A	A	A	B	C	I.L.	C	C	I.L.

613

Table 12.14 *(Continued)*

FLUID	Carbon Steel	Cast Iron	302 or 304 Stainless Steel	316 Stainless Steel	Bronze	Monel*	Hastelloy† B	Hastelloy† C	Durimet‡ 20	Titanium	Cobalt-Base Alloy 6	416 Stainless Steel	440C Hard Stainless Steel	17-4PH Hard Stainless Steel
Hydrogen	A	A	A	A	A	A	A	A	A	A	A	A	A	A
Hydrogen Peroxide	I.L.	A	A	A	C	A	B	B	A	A	I.L.	B	B	I.L.
Hydrogen Sulfide, Liquid	C	C	A	A	C	C	A	A	B	A	A	C	C	I.L.
Magnesium Hydroxide	A	A	A	A	B	A	A	A	A	A	A	A	A	I.L.
Mercury	A	A	A	A	C	B	A	A	A	A	A	A	A	B
Methanol	A	A	A	A	A	A	A	A	A	A	A	A	B	A
Methyl Ethyl Ketone	A	A	A	A	A	A	A	A	A	I.L.	A	A	A	A
Milk	C	C	A	A	A	A	A	A	A	A	A	C	C	C
Natural Gas	A	A	A	A	A	A	A	A	A	A	A	A	A	A
Nitric Acid	C	C	A	B	C	C	C	B	A	A	C	C	C	B
Oleic Acid	C	C	A	A	B	A	A	A	A	A	A	A	A	I.L.
Oxalic Acid	C	C	B	B	B	B	A	A	A	B	B	B	B	I.L.
Oxygen	A	A	A	A	A	A	A	A	A	A	A	A	A	A
Petroleum Oils, Refined	A	A	A	A	A	A	A	A	A	A	A	A	A	A
Phosphoric Acid (Aerated)	C	C	A	A	C	C	A	A	A	B	A	C	C	C
Phosphoric Acid (Air Free)	C	C	A	A	C	B	A	A	A	B	A	C	C	C
Phosphoric Acid Vapors	C	C	B	B	C	C	A	I.L.	A	B	C	C	C	C.
Picric Acid	C	C	A	A	C	C	A	A	A	I.L.	B	B	B	I.L.
Potassium Chloride	B	B	A	A	B	B	A	A	A	A	I.L.	C	C	I.L.
Potassium Hydroxide	B	B	A	A	B	A	A	A	A	A	I.L.	B	B	I.L.
Propane	A	A	A	A	A	A	A	A	A	A	A	A	A	A
Rosin	B	B	A	A	A	A	A	A	A	I.L.	A	A	A	A
Silver Nitrate	C	C	A	A	C	C	A	A	A	A	B	B	B	I.L.
Sodium Acetate	A	A	B	A	A	A	A	A	A	A	A	A	A	A
Sodium Carbonate	A	A	A	A	A	A	A	A	A	A	A	B	B	A
Sodium Chloride	C	C	B	B	A	A	A	A	A	A	A	B	B	B
Sodium Chromate	A	A	A	A	A	A	A	A	A	A	A	A	A	A
Sodium Hydroxide	A	A	A	A	C	A	A	A	A	A	A	B	B	A
Sodium Hypochloride	C	C	C	C	B-C	B-C	C	A	B	A	I.L.	C	C	I.L.
Sodium Thiosulfate	C	C	A	A	C	C	A	A	A	A	I.L.	B	B	I.L.
Stannous Chloride	B	B	C	A	C	B	A	A	A	A	I.L.	C	C	I.L.
Stearic Acid	A	C	A	A	B	B	A	A	A	A	B	B	B	I.L.
Sulfate Liquor (Black)	A	A	A	A	C	A	A	A	A	A	A	I.L.	IL.	I.L.
Sulfur	A	A	A	A	C	A	A	A	A	A	A	A	A	A
Sulfur Dioxide, Dry	A	A	A	A	A	A	B	A	A	A	A	B	B	I.L.
Sulfur Trioxide, Dry	A	A	A	A	A	A	B	A	A	A	A	B	B	I.L.
Sulfuric Acid (Aerated)	C	C	C	C	C	C	A	A	A	B	B	C	C	C
Sulfuric Acid (Air Free)	C	C	C	C	B	B	A	A	A	B	B	C	C	C
Sulfurous Acid	C	C	B	B	B	C	A	A	A	A	B	C	C	I.L.
Tar	A	A	A	A	A	A	A	A	A	A	A	A	A	A
Trichloroethylene	B	B	B	A	A	A	A	A	A	A	A	B	B	I.L.
Turpentine	B	B	A	A	A	B	A	A	A	A	A	A	A	A
Vinegar	C	C	A	A	B	A	A	A	A	I.L.	A	C	C	A
Water, Boiler Feed	B	C	A	A	C	A	A	A	A	A	A	B	A	A
Water, Distilled	A	A	A	A	A	A	A	A	A	A	A	B	B	I.L.
Water, Sea	B	B	B	B	A	A	A	A	A	A	A	C	C	A
Whiskey and Wines	C	C	A	A	A	B	A	A	A	A	A	C	C	I.L.
Zinc Chloride	C	C	C	C	C	C	A	A	A	A	B	C	C	I.L.
Zinc Sulfate	C	C	A	B	A	A	A	A	A	A	B	B	B	I.L.

*Trademark of International Nickel Co.
†Trademark of Stellite Div., Cabot Corp.
‡Trademark of Duriron Co.

A—Recommended.
B—Minor to moderate effect. Proceed with caution.
C—Unsatisfactory.
I.L.—Information lacking.

Source: General Catalog 501, 5th ed., Fisher Controls, Marshalltown, Iowa. Used with permission of Fish Controls International, Inc.

Table 12.15 Characteristics of Particles and Particle Dispersoids

Methods for Particle Size Analysis

Sieving — Electroformed Sieves — Microscope — Electron Microscope — Ultramicroscope+ — Impingers — Centrifuge — Ultracentrifuge — Turbidimetry++ — Sedimentation — Elutriation — X-Ray Diffraction+ — Adsorption+ — Permeability+ — Light Scattering++ — Nuclei Counter — Electrical Conductivity — Visible to Eye — Machine Tools (Micrometers, Calipers, etc.) — Scanners

+Furnishes average particle diameter but no size distribution.
++Size distribution may be obtained by special calibration.

Types of Gas Cleaning Equipment

Settling Chambers — Centrifugal Separators — Liquid Scrubbers — Cloth Collectors — Packed Beds — Common Air Filters — Impingement Separators — Mechanical Separators — High Efficiency Air Filters — Thermal Precipitation (used only for sampling) — Electrical Precipitators — Ultrasonics (very limited industrial application)

Terminal Gravitational Settling* [for spheres, sp. gr. 2.0]

In Air at 25°C. 1 atm. — Reynolds Number, Settling Velocity, cm/sec.
In Water at 25°C. — Reynolds Number, Settling Velocity, cm/sec.

Particle Diffusion Coefficient,* cm²/sec.

In Air at 25°C. 1 atm.
In Water at 25°C.

Particle Diameter, microns (μ)

0.0001   0.001 (1mμ.)   0.01   0.1   1   10   100   1,000 (1mm.)   10,000 (1cm.)

*Stokes-Cunningham factor included in values given for air but not included for water

PREPARED BY C. E. LAPPLE

Source: C. R. Lapple, *SRI Journal,* **5**, 94 (Third Quarter 1961).

Table 12.16 Orifice Coefficients

Coefficients of discharge (c) for circular orifices, with full contractions.

Head from center of orifice in feet	Diameters in feet					
	0.02	0.05	0.1	0.2	0.6	1.0
0.5	....	0.627	0.615	0.600	0.592	......
0.8	0.648	0.620	0.610	0.601	0.594	0.591
1.0	0.644	0.617	0.608	0.600	0.595	0.591
1.5	0.637	0.613	0.605	0.600	0.596	0.593
2.0	0.632	0.610	0.604	0.599	0.597	0.595
2.5	0.629	0.608	0.603	0.599	0.598	0.596
3.0	0.627	0.606	0.603	0.599	0.598	0.597
3.5	0.625	0.606	0.602	0.599	0.598	0.596
4.0	0.623	0.605	0.602	0.599	0.597	0.596
6.0	0.618	0.604	0.600	0.598	0.597	0.596
8.0	0.614	0.603	0.600	0.598	0.596	0.596
10.0	0.611	0.601	0.598	0.597	0.596	0.595
20.0	0.601	0.598	0.596	0.596	0.596	0.594
50.0	0.596	0.595	0.594	0.594	0.594	0.593
100.0	0.593	0.592	0.592	0.592	0.592	0.592

Coefficients of discharge (c) for square orifices, with full contractions.

Head from center of orifice in feet	Length of side of square in feet					
	0.02	0.05	0.1	0.2	0.6	1.0
0.5	....	0.633	0.619	0.605	0.597	......
0.8	0.652	0.625	0.615	0.605	0.600	0.597
1.0	0.648	0.622	0.613	0.605	0.601	0.599
1.5	0.641	0.619	0.610	0.605	0.602	0.601
2.0	0.637	0.615	0.608	0.605	0.604	0.602
2.5	0.634	0.613	0.607	0.605	0.604	0.602
3.0	0.632	0.612	0.607	0.605	0.604	0.603
3.5	0.630	0.611	0.607	0.605	0.604	0.602
4.0	0.628	0.610	0.606	0.605	0.603	0.602
6.0	0.623	0.609	0.605	0.604	0.603	0.602
8.0	0.619	0.608	0.605	0.604	0.603	0.602
10.0	0.616	0.606	0.604	0.603	0.602	0.601
20.0	0.606	0.603	0.602	0.602	0.601	0.600
50.0	0.602	0.601	0.600	0.600	0.599	0.599
100.0	0.599	0.598	0.598	0.598	0.598	0.598

Source: Hamilton Smith's *Hydraulics.*

Table 12.17 Weir Coefficients (Contracted)

Coefficients of discharge (c) for contracted weirs for use in the Hamilton Smith formula.

Effective head in feet	Length of weir in feet									
	0.66	1	2	3	4	5	7	10	15	19
0.1	0.632	0.639	0.646	0.652	0.653	0.653	0.654	0.655	0.655	0.656
0.2	0.611	0.618	0.626	0.630	0.631	0.631	0.632	0.633	0.634	0.634
0.25	0.605	0.612	0.621	0.624	0.625	0.626	0.627	0.628	0.628	0.629
0.3	0.601	0.608	0.616	0.619	0.621	0.621	0.623	c.624	0.624	0.625
0.4	0.595	0.601	0.609	0.613	0.614	0.615	0.617	0.618	0.619	0.620
0.5	0.590	0.596	0.605	0.608	0.610	0.611	0.613	0.615	0.616	0.617
0.6	0.587	0.593	0.601	0.605	0.607	0.608	0.611	0.613	0.614	0.615
0.8	......	......	0.595	0.600	0.602	0.604	0.607	0.611	0.612	0.613
1.0	......	......	0.590	0.595	0.598	0.601	0.604	0.608	0.610	0.611
1.2	......	......	0.585	0.591	0.594	0.597	0.601	0.605	0.608	0.610
1.4	......	......	0.580	0.587	0.590	0.594	0.598	0.602	0.606	0.609
1.6	......	......	......	0.582	0.587	0.591	0.595	0.600	0.604	0.607

Table 12.18 Weir Coefficients (Suppressed)

Coefficients of discharge (c) for suppressed weirs for use in the Hamilton Smith formula.

Effective head in feet	Length of weir in feet								
	0.66	2	3	4	5	7	10	15	19
0.1	......	......	......	......	0.659	0.658	0.658	0.657	0.657
0.2	0.656	0.645	0.642	0.641	0.638	0.637	0.637	0.636	0.635
0.25	0.653	0.641	0.638	0.636	0.634	0.633	0.632	0.631	0.630
0.3	0.651	0.639	0.636	0.633	0.631	0.629	0.628	0.627	0.626
0.4	0.650	0.636	0.633	0.630	0.628	0.625	0.623	0.622	0.621
0.5	0.650	0.637	0.633	0.630	0.627	0.624	0.621	0.620	0.619
0.6	0.651	0.638	0.634	0.630	0.627	0.623	0.620	0.619	0.618
0.8	0.656	0.643	0.637	0.633	0.629	0.625	0.621	0.620	0.618
1.0	......	0.648	0.641	0.637	0.633	0.628	0.624	0.621	0.619
1.2	......	......	0.646	0.641	0.636	0.632	0.626	0.623	0.620
1.4	......	......	......	0.644	0.640	0.634	0.629	0.625	0.622
1.6	......	......	......	0.647	0.642	0.637	0.631	0.626	0.623

Source: Hamilton Smith's *Hydraulics.*

Table 12.19 Friction Factors (Clean Cast Iron)
Values of friction factor (f) for clean cast-iron pipes.

Diam-eter in inches	Velocity in feet per second						
	0.5	1	2	3	6	10	20
1	0.0398	0.0353	0.0317	0.0299	0.0266	0.0244	0.0228
3	0.0354	0.0316	0.0288	0.0273	0.0248	0.0232	0.0218
6	0.0317	0.0289	0.0264	0.0252	0.0231	0.0219	0.0208
9	0.0290	0.0269	0.0247	0.0237	0.0220	0.0209	0.0200
12	0.0268	0.0251	0.0233	0.0224	0.0209	0.0201	0.0192
18	0.0238	0.0224	0.0211	0.0204	0.0193	0.0188	0.0181
24	0.0212	0.0194	0.0193	0.0187	0.0180	0.0176	0.0170
30	0.0194	0.0186	0.0179	0.0175	0.0170	0.0166	0.0161
36	0.0177	0.0172	0.0167	0.0164	0.0160	0.0156	0.0152
48	0.0153	0.0150	0.0147	0.0145	0.0143	0.0141	0.0138
60	0.0137	0.0135	0.0133	0.0132	0.0130	0.0128	0.0125
72	0.0125	0.0124	0.0122	0.0120	0.0118	0.0117	0.0117
96	0.0109	0.0107	0.0106	0.0106	0.0105	0.0104	0.0103

Table 12.20 Friction Factors (Old Cast Iron)
Values of friction factor (f) for old cast-iron pipes.

Diameter in inches	Velocity in feet per second			
	1	3	6	10
3	0.0608	0.0556	0.0512	0.0488
6	0.0540	0.0468	0.0432	0.0412
9	0.0488	0.0420	0.0400	0.0368
12	0.0432	0.0384	0.0356	0.0336
15	0.0396	0.0348	0.0324	0.0312
18	0.0348	0.0312	0.0292	0.0276
24	0.0304	0.0268	0.0252	0.0240
30	0.0268	0.0244	0.0228	0.0220
36	0.0244	0.0224	0.0208	0.0200
42	0.0232	0.0208	0.0200	0.0192
48	0.0228	0.0204	0.0196	0.0184

Table 12.21 Channel Coefficients (Chezy)
Values of coefficient (c) in Chezy formula.

Radius in Feet	Velocity in feet per second						
	1	2	3	4	6	10	15
0.5	96	104	109	112	116	121	124
1.0	109	116	121	124	129	134	138
1.5	117	124	128	132	136	143	147
2.0	123	130	134	137	142	150	155
2.5	128	134	139	142	147	155	....
3.0	132	138	142	145	150	....	....
3.5	135	141	145	149	153	....	....
4.0	137	143	148	151	....	....	....

Table 12.22　Channel Coefficients (Kutter)

Values of coefficients (c) in Kutter's formula.

Slope	n	Hydraulic radius r in feet										
		0.2	0.4	0.6	0.8	1.0	1.5	2.0	6.0	10.0	15.0	50.0
0.00005	0.010	87	109	123	133	140	154	164	199	213	220	245
	0.015	52	66	76	83	89	99	107	138	150	159	181
	0.020	35	45	53	59	64	72	80	105	116	125	148
	0.025	26	35	41	45	49	57	62	85	96	104	127
	0.030	22	28	33	37	40	47	51	72	83	90	112
	0.040	15	20	24	27	29	34	38	56	64	71	93
0.0001	0.010	98	118	131	140	147	158	167	196	206	212	227
	0.015	57	72	81	88	93	103	109	134	143	150	166
	0.020	38	50	57	63	67	75	81	102	111	118	134
	0.025	28	38	43	48	51	59	64	84	93	98	114
	0.030	23	30	35	39	42	48	52	72	78	85	100
	0.040	16	22	25	28	31	35	30	54	62	68	83
0.0002	0.010	105	125	137	145	150	162	169	193	202	206	220
	0.015	61	76	84	91	96	105	110	132	140	145	158
	0.020	42	53	60	65	68	76	82	100	108	113	126
	0.025	30	40	45	50	54	60	65	83	90	95	108
	0.030	25	32	37	40	43	49	53	69	77	82	94
	0.040	17	23	26	29	32	36	40	53	60	65	78
0.0004	0.010	110	128	140	148	153	164	171	192	198	203	215
	0.015	64	78	87	93	98	106	112	130	137	142	154
	0.020	43	55	61	67	70	77	83	99	106	110	123
	0.025	32	42	47	51	55	60	65	82	88	92	104
	0.030	26	33	38	41	44	50	54	68	75	80	91
	0.040	18	23	27	30	32	37	40	53	59	63	75
0.001	0.010	113	132	143	150	155	165	172	190	197	201	212
	0.015	66	80	88	94	98	107	112	130	135	141	151
	0.020	45	56	62	68	71	78	84	98	105	109	120
	0.025	33	43	48	52	55	61	65	81	87	91	101
	0.030	27	34	38	42	45	50	54	68	74	78	89
	0.040	18	24	27	30	33	37	40	53	58	61	72
0.01	0.010	114	133	143	151	156	165	172	190	196	200	210
	0.015	67	81	89	95	99	107	113	129	135	140	150
	0.020	46	57	63	68	72	78	84	98	105	108	119
	0.025	34	44	49	52	56	62	65	80	86	90	100
	0.030	27	35	39	43	45	51	55	67	73	77	87
	0.040	19	24	28	30	33	37	40	52	58	61	71

Table 12.23　Channel Coefficients (Bazin)

Values of coefficients (c) in Bazin's formula.

Hydraulic radius in feet	Coefficient of roughness m					
	0.06	0.16	0.46	0.85	1.30	1.75
0.2	126	96	55	36	25	19
0.3	132	103	63	41	30	23
0.4	134	108	68	46	33	26
0.5	136	112	71	50	36	29
0.75	140	118	80	57	42	34
1.0	142	122	86	62	47	38
1.25	143	125	90	66	51	41
1.5	145	127	94	70	54	44
2.0	146	131	99	76	59	49
2.5	147	133	104	80	63	53
3.0	148	135	106	83	67	57
5.0	150	140	115	93	77	65
10.0	152	144	125	106	91	79
20.0	154	148	133	117	103	92

Source: Russell, *Textbook on Hydraulics.*

Table 12.24 Tank Capacities (Vertical Cylindrical)

Diameter	Area in Sq. Ft. Cu. Ft. per 1' of Depth	U.S. Gallons per 1' of Depth	Diameter	Area in Sq. Ft. Cu. Ft. per 1' of Depth	U.S. Gallons per 1' of Depth	Diameter	Area in Sq. Ft. Cu. Ft. per 1' of Depth	U.S. Gallons per 1' of Depth
1'	0.785	5.87	6'	28.27	211.5	28'	615.8	4606.
1' 1"	0.922	6.89	6' 3"	30.68	229.5	28' 6"	637.9	4772.
1' 2"	1.069	8.00	6' 6"	33.18	248.2	29'	660.5	4941.
1' 3"	1.227	9.18	6' 9"	35.78	267.7	29' 6"	683.5	5113.
1' 4"	1.396	10.44	7'	38.48	287.9	30'	706.9	5288.
1' 5"	1.576	11.79	7' 3"	41.28	308.8	31'	754.8	5646.
1' 6"	1.767	13.22	7' 6"	44.18	330.5	32'	804.3	6016.
1' 7"	1.969	14.73	7' 9"	47.17	352.9	33'	855.3	6398.
1' 8"	2.182	16.32	8'	50.27	376.0	34'	907.9	6792.
1' 9"	2.405	17.99	8' 3"	53.46	399.9	35'	962.1	7197.
1' 10"	2.640	19.75	8' 6"	56.75	424.5	36'	1018.	7616.
1' 11"	2.885	21.58	8' 9"	60.13	449.8	37'	1075.	8043.
2'	3.142	23.50	9'	63.62	475.9	38'	1134.	8483.
2' 1"	3.409	25.50	9' 3"	67.20	502.7	39'	1195.	8940.
2' 2"	3.687	27.58	9' 6"	70.88	530.2	40'	1257.	9404.
2' 3"	3.976	29.74	9' 9"	74.66	558.5	41'	1320.	9876.
2' 4"	4.276	31.99	10'	78.54	587.5	42'	1385.	10360.
2' 5"	4.587	34.31	10' 6"	86.59	647.7	43'	1452.	10860.
2' 6"	4.909	36.72	11'	95.03	710.9	44'	1521.	11370.
2' 7"	5.241	39.21	11' 6"	103.9	777.0	45'	1590.	11900.
2' 8"	5.585	41.78	12'	113.1	846.0	46'	1662.	12430.
2' 9"*	5.940	44.43	12' 6"	122.7	918.0	47'	1735.	12980.
2' 10"	6.305	47.16	13'	132.7	992.9	48'	1810.	13540.
2' 11"	6.681	49.98	13' 6"	143.1	1071.	49'	1886.	14110.
3'	7.069	52.88	14'	153.9	1152.	50'	1964.	14690.
3' 1"	7.467	55.86	14' 6"	165.1	1235.	52'	2124.	15890.
3' 2"	7.876	58.92	15'	176.7	1322.	54'	2290.	17130.
3' 3"	8.296	62.06	15' 6"	188.7	1412.	56'	2463.	18420.
3' 4"	8.727	65.28	16'	201.1	1504.	58'	2642.	19760.
3' 5"	9.168	68.58	16' 6"	213.8	1600.	60'	2827.	21150.
3' 6"	9.621	71.97	17'	227.0	1698.	62'	3019.	22580.
3' 7"	10.08	75.44	17' 6"	240.5	1799.	64'	3217.	24060.
3' 8"	10.56	78.99	18'	254.5	1904.	66'	3421.	25530.
3' 9"	11.04	82.62	18' 6"	268.8	2011.	68'	3632.	27170.
3' 10"	11.54	86.33	19'	283.5	2121.	70'	3848.	28790.
3' 11"	12.05	90.13	19' 6"	298.6	2234.	72'	4072.	30450.
4'	12.57	94.00	20'	314.2	2350.	74'	4301.	32170.
4' 1"	13.10	97.96	20' 6"	330.1	2469.	76'	4536.	33930.
4' 2"	13.64	102.0	21'	346.4	2591.	78'	4778.	35740.
4' 3"	14.19	106.1	21' 6"	363.1	2716.	80'	5027.	37600.
4' 4"	14.75	110.3	22'	380.1	2844.	82'	5281.	39500.
4' 5"	15.32	114.6	22' 6"	397.6	2974.	84'	5542.	41450.
4' 6"	15.90	119.0	23'	415.5	3108.	86'	5809.	43450.
4' 7"	16.50	123.4	23' 6"	433.7	3245.	88'	6082.	45490.
4' 8"	17.10	128.0	24'	452.4	3384.	90'	6362.	47590.
4' 9"	17.72	132.6	24' 6"	471.4	3527.	92'	6648.	49720.
4' 10"	18.35	137.3	25'	490.9	3672.	94'	6940.	51920.
4' 11"	18.99	142.0	25' 6"	510.7	3820.	96'	7238.	54140.
5'	19.63	146.9	26'	530.9	3972.	98'	7543.	56420.
5' 3"	21.65	161.9	26' 6"	551.5	4126.	100'	7854.	58750.
5' 6"	23.76	177.7	27'	572.6	4283.			
5' 9"	25.97	194.3	27' 6"	594.0	4443.			

Table 12.25 Tank Capacities (Horizontal Cylindrical)

Contents of Tanks with Flat Ends When Filled to Various Depths

Contents in U.S. gallons per foot of length.

To ascertain the contents of a tank over one-half full: Let *h* = depth of unfilled portion. Find from the table the quantity corresponding to a depth *h*. Subtract this quantity from the contents of a full tank.

Diameter of tank inches	Full tank	Depth of liquid, in inches = *h*																			
		3"	6"	9"	12"	15"	18"	21"	24"	27"	30"	33"	36"	39"	42"	45"	48"	51"	54"	57"	60"
12"	5.88	1.15	2.94																		
18"	13.22	1.45	3.86	6.61																	
24"	23.50	1.70	4.60	8.05	11.75																
30"	36.72	1.91	5.23	9.27	13.72	18.36															
36"	52.88	2.12	5.79	10.34	15.43	20.85	26.44														
42"	71.97	2.28	6.31	11.31	16.97	23.07	29.47	35.99													
48"	94.01	2.45	6.78	12.20	18.38	25.10	32.20	39.54	47.60												
54"	118.98	2.60	7.22	13.04	19.68	26.97	34.72	42.80	51.08	59.49											
60"	146.89	2.75	7.64	13.82	20.91	28.72	37.06	45.82	54.87	64.11	73.44										
66"	177.73	2.89	8.04	14.56	22.07	30.37	39.28	48.65	58.39	68.41	78.50	88.86									
72"	211.52	3.02	8.42	15.26	23.17	31.92	41.36	51.32	61.71	72.45	83.41	94.54	105.76								
78"	248.24	3.15	8.78	15.94	24.21	33.41	43.34	53.86	64.87	76.27	87.97	99.90	111.97	124.13							
84"	287.90	3.26	9.12	16.57	25.24	34.85	45.24	56.29	67.87	79.91	92.30	104.98	117.85	130.87	143.95						
90"	330.49	3.43	9.46	17.20	26.20	36.21	47.05	58.61	70.75	83.39	96.43	109.81	123.45	137.28	151.23	165.25					
96"	376.02	3.50	9.79	17.80	27.13	37.52	48.81	60.84	73.52	86.73	100.39	114.44	128.79	143.40	158.17	173.06	188.01				
102"	424.50	3.61	10.10	18.37	28.01	39.00	50.49	62.99	76.18	89.94	104.20	118.89	133.92	149.25	164.81	180.53	196.37	212.25			
108"	476.10	3.71	10.39	18.94	28.90	40.03	52.14	65.09	78.74	93.04	107.87	123.17	138.87	154.89	171.19	187.71	204.37	221.14	238.05		
114"	530.25	3.78	10.74	19.49	29.75	41.22	53.73	67.10	81.24	96.05	111.43	127.31	143.63	160.33	177.33	194.60	212.05	229.65	247.37	265.13	
120"	587.54	3.91	10.98	20.02	30.57	42.39	55.26	69.06	83.65	98.95	114.87	131.32	148.25	165.58	183.27	201.24	219.46	237.87	256.43	275.08	293.77

Table 12.26 Contents of Standard Dished Heads When Filled to Various Depths

Contents in U.S. gallons for one head only. This table is only approximate but close enough for practical use.

To ascertain the contents of a head over one-half full: Let h = depth of unfilled portion. Find from the table the quantity corresponding to a depth h. Subtract this quantity from the contents of a full head.

Radius = Diameter

Depth of liquid, in inches = h

Diameter of head inches	Full head	3"	6"	9"	12"	15"	18"	21"	24"	27"	30"	33"	36"	39"	42"	45"	48"	51"	54"	57"	60"
12"	0.40	0.05	0.20																		
18"	1.36	0.07	0.32	0.68																	
24"	3.22	0.08	0.41	0.95	1.61																
30"	6.30	0.10	0.49	1.18	2.10	3.15															
36"	10.88	0.11	0.56	1.39	2.54	3.92	5.44														
42"	17.28	0.12	0.63	1.59	2.94	4.64	6.57	8.64													
48"	25.79	0.13	0.68	1.75	3.31	5.29	7.62	10.19	12.89												
54"	36.72	0.14	0.74	1.90	3.64	5.91	8.60	11.65	14.95	18.36											
60"	50.37	0.14	0.82	2.07	3.98	6.49	9.54	13.03	16.87	20.96	25.18										
66"	67.04	0.15	0.83	2.19	4.25	6.98	10.35	14.30	18.68	23.43	28.42	33.52									
72"	87.04	0.16	0.88	2.32	4.52	7.47	11.15	15.48	20.38	25.74	31.46	37.43	43.52								
78"	110.66	0.17	0.93	2.44	4.79	7.97	11.94	16.65	22.02	27.97	34.39	41.16	48.20	55.33							
84"	138.22	0.18	0.98	2.59	5.07	8.44	12.69	17.78	23.60	30.11	37.19	44.75	52.67	60.83	69.11						
90"	170.01	0.18	1.00	2.68	5.33	8.91	13.44	18.86	25.12	32.18	39.90	48.22	56.99	66.14	75.52	85.00					
96"	206.32	0.20	1.07	2.83	5.59	9.36	14.14	19.90	26.60	34.17	42.52	51.53	61.13	71.22	81.66	92.34	103.16				
102"	247.48	0.22	1.14	3.01	5.89	9.87	14.92	21.01	28.11	36.18	45.19	54.91	65.31	76.29	87.73	99.56	111.59	123.74			
108"	293.77	0.20	1.13	3.03	6.04	10.21	15.50	21.93	29.47	38.03	47.56	57.97	69.14	81.65	93.53	106.47	119.76	133.26	146.88		
114"	345.51	0.21	1.16	3.12	6.25	10.55	16.06	22.80	30.70	39.73	49.81	60.88	72.85	85.61	99.05	113.07	127.56	142.41	157.51	172.75	
120"	402.27	0.21	1.19	3.23	6.47	10.93	16.68	23.70	31.96	41.43	52.04	63.73	76.40	89.95	104.32	119.39	135.04	151.15	167.62	184.32	201.13

623

Table 12.27 Pipe Data (Carbon and Alloy Steel—Stainless Steel)[a]

Nominal Pipe Size (Inches)	Outside Diam. (Inches)	Identification — Steel — Iron Pipe Size	Identification — Steel — Sched. No.	Identification — Stainless Steel Sched. No.	Wall Thickness (t) Inches	Inside Diameter (d) Inches	Area of Metal Square Inches	Transverse Internal Area (a) Square Inches	Transverse Internal Area (A) Square Feet	Moment of Inertia (I) Inches[4]	Weight Pipe Pounds per foot	Weight Water Pounds per foot of pipe	External Surface Sq. Ft. per foot of pipe	Section Modulus $\left(2\,\dfrac{I}{O.D.}\right)$
1/8	0.405	...	...	10S	.049	.307	.0548	.0740	.00051	.00088	.19	.032	.106	.0043
		STD	40	40S	.068	.269	.0720	.0568	.00040	.00106	.24	.025	.106	.0052
		XS	80	80S	.095	.215	.0925	.0364	.00025	.00122	.31	.016	.106	.0066
1/4	0.540	...	...	10S	.065	.410	.0970	.1320	.00091	.00279	.33	.057	.141	.0103
		STD	40	40S	.088	.364	.1250	.1041	.00072	.00331	.42	.045	.141	.0122
		XS	80	80S	.119	.302	.1574	.0716	.00050	.00377	.54	.031	.141	.0139
3/8	0.675	...	...	10S	.065	.545	.1246	.2333	.00162	.00586	.42	.101	.178	.0173
		STD	40	40S	.091	.493	.1670	.1910	.00133	.00729	.57	.083	.178	.0216
		XS	80	80S	.126	.423	.2173	.1405	.00098	.00862	.74	.061	.178	.0255
1/2	0.840	...	...	5S	.065	.710	.1583	.3959	.00275	.01197	.54	.172	.220	.0284
		...	...	10S	.083	.674	.1974	.3568	.00248	.01431	.67	.155	.220	.0340
		STD	40	40S	.109	.622	.2503	.3040	.00211	.01709	.85	.132	.220	.0406
		XS	80	80S	.147	.546	.3200	.2340	.00163	.02008	1.09	.102	.220	.0478
		...	160	...	.187	.466	.3836	.1706	.00118	.02212	1.31	.074	.220	.0526
		XXS	...	...	.294	.252	.5043	.050	.00035	.02424	1.71	.022	.220	.0577
3/4	1.050	...	...	5S	.065	.920	.2011	.6648	.00462	.02450	.69	.288	.275	.0467
		...	...	10S	.083	.884	.2521	.6138	.00426	.02969	.86	.266	.275	.0566
		STD	40	40S	.113	.824	.3326	.5330	.00371	.03704	1.13	.231	.275	.0705
		XS	80	80S	.154	.742	.4335	.4330	.00300	.04479	1.47	.188	.275	.0853
		...	160	...	.219	.612	.5698	.2961	.00206	.05269	1.94	.128	.275	.1003
		XXS	...	...	.308	.434	.7180	.148	.00103	.05792	2.44	.064	.275	.1103
1	1.315	...	...	5S	.065	1.185	.2553	1.1029	.00766	.04999	.87	.478	.344	.0760
		...	...	10S	.109	1.097	.4130	.9452	.00656	.07569	1.40	.409	.344	.1151
		STD	40	40S	.133	1.049	.4939	.8640	.00600	.08734	1.68	.375	.344	.1328
		XS	80	80S	.179	.957	.6388	.7190	.00499	.1056	2.17	.312	.344	.1606
		...	160	...	.250	.815	.8365	.5217	.00362	.1251	2.84	.230	.344	.1903
		XXS	...	...	.358	.599	1.0760	.282	.00196	.1405	3.66	.122	.344	.2136
1¼	1.660	...	...	5S	.065	1.530	.3257	1.839	.01277	.1038	1.11	.797	.435	.125
		...	...	10S	.109	1.442	.4717	1.633	.01134	.1605	1.81	.708	.435	.193
		STD	40	40S	.140	1.380	.6685	1.495	.01040	.1947	2.27	.649	.435	.2346
		XS	80	80S	.191	1.278	.8815	1.283	.00891	.2418	3.00	.555	.435	.2913
		...	160	...	.250	1.160	1.1070	1.057	.00734	.2839	3.76	.458	.435	.3421
		XXS	...	...	.382	.896	1.534	.630	.00438	.3411	5.21	.273	.435	.4110
1½	1.900	...	...	5S	.065	1.770	.3747	2.461	.01709	.1579	1.28	1.066	.497	.1662
		...	...	10S	.109	1.682	.6133	2.222	.01543	.2468	2.09	.963	.497	.2598
		STD	40	40S	.145	1.610	.7995	2.036	.01414	.3099	2.72	.882	.497	.3262
		XS	80	80S	.200	1.500	1.068	1.767	.01225	.3912	3.63	.765	.497	.4118
		...	160	...	.281	1.338	1.429	1.406	.00976	.4824	4.86	.608	.497	.5078
		XXS	...	...	.400	1.100	1.885	.950	.00660	.5678	6.41	.42	.497	.5977
2	2.375	...	...	5S	.065	2.245	.4717	3.958	.02749	.3149	1.61	1.72	.622	.2652
		...	...	10S	.109	2.157	.7760	3.654	.02538	.4992	2.64	1.58	.622	.420
		STD	40	40S	.154	2.067	1.075	3.355	.02330	.6657	3.65	1.45	.622	.5606
		XS	80	80S	.218	1.939	1.477	2.953	.02050	.8679	5.02	1.28	.622	.7309
		...	160	...	.344	1.687	2.190	2.241	.01556	1.162	7.46	.97	.622	.979
		XXS	...	...	.436	1.503	2.656	1.774	.01232	1.311	9.03	.77	.622	1.104
2½	2.875	...	...	5S	.083	2.709	.7280	5.764	.04002	.7100	2.48	2.50	.753	.4939
		...	...	10S	.120	2.635	1.039	5.453	.03787	.9873	3.53	2.36	.753	.6868
		STD	40	40S	.203	2.469	1.704	4.788	.03322	1.530	5.79	2.07	.753	1.064
		XS	80	80S	.276	2.323	2.254	4.238	.02942	1.924	7.66	1.87	.753	1.339
		...	160	...	.375	2.125	2.945	3.546	.02463	2.353	10.01	1.54	.753	1.638
		XXS	...	...	.552	1.771	4.028	2.464	.01710	2.871	13.69	1.07	.753	1.997
3	3.500	...	...	5S	.083	3.334	.8910	8.730	.06063	1.301	3.03	3.78	.916	.743
		...	...	10S	.120	3.260	1.274	8.347	.05796	1.822	4.33	3.62	.916	1.041
		STD	40	40S	.216	3.068	2.228	7.393	.05130	3.017	7.58	3.17	.916	1.724
		XS	80	80S	.300	2.900	3.016	6.605	.04587	3.894	10.25	2.86	.916	2.225
		...	160	...	.438	2.624	4.205	5.408	.03755	5.032	14.32	2.35	.916	2.876
		XXS	...	...	.600	2.300	5.466	4.155	.02885	5.993	18.58	1.80	.916	3.424

Nom. Pipe Size Inches	Outside Diam. Inches	Identification — Steel Iron Pipe Size	Sched. No.	Stainless Steel Sched. No.	Wall Thickness (t) Inches	Inside Diameter (d) Inches	Area of Metal Square Inches	Transverse Internal Area (a) Square Inches	(A) Square Feet	Moment of Inertia (I) Inches^4	Weight Pipe Pounds per foot	Weight Water Pounds per foot of pipe	External Surface Sq. Ft. per foot of pipe	Section Modulus $\left(2\dfrac{I}{O.D.}\right)$
		. . .	. . .	5S	.083	4.334	1.152	14.75	.10245	2.810	3.92	6.39	1.178	1.249
		. . .	. . .	10S	.120	4.260	1.651	14.25	.09898	3.963	5.61	6.18	1.178	1.761
		STD	40	40S	.237	4.026	3.174	12.73	.08840	7.233	10.79	5.50	1.178	3.214
4	4.500	XS	80	80S	.337	3.826	4.407	11.50	.07986	9.610	14.98	4.98	1.178	4.271
		. . .	120	. . .	.438	3.624	5.595	10.31	.0716	11.65	19.00	4.47	1.178	5.178
		. . .	160	. . .	.531	3.438	6.621	9.28	.0645	13.27	22.51	4.02	1.178	5.898
		XXS	. . .	. . .	.674	3.152	8.101	7.80	.0542	15.28	27.54	3.38	1.178	6.791
		. . .	. . .	5S	.109	6.407	2.231	32.24	.2239	11.85	7.60	13.97	1.734	3.576
		. . .	. . .	10S	.134	6.357	2.733	31.74	.2204	14.40	9.29	13.75	1.734	4.346
		STD	40	40S	.280	6.065	5.581	28.89	.2006	28.14	18.97	12.51	1.734	8.496
6	6.625	XS	80	80S	.432	5.761	8.405	26.07	.1810	40.49	28.57	11.29	1.734	12.22
		. . .	120	. . .	.562	5.501	10.70	23.77	.1650	49.61	36.39	10.30	1.734	14.98
		. . .	160	. . .	.719	5.187	13.32	21.15	.1469	58.97	45.35	9.16	1.734	17.81
		XXS	. . .	. . .	.864	4.897	15.64	18.84	.1308	66.33	53.16	8.16	1.734	20.02
		. . .	. . .	5S	.109	8.407	2.916	55.51	.3855	26.44	9.93	24.06	2.258	6.131
		. . .	. . .	10S	.148	8.329	3.941	54.48	.3784	35.41	13.40	23.61	2.258	8.212
		. . .	20	. . .	.250	8.125	6.57	51.85	.3601	57.72	22.36	22.47	2.258	13.39
		. . .	30	. . .	.277	8.071	7.26	51.16	.3553	63.35	24.70	22.17	2.258	14.69
		STD	40	40S	.322	7.981	8.40	50.03	.3474	72.49	28.55	21.70	2.258	16.81
8	8.625	. . .	60	. . .	.406	7.813	10.48	47.94	.3329	88.73	35.64	20.77	2.258	20.58
		XS	80	80S	.500	7.625	12.76	45.66	.3171	105.7	43.39	19.78	2.258	24.51
		. . .	100	. . .	.594	7.437	14.96	43.46	.3018	121.3	50.95	18.83	2.258	28.14
		. . .	120	. . .	.719	7.187	17.84	40.59	.2819	140.5	60.71	17.59	2.258	32.58
		. . .	140	. . .	.812	7.001	19.93	38.50	.2673	153.7	67.76	16.68	2.258	35.65
		XXS	. . .	. . .	.875	6.875	21.30	37.12	.2578	162.0	72.42	16.10	2.258	37.56
		. . .	160	. . .	.906	6.813	21.97	36.46	.2532	165.9	74.69	15.80	2.258	38.48
		. . .	. . .	5S	.134	10.482	4.36	86.29	.5992	63.0	15.19	37.39	2.814	11.71
		. . .	. . .	10S	.165	10.420	5.49	85.28	.5922	76.9	18.65	36.95	2.814	14.30
		. . .	20	. . .	.250	10.250	8.24	82.52	.5731	113.7	28.04	35.76	2.814	21.15
		. . .	30	. . .	.307	10.136	10.07	80.69	.5603	137.4	34.24	34.96	2.814	25.57
		STD	40	40S	.365	10.020	11.90	78.86	.5475	160.7	40.48	34.20	2.814	29.90
10	10.750	XS	60	80S	.500	9.750	16.10	74.66	.5185	212.0	54.74	32.35	2.814	39.43
		. . .	80	. . .	.594	9.562	18.92	71.84	.4989	244.8	64.43	31.13	2.814	45.54
		. . .	100	. . .	.719	9.312	22.63	68.13	.4732	286.1	77.03	29.53	2.814	53.22
		. . .	120	. . .	.844	9.062	26.24	64.53	.4481	324.2	89.29	27.96	2.814	60.32
		XXS	140	. . .	1.000	8.750	30.63	60.13	.4176	367.8	104.13	26.06	2.814	68.43
		. . .	160	. . .	1.125	8.500	34.02	56.75	.3941	399.3	115.64	24.59	2.814	74.29
		. . .	. . .	5S	.156	12.438	6.17	121.50	.8438	122.4	20.98	52.65	3.338	19.2
		. . .	. . .	10S	.180	12.390	7.11	120.57	.8373	140.4	24.17	52.25	3.338	22.0
		. . .	20	. . .	.250	12.250	9.82	117.86	.8185	191.8	33.38	51.07	3.338	30.2
		. . .	30	. . .	.330	12.090	12.87	114.80	.7972	248.4	43.77	49.74	3.338	39.0
		STD	. . .	40S	.375	12.000	14.58	113.10	.7854	279.3	49.56	49.00	3.338	43.8
		. . .	40	. . .	.406	11.938	15.77	111.93	.7773	300.3	53.52	48.50	3.338	47.1
12	12.75	XS	. . .	80S	.500	11.750	19.24	108.43	.7528	361.5	65.42	46.92	3.338	56.7
		. . .	60	. . .	.562	11.626	21.52	106.16	.7372	400.4	73.15	46.00	3.338	62.8
		. . .	80	. . .	.688	11.374	26.03	101.64	.7058	475.1	88.63	44.04	3.338	74.6
		. . .	100	. . .	.844	11.062	31.53	96.14	.6677	561.6	107.32	41.66	3.338	88.1
		XXS	120	. . .	1.000	10.750	36.91	90.76	.6303	641.6	125.49	39.33	3.338	100.7
		. . .	140	. . .	1.125	10.500	41.08	86.59	.6013	700.5	139.67	37.52	3.338	109.9
		. . .	160	. . .	1.312	10.126	47.14	80.53	.5592	781.1	160.27	34.89	3.338	122.6

entification, wall thickness, and weights are extracted from ANSI B36 10 and B36 19. The notations STD, XS, and XXS licate Standard, Extra Strong, and Double Extra Strong pipe, respectively.

ansverse internal area values listed in square feet also represent volume in cubic feet per foot of pipe length.

Table 12.28 Properties of Saturated Steam (Pressure: in. Hg Absolute)

Absolute Pressure (in. Hg)	Temp. (°F)	Specific Volume (ft³/lb)	Absolute Pressure (in. Hg)	Temp. (°F)	Specific Volume (ft³/lb)
0.18	32.00	3306	3.75	123.1	188
0.20	34.56	2997	4.00	125.4	177
0.30	44.96	2039	4.25	127.7	167
0.40	52.64	1553	4.50	129.8	158
0.50	58.80	1256	4.75	131.8	150
0.60	63.96	1057	5.00	133.8	143
0.70	68.40	914	5.50	137.0	131
0.80	72.33	806	6.00	141.0	121
0.90	75.85	721	6.50	144.0	112
1.00	79.03	652	7.00	147.0	105
1.25	85.93	528	8.00	152.0	92
1.50	91.72	445	9.00	157.0	83
1.75	96.72	385	10.00	162.0	75
2.00	101.14	339	12.00	169.0	63
2.25	105.11	304	14.00	176.0	55
2.50	108.71	275	16.00	182.0	48
2.75	112.01	251	18.00	188.0	43
3.00	115.06	232	20.00	192.0	39
3.25	117.90	215	25.00	203.0	32
3.50	120.60	200	29.92	212.0	27

Table 12.29 Properties of Saturated Steam (Temperature)[a]

Temp. (°F) t	Press. $\left(\dfrac{lbf}{in.^2}\right) p$	Specific Volume		Internal Energy			Enthalpy			Entropy		
		Sat. Liquid v_f	Sat. Vapor v_g	Sat. Liquid u_f	Evap. u_{fg}	Sat. Vapor u_g	Sat. Liquid h_f	Evap. h_{fg}	Sat. Vapor h_g	Sat. Liquid s_f	Evap. s_{fg}	Sat. Vapor s_g
32	.08859	.016022	3305.	−.01	1021.2	1021.2	−.01	1075.4	1075.4	−.00003	2.1870	2.1870
35	.09992	.016021	2948.	2.99	1019.2	1022.2	3.00	1073.7	1076.7	.00607	2.1704	2.1764
40	.12166	.016020	2445.	8.02	1015.8	1023.9	8.02	1070.9	1078.9	.01617	2.1430	2.1592
45	.14748	.016021	2037.	13.04	1012.5	1025.5	13.04	1068.1	1081.1	.02618	2.1162	2.1423
50	.17803	.016024	1704.2	18.06	1009.1	1027.2	18.06	1065.2	1083.3	.03607	2.0899	2.1259
60	.2563	.016035	1206.9	28.08	1002.4	1030.4	28.08	1059.6	1087.7	.05555	2.0388	2.0943
70	.3632	.016051	867.7	38.09	995.6	1033.7	38.09	1054.0	1092.0	.07463	1.9896	2.0642
80	.5073	.016073	632.8	48.08	988.9	1037.0	48.09	1048.3	1096.4	.09332	1.9423	2.0356
90	.6988	.016099	467.7	58.07	982.2	1040.2	58.07	1042.7	1100.7	.11165	1.8966	2.0083
100	.9503	.016130	350.0	68.04	975.4	1043.5	68.05	1037.0	1105.0	.12963	1.8526	1.9822
110	1.2763	.016166	265.1	78.02	968.7	1046.7	78.02	1031.3	1109.3	.14730	1.8101	1.9574
120	1.6945	.016205	203.0	87.99	961.9	1049.9	88.00	1025.5	1113.5	.16465	1.7690	1.9336
130	2.225	.016247	157.17	97.97	955.1	1053.0	97.98	1019.8	1117.8	.18172	1.7292	1.9109
140	2.892	.016293	122.88	107.95	948.2	1056.2	107.96	1014.0	1121.9	.19851	1.6907	1.8892
150	3.722	.016343	96.99	117.95	941.3	1059.3	117.96	1008.1	1126.1	.21503	1.6533	1.8684
160	4.745	.016395	77.23	127.94	934.4	1062.3	127.96	1002.2	1130.1	.23130	1.6171	1.8484
170	5.996	.016450	62.02	137.95	927.4	1065.4	137.97	996.2	1134.2	.24732	1.5819	1.8293
180	7.515	.016509	50.20	147.97	920.4	1068.3	147.99	990.2	1138.2	.26311	1.5478	1.8109
190	9.343	.016570	40.95	158.00	913.3	1071.3	158.03	984.1	1142.1	.27866	1.5146	1.7932
200	11.529	.016634	33.63	168.04	906.2	1074.2	168.07	977.9	1145.9	.29400	1.4822	1.7762

Table 12.29 (*Continued*)

Temp. (°F) t	Press. $\left(\frac{lbf}{in.^2}\right)$ p	Specific Volume		Internal Energy			Enthalpy			Entropy		
		Sat. Liquid v_f	Sat. Vapor v_g	Sat. Liquid u_f	Evap. u_{fg}	Sat. Vapor u_g	Sat. Liquid h_f	Evap. h_{fg}	Sat. Vapor h_g	Sat. Liquid s_f	Evap. s_{fg}	Sat. Vapor s_g
210	14.125	.016702	27.82	178.10	898.9	1077.0	178.14	971.6	1149.7	.30913	1.4508	1.7599
212	14.698	.016716	26.80	180.11	897.5	1077.6	180.16	970.3	1150.5	.31213	1.4446	1.7567
220	17.188	.016772	23.15	188.17	891.7	1079.8	188.22	965.3	1153.5	.32406	1.4201	1.7441
230	20.78	.016845	19.386	198.26	884.3	1082.6	198.32	958.8	1157.1	.33880	1.3901	1.7289
240	24.97	.016922	16.327	208.36	876.9	1085.3	208.44	952.3	1160.7	.35335	1.3609	1.7143
250	29.82	.017001	13.826	218.49	869.4	1087.9	218.59	945.6	1164.2	.36772	1.3324	1.7001
260	35.42	.017084	11.768	228.64	861.8	1090.5	228.76	938.8	1167.6	.38193	1.3044	1.6864
270	41.85	.017170	10.066	238.82	854.1	1093.0	238.95	932.0	1170.9	.39597	1.2771	1.6731
280	49.18	.017259	8.650	249.02	846.3	1095.4	249.18	924.9	1174.1	.40986	1.2504	1.6602
290	57.53	.017352	7.467	259.25	838.5	1097.7	259.44	917.8	1177.2	.42360	1.2241	1.6477
300	66.98	.017448	6.472	269.52	830.5	1100.0	269.73	910.4	1180.2	.43720	1.1984	1.6356
310	77.64	.017548	5.632	279.81	822.3	1102.1	280.06	903.0	1183.0	.45067	1.1731	1.6238
320	89.60	.017652	4.919	290.14	814.1	1104.2	290.43	895.3	1185.8	.46400	1.1483	1.6123
330	103.00	.017760	4.312	300.51	805.7	1106.2	300.84	887.5	1188.4	.47722	1.1238	1.6010
340	117.93	.017872	3.792	310.91	797.1	1108.0	311.30	879.5	1190.8	.49031	1.0997	1.5901
350	134.53	.017988	3.346	321.35	788.4	1109.8	321.80	871.3	1193.1	.50329	1.0760	1.5793
360	152.92	.018108	2.961	331.84	779.6	1111.4	332.35	862.9	1195.2	.51617	1.0526	1.5688
370	173.23	.018233	2.628	342.37	770.6	1112.9	342.96	854.2	1197.2	.52894	1.0295	1.5585
380	195.60	.018363	2.339	352.95	761.4	1114.3	353.62	845.4	1199.0	.54163	1.0067	1.5483
390	220.2	.018498	2.087	363.58	752.0	1115.6	364.34	836.2	1200.6	.55422	.9841	1.5383
400	247.1	.018638	1.8661	374.27	742.4	1116.6	375.12	826.8	1202.0	.56672	.9617	1.5284
410	276.5	.018784	1.6726	385.01	732.6	1117.6	385.97	817.2	1203.1	.57916	.9395	1.5187
420	308.5	.018936	1.5024	395.81	722.5	1118.3	396.89	807.2	1204.1	.59152	.9175	1.5091
430	343.3	.019094	1.3521	406.68	712.2	1118.9	407.89	796.9	1204.8	.60381	.8957	1.4995
440	381.2	.019260	1.2192	417.62	701.7	1119.3	418.98	786.3	1205.3	.61605	.8740	1.4900
450	422.1	.019433	1.1011	428.6	690.9	1119.5	430.2	775.4	1205.6	.6282	.8523	1.4806
460	466.3	.019614	.9961	439.7	679.8	1119.6	441.4	764.1	1205.5	.6404	.8308	1.4712
470	514.1	.019803	.9025	450.9	668.4	1119.4	452.8	752.4	1205.2	.6525	.8093	1.4618
480	565.5	.020002	.8187	462.2	656.7	1118.9	464.3	740.3	1204.6	.6646	.7878	1.4524
490	620.7	.020211	.7436	473.6	644.7	1118.3	475.9	727.8	1203.7	.6767	.7663	1.4430
500	680.0	.02043	.6761	485.1	632.3	1117.4	487.7	714.8	1202.5	.6888	.7448	1.4335
520	811.4	.02091	.5605	508.5	606.2	1114.8	511.7	687.3	1198.9	.7130	.7015	1.4145
540	961.5	.02145	.4658	532.6	578.4	1111.0	536.4	657.5	1193.8	.7374	.6576	1.3950
560	1131.8	.02207	.3877	557.4	548.4	1105.8	562.0	625.0	1187.0	.7620	.6129	1.3749
580	1324.3	.02278	.3225	583.1	515.9	1098.9	588.6	589.3	1178.0	.7872	.5668	1.3540
600	1541.0	.02363	.2677	609.9	480.1	1090.0	616.7	549.7	1166.4	.8130	.5187	1.3317
620	1784.4	.02465	.2209	638.3	440.2	1078.2	646.4	505.0	1151.4	.8398	.4677	1.3075
640	2057.1	.02593	.1805	668.7	394.5	1063.2	678.6	453.4	1131.9	.8681	.4122	1.2803
660	2362.	.02767	.14459	702.3	340.0	1042.3	714.4	391.1	1105.5	.8990	.3493	1.2483
680	2705.	.03032	.11127	741.7	269.3	1011.0	756.9	309.8	1066.7	.9350	.2718	1.2068
700	3090.	.03666	.07438	801.7	145.9	947.7	822.7	167.5	990.2	.9902	.1444	1.1346
705.44	3204.	.05053	.05053	872.6	0	872.6	902.5	0	902.5	1.0580	0	1.0580

Source: J. H. Keenan, F. G. Keyes, P. G. Hill, and J. G. Moore, *Steam Tables*, John Wiley & Sons, Inc., New York, 1969.

[a] Symbols used: h, specific enthalpy, Btu/lb; p, pressure, lbf/in.^{2}; s, specific entropy, Btu/lb, degrees Rankine; t, thermodynamic temperature, degrees Fahrenheit; u, specific internal energy, Btu/lb; and v, specific volume, ft^3/lb. Note the use of the following subscripts: f, refers to a property of liquid in equilibrium with vapor; g, refers to a property of vapor in equilibrium with liquid; i, refers to a property of solid in equilibrium with vapor; fg, refers to a change by evaporation; and ig, refers to a change by sublimation.

Table 12.30 Properties of Saturated Steam (Pressure)[a]

Press. $\left(\dfrac{\text{lbf}}{\text{in.}^2}\right)$ p	Temp. (°F) t	Specific Volume Sat. Liquid v_f	Sat. Vapor v_g	Internal Energy Sat. Liquid u_f	Evap. u_{fg}	Sat. Vapor u_g	Enthalpy Sat. Liquid h_f	Evap. h_{fg}	Sat. Vapor h_g	Entropy Sat. Liquid s_f	Evap. s_{fg}	Sat. Vapor s_g
.08866	32.02	.016022	3302.	.00	1021.2	1021.2	.01	1075.4	1075.4	.00000	2.1869	2.1869
.10	35.02	.016021	2946.	·3.02	1019.2	1022.2	3.02	1073.7	1076.7	.00612	2.1702	2.1764
.20	53.15	.016027	1526.	21.22	1007.0	1028.2	21.22	1063.5	1084.7	.04225	2.0736	2.1158
.30	64.46	.016041	1039.	32.55	999.4	1031.9	32.56	1057.1	1089.6	.06411	2.0166	2.0807
.40	72.84	.016056	792.0	40.94	993.7	1034.7	40.94	1052.3	1093.3	.07998	1.9760	2.0559
.50	79.56	.016071	641.5	47.64	989.2	1036.9	47.65	1048.6	1096.2	.09250	1.9443	2.0368
.60	85.19	.016086	540.0	53.26	985.4	1038.7	53.27	1045.4	1098.6	.10287	1.9184	2.0213
.70	90.05	.016099	466.9	58.12	982.1	1040.3	58.12	1042.6	1100.7	.11174	1.8964	2.0081
.80	94.35	.016112	411.7	62.41	979.2	1041.7	62.41	1040.2	1102.6	.11951	1.8773	1.9968
.90	98.20	.016124	368.4	66.25	976.6	1042.9	66.25	1038.0	1104.3	.12642	1.8604	1.9868
1.0	101.70	.016136	333.6	69.74	974.3	1044.0	69.74	1036.0	1105.8	.13266	1.8453	1.9779
1.5	115.65	.016187	227.7	83.65	964.8	1048.5	83.65	1028.0	1111.7	.15714	1.7867	1.9438
2.0	126.04	.016230	173.7	94.02	957.8	1051.8	94.02	1022.1	1116.1	.17499	1.7448	1.9198
3.0	141.43	.016300	118.7	109.38	947.2	1056.6	109.39	1013.1	1122.5	.20089	1.6852	1.8861
4.0	152.93	.016358	90.64	120.88	939.3	1060.2	120.89	1006.4	1127.3	.21983	1.6426	1.8624
5.0	162.21	.016407	73.53	130.15	932.9	1063.0	130.17	1000.9	1131.0	.23486	1.6093	1.8441
6.0	170.03	.016451	61.98	137.98	927.4	1065.4	138.00	996.2	1134.2	.24736	1.5819	1.8292
7.0	176.82	.016490	53.65	144.78	922.6	1067.4	144.80	992.1	1136.9	.25811	1.5585	1.8167
8.0	182.84	.016526	47.35	150.81	918.4	1069.2	150.84	988.4	1139.3	.26754	1.5383	1.8058
9.0	188.26	.016559	42.41	156.25	914.5	1070.8	156.27	985.1	1141.4	.27596	1.5203	1.7963
10	193.19	.016590	38.42	161.20	911.0	1072.2	161.23	982.1	1143.3	.28358	1.5041	1.7877
14.696	211.99	.016715	26.80	180.10	897.5	1077.6	180.15	970.4	1150.5	.31212	1.4446	1.7567
15	213.03	.016723	26.29	181.14	896.8	1077.9	181.19	969.7	1150.9	.31367	1.4414	1.7551
20	227.96	.016830	20.09	196.19	885.8	1082.0	196.26	960.1	1156.4	.33580	1.3962	1.7320
25	240.08	.016922	16.306	208.44	876.9	1085.3	208.52	952.2	1160.7	.35345	1.3607	1.7142
30	250.34	.017004	13.748	218.84	869.2	1088.0	218.93	945.4	1164.3	.36821	1.3314	1.6996
35	259.30	.017078	11.900	227.93	862.4	1090.3	228.04	939.3	1167.4	.38093	1.3064	1.6873
40	267.26	.017146	10.501	236.03	856.2	1092.3	236.16	933.8	1170.0	.39214	1.2845	1.6767
45	274.46	.017209	9.403	243.37	850.7	1094.0	243.51	928.8	1172.3	.40218	1.2651	1.6673
50	281.03	.017269	8.518	250.08	845.5	1095.6	250.24	924.2	1174.4	.41129	1.2476	1.6589
55	287.10	.017325	7.789	256.28	840.8	1097.0	256.46	919.9	1176.3	.41963	1.2317	1.6513
60	292.73	.017378	7.177	262.06	836.3	1098.3	262.25	915.8	1178.0	.42733	1.2170	1.6444
65	298.00	.017429	6.657	267.46	832.1	1099.5	267.67	911.9	1179.6	.43450	1.2035	1.6380
70	302.96	.017478	6.209	272.56	828.1	1100.6	272.79	908.3	1181.0	.44120	1.1909	1.6321
75	307.63	.017524	5.818	277.37	824.3	1101.6	277.61	904.8	1182.4	.44749	1.1790	1.6265
80	312.07	.017570	5.474	281.95	820.6	1102.6	282.21	901.4	1183.6	.45344	1.1679	1.6214
85	316.29	.017613	5.170	286.30	817.1	1103.5	286.58	898.2	1184.8	.45907	1.1574	1.6165
90	320.31	.017655	4.898	290.46	813.8	1104.3	290.76	895.1	1185.9	.46442	1.1475	1.6119
95	324.16	.017696	4.654	294.45	810.6	1105.0	294.76	892.1	1186.9	.46952	1.1380	1.6076
100	327.86	.017736	4.434	298.28	807.5	1105.8	298.61	889.2	1187.8	.47439	1.1290	1.6034
110	334.82	.017813	4.051	305.52	801.6	1107.1	305.88	883.7	1189.6	.48355	1.1122	1.5957
120	341.30	.017886	3.730	312.27	796.0	1108.3	312.67	878.5	1191.1	.49201	1.0966	1.5886
130	347.37	.017957	3.457	318.61	790.7	1109.4	319.04	873.5	1192.5	.49989	1.0822	1.5821
140	353.08	.018024	3.221	324.58	785.7	1110.3	325.05	868.7	1193.8	.50727	1.0688	1.5761
150	358.48	.018089	3.016	330.24	781.0	1111.2	330.75	864.2	1194.9	.51422	1.0562	1.5704
160	363.60	.018152	2.836	335.63	776.4	1112.0	336.16	859.8	1196.0	.52078	1.0443	1.5651
170	368.47	.018214	2.676	340.76	772.0	1112.7	341.33	855.6	1196.9	.52700	1.0330	1.5600
180	373.13	.018273	2.533	345.68	767.7	1113.4	346.29	851.5	1197.8	.53292	1.0223	1.5553
190	377.59	.018331	2.405	350.39	763.6	1114.0	351.04	847.5	1198.6	.53857	1.0122	1.5507
200	381.86	.018387	2.289	354.9	759.6	1114.6	355.6	843.7	1199.3	.5440	1.0025	1.5464

Table 12.30 *(Continued)*

Press. $\left(\dfrac{\text{lbf}}{\text{in.}^2}\right)$ p	Temp. (°F) t	Specific Volume		Internal Energy			Enthalpy			Entropy		
		Sat. Liquid v_f	Sat. Vapor v_g	Sat. Liquid u_f	Evap. u_{fg}	Sat. Vapor u_g	Sat. Liquid h_f	Evap. h_{fg}	Sat. Vapor h_g	Sat. Liquid s_f	Evap. s_{fg}	Sat. Vapor s_g
250	401.04	.018653	1.8448	375.4	741.4	1116.7	376.2	825.8	1202.1	.5680	.9594	1.5274
300	417.43	.018896	1.5442	393.0	725.1	1118.2	394.1	809.8	1203.9	.5883	.9232	1.5115
350	431.82	.019124	1.3267	408.7	710.3	1119.0	409.9	795.0	1204.9	.6060	.8917	1.4978
400	444.70	.019340	1.1620	422.8	696.7	1119.5	424.2	781.2	1205.5	.6218	.8638	1.4856
450	456.39	.019547	1.0326	435.7	683.9	1119.6	437.4	768.2	1205.6	.6360	.8385	1.4746
500	467.13	.019748	.9283	447.7	671.7	1119.4	449.5	755.8	1205.3	.6490	.8154	1.4645
600	486.33	.02013	.7702	469.4	649.1	1118.6	471.7	732.4	1204.1	.6723	.7742	1.4464
700	503.23	.02051	.6558	488.9	628.2	1117.0	491.5	710.5	1202.0	.6927	.7378	1.4305
800	518.36	.02087	.5691	506.6	608.4	1115.0	509.7	689.6	1199.3	.7110	.7050	1.4160
900	532.12	.02123	.5009	523.0	589.6	1112.6	526.6	669.5	1196.0	.7277	.6750	1.4027
1000	544.75	.02159	.4459	538.4	571.5	1109.9	542.4	650.0	1192.4	.7432	.6471	1.3903
1100	556.45	.02195	.4005	552.9	553.9	1106.8	557.4	631.0	1188.3	.7576	.6209	1.3786
1200	567.37	.02232	.3623	566.7	536.8	1103.5	571.7	612.3	1183.9	.7712	.5961	1.3673
1300	577.60	.02269	.3297	579.9	519.9	1099.8	585.4	593.8	1179.2	.7841	.5724	1.3565
1400	587.25	.02307	.3016	592.7	503.3	1096.0	598.6	575.5	1174.1	.7964	.5497	1.3461
1500	596.39	.02346	.2769	605.0	486.9	1091.8	611.5	557.2	1168.7	.8082	.5276	1.3359
1750	617.31	.02450	.2268	634.4	445.9	1080.2	642.3	511.4	1153.7	.8361	.4748	1.3109
2000	636.00	.02565	.18813	662.4	404.2	1066.6	671.9	464.4	1136.3	.8623	.4238	1.2861
2250	652.90	.02698	.15692	689.9	360.7	1050.6	701.1	414.8	1115.9	.8876	.3728	1.2604
2500	668.31	.02860	.13059	717.7	313.4	1031.0	730.9	360.5	1091.4	.9131	.3196	1.2327
2750	682.46	.03077	.10717	747.3	258.6	1005.9	763.0	297.4	1060.4	.9401	.2604	1.2005
3000	695.52	.03431	.08404	783.4	185.4	968.8	802.5	213.0	1015.5	.9732	.1843	1.1575
3200	705.27	.04805	.05444	862.1	25.6	887.7	890.6	29.3	919.9	1.0478	.0252	1.0730
3203.6	705.44	.05053	.05053	872.6	0	872.6	902.5	0	902.5	1.0580	0	1.0580

Source: J. H. Keenan, F. G. Keyes, P. G. Hill, and J. G. Moore, *Steam Tables,* John Wiley & Sons, Inc., New York, 1969.

[a] See footnotes to Table 12.29.

Table 12.31 Properties of Superheated Steam[a]

Absolute Pressure (lb/in²) Sat. Temp. (°F)		Temperature (°F)										
		200	300	400	500	600	700	800	900	1000	1100	1200
1 (101.70)	v	392.5	452.3	511.9	571.5	631.1	690.7	750.3	809.9	869.5	929.0	988.6
	h	1150.1	1195.7	1241.8	1288.5	1336.1	1384.5	1433.7	1483.8	1534.8	1586.8	1639.6
	s	2.0508	2.1150	2.1720	2.2235	2.2706	2.3142	2.3550	2.3932	2.4294	2.4638	2.4967
5 (162.21)	v	78.15	90.24	102.24	114.20	126.15	138.08	150.01	161.94	173.86	185.78	197.70
	h	1148.6	1194.8	1241.2	1288.2	1335.8	1384.3	1433.5	1483.7	1534.7	1586.7	1639.5
	s	1.8715	1.9367	1.9941	2.0458	2.0930	2.1367	2.1775	2.2158	2.2520	2.2864	2.3192
10 (193.19)	v	38.85	44.99	51.03	57.04	63.03	69.01	74.98	80.95	86.91	92.88	98.84
	h	1146.6	1193.7	1240.5	1287.7	1335.5	1384.0	1433.3	1483.5	1534.6	1586.6	1639.4
	s	1.7927	1.8592	1.9171	1.9690	2.0164	2.0601	2.1009	2.1393	2.1755	2.2099	2.2428
14.696 (211.99)	v		30.52	34.67	38.77	42.86	46.93	51.00	55.07	59.13	63.19	67.25
	h		1192.6	1239.9	1287.3	1335.2	1383.8	1433.1	1483.4	1534.5	1586.4	1639.3
	s		1.8157	1.8741	1.9263	1.9737	2.0175	2.0584	2.0967	2.1330	2.1674	2.2003
20 (227.96)	v		22.36	25.43	28.46	31.47	34.47	37.46	40.45	43.44	46.42	49.41
	h		1191.5	1239.2	1286.8	1334.8	1383.5	1432.9	1483.2	1534.3	1586.3	1639.2
	s		1.7805	1.8395	1.8919	1.9395	1.9834	2.0243	2.0627	2.0989	2.1334	2.1663
40 (267.26)	v		11.038	12.623	14.164	15.685	17.196	18.701	20.202	21.700	23.20	24.69
	h		1186.8	1236.4	1284.9	1333.4	1382.4	1432.1	1482.5	1533.8	1585.9	1638.9
	s		1.6993	1.7606	1.8140	1.8621	1.9063	1.9474	1.9859	2.0223	2.0568	2.0897
60 (292.73)	v		7.260	8.353	9.399	10.425	11.440	12.448	13.452	14.454	15.454	16.452
	h		1181.9	1233.5	1283.0	1332.1	1381.4	1431.2	1481.8	1533.2	1585.4	1638.5
	s		1.6496	1.7134	1.7678	1.8165	1.8609	1.9022	1.9408	1.9773	2.0119	2.0448
80 (312.07)	v			6.217	7.017	7.794	8.561	9.321	10.078	10.831	11.583	12.333
	h			1230.6	1281.1	1330.7	1380.3	1430.4	1481.2	1532.6	1584.9	1638.1
	s			1.6790	1.7346	1.7838	1.8285	1.8700	1.9087	1.9453	1.9799	2.0130

		4.934	5.587	6.216	6.834	7.445	8.053	8.657	9.260	9.861
100	v	4.934	5.587	6.216	6.834	7.445	8.053	8.657	9.260	9.861
(327.86)	h	1227.5	1279.1	1329.3	1379.2	1429.6	1480.5	1532.1	1584.5	1637.7
	s	1.6517	1.7085	1.7582	1.8033	1.8449	1.8838	1.9204	1.9551	1.9882
120	v	4.079	4.633	5.164	5.682	6.195	6.703	7.208	7.711	8.213
(341.30)	h	1224.4	1277.1	1327.8	1378.2	1428.7	1479.8	1531.5	1584.0	1637.3
	s	1.6288	1.6868	1.7371	1.7825	1.8243	1.8633	1.9000	1.9348	1.9679
140	v	3.466	3.952	4.412	4.860	5.301	5.739	6.173	6.605	7.036
(353.08)	h	1221.2	1275.1	1326.4	1377.1	1427.9	1479.1	1531.0	1583.6	1636.9
	s	1.6088	1.6682	1.7191	1.7648	1.8068	1.8459	1.8827	1.9176	1.9507
160	v	3.007	3.440	3.848	4.243	4.631	5.015	5.397	5.776	6.154
(363.60)	h	1217.8	1273.0	1325.0	1376.0	1427.0	1478.4	1530.4	1583.1	1636.5
	s	1.5911	1.6518	1.7034	1.7494	1.7916	1.8308	1.8677	1.9026	1.9358
180	v	2.648	3.042	3.409	3.763	4.110	4.453	4.793	5.131	5.467
(373.13)	h	1214.4	1270.9	1323.5	1374.9	1426.2	1477.7	1529.8	1582.6	1636.1
	s	1.5749	1.6372	1.6893	1.7357	1.7781	1.8175	1.8545	1.8894	1.9227
200	v	2.361	2.724	3.058	3.379	3.693	4.003	4.310	4.615	4.918
(381.86)	h	1210.8	1268.8	1322.1	1373.8	1425.3	1477.1	1529.3	1582.2	1635.7
	s	1.5600	1.6239	1.6767	1.7234	1.7660	1.8055	1.8425	1.8776	1.9109
220	v	2.125	2.463	2.771	3.065	3.352	3.635	3.914	4.192	4.468
(389.94)	h	1207.2	1266.6	1320.6	1372.7	1424.5	1476.4	1528.7	1581.7	1635.4
	s	1.5461	1.6116	1.6651	1.7122	1.7549	1.7946	1.8318	1.8668	1.9002
240	v	1.9283	2.246	2.531	2.803	3.068	3.328	3.585	3.840	4.094
(397.46)	h	1203.3	1264.4	1319.1	1371.6	1423.6	1475.7	1528.2	1581.2	1635.0
	s	1.5329	1.6002	1.6545	1.7018	1.7448	1.7846	1.8219	1.8570	1.8904

Table 12.31 (Continued)

Absolute Pressure (lb/in²) Sat. Temp. (°F)		200	300	400	Temperature (°F) 500	600	700	800	900	1000	1100	1200
260 (404.51)	v				2.061	2.329	2.582	2.827	3.068	3.306	3.542	3.777
	h				1262.1	1317.6	1370.5	1422.7	1475.0	1527.6	1580.8	1634.6
	s				1.5896	1.6446	1.6923	1.7355	1.7754	1.8128	1.8480	1.8814
280 (411.15)	v				1.9033	2.155	2.392	2.621	2.846	3.067	3.287	3.505
	h				1259.9	1316.0	1369.4	1421.9	1474.3	1527.0	1580.3	1634.2
	s				1.5796	1.6353	1.6834	1.7268	1.7669	1.8043	1.8396	1.8731
300 (417.43)	v				1.7662	2.004	2.227	2.442	2.653	2.860	3.066	3.270
	h				1257.5	1314.5	1368.3	1421.0	1473.6	1526.5	1579.8	1633.8
	s				1.5701	1.6266	1.6751	1.7187	1.7589	1.7964	1.8317	1.8653
350 (431.82)	v				1.4913	1.7025	1.8975	2.085	2.267	2.446	2.624	2.799
	h				1251.5	1310.6	1365.4	1418.8	1471.8	1525.0	1578.6	1632.8
	s				1.5482	1.6068	1.6562	1.7004	1.7409	1.7787	1.8142	1.8478
400 (444.70)	v				1.2843	1.4760	1.6503	1.8163	1.9776	2.136	2.292	2.446
	h				1245.2	1306.6	1362.5	1416.6	1470.1	1523.6	1577.4	1631.8
	s				1.5282	1.5892	1.6397	1.6844	1.7252	1.7632	1.7989	1.8327
450 (456.39)	v				1.1226	1.2996	1.4580	1.6077	1.7524	1.8941	2.034	2.172
	h				1238.5	1302.5	1359.6	1414.4	1468.3	1522.2	1576.3	1630.8
	s				1.5097	1.5732	1.6248	1.6701	1.7113	1.7495	1.7853	1.8192
500 (467.13)	v				0.9924	1.1583	1.3040	1.4407	1.5723	1.7008	1.8271	1.9518
	h				1231.5	1298.3	1356.7	1412.1	1466.5	1520.7	1575.1	1629.8
	s				1.4923	1.5585	1.6112	1.6571	1.6987	1.7371	1.7731	1.8072
550 (477.07)	v				0.8850	1.0424	1.1779	1.3040	1.4249	1.5426	1.6581	1.7720
	h				1224.1	1293.9	1353.7	1409.8	1464.7	1519.3	1573.9	1628.8
	s				1.4755	1.5448	1.5987	1.6452	1.6872	1.7259	1.7620	1.7962
600 (486.33)	v				0.7947	0.9456	1.0727	1.1900	1.3021	1.4108	1.5173	1.6222
	h				1216.2	1289.5	1350.6	1407.6	1462.9	1517.8	1572.7	1627.8
	s				1.4592	1.5320	1.5872	1.6343	1.6766	1.7155	1.7519	1.7861
700 (503.23)	v					0.7929	0.9073	1.0109	1.1089	1.2036	1.2960	1.3868
	h					1280.2	1344.4	1402.9	1459.3	1514.9	1570.2	1625.8
	s					1.5081	1.5661	1.6145	1.6576	1.6970	1.7337	1.7682

800 (518.36)	v	0.6776	0.7829	0.8764	0.9640	1.0482	1.1300	1.2102
	h	1270.4	1338.0	1398.2	1455.6	1511.9	1567.8	1623.8
	s	1.4861	1.5471	1.5969	1.6408	1.6807	1.7178	1.7526
900 (532.12)	v	0.5871	0.6859	0.7717	0.8513	0.9273	1.0009	1.0729
	h	1260.0	1331.4	1393.4	1451.9	1508.9	1565.4	1621.7
	s	1.4652	1.5297	1.5810	1.6257	1.6662	1.7036	1.7386
1000 (544.75)	v	0.5140	0.6080	0.6878	0.7610	0.8305	0.8976	0.9630
	h	1248.8	1324.6	1388.5	1448.1	1505.9	1562.9	1619.7
	s	1.4450	1.5135	1.5664	1.6120	1.6530	1.6908	1.7261
1100 (556.45)	v	0.4532	0.5441	0.6190	0.6871	0.7513	0.8131	0.8731
	h	1236.7	1317.5	1383.5	1444.3	1502.8	1560.4	1617.6
	s	1.4252	1.4982	1.5529	1.5993	1.6409	1.6790	1.7146
1200 (567.37)	v	0.4017	0.4906	0.5617	0.6255	0.6583	0.7426	0.7982
	h	1223.6	1310.2	1378.4	1440.4	1499.7	1557.9	1615.5
	s	1.4054	1.4837	1.5402	1.5876	1.6297	1.6682	1.7040
1400 (587.25)	v	0.3175	0.4059	0.4713	0.5285	0.5815	0.6319	0.6805
	h	1193.1	1294.8	1367.9	1432.5	1493.5	1552.8	1611.4
	s	1.3642	1.4562	1.5168	1.5661	1.6094	1.6487	1.6851
1600 (605.06)	v		0.3415	0.4032	0.4557	0.5036	0.5488	0.5921
	h		1278.1	1357.0	1424.4	1487.1	1547.7	1607.1
	s		1.4299	1.4953	1.5468	1.5913	1.6315	1.6684
1800 (621.21)	v		0.2905	0.3500	0.3989	0.4430	0.4842	0.5235
	h		1259.9	1345.7	1416.1	1480.7	1542.5	1602.9
	s		1.4042	1.4753	1.5291	1.5749	1.6159	1.6534

Table 12.31 (*Continued*)

Absolute Pressure (lb/in²) Sat. Temp. (°F)		Temperature (°F)										
		200	300	400	500	600	700	800	900	1000	1100	1200
2000 (636.00)	v						0.2487	0.3071	0.3534	0.3945	0.4325	0.4685
	h						1239.8	1333.8	1407.6	1474.1	1537.2	1598.6
	s						1.3782	1.4562	1.5126	1.5598	1.6017	1.6398
2500 (668.31)	v						0.16839	0.2291	0.2712	0.3069	0.3393	0.3696
	h						1176.6	1301.7	1385.4	1457.2	1523.8	1587.7
	s						1.3073	1.4112	1.4752	1.5262	1.5704	1.6101
3000 (695.52)	v						0.09771	0.17572	0.2160	0.2485	0.2772	0.3036
	h						1058.1	1265.2	1361.7	1439.6	1510.1	1576.6
	s						1.1944	1.3675	1.4414	1.4967	1.5434	1.5848
3206.2[b] (705.34)	v							0.1591	0.1980	0.2290	0.2564	0.2816
	h							1250.6	1353.4	1434.8	1507.3	1575.1
	s							1.3506	1.4293	1.4872	1.5351	1.5773

Source: J. H. Keenan, F. G. Keyes, P. G. Hill, and J. G. Moore, *Steam Tables*, John Wiley & Sons, Inc., New York, 1969.

[a] See footnotes to Table 12.29.

Table 12.32   Metric Prefixes and Symbols

Multiplication Factor	Prefix	Symbol
1 000 000 000 000 000 000 = 10^{18}	exa	E
1 000 000 000 000 000 = 10^{15}	peta	P
1 000 000 000 000 = 10^{12}	tera	T
1 000 000 000 = 10^{9}	giga	G
1 000 000 = 10^{6}	mega	M
1 000 = 10^{3}	kilo	k
100 = 10^{2}	hecto*	h
10 = 10^{1}	deka*	da
0.1 = 10^{-1}	deci *	d
0.01 = 10^{-2}	centi*	c
0.001 = 10^{-3}	milli	m
0.000 001 = 10^{-6}	micro	µ
0.000 000 001 = 10^{-9}	nano	n
0.000 000 000 001 = 10^{-12}	pico	p
0.000 000 000 000 001 = 10^{-15}	femto	f
0.000 000 000 000 000 001 = 10^{-18}	atto	a

Source: B. D. Tapley, *Handbook of Engineering Fundamentals,* 4th ed. Copyright © 1990 by John Wiley & Sons, Inc. Reprinted with permission.

* Avoid use if possible.

Table 12.33 Length $[L]^a$

Multiply Number of → by ... to Obtain →	Centimeters	Feet	Inches	Kilometers	Nautical Miles	Meters	Mils	Miles	Millimeters	Yards
Centimeters	1	30.48	2.540	10^5	1.853×10^5	100	2.540×10^{-3}	1.609×10^5	0.1	91.44
Feet	3.281×10^{-2}	1	8.333×10^{-2}	3281	6080.27	3.281	8.333×10^{-5}	5280	3.281×10^{-3}	3
Inches	0.3937	12	1	3.937×10^4	7.296×10^4	39.37	0.001	6.336×10^4	3.937×10^{-2}	36
Kilometers	10^{-5}	3.048×10^{-4}	2.540×10^{-5}	1	1.853	0.001	2.540×10^{-8}	1.609	10^{-6}	9.144×10^{-4}
Nautical Miles		1.645×10^{-4}		0.5396	1	5.396×10^{-4}		0.8684		4.934×10^{-4}
Meters	0.01	0.3048	2.540×10^{-2}	1000	1853	1		1609	0.001	0.9144
Mils	393.7	1.2×10^4	1000	3.937×10^7		3.937×10^4	1		39.37	3.6×10^4
Miles	6.214×10^{-6}	1.894×10^{-4}	1.578×10^{-5}	0.6214	1.1516	6.214×10^{-4}		1	6.214×10^{-7}	5.682×10^{-4}
Millimeters	10	304.8	25.40	10^6		1000	2.540×10^{-2}		1	914.4
Yards	1.094×10^{-2}	0.3333	2.778×10^{-2}	1094	2027	1.094	2.778×10^{-5}	1760	1.094×10^{-3}	1

Source: B. D. Tapley, *Handbook of Engineering Fundamentals*, 4th ed. Copyright © 1990 by John Wiley & Sons, Inc. Reprinted with permission.

[a] In Tables 12.33 to 12.51, multiply a unit across the top of the table by the numerical value to obtain a unit on the left.

Table 12.34 Area $[L^2]^a$

Multiply Number of → by → to Obtain ↓

to Obtain ↓ \ of →	Acres	Circular Mils	Square Centimeters	Square Feet	Square Inches	Square Kilometers	Square Meters	Square Miles	Square Millimeters	Square Yards
Acres	1			2.296×10^{-5}		247.1	2.471×10^{-4}	640		2.066×10^{-4}
Circular Mils		1	1.973×10^5	1.833×10^8	1.273×10^6		1.973×10^9		1973	
Square Centimeters		5.067×10^{-6}	1	929.0	6.452	10^{10}	10^4	2.590×10^{10}	0.01	8361
Square Feet	4.356×10^4		1.076×10^{-3}	1	6.944×10^{-3}	1.076×10^7	10.76	2.788×10^7	1.076×10^{-5}	9
Square Inches	6,272,640	7.854×10^{-7}	0.1550	144	1	1.550×10^9	1550	4.015×10^9	1.550×10^{-3}	1296
Square Kilometers	4.047×10^{-3}		10^{-10}	9.290×10^{-8}	6.452×10^{-10}	1	10^{-6}	2.590	10^{-12}	8.361×10^{-7}
Square Meters	4047		0.0001	9.290×10^{-2}	6.452×10^{-4}	10^6	1	2.590×10^6	10^{-6}	0.8361
Square Miles	1.562×10^{-3}		3.861×10^{-11}	3.587×10^{-8}		0.3861	3.861×10^{-7}	1	3.861×10^{-13}	3.228×10^{-7}
Square Millimeters		5.067×10^{-4}	100	9.290×10^4	645.2	10^{12}	10^6	2.590×10^{12}	1	8.361×10^5
Square Yards	4840		1.196×10^{-4}	0.1111	7.716×10^{-4}	1.196×10^6	1.196	3.098×10^6	1.196×10^{-6}	1

Source: B. D. Tapley, *Handbook of Engineering Fundamentals*, 4th ed. Copyright © 1990 by John Wiley & Sons, Inc. Reprinted with permission.

[a] See the footnotes to Table 12.33.

Table 12.35   Volume $[L^3]^a$

to Obtain ↓ \ Multiply Number of → by →	Bushels (Dry)	Cubic Centimeters	Cubic Feet	Cubic Inches	*Cubic Meters*	Cubic Yards	Gallons (Liquid)	Liters	Pints (Liquid)	Quarts (Liquid)
Bushels (Dry)	1		0.8036	4.651×10^{-4}	28.38			2.838×10^{-2}		
Cubic Centimeters	3.524×10^{4}	1	2.832×10^{4}	16.39	10^{6}	7.646×10^{5}	3785	1000	473.2	946.4
Cubic Feet	1.2445	3.531×10^{-5}	1	5.787×10^{-4}	35.31	27	0.1337	3.531×10^{-2}	1.671×10^{-2}	3.342×10^{-2}
Cubic Inches	2150.4	6.102×10^{-2}	1728	1	6.102×10^{4}	46,656	231	61.02	28.87	57.75
Cubic Meters	3.524×10^{-2}	10^{-6}	2.832×10^{-2}	1.639×10^{-5}	1	0.7646	3.785×10^{-3}	0.001	4.732×10^{-4}	9.464×10^{-4}
Cubic Yards		1.308×10^{-6}	3.704×10^{-2}	2.143×10^{-5}	1.308	1	4.951×10^{-3}	1.308×10^{-3}	6.189×10^{-4}	1.238×10^{-3}
Gallons (Liquid)		2.642×10^{-4}	7.481	4.329×10^{-3}	264.2	202.0	1	0.2642	0.125	0.25
Liters	35.24	0.001	28.32	1.639×10^{-2}	1000	764.6	3.785	1	0.4732	0.9464
Pints (Liquid)		2.113×10^{-3}	59.84	3.463×10^{-2}	2113	1616	8	2.113	1	2
Quarts (Liquid)		1.057×10^{-3}	29.92	1.732×10^{-2}	1057	807.9	4	1.057	0.5	1

Source: B. D. Tapley, *Handbook of Engineering Fundamentals*, 4th ed. Copyright © 1990 by John Wiley & Sons, Inc. Reprinted with permission.

[a] See the footnotes to Table 12.33.

Table 12.36 Linear Velocity $[LT^{-1}]$[a]

to Obtain ↓ \ Multiply Number of → by	Centimeters per Second	Feet per Minute	Feet per Second	Kilometers per Hour	Kilometers per Minute	Knots[a]	Meters per Minute	*Meters per Second*	Miles per Hour	Miles per Minute
Centimeters per Second	1	0.5080	30.48	27.78	1667	51.48	1.667	100	44.70	2682
Feet per Minute	1.969	1	60	54.68	3281	101.3	3.281	196.8	88	5280
Feet per Second	3.281×10^{-2}	1.667×10^{-2}	1	0.9113	54.68	1.689	5.468×10^{-2}	3.281	1.467	88
Kilometers per Hour	0.036	1.829×10^{-2}	1.097	1	60	1.853	0.06	3.6	1.609	96.54
Kilometers per minute	0.0006	3.048×10^{-4}	1.829×10^{-2}	1.667×10^{-2}	1	3.088×10^{-2}	0.001	0.06	2.682×10^{-2}	1.609
Knots[a]	1.943×10^{-2}	9.868×10^{-3}	0.5921	0.5396	32.38	1	3.238×10^{-2}	1.943	0.8684	52.10
Meters per Minute	0.6	0.3048	18.29	16.67	1000	30.88	1	60	26.82	1609
Meters per Second	0.01	5.080×10^{-3}	0.3048	0.2778	16.67	0.5148	1.667×10^{-2}	1	0.4470	26.82
Miles per Hour	2.237×10^{-2}	1.136×10^{-2}	0.6818	0.6214	37.28	1.152	3.728×10^{-2}	2.237	1	60
Miles per Minute	3.728×10^{-4}	1.892×10^{-4}	1.136×10^{-2}	1.036×10^{-2}	0.6214	1.919×10^{-2}	6.214×10^{-4}	3.728×10^{-2}	1.667×10^{-2}	1

Source: B. D. Tapley, Handbook of Engineering Fundamentals, 4th ed. Copyright © 1990 by John Wiley & Sons, Inc. Reprinted with permission.

[a] See the footnotes to Table 12.33.
[b] Nautical miles per hour.

Table 12.37   Mass $[M]^{a,b}$

Multiply Number of → / to by / Obtain ↓	Grains	Grams	*Kilograms*	Milligrams	Ounces[c]	Pounds[c]	Tons (Long)	Tons (Metric)	Tons (Short)
Grains	1	15.43	1.543×10^4	1.543×10^{-2}	437.5	7000			
Grams	6.481×10^{-2}	1	1000	0.001	28.35	453.6	1.016×10^6	$\times 10^6$	9.072×10^5
Kilograms	6.481×10^{-5}	0.001	1	10^{-6}	2.835×10^{-2}	0.4536	1016	1000	907.2
Milligrams	64.81	1000	10^6	1	2.835×10^4	4.536×10^5	1.016×10^9	10^9	9.072×10^8
Ounces[c]	2.286×10^{-3}	3.527×10^{-2}	35.27	3.527×10^{-5}	1	16	3.584×10^4	3.527×10^4	3.2×10^4
Pounds[c]	1.429×10^{-4}	2.205×10^{-3}	2.205	2.205×10^{-6}	6.250×10^{-2}	1	2240	2205	2000
Tons (Long)		9.842×10^{-7}	9.842×10^{-4}	9.842×10^{-10}	2.790×10^{-5}	4.464×10^{-4}	1	0.9842	0.8929
Tons (Metric)		10^{-6}	0.001	10^{-9}	2.835×10^{-5}	4.536×10^{-4}	1.016	1	0.9072
Tons (Short)		1.102×10^{-6}	1.102×10^{-3}	1.102×10^{-9}	3.125×10^{-5}	0.0005	1.120	1.102	1

Source: B. D. Tapley, *Handbook of Engineering Fundamentals,* 4th ed. Copyright © 1990 by John Wiley & Sons, Inc. Reprinted with permission.

[a] See the footnotes to Table 12.33.

[b] These same conversion factors apply to the *gravitational* units of force having the corresponding names. The dimensions of these units when used as gravitational units of force are MLT^{-2}; see Table 12.38.

[c] Avoirdupois pounds and ounces.

Table 12.38   Force $[MLT^{-2}]$ or $[F]^{a,b}$

to Obtain ↓ / Multiply Number of → by	Dynes	Grams	Joules per Centimeter	Newtons or Joules per Meter	Kilograms	Pounds	Poundals
Dynes	1	980.7	10^7	10^5	9.807×10^5	4.448×10^5	1.383×10^4
Grams	1.020×10^{-3}	1	1.020×10^4	102.0	1000	453.6	14.10
Joules per Centimeter	10^{-7}	9.807×10^{-5}	1	.01	9.807×10^{-2}	4.448×10^{-2}	1.383×10^{-3}
Newtons, or Joules per Meter	10^{-5}	9.807×10^{-3}	100	1	9.807	4.448	0.1383
Kilograms	1.020×10^{-6}	0.001	10.20	0.1020	1	0.4536	1.410×10^{-2}
Pounds	2.248×10^{-6}	2.205×10^{-3}	22.48	0.2248	2.205	1	3.108×10^{-2}
Poundals	7.233×10^{-5}	7.093×10^{-2}	723.3	7.233	70.93	32.17	1

Source: B. D. Tapley, *Handbook of Engineering Fundamentals,* 4th ed. Copyright © 1990 by John Wiley & Sons, Inc. Reprinted with permission.

[a] See the footnotes to Table 12.33.
[b] Conversion factors between absolute and gravitational units apply only under standard acceleration due to gravity conditions.

Table 12.39 Pressure or Force per Unit Area $[ML^{-1}T^{-2}]$ or $[FL^{-2}]^a$

to Obtain ↓ \ Multiply Number of → by	Atmospheres[b]	Baryes or Dynes per Square Centimeter	Centimeters of Mercury at 0°C[b]	Inches of Mercury at 0°C[b]	Inches of Water at 4°C	Kilograms per Square Meter[d]	Pounds per Square Foot	Pounds per Square Inch	Tons (Short) per Square Foot	Pascal
Atmospheres[b]	1	9.869×10^{-7}	1.316×10^{-2}	3.342×10^{-2}	2.458×10^{-3}	9.678×10^{-5}	4.725×10^{-4}	6.804×10^{-2}	0.9450	9.869×10^{-6}
Baryes or Dynes per Square Centimeter	1.013×10^6	1	1.333×10^4	3.386×10^4	2.491×10^3	98.07	478.8	6.895×10^4	9.576×10^5	10
Centimeters of Mercury at 0°C[b]	76.00	7.501×10^{-5}	1	2.540	0.1868	7.356×10^{-3}	3.591×10^{-2}	5.171	71.83	7.501×10^{-4}
Inches of Mercury at 0°C[b]	29.92	2.953×10^{-5}	0.3937	1	7.355×10^{-2}	2.896×10^{-3}	1.414×10^{-2}	2.036	28.28	2.953×10^{-4}
Inches of Water at 4°C	406.8	4.015×10^{-4}	5.354	13.60	1	3.937×10^{-2}	0.1922	27.68	384.5	4.015×10^{-3}
Kilograms per Square Meter[d]	1.033×10^4	1.020×10^{-2}	136.0	345.3	25.40	1	4.882	703.1	9765	0.1020
Pounds per Square Foot	2117	2.089×10^{-3}	27.85	70.73	5.204	0.2048	1	144	2000	2.089×10^{-2}
Pounds per Square Inch	14.70	1.450×10^{-5}	0.1934	0.4912	3.613×10^{-2}	1.422×10^{-3}	6.944×10^{-3}	1	13.89	1.450×10^{-4}
Tons (Short) per Square Foot	1.058	1.044×10^{-6}	1.392×10^{-2}	3.536×10^{-2}	2.601×10^{-3}	1.024×10^{-4}	0.0005	0.072	1	1.044×10^{-5}
Pascal	1.013×10^5	10^{-1}	1.333×10^3	3.386×10^3	2.491×10^2	9.807	47.88	6.895×10^3	9.576×10^4	1

Source: B. D. Tapley, *Handbook of Engineering Fundamentals*, 4th ed. Copyright © 1990 by John Wiley & Sons, Inc. Reprinted with permission.

[a]See the footnotes to Table 12.33.

[b]Definition: One atmosphere (standard) = 76 cm Hg at 0°C.

[c]To convert height h of a column of mercury at t degrees Celsius to the equivalent height h_0 at 0°C use $h_0 = h[1 - (m - l)t/(1 + mt)]$, where $m = 0.0001818$ and $l = 18.4 \times 10^{-6}$ if the scale is engraved on brass; $l = 8.5 \times 10^{-6}$ if on glass. This assumes the scale is correct at 0°C; for other cases (any liquid), see *International Critical Tables*, Vol. 1, p. 68.

[d]$1 \text{ g/cm}^2 = 10 \text{ kg/m}^2$.

Table 12.40   Torque or Moment of Force $[ML^2T^{-2}]$ or $[FL]^{a,b}$

to Obtain ↓ / Multiply Number of → by	Dyne-Centimeters	Gram-Centimeters	Kilogram-Meters	Pound-Feet	Newton-Meter
Dyne-Centimeters	1	980.7	9.807×10^7	1.356×10^7	10^7
Gram-Centimeters	1.020×10^{-3}	1	10^5	1.383×10^4	1.020×10^4
Kilogram-Meters	1.020×10^{-8}	10^{-5}	1	0.1383	0.1020
Pound-Feet	7.376×10^{-8}	7.233×10^{-5}	7.233	1	0.7376
Newton-Meter	10^{-7}	9.807×10^{-4}	9.807	1.356	1

Source: B. D. Tapley, *Handbook of Engineering Fundamentals,* 4th ed. Copyright © 1990 by John Wiley & Sons, Inc. Reprinted with permission.

[a] See the footnotes to Table 12.33.

[b] Same dimensions as energy; more properly, torque should be expressed as newton-meters per radian to avoid this confusion.

Table 12.41 Energy, Work, and Heat $[ML^2T^{-2}]$ or $[FL]^{a,b}$

to Obtain ↓ \ Multiply Number of → by	British Thermal Units[c]	Centimeter-Grams	Ergs or Centimeter-Dynes	Foot-Pounds	Horsepower-Hours	Joules[d] or Watt-Seconds	Kilogram-Calories[c]	Kilowatt-Hours	Meter-Kilograms	Watt-Hours
British Thermal Units[c]	1	9.297×10^{-8}	9.480×10^{-11}	1.285×10^{-3}	2545	9.480×10^{-4}	3.969	3413	9.297×10^{-3}	3.413
Centimeter-Grams	1.076×10^{7}	1	1.020×10^{-3}	1.383×10^{4}	2.737×10^{10}	1.020×10^{4}	4.269×10^{7}	3.671×10^{10}	10^{5}	3.671×10^{7}
Ergs or Centimeter-Dynes	1.055×10^{10}	980.7	1	1.356×10^{7}	2.684×10^{12}	10^{7}	4.186×10^{10}	3.6×10^{13}	9.807×10^{7}	3.6×10^{10}
Foot-Pounds	778.0	7.233×10^{-5}	7.367×10^{-8}	1	1.98×10^{6}	0.7376	3087	2.655×10^{6}	7.233	2655
Horsepower-Hours	3.929×10^{-4}	3.654×10^{-11}	3.722×10^{-14}	5.050×10^{-7}	1	3.722×10^{-7}	1.559×10^{-3}	1.341	3.653×10^{-6}	1.341×10^{-3}
Joules[d] or Watt-Seconds	1054.8	9.807×10^{-5}	10^{-7}	1.356	2.684×10^{6}	1	4186	3.6×10^{6}	9.807	3600
Kilogram-Calories[c]	0.2520	2.343×10^{-8}	2.389×10^{-11}	3.239×10^{-4}	641.3	2.389×10^{-4}	1	860.0	2.343×10^{-3}	0.8600
Kilowatt-Hours	2.930×10^{-4}	2.724×10^{-11}	2.778×10^{-14}	3.766×10^{-7}	0.7457	2.778×10^{-7}	1.163×10^{-3}	1	2.724×10^{-6}	0.001
Meter-Kilograms	107.6	10^{-5}	1.020×10^{-8}	0.1383	2.737×10^{5}	0.1020	426.9	3.671×10^{5}	1	367.1
Watt-Hours	0.2930	2.724×10^{-8}	2.778×10^{-11}	3.766×10^{-4}	745.7	2.778×10^{-4}	1.163	1000	2.724×10^{-3}	1

Source: B. D. Tapley, Handbook of Engineering Fundamentals, 4th ed. Copyright © 1990 by John Wiley & Sons, Inc. Reprinted with permission.

[a] See the footnotes to Table 12.33.
[b] See the note at the bottom of Table 12.42.
[c] Mean calorie and Btu used throughout. One gram-calorie = 0.001 kilogram-calorie; 1 Ostwald calorie = 0.1 kilogram-calorie. The IT cal, 1000 international steam table calories, has been defined as the 1/860 part of the international kilowatthour (see Mechanical Engineering, Nov. 1935, p. 710). Its value is very nearly equal to the mean kilogram-calorie, 1 IT cal–1.00037 kilogram-calories (mean). 1 Btu = 251.996 IT cal.
[d] Absolute joule, defined as 10^{7} ergs. The international joule, based on the international ohm and ampere, equals 1.0003 absolute joules.

Table 12.42 Power or Rate of Doing Work $[ML^2T^{-3}]$ or $[FLT^{-1}]^{a,b}$

Multiply Number of → / by → / to Obtain ↓	British Thermal Units per Minute	Ergs per Second	Foot-Pounds per Minute	Foot-Pounds per Second	Horsepower[b]	Kilogram-Calories per Minute	Kilowatts	Metric Horsepower	Watts
British Thermal Units per Minute	1	5.689×10^{-9}	1.285×10^{-3}	7.712×10^{-2}	42.41	3.969	56.89	41.83	5.689×10^{-2}
Ergs per Second	1.758×10^{8}	1	2.259×10^{5}	1.356×10^{7}	7.457×10^{9}	6.977×10^{8}	10^{10}	7.355×10^{9}	10^{7}
Foot Pounds per Minute	778.0	4.426×10^{-6}	1	60	3.3×10^{4}	3087	4.426×10^{4}	3.255×10^{4}	44.26
Foot-Pounds per Second	12.97	7.376×10^{-8}	1.667×10^{-2}	1	550	51.44	737.6	542.5	0.7376
Horsepower[b]	2.357×10^{-2}	1.341×10^{-10}	3.030×10^{-5}	1.818×10^{-3}	1	9.355×10^{-2}	1.341	0.9863	1.341×10^{-3}
Kilogram-Calories per Minute	0.2520	1.433×10^{-9}	3.239×10^{-4}	1.943×10^{-2}	10.69	1	14.33	10.54	1.433×10^{-2}
Kilowatts	1.758×10^{-2}	10^{-10}	2.260×10^{-5}	1.356×10^{-3}	0.7457	6.977×10^{-2}	1	0.7355	10^{-3}
Metric Horsepower	2.390×10^{-2}	1.360×10^{-10}	3.072×10^{-5}	1.843×10^{-3}	1.014	9.485×10^{-2}	1.360	1	1.360×10^{-3}
Watts	17.58	10^{-7}	2.260×10^{-2}	1.356	745.7	69.77	1000	735.5	1

Source: B. D. Tapley, Handbook of Engineering Fundamentals, 4th ed. Copyright © 1990 by John Wiley & Sons, Inc. Reprinted with permission.

[a] See the footnotes to Table 12.33.

[b] The "horsepower" used in these tables is equal to 550 foot-pounds per second by definition. Other definitions are 1 horsepower equals 746 watts (U.S. and Great Britain) and 1 horsepower equals 736 watts (continental Europe). Neither of the latter definitions is equivalent to the first; the "horsepowers" defined in these latter definitions are widely used in the rating of electrical machinery. 1 Cheval-vapeur = 75 kilogram-meters per second; 1 Poncelet = 100 kilogram-meters per second.

Table 12.43 Quantity of Electricity and Dielectric Flux [Q][a]

to Obtain ↓ / Multiply Number of → by ↘	Abcoulombs	Ampere-Hours	*Coulombs*	Faradays	Stat-Coulombs
Abcoulombs	1	360	0.1	9649	3.335×10^{-11}
Ampere-Hours	2.778×10^{-3}	1	2.778×10^{-4}	26.80	9.259×10^{-14}
Coulombs	10	3600	1	9.649×10^{4}	3.335×10^{-10}
Faradays	1.036×10^{-4}	3.731×10^{-2}	1.036×10^{-5}	1	3.457×10^{-15}
Statcoulombs	2.998×10^{10}	1.080×10^{13}	2.998×10^{9}	2.893×10^{14}	1

Source: B. D. Tapley, *Handbook of Engineering Fundamentals,* 4th ed. Copyright © 1990 by John Wiley & Sons, Inc. Reprinted with permission.

[a] See the footnotes to Table 12.33.

Table 12.44 Charge per Unit Area and Electric Flux Density [QL^{-2}][a]

to Obtain ↓ / Multiply Number of → by ↘	Abcoulombs per Square Centimeter	Coulombs per Square Centimeter	Coulombs per Square Inch	Statcoulombs per Square Centimeter	*Coulombs per Square Meter*
Abcoulombs per Square Centimeter	1	0.1	1.550×10^{-2}	3.335×10^{-11}	10^{-5}
Coulombs per Square Centimeter	10	1	0.1550	3.335×10^{-10}	10^{-4}
Coulombs per Square Inch	64.52	6.452	1	2.151×10^{-9}	6.452×10^{-4}
Statcoulombs per Square Centimeter	2.998×10^{10}	2.998×10^{9}	4.647×10^{8}	1	2.998×10^{5}
Coulombs per Square Meter	10^{5}	10^{4}	1550	3.335×10^{-6}	1

Source: B. D. Tapley, *Handbook of Engineering Fundamentals,* 4th ed. Copyright © 1990 by John Wiley & Sons, Inc. Reprinted with permission.

[a] See the footnotes to Table 12.33.

Table 12.45 Current Density $[QT^{-1}L^{-2}]^a$

to Obtain ↓ / Multiply Number of → by ↘	Abamperes per Square Centimeter	Amperes per Square Centimeter	Amperes per Square Inch	Statamperes per Square Centimeter	*Amperes per Square Meter*
Abamperes per Square Centimeter	1	0.1	1.550×10^{-2}	3.335×10^{-11}	10^{-5}
Amperes per Square Centimeter	10	1	0.1550	3.335×10^{-10}	10^{-4}
Amperes per Square Inch	64.52	6.452	1	2.151×10^{-9}	6.452×10^{-4}
Statamperes per Square Centimeter	2.998×10^{10}	2.998×10^{9}	4.647×10^{x}	1	2.998×10^{5}
Amperes per Square Meter	10^{5}	10^{4}	1550	3.335×10^{-6}	1

Source: B. D. Tapley, *Handbook of Engineering Fundamentals,* 4th ed. Copyright © 1990 by John Wiley & Sons, Inc. Reprinted with permission.

[a] See the footnotes to Table 12.33.

Table 12.46 Electric Field Intensity and Potential Gradient $[MQ^{-1}LT^{-2}]$ or $[FQ^{-1}]$[a]

Multiply Number of → by → to Obtain ↓	Abvolts per Centimeter	Microvolts per Meter	Millivolts per Meter	Statvolts per Centimeter	Volts per Centimeter	Kilovolts per Centimeter	Volts per Inch	Volts per Mil	Volts per Meter
Abvolts per Centimeter	1	1	1000	2.998×10^{10}	10^8	10^{11}	3.937×10^7	3.937×10^{10}	10^6
Microvolts per Meter	1	1	1000	2.998×10^{10}	10^8	10^{11}	3.937×10^7	3.937×10^{10}	10^6
Millivolts per Meter	0.001	0.001	1	2.998×10^7	10^5	10^8	3.937×10^4	3.937×10^7	1000
Statvolts per Centimeter	3.335×10^{-11}	3.335×10^{-11}	3.335×10^{-8}	1	3.335×10^{-3}	3.335	1.313×10^{-3}	1.313	3.335×10^{-5}
Volts per Centimeter	10^{-8}	10^{-8}	10^{-5}	299.8	1	1000	0.3937	393.7	10^{-2}
Kilovolts per Centimeter	10^{-11}	10^{-11}	10^{-8}	0.2998	0.001	1	3.937×10^{-4}	0.3937	10^{-5}
Volts per Inch	2.540×10^{-8}	2.540×10^{-8}	2.540×10^{-5}	761.6	2.540	2540	1	1000	2.540×10^{-2}
Volts per Mil	2.540×10^{-11}	2.540×10^{-11}	2.540×10^{-8}	0.7616	2.540×10^{-3}	2.540	0.001	1	2.540×10^{-5}
Volts per Meter	10^{-6}	10^{-6}	10^{-3}	2.998×10^4	100	10^5	39.37	3.937×10^4	1

Source: B. D. Tapley, Handbook of Engineering Fundamentals, 4th ed. Copyright © 1990 by John Wiley & Sons, Inc. Reprinted with permission.

[a] See the footnotes to Table 12.33.

Table 12.47   Capacitance $[M^{-1}Q^2L^{-2}T^2]$ **or** $[F^{-1}Q^2L^{-1}]^a$

to Obtain ↓ / Multiply Number of → by →	Abfarads	*Farads*	Microfarads	Statfarads
Abfarads	1	10^{-9}	10^{-15}	1.112×10^{-21}
Farads	10^9	1	10^{-6}	1.112×10^{-12}
Microfarads	10^{15}	10^6	1	1.112×10^{-6}
Statfarads	8.988×10^{20}	8.988×10^{11}	8.988×10^5	1

Source: B. D. Tapley, *Handbook of Engineering Fundamentals,* 4th ed. Copyright © 1990 by John Wiley & Sons, Inc. Reprinted with permission.

[a] See the footnotes to Table 12.33.

Table 12.48 Inductance $[MQ^{-2}L^2]$ or $[FQ^{-2}LT^2]^a$

Multiply Number of → / to Obtain ↓	Abhenries[b]	*Henries*	Microhenries	Millihenries	Stathenries
Abhenries[b]	1	10^9	1000	10^6	8.988×10^{20}
Henries	10^{-9}	1	10^{-6}	0.001	8.988×10^{11}
Microhenries	0.001	10^6	1	1000	8.988×10^{17}
Millihenries	10^{-6}	1000	0.001	1	8.988×10^{14}
Stathenries	1.112×10^{-21}	1.112×10^{-12}	1.112×10^{-18}	1.112×10^{-15}	1

Source: B. D. Tapley, *Handbook of Engineering Fundamentals,* 4th ed. Copyright © 1990 by John Wiley & Sons, Inc. Reprinted with permission.

[a] See the footnotes to Table 12.33.
[b] An abhenry is sometimes called a "centimeter."

Table 12.49 Magnetic Field Intensity, Potential Gradient, and Magnetizing Force $[QL^{-1}T^{-1}]^a$

Multiply Number of → / to Obtain ↓	Abampere-Turns per Centimeter	Ampere-Turns per Centimeter	Ampere-Turns per Inch	Oersteds (Gilberts per Centimeter)	*Ampere-Turns per Meter*
Abampere-Turns per Centimeter	1	0.1	3.937×10^{-2}	7.958×10^{-2}	10^{-3}
Ampere-Turns per Centimeter	10	1	0.3937	0.7958	10^{-2}
Ampere-Turns per Inch	25.40	2.540	1	2.021	2.54×10^{-2}
Oersteds (Gilberts per Centimeter)	12.57	1.257	0.4950	1	1.257×10^{-2}
Ampere-Turns per Meter	10^3	10^2	39.37	79.58	1

Source: B. D. Tapley, *Handbook of Engineering Fundamentals,* 4th ed. Copyright © 1990 by John Wiley & Sons, Inc. Reprinted with permission.

[a] See the footnotes to Table 12.33.

Table 12.50 Photometric Units

	Common Unit	Multiply by	to Get *SI Unit*
Luminous intensity	International candle	9.81×10^{-1}	**Candela**
Luminance	Candela/in^2	1.550×10^3	**Candela / m^2**
	Candela/cm^2	1×10^4	**Candela / m^2**
	Foot · lambert	3.4263	**Candela / m^2**
Luminous flux	Candela · steradian	1.0000	**Lumen**
	Candle power (spher.)	12.566	**Lumen**
Quantity of light flux			**Lumen · sec**
Luminous exitance[a]			**Lumens / m^2**
Illuminance[b]	Lambert	3.103×10^3	**Candela / m^2**
	Foot candles	1.0764×10	**Lumens / m^2**
	Lumens per ft^2	1.0764×10	**Lumens / m^2**
	Lux	1.000	**Lumens / m^2**
	Phots	1×10^4	**Lumens / m^2**
Luminous efficacy			**Lumens / watt**

Source: B. D. Tapley, *Handbook of Engineering Fundamentals,* 4th ed. Copyright © 1990 by John Wiley & Sons, Inc. Reprinted with permission.

[a]Luminous emittance.
[b]Luminous flux density.

Table 12.51 Thermal Conductivity $LMT^{-3}t^{-1}$ [a,b]

From → To ↓ (by)	Btu·ft per h·ft²·°F	Btu·in per h·ft²·°F	Btu·in per sec·ft²·°F	Joules per m·s·°C	kcal per m·h·°C	erg per cm·s·°C	kcal per m·s·°C	cal per cm·s·°C	W per ft·°C	W per m·K
Btu·ft per h·ft²·°F	1	8.333×10^{-2}	3.0×10^{2}	5.778×10^{-1}	6.720×10^{-1}	5.778×10^{-6}	2.419×10^{3}	2.419×10^{2}	1.895	5.778×10^{-1}
Btu·in per h·ft²·°F	12	1	3.6×10^{3}	6.933	8.064	6.933×10^{-5}	2.903×10^{4}	2.903×10^{3}	2.275×10^{1}	6.933
Btu·in per s·ft²·°F	3.333×10^{-3}	2.778×10^{-4}	1	1.926×10^{-3}	2.240×10^{-3}	1.926×10^{-8}	8.064	8.064×10^{-1}	6.319×10^{-3}	1.926×10^{-3}
Joules per m·s·°C	1.731	1.442×10^{-1}	5.192×10^{2}	1	1.163	1.000×10^{-5}	4.187×10^{3}	4.187×10^{2}	3.281	1.0
kcal per m·h·°C	1.483	1.240×10^{-1}	4.465×10^{7}	8.599×10^{-1}	1	8.599×10^{-6}	3.6×10^{3}	3.6×10^{2}	2.821	8.599×10^{-1}
erg per cm·s·°C	1.731×10^{5}	1.442×10^{4}	5.192×10^{7}	1.0×10^{5}	1.163×10^{5}	1	4.187×10^{8}	4.187×10^{7}	3.281×10^{5}	1.0×10^{5}
kcal per m·s·°C	4.134×10^{-4}	3.445×10^{-5}	1.240×10^{-1}	2.388×10^{-4}	2.778×10^{-4}	2.388×10^{-9}	1	1.0×10^{-1}	7.835×10^{-4}	2.388×10^{-4}
cal per cm·s·°C	4.134×10^{-3}	3.445×10^{-4}	1.240	2.388×10^{-3}	2.778×10^{-3}	2.388×10^{-8}	10	1	7.835×10^{-3}	2.388×10^{-3}
W per ft·°C	5.276×10^{-1}	4.395×10^{-2}	1.582×10^{2}	3.048×10^{-1}	3.545×10^{-1}	3.048×10^{-6}	1.276×10^{3}	1.276×10^{2}	1	3.048×10^{-1}
W per m·K	1.731×10^{-1}	1.442×10^{-1}	5.192×10^{2}	1.0	1.163	1.00×10^{-5}	4.187×10^{3}	4.187×10^{2}	3.281	1

Source: B. D. Tapley, Handbook of Engineering Fundamentals, 4th ed. Copyright © 1990 by John Wiley & Sons, Inc. Reprinted with permission.

[a] See the footnotes to Table 12.33.
[b] International table Btu = 1.055056×10^{3} joules; and international table cal = 4.1868 joules are used throughout.

Table 12.52 Impurities in Water

U. S. Systems of Expressing Impurities

1 grain per gallon	= 1 grain calcium carbonate (CaCO₃)	per U.S. gallon of water

1 grain per gallon = 1 grain calcium carbonate ($CaCO_3$) per U.S. gallon of water
1 part per million = 1 part calcium carbonate ($CaCO_3$) per 1,000,000 parts of water
1 part per hundred thousand = 1 part calcium carbonate ($CaCO_3$) per 100,000 parts of water

Foreign Systems of Expressing Impurities

1 English degree (or °Clark) = 1 gram calcium carbonate ($CaCO_3$) per **British** Imperial gal of water
1 French degree = 1 part calcium carbonate ($CaCO_3$) per 100,000 parts of water
1 German degree = 1 part calcium oxide (CaO) per 100,000 parts of water

Conversions

	Parts CaCO₃ per Million (ppm)	Parts CaCO₃ per Hundred Thousand (Pts./100,000)	Grains CaCO₃ per U.S. Gallon (gpg)	English Degrees (° Clark)	French Degrees (° French)	German Degrees (° German)	Milliequivalents per Liter or Equivalents per Million
1 Part per million	1.	0.1	0.0583	0.07	0.1	0.0560	0.020
1 Part per hundred thousand	10.0	1.	0.583	0.7	1.	0.560	0.20
1 Grain per U. S. gallon	17.1	1.71	1.	1.2	1.71	0.958	0.343
1 English or Clark degree	14.3	1.43	0.833	1.	1.43	0.800	0.286
1 French degree	10.	1.	0.583	0.7	1.	0.560	0.20
1 German degree	17.9	1.79	1.04	1.24	1.79	1.	0.357
1 Milliequivalent per liter or 1 Equivalent per million	50.	5.	2.92	3.50		2.80	1.

Source: Reprinted from *The Permutit Water and Waste Treatment Data Book*, © The Permutit Co., Inc.

Table 12.53 Water Analysis

Conversions

	Parts per Million (ppm)	Milligrams per Liter (mg/l)	Grams per Liter (g/l)	Parts per Hundred Thousand (Pts./100,000)	Grains per U.S. Gallon (gr/U.S. gal)	Grains per British Imperial Gallon	Kilograins per Cubic Foot (kgr/ft³)
1 Part per million	1.	1.	0.001	0.1	0.0583	0.07	0.0004
1 Milligram per liter	1.	1.	0.001	0.1	0.0583	0.07	0.0004
1 Gram per liter	1000.	1000.	1.	100.	58.3	70.	0.436
1 Part per hundred thousand	10.	10.	0.01	1.	0.583	0.7	0.00436
1 Grain per U.S. gallon	17.1	17.1	0.017	1.71	1.	1.2	0.0075
1 Grain per British Imperial gallon	14.3	14.3	0.014	1.43	0.833	1.	0.0062
1 Kilograin per cubic foot	2294.	2294.	2.294	229.4	134.	161.	1.

NOTE: In practice, water analysis samples are measured by volume, not by weight, and corrections for variations in specific gravity are practically never made. Therefore, parts per million are assumed to be the same as milligrams per liter and hence the above relationships are, for practical purposes, true.

Equivalents

Water analyses may also be expressed as:

(1) Equivalents per million (epm) $= \dfrac{\text{No. of ppm of substance present}}{\text{Equivalent weight of substance}}$

(2) Milliequivalents per liter (meq/l) $=$ Equivalents per million

(3) Parts per million expressed as $CaCO_3$ $=$ No. of ppm $CaCO_3$ equivalent to No. of ppm of substance present

(4) Fiftieths of equivalents per million (epm/50) $= \dfrac{\text{No. of ppm of substance present} \times 50}{\text{Equivalent weight of substance}}$

NOTES: Numerically (1) and (2) are equal.
Numerically (3) and (4) are equal.

Source: Reprinted from *The Permutit Water and Waste Treatment Data Book*, © The Permutit Co., Inc.

Table 12.54   Sodium Zeolite Reactions[a]

Softening

$$Na_2Z + \left.\begin{matrix} Ca \\ Mg \end{matrix}\right\} \left\{\begin{matrix} (HCO_3)_2 \\ SO_4 \\ Cl_2 \end{matrix}\right. = \left.\begin{matrix} Ca \\ Mg \end{matrix}\right\} Z + \left\{\begin{matrix} 2NaHCO_3 \\ Na_2SO_4 \\ 2NaCl \end{matrix}\right.$$

Sodium Zeolite + $\left.\begin{matrix} \text{Calcium} \\ \text{and/or} \\ \text{Magnesium} \end{matrix}\right\}$ $\left\{\begin{matrix} \text{Bicarbonates,} \\ \text{Sulfates,} \\ \text{and/or Chlorides} \end{matrix}\right.$ = $\left.\begin{matrix} \text{Calcium} \\ \text{and/or} \\ \text{Magnesium} \end{matrix}\right\}$ Zeolite + $\left\{\begin{matrix} \text{Sodium bicarbonate,} \\ \text{Sodium sulfate,} \\ \text{and/or} \\ \text{Sodium chloride} \end{matrix}\right.$

(insoluble)   (soluble)   (insoluble)   (soluble)

Regeneration

$$\left.\begin{matrix} Ca \\ Mg \end{matrix}\right\} Z + 2NaCl = Na_2Z + \left.\begin{matrix} Ca \\ Mg \end{matrix}\right\} Cl_2$$

$\left.\begin{matrix} \text{Calcium} \\ \text{and/or} \\ \text{Magnesium} \end{matrix}\right\}$ Zeolite + Sodium Chloride = Sodium Zeolite + $\left.\begin{matrix} \text{Calcium} \\ \text{and/or} \\ \text{Magnesium} \end{matrix}\right\}$ Chlorides

(insoluble)   (soluble)   (insoluble)   (soluble)

COMPENSATED HARDNESS: The hardness of a water for softening by the zeolite process should be compensated when:
1. The total hardness (T.H.) is over 400 ppm as $CaCO_3$, or
2. The sodium salts (Na) are over 100 ppm as $CaCO_3$.

Calculate compensated hardness as follows:

$$\text{Compensated Hardness (ppm)} = \text{Total Hardness (ppm)} \times \frac{9000}{9000 - \text{Total Cations (ppm)}}.$$

Express compensated hardness as:
1. Next higher tenth of a grain up to 5.0 grains per gallon.
2. Next higher half of a grain from 5.0 to 10.0 grains per gallon.
3. Next higher grain above 10.0 grains per gallon.

SALT CONSUMPTION: The salt consumption with the zeolite water softener ranges between 0.37 and 0.45 lb of salt per 1000 grains of hardness, expressed as calcium carbonate, removed. This range is due to two factors: (1) the composition of the water and (2) the operating exchange value at which the zeolite is to be worked. The lower salt consumptions may be attained with waters that are not excessively hard nor high in sodium salts and where the zeolite is not worked at its maximum capacity.

Source: Reprinted from *The Permutit Water and Waste Treatment Data Book.* © The Permutit Co., Inc.
[a] The symbol Z represents zeolite radical.

Table 12.55   Salt and Brine Equivalents

1 U.S. gallon of saturated salt brine weighs 10 lb and contains 2.48 lb salt.
1 Cubic foot of saturated salt brine weighs 75 lb and contains 18.5 lb salt.
Pounds of salt × 0.405 = gallons of saturated salt brine.
1 Cubic foot of dry salt weighs from slightly less than 50 lb to somewhat over 70 lb.
(Evaporated salts are near the lower limits; rock salts are near the upper limits.)

Source: Reprinted from *The Permutit Water and Waste Treatment Data Book.* © The Permutit Co., Inc.

Table 12.56 Hydrogen Cation Exchanger Reactions[a]

Reactions with Bicarbonates

$$\begin{Bmatrix} Ca \\ Mg \\ Na_2 \end{Bmatrix} (HCO_3)_2 \; + \; H_2Z \; = \; \begin{Bmatrix} Ca \\ Mg \\ Na_2 \end{Bmatrix} Z \; + \; 2H_2O \; + \; 2CO_2$$

Calcium, Magnesium, and/or Sodium Bicarbonate + Hydrogen cation exchanger = Calcium, Magnesium, and/or Sodium Cation Exchanger + Water + Carbon dioxide

(soluble)   (insoluble)   (insoluble)   (soluble gas)

Reactions with Sulfates or Chlorides

$$\begin{Bmatrix} Ca \\ Mg \\ Na_2 \end{Bmatrix} \begin{Bmatrix} SO_4 \\ Cl_2 \end{Bmatrix} \; + \; H_2Z \; = \; \begin{Bmatrix} Ca \\ Mg \\ Na_2 \end{Bmatrix} Z \; + \; \begin{Bmatrix} H_2SO_4 \\ 2HCl \end{Bmatrix}$$

Calcium, Magnesium, and/or Sodium Sulfates and/or Chlorides + Hydrogen Cation Exchanger = Calcium, Magnesium, and/or Sodium Cation Exchanger + Sulfuric and/or Hydrochloric acids

(soluble)   (insoluble)   (insoluble)   (soluble)

Regeneration Reactions

$$\left.\begin{array}{c}\text{Ca}\\\text{Mg}\\\text{Na}_2\end{array}\right\}\text{Z} \quad + \quad \text{H}_2\text{SO}_4 \quad = \quad \text{H}_2\text{Z} \quad + \quad \left.\begin{array}{c}\text{Ca}\\\text{Mg}\\\text{Na}_2\end{array}\right\}\text{SO}_4$$

Calcium, Magnesium, and/or Sodium Cation Exchanger + Sulfuric acid = Hydrogen Cation Exchanger + Calcium, Magnesium, and/or Sodium Sulfates

(insoluble)   (soluble)   (insoluble)   (soluble)

Source: Reprinted from *The Permutit Water and Waste Treatment Data Book.* © The Permutit Co., Inc.

[a]The symbol Z represents hydrogen cation exchanger radical.

Table 12.57 Weakly Basic Anion Exchanger Reactions[a]

Reactions with Sulfuric and Hydrochloric Acids

$$D + \begin{Bmatrix} H_2SO_4 \\ 2HCl \end{Bmatrix} = D \begin{Bmatrix} \cdot H_2SO_4 \\ \cdot 2HCl \end{Bmatrix}$$

Weakly Basic Anion Exchanger	+	Sulfuric and/or Hydrochloric acids	=	Weakly Basic Anion Exchanger	Hydrosulfate and/or Hydrochloride
(insoluble)		(soluble)			(insoluble)

Regeneration Reactions

$$D \begin{Bmatrix} \cdot H_2SO_4 \\ \cdot 2HCl \end{Bmatrix} + Na_2CO_3 = D + Na_2 \begin{Bmatrix} SO_4 \\ Cl_2 \end{Bmatrix} + H_2O + CO_2$$

Weakly Basic Anion Exchanger	Hydrosulfate and/or Hydrochloride	+ Soda ash =	Weakly Basic Anion Exchanger	+ Sodium	Sulfate and/or Chloride	+ Water + Carbon dioxide
(insoluble)		(soluble)	(insoluble)		(soluble)	(soluble gas)

Source: Reprinted from *The Permutit Water and Waste Treatment Data Book,* © The Permutit Co., Inc.

Table 12.58   Codes

Literally thousands of codes exist, but several are very widely used. Their specific applicability will vary from one jurisdiction to another and in all cases should be confirmed prior to use. In addition, some jurisdictions will supplement the codes with specific local requirements. The general content of the more widely used codes is listed here.

National Electrical Code
Widely adopted as the controlling code for electrical design and construction.

1. General
2. Wiring and Protection
3. Wiring Methods and Materials
4. Equipment for General Use
5. Special Occupancies
6. Special Equipment
7. Special Conditions
8. Communications Systems
9. Tables and Examples

ANSI/ASW D1.1: Structural Welding Code, Steel
The basic code for welding of steel.

1. General Provisions
2. Design of Welded Connections
3. Workmanship
4. Technique
5. Qualification
6. Inspection
7. Stud Welding
8. Statically Loaded Structures
9. Dynamically Loaded Structures
10. Tubular Structures
11. Strengthening and Repairing Existing Structures

1995 ASME Boiler and Pressure Vessel Code
An American National Standard; in most of the United States and many other parts of the world as well, the American Society of Mechanical Engineers (ASME) Boiler and Pressure Vessel Code has been adopted by jurisdictional bodies as the governing code.

I. Power Boilers
II. Materials
  Part A—Ferrous Materials Specifications
  Part B—Nonferrous Materials Specifications
  Part C—Specifications for Welding Rods, Electrodes, and Filler Metals
  Part D—Properties
III. Subsection NCA—General Requirements for Division 1 and Division 2
  Division 1
    Subsection NB—Class 1 Components
    Subsection NC—Class 2 Components
    Subsection ND—Class 3 Components
    Subsection NE—Class MC Components
    Subsection NF—Component Supports
    Subsection NG—Core Support Structures
    Subsection NH—Class 1 Components in Elevated Temperature Service
    Appendices
  Division 2—Code for Concrete Reactor Vessels and Containments
IV. Heating Boilers
V. Nondestructive Examination
VI. Recommended Rules for Care and Operation of Heating Boilers
VII. Recommended Rules for Care of Power Boilers
VIII. Pressure Vessels
  Division 1
  Division 2—Alternative Rules
IX. Welding and Brazing Qualifications
X. Fiberglass-Reinforced Plastic Pressure Vessels
XI. Rules for Inservice Inspection of Nuclear Power Plant Components

The code is intended to provide rules for the design, fabrication and operation of these equipment and components and is extensive and extremely detailed. The code is revised triannually and intermediate annual revisions are issued as addenda to maintain currency of the code. Before

Table 12.58 *(Continued)*

use of the code, the engineer should assure himself of the code applicability to the jurisdiction (usually state) in which the item will be assembled (as distinct from fabricated) and operated as well as specific additional or differing requirements particular to that jurisdiction.

Uniform Building Code
The ICBO (International Conference of Building Officials) Building Code; widely adopted west of the Mississippi River as the controlling code for general building and construction work—it is one of a family of codes covering mechanical systems and similar building considerations—for exact jurisdictions, consult the Council of American Building Officials.

1. Administration and Terms
2. Building Planning
3. Fire Protection
4. Occupant Needs
5. Building Envelope
6. Structural Systems
7. Structural Materials
8. Nonstructural Materials
9. Building Services
10. Special Devices and Conditions
11. Standards

BOCA National Building Code
The BOCA (Building Officials and Code Administrators) National Building Code; widely adopted as the basic building code east of the Mississippi River and north of Arkansas, Kentucky, and Virginia.

1. Administration and Enforcement
2. Definitions
3. Use Group Classification
4. Types of Construction
5. General Building Limitations
6. Special Use and Occupancy Requirements
7. Interior Environmental Requirements
8. Means of Egress
9. Fireresistive Construction
10. Fire Protection Systems
11. Structural Loads
12. Foundation Systems and Retaining Walls
13. Materials and Tests
14. Masonry
15. Concrete
16. Gypsum and Plaster
17. Wood
18. Steel
19. Lightweight Metal Alloys
20. Plastic
21. Exterior Walls
22. Vertical and Sloped Glass and Glazing
23. Roofs and Roof Coverings
24. Masonry Fireplaces
25. Mechanical Equipment and Systems
26. Elevator, Dumbwaiter, and Conveyor Equipment: Installation and Maintenance
27. Electric Wiring: Equipment and Systems
28. Plumbing Systems
29. Signs
30. Precautions During Building Operations
31. Energy Conservation
32. Repair, Alteration, Addition to, and Change of Use of Existing Buildings

Table 12.58 (*Continued*)

Standard Building Code
Widely adopted for use in southern tier states east of the Mississippi.

1. Administration	19. Concrete
2. Definitions	20. Light Metal Alloys
3. Occupancy Classifications	21. Masonry
4. Special Occupancy	22. Steel
5. General Building Limitations	23. Wood
6. Construction Types	24. Glass and Glazing
7. Fire-Resistant Materials and Construction	25. Gypsum Board and Plaster
8. Interior Finishes	26. Plastic
9. Fire Protection Systems	27. Electrical Systems
10. Means of Egress	28. Mechanical Systems
11. Accessibility for People with Physical Disabilities	29. Plumbing Systems
12. Interior Environment	30. Elevators and Conveying Systems
13. Energy Conservation	31. Special Construction
14. Exterior Wall Covering	32. Construction in the Public Right-of-Way
15. Roofs and Roof Structures	33. Site Work, Demolition and Construction
16. Structural Loads	34. Existing Buildings
17. Structural Tests and Inspections	35. Reference Standards
18. Foundations and Retaining Walls	

Uniform Fire Code
Extremely broad coverage with detailed requirements included; widely adopted as the local fire code.

Part I. General	Part VI. Special Equipment
Part II. Definitions	Part VII. Special Subjects
Part III. General Provisions for Safety	Part VIII. Standards
Part IV. Special Occupancy Uses	Part IX. Appendices
Part V. Special Processes	

Table 12.59 Electromagnetic Spectrum

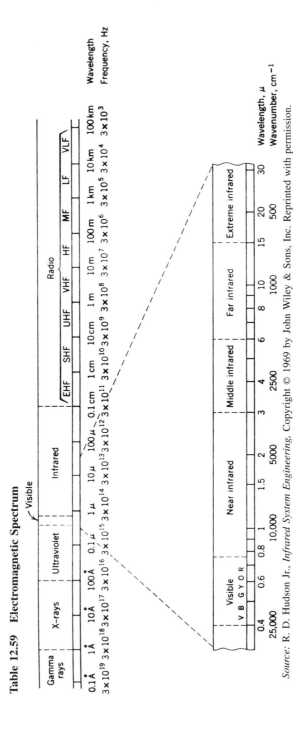

Source: R. D. Hudson Jr., *Infrared System Engineering,* Copyright © 1969 by John Wiley & Sons, Inc. Reprinted with permission.

Table 12.60 Selected Copy Editing and Proofreading Marks

Mark	Meaning	Marked in Manuscript	Set in Type
—	Set in italics (ital)	Goethe's <u>Faust</u>	Goethe's *Faust*
=	Set in small capitals (sc, esc)	<u>d</u>-glucose	D-glucose
≡	Set in capitals (cap)	<u>new york</u>	New York
/	Set in lowercase (lc)	the Ñuclear Ãge	the nuclear age
⌇	Set in boldface (bf)	A · B	**A · B**
⌐	Move left; flush left (fl, fl left)	⌐ where <u>t</u> = time.	where *t* = time.
⌐	Move right; flush right (fl right)	(25a)	(25a)
¶	Paragraph indent (para indent)	¶ The ethologist D. Morris gave his young chimpanzee Congo a pencil and studied his "drawing activity."	The ethologist D. Morris gave his young chimpanzee Congo a pencil and studied his "drawing activity." Even the first scratchings were not random:
↩	Run in	Even the first scratchings were not random: ¶ [Congo] carried in him the germ . . . of visual patterns." The	"[Congo] carried in him the germ . . . of visual patterns."
⌐¶	New line, paragraph indent	chimpanzee passed through several stages characteristic of early child art.	The chimpanzee passed through several stages characteristic of early child art.
⌐	New line, flush left		
⊔	Move down	x<u>y</u>	*xy*
⊓	Move up	xy	*xy*
⏋⌐	Center (ctr), display	⏋ x + y ⌐	*x* + *y*
⊓⊔	Transpose (tr)	Smith, A. B.	A. B. Smith
∼	Transpose (tr)	centɽe	center
#	Insert full space (#)	A B Smith etal.	A. B. Smith et al.
◡	Close up	non uniform	nonuniform
( )	Close up	1. Purpose ( ) 2. Methods 3. Results	1. Purpose 2. Methods 3. Results
⌒	Bring closer	In ~~the~~ two cases	In two cases

663

Table 12.60 *(Continued)*

Mark	Meaning	Marked in Manuscript	Set in Type
ℯ	Delete	right–hand hand page	right-hand page
⌐	Delete	right–haand page	right-hand page
(stet)	Restore deleted copy (stet)	right–hand page	right-hand page
∧	Insert copy	right‿page =hand	right-hand page
(en)	Set en fraction	½ (en)	½
(em)	Set em fraction	½ (em)	½
(shill)	Set shilling fraction (shill, sh)	½ (shill)	1/2
⊙	Set a period	J. AmChem. Soc.	*J. Am. Chem. Soc.*
⋏	Set a comma	a⋏. a₂. ...⋏a⋏	$a_1, a_2, \ldots, a_n$
⊙	Set a colon	New York⊙ Wiley, 1973	New York: Wiley, 1973
⋏	Set a semicolon	a = 2⋏ b = 5	$a = 2; b = 5$
⋎	Set a prime	A⋎	A'
⋎⋎	Set quotation marks	⋎Bromine⋎ comes from the French	"Bromine" comes from the French
=	Set hyphen	high=frequency radiation	high-frequency radiation
⊜ ⋧	Insert hyphen	unionized calcium	un-ionized calcium

Mark in Margin	Mark in Proof	Meaning
cap /	are capitalized in english titles;	Capital letter
lc /	only the first word and Proper	Lowercase letter
tr /	nouns, in French and Italain ti-	Transpose
⋏/	tles; and all nouns in German.	Insert comma
⌒/	Setting the volume number of	Delete and close up
✕	a periodical in distinctive type	Fix broken letter
⌒/	(italics or boldface) and page	Delete hyphen, close up
ℰ	numbers in ordinary type makes	Invert letter
⋎⋎	"Vol. and "p." unnecessary.	Insert quotes
=/	The year of publication is en	Insert hyphen
e/	closed in parenthesis only when	Substitute e for i
tr/	it is actually parenthetical	Transpose
eq #	information, as when, for ex-	Equalize spacing
⊏	ample, the volume and page of	Move left
stet /	a periodical are cited and the	Let type stand
wf/	year is added for the convenience	Change to right font
⁋/	of the reader. The following	Begin new paragraph
:/	illustrates a bibliographic item/	Change period to colon
sc	GAY, CHARLES M., and Charles	Set in small caps
ital /	DE VAN FAWCETT, *Mechanical and*	Set in itals
for /	*Electrical Equipment of Buildings*,	Change *of* to *for*
rom	*John* Wiley & Sons, New York,	Set in roman
⌒/ #/	second edition, p. 229 1945.	Close up; insert space

Proofreader's marks.

Table 12.61   Number of Each Day of the Year

Day of Mo.	Jan.	Feb.	Mar.	Apr.	May	Jun.	Jul.	Aug.	Sep.	Oct.	Nov.	Dec.
1	1	32	60	91	121	152	182	213	244	274	305	335
2	2	33	61	92	122	153	183	214	245	275	306	336
3	3	34	62	93	123	154	184	215	246	276	307	337
4	4	35	63	94	124	155	185	216	247	277	308	338
5	5	36	64	95	125	156	186	217	248	278	309	339
6	6	37	65	96	126	157	187	218	249	279	310	340
7	7	38	66	97	127	158	188	219	250	280	311	341
8	8	39	67	98	128	159	189	220	251	281	312	342
9	9	40	68	99	129	160	190	221	252	282	313	343
10	10	41	69	100	130	161	191	222	253	283	314	344
11	11	42	70	101	131	162	192	223	254	284	315	345
12	12	43	71	102	132	163	193	224	255	285	316	346
13	13	44	72	103	133	164	194	225	256	286	317	347
14	14	45	73	104	134	165	195	226	257	287	318	348
15	15	46	74	105	135	166	196	227	258	288	319	349
16	16	47	75	106	136	167	197	228	259	289	320	350
17	17	48	76	107	137	168	198	229	260	290	321	351
18	18	49	77	108	138	169	199	230	261	291	322	352
19	19	50	78	109	139	170	200	231	262	292	323	353
20	20	51	79	110	140	171	201	232	263	293	324	354
21	21	52	80	111	141	172	202	233	264	294	325	355
22	22	53	81	112	142	173	203	234	265	295	326	356
23	23	54	82	113	143	174	204	235	266	296	327	357
24	24	55	83	114	144	175	205	236	267	297	328	358
25	25	56	84	115	145	176	206	237	268	298	329	359
26	26	57	85	116	146	177	207	238	269	299	330	360
27	27	58	86	117	147	178	208	239	270	300	331	361
28	28	59	87	118	148	179	209	240	271	301	332	362
29	29	[a]	88	119	149	180	210	241	272	302	333	363
30	30		89	120	150	181	211	242	273	303	334	364
31	31		90		151		212	243		304		365

[a] In leap years, after February 28, add 1 to the tabulated number.

Table 12.62 Latin Terms

ca.	*circa*	about
cf.	*confer*	compare
e. g.	*exempli gratia*	for example
et al.	*et alibi* *et alii or aliae*	and elsewhere and others
etc.	*et cetera*	and others, and so forth
fl.	*floruit* *fluidus*	flourished fluid
i. e.	*id est*	that is
ibid.	*ibidem*	the same place
inf.	*infra*	below
in loc. cit.	*in loco citato*	in the place cited
op. cit.	*opere citato*	in the work cited
pass.	*passim*	here and there
per	*per*	through, by means of
q. v.	*quod vide*	which see
sc.	*scilicet*	namely
sic	*sic*	thus
sup.	*supra*	above
viz.	*videlicet*	namely
v. or vs.	*versus*	against

Table 12.63 Greek Alphabet

A	α	Alpha	N	ν	Nu
B	β	Beta	Ξ	ξ	Xi
Γ	γ	Gamma	O	o	Omicron
Δ	δ	Delta	Π	π	Pi
E	ε	Epsilon	P	ρ	Rho
Z	ζ	Zeta	Σ	σ	Sigma
H	η	Eta	T	τ	Tau
Θ	θ	Theta	Υ	υ	Upsilon
I	ι	Iota	Φ	φ	Phi
K	κ	Kappa	X	χ	Chi
Λ	λ	Lambda	Ψ	ψ	Psi
M	μ	Mu	Ω	ω	Omega

Table 12.64   Standard Electrode (Oxidation) Potentials of Some Electrodes in Aqueous Solution at 25°C

Electrode Reaction	Standard Potential (volts)	Electrode Reaction	Standard Potential (volts)
$Li = Li^+ + \epsilon$	3.045	$Ag + I^- = AgI + \epsilon$	0.151
$K = K^+ + \epsilon$	2.925	$Sn = Sn^{2+} + 2\epsilon$	0.136
$Na = Na^+ + \epsilon$	2.714	$Pb = Pb^{2+} + 2\epsilon$	0.036
$Mg = Mg^{2+} + 2\epsilon$	2.37	$H_2 = 2H^+ + 2\epsilon$	0.000
$Al = Al^{3+} + 3\epsilon$	1.66	$Cu = Cu^{2+} + 2\epsilon$	−0.337
$Zn = Zn^{2+} + 2\epsilon$	0.763	$2OH^- = H_2O + \frac{1}{2}O_2 + \epsilon$	−0.401
$Fe = Fe^{2+} + 2\epsilon$	0.440	$2Hg = Hg^{2+} + 2\epsilon$	−0.789
$Cd = Cd^{2+} + 2\epsilon$	0.403	$Ag = Ag^+ + \epsilon$	−0.799
$Ni = Ni^{2+} + 2\epsilon$	0.250	$2Cl^- = Cl_2 + 2\epsilon$	−1.360

Source: E. U. Condon and H. Odishaw, *Handbook of Physics,* © McGraw-Hill. Used with the permission of McGraw-Hill Book Company.

Table 12.65   Selected Technical Societies and Associations

American Association of State Highway and Transportation Officials, 444 North Capitol Street NW, Washington, DC 20001; (206) 624-5800, Fax (202) 624-5806

American Concrete Institute, P.O. Box 9094, Farmington Hills, MI 48333; (810) 848-3700, Fax (810) 848-3701

American Forest & Paper Association, 1111 19th Street NW, Washington, DC 20036; (202) 463-2700, Fax (202) 463-2785

American Gas Association, 1515 Wilson Boulevard, Arlington, VA 22209; (703) 841-8400, Fax (703) 841-8406

American Institute of Architects, 1735 New York Avenue NW, Washington, DC 20006; (202) 626-7300, Fax (202) 626-7421

American Institute of Chemical Engineers, 345 East 47th Street, New York, NY 10017; (212) 705-7338, Fax (212) 752-3294

American Iron and Steel Institute, 1101 17th Street NW, Washington, DC 20036-4700; (202) 452-7100, Fax (202) 463-6573

American Institute of Steel Construction, 1 East Wacker Drive, Chicago, IL 60601-2001; (312) 670-2400, Fax (312) 670-5403

American National Standards Institute, 11 West 42nd Street, New York, NY 10036; (212) 642-4900, Fax (212) 398-0023

American Petroleum Institute, 1220 L Street NW, Washington, DC 20005; (202) 682-8000, Fax (202) 685-8232

American Society for Quality Control, 611 East Wisconsin Avenue, Milwaukee, WI 53201-3005; (800) 248-1946, Fax (414) 272-8575

American Society for Testing and Materials, 1916 Race Street, Philadelphia, PA 19103-1187; (215) 299-5400, Fax (215) 977-9679

American Society of Civil Engineers, 1801 Alexander Bell Drive, Reston, VA 20191; (703) 295-6057, Fax (703) 295-6333

American Society of Heating, Refrigerating, and Air-Conditioning Engineers, 1791 Tullie Circle NE, Atlanta GA 30329; (404) 636-8400, Fax (404) 321-5478

American Society of Mechanical Engineers, 345 East 47th Street, New York, NY 10017; (212) 705-7722, Fax (212) 705-7739

American Water Works Association, 6666 West Quincy Avenue, Denver, CO 80235; (303) 794-7711, Fax (303) 795-1440

American Wind Energy Association, 122 C Street NW, Washington, DC 20001; (202) 383-2500, Fax (202) 383-2505

Automatic Identification Manufacturers, Inc., 634 Alpha Drive, Pittsburgh, PA 15238-2802; (412) 963-8588, Fax (412) 963-8753

Chemical Manufacturers Association, 2501 M Street NW, Washington, DC 20037; (202) 887-1100, Fax (202) 887-1237

Composites Fabricators Association, 1735 North Lynn Street, Arlington, VA 22209-2202; (703) 524-3332, Fax (703) 524-2303

Energy Efficiency and Renewable Energy Clearinghouse, P.O. Box 3048, Merrifield, VA 22116; (800) 363-3732, Fax (703) 893-0400

Forest Products Laboratory, U.S. Department of Agriculture, 1 Grifford Pinchot Drive, Madison, WI 53705-2398; (608) 231-9236

Hydraulic Institute, 9 Sylvan Parkway, Parsippany, NJ 07054-3802; (201) 267-9700, Fax (201) 267-9055

Institute of Electrical and Electronic Engineers, 345 East 47th Street, New York, NY 10017; (212) 705-7900, Fax (212) 705-4929

International Fire Code Institute, 9300 Jollyville Road, Austin, TX 78759-7455; (512) 345-2633, Fax (512) 343-9116

Laser Institute of America, 12424 Research Parkway, Orlando FL 32826-3274; (407) 380-1553, Fax (407) 380-5588

National Climatic Data Center, 151 Patton Avenue, Asheville, NC 28801; (704) 271-4800, Fax (704) 271-4876

National Electrical Manufacturers Association, 2101 L Street NW, Washington, DC 20037; (202) 457-8400, Fax (202) 457-8411

National Fire Protection Association, 1 Batterymarch Park, P.O. Box 9101, Quincy, MA 02269-9101; (617) 770-3000, Fax (617) 770-0700

National Fluid Power Association, 3333 North Mayfair Road, Milwaukee, WI 53222-3219; (414) 778-3357, Fax (414) 778-3361

National Institute of Standards and Technology, Quince Orchard Road, Gaithersburg, MD 20899; (301) 975-2000, Fax (301) 975-3839

National Society of Professional Engineers, 1420 King Street, Alexandria, VA 22314; (703) 684-2800, Fax (703) 836-4875

Rubber Manufacturers Association, 1400 K Street NW, Washington, DC 20005; (202) 682-4863, Fax (202) 682-4800

SAE (Society of Automotive Engineers) International, 400 Commonwealth Drive, Warrendale, PA 15096-0001; (412) 776-4841, Fax (412) 776-5760

Underwriters Laboratories, 333 Pfingsten Road, Northbrook, IL 60062; (708) 272-8800, Fax (708) 272-8129

U.S. Department of Energy Laboratory, 1617 Cole Boulevard, Golden, CO 80401; (303) 275-4099, Fax (303) 275-4091

Table 12.66 Factorials

n	n! = 1·2·3···n	1/n!	n	n! = 1·2·3···n	1/n!
1	1	1.	11	$399{,}168 \times 10^2$	0.250521×10^{-7}
2	2	0.5	12	$479{,}002 \times 10^3$	$.208768 \times 10^{-8}$
3	6	.166667	13	$622{,}702 \times 10^4$	$.160590 \times 10^{-9}$
4	24	$.416667 \times 10^{-1}$	14	$871{,}783 \times 10^5$	$.114707 \times 10^{-10}$
5	120	$.833333 \times 10^{-2}$	15	$130{,}767 \times 10^7$	$.764716 \times 10^{-12}$
6	720	$.138889 \times 10^{-2}$	16	$209{,}228 \times 10^8$	$.477948 \times 10^{-13}$
7	5,040	$.198413 \times 10^{-3}$	17	$355{,}687 \times 10^9$	$.281146 \times 10^{-14}$
8	40,320	$.248016 \times 10^{-4}$	18	$640{,}237 \times 10^{10}$	$.156192 \times 10^{-15}$
9	362,880	$.275573 \times 10^{-5}$	19	$121{,}645 \times 10^{12}$	$.822064 \times 10^{-17}$
10	3,628,800	$.275573 \times 10^{-6}$	20	$243{,}290 \times 10^{13}$	$.411032 \times 10^{-18}$

Table 12.67   Selected Physical Constants

Second	1/86,400 mean solar days
Temperature 0°C, ice point	273.15°K or 491.67°R
Avogadro number	6.02252×10^{23} molecules/mol
Boltzmann constant	1.38054×10^{-23} joules/°K
Elementary charge	1.60210×10^{-19} coulombs
Electron rest mass	9.1091×10^{-28} grams
Faraday constant (electromag)	9.64870×10^{4} coulombs/mol
Gravitational constant	6.670×10^{-8} dyne-cm²/gm²
Ideal gas constant	8.3143 joules/°K-mol
Light speed in vacuum	2.997925×10^{8} meters/second
Planck constant	6.6256×10^{-34} joule-seconds
Standard acceleration of gravity	9.80665 meters/sec²
Standard acceleration of gravity	32.1740 feet/sec²
Standard atmospheric pressure	1013250 dynes/cm²
Stefan-Boltzmann constant	5.6697×10^{-5} ergs/cm²-sec-°K⁴
Volume ideal gas at 0°C, 1 atmos	22413.6 cm³/gram mol

Table 12.68   Solar System Characteristics

Body	Mean dist. from sun 10⁶ km	Orbital period sidereal days	Axial period sidereal hr	Equatorial diameter km	Mass 10²⁴ kg	Density g/cm³	g m/sec²	Escape velocity km/sec
Sun	—	—	609 h, 6′	1.392×10^{6}	1980000	1.39	271	—
Mercury	57.85	87.97	—	4800	0.32	5.3	3.33	4.2
Venus	108.2	224.70	30 h	12400	4.9	4.95	8.52	10.3
Earth	149.6	365.26	23 h, 56′, 4.1″	12756.6	6.0	5.52	9.81	11.2
Moon	0.38*	27.32	655 h, 43′, 11″	3478	0.074	3.39	1.62	2.4
Mars	227.9	686.98	24 h, 37′, 23″	6783	0.64	3.95	3.77	5.1
Jupiter	778.3	4332.6	9 h, 50′, 30″	142600	1900	1.33	25.1	61
Saturn	1428	10759	10 h, 14′	119000	570	0.69	10.72	37
Uranus	2872	30687	10 h, 49′	51500	87	1.56	8.83	22
Neptune	4498	60184	15 h, 40′	49900	103	2.27	11.00	25
Pluto	5910	90700	16 h	12800	5.6	5	9.1	10

*Distance from Earth.

Table 12.69 Periodic Table of the Elements

Key

atomic number → 29 ← atomic weight; parentheses indicates longest lived isotope

symbol → Cu

oxidation states +1 and +2 → Copper 1,2 ← name

(Ar)3d¹⁰4s¹ → electron structure same as Ar, plus 10e⁻ in 3d and 1e in 4s orbitals

63.54

GROUP 1A	2A	3B	4B	5B	6B	7B	8		
1 1.00797 H Hydrogen ±1 $1s^1$									
3 6.939 Li Lithium 1 $1s^2 2s^1$	4 9.0122 Be Beryllium 2 $1s^2 2s^2$								
11 22.9898 Na Sodium 1 $(Ne)3s^1$	12 24.312 Mg Magnesium 2 $(Ne)3s^2$								
19 39.102 K Potassium 1 $(Ar)4s^1$	20 40.08 Ca Calcium 2 $(Ar)4s^2$	21 44.956 Sc Scandium 3 $(Ar)3d^1 4s^2$	22 47.90 Ti Titanium 2,3,4 $(Ar)3d^2 4s^2$	23 50.942 V Vanadium 2,3,4,5 $(Ar)3d^3 4s^2$	24 51.996 Cr Chromium 2,3,6 $(Ar)3d^5 4s^1$	25 54.938 Mn Manganese 2,3,4,6,7 $(Ar)3d^5 4s^2$	26 55.847 Fe Iron 2,3,4,6 $(Ar)3d^6 4s^2$	27 58.933 Co Cobalt 2,3,4 $(Ar)3d^7 4s^2$	
37 85.47 Rb Rubidium 1 $(Kr)5s^1$	38 87.62 Sr Strontium 2 $(Kr)5s^2$	39 88.905 Y Yttrium 3 $(Kr)4d^1 5s^2$	40 91.22 Zr Zirconium 4 $(Kr)4d^2 5s^2$	41 92.906 Nb Niobium 3,5 $(Kr)4d^4 5s^1$	42 95.94 Mo Molybdenum 3,5,6 $(Kr)4d^5 5s^1$	43 (98) Tc Technetium 2,4,7 $(Kr)4d^6 5s^1$	44 101.07 Ru Ruthenium 2,3,4,6,8 $(Kr)4d^7 5s^1$	45 102.905 Rh Rhodium 2,3,4,6 $(Kr)4d^8 5s^1$	
55 132.905 Cs Cesium 1 $(Xe)6s^1$	56 137.34 Ba Barium 2 $(Xe)6s^2$	57 138.91 La Lanthanum 3 $(Xe)5d^1 6s^2$	72 178.49 Hf Hafnium 4 $(Xe)4f^{14}5d^2 6s^2$	73 180.948 Ta Tantalum 4 $(Xe)4f^{14}5d^3 6s^2$	74 183.85 W Tungsten 2,4,5,6 $(Xe)4f^{14}5d^4 6s^2$	75 186.2 Re Rhenium −1,3,4,6,7 $(Xe)4f^{14}5d^5 6s^2$	76 190.2 Os Osmium 2,3,4,6,8 $(Xe)4f^{14}5d^6 6s^2$	77 192.2 Ir Iridium 2,3,4,6 $(Xe)4f^{14}5d^7 6s^2$	
87 (223) Fr Francium 1 $(Rn)7s^1$	88 (226) Ra Radium 2 $(Rn)7s^2$	89 (227) Ac Actinium 3 $(Rn)6d^1 7s^2$							

58 140.12 Ce Cerium 3,4 $(Xe)4f^1 5d^1 6s^2$	59 140.907 Pr Praseodymium 3,4 $(Xe)4f^3 5d^0 6s^2$	60 144.24 Nd Neodymium 3,4 $(Xe)4f^4 5d^0 6s^2$	61 (147) Pm Promethium 3 $(Xe)4f^5 5d^0 6s^2$	62 150.35 Sm Samarium 2,3 $(Xe)4f^6 5d^0 6s^2$	63 151.96 Eu Europium 2,3 $(Xe)4f^7 5d^0 6s^2$
90 232.038 Th Thorium 3,4 $(Rn)6d^2 7s^2$	91 (231) Pa Protoactinium 4,5 $(Rn)5f^2 6d^1 7s^2$	92 238.03 U Uranium 3,4,5,6 $(Rn)5f^3 6d^1 7s^2$	93 (237) Np Neptunium 3,4,5,6 $(Rn)5f^4 6d^1 7s^2$	94 (244) Pu Plutonium 3,4,5,6 $(Rn)5f^6 6d^0 7s^2$	95 (243) Am Americium 3,4,5,6 $(Rn)5f^7 6d^0 7s^2$

Basis: $^{12}C = 12.0000$.

Table 12.69 (Continued)

3A	4A	5A	6A	7A	He (Noble Gases)
					2 4.0026 **He** Helium 0 $1s^2$
5 10.811 **B** Boron 3 $1s^22s^22p^1$	6 12.0112 **C** Carbon ±4,2 $1s^22s^22p^2$	7 14.0067 **N** Nitrogen −3,2,3,4,5 $1s^22s^22p^3$	8 15.9994 **O** Oxygen −2 $1s^22s^22p^4$	9 18.9984 **F** Fluorine −1 $1s^22s^22p^5$	10 20.183 **Ne** Neon 0 $1s^22s^22p^6$
13 26.9815 **Al** Aluminum 3 $(Ne)3s^23p^1$	14 28.086 **Si** Silicon 4 $(Ne)3s^23p^2$	15 30.9738 **P** Phosphorus ±3,4,5 $(Ne)3s^23p^3$	16 32.064 **S** Sulfur −2,4,6 $(Ne)3s^23p^4$	17 35.453 **Cl** Chlorine ±1,3,5,7 $(Ne)3s^23p^5$	18 39.948 **Ar** Argon 0 $(Ne)3s^23p^6$

1B	2B	3A	4A	5A	6A	7A	Noble Gases
28 58.71 **Ni** Nickel 2,4,6 $(Ar)3d^84s^2$							
29 63.54 **Cu** Copper 1,2 $(Ar)3d^{10}4s^1$	30 65.37 **Zn** Zinc 2 $(Ar)3d^{10}4s^2$	31 69.72 **Ga** Gallium 3 $(Ar)3d^{10}4s^24p^1$	32 72.59 **Ge** Germanium 2,4 $(Ar)3d^{10}4s^24p^2$	33 74.922 **As** Arsenic ±3,5 $(Ar)3d^{10}4s^24p^3$	34 78.96 **Se** Selenium −2,4,6 $(Ar)3d^{10}4s^24p^4$	35 79.909 **Br** Bromine ±1,3,5,7 $(Ar)3d^{10}4s^24p^5$	36 83.80 **Kr** Krypton 0 $(Ar)3d^{10}4s^24p^6$
47 107.870 **Ag** Silver 1 $(Kr)4d^{10}5s^1$	48 112.40 **Cd** Cadmium 2 $(Kr)4d^{10}5s^2$	49 114.82 **In** Indium 3 $(Kr)4d^{10}5s^25p^1$	50 118.69 **Sn** Tin 2,4 $(Kr)4d^{10}5s^25p^2$	51 121.75 **Sb** Antimony ±3,5 $(Kr)4d^{10}5s^25p^3$	52 127.60 **Te** Tellurium −2,4,6 $(Kr)4d^{10}5s^25p^4$	53 126.904 **I** Iodine ±1,5,7 $(Kr)4d^{10}5s^25p^5$	54 131.30 **Xe** Xenon 0(+4 +6) $(Kr)4d^{10}5s^25p^6$
46 106.4 **Pd** Palladium 2,4,6 $(Kr)4d^{10}5s^0$							
79 196.967 **Au** Gold 1,3 $(Xe)4f^{14}5d^{10}6s^1$	80 200.59 **Hg** Mercury 1,2 $(Xe)4f^{14}5d^{10}6s^2$	81 204.37 **Tl** Thallium 1,3 $(Xe)4f^{14}5d^{10}6s^26p^1$	82 207.19 **Pb** Lead 2,4 $(Xe)4f^{14}5d^{10}6s^26p^2$	83 208.980 **Bi** Bismuth ±3,5 $(Xe)4f^{14}5d^{10}6s^26p^3$	84 (210) **Po** Polonium −2,4,6 $(Xe)4f^{14}5d^{10}6s^26p^4$	85 (210) **At** Astatine ±1,5 $(Xe)4f^{14}5d^{10}6s^26p^5$	86 (222) **Rn** Radon 0(+6) $(Xe)4f^{14}5d^{10}6s^26p^6$
78 195.09 **Pt** Platinum 2,4,6 $(Xe)4f^{14}5d^96s^1$							

LANTHANIDES

64 157.25 **Gd** Gadolinium 3 $(Xe)4f^75d^16s^2$	65 158.924 **Tb** Terbium 3,4 $(Xe)4f^95d^06s^2$	66 162.50 **Dy** Dysprosium 3 $(Xe)4f^{10}5d^06s^2$	67 164.930 **Ho** Holmium 3 $(Xe)4f^{11}5d^06s^2$	68 167.26 **Er** Erbium 3 $(Xe)4f^{12}5d^06s^2$	69 168.934 **Tm** Thulium 2,3 $(Xe)4f^{13}5d^06s^2$	70 173.04 **Yb** Ytterbium 2,3 $(Xe)4f^{14}5d^06s^2$	71 174.97 **Lu** Lutetium 3 $(Xe)4f^{14}5d^16s^2$
96 (245) **Cm** Curium 3 $(Rn)5f^76d^17s^2$	97 247 **Bk** Berkelium 3,4 $(Rn)5f^86d^17s^2$	98 (251) **Cf** Californium (3) $(Rn)5f^96d^17s^2$	99 (252) **Es** Einsteinium (3) $(Rn)5f^{10}6d^17s^2$	100 (253) **Fm** Fermium (3) $(Rn)5f^{11}6d^17s^2$	101 (256) **Md** Mendelevium (3) $(Rn)5f^{12}6d^17s^2$	102 (253) **No** Nobelium (3) $(Rn)5f^{13}6d^17s^2$	103 (257) **Lw** Lawrencium (3) $(Rn)5f^{14}6d^17s^2$

ACTINIDES

INDEX